Tubular
Structures V

Other books available from E & FN Spon

Aluminium Structural Analysis Recent European Advances
Edited by P.S. Bulson

Architecture and Construction in Steel
Edited by Alan Blanc, Michael McEvoy and Roger Plank

Computer Methods in Structural Analysis
J.L. Meek

Constructional Steel Design: An International Guide
Edited by P.J. Dowling, J.E. Harding and R. Bjorhovde

Constructional Steel Design: World Developments
Edited by P.J. Dowling, J.E. Harding, R. Bjorhovde and
E. Martinez-Romero

Flexural-Torsional Buckling of Structures
N.S. Trahair

Limit States Design of Steel Structures
D.A. Nethercot

Testing of Metals for Structures
Edited by F.M. Mazzolani

The Behaviour and Design of Steel Structures
N.S. Trahair and M.A. Bradford

Tubular Structures: Proceedings of Third International Symposium
Edited by E. Niemi and P. Makelainen

For details of these and other books, contact the Promotion Department,
E & FN Spon, 2–6 Boundary Row, London SE1 8HN, Tel: 071-865 0066

Tubular Structures V

Proceedings of the Fifth International Symposium
Nottingham, United Kingdom
25–27 August 1993

Edited by

M.G. COUTIE

and

G. DAVIES

Department of Civil Engineering
The University of Nottingham
University Park
Nottingham
United Kingdom

E & FN SPON
An Imprint of Chapman & Hall
London · Glasgow · New York · Tokyo · Melbourne · Madras

Published by E & FN Spon, an imprint of Chapman & Hall,
2-6 Boundary Row, London SE1 8HN, UK

Chapman & Hall, 2-6 Boundary Row, London SE1 8HN, UK

Blackie Academic & Professional, Wester Cleddens Road,
Bishopbriggs, Glasgow G64 2NZ, UK

Chapman & Hall Inc., One Penn Plaza, 41st Floor, New York NY10019, USA

Chapman & Hall Japan, Thomson Publishing Japan, Hirakawacho
Nemoto Building, 6F, 1-7-11 Hirakawa-cho, Chiyoda-ku, Tokyo 102, Japan .

Chapman & Hall Australia, Thomas Nelson Australia, 102 Dodds Street,
South Melbourne, Victoria 3205, Australia

Chapman & Hall India, R. Seshadri, 32 Second Main Road, CIT East,
Madras 600 035, India

First edition 1993

© 1993 E & FN Spon

Printed in Great Britain at the University Press, Cambridge

ISBN 0 419 18770 7

A catalogue record for this book is available from the British Library

Library of Congress Cataloging-in-Publication data available

Publisher's Note
This book has been produced from camera ready copy provided by the
individual contributors. This method of production has allowed us to
supply finished copies to the delegates at the Symposium.

♾ Printed on acid-free text paper, manufactured in accordance
with ANSI/NISO Z 39.48-1992 (Permanence of Paper).

Contents

Preface

vii

This volume consists of papers presented at the 5th International Symposium on Tubular Structures held at Nottingham, United Kingdom, 25-27 August 1993, organised under the auspices of the Department of Civil Engineering, The University of Nottingham.

INTERNATIONAL PROGRAMME COMMITTEE

Prof F M Burdekin	University of Manchester Institute of Science and Technology
Dr G Davies	Secretary IIW Subcommission XV-E, The University of Nottingham
Mr D Dutta	Chairman CIDECT Technical Commission, Mannesmann-röhren-Werke AG, Düsseldorf
Prof T H Hyde	The University of Nottingham
Prof Y Kurobane	Kumamoto University
Dr P W Marshall	Shell Oil Company, Houston
Prof E Niemi	Lappeenranta University of Technology
Prof J A Packer	Chairman IIW Subcommission XV-E, Toronto University
Prof dr J Wardenier	Delft University of Technology
Mr N F Yeomans	Chairman CIDECT Joint Working Group, British Steel Tubes and Pipes, Corby
Dr N Zettlemoyer	Exxon Production Research Company, Houston

LOCAL ORGANISING COMMITTEE

Dr G Davies and Prof D A Nethercot	Co-Chairmen, The University of Nottingham
Dr M G Coutie	The University of Nottingham
Mr E Hole	British Steel Tubes and Pipes, Corby
Prof T H Hyde	The University of Nottingham
Dr J Tolloczko	The Steel Construction Institute
Mr N F Yeomans	British Steel Tubes and Pipes, Corby
Miss Gayle Lowe	Committee Secretary, The University of Nottingham

SPONSORS

The University of Nottingham, Department of Civil Engineering
Comité International pour le Développement et l'Étude de la Construction Tubulaire (CIDECT)
International Institute of Welding: Subcommission XV-E Tubular Structures
British Steel Tubes and Pipes, Corby

CO-SPONSORS

The Steel Construction Institute
American Society of Civil Engineers; Tubular Structures Group

Preface

This is the Fifth in the series of International Symposia on Tubular Structures, devoted to the use of hollow sections in both on and off-shore situations. It follows Symposia previously held in Boston, USA (1984); Tokyo, Japan (1986); Lappeenranta, Finland (1989); and Delft, The Netherlands (1991). In response to the initial invitation for submissions, 102 synopses were received. From these the International Programme Committee in co-operation with the Local Organising Committee have been able to select - after proper review - the 75 contributions from over 10 countries to be presented at the Symposium, and collected together in this volume. The Committees feel that the papers are of a high standard reflecting the wide range of interest in this form of construction. They cover developments in design, planning, fabrication and construction of structures, prestigious international developments, and the continuing work of research and development, which are proceeding throughout the world. These papers describe the innovative use of tubular steel, while also indicating developments in the understanding of the behaviour of such structures, with refinement of design procedures.

An overview of the papers clearly displays the reasons why architects choose these sections, particularly for prestigious structures where aesthetic detail and ease of maintenance are important features. The varied use of rectangular sections in columns for multi-storey buildings in earthquake regions has generated a group of papers on the behaviour of large cold formed sections fabricated in different ways. The ductility and energy absorbing features of such members and connections receives attention. The behaviour of joints within structures, under repeated loading whether CHS or RHS is of considerable interest. Over the years the use of the finite element method of analysis for the estimation of peak stress and strain at welds has developed. Arguments as to the appropriate values of concentration factors are being rapidly rationalised in the light of significant experimental and analytical work in this field, and are linked to the development of new design recommendations. While this approach to fatigue is based on elastic analysis, significant advances are being made in strength and serviceability studies by incorporating non-linear material and large deflection geometric properties. This volume of the Symposium Proceedings includes several papers in which finite element solutions are calibrated to the results of a carefully conducted series of experiments on three dimensional behaviour of joints. The use of the FE technique in the production and assessment of new design proposals augers well for the future. Papers also describe the steady development in fire and corrosion protection, as well as probing the future with expert systems.

Free exchange of information has been an important factor in the rapid development of the subject of tubular structures, and the roles played by CIDECT and IIW Subcommission XV-E 'Welded Joints in Tubular Structures' have made a major contribution to this exchange and co-operation over the years. The friendliness and approachability of successive chairmen and committee members of these two groups has facilitated this exchange. It is hoped that the spirit of constructive criticism evident in previous Symposia will also be present in Nottingham. It is a particular pleasure to the co-chairmen and editors to have Professor Jeff Packer as the current IIW Subcommission XV-E chairman, as his interest in tubular structures was awakened when a research assistant at this University in the 1970's.

We are of course most grateful to all those authors who have contributed to the papers of this volume, and for their patience in carrying out the manuscript revisions requested by local paper reviewers and editors, in order to reach a fairly uniform standard. Mr Nick Clarke of Spons has been a great encouragement throughout.

The low level of Symposium fees to delegates and particularly to authors, has only been possible through the generosity of British Steel - Tubes and Pipes at Corby, and

CIDECT. Their optimism and foresight is particularly appreciated at a time of industrial recession.

The willingness of Mr Richard Heckels, Mr Anthony Hunt, Professor Jeff Packer and Mr Dipak Dutta to participate in the opening session of the Symposium has been a great encouragement.

As co-chairmen of the local committee and as editors we wish to express our highest admiration for the hard work and professionalism of Miss Gayle Lowe in handling both the secretarial and many of the organisational aspects of this Symposium.

Prof D A Nethercot   Co-Chairman
Dr G Davies   Co-Chairman and Editor
Dr M G Coutie   Editor

The University of Nottingham

August 1993

PART 1

PROJECTS IN TUBULAR CONSTRUCTION

Architecture, planning, design, fabrication and construction

1 UNITED KINGDOM PAVILION, EXPO '92

D. HADDEN
Ove Arup & Partners, London, UK
K. FORD
formerly Tubeworkers Ltd, Warwick, UK

Abstract
The UK Pavilion EXPO'92 is a notable example of how the use of
tubular steel elements can combine engineering efficiency in a long
span structure with clear architectural expression of the building
frame. The paper describes the evolution of the structural form
from design competition to fabrication and erection. The 'kit-of-
parts' approach, which combined precise fabrication in factory
conditions off-site with simple site assembly, was fundamental to
the successful completion of a structure made in Britain and built
in Spain.
    While the Pavilion was designed to provide exhibition space
which would be a cool haven from the intense heat of Seville in
summer, the building itself was the primary exhibit. One key
message conveyed by the Pavilion was the responsible use of energy
by passive means. Another was to demonstrate British steelwork
capability in an international setting.
    Together the designers and constructors addressed the balance
between numerous small elements giving transportation flexibility
but putting greater emphasis on site work and large prefabricated
assemblies generally connected by simple pin joints. In the final
design the detailing of the connections informed the observer of
the sequence of erection and structural action of the members.
    The overall approach to analysis of the main elements of the
structure is described. The use of overlap joints and their
influence on the global analysis is discussed together with the
treatment of seismic loads on the structure. Quality control
during fabrication and dimensional accuracy in a structure with
limited scope for site tolerance is considered.
Keywords: Tubular Structures, Exhibition Pavilions, Structural
Form, Architecture, Prefabrication, Transportability.

1 Introduction

From April to October 1992 Seville in southern Spain was the venue
for Expo'92, the first Universal Exposition to be held anywhere
since Osaka in 1970 and the first in Europe since Brussels in 1958.
    In addition to the various theme pavilions over 100 countries
built national pavilions on the 215 hectare site by the River
Guadalquivir near the heart of the historic Andalucian city from
which Columbus had embarked on his great voyages of discovery 500
years earlier.

Tubular Structures V. Edited by M.G. Coutie and G. Davies.
© 1993 E & FN Spon, 2–6 Boundary Row, London SE1 8HN. ISBN 0 419 18770 7.

Fig.1. Pavilion from north east

In 1988 the British government confirmed its intention to participate fully in Expo'92, Prime Minister Thatcher declaring that Britain's pavilion would be "quite the best". In December 1988 the Department of Trade and Industry invited selected firms of architects and designers to submit proposals for the building. Shortly afterwards Nicholas Grimshaw & Partners Ltd, Ove Arup & Partners as structural and building services engineers and cost consultants Davis Langdon & Everest were appointed as the professional team for the Pavilion which was to be funded by central government and sponsorship from industry: British Steel in particular made a major contribution.

The client's objectives and brief for the Pavilion are described elsewhere, Gardner et al (1993).

2 Principal Design Objectives

From the outset the architect and his team were determined to create a piece of 'serious' architecture and avoid the Disneyesque. The 1851 Great Exhibition Building and the Festival of Britain pavilions 100 years later were seen as more appropriate points of reference. Off-site prefabrication of major building elements, preferably in the UK, was considered essential. A responsible energy policy, and a strategy for re-use either of the whole Pavilion or components, were important.

Nicholas Grimshaw has gained his reputation as a leader of British architecture by his use of industrial techniques to create building components and his ability to exploit technology transfer.

4

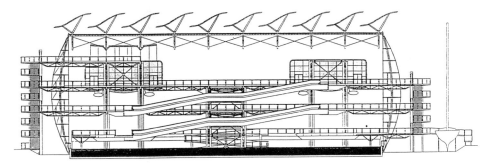

Fig.2. Longitudinal section

The design developed by the Pavilion team fully embodied this
philosophy particularly in the way that traditional methods of
coping with the hot, dry summers of Seville were adopted and
refined using modern production methods and analytical tools.
It is the essence of a Grimshaw building that how it was built
and how it works are there to be seen.   That the structure would be
exposed was never in doubt.   Steel as the primary structural
material could be fabricated in the UK thus distancing the project
from the inevitable over-heating of the local construction industry
in Seville due to the enormous volume of work for Expo'92.
Furthermore an opportunity for the British steelwork industry to
demonstrate its abilities in such an international setting was not
to be missed.
To create the unobstructed spaces called for in the brief long
span primary members were required.   Trusses were a logical
solution with tubular chords and bracings ideal for elements under
mainly axial load.   Tubular elements were also entirely compatible
with the architect's vision of a clean, efficient structure.

3 Overall Form

The Pavilion had a rectangular external envelope with internal plan
dimensions of 64.8m x 32m, and a height of 24.8m from lower ground
floor to roof. Early press articles referred to it as 'the size of
Westminster Abbey' - in cross-section at least a valid comparison.
The longitudinal axis ran north-south, with the east wall as
principal public elevation and the west wall facing various service
buildings.
The five main floor levels accomodated exhibition space,
circulation routes, plantrooms and management offices amounting to
some 6000m^2.
Only the lower ground floor occupied the whole plan area. The
partial floors above housed the principal exhibition spaces within
the protection of the envelope. Two enclosed theatre pods were
located towards the north and south end walls, with two open decks
between them to provide access to the central feature exhibit.
Visitor entry was via a bridge over an external lake onto the
concourse level. The Pavilion had no internal stairs; visitors
proceeded to and from the upper floors by inclined travelators,
each 50m long and arranged in pairs on opposite sides of the

building. From the concourse they could exit directly via a bridge at the north elevation or descend by a ramp to the restaurant below. Two 13-person hydraulic scenic lifts took VIPs from their own entrance to the VIP Suite on top of the south pod.

4 Environmental Control and the Building Fabric

Before examining the structure in detail it is worth considering the building fabric and its role in modifying the harsh climate for this, as much as any other factor, was fundamental to the total building engineering design.

Rather than treating the interior as one space with uniform conditions throughout, the arrangement of envelope and specialized areas allowed a hierarchical approach to environmental control. Visitors queuing for an hour in the sun could be discomforted by a dramatic temperature drop on entry, but clearly 200 people seated in a compact theatre would produce a significant heat gain. The solution was to provide a moderated environment within the envelope to which visitors could adjust before entering the fully conditioned inner spaces. Typical internal temperatures were 28°C on the Concourse level and 22°C in the pods and offices, but these were not fixed and reflected seasonal changes in external conditions.

The Pavilion orientation meant that direct solar radiation fell on the east wall for only 2-3 hours each day after public opening; the roof, south elevation and, in particular, west elevation were exposed to the sun's full strength for most of the day. The constructional forms of the principal surfaces of the building were in response to these conditions and extensive computer modelling was used to evaluate their effectiveness.

4.1 Roof
This was a lightweight sandwich of metal decking, insulation and waterproofing to minimize loads on the long-span roof members. At night the relatively thin insulation allowed some interior cooling. By day, curved south-facing shades on light steel frames shielded the roof from the sun and, with natural air movement across, limited its temperature rise. The shades themselves were mostly light-coloured fabric panels, although approximately $\frac{1}{3}$ of the shade area was provided by banks of photovoltaic cells which supplied power for the east water wall.

4.2 East wall
The main public face of the Pavilion was glazed for its entire length with water pouring continuously down the outside. In engineering terms the water limited the maximum glass temperature to around 24°C so that it remained below skin temperature, thus contributing to visitor comfort. Surface evaporation helped the microclimate in front of the Pavilion, providing lower temperatures for the queuing public.

4.3 West wall
Neither time nor architectural style permitted the creation of a genuine masonry wall on the west side. Instead, water-filled stacked steel tanks provided enormous mass to act as a 'thermal flywheel' which, together with the concrete concourse and lower

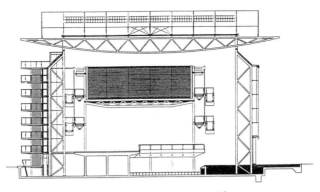

Fig.3. Transverse section

ground floor slabs, helped stabilize internal temperatures. The
wall's internal face remained almost static at c.30°C, while
external temperatures varied daily through a 25°C range.

4.4 End walls
Tensioned fabric formed the north and south skins, softening
incoming light by day and allowing these elevations to glow
spectacularly at night. To improve the thermal performance of the
south wall a layer of external fabric 'flysheets' provided shade to
the wall itself. This additional line of defence was not required
at the north end.

5 Structural design and construction sequence

The key objectives were for an assembly of identifiable parts,
conveying to the observer how they were made and how they
functioned, to be erected simply on site, with the prefabricated
elements easily transportable. The steel superstructure
demonstrated the philosophy of precise off-site fabrication by a
skilled workforce in harness with erection requiring minimum
temporary works. Many connections between steel members were made
with a single pin, thus defining the structural action while
allowing rapid assembly with potential for subsequent dismantling.
This approach contrasted with that adopted on several other
pavilions where extensive site welding was used, sometimes with a
lack of finesse in the end product.
    Structural design was generally carried out to UK Codes of
Practice with lateral forces derived from the Spanish wind load and
seismic codes.

5.1 Internal pods and decks
Following construction of the foundations and substructure the
first superstructure elements to be erected were the north and
south pods and central decks. Each comprised two identical floors,
one 5m above the other. The upper central deck was at the same
level (+12.5m) as the lower pod floors. Each floor included a
concrete slab with composite metal decking supported on the bottom
flanges of steel universal beams. The void between top of slab and
the raised floor was used for services distribution.
    The floor beams were carried by two parallel tubular steel
Warren trusses spanning 20m between a pair of columns, each of

which consisted of two thick-walled tubes battened together. The trusses acted compositely with the concrete slab and reduced in depth towards their supports. By splitting each column the visual bulk of an equivalent single member was avoided and the trusses passed between the two legs, allowing a single pin connection between supported and supporting members. Extending the upper trusses beyond the columns enabled a cantilever support system for the interconnecting walkways and travelators to be devised. Thus the access routes to the upper exhibition spaces related structurally only to the pods and decks which they served and visitors on the travelators had unobstructed views of the water wall.

The pods and central decks derived their east-west lateral stability from pinned connections to the envelope west wall at the tops of the pod columns. Independent bracing to the pods in this direction would have inhibited concourse level circulation, with moment frame action unacceptably increasing the pod column dimensions. As the pods and central decks were offset towards the west wall, linking the structures could be achieved without visual ambiguity.

North-south stability of the internal structures was provided by vertical cross-bracing between the columns to the central decks. The two pods are linked to this restraint system by the interconnecting walkways at +12.5m level. Erection of these structures was relatively straightforward and commenced with the assembly of the central deck frame. With its permanent bracing in place this frame was stable north to south. The north and south pods could be immediately stabilized in the same direction by linking them to the central frame. Temporary guys and restraint from the concourse level slab secured all three structures east to west until they could be linked to the external west wall.

5.2 External envelope

Once the main internal elements were in place, erection of the envelope could proceed. This consisted of 10 identical tubular steel frames at 7.2m spacing, each comprising two vertical wall trusses 21.7m high to support a 32m span roof truss. The latter, which had radiused lower chords, were connected to the wall trusses by single pins at each junction. Each frame acted not as a portal but as a pair of vertical cantilevers linked by the roof member. Transverse horizontal forces on the envelope were thus resisted by bending in the wall trusses and axial force in the roof truss.

Having established the principle of using trussed members early in the design process the relative merits of N-trusses and Warren trusses were compared before selecting the latter. Two main factors influenced this decision. Firstly, after examining drawings and physical models, it was felt that with N-trusses the density of members, when viewed obliquely so that a number of frames were visible, would diminish the clarity of the structural system. Secondly the symmetry of the Warren truss bracings at node points allowed more elegant detailing of these joints and facilitated the development of a bolted mid-span splice for the roof trusses.

In a case such as this where the wall truss bracings were subjected to load reversals depending on wind direction there was no penalty in member size with the Warren truss pattern and indeed the more open arrangement resulted in a 7.5% saving in steel tonnage compared to the N-truss scheme.

The connections between the truss chords and bracings were configured as overlap joints with negative eccentricities equal to the chord radii. Joint symmetry was preserved by introducing central division plates perpendicular to the chords to which the bracings were welded. This overlap geometry was chosen to enable the full load capacities of the bracing members to be utilised without being limited by the joint strengths. The moments induced in the chords by the joint eccentricity were found to have little effect on the chord section sizes.

The central bay of the envelope contained cross-bracing in both walls and across the roof to stabilize the structure longitudinally. This single stability bay, which aligned with the internal central deck system, had two important benefits. Firstly it allowed the envelope to adjust to thermal expansion and contraction without 'locking in' additional stresses. Secondly, once assembled it provided a fixed point from which erection of the rest of the envelope could proceed on two fronts without temporary works. The wall trusses north and south of the central bay were tied to it by slender horizontal tubes. Purlins spanning between the roof trusses tied them to the central bay as well as carrying the roof decking and finishes.

5.3 Analysis of main structures

For the internal pods and decks and the envelope structure analysis was generally linear elastic assuming small deflections. One exception to this was the stability analysis for the lower roof truss chords. As the trusses cantilevered beyond their supports compression in certain elements of the lower chords due to roof shading loads, thermal effects and allowances for suspended exhibits was considered. While the upper chords were restrained by the roof purlins the lower chords were restrained only at their supports and by tension bracing to top chord level at third-points along their internal spans. The analysis carried out took account of the most onerous pattern of varying axial load in the lower chords and the spring stiffnesses of the restraints to determine the buckling mode shape. An equivalent slenderness was then derived from which the member load capacity could be evaluated.

The north and south end walls, being tied horizontally to the adjacent envelope frames (see 5.4), imposed additional compressive forces on the two end roof trusses. To ensure lower chord stability in these cases diagonal tubular plan braces were included between the tips of the trusses and the restrained top chord level of the nearest inner trusses. These braces acted to constrain the lower chords buckling mode shapes and hence limit their slenderness.

Although in overall terms the main Pavilion structure was not 'continuous' a sway classification check in line with Section 5 BS 5950:Part1 was carried out. Having designated the structure as a 'sway frame', the sway amplification factors in both longitudinal and transverse directions were applied to horizontal design loads.

5.4 North and south sail walls

The design concept for the end walls to the Pavilion was to erect fabric membranes tensioned between a series of mast structures, to provide a maritime image of sail walls.

25m high and 32m wide, these walls consisted of PVC-coated

Fig.4. End wall mast

polyester textile membranes in 4m wide bays between nine masts.
These masts were bowed, spanning vertically between ground level
and the Pavilion roof.

The nine masts each consisted of a central tubular spine
stiffened by stays. The stays were held off the spine by tapered
tubular spreaders cantilevering out each side of the spine normal
to the general plane of the fabric membrane. The stays were
continuous steel rods tensioned between the ends of the spreaders
and terminating at the top and bottom of the mast by connecting
back.to the central spine. The stay rods were prestressed against
the axial stiffness of the tubular spine.

The spine, stays and spreaders formed a cable truss spanning
vertically over 25m and resisting the principal wind loading by
changing the prestress forces in the stays - one increasing and the
other decreasing according to the wind direction.

The cable truss masts did not have cross-bracing to provide
shear stiffness. Instead, the rigid cantilever connection of the
spreaders to the tubular spine resisted the shear forces. The
spine was stabilised against buckling in the plane of the mast by
the stays. Under uniform loadings the lateral forces of the
adjoining fabric panels balanced out. For asymmetric loadings the
lateral stability of the spine, and hence the stability of the mast
out of its plane was provided by horizontal tie rods running from
the ends of the spreaders across all nine masts and being anchored
back to the end frame of the main envelope structure. The
spreaders made use of their rigid connections to the central spine
to act as beams, stabilising the spine.

Due to the use of stay rods on both sides of the central tubular
spine, different loading conditions resulted in some of the stays
tending to slacken. Therefore the structural stiffness matrix
varied, requiring a non-linear analysis using the Oasys FABLON
dynamic relaxation program. Wind, prestress, self-weight and

temperature loadings were taken into account. The lightweight nature of the end walls meant that seismic loadings were not significant.

At the top of each mast the connection to the main envelope roof was articulated to permit relative vertical movement, preventing the transfer of vertical loads.

5.5 Seismic design
In addition to gravity and wind loads the Expo'92 rules required the Pavilion structure to withstand an earthquake of Intensity VII as defined in the Spanish seismic code.

The building form meant that in the north-south direction relatively straightforward equivalent static load analyses for the internal and external structures could be carried out. Both empirical rules and modal analyses of the lateral resisting frames were used to estimate the periods of vibration. In this direction the maximum horizontal seismic force was estimated to be 7.5% of the vertical load.

In the transverse east-west direction, however, the linking of the pods to the envelope made this approach inappropriate. Instead a dynamic analysis of a typical frame was therefore carried out by setting up a typical plane frame structural model with appropriate lumped masses using the Oasys GSA program. This data was then transferred into PAFEC for execution. Estimated seismic forces of 2.5% of vertical load in this direction proved less critical than wind loading.

5.6 Transportability and element size
Early on the design team and management contractor Trafalgar House Construction Management Ltd considered and discussed with leading fabricators and British Steel the balance between small member sizes for ease of transportation and larger elements which would maximise fabrication efficiency and reduce the number of site connections. Breaking the structure down into single elements which could be containerised, possibly using castings for repetitive connection details, was compared to the feasibility of transporting large fabricated assemblies from the UK to Spain.

This exercise showed that in this case it was preferable to carry out as much work as possible in the fabrication shop and that large members could be efficiently shipped hundreds of miles to site. Various possible routes were considered which involved shipping from the UK to a port in southern Spain. Due to the amount of handling involved in loading and unloading and the possibility of delays at the port or site, Tubeworkers Ltd the fabricators, decided to deliver all the loads using UK based transport and ferry services. In the early stages of the construction programme, a fleet of lorries used the ferry from Plymouth to Santander in northern Spain but as work progressed into holiday periods when the Santander ferry was fully booked the final loads went by ferry into northern France and then by road through France and Spain. The most awkward loads involving the masts for the north and south walls went via this latter route.

6 Fabrication and Connection Details

The basic design philosophy dictated the overall shape of each component and the detailed joint design was developed between the architect, engineer and the fabricator to suit the axial loads and bending moments computed from the global analysis and the fabricator's workshop techniques.

As noted earlier the components were generally fabricated as complete units, i.e. the pod columns and trusses and the main columns for the external envelope. The main roof trusses were 48 metres long and made in two pieces, joined at site with a fork and pinned splice in the tension chord and a bolted splice in the top chord. The majority of the remaining main connections were made with pins varying from 200 to 60mm diameter. To ensure a good "fit-up" between individual hollow sections at the jigging stage of each lattice component, the ends of the bracings were prepared on an automatic shaping machine to give the correct profile and welding preparation. This ensured accurate fabrication, economic use of welding consumables and the minimum distortion due to welding.

A quality control system ensured that welding procedures were prepared and proven prior to the work commencing and appropriate workshop inspection and non-destructive testing carried out to prove that the quality of workmanship and accuracy of fabrication was maintained. Non-destructive testing was carried out using ultrasonic testing of butt welds and magnetic particle inspection of fillet welds. Through thickness tested plate was incorporated at joints where tensile forces were transferred through the thickness of the plate. M.I.G. and Synergic M.I.G. welding techniques were selected for all workshop fabrication as being the most adaptable and economical for this type of construction. Welding wire compatible with Grade 50 steel was used throughout the fabrication irrespective of the grade of the parent metal (i.e. Grade 43 or 50).

The masts for the north and south walls required particular attention. The main spine was formed from 323.9 x 6.3 CHS were welded together prior to bending. The tapered arms were fabricated from bent plate, the radius ends to each plate being constant and the sloping sides reducing in depth to create the tapered unit. At the junction between the tapered arms and the spine, the CHS was stiffened by "rod inserts".

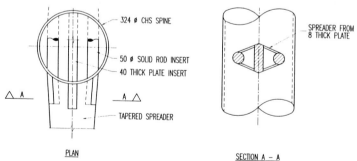

Fig.5. End wall spreader/spine connection

When each mast was fabricated and painted, the tension rods connecting the ends of the arms on the inside and outside of the curve were tensioned to a pre-determined value.

7 Construction on site

Erection of the primary elements was relatively straightforward, components being lifted into position by lorry-mounted mobile cranes. Access for the erectors was from mobile hydraulic hoists. These machines, steerable from the cradle, proved to be invaluable tools for the erection of the structure and subsequently for fixing plant and finishes.

The main pins were too large to handle easily and a special support and jacking frame was used to lift them into position and to push the pins into their final locations. For main span roof trusses each half was lifted and pinned onto the vertical wall truss and then the centre splice completed in the air.

Inevitably some damage occurred to the paint finish and at the final stage a specialist firm, using abseiling techniques, carried out the necessary "making good" of the finishes.

8 The Pavilion during and after Expo '92

Throughout the duration of Expo the Pavilion attracted excellent crowds and high praise from many quarters. In June 1992 it received an award from the Energy Pavilion at Expo for its imaginative use of solar power and water to control internal conditions. It was awarded a High Commendation in the Building Category of the British Construction Industry Awards for 1992 and a Special Commendation as an outstanding ambassador for British design and construction. Ove Arup & Partners were joint winners of the British Consultants of the Year Award from the British Consultants Bureau for their work on the Pavilion. The Pavilion won the Architectural Review award for European lighting design and the Institution of Structural Engineers awarded Ove Arup & Partners a Structural Award for work on this Pavilion and the Pavilion of the Future, also at Expo.

Following the closing of Expo'92 the Pavilion was acquired by a UK based communications company. The new owners intend to exploit fully the building's capacity for reuse by dismantling it for relocation and a new role in north London.

9 References

Gardner, I., Hadden, D., Hall, M. and Harris, T (1993) Engineering aspects of the UK Pavilion Expo'92 Seville. The Structural Engineer, 71, 37-46.

2 MELBOURNE CRICKET GROUND NEW SOUTHERN STAND ROOF – CASE HISTORY

M.J. BARKER
Mott MacDonald, London, UK
Connell Wagner, Australia

Abstract

This paper gives a case study of the redevelopment of the Melbourne Cricket Ground South Stand, with emphasis on the lightweight tubular steel roof structure. It gives some general background to the project as a whole, it's conception and the successful way the severe constraints imposed on the design team were tackled by adopting well thought out concepts and implementing them efficiently. The fast track programme coupled with the technical demands of the project required the constructor to be involved with the design team at an early stage. The paper discusses the tendering process, the design team novation and the subsequent relationship of the parties forming the construction team, which bridged the traditional divide between the designers and contractor.

Keywords: Melbourne Cricket Ground (MCG), Tube steel roof, Tolerances, Erection, Analysis, Design, Circular Hollow Section.

1 Introduction

The Melbourne Cricket Ground (MCG) holds a special place in the heart of most Australian sports fans. For the last century, the stadium has hosted some of the most memorable sporting events in the country, including the 1956 Olympic games. A continued programme of upgrading has been the hallmark of the MCG and with this in mind attention was turned to the old Southern Stand. This was built in 1936-37 at a cost (then) of £100,000. It's condition was rather poor, and as such the decision was made to completely reconstruct the stand. Plans drawn up were for a stand holding 48,000 spectators which would contain all the modern conveniences expected in a facility of this kind. The reconstruction programme was set to allow opening of the stand for the World Series Cricket Final in March 1992. For that event there would be nearly 100,000 spectators in the stadium.

The short 15 month programme required extensive use of precast and prefabricated elements. The design team needed to address the problem of the programme when evaluating design solutions for the building, and as such the programme in many cases drove the final choice of form adopted.

The structure itself comprises four tier seating plus basement and is split into 27 bays typically 12m wide.

Level 1 seats 16,000 and provides standing room for 5,500.
Level 2 seats 5,300.
Level 3 seats 3,200 and contains the corporate boxes, which have a capacity of 1,500.
Level 4 seats 16,500.

Tubular Structures V. Edited by M.G. Coutie and G. Davies.
© 1993 E & FN Spon, 2–6 Boundary Row, London SE1 8HN. ISBN 0 419 18770 7.

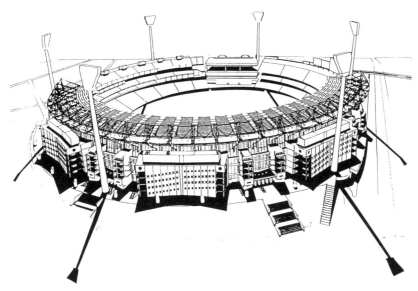

Fig.1. Melbourne Cricket Ground South Stand

The stand itself comprises the major portion of the project, however, contained with the stand reconstruction was the provision of three ramp access buildings to the rear of the stand for spectator entry and egress.

The structure is based on a general 12m grid and consists of a post tensioned concrete frame from which the steel box contilevers spring. The beam elements were formed using hollow precast shells with insitu hearts. The cantilevers are steel box beams cantilevering some 11m, tied back to the main frame by post tensioned cables laid through the box section and the hollow concrete shell beams of the main frame. The combination of the steel box and hollow shell beams proved a major success during construction and for the provision of crack free and waterproof concrete.

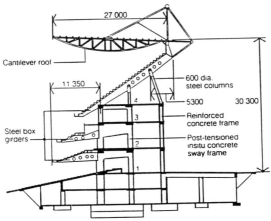

Fig.2. Typical Frame

The roof structure consists of a steel bowstring truss cantilevering 27m, which sits on a raking steel box beam. These beams form the main support for the level 4 seating plats. The light roof was designed with the aid of aeroelastic models and a comprehensive wind tunnel analysis.

The building is more comprehensive than just a grandstand. It contains, within the facility commercial offices, restaurants, function rooms, corporate boxes and security facilities. The facility can therefore cope with the range of use, 7 days a week, from single room occupancy to full capacity (48,000) spectators watching an event.

2 The Team

The Architects, Tomkins Shaw and Evans in association with Daryl Jackson were commissioned by the MCG committee, this together with a single Engineering commission to Connell Wagner Rankine & Hill. All the engineering design took place between February and September 1989, starting again in February 1990 through to early 1991. Construction, by John Holland Construction commenced in November 1990 through to March 1992. A planned phased handover of a third of the stand was completed on schedule in November 1991.

3 Design

3.1 Concept
From a very early stage in the design process, it was decided that the only form of roof applicable to this quality of stand was a free spanning cantilever. Sight restricting columns were therefore eliminated. The Architect also wished the roof to be of as light a construction as possible, so the choice of structural material was relatively straightforward. A steel truss solution was favoured and this was duly progressed by the design team. Various configurations were considered, the final arrangement emerging as a result of detailed coordinated work between the Engineer and Architect, who both held strong views on the desired form of the final roof.

The shape, detailing and overall configuration of the main truss is designed to emphasise the strong and dramatic structural statement, which a grandstand roof provides. The sweep of the stand as it wraps itself around the oval that defines the boundary of a Australian Rules football pitch, adds majesty the structure. The complex geometry also added some headaches!. The roof is constructed from CHS tubular steel elements, ranging in size from 273mm dia. for the raking ties to 168 and 114 mm dia. in the trusses, down to rod bracing of 40 and 50 mm dia.

Grade 50C tubes were chosen for the exposed structure as they provide clean, aesthetically pleasing lines, each element can be sized very efficiently, keeping the overall scale of members down to a minimum. The simple uncluttered jointing patterns and relative ease of protection application for the exposed members is also a distinct advantage over open rolled sections.

3.2 Analysis
The structural analysis of the preferred concept began with an investigation into wind effects. As the roof was to be a light structure, deadweight was at a minimum. This coupled with the natural wind "gathering" action of such a roof ensured that wind pressures would play a major part in the design of this sensitive structure.

MEL Consultants of Monash University were commissioned to undertake wind tunnel tests on models of the roof for the purpose of providing design data, which

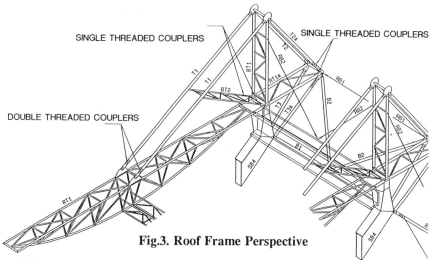

SINGLE THREADED COUPLERS

SINGLE THREADED COUPLERS

DOUBLE THREADED COUPLERS

Fig.3. Roof Frame Perspective

could be used in the final design of the roof itself. A 1:200 aeroelastic scale model of the stand and stadium was constructed and used as the base for the tests in the wind tunnel. This model was based on natural frequencies, determined by pre analysis, in the range 1.3 to 1.7 Hz. As expected the leading edge of the roof exhibited the most sensitivity. Various adaptations to the edge profile were tried in an attempt to reduce the wind effect. The method adopted was the incorporation of roof slots in radial pairs, 350mm deep at 12m centres. This simple inclusion reduced the resulting wind uplift pressures by 20%. The testing produced design wind pressure values of a triangular distribution radially over the roof, with inner edge (oval) design pressures of $2.5kN/m^2$ uplift and $0.7kN/m^2$ down pressure, reducing to zero at the outer edges.

Analysis of the structure for the various Dead, Live and Wind combinations was carried out using STRAND5 and SPACEGASS programs. Design of the members was carried out according to the Australian Standard Structural Steelwork Code AS 1250 (1981). Joint design followed the CIDECT recommendations.

3.3 Concept development

The concept of the roof had provided a very elegant lightweight structure. Careful thought was now applied to the practical aspects of the actual construction of the roof. Erection sequence, tolerances, gradual loadings of the roof during construction and temporary stability were all areas of potential problems to the constructor. The inability to solve these problems at the design stage would have rendered the roof concept worthless in practical terms. With the tight programme and the heavy penalties for delay, all aspects needed to be thoroughly investigated well before arrival on site of the trusses.

The provision of elegant pin connections at the end of such members was desirable. However the location of certain of the joints (in some cases 30m above the ground) with a potential lack of fit, made the rapid connection of the elements of paramount importance. A series of carefully placed threaded couplers were introduced in the ties and struts to allow for the adjustment of the length of these elements during construction. The method of the simple end pin and coupler provided both fabrication economies and construction efficiencies.

The couplers were designed to facilitate a 2mm adjustment for a 180° rotation.

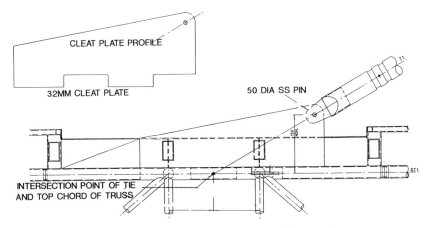

Fig.4. Truss Centre Support Joint Detail

The thread specified was of 4mm pitch standard metric thread and this provided tolerance corrections down to 1mm. The thread was detailed to allow a ± 75mm correction to the member length. The thread was sealed with a rubber ring and sealant.

The roof sheeting was underslung from the main roof structure. A fall of 1:19 from front to rear across the roof was planned, with a concrete ring gutter collecting the rainfall off the roof. The actual sheeting is underslung from a 250x150 RHS purlin section spaced at 1600mm centres. It was decided that the underside of the roof should present a flat, relatively uncluttered soffit to the spectators below. The underslinging was at first perceived to create problems for the constructor, with extensive scaffolding required to enable the sheeting to be fixed. However with a little lateral thinking, the constructor utilised a jig for the fixing of the sheeting into prefabricated sections. This operation was carried out at ground level then lifted into place by crane. The elements were generally 9 x 10.5m in size. The sheeting was screwed to the purlins in the normal way, except the process was carried out with the section inverted. The section was then turned over, lifted and the purlin ends bolted to the trusses.

The prefabricated sections were not designed to fit together. The problem of shape and tolerance of the roof members erected, would have created an impossible situation to try and match up the edges of the sections exactly. It was therefore decided that an infill section be utilised over the truss, which overlapped the fixed prefabricated sections. The infill section was fixed from the top, avoiding the need for access under the roof at high level. The whole of the roof was therefore constructed without the need for scaffolding.

The support to the main roof structure is effected by steel box girders, which primarily support the seating plats for level 4. The choice for these elements was not automatic. Early in the design, the preference for prestressed concrete beams was evident. The relatively free maintenance, consistency of material up the structure, and excellent support characteristics for the main roof structure drove the design to concrete. However the final choice was that of a box steel structure. Reasons for the adoption were:

The constructor required the beams to fit within an acceptable cranage weight window.

18

Steelwork in the open air was not considered to be a fire problem, however with the areas expected to need protection a simple sprinkler system was incorporated.
Speed of erection
Overall economy in favour of steel elements was apparent at that time.

Successful representation was made to the checking authority for the unprotected steelwork by the design team, with the help of a report from the BHP Melbourne Research Laboratory into the problem.
The utilised shape for the steel girders was a fabricated box section instead of the more usual "I" section. Reasons for this choice were:

The main roof support beams needed to be torsionally stiff. A box section provides much greater torsional rigidity than an equivalent open rolled section. Double web sections were more suitable for supporting the prestressed seating plat sections.
Painting surfaces were minimised.
The Architect much preferred the shape.

The boxes were detailed with 32 to 50 mm flange plates, with 16mm web plates. Internal stiffeners were used where necessary, and the web welds carefully detailed to minimise distortion during fabrication. The resulting elements were considered a great success both from the fabrication and construction viewpoints.
The boxes were not sealed, but painted internally, and fitted with internal condensation drainage holes. The main objections to sealing the sections were:

Temperature induced stresses in webs and diaphragms.
Condensation could not be ruled out due to the possible presence of pin holes in the weld runs. Paint protection to the internal surfaces would therefore still be necessary.
It would be impossible to determine the condition of the internal paint coat if the section were sealed.

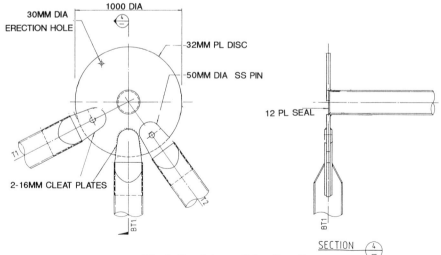

Fig.5. Roof Apex Joint Detail

3.4 Paint System

A high quality paint system was adopted for all the steelwork in the project. Hot dip galvanising was rejected as a protection due to the limitation of the dip bath and the inability to predict the potential distortion in the box girder geometry.

The following paint system was adopted, after consultation with various paint manufacturers:

1	Abrasive blast clean Class 2½
2	Inorganic Zinc Silicate 50 to 75 µm
3	Polyamide Cured Epoxy 100 to 125 µm
4	Catalised Acrylic 40 to 50 µm

The internal surfaces of the box sections were treated as 1 & 2, with polyamide cured epoxy zinc touch up to welds, except closing ones which were not treated.

3.5 Roof Truss Supports

The roof truss springing point, was the top of the box girder at level 4. The actual truss connection was facilitated by fixing the two top chords of the truss to the box section. With the requirement for prefabrication of the roof sheeting panels and purlins, accurate positioning of the main trusses was vital. A geometric analysis was carried out to determine the tolerance values of the springing point of the truss. The analysis took account of all the permissable maximum tolerance values for the various connecting elements ie, length, camber, sweep, twist etc. with a calculation of their cumulative effect.

In close collaboration with the constructor, a construction sequence was evolved. (Ref. section 4). Built into the structure were devices by which these various tolerances could be accommodated. These included:

Slotted purlin cleat holes, with friction bolts, to accommodate some of the potential movement between adjacent trusses.

Double threaded screw splices on the ties and struts making up the main support structure to the trusses. This enabled the truss tip to be accurately aligned.

Base connection to the inclined support strut of the main box girder allowed movement of the HD bolts. This allowed the springing point of the truss to be determined insitu prior to the bolts being grouted into final position.

A system of packers up to a thickness of 15mm was designed into the joint between the springing point of the box girder and the top chords of the truss. This was designed to cater for any twist of the supporting box section springing point. This packer system allowed for a possible ± 270mm horizontal movement to the tip of the truss.

4 Construction sequence

The sequence of erection of the roof was closely worked out with the constructor. With the programme constraints, (already mentioned), and the general lack of space for storage on the site, the programming for the erection sequence began with the planning of the delivery schedules for the roof elements.

Cranage was then considered, with the roof lifting requirements being programmed into the crane work schedules. The main erection sequence can be broken down into four main stages, (see over for diagram).

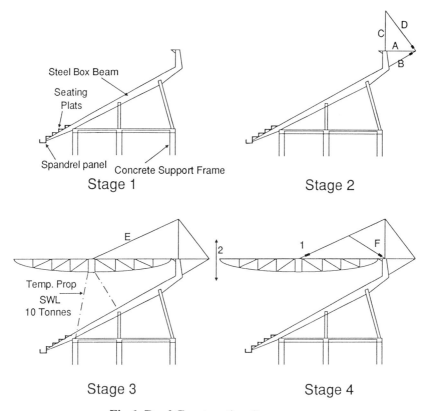

Fig.6. Roof Construction Sequence

Stage 1. The main steel box girders are erected on the concrete frame. The seating plats are fixed to the box beam, and the main precast concrete spandrel panel at the tip of the beam attached.

Stage 2. The backspan members A and B are erected. The upstand members C and D are then connected. The alignment of the assembly is then adjusted by the couplers contained in the members B and D.

Stage 3. The main truss is erected on a temporary propping system which is able to be adjusted vertically. Temporary bracing is also required to the truss to restrain it laterally, prior to the horizontal ties and the decking being connected. The main truss connection to the box beam is then made. The bolts are not tightened fully at this stage. Member E is then attached.

Stage 4. The truss is checked for position, and the necessary adjustments made with the couplers. Vertically, coupler 1 provides adjustment. 5mm movement of the coupler provides 20mm vertical movement of the tip of the truss (at 2). The lateral adjustments are also carried out using the couplers contained in the bracing members, and the bolts for the truss main connection to the box beam are finally tightened. Member F is the attached and adjusted to length using the coupler provided. The deflection expected at the truss tip under full dead load is 120mm (L/225). The walkway framing and the roof sheeting is then erected. The main bracing ties are then tensioned by their couplers.

5 Contract

The MCG committee, after discussion with the design team, adopted a Novation type of arrangement for the design team to the constructor. For this project the sequence of events were:

> The MCG commissioned a full design team, including a programming consultant, to prepare a preliminary design over a 9 month period. At the end of this period, a design/construct tender was issued to a select tender list of 6 of Australia's major contractors. The preliminary documentation contained a Brief which outlined sections of the design which could be further developed by the contractor, and those sections which were considered to be frozen.
> The tender document called for a Fixed Price and Time contract. The documents included design programmes and fees from each of the consultants for the completion of the design.
> The constructor then assumed responsibility for the design team and the preparation of the final fast track documentation to suit the individual requirements of the way the contract was to be tackled.

The MCG adopted this form of contract as it was perceived that all the advantages of a design and build contract were available without the disadvantages. It was certainly borne out that the early firming up of packages proved very desirable, and the ultimate variation of final contract sum from that quoted was in the region of 2%.

A vital aspect of this form of contract is the relationship between the design team and the constructor who "inherits" the team. On the one hand, the design team must embrace the comments made by the constructor, who is now in fact the designer's Client. The constructor must also respect the design function, have knowledge of the process, and the cost and time required for the design to be adequately completed. The team relationship for this contract was excellent and as such contributed to the overall success of the project.

6 Conclusion

The MCG Southern Stand design and construction has been a significant achievement for Australian construction. The whole process of the design, the tendering procedures, the novation of the design team to the contractor, the extensive use of prefabricated elements, the designing out of construction tolerances by intensive activity by all the team during final design, the actual construction itself; all these areas are remarkable in that the final project was completed to time and budget (A$ 140 million) in the programmed period of 15 months.

7 References

Melbourne, W.H., Davis, K. and Taylor, T.J. (1989) Aeroelastic Wind Tunnel Model Tests on the MCG Southern Grandstand Roof, Melbourne. **MEL Consultants Report 33/89.**
Peyton, J.J. and Langley, T.J. (1992) Great Southern Stand, Melbourne Cricket Ground. **Connell Wagner, Rankine & Hill.**
Standards Association of Australia (1981) The Use Of Steel In Structures. AS 1250-1981. **The Standards Association of Australia, ISBN 0 7262 2142 2.**

3 KANSAI INTERNATIONAL AIRPORT

H.A. PASSADES and P.G. DILLEY
Ove Arup & Partners, London, UK
D.E. MANGNALL
Watson Steel Ltd, Bolton, UK

Abstract
The paper describes the design, fabrication and erection of the Kansai International
Airport roof. It explains the reasons behind the form of the building and describes the
structural solutions. It also describes the various design considerations and shows how
the structure was designed for buildability. Finally, it explains how the fabrication and
delivery contract was procured and the QA/QC measures implemented.
Keywords: Airports, Arched Trusses, Curved Steel Tubes, Japan, Osaka, Shells, Tie
Rods, Toroidal Geometry.

PART 1: DESIGN

1   Introduction

As early as 1965, the Japanese authorities recognised the need for a new international
airport to serve the western regions of Japan. It was decided that the new airport would
be built offshore to avoid the nuisance of noise to residential areas and enable
unopposed 24-hour operation. The south eastern part of Osaka Bay was selected as a
suitable site and a survey of the area began in 1976.

The plans for the airport involved the construction of a 1.25km x 4.0km artificial
island, some 5km from the shore. A double level road and rail bridge would connect the
island with the mainland. Both of these huge construction programs have now been
completed. On the island are the single runway and the Passenger Terminal Building
(PTB). This is surrounded by the international and domestic cargo terminals, maintenance and night stay facilities, a fuel dump, administrative offices and the control tower.

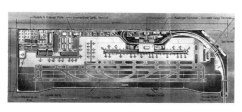

Fig.1. Plan of the island

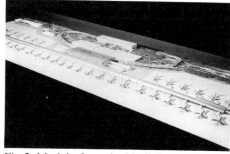

Fig.2. Model of terminal building

Tubular Structures V. Edited by M.G. Coutie and G. Davies.
© 1993 E & FN Spon, 2–6 Boundary Row, London SE1 8HN. ISBN 0 419 18770 7.

The Passenger Terminal Building was subject to an International Architectural design competition in 1988. This was won by Renzo Piano, a world renowned architect and RIBA gold medal winner, with support by engineers Ove Arup & Partners.

The winning concept was striking. The terminal roof consists of smooth flowing curves and resembles the wings of a huge aircraft, with a 'wing' span of 1.6km. Renzo was all out to recapture the sense of adventure generated by travel on the great railways of the last century.

Renzo Piano became the leader of a Joint Venture design team with Ove Arup as his consulting Structural and Services engineers. Other members of the joint venture were Japanese architects and engineers Nikken Sekkei, airport functional planners Aeroports de Paris, and local advisers on customs, immigration and quarantine facilities, Japan Airport Consultants Inc.

2   Building Form

The whole concept of the Terminal Building is based around the movement of people through it. The entire building, including the structure, light and air movement complement the logic of passenger movement.

Passengers enter the Main Terminal Building (MTB) through the canyon. This is a full height space along the whole MTB, 25m wide and 30m high. It is heavily landscaped and provides an interface between nature (outside of the building) and technology (inside). It is in this space that all vertical and longitudinal movement of passengers occurs.

Once a passenger is on the right level, his process through the MTB is linear, from landside to airside. In fact he can see the airplanes as soon as he leaves the canyon and this helps with orientation.

Once he has gone through the check-in, security checks and passport control and he has reached the airside, his movement is along one of the wings until he reaches his boarding gate. For this he can use the AGT passenger transportation system.

2.1 Geometry

The terminal building roof has a very elegant form which curves both in plan and in elevation. This is not just for aesthetic gain: its height is also controlled by the required sight lines from the control tower.

Curved geometry is normally associated with disproportionately high costs, difficulties in manufacture and difficulties in erection. We addressed these potential problems by careful control of the geometry. This consists of a mixture of toroidal and cylindrical (extrusion) geometries.

The MTB is defined by a cylindrical geometry, (i.e. a toroidal geometry with an infinite radius). This provides repetition of all structural and cladding elements.

The wing structure which makes up most of the length of the terminal building is defined by a toroidal geometry. A 2-D shape made from tangential arcs is rotated around the circumference of an inclined circle with a radius of some 16.8km. The geometry allows repetition in the vast majority of the cladding elements. Special shapes are confined to the interfaces of the toroidal geometry with the rectilinear geometry of the building frame.

Similarly, there is significant repetition in the structural elements. As a result, the fabrication and erection of the structure is simplified and savings are made in the costs of both cladding and structural elements.

 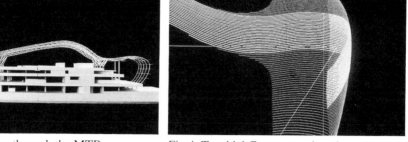

Fig.3. Section through the MTB   Fig.4. Toroidal Geometry of roof

3 Structure

3.1 Building Frame

The building frame of the PTB, designed by Nikken-Sekkei, is a conventional Japanese frame construction. It consists of box-section columns rigidly connected to I-beams with concrete floors on steel decking. This frame provides support to the roof structure.

3.2 Main Terminal Building (MTB) Roof

The sweeping shape of the MTB roof is formed by a series of 20 triangular-section, arched, triangulated tubular steel frames supported on 'cigar' splayed legs. The arch spans over 82.5m and then continues on the landside over a single vertical prop to form the canopy over the passenger drop-off area. On the airside, the truss eventually merges into the MTB wing structure.

Secondary I-beams run over and across the trusses and carry the cladding elements. In addition, they perform a very important role in absorbing energy during earthquake accelerations in a direction perpendicular to the axis of the trusses. This is achieved by the trusses rotating about their connections to the splayed bipod legs, thus causing bending and eventual plastic hinge formation in the secondaries. Under earthquake accelerations in the direction of the trusses' axis, the trusses and the bipod legs behave like big portal frames. Bracing elements in the plane of the secondaries ensure that the loads are distributed amongst all the trusses.

3.3 Wing roof

The wing roof structure is less conventional. Its shape has been selected to enable a large proportion of the vertical and lateral forces to be carried in its surface by shell action.

The structure consists of tubular curved ribs defining the shape of the roof approximately every 7.5m. The distances are approximate because the ribs follow the toroidal geometry of the roof. Each rib lies in a radial plane of the toroid at constant angular intervals. As a result they are all inclined from the vertical by varying amounts, but they all have the same basic 2-D shape with varying lengths truncated off their ends. We considered this to be crucial in order to ensure that the fabrication and erection could be carried out within reasonable cost and accuracy. This benefit was visibly demonstrated by Watson who used a single template replicated in the UK and Japan to assemble all of the ribs.

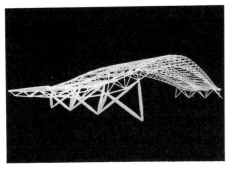 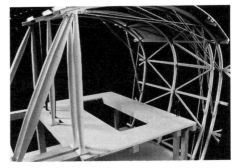

Fig.5. Computer model of MTB roof trusses   Fig.6. Model of wing roof structure

The ribs have to be stiff enough to enable the shell surface to perform properly. If this were to be achieved by pure bending action, the resulting member depths would have had to suit a span of 20m. We avoided this by stiffening alternative ribs with steel tie bars, arranged singly and in pairs to avoid clashes at crossover points. A significant proportion of the final bending moments in these ribs is due to compatibility. They can actually be reduced by reducing the bending stiffness of the elements. As a result we were able to keep the rib diameter to a maximum of 355mm in the wing. Ribs stiffened with tension bars were known in the design as 'diaphragm ribs' as they perform a function similar to the diaphragms in conventional shells.

In order to simplify assembly, we avoided prestressing the tie bars. This means that the ribs can be assembled on their sides on the ground, the bars just tightened and then the whole assembly put in position. The self weight of the structure in its final position effectively stresses the bars.

The surface of the shell is formed by Rectangular Hollow Section secondaries and bracing. The secondaries also support the cladding panels. Because the ribs lie on the toroidal geometry, the secondary and bracing members also repeat along the length of the wings, resulting in the same fabrication and erection benefits as for the ribs. This repetition extends to the cladding elements. The efficiency of the shell action is such that the size of the bracing elements is governed by the maximum allowable slenderness ratio under earthquake conditions prescribed by the Japanese code.

The airside of the ribs is connected to cantilevers of the superstructure steel frame. By controlling the stiffness of these cantilevers we were able to relieve some of the stresses induced in the roof structure by differential settlements, without compromising overall stiffness under normal conditions. Vertical support to the landside of the ribs is provided by cigar legs. Diaphragm ribs are also restrained in the across-the-wing direction by additional inclined legs.

Designing the wing roof to act like a shell and stiffening it using tie bars has resulted in a structure which is quite slender for the distances it spans. As a result it is remarkably elegant and airy, complementing the architectural concept.

4   Design considerations

The design of the terminal building was in accordance to the Building Standard Law of Japan, which refers to a permissible stress code. Because of the situation of the building on an artificial island, other design considerations were also appropriate.

The island is predicted to settle approximately 8m in the first twenty years of its life. Most of this settlement will occur in the first few months, helped by about one million sand drains that have been installed throughout the site. Even so, differential settlements are certain to occur after the building has been completed and these were one of the critical design loadcases. Once certain limits are reached, an automated jacking system installed at basement level will bring the building back to level.

Wind was another important loading to be considered since Japan suffers regular typhoons. In fact, because of the unusual aerofoil shape of the roof and the large spans involved, a wind tunnel test was considered necessary. This confirmed that the roof did not exhibit any aeroelastic behaviour and also provided us with design pressures. As expected, we had to deal with quite large uplift pressures.

Finally, there were the seismic loads to be considered. Japan lies in an area of notoriously high seismicity. Several large earthquakes have occurred there and this is reflected in their design code requirements. Two levels of earthquake need to be designed for. Level-1 represents an event statistically likely to occur once in the lifetime of the building (50 years) and this must be resisted without any damage to the structure or any substantial damage to the services or cladding. Level-2 represents an extreme event where extreme deflections and cladding damage are permitted but the structure must not collapse. The design earthquake loads are very large, and the terminal roof structure is designed to ultimately support 1.24 times its weight applied sideways!

5   Analysis

The MTB roof structure was analysed using a conventional 3-D linear analysis program.

In the case of the wing roof, structural performance relies on the geometry. As a result, analysis was carried out using FABLON, a non-linear analysis program developed in-house by Arup. This program is based on the technique of dynamic relaxation and it functions iteratively by simulating a process of damped vibrations in small time increments. The Program can also simulate buckling by taking full account of the structure's deflected shape and by modifying the bending stiffness of elements according to their axial compression. This was important in order to guard against 'shell' buckling of the wing roof surface.

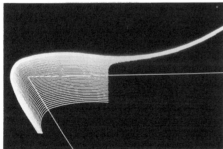
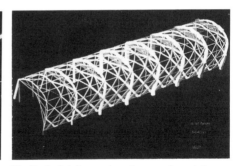

Fig.7. Effect of toroidal geometry on ribs   Fig.8. Axial forces in wing roof model

Aseismic design of the terminal roof was carried out to the equivalent static loads specified in the Building Standard Law of Japan. However, we felt that the unusual form of the structure warranted examination of its dynamic behaviour. Apart from the roof's unusual shape, the wing exhibits anisotropic and non-linear stiffness characteristics. In addition, we wanted to confirm our predictions on the effect of two specific issues. The first is the connection of two different structural forms in the MTB, the trusses and the wing; the second was the interstorey drift between levels 2 and 3 of the superstructure frame on which the airside and landside of the wing structure were supported respectively.

The dynamic analyses performed were comprehensive. Modal analyses on 3-D models of the roof structures were performed. From these, complex lumped parameter models were established to represent the roof dynamic characteristics. The final models were subjected to time-history analyses for specific actual and artificially generated seismic events.

6   Connections

Because of the need for ductility in earthquake conditions, connection design in Japan is based on the principle that the connection should not fail before the attached members have yielded. Clearly, in cases of very slender elements whose design is governed by compression (buckling), this rule becomes too onerous. In such cases it is acceptable to design the connections for the forces that would result from an analysis under the highest earthquake loads that would arise in a non-ductile structure.

Ductile connections in tubular steel require detailed consideration; more so if they are to satisfy the architectural visual requirements. Many of the node connections were becoming quite complicated and it was important not to lose sight of the force paths through them. A typical example is the node connection in the wing roof where secondaries and bracing members meet the rib and tension ties. One of the reasons for locating the plane of the secondary and bracing members outside the ribs was to try and simplify the connection detail.

In this case use was made of plates slotted through the tubes, enabling direct transfer of forces and minimising local buckling problems. Discussions with Watson gave us confidence that this type of connection is not as difficult to achieve as we first feared.

In the MTB trusses, the bracing joints to the main tubes were all overlap K joints. These have been shown experimentally to exhibit much better ductility characteristics than open K joints.

7   Fire Protection

The rules governing fire protection of structures in Japan are fairly simple and rigid. Structural elements which only support a roof must be fire protected up to a height of 4m from the floor.

Unfortunately, intumescent paint does not conform to the relevant Japanese tests and is not considered suitable as a fire protection material. Clearly, rigid application of these rules would affect the building in a very dramatic way. For this reason it was decided that we would adopt a Fire Engineering approach and get approval from a special committee of Japanese experts on the subject.

Our fire engineering calculations showed that under a design fire no structural collapse would occur until sufficient time had elapsed to allow safe escape. Even in the extreme cases where structural damage would occur after fire, this could be localised and would not cause progressive collapse. Consequently, no fire protection is necessary for public safety. In practice, an agreement was reached with the special committee that all support legs of the roof were to be fire protected using traditional materials, but the remainder of the roof structure should be fire protected using intumescent paint as a "countermeasure". This suited well the architectural requirements as the legs were to be clad anyway in order to achieve the cigar like shape. The main roof structure would have the appearance of exposed steelwork.

8   Communication of Information on Geometry

Both the 2-D and the toroidal geometries of the structure were defined very simply on drawings. However, we felt that we should provide additional information to help with the fabrication and the erection of the roof structure. As a result we prepared a geometry report with tables of calculated 3-D coordinates of some selected setting out points of the structure.

The geometry of all the wing ribs was defined in terms of the basic 2-D shape and the truncated length for each gridline location. This ensured that the overall simplicity behind the generation of the rib shapes was obvious and that the information was presented in a way which was better suited to fabrication.

PART 2: CONSTRUCTION

9   The Contract

Watson Steel Ltd were awarded the contract on 10th December 1991 to supply, fabricate and deliver the tubular roof steelwork for the wing portions of the roof; the north section being a contract with Nippon Steel Corporation (NSC) and the south section being a contract with Kawasaki Heavy Industries (KHI).

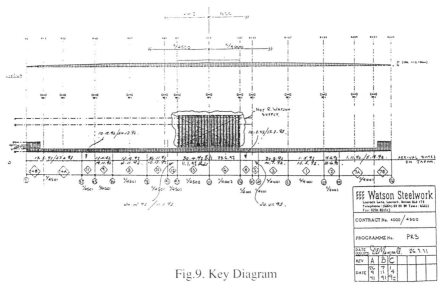

Fig.9. Key Diagram

The contract involves the supply of approximately 4000 tonnes of tubular steel-work forming a curved shell structure approx. 18m wide 20m high and 1600 metres long supported on a conventional steelwork frame with the bay centres being 7.2m.

29

The tubular steelwork for the wing roof is generally 356 ø CHS section for the main curved ribs, 250 RHS for the main rails and bracings, fabricated box sections and 300 x 250 RHS for the top and bottom rails and 30mm to 60mm dia. tie rods with architectural end fittings.

The main ribs have node features at the connection of the rails, bracings and tie rods which are manufactured from plate ranging from 32mm thick to 80mm thick, with the plate slotted through the circular tube.

The scope of the work was sectioned into fourteen stages to suit the programme of construction.

The fabrication started in December 1991 and proceeded continuously through to January 1993.

The arrival date in Japan for the first consignment was April 1992 with further shipments progressively through to March 1993.

The programming of the work in fabrication was determined to achieve the delivery dates in Japan whilst maintaining a continuous rate of production.

10 Japanese Approval

The Japanese had always been led to believe that European companies never meet cost and time constraints and that they have a high turnover of labour resulting in continual loss of experience and continuity.

One of the chief concerns was welding and at the outset Watson had to arrange to test 50 of their welders to Japanese Certification in order to gain approval to fabricate and weld steelwork destined for Japan.

A party of welding engineers from Japan came to the U.K. to witness and approve the weld tests.

In addition, Watson had to catalogue details of all Watson's personnel involved in the management and execution of the work to gain approval to the 1103 Committee in Tokyo, Japan; one of the Japanese legislative bodies.

Watson proved conclusively that they have a balanced workforce, with many personnel having more than 20 years service, with the company, and that high labour turnover is not evident at Watson.

11 Communication

Since the representatives of NSC had previous experience of working outside Japan and communicating in the English Language, they arranged their resources to primarily project manage and co-ordinate the development and progress of the work whereas KHI primarily dealt with the development of the connections, weld details and construction method.

Most communications were carried out by telefax, since this allowed for speed of communication and resolution of matters whilst maintaining a paper record of agreed details and arrangements. Telephone conversations between Watson personnel and the Japanese were kept to a minimum to avoid the difficulties in communication.

When packages had to be despatched to Japan or U.K. a courier system was used to guarantee delivery within 3/4 working days.

Both NSC and KHI despatched representatives from Japan periodically to carry out product inspection and to progress the work.

1 2   Agreement to Details

At the start of the work in September 1991 representatives of Watson visited the offices of NSC in Osaka Japan. Their task was to review the connection details and construction method with NSC and KHI personnel.

Fig.10. Computer Graphics

We found that there are some areas where we have a lot to learn from the Japanese. One good example occured where Watson, NSC and KHI were proposing alternative construction details for approval. In each instance a detailed document was prepared comparing the details shown on the contract drawings with those proposed by Watson/KHI/NSC. In addition, as revised details were agreed with the Engineer/Client copies of the contract drawings were updated as a record of the construction; these were of particular assistance in the drawing office during the preparation of the fabrication drawings.

Once the stage 1 drawings had been prepared a representative from Watson visited Japan with the drawings in order to agree the final construction prior to formal issue for approval to the Engineer. The balance of the drawings for each stage were then issued for approval by courier in line with the stage 1 drawings.

Again the Japanese produced a book of approved drawings for each of the stages as a formal record of the approved set, which was distributed to all parties.

1 3   Tie Rods

The tie rod detail indicated on the contract drawings had been specified except for the connection of the rod into the connector.

In liaison with McCalls Special Products the thread size, length and features were designed in line with the specification requirements and three full size tensile test pieces were arranged and tested at British Steel's Swinden Laboratories in Rotherham to prove the ultimate capacity of the construction, which was witnessed by the Japanese.

A common problem with tie rod connections concerns the treatment to the threaded area where one requires untreated threads with lubricant for erection, and yet treatment to those areas on completion of the work. To this end in liaison with McCalls a cover nut detail was engineered which encapsulated the thread area whilst allowing adjustment to the length of the bars. This resulted in threads being hidden from view and a simple sealant detail to complete the joint, without the need to carry out any site treatment of the threaded areas.

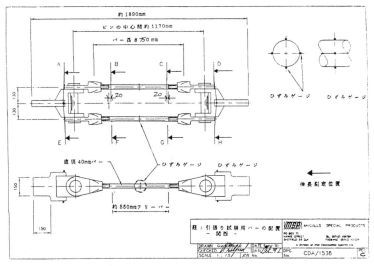

Fig.11. Tie Rod Test Assembly

The tie rods were set accurately to length, and locked off after treatment prior to delivery to Japan.

In order to ensure that each end of the tie rod was entered equally into the end fitting, set marks were provided equidistant from the end of each bar to allow verification after assembly or adjustment to length. All adjustment to length is provided by either turning the end fitting before installation, or turning the rod after installation.

14 Potential Construction Difficulties

The toroidal geometry of the roof results in every rib frame being similar, with varying lengths, due to the curvature in both plan and elevation and with each frame having progressively varied verticality and plan location, whilst allowing the repetition of node details.

The shell form of the roof dictates very high accuracy requirements since this roof is not tolerant to deviation as is the case with rectangular frameworks. In addition there is the impracticality of trial assembly of the whole of the works to verify the constructed geometry.

The varied curvature of the roof, from very large radii down to small radii, and the normal bending techniques which distort the circularity of the tubes, leads to "tailor making" joints to suit the actual distorted shape which cannot be predicted accurately.

The penetration plates and bracing stubs involved have highly complex shapes which are prone to error with normal draughting/fabrication techniques.

The checking of the accuracy of components which are 3 dimensional curved objects, all different, required considerable consideration.

In addition the distortion due to welding, since mostly full penetration butt welds were used, could also critically affect final geometry.

All the site joints are "hard" type connections either welded, pinned, or HSFG bolts in clearance holes, allowing no possibility of any adjustments on site and therefore a very high degree of control was needed on dimensions.

A further production problem was the manufacture of the 10,000 end tees for the rails and bracings which were formed from plate with full penetration butt welds with approximately 70,000 holes in 30mm plate.

15   Solutions Adopted

In view of the bending of over 600 tubes, to resolve the possible problem of cross sectional tube distortions, the use of the induction bending process was adopted which provided no loss of circularity and very accurate control of the bending radii. This enabled the pre-production of node joints to the same basic theoretical shape.

The use of the CAD system to derive the geometry, from first principles, was of paramount importance to enable the Ove Arup & Partners detailed co-ordinates to be checked and any irregularities agreed and corrected, in liaison with the Japanese Consulting Engineers.

The CAD system was also used, where beneficial, to detail the components which were derived initially from the 3D development and transposed into 2D for the drawing.

Some simple members were detailed by hand, but using the geometry co-ordinates produced by the CAD system.

The further benefits of the CAD system allowed the complex shaped plates to be down loaded direct to the cutting equipment, in electronic form, and allowed for screen nesting without any interface problems.

The geometry for the bracing stubs, being also complex, was determined using the CAD system by the preparation of wrap around templates for checking and determination of sawing angles for production.

In the factory various jig systems were developed and engineered to ensure the accuracy of the various stages of assembly.

The jigs primarily:-
(i)   checked the accuracy of the curved pipes and allowed "fine tuning" to high accuracy;
(ii)   allowed the setting out and installation of penetration plates;
(iii)   allowed the assembly of bracing and rail arms;
(iv)   allowed the introduction of bracing and rail arms onto the ribs;
(v)   allowed for the final checking of the complete assembly after welding and any final adjustments;

The CAD system was used to determine various bent plate fabrication aids which assisted the assembly of the work.

During fabrication a system of balanced welding was utilized in order to accommodate weld distortions.

To overcome the possible problems with the manufacture of end tees these were formed as long plate girders which were welded and drilled prior to sawing and cutting to size.

16   Final Checking

In order to satisfy both us and the Japanese, of the fit up, various trial erections were implemented periodically throughout the term involving the assembly of two complete bays of the structure, on nine occasions, at Watson's Lostock Works.

Once erected the whole assembly was surveyed using Laser Distomat equipment in order to determine the shape of that section of the structure and the deflection characteristics of the frames.

In the workshop every single rib frame was checked and verified against predetermined co-ordinates from the CAD system, again using the Laser Distomat equipment.

The tolerance on the position of a bracing arm in its 3D location (i.e. resultant co-ordinates of x, y and z) was only 2mm.

All dimensional checks of components were logged on record sheets against the theoretical values to form documentary evidence of the checks that had been carried out for the QA/QC package.

17   Treatment

The whole of the steelwork was shotblasted to $SA2^1/_2$ followed by the shop application of a coat of zinc rich primer and a coat of micaceous iron oxide paint (M.I.O.)

18   Packing and Shipment

Since containerisation was the only practical and available means of transport Watson decided to ship the components in standard 40 ft. containers, all with open tops and closure sheets, due to the size of the pieces, which assured the protection of the goods during transit.

Following "Construction Led" philosophy, and techniques, two complete bays of the structure were packed into three containers including rib members, rails, bracings, tie rods, pins and temporary fixings in order to simplify the processing of components at site.

The CAD system was used, along with a scale model of the components and the containers, to plan the packing of the steelwork. Since all ribs were different, detailed forward planning was required to determine the site joint locations to suit the packing requirements. In addition temporary packing steel and timbers were required to secure the permanent steelwork during transit.

The work involved the packing of over 300 containers over the period Feb 1992 - May 1992 and August 1992 - February 1993, with up to 16 containers requiring packing per week at the peak of the work.

The decision to segment the work into specific containers assisted the contract quite considerably, since it allowed for some programme sequence changes, made by the Japanese, to be implemented without disruption to the packing schedule.

19   Sitework

The work in Japan involved the site jointing of the ribs into one piece via two site welds. In order to assist in the assembly of the ribs Watson sent three workshop/erection staff to Japan to assemble the site assembly jig and to check the ribs during assembly and after welding, in conjunction with the Japanese.

Each container was thoroughly checked and itemised to verify the quality and content of the goods during unloading in Japan.

Fig.12. Photographs

4 TEES BARRAGE BRIDGE DESIGN

N.A. CARSTAIRS
Ove Arup & Partners, Newcastle-upon-Tyne, UK

Abstract
A tubular steel structure is rarely used to support a concrete
highway bridge deck. This paper describes how the design for such a
bridge has been developed from the original architect's sketch.
Decisions made during the design process are discussed, and insights
into the unusual behaviour of the structure gained from detailed
computer analysis are described.
Keywords: Bridges, Design, Analysis, Computing, Arches, Tubular
Construction, Circular Hollow Sections.

1 Introduction

Teesside Development Corporation (TDC) was set up in 1987 to
regenerate an area of 19 square miles in the North East of England,
much of it derelict or polluted by former industrial uses. Its aims
are to create jobs, attract investment, and improve the quality of
life of the local people. TDC has commissioned a barrage across the

Fig.1. Model of Architects Original Concept

Tubular Structures V. Edited by M.G. Coutie and G. Davies.
© 1993 E & FN Spon, 2–6 Boundary Row, London SE1 8HN. ISBN 0 419 18770 7.

tidal River Tees at Blue House Point, Stockton which will maintain
high water levels upstream. Associated sewer diversions will improve
water quality and so new developments will be in an attractive
riverside setting. To enhance the appearance of this flagship
project, TDC held an architectural competition. This was won by the
Napper Collerton Partnership with an entry whose centrepiece was a
tubular steel road bridge. The bridge crossed the barrage and
approaches in eight arch spans with an overall length of 160m.
Watson Hawksley Ltd (now Montgomery Watson), the consultants for the
barrage, commissioned Ove Arup and Partners to design this bridge and
associated works.

2 Concept

The architectural concept depends on the examples of the early
ironmasters. The structure consists of tubular steel arches and
circular infill members and is kept as transparent as possible to
allow the landscape design to flow past the road line. This
contrasts with the solid concrete arch structure originally
envisaged. This conceptual design had to be developed into a
structure able to carry highway loading. The client specified that
the bridge should carry the highest standard loading specified by the
Department of Transport (45 units of HB loading) so that future
industrial development will not be restricted by the capacity of the
bridge. Unusual tubular structures are more often used to support
roofs where the design load is about a tenth of this highway load.

3 Scheme Design

The main bridge crosses the barrage in four 17.5m spans and there are
approach bridges each with two identical 17.5m spans. The arches
have a rise of 5m which gives clearance to the navigation channel and
service roads.

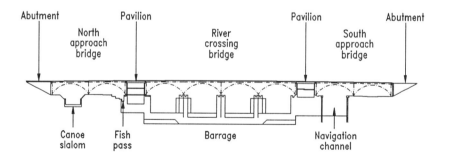

Fig.2. General Arrangement of Bridges

38

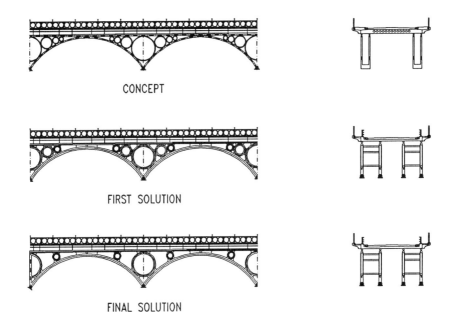

CONCEPT

FIRST SOLUTION

FINAL SOLUTION

Fig.3. Design Development

3.1 Arches and Infills

Initial calculations showed that both the arch ribs and the circular
infill members needed to be significantly larger than those shown by
the architect. Since the infill members could not be bent to the
tight radius required, a cast steel solution was proposed, with nodes
cast separately and joined with tube to tube butt welds. As a result
of these changes, a solution using touching infill members
progressively reducing in size looked much more solid than the
original concept. The architect therefore chose to reduce the number
of infill members, leaving clear space between them.

3.2 Articulation

At the piers the tubular arch members are joined together and pinned
to the supports using purpose made knuckle leaf bearings. The form
of the steelwork, with no vertical element at the piers, meant that
the bridge had to be made continuous with no movement joints. This
also reduced the risk of water containing road salts leaking on to
the bridge structure.

The combination of fixed supports and a continuous deck meant that
a change in the temperature of the bridge would cause significant
stress in the steelwork. In a normal bridge, this causes movement at
sliding bearings or bending of slender columns.

3.3 Arch Spacing

The architect had shown a pair of closely spaced arches at each side of the bridge but this arrangement would mean that all the load was carried by the inner arch. To minimise member sizes the four arches had to share the load. Introducing bearings between the deck and the arches was considered. The number required would have been expensive and would lead to future maintenance difficulties. Instead the pairs of arches were separated to give spans for the concrete deck of 2.75m and 3.6m. This allowed the transverse steelwork supporting the deck to be omitted. Horizontal bracing members were used between the pairs of arches rather than cross bracing. A decorative parapet, echoing the circular motif of the arches, was to be provided supported on shaped precast cantilevers. Since this would not meet the standards for containment of vehicles, a separate standard P2 parapet was provided beside the highway.

3.4 Buildability

At this stage initial discussions with engineers from a steelwork fabricator and foundry (with whom we had worked on previous contracts), were held. The introduction of flat plates at the joints allowed them to suggest a change from the all cast solution. The main arch ribs would be circular hollow sections bent to the required radius, and the horizontal deck support and bracing would be straight CHS sections. The infill sections would each be cast with a solid circular cross section, either with the plates as the integral part of the casting or butt welded on later.

A scheme along these lines was developed, taking the flat plates through the rolled sections to minimise punching shear problems. At this stage Tarmac Construction were appointed as main contractor for the barrage and associated works. They suggested a number of alternative methods for fabricating the circular infill members to reduce the cost of the bridge without changing the basic form. The various cross sections were sized and the contractor costed them. A

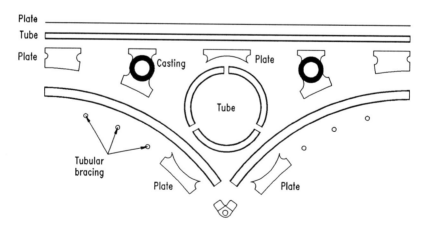

Fig.4. Final Arch Make Up

solution using special grade 100mm thick flat plate for the infills was found to be economical, but was rejected by the client as a significant departure from the original tubular concept. The larger infill members were changed from cast solid bar to segments of curved hollow section to be either cast or bent using induction bending.

4 Detailed Design

Initial analysis of parts of the steelwork showed that the bridge steelwork behaved in unusual ways. In the final design, more than half the thrust in the arch at the support was transmitted through the infill members into the deck at midspan. The proportion of vehicle load carried by a single arch depended on the position of the axles on the span. To quantify these effects, a space frame model of the whole bridge was developed.

4.1 Computer Analysis

A few years ago a model sufficiently complex to represent the bridge could only have been analysed on a main frame computer. It would have taken at least a month to define, enter and check the geometry, and the analysis would have provided a printout of numerical answers several feet high to be processed by hand. This mountain of data would give no insights into the behaviour of the structure.

Today the power of desktop computers and the programs they run make such analysis possible. A complex model with over 1000 elements can be developed in about a week using graphical data entry and copying. Post processors manipulate the results so that the critical member stresses can be printed on about a dozen sheets of paper. Above all the results may be displayed graphically.

For this bridge the final model had over 2000 elements. It was analysed on a Sun Sparc workstation using Oasys GSA analysis software. The curved elements of the bridge were modelled by short straight elements and additional moments to compensate for this inaccuracy were included in the design check. At the nodes the plates were modelled by triangulated trusses to give rigidity.

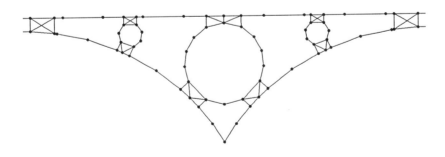

Fig.5. Basis of Computer Model

4.2 Loadcases

Which elements would be critical for the sizing of members could not
be predicted. The effects of highway loading could not therefore be
investigated using influence lines. A series of 31 loadcases was
analysed for an HB vehicle crossing half the bridge in regular
increments. Nine combinations of temperature change and support
settlement were considered giving a total of 279 loadcases. After
completion of the design, the loadcases which gave stresses within
10% of the governing stresses were examined. It was evident that the
use of this number of combinations was justified.

4.3 Design Stresses

When the initial analysis was carried out, live load effects
dominated and member sizes had to be increased. After this the
effects of changes in bridge temperature became more significant.
Further member size increases were counterproductive because an
increase in stiffness in one member increased the stresses in all the
others. It was therefore decided, once the contractor had
established that it could be procured, that grade 55 steel should be
used for the rolled tubular members. This gave an increase in
capacity without a corresponding increase in stiffness.

The proportion of dead, live and temperature loads contributing to
the critical stress in the different elements are shown in Figure 6.
A non-linear buckling analysis was used to determine the effective
length of the arch members; this gave a small reduction in limiting
compressive stress below the yield stress of the steel.

4.4 Connections

Further processing of data was needed to size the fillet welds
connecting the plates to the tubes. This calculation was done using
a spreadsheet, with data transferred automatically from the analysis
program.

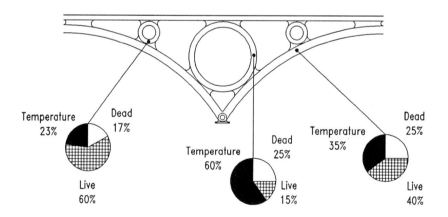

Fig.6. Make Up of Critical Member Stresses

An anomaly was found during this exercise which led to exhaustive checking of these calculations. A loadcase had been found which gave a stress of 6kN/mm in the weld connecting the smaller infill member to the top boom. However the maximum stress in the weld connecting the same member to the arch was only 2kN/mm. It was eventually established that the three components of force causing a moment on the weld were additive at the top, while one component had a relieving effect at the bottom. Detailed checks on problems of this nature are vital for a structure where it is impossible to do simple hand calculations to validate the computer analysis.

4.5 Final Steelwork Design

In addition to the strength checks described above it was necessary to do a fatigue analysis on the steelwork. Some welds were increased in size to give acceptably small changes in stress due to the passage of a standard fatigue vehicle.

The final design contains 350T of tubular steel, and 280T of plates and cast steel.

4.6 Concrete Deck Design

Design of the reinforced concrete deck is conventional. The heavy longitudinal bars, which act compositely with the steelwork, have to be provided over the full length of the deck, not just over the supports. The contractor decided to use conventional rather than permanent formwork to avoid having to thread these bars through projecting steelwork.

4.7 Corrosion Protection

The exposed steelwork is protected by a sprayed metal coating and 250μm of acrylated rubber based paint (DTp paint scheme Type 10AR). It is felt that the best protection is provided by regular inspection and maintenance, rather than by applying a special coating and then ignoring the bridge for a long time. Protection to the insides of the tubular members is provided by the fully welded construction which seals the voids against water ingress.

5 Fabrication and Erection

At the time of writing this paper, construction of the barrage has started, but steelwork fabrication and erection are still at the planning stage. It is anticipated that each basic unit will consist of two half spans of a single arch. This will be fabricated in the workshop, transported to site, and lifted into position on the barrage. Connections at the crown of the arch must be made by site welding and continuity can only be established when the air temperature is between 5 and 10°C. This restriction has been imposed to limit the stresses due to the changes in temperature to which the bridge will be subjected during its life.

6 Conclusion

Designing the Tees Barrage Bridge has been a fascinating challenge for the designers, and its construction will tax the contractor's ingenuity. Without modern computers, the job would have been much more difficult. Whilst not the cheapest solution to the problem of crossing the barrage, it will give the Teesside Development Corporation a distinctive centrepiece for their flagship project.

7 References

Design Manual for Roads and Bridges. (1992) Department of Transport, London.

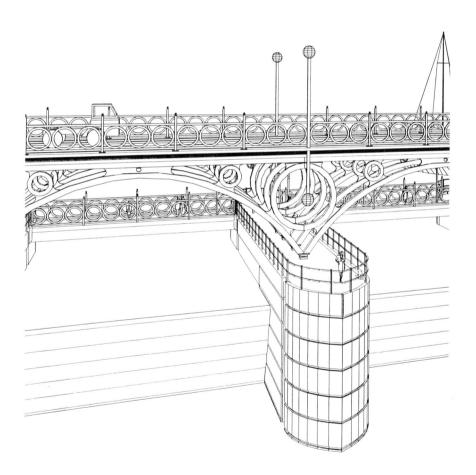

Fig.7. Architects Drawing of Final Solution

5 BRITISH AIRWAYS HEAVY MAINTENANCE HANGAR – CARDIFF WALES AIRPORT: HANGAR DESIGN

S. LUKE
Ove Arup & Partners, Cardiff, UK

Abstract
The new British Airways Hangar at Cardiff Wales Airport, code named Project Dragonfly, is a heavy maintenance facility for Boeing 747 Series aircraft and has set new standards for aircraft maintenance and building design. The project includes a 3-bay 22000m² hangar with a 6000m² mezzanine floor, a 15000m² technical support building, full access docking, overhead cranage, undercarriage lifting platforms, specialist building services, a ground run pen and 35000m² of concrete apron. This paper discusses aspects of the hangar form and design with respect to the large tubular structures, it also outlines the key issues concerning the fabrication, erection and construction of the hangar frame.
Keywords: Maintenance Hangar, Tubular Girders, Large Tubes, Hangar Structure, Tubular Connections.

1 Introduction

Project Dragonfly commenced in April 1990 following a British Airways board decision to construct their new maintenance hangar at Cardiff Wales Airport. It is to be used by British Airways Maintenance Cardiff (B.A.M.C.) to undertake heavy maintenance, routine work, repairs and modifications, their goal being to establish the 'best and most successful Boeing 747 maintenance business in the World'.

The Design Team were appointed in March 1990, work commenced on site in May 1991, and construction was completed in April 1993.

The project management for British Airways was undertaken by their Property Branch, Property Construction Group, who are involved in a vast range of construction work essential to the continued success of the airline's international operations.

Ove Arup & Partners were appointed as the Design Team Leaders for the project responsible for the Design Team Co-ordination, Contract Administration, Civil, Structural, Building Services, Architecture and Landscaping Design. The latter two elements of work were sub-contracted to the Alex Gordon Partnership and Gillespies respectively. Bucknall Austin plc were the Quantity Surveyors and Johnson Jackson Jeff Ltd the Project Planners.

Tubular Structures V. Edited by M.G. Coutie and G. Davies.
© 1993 E & FN Spon, 2–6 Boundary Row, London SE1 8HN. ISBN 0 419 18770 7.

Fig. 1. A view of the hangar prior to completion.

2 Design Influences

The hangar design study was influenced significantly by a number of
considerations relating to the appearance, its height relative to the
surrounding area and position of the 'inner transition slope' of the
'obstacle free zone' dictated by the closeness of the runways. At the
front of the hangar the height restriction is 35m above the hangar slab
and levels out at 42m over the building footprint.

The internal design constraints, outlined below, were imposed by the
Client and had a major impact on the building development :

- o Aircraft docking position to be nose-in.
- o Access tail and fuselage docking to be suspended.
- o Minimum internal volume to maximise environmental control.
- o Column free double bay, 150m wide x 90m deep.
- o Secure single bay, 75m wide x 90m deep.
- o Smooth inner surfaces to minimise dust collection.
- o Well defined flow patterns between operation areas.

The site layout was developed from a planning study and the building
design, see Fig 1, emerged from a detailed appraisal of the constraints. These and other issues relating to the building planning,
services, specialist structures and appearance are discussed in a
further paper to be published by Luke (1993).

3   Hangar Form

To establish the preferred hangar form many alternative building types
were studied from which five main scheme concepts were selected as
feasible alternatives.

The base scheme, a stepped roof, was established from an existing
138m x 83m British Airways hangar at London Heathrow. This base form
was not considered to be the most economic or visually satisfactory
arrangement for the larger Project Dragonfly configuration.

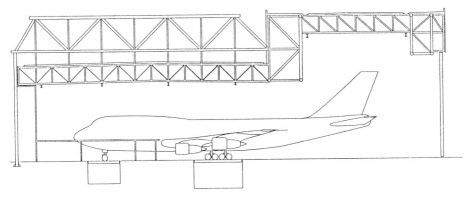

Fig. 2.   Hangar Cross-section.

Flat roof, tension assisted structures and free form schemes were
also discounted because of either cost, increased volume or conflict
with the 'inner transition zone'. Since the docking structure, which
is fundamental to the maintenance operation, is a tight fit to the
aircraft skin the control of roof deflection is critical to avoid
damage. This requirement ruled out the use of lightweight highly
responsive roof structures and the imposed load deflection limit in the
double bay was controlled to approximately 150mm for all loading conditions.

The building form selected was a modified stepped roof arrangement
incorporating distinctive exposed external triangular shaped space
trusses to support the low roof structure, see Fig 2. This structurally efficient design lowered the building profile over the aircraft
fuselage by up to 1.5m compared with the base building form to minimise
the internal volume and energy requirements. Aesthetically the fili-
gree of exposed external structure combined with the facade detailing
reduced the apparent massing of the building by softening the profile
viewed from long distances, which achieved the desired objective.

The building study also considered the most effective support building arrangement and the form selected separated the workshop and
offices to establish optimum circulation routes, see Fig 3. This was
achieved by incorporating a wide circulation street with connecting
balconies at each level and by locating staircases placed in the
central void. The roof structure over this area is fully glazed to
introduce natural light into the deep space.

4 Design Criteria

4.1 Imposed Roof Loading

The uniformly distributed site specific snow load predicted using
BS6399:Part 3(1988) Imposed Roof Loads; for Cardiff Wales Airport is
$0.27kN/m^2$ which is less onerous than the general uniformly distributed
access load. Consequently the roof was designed to withstand local
snow drifting at the roof step and a characteristic imposed load of
$0.6kN/m^2$ which would also allow access for cleaning and maintenance.

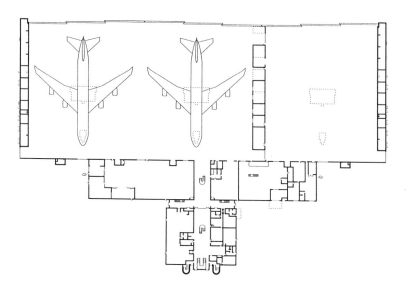

Fig. 3.   Ground Floor Plan of Hangar and Support Building.

Substantial drifting can be expected to occur against the vertical
step in the roof and BS6399:Part 3 predicts that the drifts will extend
the full length of the hangar. The characteristic loading intensity of
the snow drift varies linearly from 2.2kN/m² at the face of the vertical step to zero along a parallel line 15m from the step.

The hangar roof structure is also designed to support overhead
cranes which can operate in tandem to lift loads up to 20 tonne. The
access docking and the overhead cranes both contribute a dead load of
approximately 600 tonnes. The roof is also designed to support 280
tonnes of loading from by three two-storey high plantrooms accommo-
dating the hangar heating system. These plantrooms were incorporated
adjacent the roof step to minimise the length of ductwork to reduce
heat loss and running costs. To minimise the effect on the roof
structure they are situated close to the main tower supports.

4.2 Thermal Effects
Although the Bridges Specification BS5400:Part 2 (1978) is not directly
applicable to buildings it provided useful guidance for computing the
effective temperature of the structure influenced by daily and seasonal
fluctuations in air temperature and solar radiation.

The extent of movement in the hangar roof is influenced by the
structural articulation and centre of stiffness. In the scheme adopted
it was assumed that anchorage occurred at the main central support
between the single and double bays with movements away from this centre
on a proportional basis.

A separation joint was provided between the hangar and the support
building to isolate the effect of the hangar movements. Expansion
joints within the support building were included based on normal
building design guidelines and were incorporated at the natural weak
points in the structure articulated by the street circulation spaces.

48

The range of movement along the length of the hangar structure taking into account erection and operation was calculated to be from 90mm contraction to 130mm expansion. The largest positive temperature difference for the exposed structure generated a differential expansion of 10mm. The largest negative temperature difference generated a contraction of 25mm. These movements were taken into account to establish the secondary effects on the roof structure.

4.3 Earthquake Loads

It is perhaps noteworthy that in 1990 the British Geological Survey recorded seismic activity with its epicentre in mid-Wales. Although the earthquake was felt widely throughout the United Kingdom, it caused very little structural damage and no reported injuries. The calculated design load was found to be similar in magnitude to the notional horizontal force arising from Clause 2.4.2 of the BS5950: Part 1(1990), and as might be expected for a hangar structure the horizontal wind loads were found to be more onerous than the earthquake loading.

4.4 Wind Loading

The Code of Practice for wind loading, CP3:Chapter V, Part 2 (1972) indicates that the basic wind speed at C.W.A. for a 50 year return period is 45m/s which is equivalent to 162km/h. Amendments made to CP3:Chapter V permit the use of direction varying winds. This refinement was of great benefit for the hangar design since the doors face south east and the prevailing winds are from the west, north west to south west. Consequently the wind loading was adjusted to suit the exposure on each building elevation. The proposed new British Standard for wind loading BS6399:Part 2, 9th draft (May 1989), was available during the design process and although not used for the main design it supplemented the local wind co-efficient values given in CP3 Chapter V to assist with the building fabric and support rail development.

The wind study also indicated that the low and high level roof areas would be subject to uplift pressures with the hangar doors open or with the hangar doors closed but a 'dominant opening' such as damaged fabric, having been created.

4.5 Corrosion Protection

The identification of the corrosion environment within the hangar roof requires engineering judgement since the steelwork is exposed and will experience a range of environmental conditions depending on whether the hangar doors are open or closed. With the heating on and the doors closed, the corrosion environment is considered to be mild internal throughout with a low risk of corrosion. With the hangar doors open, whilst the roof structure remains sheltered, wind borne moisture will be blown into the hangar particularly near to the door and into the high level roof. This condition will produce a more aggressive corrosion environment than normal internal conditions.

Taking into account the design aims of the project, paint protection for the internal accessible steelwork, was selected to provide life to first maintenance in the order of fifteen to twenty years. This was achieved using a two-pack epoxy zinc phosphate primer with a two-pack epoxy micaceous iron oxide and finish coats. Where the steelwork was located behind the external cladding but concealed by inner blockwork or profiled steel single sheet lining then a bituminous paint system was specified.

5 Roof Structure

5.1 Structural Elements
The hangar roof plate is generally formed using plane frame trusses, except for the main member over the hangar door, at the roof step and the external exposed triangular structures which are space girders.

   The hangar roof structure consists of five main parts and a description of each is as follows:

(i)   The door girder 232.5m long x 9m deep x 5m wide which is 22.6m clear from the hangar floor slab to the structure.
(ii)   The spine girder is 232.5m long x 14.5m deep x 8m wide which supports the external girders and high level roof structure.
(iii) The high level trusses span 23m between door girder and spine girder and are at 7.5m centres. They provide 27.2m clearance to the hangar slab and are braced horizontally to combine with the door and spine girders to resist the horizontal wind forces. They are also braced laterally to distribute the crane loading.
(iv)   The low level roof structure is supported by six external 57m long x 7.25m deep x 7.5m wide external triangular shaped girders at 37.5m centres.
(v)   The low level roof trusses span 30m and are placed at 7.5m centres. They are connected together laterally with spreader trusses at 7.5m centres forming a two way spanning system which distributes the load; principally from the cranes. This structure provides a 16.6m clearance to the hangar floor slab.

   The main girders span 153.75m and 78.75m and are supported by lattice towers. The external girders are supported at one end by the spine girder and at the other by plane frame Vierendeel towers. The main support members are formed from tubular closed sections with diameters up to 508mm and wall thickness between 12.5m to 32mm.

5.2 Hangar Stability
Stability is provided in both the lateral and longitudinal direction by a combination of horizontal and vertical shear bracing. The location of the main stability bracing and structural elements outlined in section 5.1, are indicated in Fig 4.

   Wind loading on the gable walls is resisted by the main central towers and braced cantilever frames placed within the dividing wall between the single and double bays.

   Wind loading on the hangar doors and on the back wall of the hangar is resisted by the six support towers connected by the diagonal bracing. The wind sway movements calculated laterally were 20mm at the end towers and 75mm at the central support, not allowing for the effects of the roof cladding stiffening.

   The uplift forces generated by the overturning loads at the base of the cantilever dividing wall structure and below the main girder support towers are resisted by ground anchors.

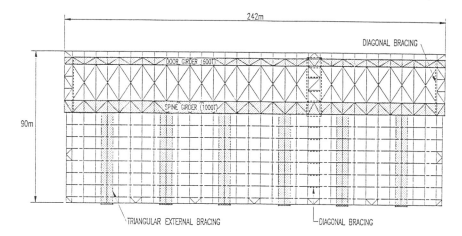

Fig. 4.   Roof Steelwork Plan Layout.

5.3 Member Design

The door and spine girders have been designed as continuous structures
to reduce the bending stresses and deflections in the large span.   The
girders incorporate tubes up to 762mm diameter with varying wall
thickness between 12.5mm - 40mm.   At the main member node locations
some joints have been strengthened using thicker walled sections,
'cans', between 16mm to 50mm thick.   The steelwork is predominantly
Grade 50C with some selected Grade 50D material at exposed locations.
The maximum ultimate bending moment in the double span door girder was
calculated at 220MNm and in the spine girder 430MNm.

From the outset the selection of roof members was influenced by
British Airways preference for closed sections to minimise dust collection and cleaning requirements.   Apart from preventing birds nesting
closed sections are structurally efficient because of their high
stiffness and ability to provide aesthetic joint configurations.

The weight of the structural steelwork in the door and spine girders
is 600 and 1000 tonnes respectively.   In the hangar roof it is 4,000
tonnes and on the project as a whole is approximately 6,000 tonnes.

To confirm the behaviour of the principle hangar roof structure a
three dimensional space frame analysis was undertaken using the OASYS
GSA programme.   The computer model was generated for the six support
towers, door and spine girder with the adjoining high level roof and
vertical wall bracing.   The number of elements was 2054 with 768 nodes
and the input data included upto 30 load cases to investigate the
effect of the varying equipment positions, wind and imposed loads.

5.4 Joint Design

The guidelines used for determining the strength of uniplanar joints
were taken from the design guide prepared by CIDECT (Draft 1990), based
on research by CIDECT and IIW (1989).   This work has received international acceptance.   The design formulae developed are also included in
British Steel's Design of Welded joints publication (1991), and also in
Eurocode 3:Part 1 (1990).

The tube to tube connections were developed on the basis of profiled fully welded joints since they are aesthetically neater and can be formed with modern computer controlled cutting machines. Other forms of joint configuration using gusset plates, ring stiffened joints, cast joints, composite joints and diaphragm stiffened chord members with grout infill were all considered but discounted because of the increased complexity, cost and effect on the appearance of the steel-work structure.

6 Procurement

Early project programming identified that placing an order for the steelwork would be a critical activity and this was targeted for early 1991. To allow the works to be tendered with substantial but prelimi-nary information a two staged approach was adopted and this enabled the contractor to advise on constructability issues prior to finalising the detailed design. The contract also included a Contractor's Design Portion element to cover specialist equipment and installation design.

Prior to the issue of the first stage tender the method of fabri-cation and availability of material was discussed with steel fabricators. Following this exercise it was considered that the industry could contribute to the cost effectiveness of the steel fabrication by allowing them to complete the fabrication joint design to reflect their experience, equipment skills and their preferred fabrication techniques. On this basis the contract documents were finalised providing member sizes, dimensions and typical load arr-angements for a fully welded hangar structure.

During the development of the fabrication information the steel fabricator, Octavius Atkinson & Sons Ltd selected by the Main Con-tractor Balfour Beatty Building Ltd, requested all secondary member connections to be bolted to reduce on-site welding. This proposal represented approximately 40% of the joints and excluded the main girder primary members and exposed triangular structures to ensure the aesthetics could be controlled. The fabricator undertook the final joint design and where the node force configuration required the joint capacities were enhanced using thick walled 'can' sections

7 Construction

The steelwork where practical was fabricated in the works using MIG and submerged arc welding techniques except for the spine and door girders. These members were brought to site with tube ends prepared for assembly using manual arc welding and bolted fabrication methods.

The site welding was undertaken by ABS Welders and at peak there were up to sixty highly skilled welders employed. Weld sizes range from small fillets to 40mm thick butt welds which in some instances required up to 50 hours continuous work to complete. Because of the critical nature of the welding an independant testing authority was resident on site throughout the operations to compliment the steel fabricators testing procedures. Weld testing was by visual, MPI and ultra-sonic methods.

Fig. 5.   View of Door Girder Erected.

The door, spine and external girders were fabricated into their
final form at ground level with built-in pre-camber. The erection
procedure for the door and spine girders provided a notable construction event since both 232.5m long structures were lifted into position
using hydraulically operated lifting equipment being supported at two
points approximately 200m apart, see Fig 5.

Each lift required approximately 10 hours to complete with the
girders being raised in 300mm incremental strokes to a height of 32m.

The initial camber built into the door and spine girders was 300mm and
220mm with the single span being formed without precamber. The predicted and actual deflection of the girders in their temporary state
between the lifting points were as follows :

	Predicted (mm)	Measured (mm)
Door Girder	760	735
Spine Girder	350	350

It is believed that the girders are the longest structural members
to be lifted by this method in Europe. The door girder was lifted on
Sunday 19 January and the spine girder on Monday 17 February 1992.

To progress the steelwork erection the high level trusses were the
next elements to be installed followed by the diagonal bracing to
provide a stable structure to allow erection of the low level roof.
Whilst this work was progressing the external girders and their support
members were also being fabricated at ground level. Following testing
and painting they were lifted into position using two mobile crawler
cranes. The low level roof and external girders were erected commencing at the braced dividing wall structure to ensure a stable arrangement in the temporary condition.

An aerial view which also indicates the exposed external girders
over the low level roof, is shown in Fig 6.

Fig. 6.   Aerial View Prior to Completion.

8 Acknowledgements

The assistance of all concerned with the project is acknowledged, in particular Mr Howard Corp who has worked on the design of the hangar roof steelwork throughout and has helped with technical aspects of this paper. I would also thank British Airways, Property Construction Group, Property Branch, for permission to publish material relating to this project and for their important contribution as the Client Project Managers.

9 References

BS5400:Part 2 (1978): Steel Concrete & Composite Bridges Specification.
BS5950:Part 1 (1990): Structural Use of Steel in Buildings.
BS6399:Part 2 9th Draft (May 1990): Wind Loading.
BS6399:Part 3 (1988): Code of Practice for Imposed Roof Loads.
British Steel Welded Tubes (1991): Design of Hollow Section Welded
    Joints, TD338.
CP3:Chapter V, Part 2 (1972): Wind Loads
CIDECT (Draft 1990): Design Guide for Hollow Section Joints under
    Predominantly Static Loads.
EC3 Eurocode No.3 (Draft November 1990): Design of Steel Structures
    Part 1 - General Notes for Building.
IIW (1989): Design Recommendations for Hollow Section Joints,
    Predominantly Static Loads IIW Doc XV-701-89;.
LUKE S.J. (1993): British Airways Maintenance Hangar, Cardiff Wales
    Airport Proc. Instn Civ Engrs (to be published).

6 NORTH POLE INTERNATIONAL BUILDING SUPERSTRUCTURE – CASE HISTORY

D.A.T. HUNTON
Mott MacDonald, London, UK
Connell Wagner, Australia
D.R. HAYNES
Mott MacDonald, London, UK

Abstract
This paper presents a case history of the development of the superstructure design for the maintenance sheds for North Pole International Railway Depot. It discusses the formation of the design team, the concept development and the design solutions adopted, which included the cladding and provision for craneage. Finally the paper discusses tendering and construction issues including use of a prototype wall panel to verify cladding details and construction procedures.
Keywords: North Pole International, Tubular Steelwork, Craneage.

1   Introduction

The Channel Tunnel Project must be considered one of the major engineering feats of this Century. North Pole International Depot is the sole facility within the United Kingdom for the cleaning, inspection, maintenance and servicing of the purpose-built international passenger trains using the Channel Tunnel. The depot combines the railway technologies of Britain, France and Belgium whose national railway companies will jointly operate the Class 373 Eurostar fleet. Up to eighteen Eurostars can be serviced nightly and during the day the European night stock and class 92 locomotives will be serviced.

The brief prepared by the British Rail Network Civil Engineer required the depot to provide for the following buildings:-

- Service Shed capable of housing six full length trains during daily maintenance.
- Repair Shed for more extensive maintenance of four half trains at any one time.
- Bogie Drop Shed for removal of bogies from the trains.
- Wheel Lathe Shed for machining wheels to correct uneven wear.
- Associated buildings for the Depot operation including Amenities and Gatehouse.

The key features that governed the development of this facility were:-

- The constraints of a site 3 kilometres long with only an average width of 70 metres.
- The complex interaction of site activites which dictated staging of construction for each building. To facilitate this the steel superstructure and cladding were combined as separate construction packages for each main building.
- The development of the design which had to be carried out over a very short period of time.

Tubular Structures V. Edited by M.G. Coutie and G. Davies.
© 1993 E & FN Spon, 2–6 Boundary Row, London SE1 8HN. ISBN 0 419 18770 7.

- The budget for the buildings which dictated that the cost of the selected design must be within 10% of a pre-established base cost using a conventional portal frame design.

The selected scheme achieved these requirements by utilising the concept of expressed triangular tubular steel frames for each of the four buildings. The triangular frames were developed to provide common solutions to wall glazing and cladding and to accommodate cranes within the building.

2   The Site

The site for North Pole International runs east-west through three London Boroughs; Kensington and Chelsea, Hammersmith and Fulham and Ealing. It is bordered on the north by the main railway line from London to the west (the old Great Western line) and to the south by Wormwood Scrubs. Mitre Bridge and the West London railway line run north-south across the centre of the site, dividing it into two sections. The service shed is located in the western part, with the other buildings on the east side.

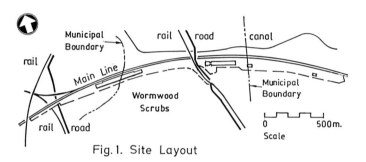

Fig. 1. Site Layout

3   The Brief

This paper is confined to the tubular steel structures adopted for the maintenance sheds namely the Service, Repair, Bogie Drop and Wheel Lathe. The Amenities building for the Service Shed, although not tubular steel, is referred to for its influence on the design.

The dimensional criteria and requirements for craneage were set by British Rail as part of a comprehensive brief for the project. This brief included the requirements for serviceability and corrosion protection of steelwork.

4   The Team

In March 1990 the design team of Mott MacDonald, lead consultant; YRM, Architect; and Mott MacDonald, structural and services consultant were given the task of developing the design of the Depot buildings, working with British Rail and their co-ordinating contractor, Kier Management.

Three weeks were permitted to produce conceptual design to meet with preliminary approval from the Client; a further five weeks were available to develop

5   The Design

5.1   Criteria

The basic criteria for each of the two main Sheds is shown in the diagrams below:

Service Inspection Shed
Clear length 413m

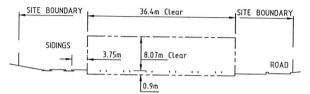

Fig.2 Crane Bay Section

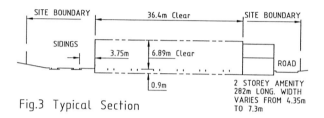

Fig.3 Typical Section

Repair Shed
Clear length 216m

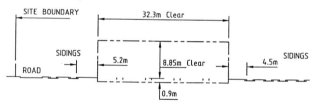

Fig.4 Crane Bay Section

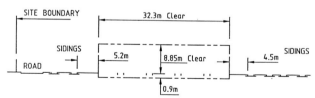

Fig.5 Typical Section

5.2   Concept

The direction of the brief and the importance of this facility in the context of the Channel Tunnel Project required that something other than a standard portal frame be used for the shed structures. However to meet budget constraints and construction programme the selected scheme had to maximise prefabrication, use repetitive elements and place a strong emphasis on buildability.

Eight alternative solutions were proposed and costed including 2 and 3 dimensional trussed portals, and portals with castellated or tapered sections.

The preferred solution was selected for the following reasons:-

-   The design concept could be arranged to suit the different span and scale of each of the four buildings, giving a consistency of appearance and design detail.

-   The architectural form introduced into the design by expressing the triangular tubular steel frames externally and internally presented an appropriate image for the facility.

-   The framing arrangement was simple and efficient in structural terms, allowing easy erection and support for cranes, which reflected well in cost comparisons. It also effectively minimised the width of site occupied by structural zones, which was a critical space planning constraint.

The design concept used externally expressed freestanding cantilever triangular steel columns at 18 metre centres. These provided both lateral and longitudinal stiffness. Diagonal members were eliminated at the base of the columns to allow pedestrian access for fire escape routes.

The columns were linked by perimeter triangular eaves trusses which provided support for the wall cladding and roof.

In the plane of the external wall both the column frame and eaves trusses were designed as vierendeel girders. These girders acted as glazing frames to provide the required 15% daylight in the external wall.

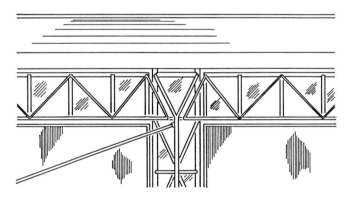

Fig.6 Part External Elevation

Internally triangular truss rafters were spaced at 18 metre centres. Triangular trusses were also used for the purlins which supported long span structural deck roof panels spanning up to 5.8 metres.

The use of triangular sections had a number of advantages for the roofs:

- The double top chord of the rafter shortened the purlin span by 2 metres.
- Additional torsional restraint such as flybracing was not required for the rafters or purlins under reverse loading due to wind.
- The horizontal truss action of the double top chord of the rafter meant that conventional roof bracing was not required.
- The double top chord of the rafter provided support for smoke vents installed above the trusses.
- The torsional stability of the rafter allowed installation of a smoke curtain under the trusses without additional bracing.
- The purlins were located and used as support for the suspended lighting system which was aligned to the rail tracks below.

At each end of the Service and Repair Sheds the roof was raised for the length of two bays (36 metres) to accommodate cranes. A simple solution was devised to achieve this using the same triangular truss philosophy for the rafter. The truss was inverted so that the double chords remained in the same plane and the bottom chord was transferred to the top as shown below.

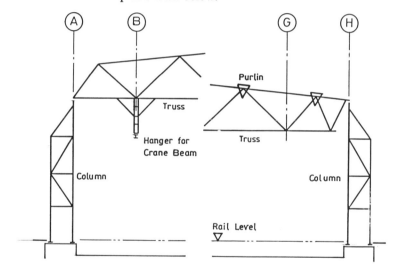

Fig.7 Rafter in Crane Bay   Fig.8 Typical Rafter

This concept had the significant advantage that no alteration was required to the columns. The columns were designed as freestanding elements so that a simple pin connection could be used at the top for the rafter connection as shown below.

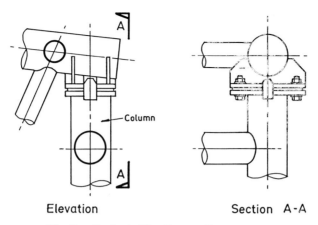

Elevation   Section A-A

Fig.9   Typical Pin Connection

In the Service Shed additional transverse triangulated trusses were added to support cranes hung from the roof structure. These trusses replaced the triangular purlins and gave a space frame appearance. The raised roof and frame configuration provided space for installation of plant platforms for servicing the building.

The maximum length of shed of 420 metres meant that thermal expansion and contraction needed to be considered. Horizontally slotted holes were provided at connection joints in the structure at 72 metre intervals.

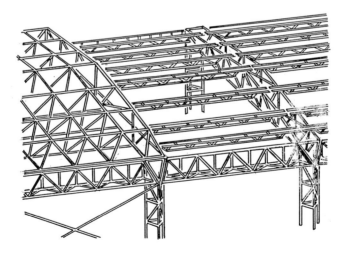

Fig.10 View Of Frame

Although this paper does not include the associated buildings, comment should be made on the Amenities building for the Service Shed. The Service Shed is open at each end to permit through movement of the trains. The Amenities building was located along one side to facilitate servicing of the trains and minimise travel distance for obtaining materials. In order to meet fire regulations it was designed as a separate structure with fire walls at the interface. This avoided the need to fire protect the shed steelwork.

5.3   Craneage

Each building had a different crane requirement. The Service Shed was unusual in that provision had to be made for three separate cranes in parallel at each end, each servicing the power car of two trainsets. As a through shed with overhead catenary electrical power supply, a complex arrangement was necessary (with control interlocks) to avoid electrical and physical conflicts. The cranes were supported from the roof structure and were designed with a central pillar projecting down between the trains bays. A horizontal jib was attached to the central pillar which could rotate through 360° to service two trains below the wire as shown below. The crane capacity was restricted to five tonnes. The adoption of transverse trusses to support these cranes provided the necessary vertical and torsional stiffness to meet the crane deflection limits.

The Repair, Bogie Drop and Wheel Lathe Sheds adopted conventional double girder electrically operated overhead travelling cranes with capacities between 10 and 40 tonnes mounted on downshop gantry beams. The gantry beams were supported on separate columns located internally within the structure adjacent to the cladding. Use of independent crane columns permitted a reduction in column spacing and gantry beam span. The external columns and eaves trusses provided the necessary lateral restraint and therefore typical columns and perimeter frames could be maintained in the crane bays. Deflection criteria were carefully examined and a detailed report prepared. The supply of cranes was not included in the shed envelope package, so design assumptions had to be clearly specified and subsequently confirmed by the crane supplier.

5.4   Tubular Steelwork

The design of the triangulated tubular steelwork, connecting elements and connections was critical structurally and architecturally. Attention to detail was of prime importance. A close working relationship was developed with the architect as these were evaluated three dimensionally. Significant use was made of CAD for modelling the building, jointing configuration, services co-ordination as well as the structural design and documentation. Availability and rationalisation of welded and seamless tube sizes, to achieve economies in fabrication, was an essential part of the design process. Detailed discussions were held with British Steel Welded Tube Division during the design development to ensure tube availability. The four buildings used approximately 700 tonnes of tube.

The trusses were designed using the STAAD3 analysis and design package. The nodes were designed using the principles set out in "Design Guide for CHS Joints" published by CIDECT and "New Development and Practices in Tubular Joint Design" by Yoshiaki Kurabane.

Typically the maximum tube size used was 219mm with the exception at the end bays where the size was increased to 324mm. Grade 50C steel was used for the tubes and grade 43C for crane support steelwork. The trusses were designed with a central splice to allow easy transportation to the site. Bolted and welded splice details were developed, however the design concept was for simple bolted connections to minimise site time during erection. The welded connections were restricted to the end bay trusses supporting the cranes.

5.5 Corrosion Protection
The design brief included a detailed specification on steel preparation, priming and painting. This was evaluated by the design team and paint suppliers and as a result a number of modifications were made. The paint coating finally specified was an epoxy micaceous iron oxide system, with a recoatable urethane top coat to provide the required finish.

5.6 Cladding System
The cladding system contained a number of interesting features. The wall cladding adopted conventional external and internal lining supported on a grid of cold formed girt sections. These panel walls were expressed as a picture frame through the connection detail to the columns, eaves truss and base of wall. The connection used longitudinal fin plates on the centre line of the tubes to achieve this separation. The glass in the columns and perimeter trusses was fitted to fin plates using neoprene gaskets. Bright steel was used for straightness and tolerance.

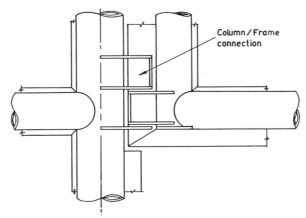

Fig.11 Column / Side Frame Connection

The roof cladding adopted a proprietary structural deck panel with external profiled sheeting, insulation and internal lining. An important feature of the roof cladding was the separation of the sheeting from the tubular steel sections. This was done to prevent corrosion from moisture entrapment and avoided holing the tube sections. Cleats were welded to the purlins for connection of the cladding above the purlins.

6 Tendering

The Sheds formed only one part of the facility at the North Pole. In order to meet the tight construction programme a series of packages were let and managed by Kier Management. The single structural steelwork and cladding packages were perceived to have advantages in staging of the work and providing the opportunity to select different contractors for each building. Since this procedure is not a common practice, only contractors with the proven capability of managing steelwork and cladding contracts were invited to tender. The packages were tendered by building contractors as well as steelwork fabricators.

7   Construction Phase

7.1   General
At the start of construction it was important to establish criteria such as

- Tube availability, fabrication set up and coating finish.
- Approval process for shop drawings.
- Tolerances for fabrication and erection.
- Methodology for erection on site.

A working relationship between the design team and contractor was set up early to facilitate quick responses to construction queries.

7.2   Fabrication
The separate contracts made for some difficulties in fabrication since a learning curve was required for each building. General fabrication followed the intent of the design drawings with only minor modification. Quality and management of the work was dependant on the contractor and this varied for each building. A prototype wall panel was required to prove cladding details and erection procedure. Separate contracts meant that this had to be repeated for each building. The prototype was erected in the fabricator's factory and it proved to be a valuable means of resolving potential problems at an early stage.

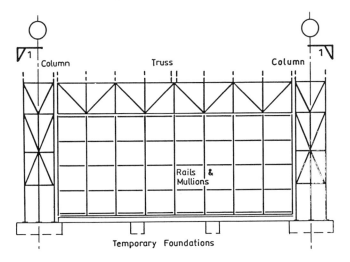

Elevation Of Steelwork

Fig. 12 Trial Assembly

7.3   Erection
The erection of the superstructure envelope followed the construction of concrete foundations. The sequence is summarised as follows:

- Erection of free standing columns commencing at one end of the building.
- Erection of perimeter frame onto the columns.
- Splicing of rafter trusses.
- Erection of rafter trusses in sequence with perimeter frames.
- Erection of roof and wall cladding.

The building was made weathertight progressively so that installation of services could commence prior to completion of the envelope. The overlapping of construction contracts presented typical problems of co-ordination. A high level of co-operation was achieved throughout the project.

8   Concluding Remarks

Traditionally railway depots have been basic functional buildings. The requirements for the Channel Tunnel Project meant that current facilities in depots were not satisfactory either to meet the servicing demands of the Eurostar Fleet or to provide the required standard of accommodation. This was made clear to the design team by British Rail at commencement. Significant emphasis was placed on image and performance of the facility.

We believe that what has been achieved has met the intent of the brief. The programme was always extremely tight but this did not restrict the quality of research investigation or the attention to detail in the development of the buildings. The use of tubular steelwork allowed the architect to achieve a strong expression of structure in these functional buildings. Commonality of detail, prefabrication and simplicity of connections were important factors in being able to meet the budget requirements and construction programme. The result has been very pleasing.

9   Acknowledgement

The authors acknowledge the assistance of the structural design engineer for the Sheds, G W Till who has provided valuable assistance in the preparation of this paper.

7 TGV STATION AT CHARLES DE GAULLE AIRPORT – PARIS

H. DUTTON and C. MAZELET
RFR, Paris, France
PETER GANNON
Watson Steel Ltd, Bolton, UK

Abstract
This paper describes the conception, design and
construction of the new TGV Station at Charles-de-Gaulle
Airport.   RFR is the structural design consultancy
acting on behalf of the client ADP/SNCF. Watson Steel
Ltd is the contractor appointed to carry out the final
design and construction of the station superstructure.
A joint venture was formed between Watson and Helmut
Fischer GmbH, Talheim, Germany, who were responsible for
the glass covering.

Keywords: Tubular Structures, Railways, Transport Interchanges,
Glazing, Lattice Girders.

1   Architectural Concept

1.1 Airport Expansion
Aeroports de Paris have adopted a strategy for the
future of the Charles de Gaulle Airport for it to
become a major central node in the newly reinforced
European Community. To meet this challenge, ADP have
decided to increase it's capacity and efficiency with a
major infrastructure development programme. Work on the
existing terminal 2, doubling its capacity, is almost
complete. A new terminal 3 is currently on the drawing
board with a view to being complete in 1996.

In conjunction with the SNCF, ADP have decided to
link the airport to the high speed TGV train system with
direct links to Great Britain,the Benelux countries and
Germany. A spur line will connect the airport to the
main line north from Gare du Nord around to the lines
going to the south of France through the Disney World
Amusement Park.

1.2 Futurist Transport Hub
Paul Andreu, ADP's chief architect and head of
Architecture and Engineering, has created a plan for the
extension of Roissy in which the train station is
located in the heart of the terminal 2 and 3

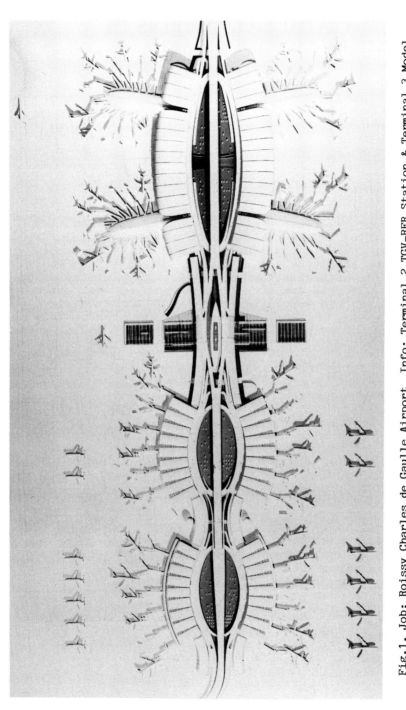

Fig.1. Job: Roissy Charles de Gaulle Airport Info: Terminal 2 TGV-RER Station & Terminal 3 Model
Architects: Paul Andreu with Pierre-Michel Delpeuch, Jean-Michel Fourcade & Anne Brison
Photographer: Michel Denance

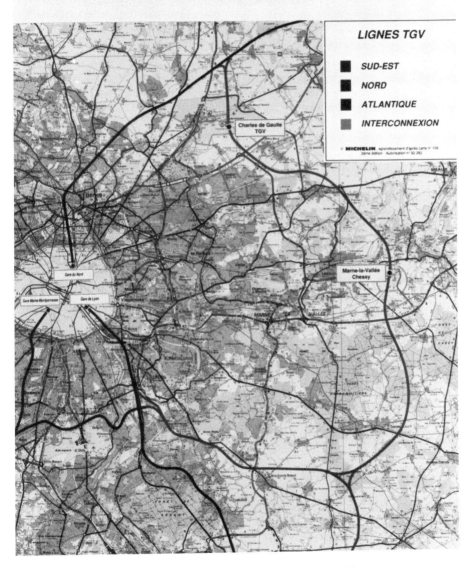

Fig.2. Map Showing TGV Network
Michelin, Reduction D'Apres Carte No.106
2EME Edition - Authorisation No. 92-282

redevelopment. It straddles the main axis of vehicular circulation at a transport hub called the 'Module d'Echanges'. This node is the interconnection point of road traffic, a small people-mover site train, the local RER suburban railway, as well as the TGV line to the air traffic system. There is also a hotel and an automatic baggage check-in and transfer terminal in this complex interchange point that recalls the ambitions of the Futurist artistic movement of Marinetti, St. Elia and others in Italy at the beginning of the century.

The railway station itself, designed by Paul Andreu, in conjunction with Jean-Marie Duthilleul of the SNCF, is underground, such that the train lines pass in a straight line under the runways and the hotel complex. The steel and glass roof structure which covers over the station is a team effort designed by the ADP team in conjunction with the late Peter Rice's Paris based office RFR. Detailed design development and construction was carried out by Watson Steel in partnership with the Helmut Fischer glass company in Germany.

The station is 500m long, determined by the length of a TGV train. It is in two halves, symmetrical about the main axis of the air terminals 2 and 3. Each half is again cut into two elements by a circular access road. The two outer elements are horizontal with a gently curved section for rain fall slope. The two inner elements are inclined with a roof slope that slowly transforms from a curved section to a flat section at the hotel end. Each element is approximately 100m long and 50m wide. The RER station covering a shorter train is approximately 300m long with a lower flatter roof.

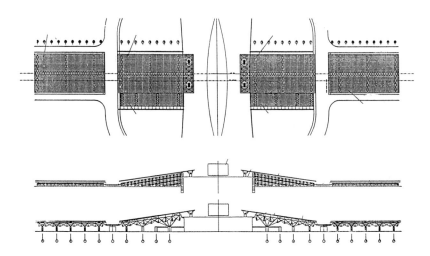

Fig.3. Plan

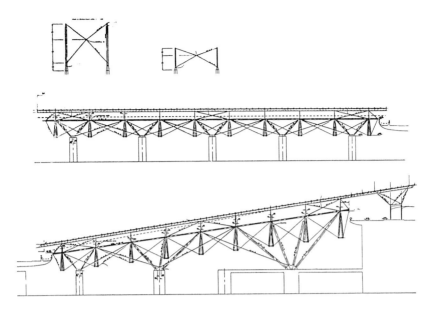

Fig.4. Longitudinal Sections

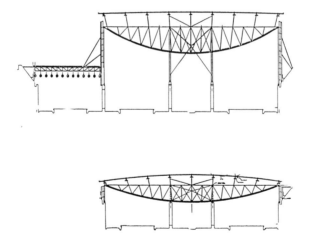

Fig. 5. Transverse Sections

1.3 Emergence

The architectural design for the steelwork is based on several simple concepts inspired by the idea of an emergence of the glass roof from underground.

First is the concept of Layers, where each successive hierarchical element of the structure is interpreted as a clearly distinct layer starting from the pylon columns, then the main transversal 'croissant' beams and finally the glass grid 'nappe' structure. The layers pull apart as the roof emerges such as to allow a view between each layer from the different levels in the Module d'Echanges complex.

Secondly, the Metamorphosis of the structure as it emerges from the underground trench, the transverse 'croissant' beams become deeper and the pylon structures become more complex, starting on paired columns, then becoming three and four column assemblies.

Thirdly, the view out toward the planes and the sky beyond from the station platform is considered particularly important given the fact that the station is underground so that the user does not feel as though he is in an underground cage.

1.4 Structural Clarity

An important notion in the general design philosophy of the office inspired by Peter Rice, is that the structures should be clearly readable and easily understood by the general public. The work can be appreciated at a greater level if the load paths are comprehensible and consequently, each member is its true size, representing the loads it can withstand and whether they are in tension , compression, or bending etc. The massive size of the station roof requires particular consideration given that all of the structure is seen at once in 'contre-jour' against the sky from a viewer on the platform. It is designed as a composition of hierarchically distinct elements superposed on each other and each with a different geometry and a different scale.

Against the backdrop of the orthogonal uniform square grid of the 'nappe' roof structure, the curved croissant beams are clearly readable, as well as the smaller inclined supporting struts and uprights. Then the pylons with their distinct inclined shape are set off as supporting the whole.

The two surfaces created by the roof glass 'nappe' and the underside of the croissant compressed tubular members dominate the longitudinal view. This is achieved by the literal interpretation of the top tension member of the croissants as a thin tie. In certain load cases, the tie would be required to resist compression. A further , third compression member was judged clumsy in the composition and so the compression case was removed by tying the croissants down to the lateral retaining walls on each side of the station.

The lateral facades are supported by cantilevered masts propped off the concrete retaining walls. These stop well below the roof edge and are on a staggered grid to that of the croissants, such that there is no confusion that they may be supporting the roof.

The smaller RER station roof is in a more conventional triangulated beam and column system with a shed roof spanning between the top and bottom members of consecutive square or rectangular section tubes, whereas in the TGV structure all of the members are in circular section members, thereby creating another hierarchical distinction between the two structures.

1.5 Glass on Steel

Both the glass for the roof and the side walls is fixed directly to the steelwork. This concept is quite different from the manner in which glass is conventionally fixed to steel using aluminium extrusions. Glass is a very strong, though brittle material. In the quest for transparency, its structural capacities have been exploited such as to dispense with the aluminium glazing bars, particularly on the side walls where a special articulated glass bolt protects the glass from local bending and stress concentrations. On the roof, the glass is laid directly on top of the steel 'nappe' grid using the Fischer patented silicone gasket system.

Finally, the view at night is also considered in the glass composition. It is treated with a white paint pattern printed and baked onto the glass called 'fritting', so that artificial light projected from the platforms is reflected back down to contribute to the general lighting of the station. Viewed from outside, the glass roof will tend to glow from the diffused light on the fritting.

1.6 Computers

The complex geometry of the structure is made possible with the use of computers. At each stage of the design and production process the computer was widely used. In the initial design, research for the form and geometry of the variable croissant beams was done on a simple 3D computer programme. Then the calculations, particularly the nonlinear analysis, were done using specially developed software. Finally, in the production phase, machines were also used to draw the multitude of complex node geometries that result.   Computers can give a far greater potential to architectural design allowing the use of complex geometries hitherto considered impossible.

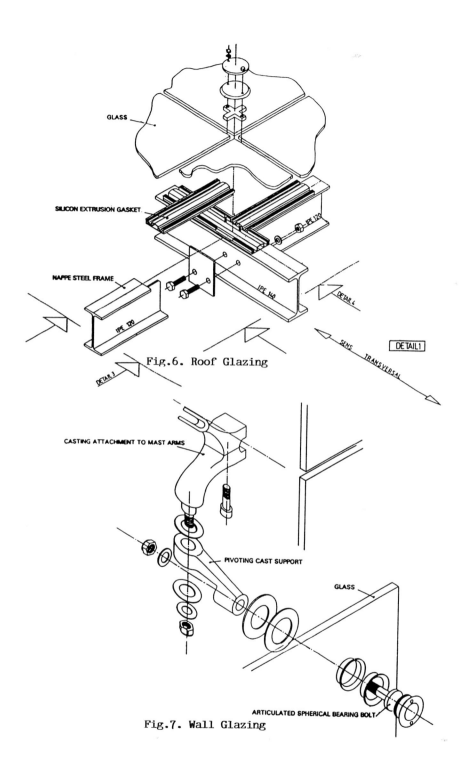

GLASS

SILICON EXTRUSION GASKET

NAPPE STEEL FRAME

IPE 120

IPE 140

IPE 120

DETAIL 4

DETAIL 3

DETAIL 1

SENS TRANSVERSAL

Fig.6. Roof Glazing

CASTING ATTACHMENT TO MAST ARMS

PIVOTING CAST SUPPORT

GLASS

ARTICULATED SPHERICAL BEARING BOLT

Fig.7. Wall Glazing

72

2   Reference Design

The station consists of three independent types of
structure:

The TGV roof structure
The Taxi rank
The <u>combined</u> facade and RER roof structure

2.1   Structural Principles

2.1.1   TGV Roof
The TGV roof structures are supported on two rows of
central concrete piers and are completely independent
from the facades.   There is no expansion joint except
at the junction with the taxi rank. They divide into
different layers of structural elements:

the pylons
the "croissant beams"
the "nappe"

The Pylons
The pylons rest on the central concrete piers.
Transversely, two pylons 11.4m apart support one
croissant.   Longitudinally, the pylons form a "fan
shaped" system of pin ended struts and ties.   The top
part of the pylon "branches" are linked to each other by
a longitudinal pin ended strut - their combined crossed
tie bars in the same vertical plane serve as bracing.

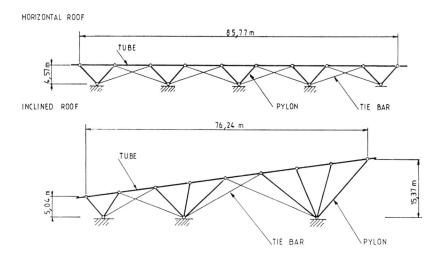

Fig.8. The Pylons

73

Transverse stability is achieved by cross bracing planes between each pair of parallel pylon branches forming a portal frame. The croissant top tie bar serves as the top member of the portal and must always remain in tension especially under transverse wind loads.

The Croissant Beams

The croissants are 47.5 m long and vary from 4 to 7m depth. They are supported at the top chord level by the pylon assemblies and have 18m long edge cantilevers on each end. They consist of a top tension member and a spindle shaped bottom vierendeel beam linked together by tubular verticals and diagonal ties. The two ends of the croissants are tied back to the ground by lateral tension members. It was decided to make the croissant stiff in the vertical plane and flexible out of plane to avoid parasitic temperature effects with the nappe in the longitudinal direction.

The top chord is naturally in tension under dead weight and other uniform downward loads. It tends to go slack under uplift or transverse wind. The transverse stability of the whole structure depends on the component of compression that this top chord can absorb under transverse wind. This is why the top chord is maintained artificially in tension by prestressing the two side ties. The bottom boom is heavily compressed, its buckling resistance is increased by the virendeel. Rotation about the top chord is prevented by tying the bottom boom to the adjacent pylon with cables.

The stiffness of the beam about a vertical axis is low which makes it sensitive to load cases of wind blowing perpendicular to the beam. Also the heavily compressed bottom boom will tend to buckle. To prevent out of plane buckling the extremities of the croissant are braced back to the stiff nappe plane.

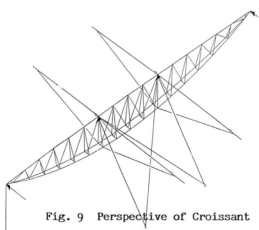

Fig. 9   Perspective of Croissant

74

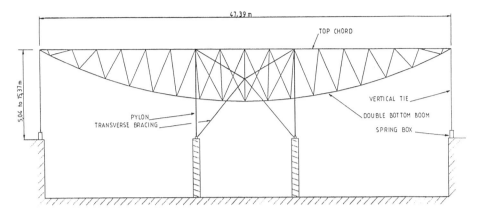

Fig.10. Transverse View of Croissant

The Nappe

The nappe "floats" on pin ended roof struts which are
spaced 6m apart on the croissant top chord (themselves
spaced 9.53m apart in the longitudinal direction). This
support system disconnects it from the rest of the
structure so that there is no expansion joint in the
nappe (except where it connects with the taxi rank). The
nappe is braced back transversely at each croissant and
longitudinally in three points to the pylons so that
some of the bracings act as security. In the plane of
the nappe there is one longitudinal braced bay and three
transverse braced bays.

The nappe consists of a continuous grillage of I and
H sections, supported by longitudinal triangular trusses
fabricated from square tubes. The original design used
pinned connections between the H sections;
Watson/Fischer's change to continuous construction
resulted in a lighter roof. The nappe is designed to
accommodate an internal gantry automatic cleaning system
which runs longitudinally along the bottom booms of the
0.6m deep triangular girders which serve as rails.

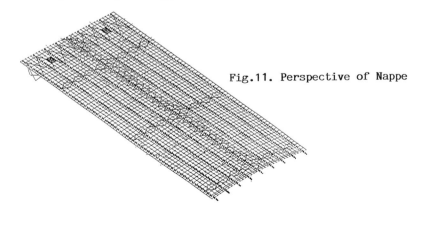

Fig.11. Perspective of Nappe

2.1.2   Facades and RER roof

The facades are completely independent from the TGV roof structure. The facade consists of pin based masts stabilised by struts which prop them back to the concrete on the east side and to the adjacent RER roof trusses on the west side. The masts and the trusses are spaced at 4.8m and the mast height varies from 0 to 15m above the RER roof level.

The glass plane is offset from the masts and suspended from them by a system of arms and cables. Each glass panel is suspended and braced individually at the corners.

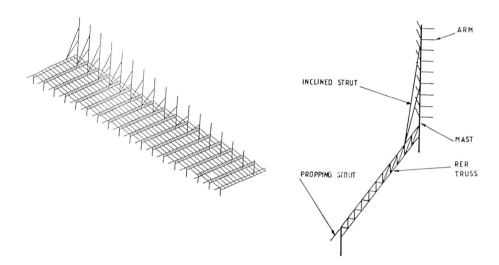

Fig.12. Perspective of
RER/Facade Structure

Fig.13. Perspective of
Mast and Truss

In plane stability of the facade

The in plane stability is achieved by four braced bays. A series of prestressed longitudinal cables link the masts back to the braced bays.

An asymmetrical wind load on the mast would induce a torsion in the mast. The longitudinal cables which are offset from the plane of the masts limit the torsional movements. By doing this the centre of rotation of the mast is moved away from the mast towards the glass plane. The horizontal loads in the longitudinal cables are transferred back to the masts by cross cables at the braced bays.

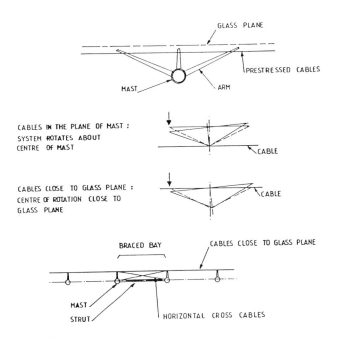

GLASS PLANE

PRESTRESSED CABLES

MAST   ARM

CABLES IN THE PLANE OF MAST :
SYSTEM ROTATES ABOUT
CENTRE OF MAST

CABLE

CABLES CLOSE TO GLASS PLANE :
CENTRE OF ROTATION CLOSE TO
GLASS PLANE

CABLE

BRACED BAY   CABLES CLOSE TO GLASS PLANE

MAST

STRUT   HORIZONTAL CROSS CABLES

Fig.14. Stability of Masts Against Torsion

2.2   Analysis
The analysis carried out by RFR determined the main
principles of the structure and member sizing.   More
exhaustive execution calculations were performed by
Watson/Fischer.   In this section we will focus on the
analysis of the croissant truss.

2.2.1   Determination of Loads
The wind loads were determined both from the French Code
and by a wind tunnel test performed at Bristol
University.   The wind turbulence was modelled in the
boundary layer wind tunnel on a 1/400 model of the
entire station for 18 different wind directions.   The
station was modelled as a closed building and the
internal pressure was calculated taking account of the
external pressure at all the openings.   The measures
included different averaging times (0.5 sec and 2 sec)
corresponding respectively to the dimensions of glass
and larger structural elements (over 10m).   To
determine with greater accuracy the local loads on the
edges, a 1/50 model of the edge of the TGV roof was also
tested.
    The SNCF performed a wind tunnel test of the inside
of the station to evaluate the train induced pressures
which turned out to be as high as 250 Mpa in the centre
part of the TGV roof.   A compilation of the results
from the different tests and the code, including an

averaging of pressures on the critical structural items, was made to determine the load cases used in the analysis.

2.2.2 Supports

The characteristics of the supports were part of the models. The concrete engineers produced a series of stiffness matrices representing the coupling of concrete movements and loads in three directions. The major parameters turned out to be the flexibility of the concrete under horizontal transverse load and were included in the analysis with different values for long term and short term load effects.

Differential settlement is a critical load case for the croissant structure as it increases the tension in the top chord and lateral ties. Load cases of imposed differential settlements were applied based on figures given by the concrete engineering office. Typical load cases were +5mm and -5mm at the two centre concrete columns or +10mm at the two columns.

2.2.3 Analysis Models

The RER roof, facades and nappe were analyzed on a linear model.

The more complex croissant beams and pylons were analyzed on a non-linear model, FABLON, developed by Ove Arup & Partners.

The program is based on the method of dynamic relaxation where the development of a structure's internal forces under an applied set of loads is followed from its initially unloaded state until equilibrium is achieved. The structure displaces incrementally according to the principle of dynamic equilibrium. Acceleration, damping and stiffness are computed at each node and tested at intervals until equilibrium is reached. The program takes into account the effects of nodal displacements on the overall geometry which enables it to model second order effects such as P-delta. The program also models the tension elements that go slack whereas in a linear model these elements go into fictitious compression and have to be manually removed. This program allowed the modelling of the out of plane buckling of the croissant.

2.2.4 Analysis of the Croissant

In plane stability

Prestressing:

As explained previously, prestressing is necessary to keep the top chord in tension. The prestress has to account for several phenomena that tend to put the top chord in compression:

- uplift
- transverse wind (the top chord serves as the top member for the portal frame)
- inclination of the roof struts towards the centre.

The value of prestress is critical as unnecessary prestress introduces extra forces in the structure.   To avoid sudden load reversals, it was decided to prestress the vertical ties so that they would remain in tension under unfactored dead load plus snow.   This choice was also made in anticipation of future fatigue calculations to limit the load variations in the lateral ties.   The initial prestress is roughly 80 KN.

Since the croissants vary in depth from 4 to 7m with vertical ties varying from 5 to 15m it was necessary to analyze three croissants: the extreme ones and the middle one.   This was sufficient to determine the member sizes of the ten different croissants.   The relative stiffness of the croissant and lateral ties varies for the different trusses.

Differential settlements and temperature effects

The truss is tied between the ground and the pylon supports.   The initial studies without spring boxes showed that the combined differential settlements and temperature effects induced high tensions in the side ties and the top chord whose section needed to be increased.   This resulted in an amplification of the parasitic effects.   To avoid this, the overall in plane stiffness of the croissant was minimised by adding spring boxes at the bottom of the vertical ties and reducing the section of the top chord and vertical ties by using high strength steel (610 Mpa ultimate tensile strength).   The spring boxes were chosen so that the combined stiffness of the vertical tie and the spring would be equal to one tenth of the stiffness that the tie would otherwise have.   Under extreme downward loads, the ties go slack which the spring boxes have to account for.

Out of Plane Stability

As already explained, the croissant is flexible out of plane.   The low stiffness (mainly achieved by the vierendeel action of the bottom boom) makes it very sensitive to loads perpendicular to the plane of the croissant.   Load cases of wind blowing on the croissant and horizontal forces applied to the top chord of the croissant were tested on the FABLON model.

Given its support conditions and 3D geometry it was not possible to make simple assumptions for the behaviour of the croissant truss and the actual buckling length of the bottom boom.   A model including the vertical struts was then used by Watson to investigate stability effects in more detail.

Dynamic Analysis
Because of their flexibility, dynamic analyses were performed on the different croissants. The vibration modes of the croissants were studied with and without vertical ties to account for the case of extreme wind when they actually go slack. In plane and out of plane vibrations were considered. The periods of vibration are lower than 0.8 sec. which is different enough from the wind period and confirms the initial 1.3 dynamic amplification factor for the wind.

Fatigue
As the tensions in the top chord and vertical ties of the croissant vary considerably under wind loads, precautions against fatigue had to be taken in the connections of those elements to avoid stress concentration. Fatigue calculations performed by Watson had to take into account acceptable fatigue life given the load variation and real wind cycle build up.

3   Final Design & Construction

3.1   Contract
The roof and walls to the station are constructed from 2000 tonnes of mainly tubular steelwork, supporting a completely glazed area of 25000m². A joint venture was set up between Watson Steel Ltd., Bolton, England, and Helmut Fischer GmbH, Talheim, Germany, to bid for the combined steel and glass package respectively.

The contract was placed in November 1991 and is due for completion early in 1994.

The interface between the two companies was decided at an early stage and resulted in the upper steel grillage, supporting the roof glazing, being supplied by Fischer.

The upper glazing grid, or nappe, is a framework approximately 1.5m square, supported above the main grid of croissant beams at 9.5m centres. The pylon lines, positioned towards the centre of the station provide support to the croissants.

A similar interface exists on the facades although the glazing system is different, being of the "bolt-on" glass type. Stainless steel brackets, supporting the glass panels, were supplied by Fischer and connected to the ends of the glazing support arms.

3.2 Final Design

The contract was placed on the reference design carried out by RFR, as previously described. A requirement of the contract, which is normal in France but unusual in Britain, is that the Contractor should carry out the final design, including both members and connections.

This requirement involved justification of the original reference design, the design of all connections, and the submission to the relevant authorities for formal approval.

Because of the complexity of the structure a close liaison was maintained with the original designers, RFR, and also with the checking authority CEP.

The conceptual design philosophy has already been described in part 2 and much of this work was repeated in the final design. This included non-linear stability checks and pre-stressing calculations for some of the tension elements.

The most sensitive areas were the top booms of the croissants and the longitudinal cable bracing system to the facades. Some fine tuning of the design was carried out to take account of variations in materials and the characteristics of supports. Also, the methods of fabrication and erection had to be considered and the effect of tolerances had to be built into the design.

The design of connections constituted a major part of the overall exercise, and because of the complexity of many of the nodes and the need to minimise details, castings were specified within the reference design. The final design of connections was carried out with recourse to various publications including Cidect, Roark and Timoshenko. Advice was also received from British Steel Tubes Division.

3.3 Construction

The main elements of the structure are shown on Figure 16. The logistics of an Anglo-German consortium building such a complex structure in France brought its own particular challenges. The constructional details finally used were a consequence of design and architectural requirements as well as the constraints dictated by fabrication, transport and erection considerations.

3.4 Pylons

The pylon legs, up to 20m long, are formed from 457 ⌀ tube, tapering down to 273 ⌀ at the top via a steel casting. This allows a welded connection to the main longitudinal member which itself incorporates castings at the croissant pick up positions.

The whole framework was trial assembled on the shop floor to ensure fit up of the site welds. For ease of

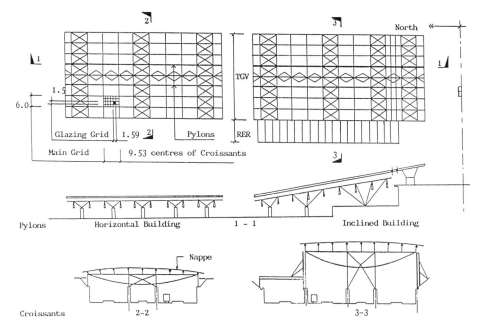

Fig.15. General Arrangement

Fig.16. Pylon Erection

site assembly the welds are positioned at the base of
the tapered cones and these were carried out on the
ground prior to erection.

Erection was facilitated by the use of pin joints
within the main longitudinal tie on either side of the
central legs.

At the base, all 4 legs are founded on a large steel
casting. A requirement of the design was that each
base should be articulated and capable of resisting
uplift. The architectural constraint was that the
joint should be completely hidden.

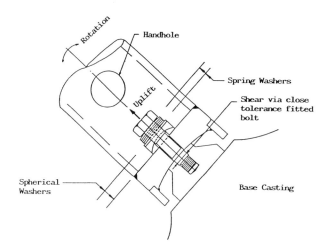

Fig.17. Pylon Bolt

The problem was solved by using a machined bolt to
locate the pylon leg onto a spherical bearing surface.
Shear transfer is achieved via a fitted shoulder, with
the upper portion of the bolt being reduced to avoid
bending stresses. Installation of the bolt was carried
out via a hand hole which was later sealed by a cover
plate.

3.5 Croissants

These are the dominant structural elements of the
building, and the most complex in terms of design and
construction.

The overall size of a croissant, at 48m long and up
to 6m deep, meant that it would have to be broken down
into smaller components for the purposes of transport
and erection.

The double CHS bottom boom was fabricated as a
vierendeel girder in lengths up to 12m and the ends
prepared for site welded connections. The tie bar top
boom was jointed at every node using threaded couplers

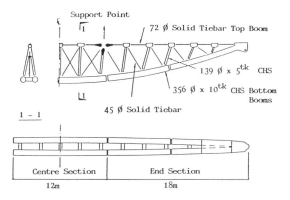

Support Point

72 ∅ Solid Tiebar Top Boom

139 ∅ x 5^{tk} CHS

356 ∅ x 10^{tk} CHS Bottom Booms

1 – 1

45 ∅ Solid Tiebar

Centre Section

End Section

12m

18m

Fig.18. Croissant Construction

Fig.19. Croissant Shop Jig

to connect either side of the struts.   These struts
were site welded at their connections to the bottom
boom, and also to the solid tie bar diagonals.
    To ensure the correct fit up of all components a
series of jigs were made for use both in the shop and on
site.
    At the croissant tips, the bottom booms converge and
it is necessary to change the detail from two tubes to a
fabricated section.
    The end section is formed from a tapered box section
with split tubes connected on either side.   The
vertical webs of the box act as backing strips to the
butt weld at the connection to the split tubes.
Submerged arc welding resulted in a smooth rounded weld
detail along the length of the section.

Fig.20. Croissant Bottom
        Boom End Fabrication

Fig.21. Croissant Bottom
        Boom Node Fabrication

An interesting feature of the design of the bottom
boom node is that the diagonal tension members do not
align with the vertical struts and this results in the
need for a load transfer system between the bottom
booms.   In some instances this is effected by the 219
diameter cross link, but when the loads become excessive
an additional gusset is required, as in Figure 21 .

The gusset is slotted into the vertical tubular strut
and welded to both faces to avoid eccentricity.   A
flush appearance is achieved on the outside by stopping
the gusset short and welding up the slot within the
tube. Another advantage of this detail is that it avoids
overstressing the main boom wall, which is subject to a
combination of axial and bending forces from the
incoming members.

A different set of parameters dictate the way in
which a typical top boom node joint is constructed.
The fact that the top boom is in tension and concern
over fatigue and brittle fracture led to the use of a
mechanical fixing rather than welding. The crotch of the
trouser leg strut member is formed by a casting,
constructed in two parts. This detail, using a spherical
upper casting, takes account of the variable angle
between the two elements and allows the use of a single
pattern for all castings. The transfer of tension from
the internal members to the top boom is effected by the
right hand coupler, with the other coupler providing
continuity on the opposite side of the node.

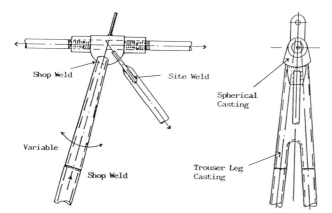

Shop Weld

Site Weld

Spherical Casting

Variable

Shop Weld

Trouser Leg Casting

Fig.22. Croissant
Top Boom
Node
Detail

Fig.23. Croissant
Support
Node

Fig.24. Croissant
Centre
Section
Erection

Fig.25. Croissant
        End
        Sections
        Erection

The most complex node on the building occurs at the connection between the croissant and the pylon.

This is again achieved by castings, the upper portion being welded into the pylon assembly and the lower section connected to the croissant.

On site, the croissants are welded up into 3 sections prior to erection. Figure 24 shows the central 12m piece connected between pylon supports. Completion of the croissant requires the erection of the two 18m long cantilever sections and the subsequent welding of the bottom booms at the splice positions.

3.6 Nappe

The nappe structure appears deceptively simple, but is in fact quite complex in terms of its structural

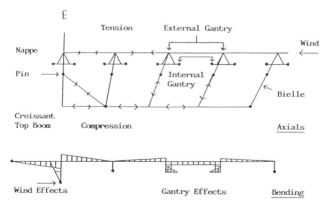

Fig.26. Nappe Design Criteria

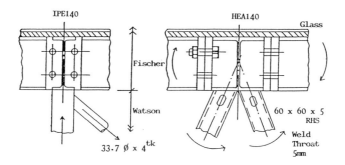

Fig.27. Nappe Connection Detail

behaviour and construction.   The basic principles of
its design are shown in figure 26 .   The requirement
to have articulated support struts and to accommodate a
maintenance gantry system above and below the glass
resulted in rigid construction transversely.   The
inclination of the supports also leads to axial forces
within the nappe members.

The torsional effects of the internal gantry are
resisted by moment connections between the triangular
maintenance rails and the incoming transverse beams.
The connection detail between the RHS verticals and the
top boom was of particular concern since it was required
to transfer this torsional effect.   Because of the need
to minimise member sizes, it was not possible to leave
a gap between the two RHS sections and the weld in this
position does not connect to the main member.   In the
absence of any suitable design method for this joint, a
series of load tests was carried out which showed that
the full moment capacity of the incoming RHS members was
realised.

In figure 27   the interface between Watson-Fischer
can be clearly seen.   The upper grillage, including the
top boom of the triangular girder, was fabricated in
Germany.   The top booms then came to England for
fabrication into the full girder, before being shipped
to France.

3.7   Facades + RER Roof
These form a completely independent structure from the
TGV roof.   The design principles have already been
explained and some aspects of the facade construction
are now described.

The 2.4m wide glass panels span between the ends of
glazing arms which themselves cantilever out from steel
masts.   Transverse stability is provided by raking
props whilst in the longitudinal direction a series of
braced bays provide restraint, with pretensioned cables
linking the intermediate masts.

Fig.28.
Erected
Nappe
Steelwork

Fig.29.
Nappe
Girder
Support

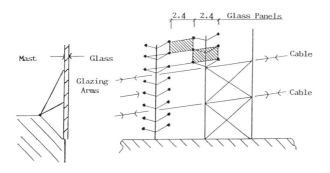

Fig.30.
Facade
Arrangement

89

The eccentricity of the glazing face, plus the possibility of asymmetric wind, causes torsion in the masts which must be transferred to the base via fitted pin connections.

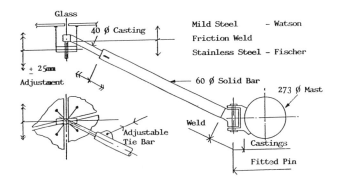

Fig.31. Glazing Arm Detail

The interface between steelwork and glazing contractor takes place at the end of the arm via a friction welded connection between stainless and mild steel. A vital feature of the detail is the adjustment which can be achieved at the end of the arm in all directions. The use of castings is again evident at both ends of the arm. A fitted pin at the lower end allows site assembly whilst providing moment fixity against wind forces.

3.8 Conclusion

This project has provided a valuable opportunity to demonstrate the versatility of tubular steelwork. The combination of tubes, tie bars and castings results in a lightweight design which complements the glazing details and achieves an overall successful solution.

The system for carrying out contracts such as this in France is also to be recommended since it involves both the original designers and the contractor in a design development situation which works to the benefit of the project as a whole.

Fig.32. General Views of TGV Station, March 1993

Fig.33. General Views of TGV Station, March 1993

PART 2

FIRE RESISTANCE

Lattice girders, columns and design

8 FIRE PERFORMANCE OF SHS LATTICE GIRDERS

M. EDWARDS
British Steel – Tubes and Pipes, Corby, UK
D.E. WAINMAN
British Steel – Technical, Rotherham, UK

Abstract
This paper describes the fire test performance of two protected SHS lattice girders and compares their behaviour to
predicted performance criteria according to British fire
test and design Standards and accepted Fire Design methods.
Recommendations are given for testing and designing such
girders and specifying their protection.
Keywords: Structural Hollow Sections, Lattice Girders, Fire
Resistance, Fire Design, Fire Tests

1 Introduction

Recently designers have begun to incorporate Structural Hollow Section (SHS) lattice girders into the framing system of
buildings and so take advantage of their high stiffness-to-
weight and strength-to- weight ratios compared to traditional rolled I-Section beams and also to use them for their
aesthetic value. Such framing systems usually require protection against fire for extended time periods.

Modern limit states design methods both at room temperature and in fire require a clear understanding of how structures and their individual members actually behave. This is
particularly important when designing against fire since
such design codes are still simplifications and also have
implicit assumptions built into them. This is true of both
the British design code, BSI (1990), and the draft Eurocode
for steel structures in fire, CEN (1992). Sufficient reserves of strength are built into the resulting design formulae to enable the designer to assess the performance of a
structural building frame in fire in terms of individual
member load ratios and resulting critical design temperatures. This enables individual member protection thickness
requirements to be assessed.

However, though most of the design guidance in both codes
is based on fire tests of individual structural members,
e.g. columns, beams and floor slabs, these have been interpreted using additional historic evidence of how such members interact when a traditional framed structure is ex-

Tubular Structures V. Edited by M.G. Coutie and G. Davies.
© 1993 E & FN Spon, 2–6 Boundary Row, London SE1 8HN. ISBN 0 419 18770 7.

posed to an actual fire. There was a need to demonstrate that the members of a relatively stiff SHS lattice girder would act in a similar way and that the same test and design criteria used for I-Section beams could be applied to SHS lattice girders.

Moreover, in a traditional building frame, protection specification and application can often be optimised since columns and beams can usually be divided into clear cut groups based on member sizes. Again, there was a need to demonstrate that similar simplifications could be applied to SHS girders.

2 Test Philosophy

Two standard fire tests were proposed, each on a protected, 4.5m. long, Centre Line noding girder using square SHS. Both girders were designed as four panel 45° Warren Girders under three point loading at the 1/4 point top chord nodes, with each load point laterally restrained in direction. Chord members were to be plastic design sections with the girder joints as strong as the bracing members.

Girder A had inner and outer bracings of different size and slenderness, but similar thermal Section Factors.

Girder B had inner and outer bracings of similar size and slenderness, but clearly different thermal Section Factors.

Each girder was protected to a uniform thickness in order to generate clear temperature differentials as required. Since tests were carried out to failure, a relatively flexible cementitious spray protection (Mandoval CP2), trowelled to a uniform thickness of 22 mm, was used rather than a board system or intumescent paint.

A light slab was cast into place on each girder top chord to serve as an independent longitudinal centre section of the furnace top. It was designed to be strong enough not to break up and fall into the furnace as each girder failed, but light enough to prevent composite action developing between a slab and a girder top chord.

2.1 Deflection Criteria
There are two traditional criteria for the failure of an I-section beam under fire test conditions:

a) a centre point deflection > l/20 mm
   i.e, for both beams, δ > 4500/20 = 225 mm

b) a rate of deflection increase > l^2/(9000.d) mm/min
   where: l is the length of the beam   = 4500 mm
          d is the overall depth of the beam   = 962 mm
so, for both beams, rate < 4500^2/(9000 x 962) = 2.33 mm/min

The tests would show whether these criteria could be applied to SHS girders.

2.2 Critical Temperature Criteria

Critical temperature criteria for elements of construction are given in Table 1 of BSI (1990) as a range of material strength reduction factors (SRF) at various strain levels. It is commonly accepted that the 0.5% level is the most accurate one for use in assessing the behaviour of steel members in fire. However, the SRF cannot always be taken as directly equal to the member load ratio (LR). The SCI (1990) Handbook to BS 5950 Part 8 states that:

For Compression Members with $l/r_y < 70$: SRF = 0.8 x LR
For Tension Members   : SRF = LR

It was intended to assess how closely these temperature criteria could be applied to SHS girder elements.

3   Girder Design and Test Load Assessment

Members sizes and material grades were intially selected by hand using a simple pin frame analysis and applied loads of 1.6 x the furnace house jack capacities. The final choice of SHS sizes and material grades is given in Table 1.

Table 1.   Lattice Girder Member Details

Member Type	SHS Size (mm)	Steel Grade	Section Factor
GIRDER A			
Inner Bracings	50 x 50 x 5.0	Grade 43C	213
Outer Bracings	90 x 90 x 5.0	"	207
Top Chord	100 x 100 x 10.0	Grade 50C	80 *
Bottom Chord	100 x 100 x 10.0	"	107 *
GIRDER B			
Inner Bracings	60 x 60 x 4.0	Grade 43C	262
Outer Bracings	60 x 60 x 8.0	"	137
Top Chord	100 x 100 x 10.0	Grade 50C	80 *
Bottom Chord	100 x 100 x 10.0		107 *

Note: * - Stated Section Factors are for 3 sided and 4 sided heating repectively.

Jack loads were then assessed more accurately using Micro-STRAN 3D, Release 3.0 (1987), a standard computer design program. Two load cases were considered:
a) The girder ultimate design load, taken as:
              1.4 x Permanent Load + 1.6 x Variable Load
b) The girder fire design/test load, taken as:
              Permanent Load + Variable Load

The permanent load comprised the girder self weight plus a uniformly distributed load of 1.928 kN/metre, due to the weight of the top slab. The variable load comprised the 3 jack loads imposed at the top chord nodes. Using common design office practice, the following assumptions were made:
- both chords were fully continuous
- bracings were pinned at each node
- material yield strengths and member geometric properties conformed to the appropriate British Standards
- top chord was laterally restrained at each load point

The resulting jack loads are given in Table 2. These were used as the imposed loads during the fire tests.

Table 2. Fire Design/Test Jack Loads

Girder	Jack Load (kN)	
	1/4 point	Centre
A	96	210
B	78	216

Girder member nominal load ratios were assessed for each load case according to BSI (1987) using the full plastic modulus of the SHS chord members. Nominal critical temperatures were then assessed to Table 1 of BSI (1990) at 0.5% strain, using the SRF/LR relationships described earlier in Section 2.2.

4 Girder Test Specimen Performance Criteria

Measured SHS properties were used to re-assess the actual member load ratios and actual critical temperatures under the imposed test loads. Since the commercial design program assumed standard material grades and dimensions, this was done by hand on a simple pro-rata basis and the knowledge that member force proportions would remain almost unchanged.

5 Girder Preparation and Instrumentation

Over sixty thermocouples were fixed to each girder, including the top slab interface. Wainman D. et al. (to be published) will give full details of the thermocouple positions.

Each top slab was then carefully cast in place. After initial curing of the slabs, each girder was sprayed with the specified fire protection. The coating was trowelled to a smooth, squared, high standard finish by Mandoval staff and left to cure for a minimum of 28 days.

6 Girder Testing

The girders were tested at the Warrington Fire Research Station.

Before each test, the specified jack loads were imposed and the resulting load point vertical deflections read using transducers, then checked against computer predictions. In both cases they agreed within 3%. Loads were then removed, the girder reloaded and all deflection transducers rezeroed. The furnace was then lit and controlled to the standard ISO temperature-time curve given by BSI (1987).

6.1 Girder B (tested 4th December 1991)
Little occured for the first 69 minutes, after which the top slab cracked transversly at mid-panel position between the loading points. After 95 minutes, further tranverse cracking of the slab occurred and the girder centre point deflection rate increased from 2 mm/min to 6 mm/min. Signs of cracking could then be seen in the protection coating on the central compression bracings. At 99 minutes the centre point deflection rate increased to 84 mm/min. Load was sustained for a further 4 minutes in order to impress an exaggerated failure pattern onto the girder.

Inspection of the cooled girder confirmed that failure had begun 95 minutes into the test by buckling of the inner compression bracings with a consequential failure of the centre sections of the compression chord (see Fig 1).

Figure 1   Girder B after Test

6.2 Girder A (tested 11th December 1991)
Little happened for the first 76 minutes, after which the top slab cracked transversly at mid-panel positions between the loading points. At 80-82 minutes cracking of the protection system became visible at the outer bracing joint to the right hand (RH) end of the tension chord. This cracking progressed steadily for the next 3 minutes. Nevertheless, the girder still remained stable until, at 86 minutes into the test, more cracking of the top slab occurred, with explosive spalling of the underside. This dislodged the already weakened protection at the RH lower chord joint and the girder centre point deflection rate increased from 2 mm/min to 6 mm/min. At 89 minutes into the test the centre

point deflection rate suddenly increased to 23 mm/min. Load was then removed to prevent further damage to the protection system and preserve the failure pattern impressed into it.

Inspection of the cooled girder confirmed that failure had begun 86 minutes into the test. Local buckling of the RH outer compression bracing had reduced the RH outer bays to a mechanism with resulting local plastic distortion of the chord SHS at three joints and a consequential failure of the RH outer tension bracing (see Fig 2). However, extensive cracking of external protection on all bracings showed that all had been at incipient failure.

Figure 2   Girder A after Test

7   Presentation of Test Data

Only selected test data is given in this paper, but full tabulated data will be published by Wainman D. et al.

8   Discussion

8.1 Failure Criteria
In both tests, each girder stayed relatively stiff with centre point deflections well below the usual beam failure criteria. Failure was relatively sudden, with beam deflection rates increasing suddenly (see Figs 3a & 3c).

8.2 Slab Behaviour
Both top slabs cracked progressively, so by the end of each test they merely retained the loading stools and contributed little additional stability to the top chords. However, they also formed a large local heat sink that markedly depressed the temperatures of each top chord (see Figs 3b & 3d).

8.2 Local Chord Behaviour at the Joints
Chord faces remained undistorted at all joints, with no signs of bracings attempting to punch through.

8.4 Temperatures of Failed Members
The measured and actual critical temperatures of selected girder members at failure are shown below in Table 3. This

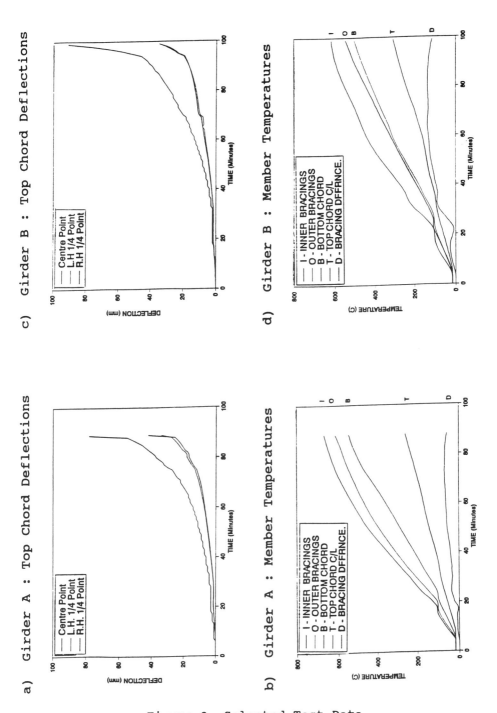

Figure 3   Selected Test Data

shows that in both girders compression bracings had failed at temperatures some 9-12°C higher than the actual critical values assessed according to BSI (1990).

However, even though tension bracings had measured temperatures substantially higher than critical, they did not fail in Girder B and only experienced consequential failure in Girder A. This implies that the methodology of BSI (1990) is conservative when predicting the performance of tension members.

Table 3 Selected Critical and Measured Temperatures

Girder Member	actual θ_{crit}	θ_{test} at failure	Temp Dif.	Failure Time
Girder A				
Centre Bottom Chord	535	538	+3	
Inner Comp. Bracing	598	621	+23	
Inner Tens. Bracing	589	621	+32	86
Outer Comp. Bracing	591	603	+12	
Outer Tens. Bracing	560	603	+43	
Girder B				
Centre Bottom Chord	533	529	-4	
Inner Tens. Bracing	583	606	+26	95
Inner Comp. Bracing	597	606	+9	

8.5 Bracing Temperature Differentials

Only a slow increase in the temperature differential between the inner and outer bracings of Girder A ocurred, reaching a maximum of 68°C. (see Fig 3b)

In contrast, the temperature differential between the inner and outer bracings of Girder B, were generated more rapidly, reaching a maximum of 136°C. (see Fig 3d)

Nevertheless, there was no noticeable difference in performance. This implies that the girders possessed sufficient flexibility and reserves of strength to absorb any internal thermal loads created by such differentials.

9 Conclusions and Recommendations

The following conclusions can be made:

SHS girders will carry their imposed loads in a highly stable manner up to failure. Test failure may then be relatively sudden. Accordingly, the most applicable criterion for defining test failure is the deflection rate.

The temperature differential created between the individual girder members and the presence of the concrete top slab did not noticeably affect girder joint strengths or the predicted performance criteria of girder members.

Drawing on these conclusions, the following simple recommendations can be made for SHS girder fire design and the specifying of their fire protection requirements:

Present BSI (1990) design procedures can be used to assess individual SHS lattice girder member load ratios and critical temperatures in fire and will give conservative results even when based on actual member properties.

Where practical, e.g. for large, deep girders used to support heavy floor loads, fire protection thicknesses can be specified in terms of the Section Factors and critical temperatures of each member group.

For shallow girders, the protection thickness requirements may be simplified by using two simple groups, chords and bracings, and taking the most critical case from each group. Resulting temperature differentials can be ignored.

Within practical limits, it may be most economic to simply specify a uniform thickness of protection based on the lowest critical temperature taken from the member group with the highest Section Factor. The resulting temperature differentials can be probably be ignored, especially in the half hour fire case.

10 References

BSI (1985), **Structural use of steelwork in building. Part 1. Code of practice for design in simple and con tinuous construction: hot rolled sections, BS 5950: Part 1:1985**, British Standards Institution, London, UK.
BSI (1987), **Fire tests on building materials and structures. Part 20. Method for the determination of the fire resistance of elements of construction (general principles), BS 476:Part 20:1987**, British Standards Institution, London, UK.
BSI (1990), **Structural use of steelwork in building. Part 8. Code of practice for fire resistant design, BS 5950:Part 8:1990**, British Standards Institution, London, UK.
CEN (1992), **Eurocode 3 : Design of steel structures - Part 1.4 Fire resistance (October 1992)**, Document No: CEN/TC 250/SC 3 N232E, European prestandard, Draft prENV 1993-1-2, European Committe for Standardisation (CEN), British Standards Institution, London, UK.
SCI (1990), **Fire Resistant Design of Steel Structures - A handbook to BS 5950:Part 8**, Steel Construction Institute, Ascot, UK.
Wainman D. et al. (to be published), **Fire Performance of SHS Lattice Girders, Test Details**, British Steel Technical, Rotherham, UK.

9 DESIGNING HOLLOW SECTION COLUMNS FOR FIRE RESISTANCE

L. TWILT
TNO Building and Construction Research, Rijswijk,
The Netherlands
M. EDWARDS
British Steel – Tubes and Pipes, Corby, UK
R. HASS
Hosser, Hass + Partner, Braunschweig, Germany
W. KLINGSCH
Universität Wuppertal, Germany

Abstract
This paper reviews some basic features of the fire design of
Structural Hollow Section (SHS) Columns. After a brief
discussion on the fire resistance concept and the associated
safety levels and assessment methods, the fire behaviour of
the relevant types of SHS columns is described in a
qualitative manner.
Keywords: Structural Hollow Sections (SHS), Columns, Fire
Resistance, Eurocodes, Fire Tests, Fire Engineering.

1 Introduction

Rectangular or circular hollow steel sections are often used
in construction because of their structural efficiency and
also because of their shape, when a visible architecture
expression is required.

Unprotected hollow sections have an inherent fire
resistance of some 15 to 30 min. When steel hollow sections
have to meet severe requirements and therefore have to
attain a significant fire resistance, additional measures
have to be taken, such as: external insulation of the steel
sections, concrete filling of the section or water cooling.

Especially in the case of concrete filled SHS columns,
additional provisions may be necessary to protect the
connections.

2 Fire Resistance

2.1 Concept

The time that a construction element can resist a fire
depends largely on the anticipated temperature development
of the fire itself. This is dependent, among other things,
on the type and amount of combustible materials present and
on the fire ventilation conditions.

In practical fire safety design, however, one single
(conventional) temperature development is used, which is
more or less representative for post flash-over fires in
buildings with relatively small compartments, such as

Tubular Structures V. Edited by M.G. Coutie and G. Davies.
© 1993 E & FN Spon, 2–6 Boundary Row, London SE1 8HN. ISBN 0 419 18770 7.

apartment buildings and offices. This is the so-called "standard fire curve", defined in ISO 834 (1975), see Fig. 1. Alternative standard fire curves are in use in the USA (Standards for Safety, 1991) and for maritime applications (IMO, 1993). Their differences from the ISO-curve are only small and of no practical significance.

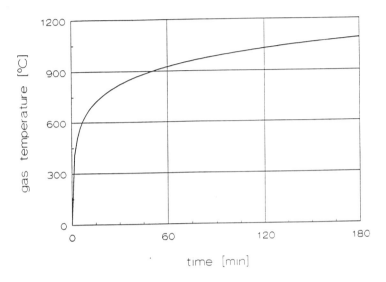

Fig. 1. The standard fire curve.

The period of time which a building component is able to withstand heat exposure according to the standard fire curve, is called the "fire resistance". To determine the fire resistance of a building component, proper performance criteria have also to be determined. These criteria are defined in relation to the anticipated function of the respective building element during fire. In general, there are three such performance criteria: stability (R), insulation (I) and integrity (E).

For building components such as columns, with only a load bearing function, the only relevant performance criterion is "stability". This criterion will, therefore, be considered hereafter.

As far as the determination of the fire resistance is concerned, there are basically two possibilities:
a: The traditional experimental approach, i.e. the determination of the fire resistance of columns on the basis of (standard) fire tests.
b: The fire engineering approach is a relatively new development that has become possible due to the recent development of computer technology.

Important factors influencing the fire resistance of columns are: load level, shape and size of the cross section and buckling length. Hence, the fire resistance of a SHS-column is not an inherent property of the column, but is influenced by a variety of design parameters. This is very important, since fire safety requirements for columns are normally expressed solely in terms of the fire resistance to be attained, and emphasises the need to consider fire resistance requirements from the beginning in a structural design.

2.2 Requirements
Safety levels in buildings referring to fully developed fires are mainly, but not exclusively, verified against standard fire tests or a numeric simulation of them. The standard fire test is not intended to reflect the temperatures and stresses that would be experienced in real fires but provides a measure of the relative performance of elements of structures and materials within the capabilities and dimensions of the standard furnaces. In general, uncertainties about structural behaviour in real fires are taken into account by making conservative fire resistance requirements.

Required levels are specified in National Codes and depend on factors such as:
- type of occupancy;
- height and size of the building;
- effectiveness of fire brigade action;
- active measures such as vents and sprinklers (but not in all countries).

An overview of fire resistance requirements as a function of the number of stories and representative for many European countries, is given in the following Table.

Table 1. Variations in required fire resistances

Type of	Requirements	Fire class
One storey	No or low requirements	Possibly up to R30
Two to three storey	No up to medium requirements	Possibly up to R60
More than three storey	Medium requirements	R60 to R120
Tall buildings	Medium requirements	R90 and more

Although quite a large variation in requirements exists, one may conclude that in most countries the required fire resistance is not beyond, say, 90 to 120 minutes. If requirements are set, the minimum value is 30 minutes (some countries however have minimum requirements of 15 or 20 minutes). Intermediate values are usually given in steps of 30 minutes, leading to a scheme of 30, 60, 90, 120 .... minutes.

2.3 Performance criteria

The fundamental concept behind all methods designed to predict structural stability in fire is that construction materials gradually lose strength and stiffness at elevated temperatures.

The loss of the yield strength of structural steel and the compression strength of concrete according to the Eurocodes is given in Fig. 2.

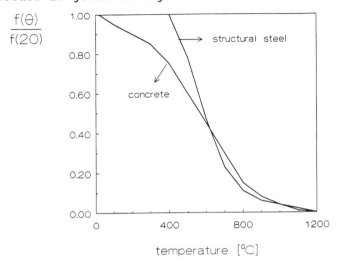

Fig. 2. Schematic material strength reduction for structural steel and concrete according to EC3 (1992), EC4 (1990).

Fig. 2 shows that there is not much difference in the relative reduction in strength of concrete and steel under high temperatures. The reason for the difference in the structural behaviour of steel and concrete elements under fire conditions is that heat propagates about 10 to 12 times faster in a steel structure than in a concrete structure of the same massiveness because the thermal conductivity of steel is higher than the thermal conductivity of concrete.

The fire resistance design of structures is normally based on the same static boundary conditions as the design under ambient temperature. In a multi-storey braced frame, the buckling length of each column at room temperature is usually assumed to be the column length between floors. However such structures are usually compartmented and any fire is likely to be limited to one storey. Therefore, any column affected by fire will lose its stiffness, while adjacent members will remain relatively cold. Accordingly, if the column is rigidly connected to the adjacent members, built-in end conditions can be assumed in the event of fire. Investigations have shown that in fire the buckling length of columns in braced frames is reduced to between 0.5 to 0.7 times the column length at room temperature, depending on

the boundary conditions (Twilt, et al. 1991). The more
conservative effective length factor (0.7) should be used
for assessing the buckling length in fire of columns on the
top floor and for the columns at the edge of a building with
only one adjacent beam. The higher reduction factor (0.5)
may be used for all other columns. The schematic structural
behaviour of columns in braced frames is shown in Fig. 3.
The above rule is also applied in EC3 (1992) and EC4 (1990).

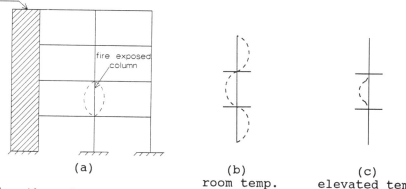

(a)

section through the building

(b)
room temp.

(c)
elevated temp.

mode of deformation

Fig. 3. Schematic structural behaviour of columns in
braced frames.

There is an increasing tendency to assess the fire
resistance of individual members or sub-assemblies by
analytical fire engineering. The Eurocodes on structural
fire design define three levels of assessments: (EC3, 1992,
EC4, 1990, EC2, 1990).
- Level 1: Design Tables and Diagrams;
- Level 2: Simple Calculation;
- Level 3: General Calculation Procedures.
"General Calculation Procedures" is the most
sophisticated and will provide, in terms of the results, the
most economic answers. The calculation procedures include a
complete thermal and mechanical analysis of the structure.
General calculation methods enable real boundary conditions
to be considered and take into account the influence of non-
uniform temperature distribution over the section and
therefore lead to more realistic failure times and
consequently, to the most competitive design. However, the
handling of such procedures is quite time-consuming and
requires expert knowledge.
For practising engineers and architects not accustomed to
handling specialised computer programmes, useful design
tools have recently been developed:
- Simple calculation procedures, which lead to a
  comprehensive design, but are limited in application
  range. They use conventional calculation procedures and
  provide normally adequate accuracy.

- Design tables and diagrams, which provide solutions on the rather safe side and allow fast design for restricted application ranges.

3 Designing unfilled SHS-columns

Calculation of the fire resistance of SHS-columns not filled with concrete comprises of two steps: (ECCS, 1985)
- the determination of the temperature development in the steel section;
- the determination of the steel temperature at which the column fails, the so-called "critical steel temperature".
The above implies the assumption of a uniform temperature distribution over both the cross section and length of the steel member.
Combining the two calculation steps gives the time at which failure of the column occurs when exposed to standard fire conditions. This time is the fire resistance of the column. The calculation scheme is illustrated in Fig. 4.

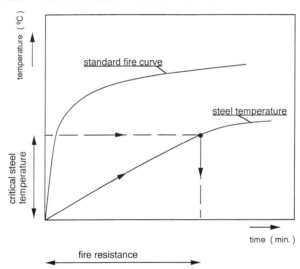

Fig. 4. Calculation scheme for the fire resistance of steel columns.

Bare steel columns (e.g. SHS-columns without external protection or concrete filling) possess only a limited fire resistance. Depending on the load level and the massivity factor, a fire resistance of 15 to 20 minutes is usually attainable; a 30 min fire resistance can only be achieved in more exceptional cases. This situation may be dramatically improved by applying thermal insulation to the column. Depending on the type and thickness of the insulation material, fire resistances of many hours can be achieved, although most requirements today are limited to 120 minutes.

4 Designing concrete filled SHS-columns for fire resistance

Structural members exposed to fire will be heated up. The temperature distribution in SHS-columns, either unprotected or externally protected, is more or less uniform. The heating behaviour of a concrete filled SHS-column will be significantly different. Because of its much higher massiveness and the combination of materials with different heat conductivity, there will be an extreme transient heating behaviour. Because of this, concrete filled SHS-columns can be designed to have a fire resistance of up to 120 min. or more without external protection. The simple calculation models for fire design, as used for unfilled SHS columns, are not appropriate here. A special fire design taking into account the transient heating and the different thermal characteristics of the various materials is necessary in these cases.

With increasing temperature, the strength and Young's modulus decrease. Thus, load bearing capacity of a structural member decreases with time while its deformation increases.

Fig. 5 explains the time dependent decrease of the resistance R of the structural member. If member resistance R has dropped to the level of the acting loading S, failure time t_u has been reached. For practical fire engineering, influence of the column slenderness has to be taken into account (EC3, 1992).

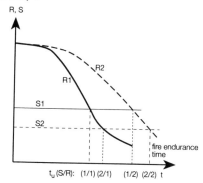

Fig. 5. Load bearing capacity of concrete filled SHS-columns as function of time.

Fig. 5 also explains, the different basic possibilities to influence the failure time t_u (= fire resistance time).

With lower load level S2 the intersection point with the column load bearing characteristic R1 will be moved from $t_{u\ (1/1)}$ to $t_{u\ (2/1)}$, thus increasing the fire resistance by overdesign.

A better method to reach a higher failure time t_u is to improve the resistance of the structural member itself by adjustment of the R1 curve to R2 by member fire design. The failure time will then increase from $t_{u\ (1/1)}$ to $t_{u\ (1/2)}$.

110

Both the above methods can be combined. Improved member design (R2) and lower utilization (S2) will lead to failure point t_u (2/2).

SHS-columns filled with concrete have a much higher load bearing capacity and longer fire resistance than unprotected empty SHS-columns. Provided the concrete is of good quality (better then C20), and the cross sectional dimensions are not too small (not less than 150 x 150 mm) a fire resistance of at least 30 minutes will be achieved. Sections with larger dimensions will have a higher fire resistance and by adding additional reinforcement to the concrete the fire resistance may be increased to over 120 minutes.

5 Designing water filled SHS-columns

Safety and reliability of the fire protection method of SHS-columns by water filling is based on two phenomena (Bond, 1975), (Hönig, et al., 1985):
- self activating;
- self controlling.
Natural circulation will be activated when heating up the water. The density of warm water is lower than the density of cold water which activates the natural circulation. This effect will be intensified when steam is formed since the mixture of water and steam bubbles has a significantly lower density than hot water only. With increasing fire severity the rate of steam production will increase too, thus forcing the cooling effect obtained by naturally activated circulation. This behaviour can be understood as a self-controlling effect, as with increasing fire severity the cooling effect itself will be intensified.

The maximum steel temperature can be estimated in a simple way as being equivalent to the boiling temperature of the contained water. The boiling temperature itself depends on the hydraulic water pressure and thus, on the static head. In addition, there will be a temperature gradient over the shell thickness of the hollow section which leads to a slight increase of the temperature of the steel surface directly exposed to the fire. However, maximum steel surface temperature will normally not reach a value where the mechanical properties of the steel will be significantly affected by temperature. Apart from situations in which extreme high values of shell thickness and water pressure apply (as for columns in highrise buildings), load bearing capacity and stiffness of a SHS-column with water filling can be assumed as independent of any fire attack as long as the natural circulation of the cooling system is working. Hence, in such cases an unlimited fire resistance may be achieved.

6 Connections and fire resistance

Since neither in practice nor during tests has failure been caused primarily by the behaviour of connections, it is not considered necessary to take additional measures for the protection of connections exposed to fire. If bolted connections are used for insulated steel members, the bolts should be as well protected as the member. This will normally lead to a local increase of insulation thickness. Concrete filled hollow sections generally fulfil fire protection requirements without further precautions. For economic reasons this construction method requires easily assembled connections between the columns and beams, which preferably should be entirely in steel. Suitably designed connections are fire resistant and can improve the fire resistance of the whole structure. The loads have to be transferred from the beams to the columns in the way that all structural components (structural steel, reinforcement and concrete) contribute to the load bearing capacity, according to their strength.

A well constructed column/beam connection should:
- provide simple installation;
- optimize prefabrication of columns and beams;
- guarantee adequate fire resistance without disturbing cladding.

If a building structure is braced (e.g. central core), the connections normally only transfer shear loads. In steel construction two different types of connections between beams and columns are used:
- Continuous beams (see Fig. 6a):
  The columns are connected to the beam by flanges, the shear forces are transferred to the columns by direct bearing. Such connections can be classified to the same fire resistance category as the composite beams and columns without further provisions.
- Continuous columns (see Fig. 6b):
  The connection of the beam to the columns is designed as a pinned joint. The shear forces are transferred to the columns by connection elements. To gain high fire resistance, these elements can either be protected, be overdesigned or specially designed, so that the forces can be transferred although the material loses its strength under fire action. See e.g. Dorn, et al. (1988).

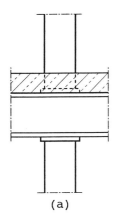 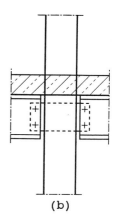

(a)   (b)

Fig. 6. Beam to column connection with continuous
beams (a) and continuous columns (b); principle
sketch.

7 The CIDECT design guide for SHS-columns exposed to fire

Within CIDECT a Design Guide for Hollow Section Columns
Exposed to Fire is under preparation (Twilt, et al. 1993).
The Design Guide, on which this paper is based, contains
quantitative information and is directed towards
architectural and engineering professions, i.e. emphasis is
on practical design aspects, rather than on scientific
details. Due attention is paid to European calculation rules
currently under way within the Eurocodes.

For a qualitative illustration of the type of information
presented in the Design Guide, refer to Fig. 7.

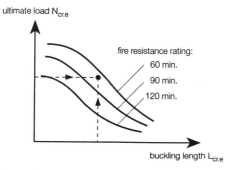

Fig. 7. Qualitative buckling curves of concrete filled SHS
columns for different fire resistance classes.

In this diagram, the collapse load of a concrete filled SHS
column is given as a function of the buckling length. For
any combination of applied load and buckling length, the

relevant fire resistant class can easily be determined. The combination indicated in Fig. 7, for example, leads to a fire resistance rating of 60 minutes. Design Guide provides practical information for unfilled and water filled SHS columns. In addition, in depth recommendations are presented on technological aspects and construction details. The Design Guide will be published during 1993.

8 References

Bond, G.V.L. (1975), **Fire and Steel Construction, Water Cooled Hollow Columns**, Constrado, Croydon (UK).

Dorn, T., Hass, R. Kordina, K. (1988), **Brandverhalten von Verbundstützen und -trägern**, Studiengesellschaft für Stahlanwendung e.V., Düsseldorf (G), project 86.

ECCS-T3 "Fire Safety of Steel Stability" (1985), **Design Manual on the European Recommendations for the Fire Safety of Steel Structures**, ECCS-T3, no. 35, first edition.

EC2 (April 1990), **Structural fire design of concrete structures**, Eurocode 2, Part 10, Luxembourg.

EC3 (October 1992), **Design of Steel Structures**, Eurocode 3, Part 1.2: Fire Resistance; Draft prENV 1993-1-2.

EC3 (April 1990), **Structural fire design of composite steel and concrete structures**, Eurocode 3, Part 4, Luxembourg.

Hönig, O., Klingsch, W., Witte, H. (1985), **Baulicher Brandschutz durch wassergefüllte Stützen in Rahmentragwerken (Fire Resistance of Water Filled Columns)**, Research Report, Studiengesellschaft für Stahlanwendung e.V., Düsseldorf (G), Forschungbericht p. 86/4.5.

IMO Resolution A.517(13) (Nov. 1983), **Recommendation on fire test procedures for "A", "B" and "F" class divisions**, International Maritime Organisation.

ISO (1975), **Fire resistance tests - Elements of building construction**, ISO 834 International Standard Organisation, first edition.

Klingsch, W. (1991), Optimization of Cross Sections of Steel Composite Columns, Proceedings of the third **International Conference on Steel-Concrete Composite Structures**, Fukuoka, Japan, p. 99-105.

Standard for Safety (1991), **Fire Tests of Building Construction and Materials**, Underwriters Laboratory UL 263 (USA).

Steel Promotion committee of Eurofer (1990), **Steel and fire safety: a global approach**, Brussels.

Twilt, L. and Both, C. (1991), **Technical Notes on the Realistic Behaviour and Design of Fire Exposed Steel and Composite Structures**, TNO Building and Construction Research, Rijswijk (NL), BI-91-069.

Twilt, L., Hass, R., Klingsch, W. (1993), **Design Guide for Hollow Section Columns Exposed to Fire**, CIDECT. Under preparation.

PART 3

BOLTED CONNECTIONS

Developments in various methods of connection

10 FLOWDRILLING FOR TUBULAR STRUCTURES

G. BANKS
British Steel Technical, Swinden Laboratories, UK

Abstract
This paper describes the Flowdrilling and Flowtapping technique for producing threaded holes in tubular steel sections. It explains the details of the thread form produced and the tensile strengths achieved with commercial quality bolts. A series of loading tests with two types of beam to column join systems both incorporating Flowdrilled holes are described
Keywords   Rolled Hollow Sections (RHS), Flowdrill, Flowtap connections and bolted connections.

1   Introduction

One of the outstanding features with the use of tubular structures has been the difficulty of direct bolting to the faces of the elements. In the past the options available have been to design a fairly complicated connection to allow the use of conventional bolts and nuts or to use a single sided bolting system which required specialised on-site tools and the necessary extra on-site supervision.
British Steel Technical has been investigating the suitability of the Flowdrill method of producing threaded holes in the wall of tubular members for use in constructional steelwork. The system was originally developed for use in the sheet metalwork industry where relatively light loads compared with structural connections are common. The recent development in the tool bit designs and the use of heavy duty drill machines has allowed the system to be used in thicker material (up to 12.5 mm to date).

2   Flowdrill Technique

The process used is one in which a hole is produced in a plate or tube wall without the metal removal associated

Tubular Structures V. Edited by M.G. Coutie and G. Davies.
© 1993 E & FN Spon, 2–6 Boundary Row, London SE1 8HN. ISBN 0 419 18770 7.

with the normal drilling process. The hole is made by rotating a carbide steel conical tool bit at high speed and pressure against the tube surface. This rapidly builds up sufficient local heating to soften the material and allows the tool to be forced through the tube wall and swage the material predominately on the inside of the parent material in the form of a truncated hollow cone on the inner surface. A greater depth of material for the production of threaded holes than that provided by the parent metal (Fig. 1) is therefore produced.

In order to produce a flat outer surface to allow components to be bolted into tight contact, tool bits can be used which incorporate a milling cutter collar. This removes any extruded metal from the outer surface once the tool has penetrated the tube wall.

A subsequent operation with a cold forming thread tap/flow tap is then performed to produce the required thread form. The rotation of the lobed shaped tap creates the thread profile without any metal removal.

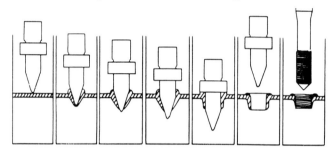

Fig. 1 Flowdrill/Flowtap Technique

For the series of investigations performed to date it was decided that the M20 thread form should be used. This was considered to be reasonably representative of the thread sizes used in the UK building industry. The drill size used for this thread size was 18.7 mm diameter as recommended by the tool bit manufacturers[1].

3   Threaded Profile Dimensional Checks

A selection of holes were produced in tube thickness of 5, 6.3, 8, 10 and 12.5 mm. The material used was from hot formed R.H.S. tubes, nominally made to BS4360 Grade 43C.

The typical profile of the various samples measured is shown in Fig. 2a along with the standard ISO metric thread form for comparison.

It can be readily seen that the Flowtap thread profile had a distinctive double crest at the top of the thread form, caused by the material being swaged into the clearance of the Flowtap tool thread profile from the Flowdrill diameter.

At the base of the thread form the grain flow, see Fig. 2b shows the continuous grain structure flowing from the parent metal into the thread profile. This detail is indicative of a sound strong thread design and is due to the cold rolling/swaging action of the flowtap tool in which no cutting of the parent metal takes place.

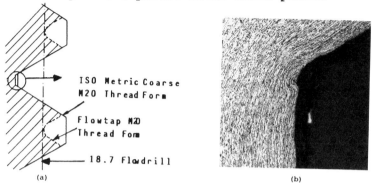

ISO Metric Coarse
M20 Thread Form

Flowtap M20
Thread Form

18.7 Flowdrill

(a)

(b)

Fig. 2 Thread Details

4   Extruded Metal Cone Profile

A sectioned and polished sample taken from each of the tube thicknesses was placed on a precision 'Shadowgraph' magnifying inspection table and the profiles of the extruded metal were measured.   Two profiles are shown in Fig. 3.   These measurements showed that the thickness of the parent metal has only a minor effect on the height of the extruded metal cone of approximately 11 to 13 mm. The greatest advantage with the thicker parent plate is that the greater diameter of the extruded cone provides increased material thickness for the formation of a full thread profile.

5   Metallurgical Examination

Hardness profiles were conducted on the above samples to check the comparable strength of the threaded portion of the hole and the parent material.   In all cases there had been a substantial increase in the strength of the material around the hole produced by the Flowdrill process.

In order to examine the reason for this increase in

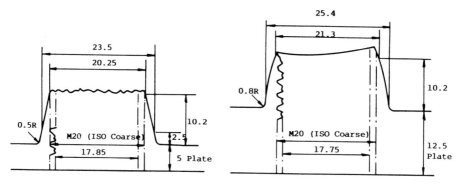

Fig. 3 Extrusion Profiles

strength a typical Flowdrill and Flowtapped sample was submitted to micrometallurgical examination.

The microstructure in the parent tube consisted of equiaxed ferrite with banded pearlite, typical of hot rolled or normalised products. The grain structure closer to the thread form was finer and showed some bainitic transformation product, indicating that the structure had been reheated above 600°C and had been partially re-austenitised at ~800°C.

The microstructure at the tip of the threaded area was generally more coarse and there was evidence of re-austenitisation and transformation to bainite.

The partial refining of the microstructure in the threaded area due to the heat generated during the Flowdrilling process had caused an increase in strength of the parent (Grade 43c) structure.

6   Direct Tension Tests

In order to obtain a useful guide to the tensile strength of Flowtapped threaded holes in various wall thicknesses a series of bolt pull-out tests has been performed.

As part of an ongoing long term investigation by British Steel Technical into the effects of fire on steel framed buildings, previous tests had been carried out on the tensile strengths of nut and bolt combinations at various temperatures. The results of some of these tests carried out at room temperature provided ideal standards for checking the comparative strength of the Flowtapped holes using bolts from the same production batch.

From the results shown on the following table it can be seen that the Flowtapped holes compared very favourably in strength with the nut/bolt assembly and only at tube wall thicknesses below 10 mm was the Flowtapped hole capacity inferior to that of the nut/bolt

Table I. (Pull Out Test)

Female Thread	Tube Thickness mm	Thread Length mm	Max. Load kN	Failure Mode
Flowdrill	5	10	143.9	Threads Stripped
Flowdrill	6.3	11	188.1	Threads Stripped
Flowdrill	8	14	214.9	Threads Stripped
Flowdrill	10	18	228.3	Bolt Failed
Flowdrill	12.5	20	228.5	Bolt Failed
Hex. Nut (GD.8)		16	227.8 (Average of 5 Tests)	Bolt Failed

* Note : All bolts used from the same production batch

assembly, and even with the thinnest tube tested (5 mm) a comparative joint strength of 63% was attained.

Generally precision grip bolts are produced to BS3692 but since building structures often use bolts in clearance holes, costs are reduced by maintaining the strength to BS3692 whilst relaxing the dimensional tolerances to suit BS4190. Note that the replacement Standards for BS3692 and BS4190 have now been issued, viz BS EN 24017/24014 and 24018.

The bolts used in these trials had been produced to tolerances which were border line for the tolerance limit of BS3692 and had therefore been supplied against the lower dimensional tolerance of BS4190.

7   Full-Scale Joint Assembly Tests

In order to provide further evidence of the suitability of Flowtapped holes for structural connections a series of beam to column connections designed in accordance with the BCSA[2] design manual was tested.

7.1 Simple Beam/Hollow Section Column Shear Connections Fig. 4

A series of tests was carried out using M20 Flowtapped holes and grade 4.6 bolts in various sizes and thicknesses of R.H.S. The thicknesses of R.H.S. ranged from 5.0 to 12.5 mm with b (breadth) to t (wall thickness) ratios between 16 and 40. In all cases the beam connections were loaded close to the column to minimise bending and rotation effects. All the tests reached the limit of the test rig (500 kN) without any failure occurring. In all cases the loads achieved were

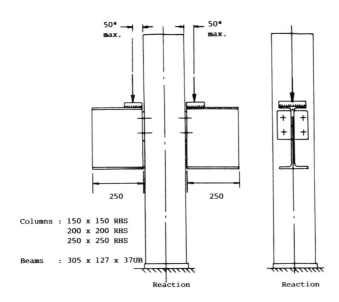

Fig. 4 Shear Connections

well in excess of the predicted ultimate shear capacity
for equivalent conventional bolt/nut systems.   The
specimens were examined after testing and the only signs
of potential failure were slight distortion of the
threaded holes in the R.H.S. and some bending of the
bolts.

7.2 Moment Carrying Beam to Column Connections – Fig. 5a
In these tests simulated bolted beam to column joints
were subjected to direct tensile loads – Fig. 5b.
    The column samples were produced from 150 mm square
R.H.S. (nominal Grade 43C) in thicknesses from 5 mm to
12.5 mm.   Cross joint type tests were carried out in a
universal tensile machine as shown in Fig. 5b.
    The results indicated that the limiting factor for
these joints was the deformation of the R.H.S. sections
and not the capacity of the Flowdrill connections. The
deformation results when compared with predicted
capacities, indicate that the IIW joint design formula[3]
for the chord face deformation of R.H.S. to R.H.S. cross
joints may be used as a design guide for predicting the
deformation criteria in these types of flowdrilled
connections.   Work is currently underway to formulate an
appropriate design method.
    Typical test results showing the amount of section
deformation induced before failure of the thread
connections occurred are shown in Fig. 6.

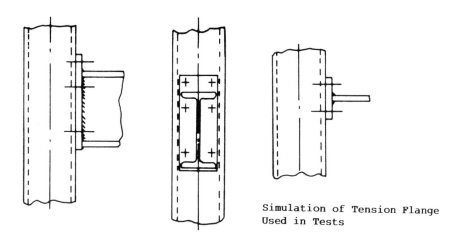

Simulation of Tension Flange
Used in Tests

Fig. 5a Extended Endplate Connection

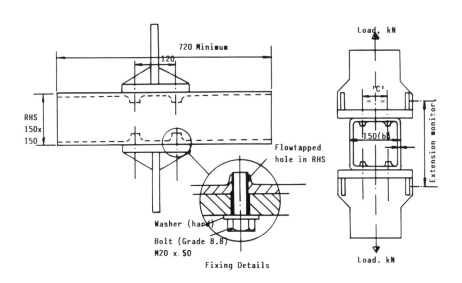

Fig. 5b Cross Joint Type Tests

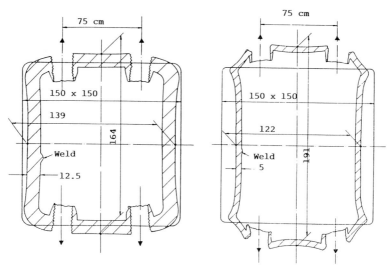

Fig. 6 Deformed RHS Sections

8   Conclusions

1.   The results and observations show that the Flowdrill
      and Flowtap processes produce sound threaded holes
      suitable for use in   constructional steelwork
      connections.
2.   The Flowdrilling and Flowtapping processes produce a
      threaded hole which has an effective thread length
      between approximately 1.8 and 3.0 times the   original
      material thicknesses.
3.   For the design of simple flexible end plate
      Flowdrilled connections (with shear loadings only)
      current analytical procedures could be   applied.
4.   For moment carrying face connections the limiting
      criterion is deformation of the hollow section and
      not failure of the Flowtapped connections.

9   References

1.   Flowdrill bv, Flowdrilling, a new manufacturing
      process - Flowdrill bv, Populierenlaan 18, 3735 LH
      Bosch en Duin. The Netherlands, October 1987.
2.   BCSA Publication 'Manual on Connections, Vol. 1 -
      Joints in Simple Construction' No. 19, 1988. Table
      4.1 (Type P2B and P3B).
3.   International Institute of Welding, 'Design
      Recommendations for Hollow Section Joints -
      Predominantly Statically Loaded', IIW Document No.
      XV-701-89, 2nd Edition, 1989.

11 BEHAVIOUR OF BLIND BOLTED MOMENT CONNECTIONS FOR HSS COLUMNS

A. GHOBARAH, S. MOURAD and R.M. KOROL
Department of Civil Engineering, McMaster University, Hamilton, Ontario, Canada

Abstract
Due to lack of access inside the hollow section column, it has been difficult to develop a practical bolted moment connection between a W-shape beam and a hollow steel column. Field welded beam-to-column connections remain the only viable method of connecting both elements. Because of problems of ensuring a dry environment and the concern about the quality of field welds, the use of hollow section columns in steel frames has been limited.
The objective of this research was to test extended end-plate bolted moment connections for hollow steel columns and to investigate the connection behaviour and assess its potential use in practice. The novelty of the connection is the use of high strength blind bolts which do not require access from inside the hollow section column. The performance of six connections was evaluated based upon the moment-rotation relationship, failure modes, stiffness and moment capacity. When compared with a similar connection using ordinary high strength A325 bolts of equivalent size, the high strength blind bolted connection was found to be very comparable. Attempts have been made to improve the performance of such a connection and to prevent the occurrence of undesirable failure modes.
Keywords: Blind Bolts, Moment Connections, HSS Columns, Extended End-Plate, Steel Frame.

1 Introduction

The use of Hollow Structural Section (HSS) columns in moment-resisting frames has been limited due to the lack of a practical bolted-type moment connection for square or rectangular columns and W-shape beams. Due to the difficulty of gaining access to the inside of the column for bolting, field welded beam-to-column joints have attracted considerable research interest, particularly in moment connections. However, the problems of ensuring dry environments and the concern about the quality of field

Tubular Structures V. Edited by M.G. Coutie and G. Davies.
© 1993 E & FN Spon, 2–6 Boundary Row, London SE1 8HN. ISBN 0 419 18770 7.

welds has limited the use of HSS sections in steel frames.
Picard and Giroux (1977) developed moment connections by
transferring the beam flange force to the hollow section
column webs or to the opposite beam flange by means of
strap angles welded to the sides of the HSS column and the
beam flanges. Dawe and Grondin (1990) tested connections
involving tension and compression plates with web clip
angles or tension plates and seat angles welded to a
doubler plate to reinforce the column walls. Meanwhile,
Tabuchi et al (1988) and Kato et al (1981) used exterior
diaphragms fitted and welded around the HSS column and the
beam flanges.

Special techniques are needed to accomodate bolted
moment connections with HSS columns. Maquoi et al (1984)
developed a beam-to-column connection by welding threaded
studs onto the walls of the HSS column. Kanatani et al
(1987) conducted an experimental study on concrete filled
hollow section columns to W-shape beams employing long
high strength bolts to be tighten from outside the
opposite flange of the column, while Kato (1989)
investigated a connection employing special circular nuts
welded to the column flange through conical holes.
Unfortunately, most of these innovative connections
involve increased fabrication and cost.

A new type of blind fasteners was developed for use in
situations where the rear side of a connection is
inaccessible. The objective of this research was to
establish their potential in practical bolted moment
connections with square columns, and to investigate joint
behaviour as determined from an experimental program. The
method of connection assemblage is to weld an extended
end-plate to the beam, and then to bolt the plate to the
flange of the HSS column in the field, as shown in Figure
1. Such a connection permits shop welding of the end-
plate and on-site bolting, hallmarks for good quality
control and economy.

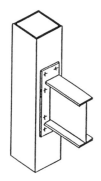
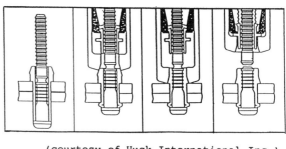

(courtesy of Huck International Inc.)

Fig.1. Typical extended   Fig.2. Installation sequence
    end-plate connection   for BOM fasteners

2 Blind fasteners

A new fastening system was recently developed by Huck International Inc., known as a Swage Lock Fastening System. By virtue of its design, it provides a consistent and high clamping force achieved by direct tensioning. The fasteners can be installed by unskilled operators using hydraluic equipment that is portable on the job site. The Swage Lock Fastening System includes blind fasteners of two basic types.

2.1 BOM fastening system
A simple schematic diagram of the Blind Oversized Mechanically Locked (BOM) fastener and its installation sequence are shown in Figure 2. These fasteners have tensile strengths about 70% that of A325 bolts, but with shear capacities about twice the A325 nominal resistance for similar diameters. Because of their high shear resistance, BOM bolts are well suitable for shear connections.

2.2 HSBB fastening system
High Strength Blind Bolts (HSBB) are still in the testing stage and have not yet been used in structural applications. A schematic representation of HSBB and the installation sequence is shown in Figure 3. As noted, the HSBB are designed for a slightly modified installation process to meet ASTM specifications for A325 bolts. Both the minimum tensile strength and clamping force of an HSBB bolt are claimed to be higher than that of an A325 bolt of similar size.

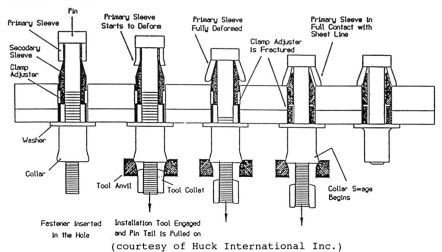

(courtesy of Huck International Inc.)

Fig.3. Installation sequence for HSBB fasteners

3 Experimental program on bolted moment connections

An experimental program was conducted on bolted end-plate moment connections of a W-shape beam and an HSS column. Five specimens using blind fasteners and one specimen using A325 were tested under monotonic loading. To compare the performance of the BOM bolts and HSBB with A325 bolts in such a connection, the first three specimens (S1, S2 and S3) were identical except for the fasteners. Details are given in Table 1. A larger column size with a thinner wall was selected for the beam-to-column connections to reveal different behaviour and failure modes of both unstiffened and stiffened column flanges. Therefore, the second set of specimens (L4, L5 and L6) utilized column size 254x254x11.13 mm and 22 mm thick end-plates, as noted in Table 1. The column flange of L5 was stiffened by a 6 mm doubler plate welded to it, while the column flange of specimen L6 was stiffened by filling the column with concrete.

Table 1. Description of tested specimens

Spec. No	Column size (mm)	End-plate (mm)	Number and Size of bolts	Remarks
S1	203x203x12.7	520x170x19	8-3/4" A325	Unstiffened
S2	203x203x12.7	520x170x19	8-19mm BOM	Unstiffened
S3	203x203x12.7	520x170x19	8-20mm HSBB	Unstiffened
L4	254x254x11.13	590x230x22	8-20mm HSBB	Unstiffened
L5	254x254x11.13	590x230x22	8-20mm HSBB	Doubler plate
L6	254x254x11.13	590x230x22	8-20mm HSBB	HSS concrete filled

3.1 Specimen fabrication

The material used for all six specimens including beams, end plates and doubler plates was in accordance with CSA G40.21-M92-300W steel with a minimum yield strength of 300 MPa. The hollow section stub columns conformed to CSA G40.21-M92 class H with a minimum yield strength of 350 MPa. For all specimens, the beam was selected to be a plastic design section, a W360x33, class 1, with its nominal plastic moment being 162 kN.m. To focus on the behaviour of the connection itself while achieving ultimate capacity, the top and bottom flanges of the beam were stiffened by a 6 mm thick cover plate at the connection. The stiffened beam was welded to the end plate by 12 mm and 6 mm fillet welds for the stiffened flange and web respectively. The nominal plastic moment of the stiffened beam was 225 kN.m. The connection configurations for S1, S2 and S3 as well as L4, L5 and L6 are shown in Figure 4. For specimen L5, the doubler plate (1000x220x6 mm) was welded all around to the column flange with 6 mm

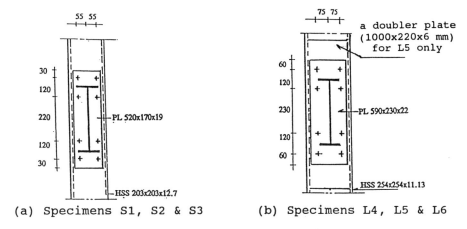

(a) Specimens S1, S2 & S3   (b) Specimens L4, L5 & L6

Fig.4. Details of connections

fillet welds and with 6 plug welds each of 12 mm diameter
(between the bolts lying along the same vertical line).
Bolt holes were drilled 2 mm larger than the nominal bolt
diameter. The column stub of L6 was filled with concrete
of nominal strength of 25 MPa, after installing the bolts.

3.2 Experimental arrangement and instrumentation

A fixture frame, especially fabricated for the tests, was
prestressed to the floor of the laboratory. A hydraulic
actuator was used to apply the load to the beam tip in a
vertical direction. The general arrangement of the set-up
is shown in Figure 5. Two angles were installed near the
end of the beam to provide lateral support. Linear Voltage
Displacement Transducers (LVDT) were used to measure beam-
tip displacement, connection rotation, out-of-plane
deformation of the column side walls and the column
flanges.

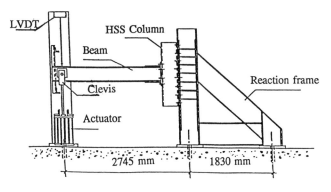

Fig.5. General arrangement of test setup

3.3 Experimental results

As noted from Figure 6, the behaviour of S3 using HSBB fasteners appears to be similar to that of S1 employing A325 bolts, with its initial stiffness slightly higher and its ultimate moment capacity slightly less. The slightly higher initial stiffness of connection S3, may be attributed to the high and consistent clamping force of the HSBB. During the later stages of loading, significant bending took place in the end-plate and in the exterior HSBB bolts on the tension side of the connection, causing partial failure of their primary sleeves. The column flange thickness of 12.7 mm seemed to be just adequate to sustain the bolt force without shearing the column flange underneath the primary sleeve. At ultimate load, a cleavage crack was initiated in S1 at the toe of the weld between the end-plate and the tension beam flange, which eventually propagated through the end-plate thickness to cause a sudden drop in load carrying capacity. Meanwhile, S3 suffered from a brittle failure in the heat affected zone of the beam flange near the end-plate. On the other hand, specimen S2 showed poor performance due to the BOM bolts' premature failure. This result was not unexpected since the nominal tensile strength of the BOM bolt is less than that of A325 by about 30%.

The maximum moment capacity sustained by specimen L4 was 245 kN.m. At failure it showed significant yielding in the column flange and at the corners of the HSS column near the bolts on the tension side. This is reflected by the large rotation capacity in Figure 6. Because of the larger thickness of end-plate and edge distance of the exterior bolts as compared to S1, S2 and S3, less bending of the end-plate was observed which resulted in minimal bending of the exterior HSBB and decreased prying action. It is evident that the column flange thickness of 11.13 mm was not adequate to resist the maximum strength of the 20 mm HSBB.

The 6 mm doubler plate in L5 resulted in a higher initial stiffness from that of L4 by about 10% and a remarkable increase in the post-elastic stiffness, thus causing a reduction in the rotation capacity, as noted in Figure 6. Its ultimate capacity was 240 kN.m at which point one of the HSBB bolts failed by stripping the swaged threads in the collar. The total thickness of the column flange and the doubler plate for this connection seems to have been adequate.

The sixth specimen of the series, L6, involved infilling of the column with concrete. This resulted in a surprising increase in the connection's initial stiffness about 3 times that without concrete (specimen L4), as shown in Figure 6. This increase is attributed to the reduction in the column flange deflection at both the tension and compression sides of the connection. Its

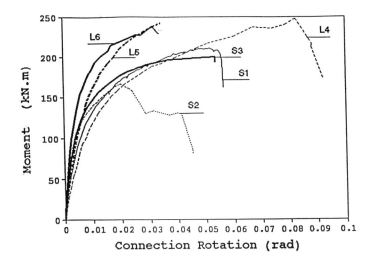

Fig.6. Moment-rotation relationships of the
tested specimens

failure mode was by shearing the column flange thickness
beneath the primary sleeves of the HSBB. The measured
moment capacity of the connection was 240 kN.m.

4 Conclusions

Based on the observations and results of the tests
conducted in this study, the following concluding remarks
can be made:

1 With the use of HSBB, a bolted end-plate connection
between W-shape beams and HSS columns is possible
2 The behaviour of the proposed bolted moment connection
using HSBB is judged to be similar to that when using A325
bolts. It appears that HSBB have a promising potential in
structural connections.
3 It was confirmed that BOM bolts will not achieve the
equivalent strength of A325 bolts.
4 Stiffening the HSS column flange with a welded doubler
plate, does not affect the initial stiffness of the
connection significantly. However, it increases the post-
elastic stiffness and improves the performance of such a
connection. In addition, the doubler plate prevents the
occurrence of undesirable shear failure of the column
flange beneath the HSBB fasteners.
5 Filling the HSS column with concrete appears to
significantly increase the initial stiffness of the
connection, thus reducing excessive rotations at such
joints in building frames.

5 Acknowledgment

The authors wish to express their gratitude to Huck International Inc. and, in particular, to P.Lowe and S.Sadri for providing the bolts, installation equipment and information necessary to undertake this study. The investigation herein described was sponsored by the Natural Sciences and Engineering Research of Canada (NSERC) through grants to McMaster University.

6 References

Dawe, J. L. and Grondin, G. Y. (1990) W-shape beam to RIS column connections. **Canadian Journal of Civil Engineering**, 17, pp 788-797.

Huck International Inc., **Industrial Fastening Systems**, Irvine, California, U.S.A

Kanatani, H., Tabuchi, M., Kamba, T. and Ishikawa, M. (1987) A study on concrete filled RHS column to H-beam connections fabricated with HT-bolts in rigid frames. **Proceedings of an Engineering Foundation Conference on Composite Construction in Steel and Concrete**, Henniker, NH, pp 614-635.

Kato, B.(1989) Bolted beam-to-column moment connection. **Proceedings of International Colloquium on Bolted and Special Structural Joints**, Moscow, volume 2, pp 29-38.

Kato, B., Maeda, Y., and Sakae, K. (1981) **Behaviour of rigid frame sub-assemblages subject to horizontal force.** Joints in Structural Steelwork, John Wiley Sones, New York, U.S.A. pp. 1.54-1.73.

Maquoi, R., Navean, X. and Rondal, J. (1984) Beam-column welded stud connections. **J. Construct. Steel Research**, 4, pp 3-26.

Mourad, S., Korol, R. and Ghobarah, A. (1993) A newly developed bolted moment connection for hollow structural columns. **Proceedings of the Annual Conference of the Canadian Socity of Civil Engineering**, Frederiction, New Brunswick, Canada.

Picard, A. and Giroux, Y. (1977) Rigid connections for tubular columns. **Canadian Journal of Civil Engineering**, 3(2), pp 134-144.

Tabuchi, M. Kanatani, H. and Kamba, T (1988) Behaviour of tubular column to H-beam connections under seismic loading. **Proceedings of the Ninth world Conference on Earthquake Engineering**, Tokyo, Japan (vol.IV), pp 181-186.

CSA-S16.1-M89. (1989) **Canadian Steel Structures for Buildings. (Limit State Design)** CSA-Standards, Ontario, Canada.

CSA-G40.21-92. (1992) **Structural Quality Steel.** Canadian Standards Association, Ontario, Canada.

12 BOLTED CONNECTIONS TO HOLLOW SECTIONS WITH THROUGH BOLTS

J. LINDNER
Technical University of Berlin, Germany

Abstract
Bolted connections to hollow sections behave differently compared with those in conventional connections, because the local support by plates, head of bolt and nut is not present. It is reported on basic tests to determine the load carrying capacity of such connections. The investigations were mainly carried out with respect to different distances of fasteners regarding the failure modes of hole bearing and failure of net section. The results are presented with regard to the parameters investigated and statistical evaluations deal with the application of the design methods used for conventional connections.
Keywords: Connections, Bolts, Bolt-distances, Tests, Statistical Evaluation

1 Introduction

For the connection of hollow sections welded connections mainly are in use. Therefore for welded truss connections intensive research has been made resulting in an international consensus for the limit state design of statically loaded welded connections involving hollow section members in planar trusses. Several national codes as well as Eurocode 3 have adopted this work. Design aids in the form of charts giving connection efficiency allow easy design.

On the other hand bolted connections are widely used in steel structures especially if static loads only have to be considered. This type of connection especially allows simple site connections.

Connections between open and tubular sections are compiled by Saidani and Nethercot (1992). For the construction of medium or long-span trusses of hollow sections field splices are used with shop welded truss segments being field-bolted together on side, see

Tubular Structures V. Edited by M.G. Coutie and G. Davies.
© 1993 E & FN Spon, 2–6 Boundary Row, London SE1 8HN. ISBN 0 419 18770 7.

Wardenier (1992). Usually flange plate connections are
used, see Fig.1.

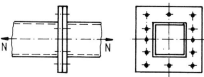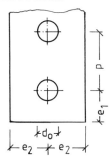

Fig.1. Bolted flange-plate   Fig.2. Bolted gusset-plate
        connection   connection

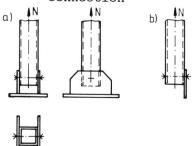

Fig.3. Connections for axial loaded   Fig.4. Distance
        elements, a) symmetric connec-   of fasteners
        tion, b) unsymmetric connection

For rectangular hollow sections gusset-plate
connections as shown in Fig.2 are also possible, which had
been studied by Birkemoe and Olynyk (1980). Following the
type of connection used for open cross sections the bolts
connect the gusset-plate and one flange of the hollow
section.

At the Technical University of Berlin additional
investigations were carried out, Lindner et al. (1992),
where through bolts are used, see Fig.3. Here the type of
connection changes compared to that of Fig.2. In this case
the local support by plates, head of bolt and nut is not
present and therefore the behaviour of these connections
differs from that of conventional connections.

The investigations were inspired by proposals of
Buchholz (1987) for the calculation of bolted connections
for hollow sections.

2   Tests

2.1 Parameter
The connections investigated here are loaded by axial
forces in the member only and therefore the bolts are
subjected to forces perpendicular to their axis.

The main objective was to investigate the load carrying capacity with regard to hole bearing and tension failure of plate, not shear failure of the bolts themselves. Therefore the end distance e_1 and the distance between bolts p are varied, see Fig.4, and Table 1.

2.2 Test specimens and implementation
The test specimens were chosen in such a way that either bearing failure or tension failure of plate occur.

The rectangular hollow sections were 350 mm long, steel grade RSt 37-2. High strength grade 10.9 bolts due to DIN 6914 were used with zinc corrosion protection. The actual dimensions of sections and bolts were measured and used for the evaluations as well as the actual mechanical properties.

Table 1. Survey on tests

test series	A	B	C	D	E	F	G	H
section	60x40x4	40x40x4	70x70x4	70x40x3	60x40x4	70x40x4	70x40x3	40x40x4
type		1 bolt - symmetric				2 bolts-unsymmetric		1 bolt-symm.
e_2/d_o	1.67	1.11	1.94	1.94	1.67	1.94	1.94	1.11
e_1/d_o	1-6	1-6	1-6	1-6	3	3	3	1-6
p/d_o	-	-	-	-	2-4	2-4	2-4	-
grade				St 37				
bolt				M 16				
d_o [mm]				18				
number of tests	20	10	10	11	12	6	12	20

The mechanical properties were determined from tensile coupon tests due to DIN 50 125, loading rate $\Delta\sigma$ approximately 10 N/mm²min. The following mean values were found:
- yield strength   300 up to 361   N/mm²,
- static yield strength   276 up to 341   N/mm²,
- ultimate tensile strength   403 up to 473   N/mm².

All test specimens were loaded by a test equipment which was similiar to that of a tensile coupon test. Load deflection curves were determined.

3 Test results and evaluation

3.1 General
The ultimate load which was reached in the tests depended on the loading rate. Therefore some of the tests were

stopped from time to time for 10 minutes and the decrease
of load was measured, which reached approximately $\Delta = 5\%$.
In the evaluations reduced test results of $F(1-\Delta)$ were
used.

3.2 Tension failure of plate
The failure occured in test series E and F at the first
bolt after hole elongation.
   For comparison a failure strength following eq. (1) was
calculated.

$$\overline{\sigma}_n = \frac{F(1 - \Delta)}{f_u \, A_n} \tag{1}$$

where
F ultimate load,
f_u ultimate tensile strength,
A_n net section.

   The values of $\overline{\sigma}_n$ from 18 tests are between 0.995 and
1.055.

3.3 Bearing capacity
As found in other tests before (Scheer (1987)) the failure
modes shear failure and bending failure occured and in
several cases a combination of both were present.

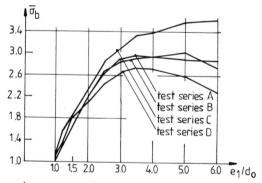

Fig.5. Bearing capacity with regard to edge distance e_1 in
the direction of load transfer

End distance e_1
   This most important value was varied between $e_1 = 1.0 \, d_0$
up to $6.0 \, d_0$. The results expressed as relative bearing
strength due to eq. (2) can be seen from Fig.5.

$$\overline{\sigma}_b = \frac{F(1 - \Delta)}{f_u \, t \, d} \tag{2}$$

136

where
t   plate thickness,
d   bolt diameter.

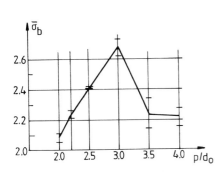
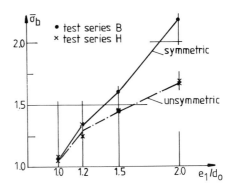

Fig.6. Load carrying capacity   Fig.7. Capacity for
       with regard to centre-   symmetric and
       to-centre distance p   unsymmetric
       in load direction   connections

Fig.5. shows that the bearing capacity increases
linearly dependent on e_1/d_0, but levels off in the region
of $e_1/d_0 = 2.5$ up to 3. For great distances a decrease can
partly be seen. This occured due to local buckling. The
factor of $e_1/d_0 = 3.0$ which is used in Eurocode 3 and DIN
18 800 part 1 (1990) as an upper limitation for
calculation is not a sufficient limitation for the type of
connection investigated here.

End distance e_2
     The variation of this parameter was possible only due
to the hollow sections investigated here for $e_2/d_0 = 1.11$,
1.67 and 1.94. The dependency of $\overline{\sigma}_b$ on e_2/d_0 is small.

Bolt distance p in load direction
     The results of test series G are shown in Fig.6. The
bearing capacity increases with increasing values of p/d_0
but decreases if the value of p/d_0 is greater than 3.0.
Local instability is even more important in this case than
with regard to the end distance e_1.
     Tests with one or two bolts are compared also. The
result shows that the bearing capacity of two bolts is
approximately 1.8 times that of one bolt instead of the
value of 2.0 which is expected due to linear addition of
the capacities.

3.4 Unsymmetric connection
These test specimens behaved in another way than the
symmetric ones. All specimens with $e_1/d_0 > 2$ failed by

rupture of bolt, bearing failure modes occured for small values of e_1 only. The relative bearing capacity can be seen from Fig.7.

4 Statistical evaluations

4.1 General

The evaluation method of Eurocode 3 is used.

The variation coefficients were taken from literature. Due to Wardenier (1992):

section width, section depth	v_B, v_H	=	0.005 ,
section thickness	v_t		0.05 .

Due to Snijder et al. (1989):

ultimate strength sections	v_{fu}		0.07 ,
hole diameter	v_{d0}		0.005 ,
distances e_1, e_2, p	v_e		0.005 .

4.2 Failure in the net section

The strength function of EC 3 is given by eq. (3a) which was additionally changed to eq. (3b).

$$V_u = 0,9 \ A_n \ f_u \qquad\qquad (3a)$$

$$V_u = 1,0 \ A_n \ f_u \qquad\qquad (3b)$$

The evaluation leads to the results of Table 2. It can be seen that the partial safety factor of 1.25 which is assumed in EC 3 and DIN 18 800 Part 1 (11.90) is reached better by eq. (3b).

Table 2. Statistical evaluation regarding failure of net section

tests	eq.	k_s	k_d	ρ	$\overline{b}$	V_δ	r_k	r_d	γ_M	ΔK	γ_M^*
18	(3a)	1.950	3.540	0.903	1.195	0.074	0.954	0.796	1.199	0.909	1.090
18	(3b)	1.950	3.540	0.903	1.075	0.074	0.859	0.716	1.199	1.010	1.211

4.3 Bearing capacity

By Snijder et al. (1989) eq. (4a) is used for connections with one bolt and eq.(4b) for connections with two bolts.

$$F_{bs} = 2,5 \ f_u \ d \ t \ \alpha \qquad\qquad (4a)$$

$$F_{bs} = 2,5 \ f_u \ d \ t \ [\alpha_1 + (n-1) \ \alpha_2] \qquad\qquad (4b)$$

where

α, α_1 the smaller value of $e_1/3d_0$ or 1.0 ,

α_2 the smaller value of $e/3d_0 - 0{,}25$ resp. 1.0 .

In addition eq. (4b) was evaluated now using α_2^* instead of α_2 where

$$\alpha_2^* = e/3d_0 \qquad\qquad \text{for } e \leq 3\ d_0 \qquad\qquad (5a)$$

$$\alpha_2^* = 2 - e/3d_0 \qquad\qquad \text{for } e > 3\ d_0 \qquad\qquad (5b)$$

The results can be seen from Table 3. Tests with one bolt are considered sufficiently by the strength function. Test results with greater end distance $e_1 > 3\ d_0$ are not yet sufficiently covered as well as tests for two bolts where too few test results are available.

Table 3. Statistical evaluation regarding failure of hole bearing

tests	eq.	k_s	k_d	ρ	$\bar{b}$	V_s	r_k	r_d	γ_M	ΔK	γ_M^*
all 63	(4)-(5)	1.780	3.250	0.961	1.262	0.136	0.942	0.746	1.263	0.921	1.163
1b 51	(4b)	1.800	3.290	0.964	1.303	0.122	0.992	0.797	1.245	0.874	1.088
1b 20 $e_1/d_0 > 3$	(4a)	1.930	3.510	<u>0.875</u>	1.239	0.128	0.922	0.729	1.265	0.941	1.190
2b 12	(4b)	2.050	3.700	<u>0.134</u>	1.164	0.154	0.813	0.614	1.324	1.068	1.413
2b 12	(4b)+(5)	2.050	3.700	<u>0.869</u>	1.068	0.084	0.863	0.718	1.202	1.005	1.2008

<u>   </u> insufficient correlation

5 Application

With regard to the fact that the investigations reported here are of limited extent the results should be used only in the range of the investigated parameters:
- relation section thickness/bolt diameter> 0.18 ,
- bolt diameterM16,
- end distance e_1 = 1.0 d_0 up to 6.0 d_0,
- distance between boltsp = 2.0 d_0 up to 4.0 d_0,
- steel gradeSt 37/St 52.

An extention is given by investigations of Mang [6] but statistical evaluations are not kwown for the time being.

6 Acknowledgement

These investigation were sponsored by the "Deutsche Forschungsgemeinschaft (DFG)" financially. Furthermore the steel producer "Mannesmannröhren-Werke AG" contributed the hollow sections and "Peiner Umformtechnik GmbH" the high strengh bolts. At the Technical University of Berlin Mr. Harms and Dipl.-Ing. Müller conducted the tests and Prof. Bamm supervised the project. This help was gratefully acknowledged.

References

Bijlaard, F.S.K., Sedlacek, G. and Stark, J.W.B. (1988) Procedure for the determination of design resistance from tests. Background report to Eurocode 3, Report B1-87-112, Rijswijk (Z.H.).

Birkemoe, P.C. and Olynyk, P.N. (1980) Bolted truss connections of rectangular steel tubes. Re-print 80-053, ASCE Spring Convention, Portland, Oregon.

Buchholz, E. (1987) Schraubenverbindungen bei Rechteck-Hohlprofilen (Bolted connections for hollow sections). Stahlbau 56, 239-245.

DIN 18 800 Teil 1 (1990) Stahlbauten, Bemessung und Konstruktion (steel structures, design and construction).

Eurocode 3 (1992) Design of Steel Structures - General Rules and Rules for Buildings.

Lindner, J., Bamm, D. und Müller, W. (1992) Normalkraftbeanspruchte geschraubte Anschlüsse an Hohlprofilen. Stahlbau 61, 371-375.

Mang, F. (1992) Hollow section construction with simple bolted connection. Cidect-Programm 6E, Univ. Karlsruhe, 1. Zwischenbericht.

Scheer, J., Maier, W., Klarhold, M. und Vajen, K. (1987) Zur Lochleibungsbeanspruchung in Schraubenverbindungen (To bearing stress utilisation in bolted connections). Stahlbau 56, 129-136.

Saidani, M., and Nethercot, D.A. (1992) Beam to column connections between tubular and open sections, Cidect report No 5Ay-6/92, international report SR91039, Nottingham, 1992.

Snijder, H.H., Ungermann, D., Stark, J.W.B., Sedlacek, G., Bijlaard, F.S.K. and Hemmert-Halswick, A. (1989) Evaluation of test results on bolted connections in order to obtain strength functions and suitabel model factors - Part A: Results, Part B: Evaluations. Background documentation to Eurocode 3, Part 1, Documents 6.01 and 6.02, Report BI-88-087, Rijswijk (Z.H.).

Wardenier, J. (1982) Hollow section joints. Delft University Press.

Wardenier, J. and Packer, J.A. (1992) Connections between hollow sections. In: Constructional Steel Design, An International Guide, London/New York, Elsevier Applied Science.

PART 4

COLD FORMED HEAVY GAUGE SECTIONS

Effect of production method
on performance

13 DEFORMABILITY OF COLD FORMED HEAVY GAUGE RECTANGULAR HOLLOW SECTIONS: DEFORMATION AND FRACTURE OF COLUMNS UNDER MONOTONIC AND CYCLIC BENDING LOAD

M. TOYODA
Department of Welding and Production Engineering,
Osaka University, Japan
Y. HAGIWARA
Steel Research Laboratories, Nippon Steel Corporation, Futtsu, Japan
H. KAGAWA
Steel Products Laboratory, NKK Corporation, Kawasaki, Japan
Y. NAKANO
Iron and Steel Research Laboratories,
Kawasaki Steel Corporation, Chiba, Japan

Abstract
A series of bending tests was carried out on cold formed columns with
400 to 490MPa class tensile strength steels in order to investigate
fracture behaviour at column corner and assure the safety against
large plastic deformation. Three types of quarter part column specimens were used: Type 1 specimen has a machined notch at the column
corner, Type 2 specimen has a notched weld bead on the column corner
and Type 3 specimen has welded joint which simulates beam–column
connection.
   Monotonic (Type 1 and 2) and cyclic loading (Type 3) was applied in
four point bending at the test temperature of 0℃ or –10℃. In the
monotonic loading tests, some of Type 2 specimens fractured in brittle
manner. While, none of Type 1 specimens showed brittle fracture under
the condition of large plastic strain more than 8%. In the cyclic
loading tests, specimens which welded with a large heat input showed
brittle fracture. On the other hand, in the case of specimens which
welded with a proper heat input, brittle fracture took place in only 3
specimens out of 27. The test results were analyzed based on fracture
mechanics approach.
Keywords: Cold formed column, Column corner, Beam–column connection,
Monotonic and cyclic loading, Brittle fracture, Fracture mechanics,
Fracture toughness, Rectangular hollow sections, Tubular structures.

1 Introduction

Cold formed rectangular columns are widely used in building structures. There are two types of cold formed columns: one is made by roll
forming steel pipe (RF column) and another is manufactured by press
bending (PB column). The strain hardening capacity and the toughness
at the corner part of cold formed columns decrease due to strain
aging. The anti-seismic design for building construction requires
sufficient deformability in structural members before clear loss of
loading capacity, caused by local buckling or brittle fracture.
   In order to assure the safety usage of RF and PB columns as
building structural members, it is important to study the brittle
fracture characteristics on their corners. A series of quarter part
column tests has been carried out by monotonic and cyclic four point
bending. The test results were evaluated on the basis of fracture
mechanics approach.

Tubular Structures V. Edited by M.G. Coutie and G. Davies.
© 1993 E & FN Spon, 2–6 Boundary Row, London SE1 8HN. ISBN 0 419 18770 7.

2 Experiment

2.1 Column material

The columns used in the present study are four kinds of RF and three kinds of PB columns. The thickness of RF columns varies from 16mm to 22mm with 400MPa tensile strength, and 22mm with 490MPa in strength. With respect to PB columns 22mm to 40mm thick ones with 490 MPa tensile strength are used. The chemical compositions and

Table 1. Chemical compositions and mechanical Properties of cold formed column tested

Cold Formed Column	Steel Grade	Thickness mm	Chemical composition (wt%)							Column corner		
			C	Si	Mn	P	S	Al	N	YP MPa	TS MPa	YR %
A	RF	SS400	16	0.22	0.20	0.70	0.017	0.004	0.040	0.0025	497	546	91
B			19	0.16	0.22	0.71	0.011	0.003	0.042	0.0040	480	495	97
C			22	0.16	0.22	0.85	0.019	0.003	0.043	0.0036	516	533	97
D		SM490	22	0.07	0.21	1.18	0.017	0.004	0.032	0.0030	600	625	96
E	PB		22	0.15	0.43	1.46	0.022	0.004	0.03	0.0023	566	643	88
F			25	0.15	0.44	1.46	0.022	0.004	0.03	0.0021	548	612	90
G			40	0.19	0.48	1.52	0.025	0.005	0.03	0.0017	607	687	88

YP: Yield point, TS: Tensile strength, YR: Yield to tensile strength Ratio (=YP/TS)

mechanical properties are shown in Table 1. It is seen that the column corners have higher yield to tensile strength ratio (YR).

2.2 Column quarter part bend specimen

Three types of quarter part column specimens were used; Type 1 specimen has a machined notch at the column corner which simulates "bruise" during construction, Type 2 specimen has a notched weld bead on the column corner and Type 3 specimen has welded joint which simulates beam-column connection. The Type 2 specimens were developed in order to investigate the brittle fracture behaviour on the corner portion although such severe defects do not exist in the actual structure.

Figure 1 shows the specimen geometry and the notch configuration. The Type 1 and 2 specimens have diaphragms inside the corner and a notch or a bead notch is machined in the middle of the specimen. The monotonic four point bend loading was applied in these specimens.

The cyclic loading test was carried out for Type 3 specimens, which have diaphragms penetrating column. Two different welding conditions were used for column-corner-diaphragm welding: one is a large heat input welding (120~150kJ/cm) and another is a conventional heat input welding (24~34kJ/cm). The full reversal cyclic loading was applied under the strain control condition at the mid-specimen. The strain amplitude incrased gradually to 32 times of the yield strain, as shown in Figure 2.

3 Experimental results and discussions

3.1 Monotonic bend test on quarter part of column corner

The Type 1 specimen has a machined notch of 1mm depth on the column corner which simulates the "bruise" during construction although the size and geometry are much more severe with respect to brittle fracture. For column A to G, 12 tests were carried out at 0℃ or -10℃. None of them showed brittle fracture. Figure 3 shows the relation between the maximum strain value attained in the tests and the plate thickness of columns. The maximum strain is more than 8% for most tests without brittle fracture, which seems to be sufficient value to

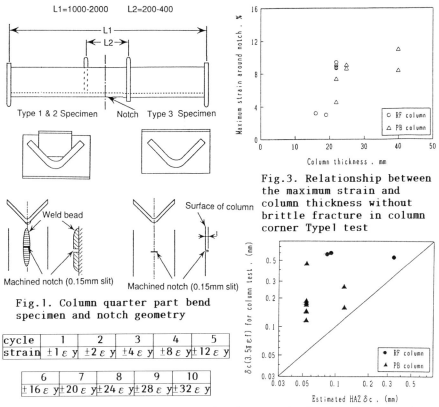

L1=1000-2000   L2=200-400

Type 1 & 2 Specimen   Notch   Type 3 Specimen

Weld bead

Surface of column

Machined notch (0.15mm slit)   Machined notch (0.15mm slit)

Fig.1. Column quarter part bend specimen and notch geometry

cycle	1	2	3	4	5
strain	$\pm 1\varepsilon_y$	$\pm 2\varepsilon_y$	$\pm 4\varepsilon_y$	$\pm 8\varepsilon_y$	$\pm 12\varepsilon_y$

6	7	8	9	10
$\pm 16\varepsilon_y$	$\pm 20\varepsilon_y$	$\pm 24\varepsilon_y$	$\pm 28\varepsilon_y$	$\pm 32\varepsilon_y$

Fig.2. Cyclic loading pattern of Type 3 quarter part column test (ε_y: yield strain)

Fig.3. Relationship between the maximum strain and column thickness without brittle fracture in column corner Type1 test

Fig.4. Relationship between δc obtained from column corner Type2 test and estimated HAZ δc

cause large plastic deformation in a column under suspected huge earthquake. Therefore, it can be concluded that "bruise" defects do not cause any harmful effect on the deformability of cold formed columns.

Brittle fracture behaviour on column corner was investigated using Type 2 specimens having a bead notch. For RF column A to D, 14 tests were carried out, while 11 tests were conducted for PB column E to G. Brittle fracture took place in 4 specimens out of 14 for RF columns and 9 specimens out of 11 for PB columns. In fractured specimens, stable crack growth was observed from a notch into the weld bead and brittle fracture occured when the ductile crack reached fusion line of the weld bead. Critical strain near a bead notch at fracture was 6~8% for RF columns and 2~5% for PB columns.

The test results showed brittle fracture are analyzed based on the CTOD criterion adopted in WES 2805 given by the following equation.

$$\delta = 3.5\varepsilon\,\bar{a} \qquad\qquad (1)$$

In this case, a crack size $\bar{a}$ is determined by assuming a ductile stable crack at onset of brittle fracture as an initial crack. The

145

results are shown in Figure 4 by comparing δc values obtained from column quarter part bend test by using Eq.(1) with the weld bead heat-affected-zone (HAZ) CTOD toughness, which is estimated from the chemical compositions and hardness of column materials using Aihara's method (1992). The tendency is nearly the same between them, although the estimated HAZ CTOD is relatively low. The discrepancy can be caused by the fact that the plastic constraint is relatively low for the column test because of a shallow notch (1~3mm depth) and in column test fracture occured by the condition of $\delta \geq \delta$c, that means the results could estimate higher critical CTOD values than the actual ones.

3.2 Cyclic bend test on quarter part of column corner

None of 6 RF column specimens tested showed brittle fracture. In the case of PB column specimens welded with conventional heat input, 1 specimen out of 15 with 25mm thickness and 2 specimens out of 6 with 40mm thickness showed brittle fracture. On the other hand, all 3 PB column specimens welded with large heat input showed brittle fracture. Examples of load-strain curves under cyclic loading are shown in Figure 5. In 5 specimens out of 6 brittle fractured ones, brittle fracture occured from ductile crack which initiated at weld toe and propagated to fusion line during loading cycles. In another one specimen, brittle fracture initiated from weld metal crack.

Table 2 shows strains and ductile crack sizes of brittle fractured specimens except for one which fractured from weld metal crack. Tensile strain in each cycle and final tensile strain ε f are defined as shown in Figure 6. Accumulated tensile plastic strain $\Sigma \varepsilon$ p is defined as the summation of tensile plastic strain throughout loading cycles. All strain values listed in Table 2 are those measured by strain gauges mounted near the weld toe. 2c and a are defined as a ductile crack length and depth at brittle fracture initiation point, respectively.

Figure 7 shows the relationship between accumulated tensile plastic strain and the maximum ductile crack depth at fracture (solid symbols) or at final unloading for non brittle fractured specimens (open symbols). It is revealed that ductile crack initiates earlier and it propagates more quickly in the specimens welded with larger heat input and in the thicker ones.

A length of ductile crack growth during loading cycle was measured

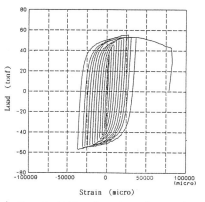 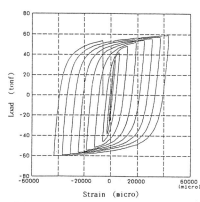

(a) Non brittle fractured specimen   (b) Brittle fractured specimen
Fig.5. Example of load-strain curves

Specimen No.	Plate thickness (mm)	Heat input (kJ/cm)	Accumulated tensile plastic strain $\Sigma \varepsilon p(\%)$	Final tensile strain $\varepsilon f(\%)$	Ductile crack size Length 2c (mm)	Ductile crack size Depth at initiation point a (mm)	Ductile crack size Equivalent crack size $\bar{a}$ (mm)
15-2-1	25	24~34	12.9	3.11	100	1.0	1.4
15-B-1	25	120~150	12.8	2.13	110	2.9	4.4
15-B-2	25	120~150	8.2	3.53	66	1.2	1.6
15-B-3	25	120~150	13.4	4.10	113	1.1	1.5
22-1-3	40	24~34	14.7	3.01	77	2.1	2.9

Table 2. Strain and ductile crack sizes of brittle frac-tured specimens

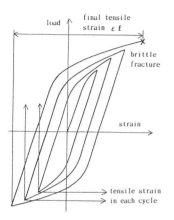

Fig.6. Definition of tensile strains

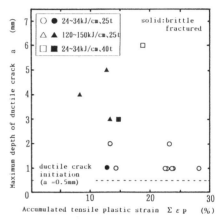

Fig.7. Relationship between accumu-lated tensile plastic strain and maximum ductile crack depth

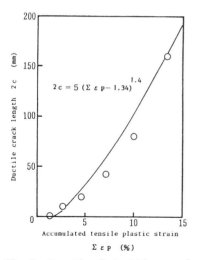

$$2c = 5 (\Sigma \varepsilon p - 1.34)^{1.4}$$

Fig.8. Growth of ductile crack length during cyclic loading (15-B-3)

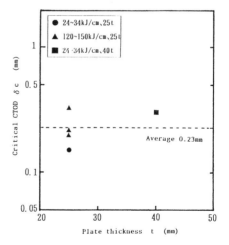

Fig.9. Estimated critical CTOD of brittle fractured specimens

in specimen 15-B-3. The results are showm in Figure 8. The relationship between ductile crack length 2c and accumulated tensile plastic
strain $\Sigma \varepsilon$ p is approximately expressed as the equation shown in the
figure.
　Brittle fracture initiation was evaluated based on fracture mechanics. Critical CTOD δ c of each brittle fractured specimen was estimated by substituting final tensile strain ε f and equivalent crack
size $\bar{a}$ into Eq.(1). Estimated results are shown in Figure 9. δ c
values of the brittle fractured specimens are in the range of 0.15~
0.32mm, and the average value is δ c=0.23mm.

4 Brittle fracture evaluation

4.1 Method for evaluation
Evaluation of Brittle fracture under cyclic loading was carried out as
schematically shown in Figure 10. The right hand side curve demonstrates the ductile crack initiation and growth resistance in terms of
$\Sigma \varepsilon$ p, and the left hand side curve shows the critical condition for
the onset of brittle fracture estimated by Eq.(1) using tensile strain
in each cycle. In this method, the δ c value is assumed as constant
regardless of the amount of cumulative plastic strain for simplifi-
cation of the evaluation. The brittle fracture initiates when the com-
bination of ε and $\bar{a}$ satisfies the relationship of δ c$\leq$3.5ε $\bar{a}$. Substituting $\bar{a}$ value at the onset of the fracture into the resistance curve
of the right side in Figure 10, the $\Sigma \varepsilon$ p value at the brittle fracture can be estimated. As shown, brittle fracture behaviour depends on
the cyclic loading pattern
and brittle fracture can
take place with larger value
of $\Sigma \varepsilon$ p for the frequently
load reversed case.

4.2 Evaluation of quarter part column test result
Using crack growth equation
similar to that in Figure 8,
ductile crack initiation
and growth resistance
curves were approximated
for 3 specimen groups; the
1st group is 25mm thick
specimens welded with
large heat input, the 2nd
group is 40mm thick specimens welded with conven-
tional heat input and the
3rd group is 25mm thick
specimens welded with conventional heat iuput. The
results are shown in Figure
11 as solid lines, where
strain values measured by
strain gauges are converted
into those at weld toe by
multipling by 0.9 according

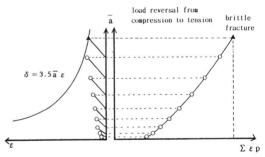

(a) Frequently load reversed case

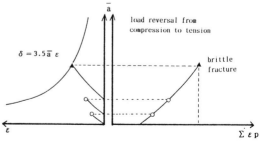

(b) Infrequently load reversed case
Fig.10. Evaluation method of brittle
fracture initiation under cyclic loading

to finite element analysis by Kamura et al. (1993).

The $\Sigma \varepsilon$ p value at brittle fracture initiation was estimated for each specimen group on the basis of the method shown in Figure 10, using δ c value of 0.23mm and strain reversal condition in Figure 2. The results are plotted by "x" in Figure 11. Solid symbols in the figure correspond to brittle fractured specimens, while open symbols correspond to non brittle fractured specimens.

4.3 Evaluation of full scale column test result

Cyclic bending test results of full scale column specimens are reported by Kikukawa et al.(1993). In those tests, 4 specimens showed brittle fracture. Brittle fracture of 2 specimens initiated from weld metal crack, while in the other 2 specimens, brittle fracture occured from ductile crack which initiated at weld toe. Brittle fracture initiation of the latter 2 specimens was estimated using the same method described in 4.1. The results are shown by "x" plots in Figure 12. In this figure, solid lines are the same as those in Figure 11.

4.4 Comparison among estimated and experimental results

Comparing Figure 11 with Figure 12, it is noticed that estimated $\Sigma \varepsilon$ p values at brittle fracture of full scale column specimens are less than those of quarter part column specimens. This difference is attributable to the difference in frequency of load reversal similar to that in Figure 10(a) and (b). The frequency of load reversal in full scale culumn tests is fewer than that in quarter part column tests.

Figure 13 shows comparison between estimated values of accumulated tensile plastic strain at brittle fracture initiation and experimental values. It is confirmed that estimated values agree relatively well with experimental values.

The proposed evaluation method, however, has some ambiguity in the estimation process of brittle fracture initiation. For example, critical CTOD δ c value used were directly calculated from results of the quater part column tests. In order to increase the accuracy of

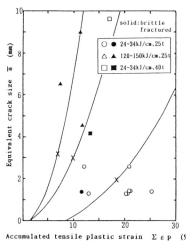

Accumulated tensile plastic strain $\Sigma \varepsilon$ p (%)

Fig.11. Brittle fracture estimation of quarter part column specimens under cyclic loading

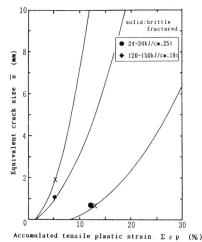

Accumulated tensile plastic strain $\Sigma \varepsilon$ p (%)

Fig.12. Brittle fracture estimation of full scale column specimens under cyclic loading

brittle fracture estimation of column
specimens, it may be necessary to use
δ c value obtained by standard CTOD
tests using small scale specimens
machined from welded joint of column.
This subject remains as for future
studies.

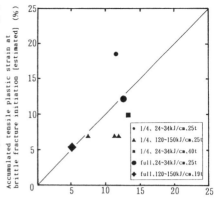

5 Conclusions

A series of bending tests was carried
out on cold formed columns with 400
to 490MPa class tensile strength
steels. Results are summarized as
follows:

Fig.13. Comparison between estimated
values of accumulated tensile plastic
strain at brittle fracture initiation
and experimental values

(1) In the monotonic loading test on
quarter part column specimen with
machined notch which simulates
"bruise" during construction, none
of specimens showed brittle fracture
under the condition of large plastic
strain more than 8%. Therefore, it
can be concluded that "bruise" defects do not cause any harmful effect
on the deformability of cold formed columns.
(2) In the cyclic loading test on quarter part column specimen with
weld joint which simulates beam-column connection, all specimens
welded with large heat input showed brittle fracture. On the other
hand, in the case of specimens welded with proper heat input, brittle
fracture took place in only 3 specimens out of 27. In the brittle
fractured specimen, it was observed that brittle fracture occured from
ductile crack which initiated at weld toe.
(3) Cyclic loading test results were analyzed based on fracture
mechanics approach. The ductile crack initiation and growth resistance
was evaluated as a function of accumulated tensile plastic strain. The
onset of brittle fracture after ductile crack growth was estimated
using CTOD criterion. As a result, condition of brittle fracture
initiations of cold formed column were reasonablly evaluated.

References

Okamoto,K., Aihara,S., Yoshie,A., Hasegawa,T., Kibe,M., Makino,H.,
    Kawasaki,H., Doi,N., Tsuzuki,T., Chijiiwa,R. and Saito,N., (1992)
    High strength offshore structural steel with high weld HAZ
    toughness produced by TMCP, Proc. PACRIM WELDCON'92-Darwin.
Kamura,H., Ito,S., and Okamoto,H., (1993) Deformability of cold formed
    heavy gauge rectangular hollow sections: Nonlinear FEM analysis on
    stress and strain behaviour, Tubular Structures, 5th International
    Symposium, Nottingham, 1993, E and FN Spon.
Kikukawa,S., Okamoto,H., Sakae,K., Nakamura,H. and Akiyama,H., (1993)
    Deformability of cold formed heavy gauge rectangular hollow
    sections: Experimental Investigation, Tubular Structures, 5th
    International Symposium, Nottingham, 1993, E and FN Spon.
The Japan Welding Engineering Society Standard, (1980) Method of
    assesment for defects in fusion welded joints with respect to
    brittle fracture, WES 2805.

14 DEFORMABILITY OF COLD FORMED HEAVY GAUGE RECTANGULAR HOLLOW SECTIONS – EXPERIMENTAL INVESTIGATION

S. KIKUKAWA, H. OKAMOTO, K. SAKAE,
H. NAKAMURA, H. AKIYAMA
Kawasaki Steel Corporation, Tokyo, Japan

Abstract
Two types of cold formed heavy rectangular hollow sections (RHSs) were tested in order to investigate their deformability. Welded RHSs were also tested for the comparison sake. Stub column tests were conducted for 46 specimens and cyclic bending tests were also conducted for 17 models which had beam-to-column connections in the mid-span. Major findings are:
1)The maximum strength of RHS can be estimated by the yield stress of the straight side of the section.
2)Welded RHS shows the best ductility in comparison with press formed RHS and roll formed RHS.
3)In the case of press formed RHS,the end of which is bevelled and welded to the beam, weld heat input has to be controlled within an adequate range in order to ensure satisfactory mechanical properties for the joint. The toe of the welded connection at the corner of RHS also has to be finished smoothly in order to avoid the stress concentration which may lead to the brittle fracture.
Keywords: Cold Forming, Rectangular Hollow Section,Ductility, Fracture, Buckling Strength, Tubular Joints, Earthquakes.

1 Introduction

Steel rectangular hollow sections(RHSs) are widely used in Japan by more than 1 million tons a year. These are so called "cold formed columns", which are manufactured by roll forming process or press bending process, and mainly used for buildings. Cold formed RHSs can be manufactured up to the size of 550 mm width*22mm wall thickness for the roll forming, and 1000 mm width*40 mm wall thickness for the press bending. Roll formed RHS is manufactured in the continuous ERW line where hot coil is roll-formed into round section and then into square section by sizing process. This means that they are enevitably subjected to the plastic cold-working at both their straight sides and 4 corners. Press formed RHS is manufactured from the steel plate by bending 4 corners of the section, and consequently receives the plastic deformations

Tubular Structures V. Edited by M.G. Coutie and G. Davies.
© 1993 E & FN Spon, 2–6 Boundary Row, London SE1 8HN. ISBN 0 419 18770 7.

at only these areas. There is concern that cold-working or plastic deformation might deteriorate the toughness, the elongation and the buckling strength of RHS. These should however have the capacity for stable plastic deformation when subjected to severe earthquake loadings.

This paper investigates the strength, ductility and the ultimate behavior of cold formed RHSs to full scale.

2 Experimental study

2.1 Stub column tests

A total of 46 column specimens composed of 11 roll formed RHSs, 21 press formed RHSs and 14 welded RHSs were tested(see Table 1). The mean value of the ratio of r(outside corner radius) to t(wall thickness of the section) was 2.7 for roll formed RHSs and 3.4 for the press formed RHSs. Press formed RHSs included 2 types for the number of welded seam. One was the group of RHSs which had one welded seam(specimen No.13, 16, 19, 23, 26 and 29) and the other was those having 2 welded seams(the rest of 21 specimens). A further parameter for press formed RHS was the outside corner radius. Specimen No.32 had larger r/t of 4.3 than other press formed RHSs. Welded RHSs were manufactured by welding 4 steel plates at the corner of the section and resulting in 4 welded seams. The materials used for press formed RHSs and welded RHSs were the same 490MPa class steels(JIS G 3106 SM 490A). For the roll formed RHS, both 400 MPa and 490MPa class steels(JIS G 3466) were used.

Tests were conducted using a 2000 tonf structure testing machine with spherical bearing plattens and specimens were axially compressed. Ratio of specimen length to the column width was 3 to 1.

2.2 Bending tests

Table 2 and Figure 1(a) show specimens which had beam-to-column connections at mid-span. A total of 17 models containing 5 roll formed RHSs, 9 press formed RHSs and 3 welded RHSs were tested. A column model was simply supported at both ends and subjected to cyclic load at beam-to-column connection. Column ends were bevelled and welded to a beam block except No.10 and No.11. Beam flanges were conversely bevelled and welded to a column for these 2 specimens.

Welding was done by GMAW(JIS Z 3312 YGW 11, 1.2φ welding wire was used) and the edge preparation was as shown in Fig.1(b). All the straight sides of column section were welded in flat position with heat input of about 2.0 kJ/mm.

Three types of welding procedure were adopted for the corner of RHS:

I   Vertical position, Multi-pass, Heat input 2.6 ~ 3.4 kJ/mm for specimen No.11,13,15 and No.17.

II   Vertical position, Single-pass, Heat input 12.0 ~

Table 1. Stub column test

No.	RHS	B*t (mm)	r (mm)	σy (MPa)	σcr (MPa)	εcr (*10^-2)	S	μ	$1/\alpha$
1		400.2*8.5	22.3	367	347	0.30	0.95	1.68	0.25
2		400.5*11.6	30.3	353	377	0.66	1.07	3.84	0.49
3		400.3*15.7	41.1	383	429	1.29	1.12	6.93	0.83
4		400.0*18.9	52.9	456	505	1.83	1.11	8.27	1.01
5		399.0*21.7	58.9	470	540	2.27	1.15	9.96	1.29
6	Roll	300.8*18.9	48.5	442	509	3.55	1.15	16.54	1.84
7	formed	401.0*12.0	30.0	497	487	0.52	0.98	2.14	0.37
8		399.7*16.1	37.4	450	482	1.17	1.07	5.33	0.74
9		399.9*18.7	52.0	539	608	1.59	1.13	6.07	0.83
10		400.5*21.8	61.9	528	625	1.91	1.18	7.44	1.16
11		350.3*18.3	51.1	545	635	2.07	1.17	7.81	1.03
12		454.1*8.9	25.6	297	257	0.20	0.86	1.39	0.27
13		448.3*11.7	39.7	292	288	0.33	0.99	2.30	0.48
14		450.1*11.9	49.7	282	279	0.39	0.99	2.81	0.51
15		452.8*15.7	53.5	273	311	0.61	1.14	4.65	0.91
16		448.9*18.5	68.3	277	392	1.59	1.42	11.86.	1.26
17		452.5*18.6	65.3	266	374	2.01	1.41	15.56	1.31
18		453.8*21.6	74.9	268	389	2.70	1.46	20.75	1.74
19		450.7*24.7	76.3	344	443	3.36	1.29	20.08	1.80
20		451.5*24.6	79.5	269	419	4.39	1.56	33.68	2.28
21	Press	400.7*24.6	76.1	267	455	5.19	1.71	40.04	2.91
22	formed	452.7*8.8	20.8	360	304	0.27	0.84	1.53	0.22
23		449.6*11.7	39.9	340	326	0.29	0.96	1.75	0.41
24		451.1*11.8	48.4	355	337	0.28	0.95	1.63	0.40
25		453.1*15.8	53.0	323	363	0.59	1.12	3.74	0.78
26		450.4*18.8	63.9	325	431	1.36	1.33	8.65	1.11
27		453.5*18.9	64.5	320	407	1.63	1.27	10.50	1.12
28		454.1*22.2	80.5	383	458	2.33	1.19	12.50	1.28
29		402.1*21.8	64.1	393	502	2.39	1.28	12.53	1.54
30		402.2*21.8	77.8	378	507	3.10	1.34	16.91	1.60
31		399.9*24.7	89.3	366	521	4.54	1.42	25.58	2.15
32		400.5*24.7	107.3	369	536	4.93	1.46	27.51	2.12
33		448.7*8.9		303	231	0.16	0.76	1.11	0.27
34		447.8*11.7		291	279	0.31	0.96	2.20	0.48
35		449.5*15.8		280	289	0.40	1.03	2.94	0.91
36		449.6*18.5		280	291	1.47	1.04	10.81.	1.24
37		450.1*21.5		279	302	2.24	1.08	16.53	1.68
38		450.2*24.4		277	353	4.02	1.28	29.94	2.19
39	Welded	400.5*24.6		277	386	5.04	1.40	37.50	2.81
40		449.5*8.9		366	233	0.14	0.64	0.79	0.22
41		449.7*11.8		342	329	0.29	0.96	1.74	0.41
42		448.5*15.9		337	346	0.46	1.02	2.80	0.77
43		450.3*18.8		331	334	0.56	1.01	3.45	1.08
44		450.3*21.9		397	377	0.53	0.95	2.73	1.23
45		400.5*22.0		397	386	1.96	0.97	10.15	1.56
46		400.5*24.7		373	451	4.66	1.21	25.76	2.10

B: Column width, t: Wall thickness
r: Outside corner radius of RHS
σy: Yield stress obtained from the straight side of RHS
σcr: Maximum stress, εcr: Average strain at σcr
$S = \sigma cr/\sigma y$, $\mu = \varepsilon cr/\varepsilon y$ ($\varepsilon y = \sigma y/E$), $\alpha = \varepsilon y(B/t)^2$

15.0 kJ/mm for specimen No.1 ~ No.8.
    III Flat position, Multi-pass, Heat input 1.5 ~ 2.5 kJ/mm
       for specimen No.12 and No.14.

For welding procedure III in the rotating jig the welding was
carried out continuously, the weld toe of the joint was
finished smoothly.

Table 2. Bending test specimens

No.	B*t*r (mm)	Beam-to-column Connection 1)	Manufacturing process	σ y of the column 2)	Welding procedure for the corner of the column	Heat input(kJ/mm)
1	400.0*11.3*29.0	C		375	Vertical position	
2	400.5*15.3*38.0	C		389	Vertical position	
3	400.0*21.7*57.5	C	Roll formed	466	Vertical position	12.0-15.0
4	450.0*18.3*51.5	C		340	Vertical position	
5	450.0*18.1*47.0	C		410	Vertical position	
6	449.5*15.5*40.5	C		360	Vertical position	
7	449.0*18.4*55.0	C	Press formed	372	Vertical position	12.0-15.0
8	450.0*24.4*69.5	C		335	Vertical position	
9	450.0*18.6	C		332	Vertical position	
10	450.0*24.7	B	Welded	343	Vertical position	1.5-2.5
11	449.7*19.1*63.0	B		390	Flat position	2.6-3.4
12	449.7*19.1*63.0	C		390	Flat position 3)	1.5-2.5
13	449.5*19.1*66.0	C	Press formed	391	Vertical position	2.6-3.4
14	449.1*25.0*87.0	C		375	Flat position 3)	1.5-2.5
15	448.1*25.0*89.0	C		370	Vertical position	2.6-3.4
16	449.1*25.3	C	Welded	401	Flat position	1.5-2.5
17	799.0*40.3*115.5	C	Press formed	374	Vertical position	2.6-3.4

1) C: Column was bevelled and welded to the beam

    B: Beam was bevelled and welded to the column

2) σ y means the yield stress. Data were obtained from the straight side of the column

3) Rotating positionner was applied

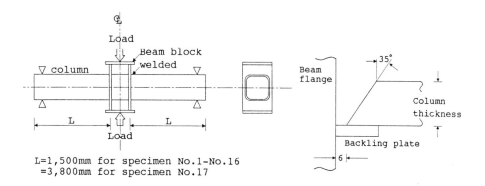

L=1,500mm for specimen No.1-No.16
  =3,800mm for specimen No.17

a) Test specimen   b) Weld edge preparation

Fig. 1. Bending test specimen details

154

3 Results and discussion

3.1 Strength and deformability at local buckling

Test results are shown in Table 1, where σcr and εcr are the mean buckling stress and corresponding strain,respectively. The latter was obtained from the average deformation divided by the height of specimen.
S is defined as the ratio of σcr to σy. It is clear from the Table that S is influenced by the column manufacturing process:

```
Roll formed RHS    --- S = 0.95 ~ 1.18
Press formed RHS ---      0.84 ~ 1.71
Welded RHS         ---      0.64 ~ 1.40
```

It is considered that the reason for the smaller S for roll formed RHSs is due to the higher yield ratio of their materials comparing with other processes.
The parameter μ is defined as $\mu = \varepsilon cr/\varepsilon y$ ($\varepsilon y = \sigma y/E$), μ measures the deformability of RHSs. Then, μ can be treated as the function of the buckling parameter α as follows:

$$\mu = a(\ 1/\alpha \ )^{b} \tag{1}$$
$$\text{where } \alpha = \varepsilon y(\ B/t \ )^{2}$$

Figure 2 shows all of the stub column test results(open symbol) which are approximated by Eq.1. Several data of welded RHSs were less than the equation. It is considered that the buckling strength of those RHSs were almost equal to σy, and the buckling deformation of the column had occured earlier than the hardening of the material occured.

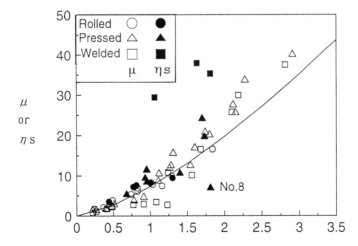

Fig. 2. Deformability vs. buckling parameter

3.2 Ductility of column member

Figure 4 shows typical relationships between the moment and the relative rotation θ of the joint which were obtained from cyclic bending tests.

Buildings must be able to absorb the energy of earthquake loadings through plastic deformation of the frames. Consequently, ductility of the members should be estimated from the total energy $E\eta s$ obtained from skeleton curve(see Fig.3) or accumulated energy integrated for each cycle of load-resistance curves. The values of $E\eta s$ are shown in Table 3, where Mp means full plastic moment calculated from plastic section modulus and the yield stress of the straight side of the column section. θp is analytically calculated value which corresponds to Mp. $E\eta s$ satisfies the relation $E\eta s \geq 6.0$ which is required for the structural rank I (in the case of B/t $\leq$ 23, AIJ(1990))except specimen No.1. Consequently, it can be concluded that RHS manufactured by cold forming process have enough ductility for practical use.

Accumulative plasticity ηs can also be obtained from cyclic bending tests as shown in Table 3. Figure 2 shows the comparison between μ(open mark) and ηs(solid mark). It is clear that μ could approximate the lowest ηs, and it is also clear that specimen No.8 has less ductility than other specimens.

Table 3. Bending test results

No.	$E\eta s$	ηs	Mmax (kN·m)	Mp	Mmax/Mp	$1/\alpha$
1	5.1	3.5	1048	926	1.13	0.4
2	12.7	7.2	1541	1262	1.22	0.7
3	11.4	9.4	2500	2015	1.24	1.3
4	19.1	8.1	2133	1640	1.30	1.0
5	17.5	7.6	2637	2014	1.31	0.8
6	11.0	5.3	1839	1499	1.23	0.6
7	8.1	9.4	2295	1784	1.29	0.9
8	10.8	6.9	3020	2048	1.47	1.8
9	58.8	29.4	2373	1727	1.37	1.0
10	50.3	35.3	3438	2303	1.49	1.8
11	13.4	11.5	2297	1919	1.20	0.9
12	13.7	11.4	2372	1919	1.24	0.9
13	10.2	8.4	2398	1915	1.25	0.9
14	32.9	24.1	3188	2260	1.41	1.7
15	27.7	19.6	3168	2209	1.43	1.7
16	47.3	37.9	3462	2718	1.27	1.6
17	14.7	10.6	16117	11395	1.41	1.4

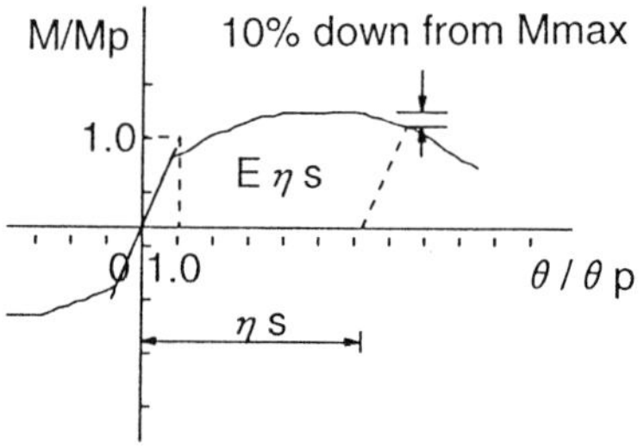

Fig. 3. Ductility factor obtained from skeleton curve

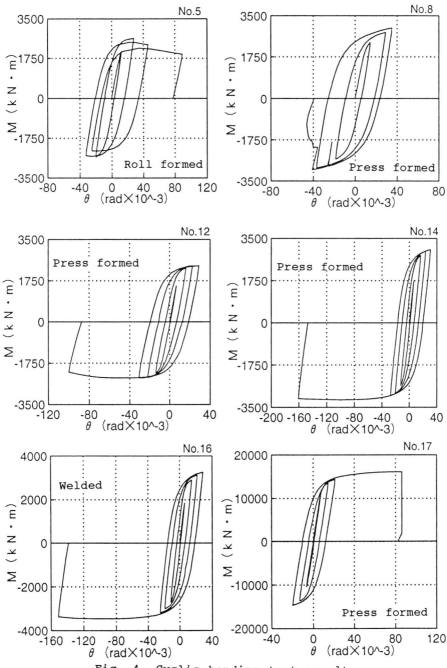

Fig. 4. Cyclic bending test results

3.3 Ultimate behavior

The maximum strength Mmax which was obtained from the bending tests exceeded Mp of the specimen as shown in Table 3. Figure 5 shows the relationships between Mmax/Mp and $1/\alpha*1/YR$, where YR is yield ratio of the material. Mmax could be approximated by a linear function of the products $1/\alpha*1/YR$.

Table 4 shows the failure mode of bending test specimens. Thirteen specimens failed by local buckling of the flanges and 4 specimens were fractured from ductile cracks, which had occured in the welded joint at the corner of the section and propagated through column flanges. Cracks occured at the toe of the joint welds for specimen No.7 and No.15. For specimen No.8 and No.17, it is considered that cracks occured in the weld metal at first, then propagated rapidly through the wall thickness. Since three of the four fractured specimens (except No.8) were locally buckled and broken after attaining the peak of M-θ curves, they showed sufficient ductility.

Ductile cracks which had occured at weld toes were observed on 7 specimens containing 3 welded RHSs and 4 press formed RHSs. Cracks did not propagate in welded RHSs, while 2 types of phenomenon were observed in press formed RHSs; one was the same as welded column and for the other the crack continued to grow. It is suggested that several differences in mechanical properties of joints, profile of weld toe and strain concentration to the corner of the section produced these different phenomena.

At first, it must be emphasized that weld heat input was much different between welded RHS specimens and press formed RHS specimens. When heat input increases, grain size in both weld metal and HAZ also grows, which results in the decrease of mechanical properties for both strength and toughness, Gooch, T.G.(1989).

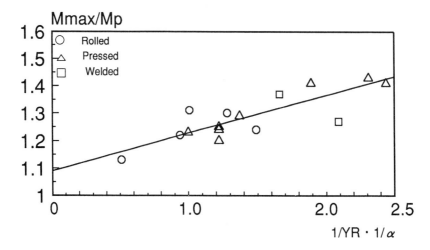

Fig. 5. Bending strength of RHS

When welding is conducted in the vertical position, the profile of the weld toe tends to be finished rougher than for the flat position. The strain concentration factor for the press formed RHS consequently tends to be larger than for the welded RHS when vertical welding is adopted. Figure 6 shows the macrographs of the profile of these welded joints. As shown in this figure, the radius r = 7 mm and the flank angle θ = 18 deg. were observed for specimen No.12(welded in the flat position), while r = 1.5 mm and θ = 39 deg. were observed for specimen No.13 which was welded in the vertical position. It is therefore preferable to control the heat input in adequate range or to finish the weld toe smoothly in order to prevent the brittle fracture of press formed RHS. As for the strain concentration in weld toe, the roll formed RHS is less than other RHSs,Kamura, H.(1993). Moreover, the buckling phenomenon governs the maximum strength of roll formed RHS. Due to these two reasons, roll formed RHS did not show any fractures from weld toe.

Table 4. Ultimate behaviors of column specimen

No.	Welding process for the corner 1)	Column	Failure mode	Ductile cracks
1		Roll formed	Locally buckled	None
2		Roll formed	Locally buckled	None
3		Roll formed	Locally buckled	None
4	II	Roll formed	Locally buckled	None
5		Roll formed	Locally buckled	None
6		Press formed	Locally buckled	None
7		Press formed	Fractured	Observed
8		Press formed	Fractured 2)	None
9	III	Welded	Locally buckled	Observed
10	III	Welded	Locally buckled	Observed
11	I	Press formed	Locally buckled	None
12	III	Press formed	Locally buckled	None
13	I	Press formed	Locally buckled	Observed
14	III	Press formed	Locally buckled	Observed
15	I	Press formed	Fractured	Observed
16	III	Welded	Locally buckled	Observed
17	I	Press formed	Fractured 2)	None

1) I : Vertical position, acceptable heat input
   II : Vertical position, excessive heat input
   III : Flat position, acceptable heat input
2) Cracks were considered to start from the defects in weld metal.

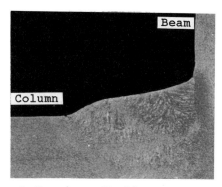
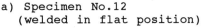
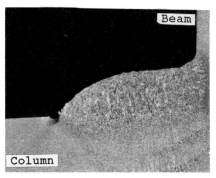

a) Specimen No.12
(welded in flat position)

b) Specimen No.13
(welded in vertical position)

Fig. 6. Macrographs of weld toe (scale of double size)

4 Conclusions

Major findings obtained through both stub column tests and
cyclic bending tests on cold formed RHSs are as follows:
1)The maximum strength of RHS can be estimated by the yield
stress of the straight side of the section.
2)Welded RHSs showed the best ductility in comparison with
press formed RHSs and roll formed RHSs. Cold formed RHSs,
however, have a sufficient ductility as columns for
buildings. Their ductility satisfies the requirements the
design guide(AIJ(1990)).
3)In the case of press formed RHS,the end of which is
bevelled and welded to the beam, weld heat input has to be
controlled in the adequate range in order to get enough
mechanical properies of the joint. The toe of the welded
connection at the corner of RHS also has to be finished
smoothly in order to avoid the stress concentration which may
lead to brittle fracture.

5 Referances

Architectural Institute of Japan(1990), **Ultimate Strength
and Deformation Capacity of Buildings in Seismic Design
(1990)**, Maruzen Inc.,Tokyo, Japan
Gooch, T.G.,(1989), Grain size and morphology in nickel-
chromium alloy weld metal, **Report 405/1989**, The Welding
Institute, Cambridge,UK.
Kamura, H., Ito, S., Okamoto, (1993), Deformability of cold
formed heavy gauge rectangular hollow sections-Non linear
FEM analysis on stress and strain behavior, to be
published on **5th Int. Symp. on Tubular Structures, E and
F.N. Spon.**

PART 5

COLUMN BEHAVIOUR

Effect of forming
the hollow section

15 ULTIMATE STRENGTH AND POST-BUCKLING BEHAVIOUR OF CHS COLUMNS – A COMPARISON BETWEEN COLD-FORMED AND HOT-FINISHED SECTIONS

K. OCHI
Kumamoto University, Japan
B.S. CHOO
University of Nottingham, UK

Abstract
Tests were carried out on 36 full scale and 9 stub columns made of steel circular hollow sections (88.9 mm diameter x3.2 mm thick). The effects of parameters such as the production process, slenderness ratio and steel grade on the ultimate compressive strength were examined. A simple theoretical approach based on the modified Ramberg-Osgood stress-strain curves for predicting the ultimate strength of tubular columns is presented. The theoretical predictions are compared with experimental data. The implications of the use of these theoretical methods for design purposes is discussed. A comparison of the experimental ultimate strength with the design criteria of Eurocode 3 (EC3) shows that the code requirements may not conservative for predicting the maximum load of cold-formed high grade steel column.
Keywords: Circular Hollow Sections, Flexural buckling, Ultimate strengths, Post-buckling, Columns.

1 Introduction

The ultimate strength and post-buckling behaviour of centrally loaded columns have been the subject of intensive study, for example, see Galambos (1988). These studies which covered various cross-sections and took into account the manufacturing process (i.e. hot-rolling or cold forming) have enabled column design curves to be developed. However, it should be noted that circular hollow sections are usually manufactured using the electric resistance welding procedure for both cold formed (CF) and hot finished (HF) production processes. These two processes have been shown by Narayanan (1982) to affect the material and induce different levels and patterns of residual stresses in the product. The effects of these changes on the behaviour of columns constructed from circular hollow sections have still to be fully understood, see Kato (1985) and Davison (1983)

The aim of this paper is to compare the behaviour of cold formed and hot finished circular hollow sections as obtained from a number of tests, against established structural design criteria.

Tubular Structures V. Edited by M.G. Coutie and G. Davies.
© 1993 E & FN Spon, 2–6 Boundary Row, London SE1 8HN. ISBN 0 419 18770 7.

2 Experimental Study

2.1 General

An experimental programme consisting of column tests, stub column tests, and tensile coupon tests, was designed to evaluate the difference between the behaviour of CF steel columns and HF steel columns. The column lengths were selected to give slenderness ratios of L/r=46, 69 and 92 where L is the length between pin supports and r is the radius of gyration. All the columns were selected from tubes with the same nominal cross-section. Two grades of steel (43C and 50C) were used and two manufacturing process (CF and HF) were selected to provide a direct comparison of heat treatment effects.

2.2 Tensile coupon tests

Typical stress-strain curves obtained for the test specimens are shown in Figure 1. Because of the non-linearity of the stress-strain curves for the CF sections, 0.2 per cent proof stress was used for CF sections. It can be seen that coupons taken from HF specimens exhibit a stress-strain behaviour which is close to that of the idealized linear elastic-plastic model. On the other hand, coupons taken from CF sections displayed varying degrees of non-linearity in their stress-strain behaviour. Yield strength and ultimate tensile strengths obtained from coupon testing are summarized in Table 1. It can be seen that yield strengths of CF tubes are higher than those of HF tubes.

Table 1. Summary of material properties

Name	Forming	Steel grade	Tensile test σ_y (MPa)	σ_u (MPa)	EL (%)	Stub column test σ_y (MPa)
HF43	Hot finished	43C	311	485	38.6	287
HF50		50C	389	544	36.3	378
CF43	Cold formed	43C	390	489	35.3	334
CF50		50C	428	496	33.6	370

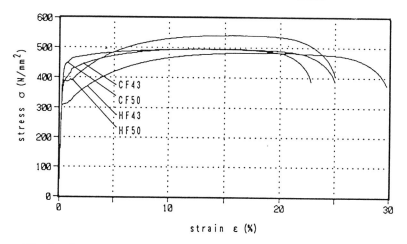

Fig.1. Stress-strain curves of tensile coupon tests.

2.3 Stub column tests

Mid-height strain measurements were made on 4 orthogonally opposite locations of the specimens using strain gauges, and displacement measurements were made using 50 mm gauge length potentiometers.

In all tests the load carrying capacity of the specimen decreased with the formation of local buckles. Averaged stress strain curves for HF and CF tube are shown in Figure 2. It can be seen that both HF and CF stub columns display varying degrees of non-linearity in their stress strain behaviour, It is probable that these stub columns were not free from residual stress, and hence the non-linearities are different for the various coupon tests. The strain at ultimate strength for HF tubes are in general different from those of the CF tubes. In general, the yield strengths obtained for CF tubes were higher the those for HF tubes.

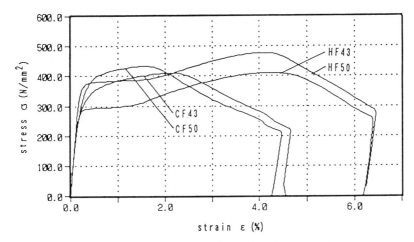

Fig.2. Stress-strain curves of stub column tests.

2.4 Full size column tests

Roller type column end fixtures were attached to specially made adaptors for the Man column test machine in the Structures Laboratory in the Civil Engineering Department at Nottingham University. The columns were placed at the centre of the loading platens while the roller plates were brought into uniform contact by means of adjustment devices contained in the roller fixtures.

Typical load versus axial deformation curves for HF43 and CF43 columns are shown in Figure 3 (a) and (b) respectively. The ultimate strengths and load deformation behaviour as affected by heat treatment is distinctly demonstrated in these figures.

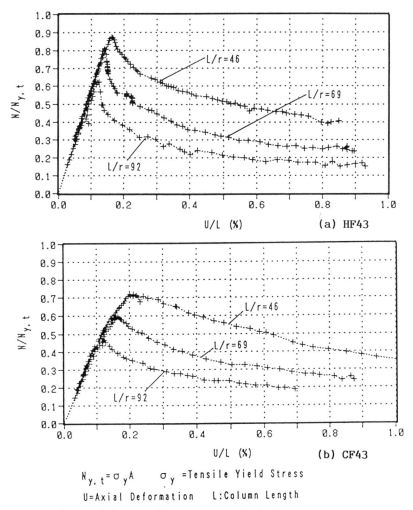

$N_{y,t} = \sigma_y A$ σ_y =Tensile Yield Stress

U=Axial Deformation   L:Column Length

Fig.3. Load-deformation curves of full scale column tests.

3 Evaluation of results

3.1 Ultimate strength

The data obtained from tests on stub and full-sized column specimens
were used to check the accuracy of various column models for
predicting column behavior of both HF and CF sections. Typical
comparisons of the stress-strain curves generated by the tangent
modulus model, see Davison (1983), with results from stub column
tests are shown in Figure 4. It can be seen that stress-strain curves
generated by the tangent modulus model fit the stub column test
results reasonably well.

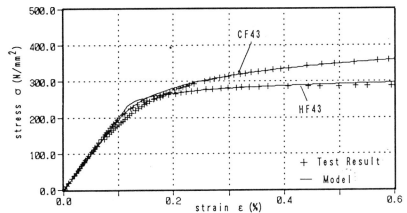

Fig.4. Comparison of stub column test stress-strain curves with Ramberg-Osgood stress-stain curves.

Since the tangent modulus theory is able to take into account the non-linear stress-strain behaviour of the column material, it was used to investigate the difference between the strengths of HF and CF columns.

Plots of experimental ultimate strength versus slenderness ratio for 43C specimens (HF and CF) are compared with those obtained using the model tangent modulus theory in Figure 5. The test results for CF sections are seen to be in close agreement with the tangent modulus curve. The tangent modulus curve for HF sections lies along the lower bounds of the test results.

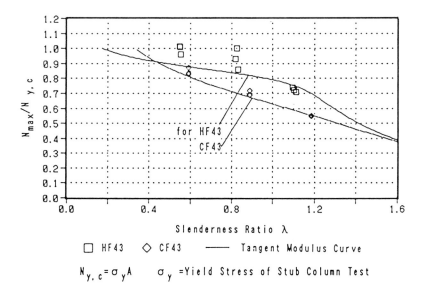

Fig.5. Comparison of column strengths with predictions using tangent modulus models.

167

The column test results (normalized using the stub column yield stress) are plotted in Figure 6. The test results of HF columns lie well above Curve a (Eurocode 3) and the results of CF columns are in general below Curve a but above Curve b.

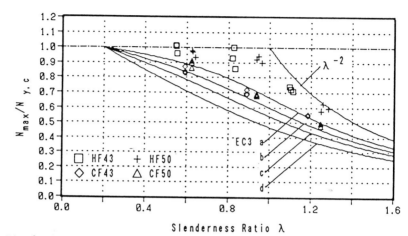

Fig.6. Comparison of column strengths. (normalized on stub column properties)

The column strengths, normalized using the coupon test yield stress, are shown in Figure 7. It can be seen that the test results for the HF columns lie above Curve a and that the results for the CF columns lie between Curve a and Curve d.

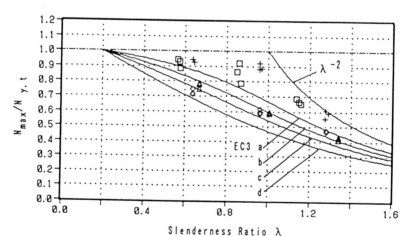

Fig.7. Comparison of column strengths. (normalized on tensile coupon properties)

Another comparison of the data using nominal properties for normalization purposes is presented in Figure 8. The results for HF columns exceed Curve a whilst the results for CF columns exceed Curve

168

b. By using the nominal yield values to normalize the data, the effects of cold working during the manufacturing process are neglected and it can be seen from Figure 8 that the test results for the CF columns lie above Curve b.

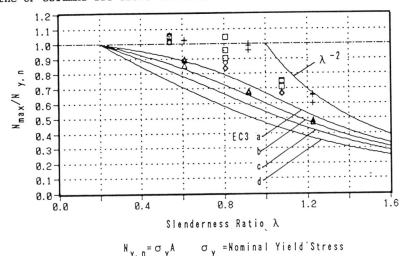

$N_{y,n} = \sigma_y A$ σ_y =Nominal Yield Stress

Fig.8. Comparison of column strengths. (normalized on nominal properties)

3.2 Post-buckling behaviour
The post-buckling behaviour under cyclic load for CF columns has been established by fitting multiple linear lines by Kato and Akiyama (1977) and Sherman (1980). Using these models, the load deformation curves for the test specimens describe above, are shown in Figure 9. A comparison of Figure 3 and 9 shows that the test data fits the numerical model fairly well, however, the multiple linear line model is not able to represent the post buckling behaviour of HF columns.

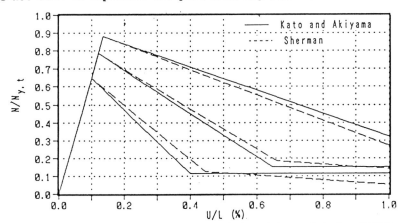

Fig.9. Load-deformation curves. (Algorithm model for HF43)

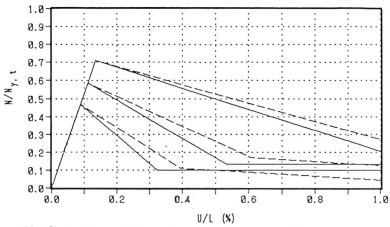

Fig.9. Continued (Algorithm models for CF43)

4 Conclusions

The results from tests on 36 tubular columns are presented. The effects of parameters such as the production process, slenderness ratio and steel grade on ultimate compressive strengths are examined. The tangent modulus theory based on stress-strain behaviour of stub columns was found to be capable of predicting the experimental ultimate strength. In general, the normalized strength test results for cold formed columns lie between Curves a and c as given in EC3 whilst test results for HF columns lie above Curve a. Hence it would appear that Curve a is an appropriate design curve for HF columns.

5 References

Davison, T. A. and Birkemoe, P. C. (1983) Column Behaviour of Cold-Formed Hollow Structural Steel Shapes, **Canadian Journal of Civil Engineering**, Vol 10, No 1,1983,pp. 125-141.
Galambos, T.V. (1988) **Guide to stability design criteria for metal structures**, 4th edition, John Wiley & Sons.
Kato, B. and Akiyama, H. (1977) Hysteretic Behaviour of Brace Frames, Transactions of AIJ, Vol 260, pp. 99-107.
Kato, B, and Aoki, H. (1978) Residual Stresses in Cold Formed Tubes, Journal of Strain Analysis, 13(3), pp. 1-2.
Kato, B. and Lee, M. (1985) Column Strength of Cold-formed square and Circular Hollow Section Members, **Journal of Structural Engineering**, Vol.31B, pp. 135-142.
Narayanan, R. (1982) **Axially compressed structures**, Stability and Strength, Applied Publishers.
Sherman, D. R. (1980) Interpretive Discussion of Tubular Beam-column Test Data, **Report**, Department of Civil Engineering, University of Wisconsin-Milwaukee.

16 COLUMN BEHAVIOUR OF PRESS-FORMED RECTANGULAR HOLLOW SECTIONS

H. NAKAMURA and N. TAKADA
Nittetsu Column Co. Ltd, Yokohama, Japan
M. YAMADA
Faculty of Engineering, University of Tohoku, Sendai, Japan

Abstract
This paper presents the results of experimental studies on the
structural behaviour of press-formed rectangular hollow sections
(hereafter, press-formed RHS) as columns. The section dimension of
the RHS used in the experiments was 350mmx350mm and the steel grades
were SM490A and SS400 as specified in JIS G3106 and JIS G3101
respectively. Two series of experiments were carried out.
The first series was to clarify the influence of cold-forming on the
mechanical properties of the steel sections. The second series was
to clarify the deformation capacity of the beam-columns subjected to
monotonic bending and cyclic bending. In the monotonic bending tests
RHS fabricated with four plates by welding (hereafter, built-up RHS)
were included for comparison. The experimental studies showed that
the degree of influence from cold-forming on the mechanical properties of RHS sections is low from a practical viewpoint and that the
press-formed RHS have sufficient deformation capacity.
Keywords: Rectangular hollow section, Press-formed RHS, Bending test
Deformation capacity, Column strength, Cyclic bending test.

1 Introduction

Cold-formed square and rectangular hollow sections (hereafter, cold-
formed RHS) are widely used for columns for buildings in Japan, with
the annual usage amounting to over one million tons.

Cold-formed RHS are produced by two manufacturing methods:
the ERW method, and the press-forming and welding method. Large size
cold-formed RHS, for example, over 550x550mm wide sections, are
mainly produced by press-forming and welding, and the market share
of such RHS is estimated at about 20%.

This paper presents the results of experimental studies on the
structural behaviour of the press-formed RHS as columns. The section
dimension of RHS used in the tests was 350x350, their wall-thicknesses were from 9mm to 22mm, the steel grades were SM 490A and
SS400 as specified in JIS G3106 and JIS G3101 respectively, and the
nominal outside corner radius of the RHS was three times the wall-
thickness.

The RHS used for test specimens were manufactured to the standards of JSS II-10 'Standard for Cold-Formed Carbon Steel Square

Tubular Structures V. Edited by M.G. Coutie and G. Davies.
© 1993 E & FN Spon, 2–6 Boundary Row, London SE1 8HN. ISBN 0 419 18770 7.

and Rectangular Hollow Sections' issued by JSSC. (JSSC:Japanese Society of Steel Construction). This paper first describes the manufacturing process and characteristics of the press-formed RHS.

It then presents the test results of the tensile and plastic bending tests on wide steel plates, which were carried out to obtain basic data on the influence from the cold-forming on the mechanical properties of the press-formed RHS. These results clarified that the influence could be evaluated by the ratio of the corner area to the total area of the flat parts.

Thirdly, test results of monotonic and cyclic bending tests are presented. These tests were carried out on the beam-column specimens with five different width-thickness ratios to clarify the column strength and deformation capacity. In order to evaluate the influence of the steel grade and the fabrication methods, press-formed RHS with SS400, and RHS fabricated with four plates of SM490 by welding were included in the monotonic bending tests. The loading points and supporting points of the beam-column test specimens were stiffened with inner-diaphragm plates.

Finally, based on these test results, the structural behaviour of the press-formed RHS, i.e. the strength and deformation capacity (or ductility), and geometric similarity between RHS with different sizes are discussed.

2 Characteristics of press-formed RHS

2.1 Manufacturing process
The manufacturing process of press-formed RHS is as follows. First, a steel plate is "gas-cut" and then it is formed into a channel by press-forming at room temperature. A pair of the channels are then welded together into a box section. Then, after passing the inspection, the press-formed RHS are finished.

2.2 Available Sizes
The largest size of press-formed RHS manufactured in Japan is 1000mm in the outside dimension and 40mm in wall-thickness. Available width -to-thickness ratios (hereafter, B/t ratio) range from 15 to 50.

2.3 Features of press-formed RHS
At the corners of press-formed RHS, two kinds of cold-formed zones exsist, i.e. one with circumferential tension strain on the outside surface and the other with circumferential compression strain on the inside. The flat parts of the section are not subjected to cold-forming and, hence, their mechanical properties are completely the same as those of the mother steel plates. Examples of the strain distribution on the inside and outside surfaces at the corners are shown in Fig.1. These are measurements by wire strain gauges and show the circumferential strain after the press-forming.

The ratio of the area of cold-formed parts to that of the whole cross section depends on the B/t ratio and varies from 10% to 30% within the range of the available sizes. This fact indicates that the press-formed RHS are composed of 70% to 90% mother steel and 10% to 30% cold-formed steel. The maximum strain in the cold-formed

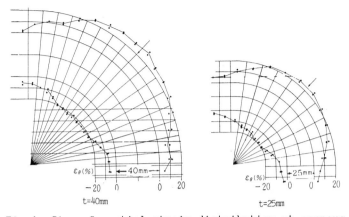

$\varepsilon_\theta(\%)$ ←—40mm—→

−20 0 0 20

t=40mm

$\varepsilon_\theta(\%)$ ←25mm→

−20 0 0 20

t=25mm

Fig.1. Circumferential strain distribution at corners

corners exists on the outside surface and its value is about 25%.

3 Mechanical properties of the corners and their influence

3.1 Mechanical properties of the corners

The figure of the coupons for tensile test of the corners and the
flat parts is shown in Fig. 2. The results of the coupon tests are
given in Table 1. The chemical composition of the steel used for
the tests is almost the same as that of the SM490A steel used for
the other test series, as referred to in Table 3.

From the test results, it is indicated that the cold-forming
causes an increase in the yield strength to an average of 1.5 times,
and in the tensile strength to 1.2 times, and it causes a decrease
in the elongation to an average of 0.55 times within the wall-thickness range from 12mm to 40mm.

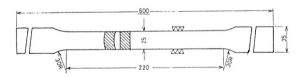

Fig.2. Configuration of tensile test coupons

3.2 Wide plate tensile test

The tensile tests on the wide plates were carried out on the basis
of the authors' idea that the cold-formed RHS had to be treated
as a whole section composed of the flat mother steel parts and the
cold-formed corners and the structural behaviour must be clarified.

The wide test coupons of steel plates, 25mm or 40mm thick, were
made by first press-forming flat steel plates into angle sections
and then flattening them again into flat plates, which had the cold-

worked zone in the centre. The ratios of the area of cold-worked zone to that of the whole section were varied from 0% (=not bent) to 100%. As an example, the shape and the test results for DR-1 coupon are shown in Fig.3. Elongation was measured by wire strain gauges of 50mm gauge length. The test results are given in Table 2. Fig.4 shows the specimen ABR-1 after the test.

From the test results of Nos.ASs, BS-1 and CS-1, it would appear that the elongation of the coupon without any cold-working is little affected by the width. Therefore, it is understood that the differences of the elongation values of Nos.ARs, BRs, CRs and DRs, which have nearly the same thicknesses but different widths, are due to the influence from cold-forming. It is clear that the wider the width, the more closely the elongation approaches that of coupons free from cold-working (i.e. mother steel plates).

Table 1. Ratio of mechanical properties at corner and flat parts
(Ratios given by corner/flat)   (Steel grade: SM490A)

Thick-ness (mm)	Number of samples	Yield strength ratio		Ult. strength ratio		Elongation ratio	
		Ave.	Range	Ave.	Range	Ave.	Range
12	2	1.47	1.44- 1.50	1.20	1.16- 1.23	0.50	0.49- 0.50
16	1	1.62	——	1.22	——	0.54	——
19	2	1.54	1.52- 1.55	1.18	1.17- 1.18	0.56	0.56- 0.56
22	2	1.56	1.50- 1.62	1.19	1.18- 1.20	0.54	0.52- 0.55
25	21	1.50	1.43- 1.57	1.17	1.15- 1.19	0.55	0.48- 0.67
28	19	1.52	1.35- 1.61	1.19	1.16- 1.24	0.54	0.44- 0.62
32	21	1.47	1.38- 1.56	1.16	1.14- 1.19	0.58	0.48- 0.69
40	9	1.50	1.45- 1.52	1.17	1.16- 1.19	0.57	0.52- 0.61

Table 2. Results of wide plate tensile tests   (Steel grade: SM490A)

Specimen nummber	Thick-ness (mm)	Width (mm)	Yield strgth. (N/mm^2)	Ult. strgth. (N/mm^2)	Elonga-tion (%)	Ratio of corner area (%)
AS-1	24.48	24.91	422	563	27.1	0
AS-2	24.49	24.98	439	562	26.6	0
AR-1	23.47	24.64	——	646	12.9	100
AR-2	23.17	24.90	626	648	12.9	100
BS-1	24.60	98.05	415	557	29.7	0
BR-1	22.95	98.84	562	635	18.0	100
BR-2	22.83	97.89	573	641	19.4	100
BR-3	23.01	97.85	580	647	18.5	100
CS-1	24.4	196.0	420	570	30.2	0
CR-1	24.4	195.9	472	584	22.3	50
CR-2	24.4	195.8	461	587	24.2	50
DR-1	24.4	266.8	462	580	24.3	37
DR-2	24.4	266.8	455	587	23.6	37
AAS-1	39.4	25.0	362	533	29.2	0
AAR-1	38.0	25.0	546	628	16.8	100
ABR-1	37.5	240.0	460	572	19.5	65

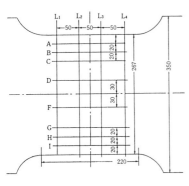

Table (b) Elongation within gauge length of 50mm:

	L₁ 27.5%	L₂ 49.9%	L₃ 22.9%
A	26.0	49.5	21.6
B	23.8	51.2	18.9
C	20.5	54.0	18.1
D	20.1	54.3	18.0
E	20.5	54.1	18.5
F	23.4	51.7	20.4
G	25.4	53.4	22.9
H	26.3	54.7	24.1

Table (c) Elongation within gauge length of 150mm:

	33.4%
A	32.4
B	31.3
C	30.9
D	30.8
E	31.0
F	31.8
G	33.9
H	35.0

(a) Measured positions

(b) Elongation within gauge length of 50mm

(c) Elongation within gauge length of 150mm

Fig.3. Elongation distribution in No.DR-1

(a) Appearance after test   (b) Fractured section

Fig.4. Deformation of specimen No.ABR-1

As shown in Fig.3(b) for No.DR-1 with the ratio of corner area of 37%, the elongation of the corner was almost the same as that of the flat parts, and, as the whole section the specimens showed an excellent ductility. The section of the broken part in No.ABR-1 was ductile as shown in Fig.4(b).

3.3 Wide plate bending tests

Bending tests were carried out on the coupons similar to those of the above tensile tests.   The width of the coupons was 200mm and their wall-thicknesses were 25mm or 40mm.   They were placed on the testing bed so that the line of the cold-worked zone was positioned tranversely to the bending line. In that condition the coupons were subjected to the planar strain condition. The coupons were bent so that the calculated value of the surface strain became 25%. As shown in Fig.5, for example, they did not cause any visual cracks and showed excellent ductility.

3.4 Influence of cold-forming on mechanical properties of RHS

From the results of the above two test series, it can be concluded that the degree of the influence of cold-forming on the

Fig.5. Section of 40mmx200mm wide plate coupon after bending

mechanical properties of RHS section is quite low. It is expected
that the structural behaviour of RHS members may be only minutely
affected by such a forming process inspite of the changes of the mechanical properties at the corners, the increase of yield strength
and tensile strength, and the decrease of the elongation property.

4 Bending tests on beam-columns

4.1 Testing procedures
Monotonic and cyclic bending tests were carried out on beam-column
specimens of 350mmx350mm press-formed RHS with five different width-
thickness ratios. The purpose for the tests was to evaluate the
strength and deformation capacity of these sections. In order to
clarify the effects of the steel grade, monotonic bending tests were
carried out on the specimens with the same B/t ratios with another
steel grade of SS400 to JIS G3101. To clarify the influence from the
manufacturing method, monotonic bending tests were also carried out
on the specimens of built-up RHS with the same steel grade, SM490A,
and with the same B/t ratios.

The loading points and supporting points of the beam-column test
specimens were stiffened with inner-diaphragm plates. The configuration of the test specimens is illustrated in Fig.6. The mechanical
properties and chemical compositions of the steel are given in
Table 3, and the dimension and the basic mechanical properties of
the specimens are given in Table 4.

A 500tonf(=4900kN) capacity Amsler testing machine was used for
the monotonic bending tests.   Deflection was measured by electric

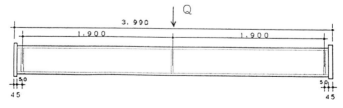

Fig.6. Configuration of bending test specimens

dial gauges at the centre of the specimens, where two gauges were set on a bar arm welded to each side of the specimen. This is to offset possible unsymmetrical deflection which tends to occur in the plastic deformation range. The strains were measured at three points 100mm apart from the loading point; at the central point of the upper flange, at a corner point in the upper flange, and at the intermediate point of these two points.

As shown in Fig.7, in the case of the cyclic bending tests, the test specimen was set in a strong frame in which the forces were self-balanced, and cyclic bending forces were applied at the centre of the specimen using 300tonf(=2940kN) hydraulic jacks (so called 'three-points loading'). Deflection and strain were measured in the same way as in the monotonic bending tests. The loading in each cycle was so controlled that when the deflection at the centre reached a pre-determined value, loading was stopped and unloading started : in the 1st cycle, deflection limits of $\pm 1 \delta p$ (δp : Centre deflection coresponding to the full plastic moment Mp), and in the 2nd cycle, deflection limits of $\pm 2 \delta p$, the limits were increased with the increment of δp .

Table 3. Mechanical properties and chemical compositions (t=9-22mm)

Steel grades	Mechanical properties			Chemical compositions (wt %)				
	Yield strght. (N/mm^2)	Ult. strght. (N/mm^2)	Elon-gation (%)	C	Si	Mn	P	S
SM490A	353 - 421	529 - 549	22.0 -28.0	0.16 -0.18	0.36 -0.40	1.40 -1.43	0.012 -0.027	0.003 -0.005
SS400	294 - 333	451 - 460	31.0 -32.0	0.11 -0.13	0.20 -0.24	0.95 -0.99	0.017 -0.023	0.003 -0.005

Table 4. Dimensions of bending test specimens

Specm. number	B (mm)	t (mm)	B/t	r (mm)	A (cm^2)	Ze (cm^3)	Zp (cm^3)	My $(kN*m)$	Mp $(kN*m)$	δp (mm)	Py (kN)	Pp (kN)
C509B	350	9	38.9	27	119.3	1300	1509	497	578	18.0	524	608
C512B	350	12	29.2	36	156.1	1660	1949	651	765	16.9	685	805
C516B	350	16	21.9	48	202.8	2086	2488	778	927	16.2	819	976
C519B	350	19	18.4	57	236.1	2367	2856	859	1037	15.9	904	1090
C522B	350	22	15.9	66	267.9	2614	3194	923	1128	15.7	972	1187
C409B	350	9	38.9	27	119.3	1300	1509	408	474	13.4	430	498
C412B	350	12	29.2	36	156.1	1660	1949	553	650	12.9	583	685
C416B	350	16	21.9	48	202.8	2086	2488	614	732	12.8	646	771
C419B	350	19	18.4	57	236.1	2367	2856	696	840	12.9	733	885
C422B	350	22	15.9	66	267.9	2614	3194	769	939	13.1	809	989
B512B	350	12	29.2	0	162.2	1767	2056	693	807	16.8	730	849
B516B	350	16	21.9	0	213.8	2275	2677	847	997	16.1	892	1050
B522B	350	22	15.9	0	288.6	2967	3550	1047	1253	15.5	1102	1319

Notes: B: Width, t: Wall-thickness,   r: Outside corner radius
       A: Whole area of section,   Ze: Elastic section modulus
       Zp: Plastic section modulus, My: Yield moment, Mp: Full plastic moment
       δp: Centre deflection at Mp, Py: Load at My,   Pp: Load at Mp

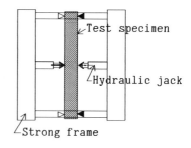

Note : Testing system

Test specimen

Hydraulic jack

Strong frame

Fig.7. Testing apparatus of cyclic bending

4.2 Influence of width-thickness ratio

The results of the monotonic bending tests are given in Table 5 and
the Q-δ curves are shown in Figs.8 and 9. From these test results,
it can be understood that the column strength and deformation capacity increased in accordance with the decrease of the B/t ratio.
In the case of the specimens with small B/t ratios, the bending
strength continued to increase up to the maximum even after local
buckling took place in the flanges, and the load began to gradually
decrease due to the occurence of local buckling in the webs.
However, for specimens Nos. C519B and C522B, the maximum strength
could not be reached because of the limitation of the measuring
system, and thus, obvious local buckling did not occur. In these
cases, the deflection angle (hereafter, defined as δc /L, where δc
is the centre deflection and L is the half length of the bending
span) could reach 7/100 at the final stage of the loading, and the
ductility factors were nearly 9.0.
    The current design standards for tubular structures of the Arch-
tectual Institute of Japan (1990) set forth the required ductility
factors for RHS columns as follows:
    For B/t$\leq$27 : Ductility factor $\geq$7
    For B/t$\leq$32 : Ductility factor $\geq$2.5
When comparing the test results with the required AIJ values,it can
be concluded that the press-formed RHS have sufficient deformation
capacity.

4.3 Influence from steel grade

The test results of the specimens for steel grade SS 400 are given
in Table 5 and Fig.8. A comparison of the Q - δ curves after
yielding given in Figs.7 and 8 showed that there isn't significant
difference between SM490A and SS400 results. From this fact,it is
clear that the structural behaviour of the press-formed RHS is
mainly determined by the geometric dimensions of the section.

4.4 Differences between monotonic and cyclic bending tests

The test results for the cyclic bending are given in Table 6 and Fig.9. Specimens Nos.C519BC and C522BC could not reach the maximum loads because of the limitation of the measuring system.

Comparison between the results in Table 5 and Table 6 indicates that the maximum strength of both series is nearly equal. On the other hand, the deflection at the maximum strength in the cyclic bending tests were smaller than those of the monotonic bending tests. It is thought that the difference was caused by the strain-hardening effect due to cyclic loading.

A skelton curve for No.C516BC is shown in Fig.8 together with that for No.C516B. Here, it is seen that the skelton curve and monotonic curve almost coincide with each other. Similar results were also reported by Kato (1982).

Taking into account the characteristics of cyclic loading, it can be noted that the deformation capacity under cyclic bending is about two times the deformation capacity under monotonic bending.

4.5 Comparison with built-up RHS

As previously decribed, the test specimens of built-up RHS were fabricated from the same SM490A plates as those used for the press-

Table 5. Test results for monotonic bending

Specimen number	Pmax (kN)	δ pmax (mm)	Pmax /Pp	δ pmax / δ p	δ pmax /L (rad)
C509B	588	32	0.98	1.7	0.017
C512B	883	80	1.10	4.7	0.042
C516B	1245	145	1.28	9.0	0.076
C519B	1627	$\geq$150	1.49	$\geq$9.4	$\geq$0.079
C522B	$\geq$ 1804	$\geq$135	$\geq$1.52	$\geq$8.6	$\geq$0.071
C409B	509	25	1.02	1.9	0.013
C412B	716	75	1.04	5.2	0.039
C416B	1010	160	1.31	12.5	0.084
C419B	1392	$\geq$160	1.56	$\geq$ 12.4	$\geq$0.084
C422B	$\geq$ 1442	$\geq$140	$\geq$1.48	$\geq$ 10.7	$\geq$0.074
B512B	882	71	1.04	4.2	0.037
B516B	1255	145	1.20	9.0	0.076
B522B	$\geq$ 1824	$\geq$135	$\geq$1.53	$\geq$8.7	$\geq$0.071

Notes: Pmax : Load at max. moment, δ pmax: Deflection at Pmax
       Pmax /Pp : Post yield strength ratio
       δ pmax/ δ p : Ductility factor at Pmax
       δ pmax /L : Deflection angle at Pmax

Table 6. Test results for cyclic bending

Specimen number	Pmax (kN)	δ pmax (mm)	Number of cycle	Notes:
C509BC	598	28	+3	Pmax : Load corresponding to max. moment.
C512BC	892	60	+4	
C516BC	1255	70	+4	
C519BC	$\geq$1471	$\geq$ 77	$\geq$ -5	δ pmax: Deflection at Pmax.
C522BC	$\geq$1657	$\geq$ 75	$\geq$ -5	

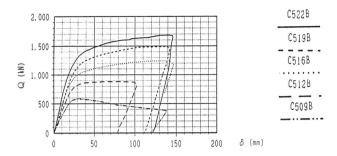

Fig.7. Q-δ curves for SM490A specimens

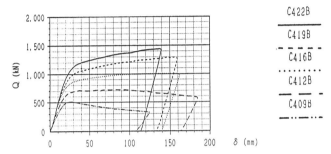

Fig.8. Q-δ curves for SS400 specimens

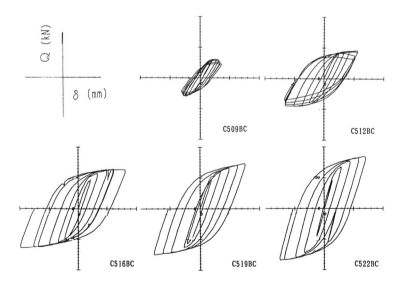

Fig.9. Q-δ curves for cyclic bending tests

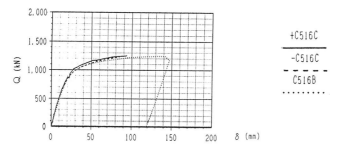

Fig.10. Skelton curves for different loading methods

formed RHS. Therefore, it is thought that the differences in the test results are mainly due to the differences in the manufacturing methods.

From the test results shown in Table 5 and Fig.7, it can be seen that the strength and ductility factors at the maximum strength and deflection angle of the press-formed RHS are almost the same as those of the built-up RHS. When comparing the properties of both sections in the actual design, it is neccessary to take into account that the section area and the elastic and plastic section modulus of the press-formed RHS are smaller than those of the built-up RHS.

4.6 Geometric similarity

Watanabe, et al.(1992) carried out bending tests on beam-columns of press-formed RHS with a width of 800mm and a thickness of 40mm, i.e B/t=20. The loading point of the specimen was stiffened with an inner-diaphragm. Some of the major test results are as follows:

Pmax / Pp   (Post yield strength ratio)   :   1.34
Ductility factor at maximum strength   :   13.5
Deflection angle at maximum strength   :   10/100

The test results of the press-formed RHS as given in Table 5 were almost similar to the above results with the same B/t. Although the number of specimens of such a large section is limited, it can be practically said that the post yield strength ratio and the ductility factor are mainly determined by B/t ratios; that is, the scale effect is fairly small. A similar fact was reported by Wallace and Krawinkler (1989).

5 Conclusions

The serial tests were carried out in order to clarify the structural behaviour of the press-formed RHS with a large outside dimension of 350mm and with wall-thicknesses ranging from 9mm to 22mm. Based on the analysis of the test results, the following are concluded.

From the tensile and bending tests of the wide plates,

(1)in the case of coupons with only mother steel, the elongation
   property are little affected by the width.
(2)In the case of wide tensile coupons including a cold-worked zone,
   the ratio of the cold-worked area to the whole section affects
   the elongation property. When the ratio is small, for example,
   less than 30%, the behaviour of the coupon is little affected by
   the cold-working.
(3)From the above conclusions, it is foreseen that the behaviour of
   press-formed RHS beam-columns, which contain 10% to 30% cold-
   formed area in the section, is little affected by the cold-forming, and that press-formed RHS have a sufficient ductility.

From the monotonic bending and cyclic bending on the press-formed
RHS and the built-up RHS,

(4)the structural behaviour, i.e. the strength and ductility, of RHS
   under bending is mainly determined by the B/t ratio of the
   compressive flange.
(5)The post yielding behaviour of the press-formed RHS with SM490 is
   almost the same as that of the press-formed RHS with SS400.
   The influence of the steel grade on the deformation capacity
   is small.
(6)The influence from the manufacturing method on the deformation
   capacity is small.
(7)The scale effect on the post yielding strength and on the deformation capacity are fairly small.

From the above conclusions, it can be concluded that the press-
formed RHS possess sufficient strength and deformation capacity
for use as columns in steel buildings.

6 References

AIJ (1990) Ultimate strength and deformation capacity of building
   in seismic design. **Archtechtual Institute of Japan,**
   Tokyo (in Japanese).
Kato, B. (1982) Beam-to-column connection reaserch in Japan. **J.**
   Struct. Div., ASCE , 108, 343-360.
Wallace, B. J. and Krawinkler, H. K. (1989) Small-scale model tests
   of structural steel assemblies.**J. Struct. Engrg., ASCE,**
   115,1999-2015.
Watanabe, K. et al. (1992) Bending tests of thick and large size
   press-formed rectangular hollow section columns. **Summaries of**
   technical papers of AIJ Annual meeting., Aug., 1992, 1285 -
   1286 (in Japanese).

17 STABILITY OF AXIALLY COMPRESSED DOUBLE-TUBE MEMBERS

S. KUWAHARA, M. TADA and T. YONEYAMA
Osaka University, Japan
K. IMAI
Kawatetsu Steel Products Corporation, Kobe, Japan

Abstract
The double-tube member presented here consists of an outer circular tube and an inner circular tube, the latter being shorter than the former. The outer tube resists axial forces and the inner tube restrains it from buckling by the bending resistance. The results obtained from axial loading tests have confirmed the lateral stiffening effect of the inner tube, as follows. The inner tube with suitable strength and stiffness prevents buckling of the member and leads to the remarkable axial deformation of the member without any decrease of loading on it.
Keywords: Buckling, Stiffening, Steel Tube, Tubular Construction.

1 Introduction

The double-tube member presented here consists of an outer circular tube and an inner circular tube, the latter being shorter than the former, not to subject to axial forces. The outer tube resists axial forces and the inner tube restrains it from buckling by the bending resistance, as shown in Figure 1. The feature of this stiffening method is that the stiffening member is parallel with the stiffened member and that the compressive member is stiffened all along its length, and not discretely.

The purpose of this paper is to obtain the criterion to prevent buckling of such double-tube column members, and to evaluate the authenticity of the criterion from the results of axial loading tests by the numerical analysis. In

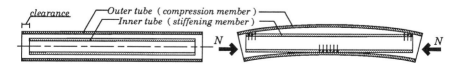

Fig.1. Double-tube member

Tubular Structures V. Edited by M.G. Coutie and G. Davies.
© 1993 E & FN Spon, 2–6 Boundary Row, London SE1 8HN. ISBN 0 419 18770 7.

addition the test results of a braced frame incorporating double-tube member is reported.

2 Criterion to prevent buckling of double-tube members

It is assumed that the material of the outer tube is elastic-perfectly plastic. Figure 2(a-c) shows the stages of loading for the double-tube member whose ends are supported by pins. The state when the axial load reaches the yield axial force of the outer tube N_y losing its bending strength is taken into consideration in (c). The equilibrium equation at such a point of the moment applied to inner tube at midspan is obtained as follows.

$$_iM_c = N_y (v + e + v_0) \tag{1}$$

where $_iM_c$ denotes the moment that acts at midspan of the inner tube, v and v_0 denote the deflection of the inner tube and the initial deflection that represents several imperfections, respectively, and e denotes the gap between the outer tube and the inner tube. To prevent buckling of the double-tube and to increase plastic axial deformation of the outer tube without any decrease of loading, the follow inequality must be satisfied.

$$_iM_c \geq {_iM_b} \tag{2}$$

Here the stiffening force is assumed to be an uniformly distributed load q. To substitute the moment - elastic deflection relation of the inner tube and eq.(1) into inequality (2), the criterion to prevent buckling of the double-tube member is obtained as follows.

$$m_b \, 1 \geq (e + v_0) \, / \, (1 - C_{sf}/k_b) \tag{3}$$

where $C_{sf} = 5/48$, $m_b = {_iM_b} \, / \, N_y \, 1$, $k_b = {_iE} \, {_iI} \, / \, N_y \, 1^2$, 1

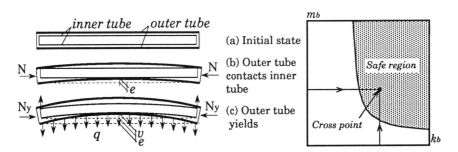

Fig.2. Stiffening mechanism

Fig.3. Criterion to prevent buckling

184

denotes the length of the double-tube and $_iE$ $_iI$ denotes the flexural stiffness of the inner tube. As $_iM_b$ is required to be defined as the maximum moment where the stiffness of inner tube can be maintained, $_iM_b$ is defined as the yield moment of the inner tube.

Inequality (3) represents the hatched region (hereafter, denoted as safe region) in Figure 3. When the cross point of the strength and the stiffness of inner tube is in the safe region, the double-tube member is stable.

3 Axial loading tests

3.1 Specimens and test procedure
Axial loading tests for double-tube members were conducted. Single-tube members with no inner tube inserted were also tested for comparison. A typical specimen is shown in Figure 4 and the list of specimens in Table 1. Outer tubes are cold-formed tubes whose section is $\phi 76.3 \times 2.8$. End plates fastened to the pins are welded at the ends of the outer tube. The section of the inner tube is $\phi 70.0 \times 5.0$, and the length is 30mm less than that of the outer tube, in order not to subject the inner tube to the axial force. The inner tube is only inserted into the outer tube without being fixed. The magnitude of the gap between the two tubes is 0.7mm.

In Figure 5, the $m_b l - k_b$ relation obtained from inequality (3) is drawn and the cross points of the strength and the stiffness of the inner tube, inserted into specimens of

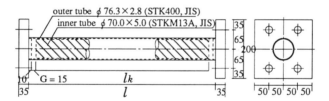

outer tube $\phi 76.3 \times 2.8$ (STK400, JIS)
inner tube $\phi 70.0 \times 5.0$ (STKM13A, JIS)

Fig.4. Test specimen

Table 1. List of specimens

Designation		l (l_k) (mm)	Loading condition
Double-tube	Single-tube		
M60D	M60	1560 (1510)	monotonic
M75D	M75	1950 (1900)	
M90D	M90	2340 (2290)	
M105D	M105	2730 (2680)	
A60D	A60	1560 (1510)	cyclic
A75D	A75	1950 (1900)	
A90D	A90	2340 (2290)	
A105D	A105	2730 (2680)	

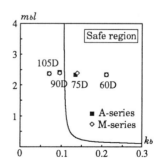

Fig.5. $m_b l - k_b$ relation

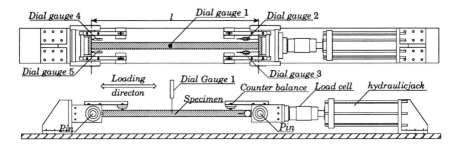

Fig.6. Test setup

double-tube members, are plotted. From Figure 5, it is found
that the plots for specimens M60D, M75D, A60D, M75D are in
the safe region.

The test setup is shown in Figure 6. The ends of a
specimen are fastened to the pins which permit the rotation
in the vertical plane and the specimen is loaded
monotonically or cyclically in axial direction by a hydraulic
jack. As the rotational centres of the pins are placed at
the inside faces of end plates, the buckling length is the
distance l between the inside faces of end plates. The
rotational friction of the pins is very small, and the moment
which is developed by the weight of end plates and pins is
cancelled by the counter balances. Axial load is measured by
a load cell. Axial displacement and lateral deflection at
the midspan are also measured.

3.2 Test result
The results of monotonic loading are reported. The load –
axial deformation relations are shown in Figure 7. The
ordinates show the axial force N divided by the yield load of
outer tube N_y, and the abscissas shows the axial deformation
u divided by the elastic axial deformation u_y when the yield
axial load is reached. The load – central deflection
relations are shown in Figures 8. The ordinates are the same
as previously, and the abscissas show the deflection v
divided by the length of specimen l. Compression is taken as
positive. The solid line shows the result for double-tube
member, and the dotted line for the single-tube. The sign▼
denotes the maximum load level recorded, and◆ denotes the
development of local buckling. ▽ indicates the rise of load
caused by the contact of the inner tube with end plates.

All single-tube members specimens buckled. And specimen
M60 developed local buckling at its mid span. The value of
maximum strength and axial deformation capacity for double-
tube members was larger than that for single-tubes which have
the same slenderness ratio. Specimens M60D,M75D didn't
buckle in elasto-plastic range, and the remarkable axial
deformation was obtained without any decrease of loading. At
the final phase, these specimens developed local buckling at

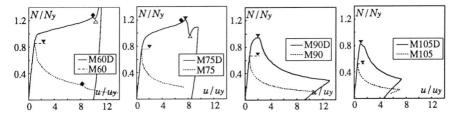

Fig.7. Load - Axial deformation (monotonic loading)

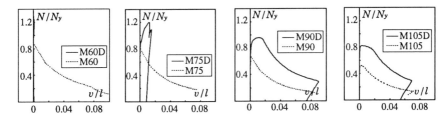

Fig.8. Load - Midspan deflection (monotonic loading)

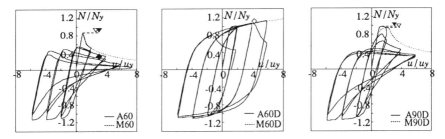

Fig.9. Load - Axial deformation (cyclic loading)

the end of the outer tube and load decrease. On the other hand, specimens M90D,M105D buckled at an earlier phase. These results correspond well to those which are predicted from inequality (3).

The results of cyclic loading are also shown in Figure 9, while monotonic loading results shown as dotted lines are added.

Specimens A60D,A75D which satisfy the above criterion didn't buckle and stable hystereses were obtained. Specimen A90D,A105D buckled at the loop whose amplitude is u/l = 1/300, after which, its energy dissipation capacity deteriorated rapidly as the number of loading cycle increased.

3.3 Comparison of analysis results and test results
The behaviour of double-tube members were also analysed by the one dimensional finite element method. The solution of

non-linear stiffness equation is the incremental perturbation method. The section of tubes is divided into six elements each area of which is concentrated to a point, as shown in Figure 10(a). Figure 10(b) shows an analytical model of double-tube members. An outer tube whose initial deflection at midspan is 1/10000 is tied to an inner tube by eight connectors. In order to simulate the contact between the inner tube and the outer tube, the connectors are assumed to have the load N - axial deformation u relation as shown in Figure 10(c). The stress - strain curve of the outer tube is modeled in four segments and that of the inner tube is bi-linear.

Figure 11 show the results of both analyses and tests for specimens M75D,M75, M90D,M90, and M105D,M105. Analysis

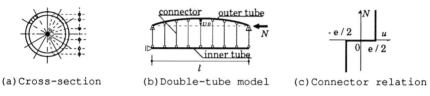

(a) Cross-section   (b) Double-tube model   (c) Connector relation

Fig.10. Analytical model of double-tube

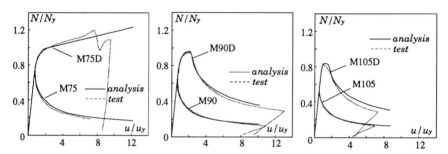

Fig.11. Comparison analysis-result with test-result

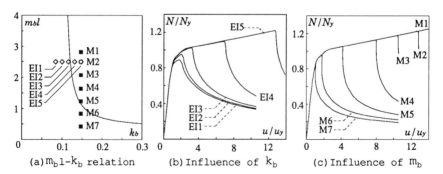

(a) $m_b l$-k_b relation   (b) Influence of k_b (c) Influence of m_b

Fig.12. Influence of stiffness and strength of inner tube

188

results correspond to test results well. And each analysis result represents the characteristics of double-tubes whether they satisfy the criterion or not.

To evaluate the applicability of the criterion in detail, various analyses are conducted. The models are grouped in two types. One group in which the stiffness of inner tubes is varied is named the EI-series; the other group where the strength is varied is named the M-series. The cross point of the strength and the stiffness of inner tubes for each model are plotted as shown in Figure 12(a), which shows that EI3 and M5 are on the boundary of the safe region. The results of analyses are shown in Figure 12(b),(c). It is obvious that the models which satisfy the above criterion show the remarkable axial plastic deformation without any decrease of loading.

4 Horizontal loading of K-braced frames

The specimen and the test setup are shown in Figure 13. The brace members consist of double-tube members which satisfy the above criterion. The ends of the bracing members are connected to the frame with pins. The inner tube is 20mm shorter than the outer tube, and is fixed at mid-span with a spot welding.

The specimen is supported with two pins at the bottom of the columns, and is loaded cyclically through the loading beam which is connected to the top of the columns at two points with pins.

The result of the test is shown in Figure 14, and that for the frame with single-tube braces is shown in Figure 15, which proves the effectiveness of this frame. ▼ denotes the development of local buckling, and ◆ denotes the development of tearing at midspan of the tensile brace member.

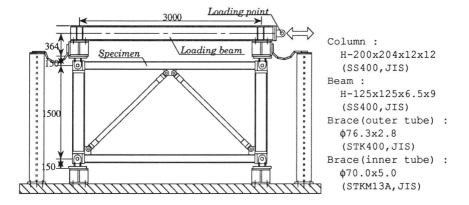

Column :
    H-200x204x12x12
    (SS400,JIS)
Beam :
    H-125x125x6.5x9
    (SS400,JIS)
Brace(outer tube) :
    φ76.3x2.8
    (STK400,JIS)
Brace(inner tube) :
    φ70.0x5.0
    (STKM13A,JIS)

Fig.13. Specimen and test setup

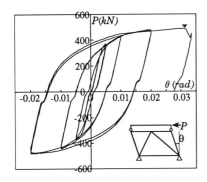 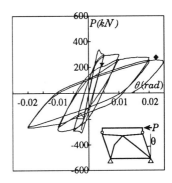

Fig.14. P - θ relation for the Fig.15. P - θ relation for the
frame with double-tube braces   frame with single-tube braces

The double-tube members used for brace members didn't
buckle, and the frame with the double-tube members showed the
stable hysteresis loops. At the final phase, the compressive
brace member developed local buckling at the end of it, and
many local buckling waves all along its length were observed.
On the other hand, the single-tube brace members buckled, and
the frame with single-tube brace members deteriorated rapidly
in its strength as the number of load cycle increased.

5 Conclusion

In this paper, the criterion to prevent buckling of double-
tube members was obtained, and the authenticity of the
criterion was evaluated experimentally and analytically.
The primary conclusions obtained are as follows.
1) The result of axial loading tests of double-tube shows
that the magnitudes of the strength and axial deformation
capacity of double-tube members exceed those of single-tubes.
2) The double-tube members which satisfy the criterion
(inequality (3)) don't buckle and show stable hysteresis
loops.
3) The axial deformation capacity of double-tube members
which satisfy the criterion is dominated by the development
of local buckling.
4) It is confirmed that the analytical model of double-tubes
in Figure 10(b) is useful to understand the behaviour of
double-tube members.
5) The result of horizontal loading of the frame braced with
double-tube members shows the stable hysteresis loops.

Acknowledgments

The authors are grateful to Dr. Kozo Wakiyama and Dr. Kazuo Inoue of Osaka University for their advice and information and to Graduate students of Osaka University Mr. Yukio Yuasa and Mr. Yoshitaka Yamashita for their help in the experiments.

References

Ishida, S. and Morisako, K. (1989) Application of incremental perturbation method to one dimensional combined materially and geometrically nonlinear finite element method (FERT-P). **Annual Trans. of Architectural Institute of Japan.** 397, 73-82 (in Japanese)

PART 6

COMPOSITE CONNECTIONS

Concrete-filled joints
and columns

18 CONNECTION OF STEEL BEAMS TO CONCRETE-FILLED TUBES

H. SHAKIR-KHALIL
Civil Engineering Department, University of Manchester, UK

Abstract
Tests have been carried out on twentyeight connection specimens. The specimens consisted of steel beams connected to columns made of concrete-filled steel tubes. The results reported here are for a series of eight specimens in which the columns were made of a steel rectangular hollow section 150x150x5 RHS. The beams were connected to the columns by means of 'finplates' which were welded to the middle of the side wall of the hollow section. In each test, loads were applied to both column and beams in order to simulate actual column loading in structures. The beams were loaded symmetrically, and the beam load was kept as a proportion of the column load. The columns were about 3m long, and thus represented a floor height in typical structures.
Keywords: Concrete-Filled, Steel, Rectangular Hollow Section, Tube, Finplate, Connection, Composite Structure.

1 Introduction

The work reported here forms part of an experimental investigation into the behaviour of composite connections. The project resulted from previous research on composite columns in which several types of concrete-filled columns have been studied. British Steel, Welded Tubes (BS-WT), Corby, England, have been actively involved in the work and provided all the steel hollow sections which were required for the experimental work. Some of the results of the investigations on the behaviour of concrete-filled rectangular hollow section columns have already been published, Shakir-Khalil et al. (1989, 1990, 1991c, 1992a).

The current study on composite connections is financed by a 3-year grant from the Science and Engineering Research Council (SERC), England. British Steel provided all the test specimens required for the investigation, and also supplied a purpose-built test rig. Both test rig and specimens were designed by British Steel in collaboration with the author. The project included a comprehensive preliminary study of bond strength in concrete-filled tubes, the results of which have been

Tubular Structures V. Edited by M.G. Coutie and G. Davies.
© 1993 E & FN Spon, 2–6 Boundary Row, London SE1 8HN. ISBN 0 419 18770 7.

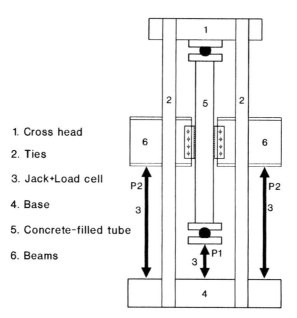

1. Cross head

2. Ties

3. Jack+Load cell

4. Base

5. Concrete-filled tube

6. Beams

Fig.1. Test Rig and Specimen

published, Shakir-Khalil (1991a, 1991b). The columns in the first series of eight connections in this project were made of concrete-filled circular hollow sections, and the test results have already been published, Shakir-Khalil (1992b).

2 Test Rig

A schematic view of both test rig and test specimen is shown in Fig.1. The test rig consists of four vertical rectangular hollow section ties which are connected to a cross-head at the top and a base at the bottom. The base is securely bolted to the laboratory strong floor. Hydraulic jacks of capacities 3000kN and 250kN apply the required loads on column and beams respectively. The side jacks are kept in the same position, and the eccentricity of load P_2 is controlled by moving the load cells on a loading platform through which the beam loads are applied. The test rig is provided with no lateral bracing as it is symmetrically loaded. The test specimens were however braced at the column mid-height with a view to eliminating the in and out-of-plane transverse displacements at that level of the test specimens.

3 Test Specimens

Eight specimens were tested in this series, and Fig.1 shows the details of the specimens. The test specimens consist of 150x150x5 RHS columns which are 2.8m

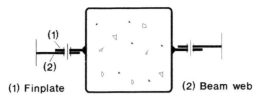

(1) Finplate   (2) Beam web

Fig.2. RHS with Finplates

long, and are provided with 15mm thick circular end plates which are welded to the column ends. The lower end plate is solid, whereas the top one is provided with a 140mm diameter hole to enable the casting of the concrete. The columns are provided at mid-height with 360mm long, 100x10mm finplates, Fig.2. The plates are welded to the column sides, and each finplate has four 22mm diameter holes in order that the side beams may be connected to the column by grade 8.8, 20mm diameter bolts, Fig.1. When testing, the column end plates are bolted to the 40mm thick loading plates of the test rig. The loading plates are provided with roller supports to allow the column freedom of rotation in the plane of loading.

Table 1. Details of Specimens and Material Properties

Specimen	B1	B2	B3	B4	B5	B6	B7	B8
Nails	--	12	--	12	--	12	--	12
e (mm)	125	125	125	125	250	250	250	250
P_1/P_2	8	8	5	5	8	8	5	5
f_{cu} N/mm^2	38.3	41.8	38.1	46.1	40.8	39.9	39.6	44.2
f_{cy} N/mm^2	31.8	34.4	32.3	39.8	34.5	33.2	33.1	35.1
f_{sd} N/mm^2	328	322	340	330	328	330	327	327
f_{ult} N/mm^2	471	477	477	470	474	472	473	472
E_s kN/mm^2	207	206	209	209	205	205	205*	210
$\mu\epsilon_y$	1600	1608	1680	1600	1600	1656	N/A	1600

(*) Assumed value based on BS 5950.

The beam loads were applied at either 125mm or 250mm from the face of the column, i.e. column eccentricities of 200mm and 325mm respectively. The beam and column loads were increased proportionately in the ratios of either 1:8 or 1:5. The details of all eight specimens are given in Table 1. As seen from the table, four of the specimens were provided with 12 'Hilti' nails, Hilti (1990), in an effort to study the effect of these shear connectors on the connection behaviour. Four of the 3.7mm diameter, 62mm long nails were provided at the top, bottom and middle of the finplates. They were placed symmetrically on both sides of the RHS sides to which the finplates were welded. When being assembled, the beams were bolted to the finplates and only hand tightened. As the beam loads were applied in the opposite

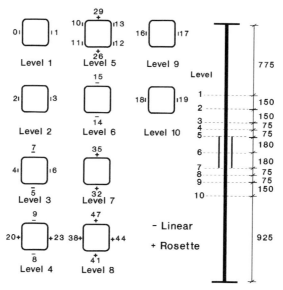

Fig.3. Strain Gauge Positions
Connections B5 - B8

direction to the self weight of the beam, this caused slip to occur in the beam-column connections during the course of the tests.

When being tested, and up to the specimen failure loads, no yield in the steel section was visible in the lower half of the specimens, and the recorded strains were well within the elastic values. The material properties of steel were therefore obtained from coupons cut from the lower ends of the columns. Table 1 gives the material properties of all specimens in which the concrete cube strength, f_{cu}, and cylinder strength, f_{cy}, are the average results obtained from testing four 100mm cubes and two 150mm solid cylinders respectively.

4   Instrumentation

Displacement transducers (DT) and electric resistance strain gauges (ERS) were used for all specimens. The displacement transducers enabled the recording of the in and out-of-plane displacements, and also the column shortening and end rotations. The locations of the strain gauges are shown in Fig.3. The strain measurements were carried out in both the longitudinal and transverse directions, and strain gauges were located at several levels above and below the finplate position. The strain measurements were carried out over distances equal to three times and twice the lateral dimension of the steel hollow section above and below the finplate positions respectively.

Data logging equipment was used to record and store the strains, displacements and load cells' outputs at each load increment. During the tests, the strain readings of three strain gauges in the vicinity of the connection were plotted on the monitor throughout the test with a view to detecting the response of the connection to the increase in the applied loads. This also helped to decide on the load increments as the failure load was approached. Early yield could also be observed through the formation of 'Luder' lines in the mill scale on the surface of the steel hollow section.

5 Test Results

The predicted failure loads and also the test results are summarised in Table 2. The predicted failure loads, N_p, of the test specimens were calculated on the basis of the British Standards. They are based on the column buckling length of 2.97m despite the fact that the columns were laterally supported at their mid-height. The predicted failure loads, N_p, are based on the squash load of the stub column, N_u, which was calculated on the basis of the experimental properties of steel and concrete. Failure loads of specimens, N_c, were also predicted on the basis of connection strength. Their values were taken as the least of the failure loads of the finplate, bolts, weld and yieldline of the side wall of the steel section.

The experimental values of the squash loads could not be established as no cut-offs from the column sections were available for such tests to be carried out. In calculating the predicted squash loads, N_u, the material partial safety factors of both steel and concrete, γ_{ms} and γ_{mc} respectively, were taken equal to unity in order that the calculated values may be used as the basis for predicting the column failure load. The N_u values were therefore calculated in accordance with BS 5400 as given by:

$$N_u = f_{sd} A_s + 0.67 f_{cd} A_c \tag{1}$$

where the terms are as defined in the notation.

The displacement transducers' recordings indicated that practically up to the failure loads, the columns suffered very little shortening, end rotation or in and out-of-plane displacements. Hence, these experimental results are not presented here.

The strain measurements showed that the average strains across the column section at levels 1-4 were practically identical. This indicated that the beam load was transferred to the column composite section within half a RHS depth above the finplate position. Figs.4 and 5 show the respective average longitudinal strains in the vicinity of and below the level of the finplate position for specimen B7. These figures are typical of the response of the rest of the test specimens in this series. Yield was recorded at level 7 at the lower end of the finplate on the steel section sides parallel to the finplates. Luder lines were also observed to form in the mill scale of the steel section between levels 7 and 6 when the recorded strains were no more than 60% of the yield strain of steel. This is probably due to the combined effect of the compressive force on the column, the transverse tensile strains due to the moment on the connection, and also the residual compressive strains which resulted from welding the finplates in position.

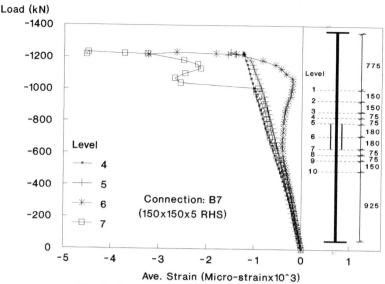

Fig.4. Ave. Strains: B7, Levels 4 - 7

Table 2. Results of Connection Tests

Specimen	B1	B2	B3	B4	B5	B6	B7	B8
N_u (kN)	1451	1479	1483	1559	1484	1478	1465	1525
N_p (kN)	1168	1188	1194	1247	1192	1188	1178	1223
N_c (kN)	1946	1910	1412	1370	973	979	679	679
N_e (kN)	1323	1408	1354	1385	1273	1197	1230	1260
R_1	--	1.04	--	0.97	--	0.94	--	0.98
R_2	1.06	1.18	1.09	1.08	--	--	--	--

NB: R_1 and R_2 ratios are based on N_e/N_u in order to allow for the variation in the material properties.

It can be seen from Fig.5 that the strains at level 8 increased at a rate which is relatively higher than those at other levels. No logical explanation could be found for this behaviour during the progress of the test. However, it was later realized that this was mainly a localised effect, and resulted from the finplates bending the walls of the steel section to which they were attached. The strain recordings at level 8 showed that the strains in the finplate sides of the RHS (ERS 38&44, Fig.3) were much higher than those recorded in the other sides (ERS 41&47, Fig.3). Towards the end of the tests, the local deformation in the sides to which the finplates were welded was clearly visible in the region below the finplate position. At the end of the tests on

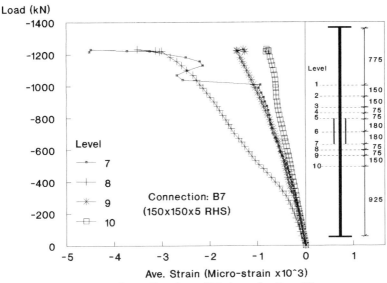

Fig.5. Ave. Strains; B7, Levels 7 - 10

specimens B5-B8 which were subjected to the large eccentricities, a yieldline type of failure was seen to have formed, and the localized effect of this type of plastic behaviour was clearly identified on the RHS side to which the finplates were connected.

As can be seen from Table 2, the predicted failure loads of the specimens which were subjected to the small eccentricity, B1-B4, were governed by the column failure load, N_p. The predicted failure loads of specimens B5-B8 were on the other hand due to connection failure, N_c. This was found to be the case in the tests. Specimens B1-B4 failed by the upper part of the column suffering large deformations, and Luder lines were clearly visible over the upper half of the column length. Some of the specimens developed a local bulge at the top of the column as the experimental load approached the squash load, N_u, of the composite section.

Specimens B5-B8, however, exhibited relatively large beam-column rotations, and failed at loads consistently lower than those of specimens B1-B4. Their failure loads are seen from Table 2 to be considerably larger than the connection failure loads, N_c, as obtained from the yieldline prediction. This is probably due to the fact that as the finplate pulled out the section wall, the outward pull caused considerable deformation in the RHS wall, and a higher resistance developed through a membrane mechanism. All specimens B5-B8 eventually failed by one of the finplates pulling out of the steel section wall by tearing the section wall at the toe of the finplate connection.

On opening the steel sections at the finplate level after the tests had been completed, none of the specimens showed that crushing of the concrete core had taken place.

6   Conclusions

Tests were carried out to investigate the moment carrying capacity of finplate connections to concrete-filled steel RHS columns. They were also conducted with a view to investigating the beam load transfer to the composite column section.

The tests have shown that the connections did not perform as well as those connected to concrete-filled circular hollow section columns. Under small eccentricities, failure resulted from overall collapse of the upper part of the column section. However, under large eccentricities beams showed relatively large rotations caused by distortions in the column sides. Under large eccentricities, the yieldline failure of the finplate side of the RHS should therefore be investigated. However, once large out-of-plane deformations take place, the failure mechanism alters to that of a membrane mechanism. This results from the concrete core preventing a fully developed yieldline type of failure from forming by preventing the tube corners from moving inwards closer to each other. The connections exhibited rotations larger than those for circular hollow section columns reported elsewhere, but in practice finplate connections are mainly designed as simple connections subjected to small eccentricities. The tests have however shown that, as a result of the pinching effect of the moment connection, the beam loads are transferred to the composite column section within a distance no longer than a column depth.

7   Notation

A_c , A_s	Areas of concrete and steel section
B, D, t	Breadth, Depth and wall thickness of steel hollow section
E_c , E_s	Elastic moduli of concrete and steel
e , e_1	Beam load eccentricity from face and centre of column
f_{cd} , f_{sd}	Design strengths of concrete and steel, given by the respective characteristic strength divided by the material partial safety factor γ_m
f_{ck} , f_{sk}	Characteristic strengths of concrete and steel, the former taken as $0.67f_{cu}$ in accordance with BS 5400
f_{cu} , f_{cy}	Characteristic 28 day cube and cylinder strengths of concrete
f_{ult}	Ultimate strength of steel
L_{ex} , L_{ey}	Effective buckling lengths of column about appropriate axis
N_c	Predicted failure load on basis of local failure at connection
N_e	Experimental failure load
N_p	Predicted failure load on basis of column failure
N_u , N_{ue}	Predicted and experimental squash loads of stub column
P	Total load on upper part of column $(P=P_1+2P_2)$
P_1 , P_2	Loads applied by middle and each side jack respectively
R_1	Load ratio of connections with and without nails
R_2	Load ratio of specimens with small and large eccentricities
γ_{mc} , γ_{ms}	Material partial safety factors of concrete and steel, given respectively by 1.5 and 1.1 in accordance with BS 5400
$\mu\epsilon_y$	Experimental yield strain of steel (in micro strain x 10^3)

8 Acknowledgements

The work reported here is part of an experimental programme financed by a 3-year research grant from the Science and Engineering Research Council (SERC). The test rig and manufactured test specimens were provided by British Steel, Welded Tubes (BS-WT), Corby, England, UK. The author wishes to acknowledge the assistance received from, and the technical discussions carried out with, the BS-WT personnel. Thanks go in particular to Mr Mike Edwards of BS-WT for his continuous and unfailing support. The author also wishes to thank Mr M A Mahmoud, the project's research assistant, for his endless assistance throughout the project.

9 References

BS 5400: Part 5 (1979) Steel, Concrete and Composite Bridges. Code of practice for the design of composite bridges. **British Standards Institution.**

BS 5950: Part 1 (1985). Structural use of steelwork in building. Code of practice for design in simple and continuous construction: hot rolled sections. **British Standards Institution.**

Hilti (Gt. Britain) Ltd. (1990) **Direct Fastening (Specifier's Guide)**, 35 Washway Road, Sale, Cheshire, England, UK.

Shakir-Khalil, H. and Zeghiche, J. (1989) Experimental behaviour of concrete-filled rolled rectangular hollow section columns. **The Structural Engineer**, Oct. 1989, Vol. 67, No. 19, pp. 346-353.

Shakir-Khalil, H. and Mouli, M. (1990) Further tests on concrete-filled rectangular hollow section columns. **The Structural Engineer**, Oct. 1990, Vol.68, No.20, pp. 405-413.

Shakir-Khalil, H. (1991a) Bond strength in concrete-filled steel hollow sections. Proceedings, **International Conference on Steel and Aluminium Structures.** Volume of Composite Steel Structures, pp. 157-168. Singapore, May 1991.

Shakir-Khalil, H. (1991b) Push-out tests on concrete-filled steel hollow sections. Proceedings, **International Symposium on Tubular Structures**, Delft, The Netherlands, June 1991. pp. 402-411.

Shakir-Khalil, H. (1991c) Tests on concrete-filled hollow section columns. Proceedings, ICCS-3; **Third International Conference on Steel-Concrete Composite Structures**, Fukuoka, Japan, Sept. 1991. pp. 89-94

Shakir-Khalil, H. (1992a) Columns of concrete-filled rectangular hollow sections. **Engineering Foundation Conference on Composite Construction; Composite Construction II.** Potosi, Missouri, USA. June 1992. Accepted to appear in a special ASCE publication.

Shakir-Khalil, H.(1992b) Full-scale tests on composite connections. **Engineering Foundation Conference on Composite Construction; Composite Construction II.** Potosi, Missouri, USA. June 1992. Accepted to appear in a special ASCE publication.

19 BUCKLING AND POST-BUCKLING STRENGTHS OF CONCRETE-FILLED SQUARE TUBES

B. TSUJI, E. TOMOYUKI, H. ENDHO and T. NISHINO
Kobe University, Japan

Abstract
Flexural buckling and post-buckling strengths of a concrete-filled square steel tubular column will be influenced by the interactions between the steel tube and filled-concrete. But the experimental data available are very few. In the present study, buckling tests of the concrete-filled square steel tubes were carried out for short or relatively short columns. Column curves are obtained using the tangent modulus equation. The tangent bending stiffness is estimated from the tangent axial stiffness obtained by the stub column tests. The column curves obtained agreed well with experimental results.
Keywords: Concrete-Filled Square Tube, Buckling Strength, Post-Buckling Strength, Column Curves.

1 Introduction

A concrete-filled steel tube is subjected to complex interaction between the concrete and steel. For example, tri-axial stress state is given to the concrete because of the confining provided by the steel tube, and local buckling of the steel tube is retarded because of the concrete filled inside. Experimental data available are very few, especially those of the concrete-filled square tubular columns (Shakir-Khalil, 1991). In the study presented, buckling tests of the concrete-filled square steel tube were carried out for short or relatively short columns. Axial compression tests for stub columns were also carried out. Buckling strength was estimated using tangent modulus theory, assuming that the bending tangent stiffness can be given as a function of the axial tangent stiffness. The estimated results were compared with the experimental ones.

2 Specimens and material properties

The specimens tested are shown in Table 3 and 4. A cold-formed square steel tube was made of mild steel (STKR400) and had a width (B) of 100mm and a thickness (t) of 2.3mm or 3.2mm. Three values for the aspect ratio (L/B) were selected for the stub column test and six values for the aspect ratio for the buckling test, where L is the

Tubular Structures V. Edited by M.G. Coutie and G. Davies.
© 1993 E & FN Spon, 2–6 Boundary Row, London SE1 8HN. ISBN 0 419 18770 7.

length of the column. Mix proportion of the concrete is shown in
Table 1. Material properties of the steel tubes obtained from stub
column tests and those of the concrete obtained from cylinder tests
are shown in Table 2.

3 Axial compression test of short concrete-filled tubes

3.1 Test procedure
As shown in Table 3, sixteen stub column specimens were tested.
Fig.1 shows the test set-up. The specimen was set on the hydraulic
testing machine. The axial contraction was measured by dial gage
type transducers and the axial strain distributions by wire strain
gages mounted on the steel tube.

3.2 Test results
Axial force (P) vs. axial displacement (ΔL) relationships obtained
are shown in Fig.2. As can be seen, after the attainment of the
maximum strength (Pmax), axial resistance deteriorates due to the
local buckling of the steel tube. As the axial deformation increases, the axial resistance increases gradually again in the relatively large deformation range. Maximum strengths, which increase
as the aspect ratio decreases, are 0.98 to 1.16 times as large as the
superposed axial resistance of the concrete and steel tube (Po).
Minimum resistances in the plastic range (Pmin) are 62% to 75% of the
maximum strength for the tubes with width-to-thickness ratio (B/t) of
43 (t=2.3mm) and 71% to 78% for the tubes with B/t ratio of 31
(t=3.2mm), as is shown in Table 3.
    Fig.3 shows the axial force vs. axial strain (ε) relationships.
The axial strain was the one measured by the wire strain gages mounted at the mid section of the steel tube. The dotted line in the
figure shows the superposed axial
resistance. The strain at which
the maximum strength was attained
increases as the width-to-thickness
ratio of the steel tube or the
aspect ratio (L/B) of the specimen
decreases. The slope of the
decreasing branch also increases
as the aspect ratio increases.
    Fig.4 shows the axial force vs.
axial tangent stiffness relationship. The tangent stiffness is
estimated as the ratio of the axial
force increment to the axial strain

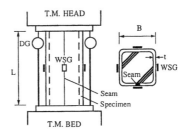

Fig.1. Stub-column test
apparatus

Table 1. Mix proportion of concrete

Water-cement ratio(%)	Constituents (N/m^3)		
	Cement	Sand	Aggregate
48.0	3800	6580	9790

Table 2. Material properties

Concrete Fc(MPa)	Steel tube σ_y(MPa)	
	t=2.3mm	t=3.2mm
43.4	311	406

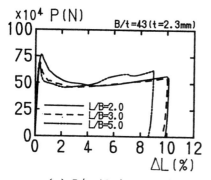
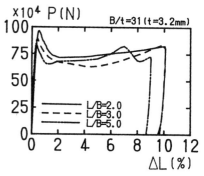

(a) B/t=43 (t=2.3mm)   (b) B/t=31 (t=3.2mm)
Fig.2. Axial force vs. axial displacement relationship

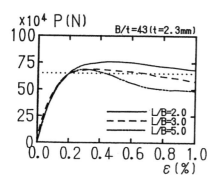
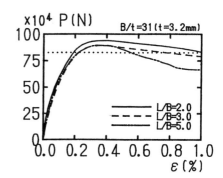

(a) B/t=43 (t=2.3mm)   (b) B/t=31 (t=3.2mm)
Fig.3. Axial force vs. axial strain relationship

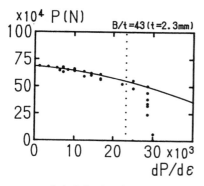
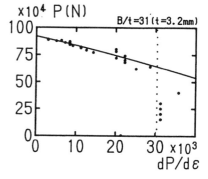

(a) B/t=43 (t=2.3mm)   (b) B/t=31 (t=3.2mm)
Fig.4. Axial force vs. axial tangent stiffness relationship

increment obtained at each step of the experiment.

4 Buckling test of centrally loaded tubes

4.1 Test procedure
As shown in Table 4, eighteen specimens, including steel tubular columns without filled-concrete, were tested. Fig.5 shows the test set-up. A 200tonf hydraulic testing machine was used. The pin end condition was adopted using the cylindrical surface bearing. Axial and lateral displacements were measured by dial gage type transducers, and axial strain distributions by wire strain gages mounted on the surface of the steel tube at the center section and the sections near the ends of the columns.
　　Each specimen was aligned such that the axial strains were essentially uniform over the mid-height cross section and the sections near the ends of the tube, so as to realize the centrally loaded condition. After the specimen was aligned in the testing machine, the centrally loaded tests were started.

4.2 Test results
Fig.6 shows the axial force vs. axial displacement relationship of the concrete-filled tubes. As can be seen, the maximum strength increases as the aspect ratio (L/B) decreases, and the descending branches of the curves have slopes increasing with increasing aspect ratio. Table 4 shows the maximum strength and the axial displacement at which the maximum strength was attained. To examine the slope at the descending branch, the ratios of the strength at the axial contraction of 1% (P_1) to the maximum strength are also shown in Table 4.
　　The centrally loaded tubes with or without filled-concrete buckled in the following manner: for the short tubes with large width-to-thickness ratio, local buckling of the steel tube appeared first, then the overall buckling occurred (type A), and for relatively short tubes with small width-to-thickness ratios, overall buckling occurred first, then the local buckling of the steel tube appeared (type B). In the former case, the maximum deflection due to the overall buckling was observed at the locally buckled portion, which appeared at random places between the pin supports. In the latter case, local buckling appeared at the center of the columns, where the maximum lateral deflection occurred due to the overall buckling. The collapse modes of each specimen are shown in Table 4. Typical strain trajectories at the center section of the columns are shown in Fig.7. For the short columns (L/B=7.5), strain reversal was observed at relatively large compressive

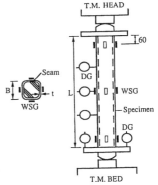

Fig.5. Buckling test apparatus

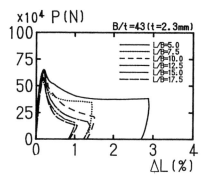
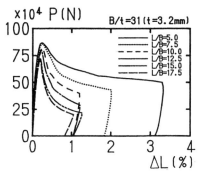

(a) B/t=43 (t=2.3mm)   (b) B/t=31 (t=3.2mm)
Fig.6. Axial force vs. axial displacement relationship

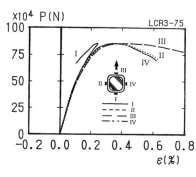
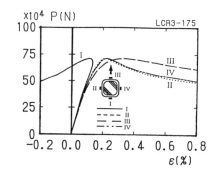

(a) L/B=75 (t=3.2mm)   (b) L/B=175 (t=3.2mm)
Fig.7. Axial strain trajectory

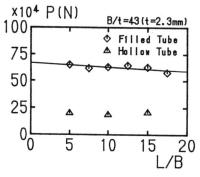
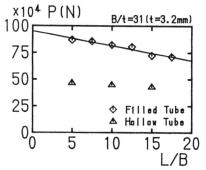

(a) B/t=43 (t=2.3mm)   (b) B/t=31 (t=3.2mm)
Fig.8. Buckling strength

strain, whereas for the relatively short columns (L/B=17.5), strain reversal was observed during the small compressive strain range.

Maximum strengths obtained (Pmax) are plotted against the aspect ratios of the columns (L/B) in Fig.8. Maximum strength decreases gradually as the aspect ratio increases, but the decrease in strength is not so severe in the cases of the specimens tested here. To examine the post-buckling behavior, the values P1/Pmax were shown in Table 4, as mentioned above. Post-buckling resistance decreases rapidly as the aspect ratio of the column increases. The effect of the width-to-thickness ratio of the steel on the post-buckling behavior is small. The axial deformations (ΔL) at which the maximum strength was attained are also shown in Table 4. The axial deformation decreases as the width-to-thickness ratio of the steel tube or aspect ratio of the column increases. The axial deformations of the hollow tubes are larger than those of concrete-filled tubes at their maximum strength.

5 Estimation of the buckling strength

Buckling and post-buckling strengths of the concrete filled tubes are affected by the interaction between the filled-concrete and steel tube. To estimate these behaviors analytically is very difficult in the case of the concrete-filled square tubes, because the interaction between the concrete and steel varies along the circumferential direction of the tube. The tangent modulus equation is usually used to obtain the buckling strength (P_{cr}) of the centrally loaded column.

$$P_{cr} = \pi^2 (EI)_t / L^2 \tag{1}$$

According to the tangent modulus theory, the bending tangent stiffness $(EI)_t$ can be obtained by summing the tangent stiffness of the filled concrete $(E_c)_t I_c$ and the steel tube $(E_s)_t I_s$ assuming that the state of stress is homogeneous:

$$(EI)_t = (E_c)_t I_c + (E_s)_t I_s \tag{2}$$

where E_c and E_s are the tangent moduli of the concrete and steel, and I_c and I_s the moment of inertia of the concrete and steel cross section, respectively. We assume here that the following equation exists between the bending tangent stiffness $(EI)_t$ and axial tangent stiffness $(EA)_t$:

$$(EI)_t / (E_c I_c + E_s I_s) = [(EA)_t / (E_c A_c + E_s A_s)]^\alpha \tag{3}$$

where A_c and A_s are the cross sectional areas of the concrete and steel tube, respectively. The relations between the axial tangent stiffness and the axial force obtained by the experiment were already shown in Fig.4. Assuming these relations adopt the quadratic form as:

$$P = a(dP/d\varepsilon)^2 + b(dP/d\varepsilon) + c \tag{4}$$

Table 3. Stub column tests

Specimen	L(mm)	t(mm)	Pmax(N)	Pmax/Po	Pmin(N)	Pmin/Pmax
SCR2-20a	200	2.09	745	1.16×10^3	461×10^3	0.618
20b	200	2.04	735	1.15	511	0.695
20c	200	2.07	773	1.20	492	0.636
SCR2-30a	300	2.05	675	1.05	504	0.747
30b	300	2.03	677	1.06	462	0.683
30c	300	2.02	667	1.04	419	0.628
SCR2-50a	500	2.04	680	1.06	461	0.678
50b	500	2.08	674	1.04	492	0.731
SCR3-20a	200	3.05	943	1.11	698	0.740
20b	200	3.05	922	1.08	701	0.761
20c	200	3.05	908	1.07	649	0.715
SCR3-30a	300	3.05	874	1.03	616	0.705
30b	300	3.04	863	1.01	667	0.773
30c	300	3.04	870	1.03	682	0.784
SCR3-50a	500	3.06	830	0.98	639	0.771
50b	500	3.05	878	1.04	653	0.744

Pmax: Maximum strength
Po : Superposed strength of the concrete and steel tube
Pmin: Minimum strength in the plastic range

Table 4. Buckling tests

Specimen	L(mm)	t(mm)	Pmax(N)	P_1(N)	P_1/Pmax	ΔL(%)	Mode
LCR2- 50	500	2.04	640×10^3	401×10^3	0.67	0.195	A
75	750	2.04	608	351	0.58	0.145	A
100	1000	2.08	620	261	0.42	0.206	B
125	1251	2.04	634	191	0.30	0.211	B
150	1500	2.08	618	154	0.25	0.201	B
175	1750	2.09	567	134	0.24	0.178	B
LCR3- 50	500	3.05	852	622	0.73	0.232	A
75	750	3.06	838	512	0.61	0.249	A
100	1000	3.05	806	418	0.52	0.246	A
125	1250	3.04	788	319	0.41	0.219	B
150	1500	3.05	708	238	0.34	0.196	B
175	1750	3.06	697	203	0.29	0.188	B
LSR2- 50	500	2.04	196	79	0.40	0.384	A
100	1000	2.03	183	42	0.23	0.231	A
150	1500	2.07	203	—	—	0.200	B
LSR3- 50	500	3.06	455	311	0.68	0.375	A
100	1000	3.06	431	102	0.24	0.329	B
150	1500	3.04	419	—	—	0.238	B

Pmax: Maximum strength
P_1 : Resistance at the axial contraction of 1%
ΔL : Axial displacement at the maximum strength

The coefficients a to c were obtained using the least-squares approximation. The estimated results are shown in Fig.4 as a solid line
in each case. The powers in the Eq.(3) were obtained as $\alpha=3$ for the
columns with B/t ratio of 43 (t=2.3mm) and $\alpha=2$ for the columns with
B/t ratio of 31 (t=3.2mm), by trial and error method. The column
curves obtained by the simple method described above agreed well with
the experimental results.

6 Conclusions

In this study, stub column compression tests and buckling tests of
the concrete-filled square steel tubes were performed. Maximum
strengths were compared with the tangent modulus buckling loads,
assuming the tangent bending stiffness is an algebraic function of
the tangent axial stiffness. Major findings obtained from this study
can be summarized as follows:
(1)Axial force vs. tangent axial stiffness relationships of the stub
   columns can be expressed as a quadratic form like Eq.(4).
(2)Two types of buckling behavior were observed:
   Type A: For short columns with large width-to-thickness ratio,
           local buckling appeared first, then overall buckling
           occurred.
   Type B: For relatively short columns with small width-to-thickness
           ratio, overall buckling occurred first, then local
           buckling occurred.
(3)Buckling strengths can be estimated using the tangent modulus
   theory, assuming that the Eq.(3) exists between the tangent
   bending stiffness and the tangent axial stiffness.
(4)The slope of the descending branch of the axial force vs. axial
   contraction relationship, after the maximum strength, increases as
   the aspect ratio of the columns increases.

7 References

Shakir-Khalil, H. (1991) Tests on concrete-filled hollow section
   columns. Proc. ICCS-3, pp. 89-94.

20 LOCAL BUCKLING OF RECTANGULAR HOLLOW SECTIONS

H.D. WRIGHT
Department of Structural Engineering,
University of Strathclyde, Glasgow, UK

Abstract
The capacity of tubular members to carry axial compression and
bending forces may be limited by the buckling capacity of the
constituent plates. These plates normally buckle in a free wave
form about their own plane. By filling the tube with a rigid
material it is possible to restrain the plates to buckle only out of
their plane with a much reduced wavelength. This increases the
buckling resistance of the plates and, therefore, the section
capacity. This paper describes a theoretical derivation of plate
buckling strengths for both the pre- and post yield condition. The
method is then used to show that the capacity of certain commonly
rolled tubular sections may be improved by being filled with a rigid
material such as concrete. Suggestions are also made as to possible
economic limits regarding the wall thicknesses of hot rolled and
cold formed tubes.
Keywords: Buckling of plates, Section classification, Tube
dimensions, Filled tubes.

1 Introduction

The capacity of tube elements to carry axial compression and bending
forces is often limited by the stress that is likely to cause
buckling in the compressed sections. This is usually most critical
in the flange plates and may occur at any stress up to the ultimate
strength of the steel depending upon the slenderness of the plate.
It is usually assumed that once buckling occurs the flange, and
therefore the element, fails.

In order to ensure that buckling does not occur Codes of Practice
such as BS 5950 Part 1 (1990) limit the slenderness of the plate.
This is usually effected by quoting limiting breadth to thickness
(b/t) ratios. b/t ratios are normally given for the plate assuming
buckling occurs at yield stress, at a certain plastic strain
following yield and at the onset of strain hardening. This relates
to the possibility of the member being class 3 (semi-compact), class
2 (compact) or class 1 (plastic) respectively. Values quoted in
current codes; BS 5950 Part 1 (1990), EC3 (1992) and AISI (1986)

Tubular Structures V. Edited by M.G. Coutie and G. Davies.
© 1993 E & FN Spon, 2–6 Boundary Row, London SE1 8HN. ISBN 0 419 18770 7.

vary considerably and it must be assumed that the derivation of these values has, to some extent, been empirical.

The likelihood of flange buckling is dependent not only upon the b/t ratio but also on the boundary conditions prevailing. It is normally assumed that the buckle, once generated will take the form of intersecting sine waves as shown in Fig. 1a. Constraining the flange to buckle in a shorter waveform increases the limiting stress at which buckling first occurs and increases the allowable b/t ratio. A simple way of doing this is to fill the tube with concrete or similar rigid substance. The flange can then only buckle in the form of intersecting cosine waves as shown in Fig. 1b.

Fig.1a. Unrestrained plate buckle

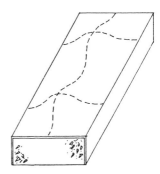

Fig.1b. Restrained plate buckle

Whilst analytical models of the buckling capacity of plates deforming into free sine waves are given in many elementary texts such as Timoshenko and Gere (1982) solutions for plates deforming into cosine and mixed sine cosine forms are not, generally, available. It is also difficult to find non-numerical methods used to determine buckling capacities for plates stressed in the post yield zone. Haaijer (1957) and Haaijer and Thurlimann (1958) presented a solution based on orthotropic plate theory that would appear to be the basis of the post-yield b/t ratios quoted in many Codes of Practice.

In this paper a method is presented for derivation of b/t ratios that limit buckling in the pre- and post yield stages of plate loading. This will be used to determine the capacity of tubes filled with a rigid material such as concrete. It will be shown that the ductility of several commonly used hot rolled sections can be improved by this means and that the rolling or cold-forming of more slender tubes could be economical.

2 Derivation of b/t ratios

The method, used here, to determine the stress at which plate buckling first occurs is the energy method. The energy associated with deforming the plate into an assumed wave form is equated to the energy associated with the axial load and shortening of the plate in its own plane. Two sets of boundary conditions are assumed: The plate is free to deform on either side of its own plane as shown in Fig. 1a. or the plate is restrained to buckle only out of its plane as shown in Fig 1b. The method allows the b/t ratios for a particular yield stress σ_y to be obtained directly.

For the free plate:

$$b/t = \left\{ \frac{\pi^2}{12 \sigma_y} \left(\frac{D_x}{\alpha^2} + 2D_{xy} + D_y + 4G \right) \right\} \tag{1}$$

where: $\alpha = \left(\frac{D_x}{D_y} \right)^{0.25}$

For the restrained plate:

$$b/t = \left\{ \frac{\pi^2}{48 \sigma_y} \left(\frac{16D_x}{\alpha^2} + 8D_{xy} + 3D_y \alpha^2 + 16G \right) \right\} \tag{2}$$

where: $\alpha = \left(\frac{16D_x}{3D_y} \right)^{0.25}$

Solutions for plates with other boundary conditions are given in another paper by the author Wright (1993).

The b/t ratio can be seen to be expressed in terms of the plate stiffnesses D_x, D_y, D_{xy} and G. For the case when buckling occurs prior to the steel yielding these are the elastic stiffnesses. Residual stresses and imperfections present in the plates following manufacture may cause premature pre-yield buckling. It is assumed here that these stresses and deformations may be compensated for by assuming that the stress present in the plate is 1.5 times that actually applied. The theoretical elastic Euler buckling curve is compared to that likely to occur when residual stresses and imperfections are present in Fig. 2. Also shown is the effect of assuming that the elastic stress is 1.5 times the yield stress.

The residual stresses, and to some extent the imperfections, are higher in welded sections than those found in rolled sections. To counter this the b/t ratios for welded plates buckling at the yield stress may be evaluated using a 1.75 rather than a 1.5 factor as previously suggested.

In both rolled and welded plates the residual stresses cause negligible plastic strain once yield is achieved and are therefore not considered for class 1 and 2 plates.

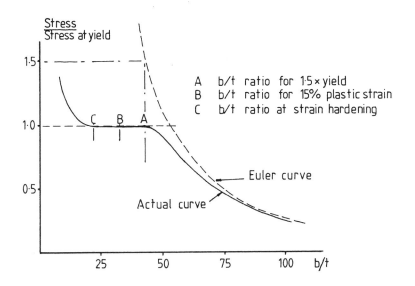

Fig.2. Buckling curve for simply supported plate

For plates buckling in the strain hardening region the stiffnesses are derived using the incremental theory of plasticity with the second invariant of the deviatoric stress tensor as the loading function. This was first presented by Haaijer (1957). Fig. 2 shows the approximate buckling curve as the stress changes from yield to ultimate.

Between these two limits, parts of the plate are yielding and parts are still elastic. The elastic portions are considerably more stiff than the yielded sections. Two assumptions may be made here: firstly the strain in the plate may be approximated by the proportion of plate that has yielded and secondly the buckling capacity of the plate is equal to a plate of equal dimensions to the proportion of the yielded plate. Consequently the buckling capacity of a plate strained, say 15%, beyond its elastic strain is equal to the buckling capacity of a plate where the wavelength of the buckle is equal to 15% of the length of the elastic buckle wavelength.

In practice this causes some discontinuity problems because a plate with an aspect ratio just less than the elastic buckling limit would, theoretically, have almost an infinite buckling resistance as the wavelength would be very small. To overcome this the wavelength of the plate is back calculated from that associated with the yield condition that has been modified for residual stresses and imperfections, point A in Fig. 2, and then increased by 15%.

For plates with b/t ratios greater than the elastic critical limit buckling may occur at stresses less than yield. Either effective width or effective thickness methods may be used to determine their resistance.

Consequently b/t ratios for plates buckling just as yield stresses are reached (class 3), plates buckling when the strain reaches an increased proportion of the elastic yield strain (class 2) and plates buckling when the strain hardening region is reached (class 1) can all be evaluated. These points are shown diagrammatically in Fig. 2. as A, B, and C.

3 Application to tubular sections

BS 5950 Part 1 (1990) provides a comprehensive table of b/t ratios for plates buckling as a free wave. For internal flanges, typical of tubes, class 1 sections must have a b/t ratio less than 26, class 2 less than 32 and class 3 less than 39. The method proposed in the previous section provides less conservative values of 32, 39, and 42 respectively which accord more closely with the EC3 (1992) values of 30, 35, and 39 and the AISI (1986) values of 33 and 44 (no class 2 values are given in the American code).

Table 1 shows those sections, currently rolled by British Steel, that are not in class 1 according to the Author's method. Also shown

Table 1. Classification of British Steel Hot Rolled tubular sections

Section size	Class	Modified class	Comment
250 x 250 x 6.3	2	2	borderline
400 x 400 x 10	2	2	borderline
450 x 250 x 10	3	2	minor axis
450 x 250 x 12.5	2	1	minor axis
400 x 200 x 10	2	1	minor axis
300 x 200 x 6.3	3	2	minor axis
300 x 200 x 8	2	1	minor axis
250 x 150 x 6.3	2	1	minor axis
250 x 100 x 5	2	1	minor axis
120 x 60 x 3.6	2	1	minor axis
100 x 60 x 3	2	1	minor axis
100 x 50 x 3	2	1	minor axis

is the effect of filling the tube with a rigid material such as concrete. It can be seen that all currently rolled sections will carry yield stresses in each axis before buckling. In two cases constituent plates are in class 3 and would therefore be incapable of showing any ductility following the achievement of yield. Filling these sections with a rigid material such as concrete raises their class one level and would allow some redistribution of moments in frames or continuous beams formed with these elements. The ductility of other sections can be improved from class 2 to class 1 by infilling the section. This would allow full plastic analysis to be used for frames that utilise filled elements such as these.

It may also be noted that the vast majority of rolled hollow sections are class 1, many with flange plates having b/t ratios considerably less than 26, the BS 5950 Part 1 limit for class 1 slenderness. This would seem to indicate that there is potential for more slender sections. Reducing the thickness of sections to the limiting b/t ratios quoted in this paper could prove economical. Very slender sections could then be used for situations when yield stress is the limiting factor and the same section filled with concrete could then provide a more ductile response for situations where plastic behaviour is expected. It may be that these slender sections are required only to resist stresses lower than yield during construction but are expected to resist yield stresses with some ductility at the ultimate limit state The construction could

Table 2. Potential square hollow section sizes

Section size	Wall thickness Class 1	Class 2	Class 3
20 x 20	0.6	0.5	0.5
25 x 25	0.76	0.63	0.6
30 x 30	0.91	0.75	0.7
40 x 40	1.21	1	0.93
50 x 50	1.51	1.25	1.2
60 x 60	1.81	1.5	1.4
70 x 70	2.12	1.75	1.63
80 x 80	2.42	2	1.86
90 x 90	2.73	2.25	2.1
100 x 100	3	2.5	2.3
120 x 120	3.6	3	2.8
150 x 150	4.5	3.75	3.5
180 x 180	5.5	4.5	4.2
200 x 200	6.1	5.0	4.8
250 x 250	7.6	6.25	5.8
300 x 300	9.1	7.5	7.0
350 x 350	10.6	8.75	8.1
400 x 400	12.12	10	9.3

be carried out with slender hollow sections that are later filled with concrete to satisfy the ultimate limit state.

Table 2 shows potential wall thicknesses for square hollow sections of between 20 and 400mm size. It is clear that many of the wall thicknesses suggested in this table cannot be hot rolled and would be better fabricated from cold forming strip steel. If cold forming is employed stiffening ribs could be rolled into the walls and much greater section sizes could be used allowing the rigid material to also carry much of the direct stress in the section.

4 Experimental validation

The analytical method proposed here has not, so far, been validated experimentally. Ideally the experiments would model very closely the boundary conditions used in the analysis but there are considerable difficulties in modelling these. The analysis assumes hinged longitudinal edges but tests on concrete filled tubes would be likely to give a boundary condition somewhere between the hinged and fixed state. This would possibly mean that the tests on concrete filled tubes would give less onerous b/t ratios than those presented in this paper.

5 Conclusions

This paper has described an analytical method that allows the buckling capacity of plates to be determined for boundary conditions that include intimate contact of the plate with a rigid medium. This method includes the buckling behaviour of plates that are strained beyond yield and allows b/t ratios to be determined for class 1, 2 and 3 plates.

The application of the method to tubes shows that currently available hot-rolled tubes generally comprise plates of quite stocky proportions that are normally in class 1 or class 2. Filling tubes with a rigid medium such as concrete will allow much more slender plates to be used. Recommendations as to limiting plate slendernesses show that filled cold formed tubes may well be structurally efficient in design situations where plastic section behaviour is required.

6 References

American Institute of Steel Construction. (1986) Load and Resistance Factor Design Specification for Structural Steel Buildings. **AISI** Chicago USA.
British Standards Institution. (1990) Structural use of Steelwork in Building Part 1 Code of Practice for Design in Simple and Continuous Construction: Hot Rolled sections. **B.S. 5950: Part 1:** London.

European Commission for Standardisation. (1992) Design of Steel Structures Part 1.1 General Rules and Rules for Building. **Eurocode 3 ENV 1993-1-1.** Brussels.

Haaijer, G. (1957) Plate Buckling in the Strain Hardening Range. **Procedings of the American Society of Civil Engineers, Engineering Mechanics Division** Vol 83 No EM2 pages 12121-121247.

Haaijer, G. and Thurlimann, B. (1958) On Inelastic Buckling of Steel. **Procedings of the American Society of Civil Engineers, Engineering Mechanics Division** Vol 84 No EM2 pages 15811-158148.

Timoshenko, S.P. and Gere J.M. (1982) **Theory of Elastic Stability.** McGraw-Hill International.

Wright, H.D. (1993) Buckling of Plates in Contact with a Rigid Medium. **The Structural Engineer** (in press).

21 CONCRETE-FILLED RECTANGULAR HOLLOW SECTION K CONNECTIONS

J.A. PACKER
University of Toronto, Canada
W.W. KENEDI
Ministry of Transportation of Ontario, Toronto, Canada

Abstract
Previous research has established the behaviour of concrete-filled Rectangular Hollow Section (RHS) T, Y and X connections with the branch loaded in either tension or compression. In this paper, tests are described on 45° gap K connections with the chord member alternately empty or filled with concrete. All (five) unfilled gap K connections showed ultimate strengths that compared well with the limit states design equations proposed by the International Institute of Welding (IIW) and the Comité International pour le Développement et l'Etude de la Construction Tubulaire (CIDECT). The results were somewhat conservative (safe), especially for the connections with a larger web member to chord width ratio. The concrete-filled K connections had a stiffness greater than, or equal to, that of their unfilled counterparts and thus it is felt that they could be designed on the basis of their ultimate strengths, as is done for unfilled K connections. For the limit states design resistance of the tension web member, the existing IIW/CIDECT effective width and punching shear design equations should be used. For the compression web member, the connection resistance can be determined from the bearing capacity of the confined concrete.
Keywords: Rectangular Hollow Sections, Connections, Joints, Concrete-Filling, Trusses.

1 Introduction

RHS steel members are now frequently chosen as the structural sections for welded trusses, but considerable expertise is needed to design unstiffened connections. There are many occasions on which just one - or a few - of the connections in a truss need to be reinforced. If the same (or perhaps just two) web members are used throughout a simply-supported truss, then the connections near the ends of the truss will be critical where the web member forces are highest. In this case these end connections can be reinforced rather than substituting a thicker chord member. Sometimes remedial reinforcement of connections is necessary because an RHS truss design has been performed based on member forces only, disregarding the connections, with the steel already having been ordered. Reinforcement typically involves plating the connecting face or the sides of the chord member but a novel alternative is for the fabricator to fill a length of the critical chord member (or members) with concrete. With short span trusses all the chord member could be filled with concrete, but with long span trusses - which can be assembled with bolted flange plates in the chords to facilitate transportation and erection - only a part of the chord need be filled with concrete as shown in Fig. 1. Concrete-filling is particularly appealing for architecturally-exposed steelwork and complete concrete-filling of a truss enhances its fire resistance too.

Many studies have been performed on concrete-filled hollow section steel members as columns, beam-columns, beams, and beam-to-column moment connections. Some tests have also been performed on grouted pile-to-tubular leg connections for the offshore industry. However, no attention has been given to the concept of full or partial concrete-filling of RHS trusses for the onshore

Tubular Structures V. Edited by M.G. Coutie and G. Davies.
© 1993 E & FN Spon, 2–6 Boundary Row, London SE1 8HN. ISBN 0 419 18770 7.

construction market. A particular study which is of relevance to RHS concrete-filled chord K (or N) connections, such as shown in Fig. 1, is that by Packer and Fear (1991) on RHS X-connections under transverse compression. It was shown that concrete-filling of RHS greatly enhances their performance under transverse compression since the RHS provides some confinement to the concrete which thereby allows it to reach bearing capacities greater than its crushing strength (as determined by cylinder compression tests).

For limit states design, the factored resistance of a concrete-filled RHS **compression-loaded X connection**, N_1^* could be taken as (Packer and Fear 1991):

$$N_1^* = \phi_c f_c' \frac{A_1}{\sin\theta_1} \left(\frac{A_2}{A_1}\right)^{0.5}$$ (1)

A_2 should be determined by dispersion of the bearing load at a slope of 2:1 longitudinally along the chord member, as shown in Fig. 2. The value of A_2 is limited by the length of concrete and $(A_2/A_1)^{0.5}$ cannot be taken greater than 3.3.

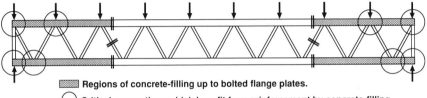

■ Regions of concrete-filling up to bolted flange plates.

◯ Critical connections which benefit from reinforcement by concrete-filling.

Fig. 1. Partial filling of RHS chord members in the critical connection regions.

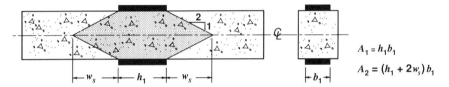

$$A_1 = h_1 b_1$$
$$A_2 = (h_1 + 2w_s) b_1$$

Fig. 2. Recommended method for determining bearing capacity of concrete-filled RHS in transverse compression (shown for 90° branch).

The following were also recommended by Packer and Fear (1991) for general design application of Eqn. (1):

$$h_0/b_0 \leq 1.4 \quad \text{(limit of experimental verification)}$$ (2a)

$$L_c \geq h_1/\sin\theta_1 + 2h_0$$ (2b)

It was subsequently recommended by Packer (1992) that, for limit states design, the factored resistance of a concrete-filled RHS **compression-loaded T or Y connection**, N_1^*, also be determined from Eqn. (1) but with a slight modification. A_2, the dispersed bearing area, should in this case be calculated assuming a stress distribution longitudinally at a slope of 2:1 through the entire depth of the chord, not just to the centreline as shown in Fig. 2. Thus, on Fig. 2 the dispersed bearing area, A_2, would be adjusted (for an inclined branch) to:

$$A_2 = (h_1/\sin\theta_1 + 4h_0) b_1$$ (3a)

Similarly, the limit of validity for L_c would need to be adjusted to:

$$L_c \geq (h_1/\sin\theta_1) + 4h_0 \tag{3b}$$

Some further comparative tests have been conducted by Kenedi (1991) on six RHS T connections with the chord member either empty or filled with concrete, but with the branch loaded in tension. He concluded that concrete-filled RHS **tension-loaded T, Y or X connections** should still be designed by the existing IIW (1989) or CIDECT (Packer et al. 1992) limit states design equations for **unfilled** T, Y and X connections.

2 Experiments on RHS Concrete-Filled Gap K Connections

Eleven RHS gap K connections were fabricated with each web member making a 45° angle to the chord. The critical connection parameters β, b_0/t_0 and gap size were varied for specimens with and without concrete in the chord. All of these test specimens were eventually tested with one web member in tension, the other in compression and the chord in tension. The limits of validity for the (unfilled) connection design formulae given by IIW (1989)/CIDECT (Packer et al. 1992) were obeyed to ensure the applicability of these design rules. With the connection eccentricity (e) falling within the IIW limits of validity, i.e. $-0.55 \leq e/h_0 \leq 0.25$, the secondary bending moments associated with noding eccentricities could be neglected with regard to their influence on connection behaviour and strength. Measured geometric and mechanical properties of the specimens are given in Table 1. Five K connections (K1 to K5) were fabricated for testing without concrete in the chord, and identical counterparts (K1C to K5C) were fabricated for testing with concrete in the chord. Specimen K1X was also concrete-filled, but in an alternate manner to the others.

Table 1. Measured Properties of RHS K Connections

Test No.	Web Members (h x b x t) mm x mm x mm	f_{y1}, f_{y2} (MPa)	A_1, A_2 (mm²)	f_c' (MPa)	Chord Member (h x b x t) mm x mm x mm	f_{y0} (MPa)	A_0 (mm²)
K1, K1C	51.1 x 50.9 x 6.32	443	1092	42.7	178.1 x 127.2 x 4.74	400	2703
K1X	As above			31.4	As above		
K2, K2C	51.1 x 50.9 x 6.32	443	1092	42.7	178.1 x 127.2 x 4.86	393	2745
K3, K3C	89.3 x 88.9 x 7.60	426	2393	42.7	178.1 x 127.2 x 4.74	400	2703
K4, K4C	89.3 x 88.9 x 7.60	426	2393	42.7	178.1 x 127.2 x 4.74	400	2703
K5, K5C	89.3 x 88.9 x 7.60	426	2393	42.7	177.9 x 127.1 x 7.54	391	4245

All RHS were cold-formed stress-relieved (Class H) with a nominal yield stress of 350 MPa. The measured yield stresses recorded in Table 1 were the average values from two longitudinal coupons taken from the "flats" of the RHS, on two adjacent sides but away from the weld bead. Cross-sectional areas were determined by weighing a prescribed length of RHS and using a density of 7850 kg/m³ for the steel (CSA 1992). The concrete crushing strengths (f_c') each represent an average of four 102 mm diameter x 203 mm long cylinder tests which were performed at approximately the same time as the concrete-filled connection tests. The concrete test cylinders were cured at room temperature inside a plastic bag, to represent conditions comparable to the confined concrete in the K connections. For the actual K connections, the concrete was dropped into the chord member while it was vertical (K1C to K5C) and then compacted by rodding. Sufficient concrete was added to give about 1000 mm of concrete centred about the connection. For K1X the concrete was placed with the chord in a horizontal position, by pushing a sheet metal "boat" full of concrete into the chord member and then squeezing the ends of the "boat" together by means of a threaded rod passing through the assembly.

This resulted in the connection region being more loosely filled with concrete - relative to vertical filling of the chord member - but it was thought to be potentially closer to the degree of concrete compaction which might be achieved in practice. The welds on all tests specimens were proportioned such that they would be non-critical (Packer and Henderson 1992).

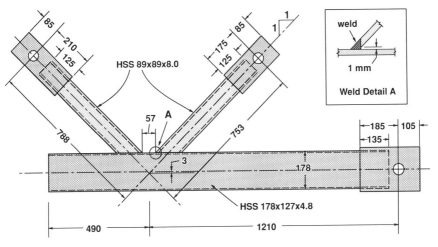

Fig. 3. Dimensions of Connection K3

Fig. 4. Typical K connection (white) with loads applied by a
universal testing machine and testing jig.

The dimensions of a typical connection are shown in Fig. 3. By means of holed plates welded to the ends of the connection members, the specimen was pinned to a specially designed testing jig. This triangular self-reacting jig was then used to convert an applied tension from a universal testing machine into a tension force in the chord member and compression force in one of the web members. A specimen under test is shown in Fig. 4. The large weight of the test jig caused bending of the vertical (tension) web member so large counterweights were added to eliminate this bending as much as possible (see Fig. 4). Each connection was loaded to failure in this manner, with displacement transducers recording the web member displacement into (or out of) the chord face and strain gauges measuring the axial and bending strains in each member. A pair of connections, in which one chord was empty and the other concrete-filled, are shown after failure in Fig. 5.

Fig. 5. A comparison of connection failure modes with the chord
empty (K3) and with the chord filled with concrete (K3C).

3 Discussion of K Connection Test Results

The yield and ultimate strengths of the K connections are given in Table 2. All of the unfilled connections failed at the compression web member while the concrete-filled connections failed at the tension web member. This marked difference in connection behaviour in the empty versus filled state is evident in Fig. 5. The connection ultimate load is reported for both the compression and tension web members since the tension web member force is always slightly higher due to the self-weight of the testing jig, specimen and counterweight. The connection yield load is determined from the load versus web member axial displacement graphs. The double tangent method is used whereby a tangent is drawn to the curve in the elastic range and another tangent is drawn to the curve after some yielding has occurred with the intersection of these lines giving the yield load. The smaller of the tension and compression web member yield loads is reported.

The five pairs of gap K connections were designed to investigate a wide range of width ratio, gap ratio and chord slenderness parameters. When the predicted ultimate strength is calculated for the **unfilled** K connections (IIW (1989), Packer and Henderson (1992) or Packer et al. (1992)), the governing failure mode in all cases is predicted to be that of chord face plastification, which agrees perfectly with the test results. Thus, the predicted connection limit states resistance (N_1^*) is compared with the actual connection ultimate strength (N_{1U}) in Table 2. (N_{1U} is used rather than N_{2U} because $N_{1U} < N_{2U}$ in all tests and $N_1^* = N_2^*$). The connection resistance is given by:

$$N_{1,2}^* = 8.9 f_{y0} t_0^2 \beta \gamma^{0.5} f(n)/\sin\theta_{1,2} \qquad (4)$$

The mean ratio of the connection ultimate strength to the design strength in Table 2, for the **unfilled** K connections, is 1.60 with a very good coefficient of variation (COV) of only 10.4%. There are two main reasons for this ratio appearing to be overly conservative. Firstly, Eqn. (4) has a built in resistance factor of approximately 0.8 to account for the variations in the test results from which the formula was derived, and for the material property variations. Second is the fact that this formula is a simplified one which assumes the use of nominal but adequate fillet welds. The five K connections fabricated here were designed so that the welds had a throat size larger than the thickness of the web member wall, so that they would be non-critical, but this does increase the effective β value and hence amplify the connection strength.

Table 2. Yield Strengths, Ultimate Strengths and Failure Modes of Connections

Test No.	β	b_0/t_0	g'	N_Y (kN)	N_{1U} (kN)	N_{2U} (kN)	Observed Failure Mode	$\dfrac{N_{1U}}{N_1{}^*}$	$\dfrac{N_{2U}}{N_2{}^*}$
K1	0.40	26.8	0.90	106	260	264	CP	1.569	
K1C	0.40	26.8	0.90	206	361	368	PS		1.110
K1X	0.40	26.8	0.90	138	370	371	PS		
K2	0.40	26.8	0.30	107	224	247	CP	1.326	
K2C	0.40	26.8	0.30	148	332	362	PS		1.082
K3	0.70	26.8	0.45	409	459	466	CP	1.586	
K3C	0.70	26.8	0.45	619	814	822	PS		1.419
K4	0.70	26.8	0.15	422	529	551	CP	1.828	
K4C	0.70	26.8	0.15	662	815	849	PS		1.465
K5	0.70	16.9	0.45	790	969	971	CP	1.706	
K5C	0.70	16.9	0.45	919	1161	1166+	CW		1.263+

Note: 1. Failure modes are described by Packer and Henderson (1992) or Packer et al. (1992). CP = Chord face plastification. PS = Punching shear, or tearing of the chord face at the base of the tension web member (For examples, see Fig. 5). CW = Compression web failure.

An inspection of the four basic failure modes for RHS gap K connections, (i.e. chord face plastification due to the "push-pull" action of the web members, shearing of the whole chord section through the gap, "effective width" or premature failure of one of the web members, and punching shear of the chord face around one of the web members, as described in Packer et al. (1992) or Packer and Henderson (1992)), will reveal that the first two will not occur with **concrete-filled** chords, thereby highlighting the latter two as being pertinent to concrete-filled K connections. Moreover, with the compression web pressing on an almost rigid foundation of concrete the "effective width" and "punching shear" failure criteria can only be applied to the tension web member. For the compression web, the connection strength would appear to be limited by bearing failure of the concrete in a manner similar to concrete-filled, compression-loaded T or Y connections. Thus, the limit states resistance of a **concrete-filled** K (or N) connection can be determined from the following criteria:

$$N_1{}^* = \phi_c f_c' A_1 \left(\frac{A_2}{A_1}\right)^{0.5} / \sin\theta_1 \tag{1}$$

with A_2 calculated from Eqn. (3a) and subject to the condition of Eqn. (3b),

$$N_2{}^* = f_{y2} t_2 (2h_2 - 4t_2 + b_2 + b_e) \tag{4}$$

$$\text{where } b_e = \frac{10}{b_0/t_0} \cdot \frac{f_{y0} t_0}{f_{y2} t_2} \cdot b_2 \quad \text{but} \le b_2, \text{ and} \tag{5}$$

$$N_2{}^* = \frac{f_{y0} t_0}{\sqrt{3}} (2h_2 / \sin\theta_2 + b_2 + b_{ep})/\sin\theta_2 \tag{6}$$

225

where $b_{ep} = \dfrac{10}{b_0 / t_0} \cdot b_2$ but $\leq b_2$ $\qquad\qquad$ (7)

The effective width terms b_e and b_{ep} should still apply to concrete-filled K connections as the concrete, in the chord only, would likely have little influence on the uneven load distribution around the **tension** web member at the joint to the chord.

Equations (1), (4) and (6) were used to determine the governing failure mode and predicted ultimate strength of each concrete-filled connection. Eqn. (1) did not govern for any connections which agrees with the observed test failures around the base of the tension web member. Punching shear (Eqn. (6)) governed for specimens K1C, K2C, K3C and K4C - which was also observed in the experiments (see Table 2) - and "effective width" governed for specimen K5C. Connection failure of K5C was not actually achieved by testing due to buckling of the compression web member away from the connection. The connection ultimate strength to design strength ratios are shown in Table 2, the mean of which is 1.27 with a COV of 12.3%. (K1X has not been included as it is effectively a duplicate of K1C). These results demonstrate that the proposed design method for concrete-filled gap K connections is representative of their behaviour and conservative for implementation.

4 Conclusions

For the range of connection parameters studied, gap K connections with concrete-filled chords were found to have superior connection yield strengths and ultimate strengths relative to their unfilled counterparts. Also, concrete-filling of such connections has been found to produce a significant change in connection failure mode. The limit states design resistance of gap K connections with concrete-filled chords can be determined from Eqns. (1), (4), (5), (6) and (7). On the basis of only two comparative tests (K1C and K1X), it was shown that the design of a gap K connection with concrete in the chord is unaffected by a moderate amount of shrinkage of the concrete away from the chord walls, thereby making this filling procedure an option for any fabricator.

5 Symbols

A_1 $\qquad$ Bearing area over which a transverse load is applied $= (h_1/\sin\theta_1)\, b_1$

A_2 $\qquad$ Dispersed bearing area

$\qquad\qquad = (h_1/\sin\theta_1 + 2w_s)\, b_1$ for X connections (see Fig. 2)

$\qquad\qquad = (h_1/\sin\theta_1 + 4h_0)\, b_1$ for T and Y connections

$\qquad\qquad$ but $(A_2/A_1)^{0.5} \leq 3.3$

b_e $\qquad$ Effective width of a web member (Eqn. (5))

b_{ep} $\qquad$ Effective punching shear width (Eqn. (7))

b_i $\qquad$ External width of RHS member i (90° to plane of truss or frame)

e $\qquad$ Noding eccentricity at a truss panel point (positive towards the outside of the truss)

f_c' $\qquad$ Crushing strength of concrete at 28 days, by cylinder tests

f_{yi} $\qquad$ Yield stress of member i

$f(n)$ $\qquad$ Function which incorporates the influence of axial stresses in compression chords

g' $\qquad$ Gap between web members on the chord face, divided by chord width

h_i $\qquad$ External depth of RHS member i (in the plane of truss or frame)

i $\qquad$ Subscript to denote member of connection; $i = 0$ designates chord; $i = 1$ refers to the branch for T, Y and X connections, or it refers to the compression web member for K and N connections; $i = 2$ refers to the tension web member for K and N connections.

L_c $\qquad$ Length of concrete in RHS chord member

N_i^* $\qquad$ Connection resistance, expressed as an axial force in member i

N_{iU} $\qquad$ Ultimate force in member i

N_Y	Yield strength of the connection, expressed as a web member force
t_i	Thickness of RHS member i
w_s	Dispersion length in concrete-filled X connections (see Fig. 2)
β	Width ratio between branch or web members and chord (excluding weld size)
	$= b_1/b_0$ for T, Y and X connections
	$= (b_1 + b_2 + h_1 + h_2)/4b_0$ for K and N connections
ϕ_c	Resistance factor for concrete in bearing (Inverse of partial safety factor, γ_M)
	$= 0.6$ in Canada
γ	Half width to thickness ratio of chord $= b_0/2t_0$
θ_i	Included angle between web member i and chord

6 References

Canadian Standards Association (1992). **General requirements for rolled or welded structural quality steel, CAN/CSA-G40.20-M92.** CSA, Toronto, Canada.

International Institute of Welding Subcommission XV-E (1989). **Design recommendations for hollow section joints - predominantly statically loaded, 2nd. ed.** IIW Doc. XV-701-89, IIW Annual Assembly, Helsinki, Finland.

Kenedi, W.W. (1991). **Concrete-filled HSS joints.** Master of Applied Science thesis, University of Toronto, Canada.

Packer, J.A., Wardenier, J., Kurobane, Y., Dutta, D., and Yeomans, N. (1992). **Design guide for rectangular hollow section (RHS) joints under predominantly static loading.** CIDECT (ed.) and Verlag TÜV Rheinland GmbH, Köln, Federal Republic of Germany.

Packer, J.A. (1992). Design criteria for concrete-filled HSS joints. **CIDECT Report No. 5AV-26/92,** University of Toronto, Toronto, Canada.

Packer, J.A., and Henderson, J.E. (1992). **Design guide for hollow structural section connections.** Canadian Institute of Steel Construction, Toronto, Canada.

Packer, J.A., and Fear, C.E. (1991). Concrete-filled rectangular hollow section X and T connections, in **Proceedings of the 4th. International Symposium on Tubular Structures** (eds. J. Wardenier and E. Panjeh Shahi), Delft University Press, Delft, The Netherlands, pp. 382-391.

7 Acknowledgements

Financial support for this project was provided by the Comité International pour le Développement et l'Etude de la Construction Tubulaire (CIDECT Program 5AV), the University Research Incentive Fund of the Government of Ontario (URIF Award TO 14-013), Ipsco Inc. and the Natural Sciences and Engineering Research Council of Canada (NSERC). The Hollow Structural Sections used in this project were provided by Stelco Inc.

PART 7

BEAM-COLUMN FRAME CONNECTIONS

Method of joint formation, behaviour

22 FIN PLATE CONNECTIONS BETWEEN RHS COLUMNS AND I-BEAMS

N.D. JARRETT
Building Research Establishment, Watford, UK
A.S. MALIK
The Steel Construction Institute, Ascot, UK

Abstract
This paper describes a series of tests and design recommendations for fin plate connections between rectangular hollow section (RHS) columns and I-beams. The tests were carried out at the Building Research Establishment in collaboration with the Steel Construction Institute on behalf of British Steel, tubes and pipes. A tensile force was applied to the beam. The maximum connection tensile forces and failure modes were determined with different axial loads applied to the column. The tests were carried out to ensure that standard connections satisfy the structural integrity requirements given in BS5950 Part 1. The influence of the column section size and column axial load on the maximum tying force and connection failure mode was investigated.
Keywords: Tubular Structures, Fin Plate, Rectangular Hollow Section, Column-Beam Joints, I-Beam, Tensile Force, Structural Integrity, British Standards.

1 Introduction

SCI and BRE have recently completed a study on connection design to maintain building stability requirements given in BSI (1990) in the event of an explosion or vehicle impact by Jarrett et al (1993). A series of partial collapses in the sixties, including Ronan Point, resulted in regulations for structures in the UK to have a minimum robustness to resist accidental loading. In steel framed buildings one method of developing robustness is to tie all the beams and columns together to reduce the sensitivity of buildings to disproportionate collapse. Beam to column connections must be capable of transferring a horizontal tying force. This paper describes a series of 15 tests carried out to determine the maximum horizontal tying force and failure modes of fin plate connections between I-beams and RHS columns and follows an extensive research programme to develop design guidelines for fin plate connections by Jarrett et al (1990) and published by SCI (1993).

Tubular Structures V. Edited by M.G. Coutie and G. Davies.
© 1993 E & FN Spon, 2–6 Boundary Row, London SE1 8HN. ISBN 0 419 18770 7.

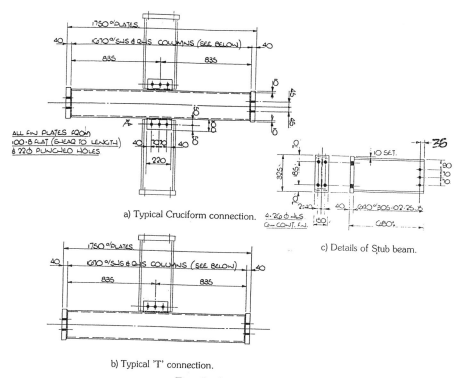

a) Typical Cruciform connection.

c) Details of Stub beam.

b) Typical 'T' connection.

Fig. 1 Specimen details

2 Test apparatus and procedure

2.1 Specimen description

Test specimen details are shown in Figure 1. an inverted tee arrangement simulated the behaviour of a connection to an external column and a cruciform arrangement was used to investigate connections to an internal column. The system used for testing the tee joint is shown in Figure 2, a similar modified system was used for the cruciform joint. Three different sizes of RHS columns were used (250x250x10, 250x150x5 and 250x150x10). Only one size I-Beam and Fin Plate was used (305x102x25 UB and a 220x100x8 plate with three M20 bolts). All steel was specified as grade 43C. Bolts were M20 grade 8.8 in 22mm diameter clearance holes. The lengths of the beam and column in each test specimen were selected to ensure that a realistic stress pattern developed at the connection and to prevent the test rig from restraining distortion of the column around the connection. The complete test series is summarised in Table 1. Holes in the fin plates were punched in line with current UK practice. All the specimens were assembled before placing in the test rig. No attempt was made to centralise the bolts in the bolt holes. Bolts were preloaded to a nominal torque of 150 Nm (approximately, 35 kN tension).

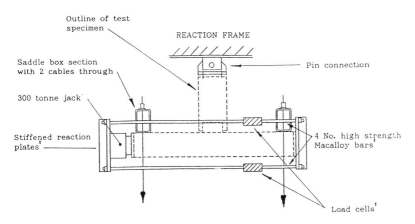

Fig. 2 Test apparatus - Tee specimen

Table 1. Summary of test series and results

Test No.	Spec. Type	RHS Size	Connection Axis	RHS Axial Load (kN)	Max. Tie Force (kN)	Failure Mode	Deflection at Failure (mm)	Strengther.ed Stub Beams
1	C	250x250x10	-	NONE	307	a	47	NO
2	C	250x150x5	MINOR	NONE	146	b	45	NO
3	C	250x150x5	MAJOR	NONE	147	b	33	NO
4	T	250x150x10	MINOR	NONE	349	a	50	NO
5	T	250x150x5	MINOR	NONE	219	b	46	NO
6	T	250x150x5	MAJOR	NONE	236	b	44	NO
7	T	250x250x10	-	NONE	443	c	60	YES
8	T	250x150x10	MAJOR	NONE	425	c,d	39	YES
9	C	250x150x5	MINOR	790	165	b,e	78	NO
10	C	250x150x5	MAJOR	650	173	b,e	45	NO
11	C	250x250x10	-	1630	295	d,e	50	YES
12	T	250x150x5	MAJOR	555	166	b,e	37	NO
13	T	250x150x5	MINOR	560	161	e	42	NO
14	T	250x250x10	-	1565	336	a	46	NO
15	T	250x150x10	MINOR	1360	347	c,d,e	43	YES

NOTE: 305x102x25UB beam size throughout
FAILURE MODES:
a - beam web pull out
b - fracture of the RHS around the weld
c - bolt shear
d - fin plate fracture
e - buckling of RHS

SPECIMEN MODES:
T - Tee section
C - Cruciform section

2.2 Test set-up, loading and instrumentation

The test rig was fabricated in BRE's workshop. Figure 2 illustrates the main features of the test set-up. In both cases the RHS column was placed horizontally in the test rig. The same arrangement was used for imposing column axial load for both specimen types. Loading plates were connected by Macalloy bars and load was applied to both plates by a 3000kN hydraulic ram placed between the specimen and one of the plates.

Load was applied to the bottom beam through an arrangement of spreader beams and Macalloy bars by a 1000 kN hydraulic ram reacting against the underside of the laboratory floor. The system was load controlled.

Instrumentation was selected to measure the applied load and the load displacement characteristic of each connection. The tying force and column axial load were measured by load cells attached to the Macalloy bars. Each load cell was calibrated before use in a 2500 kN Avery testing machine.

Two LVDT's were fixed to the specimen to measure vertical displacement either side of the connection.

2.3 Test procedure

The series of tests without column axial load was completed before any of the tests with axial load, the tying force applied to specimens was increased steadily until failure of the connection. The loading procedure for specimens with axial load varied throughout the test series. The initial order was (1) A tying force equal to half the maximum tying force measured in the tests without axial load was applied to the specimen, (2) the axial load was applied up to the maximum design value (equal to the nominal squash load of the column), (3) The tying force was increased until connection failure.

Early tests showed that columns with a high axial load could not support large tying forces and so successive tests were carried out with a column axial load equal to the service load. The reduced axial load is in line with current standards which state that accidental loads shall be combined with the service load. Service load for the columns was determined by dividing the column nominal squash load by 1.5. The test procedure was reduced to two steps consisting of applying the full axial load followed by the tying force.

3 Test results and analysis

Specimen numbers 1 to 8 were loaded to failure under tying force only. In test numbers 8 to 15, specimens were subjected to a tying force and axial column load. Figure 3 shows the tying force versus connection displacement characteristics for identical specimens tested with and without axial load. All the connections demonstrated a linear relationship initially followed by a gradual decrease in stiffness. Table 1 summarises the results and shows the following failure modes observed in the test series:

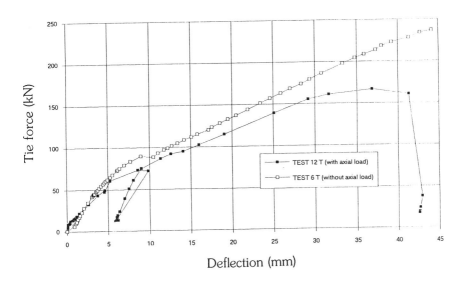

Fig. 3 Graph showing tying force versus connection deflection

.   Fracture of the RHS around the fin plate weld (figure 4).
.   Bolt pulling through the beam web.
.   Bolt shear.
.   Fracture of the fin plate at the bolt row.
.   Fracture of the fin plate starting at the weld combined with shear of one bolt.
.   Local buckling of the RHS.

All the specimens tested were able to carry tying forces more than double the minimum requirement of 75kN specified in BS5950. Fracture of the RHS wall around the fin plate weld was observed in all specimens with 5mm thick RHS sections and was the lowest failure load. Failure due to bolts pulling through the beam web was evident in Specimens 1 and 4 where the RHS wall thickness was 10mm. The same failure mode was prevented in tests 7 and 8 by strengthening the beam webs with 6mm plate. Test 7 failed due to bolt shear and test 8 failed due to a combination of shear of one bolt and plate fracture at the bolt row.

The combined influence of axial load and tying force resulted in local buckling of the RHS column sections in tests 9 to 15. Figure 4 shows an example of a local column buckle. The specimen is grossly deformed on all sides of the RHS. Table 1 includes the tying force and axial load corresponding to local buckling of the RHS. Local buckling occurred as the tying force increased after loading the column to full axial load. Tests 9, 10 and 11 were carried out with axial loads higher than service load. The axial load equalled the column squash load in test 9 and was reduced to 80% and 70% of the squash load for tests 10 and 11

Fig. 4 Failure due to fracture and local buckling of the RHS

respectively. The tying force required to buckle the column increased as the column axial load decreased.

Maximum connection deflections varied between 30mm and 80mm and were primarily due to inelastic deformation of the RHS wall. Yield of the RHS was followed by extensive rotation of the plastic hinges. The tying force resulted in significant in-plane tensile forces in the RHS wall.

A comparison of the results of tee and cruciform specimens with 5mm RHS's and the same connection detail (tests 2 and 5, 3 and 6, 11 and 15) shows that connections to tee sections carry higher tying forces than connections to cruciform sections. The reduced tying forces of the cruciform specimens may be attributed to two factors. The bending stresses resulting from loads applied to two opposite flanges of the cruciform specimens are superimposed and result in earlier failure than similar tee specimens loaded on one flange only. The support condition used to test the tee specimens resulted in global bending of the RHS and a non-uniform stress distribution along the connection weld. Fracture was first observed at the end of the plates.

Table 1 shows that the axial load appears to influence the maximum tying force of connections but only connections that failed by either fracture of the RHS or fracture of the fin plate. The influence appears to vary through the test series. The maximum tying force measured for specimens with axial load was less for some specimen types than the maximum tying force of similar specimens without axial load. Examples are tests 7 and 8 compared with test 15. Increased curvature of the RHS due to the axial load may have resulted in higher stress at the ends of the fin plate and fracture at a lower load than specimens without axial load. A similar phenomenon was observed in tests 6 and 12, see Figure 3. In contrast, the maximum tying force for connections to cruciform specimens with 5mm RHS wall thickness (tests 2 and 3 compared with tests 9 an 10) was higher for specimens with axial load than specimens without axial load.

In test 13, local buckling resulted in a drop in both the axial load and tying force. Increased pumping of the load jacks deformed the column but did not increase the load and it was not possible to fail the fin plate.

The maximum allowable tying force of a connection is that which can be carried without impairing the stability of a structure. For columns loaded to service load, axial load started to drop off (indicating local buckling) at between 80 and 100% of the maximum tying force. The tying force corresponding to local buckling for test 12 is shown in Figure 3. The maximum allowable tying force is the force corresponding to local buckling and is given in brackets in Table 1.

4 Development of Design Rules

Design rules for fin plate connections to I-columns subject to a tying force can be adapted for fin plate connections to RHS columns. The rules satisfy the structural integrity requirements of BS 5950 Part 1.

Table 2 gives the predicted ultimate loads of the test connections. Rules given in SCI (1993) were used for calculating (i) tension capacity of fin plate (ii) tension capacity of beam web (iii) bearing capacity of beam web and fin plate and (iv) capacity of column web. Simple yield line analysis results ignoring bending of the RHS at the ends of the fin plates are also included. Kapp (1974) gives guidance for the design of I-column webs based on an improved yield line analysis but still neglecting the membrane strength of the web. The web is considered as simply supported or fixed at boundaries with the flanges. Kapp's rules are used to calculate the strength of the RHS column flanges and the results are included in Table 2. The axial load applied to the column was not included in any of the calculations.

5 Comparison of Theory and Test Results

Table 3 compares the predicted failure loads and the test results. Column 2 compares the experimental failure tie force with the theoretical tie force assuming the same failure mode. Column 3 compares the experimental failure tie force and the lowest predicted failure tie force. In every case the simple yield line analysis ignoring end effects predicted the lowest tie force.

Table 2. Theoretical failure loads

Test No	Check 8* Tension Fin Plate Capacity kN	Check 9* Tension Capacity Beam Web kN	Beam Web Pull Out	Check 10** Bearing Capacity Beam Web kN	Fin Plate kN	Check 11* Tie Capacity of RHS Wall kN	Ultimate Bolt Shear Capacity kN	Yield line analysis of RHS Wall Simply Edges kN	Fixed Edges kN
1	548	420 (393)	393 (344)	191 (167)	330	165	371	210	330
7	548	854 R (799)	–	389 R (340)	330	165	371	210	330
11	548	854 R (799)	-	389 R (340)	330	165	371	210	330
14	548	420 (393)	393 (344)	191 (167)	330	165	371	210	330
4	548	420 (393)	393 (344)	191 (167)	330	156	371	198	311
8	548	854 R (799)	-	389 R (340)	330	262	371	257	427
15	548	854 R (799)	-	389 R (340)	330	156	371	198	311
2	548	420 (393)	393 (344)	191 (167)	330	38	371	49	77
3	548	420 (393)		191 (167)	330	60	371	62	101
5	548	420 (393)		191 (167)	330	38	371	49	77
6	548	420 (393)		191 (167)	330	60	371	62	101
9	548	420 (393)		191 (167)	330	38	371	49	77
10	548	420 (393)		191 (167)	330	60	371	62	101
12	548	420 (393)		191 (167)	330	60	371	62	101
13	548	420 (393)		191 (167)	330	38	371	49	77

Notes : R - Reinforced web
* Checks 8, 9, 11 based on SCI (1993) values calculated using the measured ultimate tensile strengths.
** Check 10 based on SCI (1993) values calculated using bearing strength of 825 N/mm². This is not a failure mode.
( ) Based on values for edge distance of 35mm. Others are based on edge distance of 40mm Kapp (1974) values calculated using the measured ultimate tensile strengths.

Table 3. Comparison of theoretical and experimental failure loads

Test No	Experimental Failure Loads & Modes kN		Similar failure Check Experimental	Lowest predicted Experimental	Simple Experimental	Fixed Experimental
					Yield line analysed	
1	307	a	(1.12)a	.54	.68	1.08
7	443	c	(.84)c	.37	.47	.75
11	295	d,e	(1.86)d	.56	.71	1.12
14	336	a	(1.02)a	.49	.63	.98
4	349	a	(.99)a	.45	.57	.89
8	425	c,d	(.87)c	.62	.61	1.01
15	347	c,d,e	(1.07)c	.45	.57	.90
2	146	b		.26	.34	.53
3	147	b		.41	.42	.69
5	219	b		.17	.22	.35
6	236	b		.25	.26	.43
9	165	b,e		.23	.30	.47
10	173	b,e		.35	.36	.58
12	166	b,e		.36	.37	.61
13	161	e		.24	.30	.48

Failure of the RHS sections with a wall thickness equal to 5mm (Tests 2,3,5,6,9,10,12 and 13) was generally due to fracture and buckling of the RHS wall. The predicted methods significantly underestimate the actual failure load. The large difference is due to neglect of the membrane action in the RHS although embrittlement from fast cooling after welding must also be considered. Design procedures are currently being developed to take these factors into consideration.

Failure modes of specimens with a 10mm RHS wall thickness (Tests 1,4,7,8,11,14,15) included beam web pull out, bolt shear and fin plate fracture.

For tests 1 and 14 where the failure mode was web pull out, the predicted values are higher than the tests. The test failure load may have been low because the measured edge distance to the bolt hole was up to 5mm lower than the specified edge distance and was further reduced by bearing deformation of the web and fin plate.

For test 11 the predicted failure load grossly overestimates the experimental failure load. The failure modes were a combination of fin plate fracture just above the weld and buckling of RHS. The tests showed that high local stresses are developed at the ends of the plate along the weld line causing premature failure.

6 Recommendations

Design methods are being developed to include the effect of membrane action on the load carrying capacity of the RHS wall. These will be published in due course.

Until the publication of improved design guidelines, the procedures given in the published literature by Kapp (1974) and SCI (1993) can be used and will result in a safe but conservative design. The yield line analysis given by Kapp (1974) assuming fixed edges is the least conservative method for predicting the RHS capacity and should be adopted until improved guidance is given.

7 Acknowledgements

Support from British Steel, tubes and pipes for the work described in the paper and fabrication of the test specimens by Allot Bros and Leigh Ltd is gratefully acknowledged. The authors also acknowledge the valuable contributions from R.Grantham, J. Rackham, J. Baptiste and A. Parker.

8 References

BSI (1990), Structural use of steelwork in building: Part 1: Code of practice for design in simple and continuous   construction: hot rolled sections, BS 5950, British Standards Institution, London.

Jarrett, N.D. and Grantham, R.I. (1993) Robustness tests on fin plate connections, BRE client report, Building Research Establishment, Watford.

Jarrett, N.D. (1990) Tests on beam/column web side plate connections, BRE client report CR 54/90, Building Research Establishment, Watford.

SCI (1993), Joints in simple construction volume 1: Design methods second edition, Steel Construction Institute.

Kapp, R.H. (1974) Yield line analysis of a web connection in direct tension, Engineering Journal, American Institute of Steel Construction (second quarter/1974).

23 FRAMED CONNECTIONS TO HSS COLUMNS

D.R. SHERMAN
University of Wisconsin–Milwaukee, Milwaukee,
Wisconsin, USA

Abstract
This paper discusses several methods for making simple field connections between HSS columns and either wide-flange or HSS beams. The connections are adaptations of standard field connections: pair of framing angles, single angle, tee stub, angle seat, shear tab and end plate. A testing program is currently in progress to determine if there are unique characteristics of the HSS column that influence the strength or the rotational properties of the connections as opposed to their use with wide-flange columns.

The paper will also present some unique methods of connecting to HSS columns. One of these is the use of flow-drilled holes, where a hole is essentially melted through the wall of the HSS producing an internal upset. After roll threading, a blind bolt connection is obtained. Test data shows that this type of bolting can produce satisfactory shear and tension capacities.
Keywords: Tubular Columns, Simple Connections, Bolting, Shear Tabs.

1 Introduction

Square and rectangular hollow structural sections (HSS) are being used with increasing frequency as columns in building construction. To a large extent this is due to their structural efficiency and their aesthetic appearance when exposed. The beams that frame into the HSS columns are usually the familiar wide-flange sections. A major consideration in the design of this type of framing system is the field connection between the beams and columns. The most economical beam to column connection is a simple framing connection that transmits the beam shear with a minimal moment. These simple connections are widely used in braced frame construction.

Common practice for simple field connections is to shop weld connecting elements to the beam or column and complete the connection in the field with bolting. A variety of standard simple framing connections have been developed over the years for use with wide-flange columns. These include:

Double angles
Tees
Angle beam seats
Shear tabs or wing plates
Single angles
End plates

Tubular Structures V. Edited by M.G. Coutie and G. Davies.

Except for the beam seat, the connecting element is attached to the web of the beam. These connections can be used with tubular beams if a structural tee end cap is welded to the beam so that the stem functions in a similar manner to the web of a wide-flange.

Most of these simple connections have been adopted for use with HSS columns, although in some cases concern has been expressed as to how they might perform with an HSS instead of a wide-flange column. In fact, very few tests of any of the connection types exist to verify suitable behavior and strength with HSS columns.

There are three major categories of concern in using these simple connections with HSS columns.

Will the HSS limit the capacity of the connection?
Will local distortion of the HSS impair the capacity of the column?
Will the stiffness of the HSS cause the connection to behave in a semi-rigid manner rather than as a simple shear connection?

In the following sections of the paper, these questions will be addressed for three categories of simple connections.

Connecting elements welded near the sides of the HSS.
Connecting elements welded at the center of the HSS wall.
Bolting to the HSS.

Much of the presentation is discussion, since a project involving the testing of various connections has only recently been initiated. In the case of the shear tab and the bolted connections, supporting data is presented.

2 Connections with side welds

The double angle, tee and seat connections are attached to the column with vertical welds at the two edges of the connecting elements. As shown in Figure 1, if the connecting elements are not as wide as the HSS, the welds are on the flat wall of the section near the corner. On the other hand, if the elements are as wide as the HSS, flare bevel groove welds would be required.

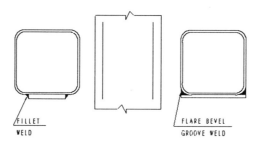

Fig. 1. Connections with two vertical side welds

Local distortion of the face of the HSS is not a strength consideration and since any moment in the connection is small, the side wall compression is not a critical factor. The primary additional strength consideration in using these connections with HSS columns is that the welds are in the vicinity of the corners that have been extremely cold-worked in many HSS manufacturing processes. In the U.S., the AWS Structural Welding Code (1992) contains a footnote cautioning about the use of cold-formed HSS in dynamic situations where low-temperature notch-toughness properties may be important. Although corner conditions have not been a problem in direct tube-to-tube truss connections, in simple connections all of the force transfer is through welds near or at the corners.

The other consideration in using these connections with HSS columns is the rotational flexibility. In practice, these connections are considered as simple hinges in the analysis of a structure. When used with wide-flange columns, the welds are toward the flexible edges of the flanges that are not directly supported by the web, while in the HSS the side walls provide stiffness to the tube face at the welds. Since most of the connection distortion is typically assumed to take place in the connecting elements, the flexibility should still exist with HSS columns. However, in the event that these connections are stiffer when used with HSS, it is important to determine if the moment-rotation characteristics require that the structure be analyzed using semi-rigid connections.

It is not anticipated that either the strength or the flexibility considerations will be critical factors with HSS columns. However, a testing program has been initiated to verify this.

3 Welds in the center of the HSS face

The two types of connections that have welds near the face of the HSS are the shear tab and the single angle. The corresponding vertical weld patterns are shown in Figure 2.

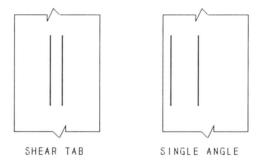

SHEAR TAB   SINGLE ANGLE

Fig. 2. Shear tab and single angle welds

A test program on shear tab connections was reported by Sherman (1991). These tests showed that distortion of the shear wall was not a critical limit state. This was due to the self limiting nature of the end rotation of the simply supported beam, which was less than the distortion required to develop a yield line mechanism. The potential critical limit state in the tube wall was punching shear. A simple expression to prevent punching shear failure can be obtained by keeping the tensile yield force on a unit length of the shear tab less than the shear fracture capacity through the HSS wall on the two potential fracture planes adjacent to the welds.

$$F_{y(tab)}t_{tab} \leq 1.2F_{u(RHS)}t_{RHS} \tag{1}$$

In this equation, the shear fracture strength of the HSS is assumed to be 0.6 times the tensile fracture strength.

The same paper by Sherman (1991) also discussed the possible reduction in the strength of the HSS column due to the local distortion of the wall at the connection. It was concluded that no significant column strength reduction occurred between shear tab connections or through-plate connections that reinforces the walls. However, this conclusion was based on only four tests using HSS with a width/thickness ratio of 16. Recently Haslam (1993) reported similar column tests with width/thickness ratios of 29 and 40. This study with eight tests included symmetric connection on both sides of the HSS and unsymmetric connections on just one side. The test setup is shown in Figure 3.

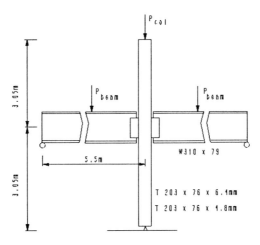

Fig. 3. Setup for shear tab column tests

In these tests, the beams were loaded to about 70% of the connection capacity and then a load was applied to the top of the column until a column buckling failure occurred in the lower portion.

Table 1 presents the column strengths as ratios of the maximum experimental load divided by the yield load given by area times the static yield strength from a tension coupon taken from the wall of the HSS. The nondimensional wall slenderness of the HSS is defined as

$$\alpha = \frac{\sqrt{E/F_y}}{b/t} \tag{2}$$

In the U.S., a thin walled tube is defined as one having α less than 0.67.

Table 1. Column strengths in shear tab vs. through-plate tests

b/t	α	connection	symmetric test	unsymmetric test
29	0.89	through-plate	0.63	0.42
		shear tab	0.61	0.46
40	0.60	through-plate	0.58	0.42
		shear tab	0.45	0.42

The symmetric tests failed with sudden buckles while the unsymmetric tests failed gradually in bending.

The conclusion from Table 1 is that shear tab connections used with HSS columns that are not thin-walled will develop the same column strength as those where the wall is reinforced with a through-plate. With thin-walled HSS, shear tabs may have a detrimental effect on the axial column capacity. For connections on only one side of the HSS column, there is no strength reduction for using shear tabs.

The simple connection made with a single angle is more complicated in that it introduces an eccentricity. However, the wall distortion of the HSS should be less than for the shear tab since some of the force is transmitted to the weld near the side wall. Therefore, the yield mechanism will not develop and the HSS column strengths should not be affected to the same extent as for a shear tab. However, it is possible that the punching shear may be more critical. The side weld should not be a problem since it is not as severely loaded as the connections discussed in the preceding section.

4 Bolting to the HSS

The end plate connection differs from the others in that the plate is welded to the beam and should be bolted to the columns as shown in Figure 4.

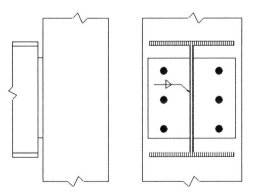

Fig. 4. End plate shear connection with bolting to HSS

This type of connection has some inherent stiffness and may be semi-rigid rather than simple. Tube wall distortion should not be a strength limit state and, of course, there are no concerns regarding welding to the HSS.

The primary concern is bolting to the HSS. Through bolts are unsatisfactory since they cannot be fully tightened (as is desirable for bolts that are subject to some tension) due to the lack of stiffness in the HSS wall. Since the inside of the HSS is inaccessible, a special fastening technique is required. One possibility is a special bolt with a rectangular head that fits into a rectangular slot in the HSS wall and, when turned 90°, provides bearing against the inside of the wall. Care must be taken to insure that the bolt retains it rotational position as it is tightened. Another possibility is a "flow-drilled" hole that replaces the nut in the bolted assembly. An evaluation of flow-drilled connections has been made by Sherman (1989).

FLOWDRILLING (1987) describes the process where a hole is forced through a flat plate by a carbide conical tool rotating at sufficient speed to produce high rapid heating which softens the material in a local area. The material that is displaced as the tool is forced through the plate forms a truncated hollow cone (bushing) on the inner surface and a small upset on the outer surface. Tools can be obtained with a milling cutting collar so that the material on the outer surface is removed, producing a flat surface allowing parts to be brought into tight contact. A cold-formed tap is then used to roll a thread into the hole without any chips or removal of material.

It is recommended that flow-drilled holes can be produced in plate thicknesses up to half the hole diameter. Hence the ranges of common bolt sizes and HSS thicknesses are compatible. In these thicknesses, the resulting thread length in a flow-drilled hole is approximately equal to the depth of heavy hex nuts. In addition, the flow-drilling produces a local hardening of the plate material that results in properties similar to a nut. The combinations of HSS thicknesses and bolt diameters that were evaluated are indicated in Table 2.

Table 2. Material thickness/bolt diameter combinations

Material Thickness (mm)	Bolt diameter (mm)				
	12.7	15.9	19.1	22.2	25.4
4.8	X	X			
6.4	X	X	X		
7.9		X	X	X	
9.5			X	X	X
12.7					X

The bolts were purchased in the U.S. using customary units for size and the tools were supplied in metric sizes. In addition, there is some uncertainty in predicting the final hole size since cooling takes place after the tool is forced through the material. This presented a problem with the 22.2 mm diameter bolts where the hole was oversized and the final threads were poorly developed. In all other sizes, threads comparable to typical nuts were produced. The threaded holes were produced in all size combinations with virtually no permanent distortion of the HSS walls.

The test program consisted of shear and tension tests. In the shear tests, bolts were tested in both the snug tight and fully tightened conditions while in the tension tests, all bolts were fully tightened by the turn-of-the-nut method (although it was the bolt head that was actually turned). Three specimens of each type were tested. The bolts were high strength A325 that have a specified proof load of 635MPa and tensile strength of 825MPa. The results of the tests are shown in Figure 5.

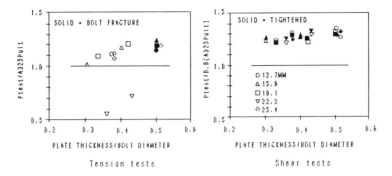

Fig. 5. Flow-drilled test results

In the tension tests, there were three failure modes by fracture of the bolt and all the other failures were the result of stripping the threads in the hole. However, except for the 22.2 mm bolts that had the poor threads due to the oversize holes, all test loads exceeded the specified tension for the fastener. There is a tendency for the tension capacity to decrease as the ratio of plate thickness to bolt diameter decreases.

The shear strength data have been normalized by 0.6 times specified bolt tensile strength. This is the ratio between shear and tension capacities when threads are in the shear plane, which will always be the case with flow-drilled holes. In all cases, including the 22.2 mm bolts with poorly threaded holes, the test strength exceeded the required capacity. There is little dependence on the ratio of plate thickness to bolt diameter and no distinction can be made between snug tight and fully tightened bolts.

Based on these tests, flow-drilling has the potential for use in blind bolting to the HSS columns. However, there are practical considerations in that the fabricator must have drilling equipment with high rotational speeds and the cost of preparing the hole is higher than normal.

5 Conclusions

Many simple beam to column connections that are used with wide-flange columns can also be used with HSS columns to produce clean and efficient connections. In some cases there are considerations regarding whether the HSS will cause the strength or behavior of the connection to be different than with a wide-flange column. A small amount of data exists for some connections and a study is underway to investigate the other connection types.

6 References

Flowdrill bv (1987) Flowdrilling: a new manufacturing process. Utrecht, Netherlands.

Haslam, C.A. (1993) The capacities of tubular columns due to shear tab beam connections. M.S. thesis, **University of Wisconsin–Milwaukee.**

Sherman, D.R. (1989) Evaluation of flow-drilled holes for bolted connection in structural
tubes. **CE DepartmentReport, University of Wisconsin–Milwaukee.**

Sherman, D.R. (1991) The design of shear tabs welded to HSS columns. **Tubular
Structures, 4th International Symposium, Delft.** 354-363.

248

24 DESIGN FORMULAE FOR CHS COLUMN-TO-BEAM CONNECTIONS WITH EXTERIOR DIAPHRAGMS

T. KAMBA and H. KANATANI
Kobe University, Japan

Abstract
In the Architectural Institute of Japan (AIJ) "Recommendations for Design and Fabrication of Tubular Structures in
Steel (1990)", design formulae for calculating the resistance of moment resisting CHS column connections with
exterior diaphragms are presented. However, it is very difficult for those who are not familiar with the AIJ recommendations to understand the design formulae.
    The purposes of this paper are to investigate the design
formulae for the resistance of connections with exterior diaphragms in more detail, and to explain them so that they
can be understood easily. The authors illustrate this point
clearly by re-evaluating the results of Rink et al.(1991), and
give a brief English commentary of Section 4.4.5 "Tubular
Column-to-Beam Connections" in the recommendations.
Keywords: CHS Column-to-Beam Connection, Local Resistance, Exterior
Diaphragm, Design Formulae, AIJ Standards

1 Introduction

The AIJ has published two different types of design recommendations for steel structures. One is a working stress
design edition which is a supplemental design guide to the
present national building code, and the other is a limit state
design edition which is the "Standard for Limit State Design
in Steel Structures (Draft)"(1990). Both of them are the same
as to the basic design concept, but resistance factors used in
the design formulae are somewhat different. The design formulae in this paper are obtained from "Recommendations for
Design and Fabrication of Tubular Structures in Steel"(1990)
(henceforth called "Tubular Recommendations") which belong
to the working stress type of AIJ recommendations.
    For hollow section connections, it is very difficult to
restrain perfectly the local flexural deformation of a tube
wall by the out-of-plane force from the beam flanges. In
order that tubular connections satisfy the condition of being moment resisting connections capable of sustaining the
ultimate beam load, the column wall must generally be stiff-

Tubular Structures V. Edited by M.G. Coutie and G. Davies.
© 1993 E & FN Spon, 2–6 Boundary Row, London SE1 8HN. ISBN 0 419 18770 7.

ened to prevent any local failure prior to developing the required plastic moment at the beam end. The Tubular Recommendations provide some requirements for the design of full-strength, welded, moment resisting connections, where I-beams are rigidly framed to tubular columns, by specifying various connection details as follows:
(a) welding diaphragms within the columns at the beam flange levels (**through diaphragms** or **internal diaphragms**);
(b) enlarging the flange width at the beam ends or fitting plate diaphragms around the column for each beam flange (**exterior diaphragms**);
(c) filling the tubular column with concrete.
   The purposes of this paper are to present an English text of Section 4.4.5 "Tubular Column-to-Beam Connections" of the Tubular Recommendations (AIJ 1990), and to investigate the design formulae for the resistance of connections with exterior diaphragms in more detail. Hopefully, this explanation will enable them to be understood easily.

2 Design Formulae

Table 1 gives the English text of Section 4.4.5 "Tubular Column-to-Beam Connections" and Table 4.4 in the Tubular Recommendations. When the connections are stiffened by exterior diaphragms, the local resistance of the CHS and RHS connections may be obtained from the design formulae shown in Table 4.4(a) and 4.4(b), respectively.
   The coefficients in the formulae in Table 4.4(a) have been determined such that the connection resistance (in the long term) is almost half the local strength from the test results.
   Rink et al.(1991) have presented a comparison between the calculated load/deflection behaviour of CHS column connections with diamond plates, which were a kind of exterior diaphragm, by the elastic-plastic finite element method, and the design formula. Table 2 shows their theoretical specimens and calculated results, where M_u is the ultimate load and $_bM_p$ is the beam plastic moment. M_{AIJ} in Rink et al. (1991) is the local strength of a connection estimated by Eqn.(1), which is obtained from twice the design formula (Table 4.4(a)) and substituting the actual yield stress of column, $_p\sigma_y$, into F_2. M'_{AIJ} is obtained by the authors by recalculating the value of the tapered flange width, $B'_f{}^*$, correctly and then using Eqn(2).

$$P_u = 2 \cdot P_a = (6.56\frac{B'_f}{D} + 2.86)\, t\sqrt{t_s \cdot (t + h_s)} \cdot {}_p\sigma_y \tag{1}$$

$$M_{AIJ} = P_u \cdot h_b = P_u \cdot (H - t_f) \tag{2}$$

250

The column diameter, D, the column thickness, t, the diaphragm thickness, ts, the height of the beam, H, the beam flange thickness, tf, and the beam web thickness, tw, of each specimen are constant, and those values are shown under Table 2.

Table 1. Translation of Section 4.4.5 "Tubular Column-to-Beam Connections" in the Tubular Recommendations (AIJ 1990)

---

(1) **Moment connections**, subjected to combined axial, bending, and shear stresses, shall be proportioned to satisfy the following requirements:

(a) Joint panels, which are the parts surrounded by columns and beams, shall be proportioned to have sufficient load carrying capacity for shear force arising therein under lateral loading.

When the joint panel is adequately stiffened at the beam flange levels, the shear strength will be adequate providing the following is satisfied:

$$_pM_y \geq {_bM^L} + {_bM^R} \qquad (12)$$

where $_pM_y$ = moment resistance of the joint panel under wind or seismic loading = $4/3 \cdot F_s \cdot V_e$

F_s = shear yield stress = $F_y/\sqrt{3}$

F_y = minimum specified yield stress for the material in tension

V_e : effective volume of the joint panel

For circular hollow sections $\quad V_e = \dfrac{\pi}{2} \cdot h_b D_m t_p$

For square hollow sections $\quad V_e = \dfrac{16}{9} \cdot h_b D_m t_p$

h_b : distance between upper and lower flange centres

D_m : mean diameter (or width) of the column = $D - t_p$

t_p : thickness of the column at the connection

$_bM^L, {_bM^R}$: bending moments of the left and right, respectively, of the connection at the beam ends (column face).

(b) Connections shall be designed to prevent local failure of the column wall by concentrated forces from the beam flanges. Reinforcement of the connections shall be according to the following principles:

(i) When connections are **stiffened by through diaphragms or inner diaphragms**, the thickness of diaphragms shall be not thinner than the beam flange thickness and the ratio of column diameter to diaphragm thickness should not exceed $730/\sqrt{F_y}$, with F_y in MPa. Welds to the diaphragms and tubes shall be groove welds, and the strength of welds shall be equivalent to the parent metal.

(ii) When connections are stiffened by **exterior diaphragms**, the corners of diaphragms shall be finished smoothly to avoid excessive stress concentrations. The local resistance (in the long term) of such connections may be obtained from the equations shown in Table 4.4.

(2) **Pinned connections** at the beam ends, fabricated with through gusset plates and such like, shall conform to Clause 4.4.4.

(3) **Concrete-filled column-to-beam connections** may be designed by taking account of the reinforcing effects of the concrete for shear resistance of the connections and for local failure of the tube walls.

Table 1 (Continued)

Table 4.4 (a) Factored connection resistance of CHS column (Exterior diaphragm)

Shape of exterior diaphragm	Connection Resistance P_a	Range of Validity
(diagrams of exterior diaphragm shapes with $\alpha=45°$, $\theta \le 30°$, hs, t, $B'f$, ts, $r \ge 10$ mm, Bf, D, P)	$P_a =(3.28\dfrac{B'f}{D}+1.43)t\ \sqrt{ts\cdot(t+hs)}\cdot F_2$ for $\sqrt{2}(D/2+hs)\ge D$, $B'f=D$ for $\sqrt{2}(D/2+hs)< D$, $B'f$ to be determined from connection details as a tangent to the tube	$15 \le D/t \le 55$ $\dfrac{B'f}{2ts}\le\dfrac{237}{\sqrt{F_1}}$

Table 4.4 (b) Factored connection resistance of RHS column (Exterior diaphragm)

Shape of exterior diaphragm	Connection Resistance P_a	Range of Validity
(diagram of RHS column exterior diaphragm with $\alpha=45°$, $\theta \le 30°$, $r \ge 10$ mm, D, hs, ts, P)	$P_a = 1.48(\dfrac{t}{D}\cdot\dfrac{ts}{t+hs})^{2/3}(\dfrac{t+hs}{D})D^2\dfrac{F_1}{Y}$ where, Y is the yield ratio $(=F_y/F_u)$ $Y = 0.57$ for mild steel ($F_u =400$MPa) $Y = 0.66$ for high-strength steel ($F_u =490$MPa)	$17 \le D/t \le 67$ $hs/D \le 0.4$ $0.75 \le ts/t \le 2.0$ $\dfrac{D/2+hs}{ts} \le \dfrac{237}{\sqrt{F_1}}$ $\theta \le 30°$

Remark : When beam flange forces are induced by temporary loading, connection resistance in the short term may be increased one-half above the values.

Notations: Bf =flange width, D =external diameter or width of tube, F_1 =specified minimum yield stress of diaphragm in MPa, F_2 =specified minimum yield stress of tube in MPa, hs =edge distance of exterior diaphragm, P =connection resistance= bM/hb, P_a =connection resistance in the long term, t =thickness of tube, ts =thickness of exterior diaphragm

Table 2. Theoretical specimens and results in Rink et al.(1991)

Specimen	Bf (mm)	hs (mm)	B'f (mm)	B'f* (mm)	Mu (kNm)	bMp (kNm)	Mu/bMp	MAIJ (kNm)	Mu/MAIJ	M'AIJ (kNm)	Mu/M'AIJ	r (mm)
HB1	300	70	806	639	4589	4040	1.14	4795	0.99	4027	1.17	863
HB2	300	120	877	764	5348	4040	1.32	6207	0.89	5503	1.00	984
HB3	300	200	990	980	6363	4040	1.58	8346	0.79	8039	0.82	1177
HB4	400	70	806	645	4741	5073	0.93	4795	1.02	4048	1.21	693
HB5	400	120	877	766	5504	5073	1.08	6207	0.91	5511	1.03	813
HB6	400	200	990	980	6466	5073	1.27	8346	0.80	8039	0.83	1007
HB7	500	70	806	653	4940	6106	0.81	4795	1.06	4081	1.25	522
HB8	500	120	877	768	5668	6106	0.93	6207	0.94	5524	1.06	643
HB9	500	200	990	980	6697	6106	1.10	8346	0.83	8039	0.86	836

Constant dimensions (mm) : D=1000,t=30,ts=30,H=1000,tf=30,tw=12

Some of their conclusions were as follows: (a) increasing
the width of the stiffening and the thickness of the tubular
column increased the ultimate strength of the connection
considerably, but not as much as would be suggested by the
design formulae; and (b) the design formulae for the ultimate load gave an overoptimistic result in almost all cases.
However, it appears that they have misused the design
formulae because the expressions were unclear.

2.1 Flange width

It is very important to calculate the width of the tapered
flange, B'f, exactly from the connection details, because it
affects the local resistance of connections directly. In
Table 4.4(a) of the recommendation, there is an upper limit
of D imposed on B'f.

For $\sqrt{2}(D/2+hs) \geq D$, B'f =D   (3)

For $\sqrt{2}(D/2+hs) < D$, B'f =B'f from connection details.   (4)

In Rink et al.(1991), the tapered flange width was calculated using B'f=$\sqrt{2}$(D/2+hs) regardless of whether B'f was
smaller than D or not and M$_{AIJ}$ was calculated using H instead of
hb=H-tf, so it consequently overestimated M$_{AIJ}$. For specimens
with hs=70mm and 120mm, the real flange width, B'f*, is much
smaller than those of Rink et al.(1991) (Table 2).

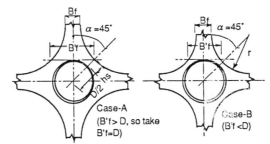

Fig.1. Shapes of Diamond Plates for Rink et al.(1991)

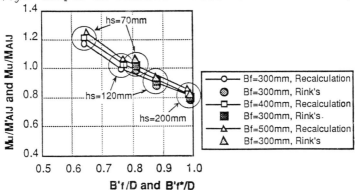

Fig.2. Mu/M$_{AIJ}$ (Mu/M'$_{AIJ}$) − B'f/D (B'f*/D) relationships

Figure 2 shows the relationship between Mu/M$_{AIJ}$ (Mu/M'$_{AIJ}$) and B'f/D (B'f*/D) for each specimen. It can be seen that: (1) a difference in B'f affects the estimated strength of the connections, (2) the design formula estimates the strength of the connections well with hs=70mm and 120m (Mu/M'$_{AIJ}$=1.0~1.25), and (3) the larger the tapered flange width, the smaller the ratio of Mu/M$_{AIJ}$ or Mu/M'$_{AIJ}$.

2.2 Failure mechanism

As mentioned above, the larger the flange width, the smaller the ratio of Mu/M'$_{AIJ}$. That is, increasing the width of the stiffening would not increase the ultimate strength of the connection as much as would be suggested by the design formula. The Mu/M'$_{AIJ}$ ratio of specimens with hs=200mm is approximately 0.8 and the AIJ design formula overestimated those strengths. The AIJ design formula is based on an assumption of local connection failure, caused by the normal force of beam flanges (i.e. local collapse of tubes, fracture of diaphragms, flange local buckling, etc.). However, all Rink et al's specimens failed by local diaphragm buckling under compression from the I-beam flange, and this failure mode is not covered by the AIJ recommendations when the diaphragm width-to-thickness ratio is in the range used in HB3, HB6 and HB9 (B'f*/(2ts)=16.3).

2.3 Limiting values for diaphragm width-to-thickness ratio

The width-to-thickness ratios, B'f*/(2ts), of the diamond plates at the column surface of Rink's specimens were 10.7~10.9 (hs=70mm), 12.7~12.8(hs=120mm) and 16.3(hs=200mm). However, the limiting width-to-thickness ratio of the flange plate in the AIJ "Design Standard for Steel Structures"(1973), the AIJ "Recommendation for the Plastic Design of Steel Structures" (1975) and the Building Center of Japan "Guidelines to Structural Calculation under the Building Standard Law" (1990) is 12.6 (=237/$\sqrt{Fy}$; Fy=355 MPa), 8.5 and 8, respectively. Since the value of 12.6 in the AIJ "Design Standard for Steel Structures"(1973) is not strict, the value of 8 or less is usually used as the limiting width-to-thickness ratio of the flange plate in structural steel design in Japan.

3 Commentary on exterior diaphragms

Here, the commentary on the exterior diaphragm, Part (1)i(b)(ii) in Section 4.4.5 of "Tubular Column-to-Beam Connections", is described. There are various joint details for reinforcement against local failure of connections. The most simple forms of connection reinforcement, using exterior diaphragms, are shown in Fig.3. The stress concentration points can be seen at the circled parts in the figure. If a crack developed at this location, premature fracture of the connection may result before the connection underwent large plastic deformation. To avoid this type of local failure, the following requirements are effective: (1) making no sharp angles at the diaphragms, (2) making no weld joints at the circled parts in Fig.3 and (3) when necessary, paying attention to the weld ends (such as using end tabs).

For the exterior diaphragm, which is the simplest form of stiffening against local failure, extensive studies have been carried out and the local strength of such connections has been identified. The design formulae based on those studies have been presented in Table 4.4. The coefficients in the formulae have been determined such that the connection resistance (in the short term) is almost equal to the yield resistance obtained from the test results.

The design formulae for CHS column connections shown in Table 4.4.(a) apply to Figure 3(a) and (b). However, these formulae can be also applied to the connection types shown in Figure 3(c) and (d) by setting B'f =Bf, and to a no reinforcement connection which flanges are directly welded to a column by setting hs =0.

Although the quoted experimental formulae were obtained by test results in which the flange force was in tension, they may be applicable to connections in which the flange force is in compression. The local resistance of connections which are subjected to a compression force from the flange usually decreases compared with a tensile force, so local failure of a real connection in a moment resisting frame would be usually caused at the compression flange level. The reasons are the out-of-plane deformation of the tube wall and local buckling at the tapered flange. Results have been reported in which the compressive strengths of simplified connection

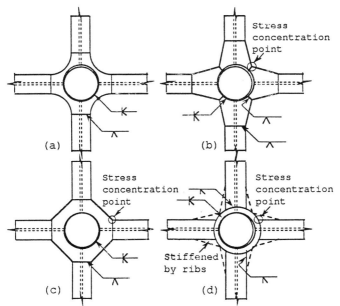

Fig.3. Reinforcement by exterior diaphragms

tests, with large diameter-to-thickness ratios (D/t=70 and 100), were about 70% of the tensile strengths. However, for typical moment connections the difference between the absolute value of compressive and tensile stress in the upper and lower beam flanges is not very much, and the depth of the beam is almost one or two times the diameter of the column. Hence, the compression strength of connections is not significantly decreased as the simplified compression tests would suggest because of the interaction of the tensile and compressive local deformations.

For CHS column connections, when the sign of the bending moment of the right and left beams is the same under lateral loading, the local strength of a connection may increase compared with the case of bending moments of the opposite sign. But, here a clear distinction is not drawn between them.

In order to increase both the local resistance of the connection and the shear resistance of the joint panel, it is most effective to increase the thickness of the column at the connection. It is very difficult, however, to restrict the local deformation by increasing the column thickness (with the range of validity $15 \leq D/t \leq 55$), so exterior diaphragms are usually used except for light beams.

4 Conclusions

1. The design formula estimates the theoretical strength of connections with $B'f^*/(2ts) \leq 12.8$ well.
2. However, the necessary care to be taken in using the design formulae is as follows:
   (a) evaluating the width of the tapered flange, $B'f$, correctly;
   (b) noting the limiting value for the width-to-thickness ratio of the diaphragm; and
   (c) avoiding stress concentrations.

5 References

AIJ, **Standard for Limit State Design in Steel Structures** (Draft), (1990) Tokyo, Japan

AIJ, **Recommendations for Design and Fabrication of Tubular Structures in Steel**, (1990) Tokyo, Japan (in Japanese)

AIJ, **Design Standard for Steel Structures**, (1973) (in Japanese)

AIJ, **Recommendation for the Plastic Design of Steel Structures**, (1975) Tokyo, Japan (in Japanese)

Building Center of Japan, **Guidelines to Structural Calculation under the Building Standard Law**, (1990) Tokyo, Japan (in Japanese)

Rink, H.D., de Winkel, G.D., Wardenier, J. and Puthli, R.S., Numerical Investigation into the Static Strength of Stiffened I-Beam-to-Column Connections", (1991) in **4th. International Symposium on Tubular Structures**, Proceedings, Delft, The Netherlands, pp.461-470.

25 DESIGN OF SHS COLUMN-TO-H BEAM CONNECTIONS WITH UNEQUAL WIDTH COLUMNS OF UPPER AND LOWER STOREYS

M. TABUCHI, H. KANATANI, T. TANAKA
Kobe University, Japan
S. QIN
Fujiki Komuten Co. Ltd, Tokyo, Japan

Abstract
In this paper, for simple fabrication of SHS column-to-H beam connections with unequal column width of upper and lower stories, a simple connection detail which is composed of a joint panel with the same width as a lower column and a pair of diaphragm plates is examined. In this type of connection, the upper diaphragm must be proportioned to be able to transfer the strength of the upper column by its out-of-plane bending resistance.

A series of tests on the connections with various details, including the reinforcement of the diaphragm plate, is carried out to investigate the behaviour of the connections. Plastic analysis based on the yield line theory is tried to estimate the strength of the connections.

Keywords: Beam-to-column connection, SHS, Through diaphragm, Yield strength, Yield line theory.

1 Introduction

Moment resisting steel building frames consisting of SHS columns are widely used in Japan. According to the aseismic design criterion, every connection in moment resisting frames must have sufficient strength for carrying full plastic moments of members jointed there and appropriate rigidity.

When SHS column width of an upper story is smaller than that of a lower story in those frames, some complicated problems occur in fabrication to satisfy the above requirement for the connections. Commonly, a detail with the tapered joint panels built up by four steel plates (Fig. 1 (a)) is used for smooth transference of the upper column stresses. Such a type of connection, however, is laborious and leads to an increase in the cost.

Here, a simple connection detail shown in Fig. 1 (b), which is composed of a joint panel with the same width as the lower column and a pair of diaphragm plates, is examined. In this type of connection, the upper diaphragm must be proportioned to be able to transfer the strength of the upper column by its out-of-plane bending resistance.

As to the behaviour of the connection, a forerunning study has been performed by Imai et al. (1987) and recently a series of tests on the connections with larger dimension of the members than those in this study have been carried out by Hashimoto et al. (1992).

In this study, a series of tests on the connections with various details, including the reinforcement of the diaphragm plate, is carried out to investigate the behaviour of the connections. And, the plastic analysis based on the yield line theory is tried to estimate the strength of the connections.

Tubular Structures V. Edited by M.G. Coutie and G. Davies.
© 1993 E & FN Spon, 2–6 Boundary Row, London SE1 8HN. ISBN 0 419 18770 7.

2 Test

Twentysix specimens were prepared for this study. Figure 2 shows test specimens. Each specimen is composed of an upper column, an upper diaphragm and a joint panel which is called lower column in later discussions, because they have the same dimensions. Each specimen has two rib plates (PL 9) simulating beam webs for preventing premature local buckling of the lower column.

The upper column and the lower one are welded to the diaphragm plate by a single bevel groove weld.

Test variables are as follows:

(1) Difference between upper column width, d, and lower one, D.
(2) Thickness of diaphragm plate, t_d.
(3) Eccentricity between upper column and lower one.
(4) Reinforcement of diaphragm plate.

The details of the specimens are shown in Table 1. As to the eccentricity between the upper and the lower columns, four types of specimens were investigated (Fig. 2 (a)).

Type 1: No eccentricity (e=0).
Type 2: Intermediate type of Type 1 and Type 3 (e=(D-d)/4).
Type 3: Each one side of upper and lower columns is trued up (e=(D-d)/2).
Type 4: Each one corner of upper and lower columns is trued up.

Four specimens with reinforced diaphragm (designated by R in specimen name) were also tested. SHS with smaller width than that of the lower column by 50 mm and 150 mm in length was welded to the back of the diaphragm plate (Fig. 2 (b)).

The columns of all specimens are cold-formed square hollow sections (STKR 400). The materials of the diaphragm plates are SS400 . The mechanical properties of the materials are shown in Table 2.

The test set-up is shown in Fig.2 (c). The specimen was fixed on the testing bed by tension rod and the load was applied at the end of the upper column by a hydraulic jack. The displacement at the end of the upper column was measured by a displacement transducer.

3 Test results

Figures 3 (a) - (i) show the load-deflection curves of all specimens. The test results are summarized in Table 3. The yield load, P_y, and the ultimate load, P_u, are defined as follows:

P_y is the load when stiffness of the specimen becomes one third of the initial stiffness.

P_u is the load when the rotation angle of the specimen ($=\delta/L$) becomes 1/20 rad..

Three failure modes were observed in the tests.

Mode 1: Flexural failure of diaphragm.
Mode 2: Local buckling of upper column.
Mode 3: Crack at column corner or weld.

When D-d=100mm, specimens failed according to Mode 1 independent of the diaphragm thickness, except for a specimen BMC100-22 which failed according to Mode 2.

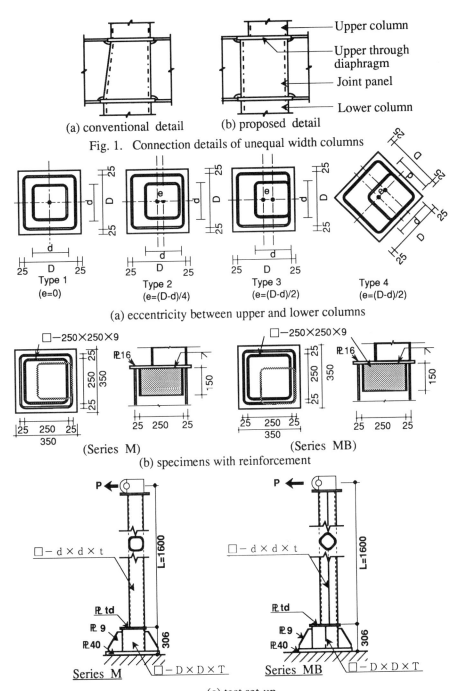

Upper column

Upper through diaphragm

Joint panel

Lower column

(a) conventional detail (b) proposed detail

Fig. 1. Connection details of unequal width columns

Type 1 (e=0)

Type 2 (e=(D-d)/4)

Type 3 (e=(D-d)/2)

Type 4 (e=(D-d)/2)

(a) eccentricity between upper and lower columns

□−250×250×9

PL16

(Series M)

□−250×250×9

PL16

(Series MB)

(b) specimens with reinforcement

P

□−d×d×t

L=1600

PL td

PL 9

PL 40

Series M □−D×D×T

P

□−d×d×t

L=1600

PL td

PL 9

PL 40

Series MB □−D×D×T

(c) test set-up

Fig. 2. Test specimens

259

Even in specimens with D-d=50mm, Mode 1 occurred, when the diaphragm thickness became small. In the specimens governed by Mode 1, the ultimate strength, P_u, did not reached the ultimate strength of the upper column, P_{cp}. On the other hand, in the specimens governed by Mode 2 or Mode 3, the values of the P_u/P_{cp} ratios are more than 1.0.

Four specimens with asterisk marks at P_{max} value in Table 3 showed small deformation capacity (Fig. 3(i)), because these specimens had slag inclusions at the corner weld of the columns.

When D-d=100mm, specimens with the reinforcement can transfer the ultimate strength of the upper column. This reinforcement method is easy for fabrication and has sufficient effect in improving the strength of the connection.

4 Analytical predictions

Plastic analysis based on the yield line theory is tried to estimate the strength of the specimens with Mode 1 type of failure. Three mechanisms correspond to the connection type shown in Fig. 2 (a) are assumed.

4.1 Mechanism for Type 1 and Type 2 connections
Mechanism 1 shown in Fig. 4(a) corresponds to the connection of Type 1 and Type 2 . In this mechanism, yield lines indicated by dashed lines are formed on the diaphragm plate whose full plastic moment per unit length is M_{do} ($=\sigma_{yd} t_d^2/4$). The full plastic moment of the yield lines indicated by solid lines takes a smaller value between M_{do} and M_{lo} ($=\sigma_{yl} T^2/4$) which is the full plastic moment per unit length of the lower column wall. A variable x is determined to give the minimum failure load of the connection. The failure load is given by Eq. (1).

$$P_{cu} \cdot L = \left\{ \left(M_{do} + M_{lo}\right)\left(\frac{b}{l_2} + \frac{a - x + l_2}{b - a}\right)(a - x) + b \cdot M_{do} \right\} + \left\{ \left(M_{do} + M_{lo}\right)\left(\frac{b}{l_1} + \frac{x + l_1}{b - a}\right) + a \cdot M_{do} \right\}$$
(1)

in which $x = \frac{1}{2}a - \frac{1}{4}\left\{ (l_1 - l_2)\frac{(1/l_1 - 1/l_2) \cdot D}{b - a} \right\}$

L is the distance between the surface of the diaphragm plate and the end of the upper column (=1600mm).
For calculation of the strength, it is assumed that the yield lines indicated by solid lines are formed in the centre of the wall thickness of the lower column and the yield line 3 is formed along the toe of the weld of the upper column.

4.2 Mechanism for Type 3 connection
Mechanism 2 shown in Fig. 4 (b) corresponds to the connection of Type 3. Yield lines indicated by dashed lines and solid lines have the same meaning as the mechanism 1. The failure load is given by Eq. (2).

$$P_{cu} \cdot L = \left(M_{do} + M_{lo}\right)\left\{ \frac{b}{l_2} + \frac{(a + l_2)a}{b - a} + 2b \cdot M_{do} \right\}$$
(2)

The location of the yield lines is the same as the mechanism 1.

4.3 Mechanism for Type 4 connection
Mechanism 3 shown in Fig. 4 (c) corresponds to the connection of Type 4. In this case, even if sufficient yield lines are formed on the diaphragm plate, the connection is not yet kinematically admissible, because vertical displacement of two sides of the upper column (DE,EF) is restricted by the lower column. Therefore, it is assumed that a kinematically

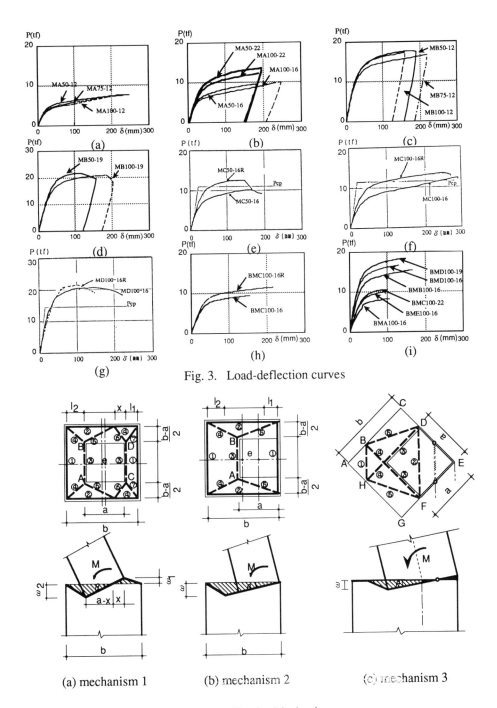

Fig. 3. Load-deflection curves

(a) mechanism 1   (b) mechanism 2   (c) mechanism 3

Fig. 4. Mechanism

admissible state may occur when the diaphragm plate failure is accompanied by failure of an angle shape composed of two sides of the upper column (a section DEF). The failure load is obtained from the sum of failure loads of the diaphragm, M_{pd}, and the full plastic moment of the angle shaped section, M_{pa}.

$$P_{cu} \cdot L = M_{pd} + M_{pa} = \frac{4\sqrt{6}}{3} M_{do} \frac{b^2}{b-a} + \frac{\sqrt{2} \cdot N_{uo} \cdot (d-t)^2}{4} \tag{3}$$

where N_{uo} is the yield force of the upper column per unit length ($= \sigma_{yu} t$). For calculation of Mpd, it is assumed that the yield line 3 is formed along the toe of the weld and points B, D, F and H are located at the inner surface of the lower column.

5   Comparison between test and calculated results

Table 4 shows the calculated failure load, P_{cu}, of all specimens with Mode 1 type of failure. The experimental value, Py, and the ratios, Py/P_{cu}, are also indicated. For calculation of the failure load, the yield stress obtained from the tensile coupon tests shown in Table 2 is used.

In series M, Eq. (1) leads to underestimation of the strength of the connections of Type 1 and 2, and Eq. (2) leads to overestimation of the strength of the connections of Type 3. The mean and the coefficient of variation of 16 values of Py/Pcu ratios are 1.02 and 0.13 respectively. These results indicate that Eqs. (1), (2) and (3) are suitable for predicting the yield strength of the connections with sufficient degree of accuracy.

6   Conclusions

The following points are the main conclusions of the present study:

1) Specimens with large differences of the width between upper and lower columns failed by flexural failure of the diaphragm plate. Even in specimens with small differences of the column width, this failure mode occurred, when the diaphragm thickness was small.
2) For this failure mode, Eqs. (1)-(3) based on the yield line theory are suitable for predicting the yield strength of the connections.
3) It is useful for reinforcement of the diaphragm to weld SHS with smaller width than the lower column by 50mm to the back of the diaphragm plate.

7   Acknowledgement
The authors wish to thank Nippon Steel Metal Products Co.,Ltd. for financial support in carrying out the experimental work.

8   References
Imai, K. and Nagata, T. (1978) On the column joints with unequal width SHS, **Summaries of Technical Papers of Annual Meeting AIJ**, pp.1305-1306 (in japanese).

Hashimoto, K. Tanuma, Y. Nagai, M. Fujisawa, K. Yamada, N. and Sugawara, H. (1992) Strength of beam-to-column connections with diaphragm using unequal square hollow section for columns (part 2-4), **Summaries of Technical Papers of Annual Meeting AIJ**, pp.1565-1570 (in japanese).

Table 1. Test specimens

Specimen	Upper column d×d×t	Lower column D×D×T	Diaphragm t_d	Eccentricity e	Connection type	Material (see Table 2)
MA50-12	□-200×200×9	□-300×300×9	12	0	1	1,7,11
MA75-12			12	25	2	1,7,11
MA100-12			12	50	3	1,7,11
MA50-16			16	0	1	1,7,12
MA100-16			16	50	3	1,7,12
MA50-22			22	0	1	1,7,17
MA100-22			22	50	3	1,7,17
MB50-12	□-250×250×9	□-300×300×9	12	0	1	4,7,11
MB75-12			12	12.5	2	4,7,11
MB100-12			12	25	3	4,7,11
MB50-19			19	0	1	4,7,15
MB100-19			19	25	3	4,7,15
MC50-16	□-200×200×9	□-300×300×12	16	0	1	2,9,13
MC50-16R			16	0	1	2,9,13
MC100-16			16	50	3	2,9,13
MC100-16R			16	50	3	2,9,13
MD100-16	□-250×250×9	□-300×300×12	16	25	3	5,9,13
MD100-16R			16	25	3	5,9,13
BMC100-16	□-200×200×9	□-300×300×12	16	50	4	3,10,14
BMC100-16R			16	50	4	3,10,14
BMC100-22			22	50	4	3,10,18
BMA100-16	□-200×200×9	□-300×300×9	16	50	4	3,8,14
BMD100-16	□-250×250×9	□-300×300×12	16	25	4	6,10,14
BMD100-19			19	25	4	6,10,16
BMB100-16	□-250×250×9	□-300×300×9	16	25	4	6,8,14
BME100-16	□-200×200×9	□-250×250×9	16	25	4	3,8,14

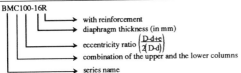

BMC100-16R
- with reinforcement
- diaphragm thickness (in mm)
- eccentricity ratio $\left(\dfrac{D-d+e}{2(D-d)}\right)$
- combination of the upper and the lower columns
- series name

Table 2. Mechanical properties

No.	Member	σ_y (tf/cm^2)	σ_u (tf/cm^2)	El. (%)
1		3.92	4.77	43.0
2	□-200×200×9	3.90	4.73	33.0
3		3.61	4.64	41.6
4		3.93	4.75	46.5
5	□-250×250×9	3.05	4.05	40.3
6		3.41	4.60	43.0
7	□-300×300×9	4.04	5.12	47.0
8		3.55	4.67	41.9
9	□-300×300×12	2.99	4.35	39.3
10		3.65	4.93	47.8
11	PL=12	2.83	4.48	49.6
12	PL=16	2.80	4.48	35.6
13		3.01	4.38	43.4
14		2.68	4.47	47.2
15	PL=19	2.76	4.60	37.3
16		2.77	4.57	48.9
17	PL=22	2.73	4.59	30.1
18		2.50	4.42	51.6

263

Table 3. Test results

Specimen	Py (tf)	Pu (tf)	Pmax (tf)	Pcp (tf)	Ko (tf/cm)	Kco (tf/cm)	Ko/Kco	Pu/Py	Pu/Pcp	Pmax/Pcp	Failure Mode
MA50-12	4.13	6.08	6.61	11.56	2.67	4.99	0.54	1.47	0.53	0.57	1
MA75-12	3.75	5.58	7.17	11.56	2.33	4.99	0.47	1.49	0.48	0.62	1
MA100-12	3.63	5.21	6.84	11.56	2.01	4.99	0.40	1.44	0.45	0.59	1
MA50-16	5.81	7.92	9.45	11.56	2.86	4.99	0.57	1.37	0.69	0.82	1
MA100-16	5.21	7.11	10.01	11.56	2.69	4.99	0.54	1.37	0.62	0.87	1
MA50-22	9.15	11.57	13.69	11.56	3.66	4.99	0.73	1.26	1.00	1.18	1
MA100-22	8.25	10.57	12.55	11.56	3.24	4.99	0.65	1.28	0.91	1.09	1
MB50-12	12.31	16.15	17.70	18.64	6.30	8.66	0.69	1.31	0.87	0.95	1
MB75-12	12.21	15.94	17.32	18.64	6.00	8.66	0.73	1.31	0.86	0.93	1
MB100-12	11.25	14.40	16.64	18.64	5.55	8.66	0.64	1.28	0.77	0.89	1
MB50-19	17.25	21.24	21.60	18.64	7.00	8.66	0.81	1.23	1.14	1.16	3
MB100-19	15.21	19.25	20.80	18.64	5.91	8.66	0.68	1.27	1.03	1.12	2
MC50-16	6.41	8.84	10.35	11.51	3.59	5.25	0.68	1.38	0.77	0.90	1
MC50-16R	9.14	11.90	12.87	11.51	3.70	5.25	0.70	1.30	1.03	1.12	2
MC100-16	6.28	8.57	12.86	11.51	3.11	5.25	0.59	1.36	0.74	1.12	1
MC100-16R	9.64	11.29	13.99	11.51	3.40	5.25	0.65	1.17	0.98	1.22	3
MD100-16	13.68	19.55	21.06	14.47	6.95	9.38	0.74	1.43	1.35	1.46	3
MD100-16R	18.85	20.96	21.16	14.47	6.72	9.38	0.72	1.11	1.45	1.46	3
BMC100-16	6.97	8.60	9.34	10.24	2.59	5.02	0.52	1.23	0.84	0.91	1
BMC100-16R	8.24	10.05	11.42	10.24	2.73	5.02	0.54	1.22	0.98	1.12	2
BMC100-22	8.36	10.40	10.55*	10.24	3.35	5.02	0.67	1.24	1.02	1.03	2
BMA100-16	6.70	7.95	8.18*	10.24	2.28	4.85	0.47	1.19	0.78	0.80	1
BMD100-16	12.29	15.88	16.69*	15.45	6.38	8.93	0.71	1.29	1.03	1.08	3
BMD100-19	13.46	17.29	18.36	15.45	6.64	8.93	0.74	1.28	1.12	1.19	3
BMB100-16	11.07	14.16	15.41	15.45	5.28	8.41	0.63	1.28	0.92	1.00	1
BME100-16	8.17	9.83	10.13*	10.24	3.16	4.33	0.73	1.20	0.96	0.99	1

* : Poor welded
Py : Yeild strength
Pu : Ultimate strength
Pmax : Maximum strength
Pcp : Calculated ultimate strength of the upper column
Ko : Initial rigidity
Kco : Calculated initial rigidity

Table 4. Comparison between test and calculated strength for Model type failure

Specimen	Py (tf)	Pcu (tf)	Py/Pcu
MA50-12	4.13	3.56	1.16
MA75-12	3.75	4.11	0.91
MA100-12	3.63	4.26	0.85
MA50-16	5.81	5.16	1.13
MA100-16	5.21	6.05	0.86
MA50-22	9.15	8.29	1.10
MA100-22	8.25	9.56	0.86
MB50-12	12.31	10.12	1.22
MB75-12	12.21	12.60	0.97
MB100-12	11.25	11.98	0.94
MC50-16	6.41	6.03	1.06
MC100-16	6.28	7.10	0.88
BMC100-16	6.97	5.76	1.21
BMB100-16	11.07	11.27	0.98
BME100-16	8.17	7.58	1.08

26 STRENGTH OF BEAM-TO-COLUMN CONNECTIONS WITH DIAPHRAGMS IN COLUMNS CONSTRUCTED OF UNEQUAL SQUARE HOLLOW SECTIONS

N. YAMADA, K. FUJISAWA, M. NAGAYASU
Kawasaki Steel Corporation, Tokyo, Japan
Y. TANUMA, K. HASHIMOTO
Hokkaido Institute of Technology, Japan
K. MORITA
Chiba University, Japan

Abstract
A connection with an upper-column diaphragm of increased thickness was proposed for connections of upper and lower columns with unequal outer dimensions. Compression, bending, and subassemblage tests were conducted. It is found that the yield strength of such connections can be estimated with good accuracy using yield line theory, and that connections with structural performance equivalent to that in the conventional method can be obtained even under combined axial loading and bending if the thickness of the diaphragm is such that its strength exceeds the nominal yield strength of the upper column.
Keywords: Diaphragm, Yield Line Theory, Beam-to-Column Connection.

1 Introduction

Multi-story steel structures are designed with upper columns of decreasing size relative to that of the lower columns. Two types of detail are conceivable for beam-to-column connections of H-shape beams and square hollow section columns in this type of construction. In the method commonly used in Japan, the connection takes the form of a truncated pyramid (termed the "pyramid type" below). In the second method, called the butt plate connection, the upper column is welded directly to the diaphragm of the connection (Fig. 1). The more economical of the two methods is the butt plate connection with a heavy diaphragm, but the structural performance of this connection has not been clarified yet. For this reason, this paper discusses a method of evaluating the yield strength of butt plate connections and introduces a method of determining the thickness of the butt-plate diaphragm, based on an experimental confirmation of structural performance.

Tubular Structures V. Edited by M.G. Coutie and G. Davies.
© 1993 E & FN Spon, 2–6 Boundary Row, London SE1 8HN. ISBN 0 419 18770 7.

2 Theory

2.1 Compression strength and bending strength
In butt plate connections, three types of joining the
upper and lower columns are possible, as shown in Fig. 1.
Collapse mechanisms were assumed for each of the three
types as illustrated in Fig. 2. The yield line
corresponding to the outer dimensions of the lower column
was formed on the lower column, while the other yield
lines were formed on the diaphragm. External and internal
virtual works for each of the collaspe mechanisms are
shown in Table 1. The compression strength and bending
strength were then obtained using yield line theory.

2.2 M-N interaction
In beam-to-column connections, bending moment acts on the
connection simultaneously with axial loading.

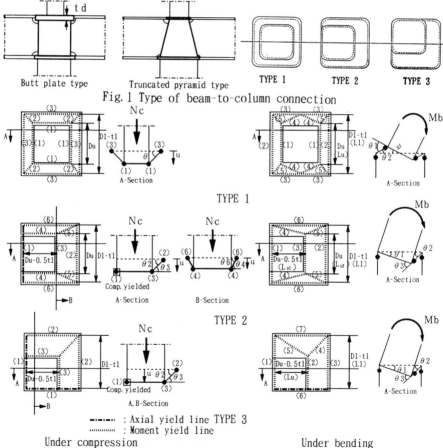

Fig. 1 Type of beam-to-column connection

Fig. 2 Yield line mechanism

Table. 1 External and internal virtual work

Collaspe mechanism	External work		Internal work
Comp.	P_c u	TYPE 1	$4(m_b(L_u \theta_1 + L_2 \theta_2)+m_a L_1 \theta_3)$
		TYPE 2	$0.5(L_1+L_u)t_1 \sigma_{1y} u + m_a L_1(\theta_2 +2\theta_6)$ $+m_b(L_{u2} \theta_3 +2L_{u1} \theta_4 +2L_5 \theta_5)$
		TYPE 3	$(L_1+L_u)t_1 \sigma_{1y} u +2m_a L_1 \theta_2 + m_b(2L_u \theta_3 + L_4 \theta_4)$
Bend.	$M_b \phi$	TYPE 1	$2m_a L_1(\theta_2 + \theta_3)+2m_b(L_u \theta_1 +2(L_4 \theta_4 + L_5 \theta_5))$
		TYPE 2	$m_a L_1(\theta_1 + \theta_2)+m_b(L_u \theta_3 +2(L_4 \theta_4 + L_5 \theta_5))$ $+2m_a' L_1 \theta_6$
		TYPE 3	$m_a L_1(\theta_1 + \theta_2)+m_a' L_1 \theta_7 + m_b(L_u \theta_3 + L_4 \theta_4 + L_5 \theta_5)$ $+0.5 L_1 t_1 \sigma_{1y}$

$$m_a=\frac{t_1^2 \sigma_{1y}}{4} - \frac{1}{4\sigma_{1y}}\left(\frac{P_c}{4L_1}\right)^2 \qquad \text{(Lower column under compression)}$$

$$m_a=\frac{t_1^2 \sigma_{1y}}{4} - \frac{1}{4\sigma_{1y}}\left(\frac{3M_b}{4L_1^2}\right)^2 \qquad \text{(Lower column flange under bending)}$$

$$m_a' =\frac{t_1^2 \sigma_{1y}}{4} - \frac{1}{4\sigma_{1y}}\left(\frac{3M_b}{4\sqrt{3} L_1^2}\right)^2 \qquad \text{(Lower column web under bending)}$$

$$m_b=\frac{t_d^2 \sigma_{dy}}{4}$$

A bending (M)/axial (N) interaction relationship for the connections is therefore introduced in order to evaluate the strength of connections subjected to combined bending and axial loads.

TYPE 1
As shown in Fig. 3, three modes of collapse mechanism were assumed, corresponding to the degree of axial loading acting on the connection. External and internal virtual works were calculated for each of the collapse mechanisms. The M-N interaction of each connection was obtained in terms of K on the basis of the yield line theory. When the value of K in each of the modes is selected so as to minimize the M-N interaction, the M-N interaction relationship can be expressed as follows.

CASE 1
$$\frac{2}{L_u}\left(M-\frac{4L_1 L_u}{L_1-L_u}m_a-\frac{2L_1(L_1+L_u)}{L_1-L_u}m_b\right)+\left(N-\frac{4L_u}{L_1-L_u}m_b\right)=0$$

CASE 2
$$\frac{2}{L_1}\left\{M-\frac{2(L_1^2-L_1 L_u+2L_u^2)}{L_1-L_u}m_b\right\}+\left\{N-\left(\frac{8L_1}{L_1-L_u}m_a+\frac{8L_1-4L_u}{L_1-L_u}m_b\right)\right\}=0$$

CASE 3
$$N-\frac{8L_1}{L_1-L_u}(m_a+m_b)=0$$

Where,

> m_a:Full plastic moment per unit length of the yield
> line formed on the outer dimensions of the lower
> column
> m_b:Full plastic moment per unit length formed on the
> diaphragm

TYPE 2 and 3
The M-N interactions for Types 2 and 3 are presumed to be
a linear relationship between the yield strength for
collapse due to compressive loading only (0, N_c), and the
yield strength for collapse due to bending moment only (0,
M_b).

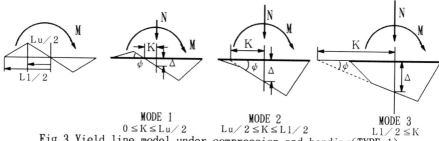

MODE 1   MODE 2   MODE 3
$0 \leq K \leq L_u / 2$ $L_u / 2 \leq K \leq L_1 / 2$ $L_1 / 2 \leq K$
Fig. 3 Yield line model under compression and bending(TYPE 1)

3 Test programme and results

3.1 Test programme
To verify the strength equations, compression, bending,
and subassemblage tests were performed with model beam-to-
column connections. The difference in the outer
dimensions of the upper and lower columns is 150 mm. The
experimental variables were selected in order to
investigate the effect on connection strength of: (1) the
thickness of the diaphragm, ($t_d=16,32,50$ mm) (2) the type
of offsetting the column axes, and (3) the axial loading.
    In addition, one pyramid type test model was fabri-
cated for the subassemblage test, along with the six butt
plate type specimens, so as to confirm that the butt plate
connection offers the same structural performance as the
pyramid type. The six butt plate specimens comprised upper
and lower columns which differed in outer dimensions by
150 mm, in combination with a diaphragm either 16 mm or 50
mm in thickness for each of the three connection types described in Section 2.1. In the compression and bending
tests, monotonic loading was applied until collapse of the
test specimen. In the subassemblage tests, a fixed axial
load was given to the column, while cyclic reversal loading was applied to both beam ends. The configuration of
the various specimens is shown in Fig.4.

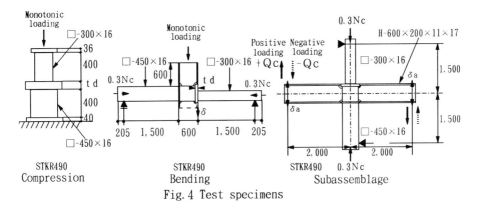

Fig. 4 Test specimens

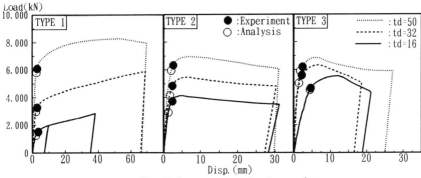

Fig. 5 Compression test results

3.2 Compression test

Fig. 5 shows the load/deformation relationship. The diaphragm deformation (u) illustrated in Fig. 2 was measured by using a displacement gauge located at the diaphragm immediately below the upper column.

The analytic results obtained from the yield line theory are indicated by open circles. The black circles show the load at the point when tangential rigidity reached one-third of its initial value, which was defined as the experimental yield load. It is possible to estimate experimental results with good accuracy by using compression strength values for the beam-to-column connection obtained by means of yield line theory.

3.3 Bending test

Fig. 6 shows the relationship between the bending moment at the diaphragm position and the rotational angle of the upper column (δ/1500) for the specimens. The bending strength of the beam- to-column connection according to yield line theory gave an accurate indication of the experimental results.

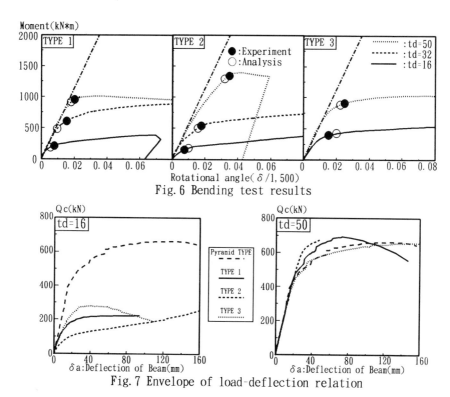

Fig. 6 Bending test results

Fig. 7 Envelope of load-deflection relation

3.4 Subassemblage test

The relationship between storey shear force acting on the upper column and the amount of accumulated deformation in the entire test specimen, as measured at the beam ends, is shown by the envelope curve in Fig.7.

When the 16mm diaphragm is used, structural performance equal to that of the pyramid type connection cannot be found in butt plate test specimens which are designed to collapse at the connection. This is because the diaphragm yields at an early point, after which only the diaphragm deforms. In contrast, with the 50 mm diaphragm, final failure in all the butt plate connection specimens took the form of local buckling of the upper column.

4 Discussion

4.1 M-N interaction

Fig. 8 presents a comparison of the M-N interaction for the connections obtained from the yield line theory and the experimental values. The figure also shows the actual M-N interaction for the upper column as calculated from material test results. The experimental results enclosed in circles are from the subassemblage test.

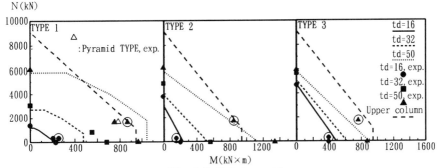

Fig. 8 M-N interaction

The experimental values for all the connection types and diaphragm thicknesses were estimated with good accuracy from the M-N interaction obtained analytically. Moreover, if the diaphragm thickness is selected so that the strength of the diaphragm exceeds the nominal yield strength of the upper column, it is possible to design a connection with the yield strength higher than that of the upper column.

4.2 Flexural rigidity

As shown in Fig. 9, the butt plate bending test specimen was divided into two models, and flexural regidity (K_b) was calculated by applying the parallel spring model of rigidity K1 of Model 1 and rigidity K2 of Model 2. Rigidity K2, which results from the deformation of the diaphragm, was calculated for a diaphram element fixed at both flanges of the lower column, with a concentrated load acting on the point at which a unit length section of the diaphragm is attached to the upper column. The calculation model for K2 is shown in Fig. 10. The calculation for the pyramid type connection assumes that flexural rigidity (K_p) is given by the arithmetical mean of the upper and lower columns.

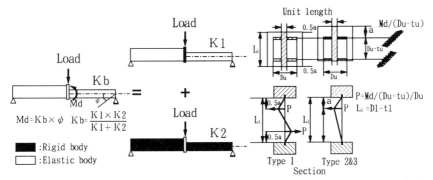

Fig. 9 Flexural rigidity model   Fig. 10 Calculation model for K2

Table 2 shows the comparative results of a flexural rigidity of bending tests. The analytic values provided an accurate estimate of the flexural rigidity of butt plate specimens whose strength exceeds the nominal yield strength of the upper column. The predictive values obtained with the flexural rigidity models of the butt plate and pyramid type connections were substantially equal.

Table. 2 Flexural rigidity of bending tests

	The thickness of diaphragm	Butt plate type Exp. (Ke)	Butt plate type Cal. (Kb)	Pyramid type Cal. (Kp)	Ke/Kb	Kb/Kp
TYPE 1	16	4.33E+05	5.10E+04	7.27E+05		
	32	5.50E+05	2.91E+05		1.89	0.40
	50	5.69E+05	5.79E+05		0.98	0.80
TYPE 2	16	2.79E+05	5.18E+04	7.27E+05		
	32	5.03E+05	2.94E+05		1.71	0.40
	50	5.05E+05	5.79E+05		0.87	0.80
TYPE 3	16	4.06E+05	5.18E+04	7.27E+05		
	50	6.30E+05	5.79E+05		1.09	0.80

5   Conclusion

Beam-to-column connections of the butt plate type were evaluated for the yield strength and rigidity, with the following conclusions.

(1)   The yield strength of test specimens subjected to pure compression, pure bending, and combined bending and axial loading can be estimated with good accuracy using yield line theory.

(2)   An evaluation of the rigidity of test specimens of butt plate and pyramid type beam-to-column connections gives an accurate estimate of the flexural rigidity of test specimens whose yield strength exceeds the nominal yield strength of the upper column.

(3)   In butt plate connections, structural performance equal to that of pyramid type connections can be secured if the thickness of the diaphragm plate is set so that the strength of the beam-to-column connection exceeds the nominal yield strength of the upper column.

6   References

Architectural Institute of Japan (1990) **Recommendation for the Design and Fabrication of Tubular Structures in Steel**

G. Davies and E. Panjehshahi (1984) Tee joints in rectangular hollow sections (RHS) under Combined Axial Loading and Bending, **7th International Symposium on Steel Structures**, Gdansk, Poland, P61~68

PART 8

ECONOMIC APPRAISAL AND DESIGN

Optimisation and expert systems

27 MINIMUM COST NODAL CONNECTIONS FOR THE CUBIC SPACE FRAME

L.A. KUBIK
ASW-CUBIC Structures Ltd, Westwood Business Park, Coventry, UK
S. SHANMUGANATHAN
Department of Civil and Structural Engineering,
The Nottingham Trent University, UK

Abstract
The fully welded nodal connections in the CUBIC Space Frame are required to transfer bending moment, shear and axial forces between the chords and the vertical posts. A number of connection details have been developed and used successfully in practice. Both strength and stiffness criteria must be satisfied, but it is also important that the connection details can be manufactured cost effectively. After introducing a number of refined yield line mechanisms for determining stiffener sizes, consideration is given to the factors which influence cost. Initial cost comparisons are then included for a commonly used stiffener detail.
Keywords: Connections, Cost, Optimization, Space frame.

1 Introduction

The CUBIC Space Frame is a modular, 3-dimensional structure, in which the members are required to resist combinations of shear force, bending moment and axial force to support the applied loads [see Kubik and Kubik (1991), Kubik (1991) and Kubik (1993)]. It comprises upper and lower grids of chords connected together by vertical posts (Fig 1); there are no diagonal bracing members.

In a typical structure universal beams and columns are used for the top and bottom chords and structural hollow sections are used for the connecting posts. Modular construction is achieved by cutting the chord grids mid-way between the posts and introducing site bolted splices.

In May 1991 a programme of research into optimization of the CUBIC Space Frame was started at Nottingham Trent University, with minimum cost as the ultimate objective [Shanmuganathan and Kubik (1993)]. The number and types of connection details used have a major influence on the overall cost of the structure. Therefore, a sep-

Tubular Structures V. Edited by M.G. Coutie and G. Davies.
© 1993 E & FN Spon, 2–6 Boundary Row, London SE1 8HN. ISBN 0 419 18770 7.

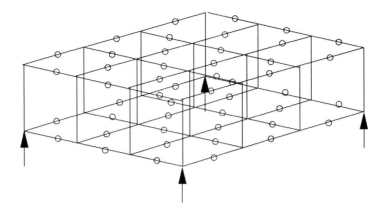

Fig. 1. CUBIC Space Frame showing splices at mid-chords

arate study was initiated to (a) identify the possible connection details for the CUBIC Space Frame and (b) investigate the factors that influence the cost of these connection details. This paper describes the initial work done in the cost study for the nodal connections.

A variety of joint details are being considered in the study, although for this paper attention will be focused on the detail shown in Fig 2.

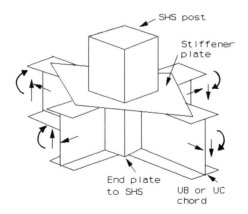

Fig. 2. Nodal joint detail for a diamond stiffener plate

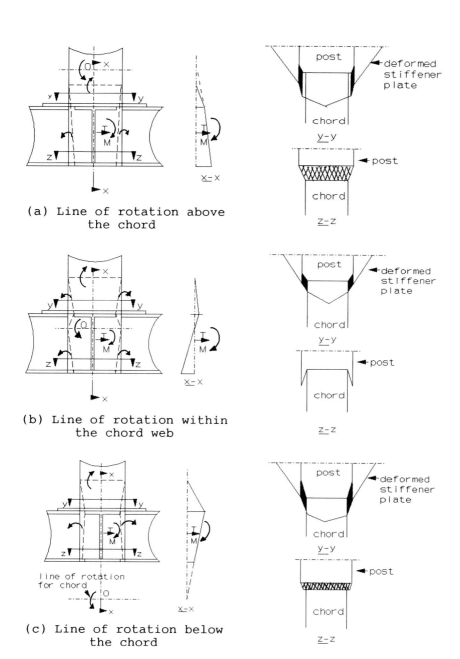

(a) Line of rotation above
the chord

(b) Line of rotation within
the chord web

(c) Line of rotation below
the chord

Fig. 3. General forms of yield line pattern used in
joint analysis

2 Failure mechanisms

The factors which influence the design of CUBIC Space
Frame nodal joints were discussed in a previous paper by
Kubik and McConnel (1989) and a number of yield line
patterns were identified. These yield line patterns
enabled joint capacities to be determined rapidly for a
variety of configurations and incident forces, permitting simple interaction diagrams to be produced.

Since then, more general forms of the yield line patterns have been developed, as shown in Fig 3, which
reflect the gradual transition from one yield line mechanism to the next depending upon the location of the
line of rotation. These more general yield line patterns
result in a slight rounding of the corners of the interaction diagram for a typical joint detail, as shown in
Fig 4. The present study uses these general forms of
yield line pattern.

Appropriate computer software has been developed to
simplify the design process. This software also includes
automatic checks on the joint capacity due to interactions between the forces and moments at the node from
each of the incident members and a check on biaxial
stresses in the end and stiffener plates.

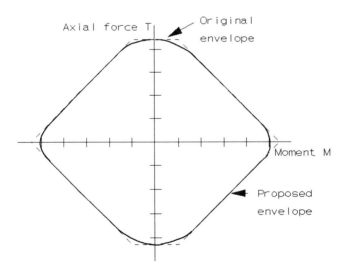

Fig. 4. Interaction diagram for a typical joint detail

3 Factors influencing cost

The components for a typical nodal joint are shown in Fig 2, and comprise up to 4 chords, an SHS post, an end plate (or internal stiffener to the SHS) and a stiffener plate (or external stiffener to the SHS). For given chords and post, the end plate size is dependent upon the post size and thickness and the chord flange thickness and width. The main variables which therefore influence the nodal fabrication cost are the stiffener plate size and thickness, and the associated cutting, handling and welding.

Typically the cost of the cutting and welding process is related to the time involved multiplied by an hourly rate [Firkins and Hemphill (1991), Technical General Secretariat (1981) and Kenyon (1982)]. Handling costs are allowed for by increasing the basic process times by a multiplier dependent on the fabrication process. Material costs are related to the weight of material, including an appropriate allowance for wastage, multiplied by the cost per tonne.

4 Initial cost comparisons

Preliminary cost studies have been undertaken using three different stiffener sizes and thicknesses for the joint detail shown in Fig 2. For these three cases 203x133x25kg/m UB chords have been used and 200x200x6.3 SHS post with similar applied forces, assuming grade Fe510 steel. The relative costs derived using published data [Firkins and Hemphill (1991), Technical General Secretariat (1981) and Kenyon (1982)] are summarised in Table 1. It should be appreciated that a considerable amount of additional work remains to be done for a variety of joint types and details, with checks on the sensitivity to the assumed fabrication times and costs, before any final conclusions can be drawn.

Table 1. Cost ratio for different stiffener sizes

Item	Stiffener size (mm)		
	382x382x10	354x354x15	332x332x20
Material	1.00	1.27	1.48
Cutting	1.00	1.07	1.14
Welding	1.00	0.90	1.46
Combined cost	1.00	1.06	1.40

From Table 1 it can be seen that the material cost increases as the plate thickness increases, whereas the cutting cost is little affected. However, the welding cost decreases initially as the plate thickness increases from 10 to 15mm, and then increases as the plate thickness increases from 15mm to 20mm. This apparent anomaly is simply due to the decision to adopt discrete weld sizes as used in practice. As a result the same weld size is necessary for both the 10mm and 15mm thick stiffeners, whereas a larger weld size is required for the 20mm thick stiffener. The stress level in the weld for the 10mm thick stiffener is lower than that in the weld for the 15mm thick stiffener, but not sufficiently low to permit a smaller weld size. Consideration could be given to the benefits of reducing the weld length and therefore increasing the weld stress by adopting a truncated stiffener detail, such as that shown in Fig 5.

Looking at combined costs for material, cutting and welding, it is apparent that there is only a small cost difference between the 10mm and 15mm thick stiffener, with the 10mm having the lowest cost for the assumed cost parameters. The 20mm thick stiffener is, however, significantly more expensive due to the increased material and welding costs.

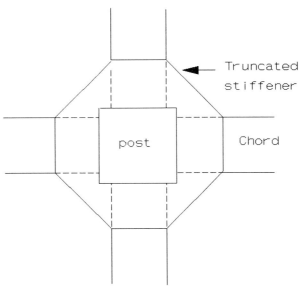

Fig. 5. Truncated stiffener for the nodal joint

5 Conclusions

A number of revised yield line mechanisms have been described for CUBIC Space Frame joints which refine those previously proposed. These mechanisms have been used in the software which has been developed as part of a cost optimization study to identify and then produce preliminary costs for a range of acceptable joint details. For the diamond shaped nodal stiffener it appears that thinner stiffeners are more cost effective than thicker stiffeners, subject to fully utilising the available weld strength. In this context advantage may be gained from using truncated diamond stiffeners (Fig 5).

6 Acknowledgements

The research into optimization of CUBIC Space Frame structures is jointly funded by ASW-CUBIC Structures Ltd and Nottingham Trent University, to whom the authors are grateful.

7 References

Firkins, A and Hemphill, D. (1991) Fabrication cost of structural steelwork, **Steel Construction Today,** May, 117-124.

Kenyon, W (1982) The cost of Fabrication, **Welding and fabrication technology,** chapter 25, 200-206.

Kubik, L.A. and McConnel, R.E. (1989) The behaviour of CUBIC Space Frame joints, **Proceedings of the International Symposium on Tubular Structures** (edited by Niemi, E.), Lappenranta, Finland, 410-417.

Kubik, M.L. & Kubik L.A. (1991) An introduction to the CUBIC Space Frame, **International Journal of Space Structures,** Issue 6, 1.

Kubik, L.A. (1991) The CUBIC Space Frame in theory and in practice, **Steel Construction Today,** March, 59-63.

Kubik, L.A. (1993) The first 10 years of the CUBIC Space Frame, paper submitted for **the Fourth International Conference on Space Structures,** Sept 6-10.

Shanmuganathan, S. and Kubik, L.A. (1993) Optimization of CUBIC Space Frame Structures, paper submitted for **the Fourth International Conference on Space Structures,** Sept 6-10.

Technical General Secretariat (1981) **European Recommendations for Steel Construction,** 211-214.

28 EXPERT SYSTEM FOR THE DESIGN AND FABRICATION OF TUBULAR STRUCTURES

M. HEINISUO
Tampere University of Technology, Finland
M. LAURILA
KPM-Engineering Ltd, Tampere, Finland

Abstract
The paper deals with the first stage of an expert system
for steel structures. A link between a CAD-system and a
database-program is developed. The formulation is done so
that the program can be used by fabricators and designers
to estimate the masses and costs of the new project. The
completed or preliminary designs are stored into the
database and can be processed by the system to utilise all
the possibilities that the used general database program
can offer. The system can be used for the plane frame
structures that involve tubular trusses, beams and
columns.
Keywords: Knowledge base, Expert system, Design, Lattice
girders, Hollow sections

1 Introduction

CAD-systems developed recently can be used very
efficiently for the preliminary design of the project. The
instigator of the project is usually someone apart from
the designer (usually the fabricator of the structure,
contractor) who has the knowledge of the structure that
the designer may not have. These are e.g. the prices of
the materials of the day, stories available,
transportation and erection costs. On the other hand if
the designer has worked for some years and has used the
same program all the time then he will have many completed
or preliminary designs stored in the memory of the
computer. One way to use the database-program is to use it
as a library. This is very important when the number of
completed designs is large.
   One idea in the present study was to combine the
databases of the contractor and the designer. The database
of the contractor in the present version is very modest.
It was decided that only that the unit price list would be
included in the database and it could be changed by the
user of the system. It is well-known for example that the
erection prices are determined by the contractors in many

Tubular Structures V. Edited by M.G. Coutie and G. Davies.
© 1993 E & FN Spon, 2–6 Boundary Row, London SE1 8HN. ISBN 0 419 18770 7.

different ways. Hence, the various combinations offered by the general database program are available to the user.

Another aim of the study was to try to select the essential parameters of the problem. The need was to find out some ways to estimate the price of the new project by using the completed works stored on the database. One important aspect is to use the graphics of the structure so that quite wrong decisions can be avoided. The authors had learnt this fact from a previous project completed for wooden nail-plate trusses. Some other ideas have also been derived from that project.

No similar system was found from the literature. The contractors have their own cost-estimation programs that are usually based on the unit prices for specific building systems, masses of the buildings etc. It is believed that this kind of expert system for steel frames can serve as a part of the cost-estimation for new project.

2 The CAD-system used in the study

The CAD-program is described in detail in other papers bu Heinisuo et al (1991) and Heinisuo (1992). The designer creates first the geometry of the frame. From the geometric model the designer can output the drawings and other documents. From the CAD-program links have been built to the cutting machines by Lehtinen (1991) and to Auto-Cad. The next stage of the project may be the data transfer program from the CAD-program to the final database of the contractor. This has not yet been able to do because the databases of the contractors in Finland are not available.

The calculation model is produced automatically from the geometric model. The calculation model is a so called 2,5-dimensional elastic model. The strength of bars is checked according to the Finnish codes of practise by Teräsrakenteet, RIL (1988).

The program also checks the capacity of the joints. The capacity check is in most cases done by using the plastic theory. Also the second order effects are taken into account approximately in the joints which can increase the second order effect (see Treiberg (1987)). This is done by using the result of the eigenvalue problem of the 2D-analysis. The design of joints in tubular trusses is programmed accord- ing to the equations of Wardenier by Wardenier (1982) and the joints of beams and columns follow mainly Ref. by Treiberg (1987). The column base design is performed by using the Swedish regulations by Zoetemeijer (1983).

When all the design criteria are accepted then the design is performed. This holds especially for the final design. In the preliminary stage of the project the time

of the design is often limited to some hours (or even
minutes today) and then all the stages of the design
cannot be carried out although the program is very quick
and easy to use.

At this stage of the project the 2D calculation model
was chosen as the source of the database of the expert
system. There was one main reason for the use of this
model. The same 2D calculation process has been used for a
long time for the design purposes of steel frames so that
there is enough data for the database.

3 The link between the CAD-program and the expert system

The masses of the members are concerned in the following.
At this stage of the project the fittings of the structure
(usually 5-10 % of the whole mass of the project) are not
considered. Also the painting areas of these parts are
neglected.

A general purpose program PARADOX is used in this
study. It is chosen because it is easy to use and
available for micro-computers. When building the link
between the database for PARADOX and the CAD-program it
must be bourne in mind that the database-program must be
easy to use. So, some of the work done by the final user
of the program must be done beforehand using the transfer
program.

It must be noted also, that all the designs in the
database are not completed projects. Some designs are made
only for the order stage of the project, or they are some
studies where the design is not totally finished. In spite
of this these designs are left in the database in order to
get a wider sample for cost estimation.

The four different objects which are present are:
columns or beams, top and lower chords and bracings of the
trusses. The other data transferred from the CAD-program
to the database (most of the following parameters are
calculated from the geometric data transferred) can be
seen in Figure 1.

The data presented in Figure 1 are in the form view.
The same data can also be presented in the table view and
one can work also in that form. The main parameters that
were tested are shown in Figure 1. Most of the parameters
are explained by the names of the parameters. It was
realized recently, that all the parameters were not needed
and that some of them are only informative. When the first
outputs of the system were considered it was found that
some more partitioning of the structures was needed.

```
Project:              **** TR-MIKRO *****           Date  2.01.91  Record  21
AURAJOKI OY / SALON TEHDAS
─────────────────────────────────────────────────────────────────────────────
Type F2 Strength 340 Mpa    Column      3   ┌Directory and filenames ─────────
Length          37560 mm    Topchord    2   │102_PPTH          /TARJOUS
Lowerc.length   18405 mm    Lower       2   │07_AURAJOKI
Span            18675 mm    Other      34   │
Height           8750 mm    Beam        0   │
Free height      6990 mm    Sum        41   │
Left 7650 Right 7650 mm     Element   100   │
Topchord slope  3.35 1:17.1 Joint      61   │
Lower            0   1:0    Support     3   └──────── Amount:── /unit ────────
┌Loading:─────────────────────────────────  │
│Distance         6000   mm    Loadcase  5   │Steel   3047 kg  5.00    15235
│Liveload         1.6    kN/m2  combi    4   │Paint     83 m2  1.00       83
│Deadload Top     .5     Materfactor 1.5    │TOTAL COST               15318
│         Lower   .2     Loadfactor  1      │
│Windload         .6     Displ 0    T  0 °  │Floor area        225  m2
│                    ─X──────────Y─         │Volume           1575  m3
│Baseload sum     33          518           │
│Real reaction    33          508           └
─────────────────────────────────────────────────── Average 2470kg 68.6m2
```

Fig.1. Data transferred from the CAD-system.

It was decided that the following designation would be used for the various types of frames:
A: archs, B: beams, C: columns,
F: frames (only beams and columns),
F1:   one-bayed frames (columns and trusses),
F2:   two-bayed frames, Fm: multi-bayed frames
O: others, M: multistory frames, T: trusses
The most important parameters are the graphics of the designs. The different loading cases are also transferred in graphical form to the database. Some of the aforementioned parameters are shown in the Figures 2 and 3 which display two examples in the graphical form.

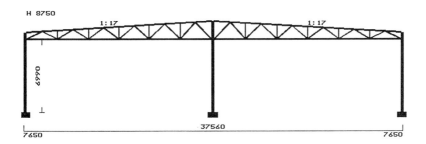

07_AURAJOKI   6000c/c live 1.6 dead .50   3047kg   83m2

Fig.2. Graphical form of the frame transferred.

285

Loadcase 1/5

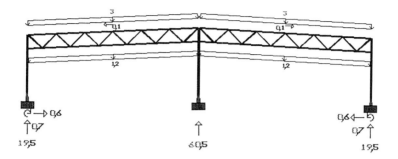

Fig. 3. An example of the loading data transferred in the graphical form.

4 The database program

The database program is written by using PAL, the Paradox Application Language. All the facilities of Paradox are available. This environment seemed to be good enough for this problem. One can give different selection criteria etc. In the present study only the price list (Figure 4) is from the database of the contractor. This price list can be changed by the user or new ones can be created easily. As seen from Figure 4 the price list is very simplified. All the other data are transferred from the CAD-program.

PROFIL	MATERIAL			Fabrication	Transportation		
	Fe360	Fe450	Fe510		Painting		Erection
	/kg	/kg	/kg	/kg	/m2	/kg	/kg
SHS (RHS) t <=6.3mm							
truss	3.20	3.20	4.00	.50	1.00	.10	1.00
beam/column	3.30	3.30	4.10	1.00	1.00	.20	1.50
SHS (RHS) t > 6.3mm							
truss	3.90	3.90 ◄	4.80	.50	1.00	.10	1.00
beam/column	4.00	4.00	4.00	1.00	1.00	.20	1.50
HOT ROLLED H<=300	3.16	3.16	3.80	1.00	1.00	.20	1.50
HOT ROLLED H> 300	3.40	3.40	3.80	1.00	1.00	.20	1.50
WELDED	2.50	2.50	2.67	1.50	1.00	.20	1.50

UNIT   PRICES

Pricelist   1/3
RAUTARUUKKI

AVERAGE   MATERIAL+WORK   5.00 /kg   PAINTING   1.00 /m2

Fig.4. Price list of the expert system.

5 Examples

As an example Figure 5 shows the consumption of the material of one database (1000 designs, kg <= 5000). It can be seen that there is no correlation in this space.

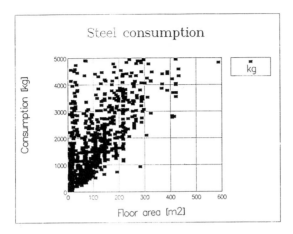

Fig.5. Consumption of the materials of the database (1000 designs).

The sample can be limited e.g. by collecting only two-bayed frames from the whole sample. By using moreover the selection criteria FreeH <= 3000 and Height <= 10000 the result can be seen in Figure 6.

Figures 7 and 8 show the painting areas of the same sample as used in Figure 6. and the price list used is shown in Figure 4.

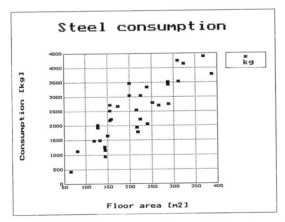

Fig.6. Material consumption of two-bayed frames (36 samples).

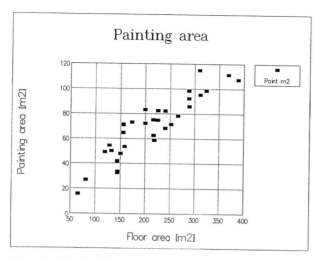

Fig.7. Painting area of the two-bayed frame
(36 samples).

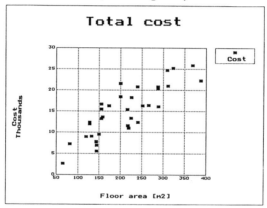

Fig.8. Material cost of the two-bayed frame
(36 samples).

6 Conclusions

An expert system can be used to estimate at least the
lower bound of the materials and prices very generally. By
using the selection criteria of the general database
program one can estimate the price of a new project very
accurately if the database is wide enough. The same
conclusion can be drawn from the other cases calculated by
the system.

The most promising fields to enlarge the system are to
take into account the fittings, bolts and welds. Also, the
3D handling of the whole structure is important. The
general database of the others parties than the designer
must be created.

7 Acknowledgments

The financial support of Rautaruukki Oy is gratefully acknowledged.

8 References

Heinisuo M., Möttönen A., Paloniemi T., Nevalainen P. (1991) **Automatic design of steel frames in a cad-system**, Proceedings of the fourth Finnish Mechanics Days, Lappeenranta University of Technology.

Heinisuo M. (1992) **Teräsrakennerungon cad-suunnitteluohjelmisto**, The Finnish Civil Engineering & Construction Journal 2/92 (in Finnish).

Lehtinen H. (1991) **Teräsprofiilin geometriatietojen siirto katkaisulaitteelle**, Diploma thesis, Tampere University of Technology, (in Finnish).

Teräsrakenteet, RIL 173-1988 (in Finnish).

Treiberg T. (1987) **Pelarfot**, Stålbyggnadsinstitutet In Swedish).

Wardenier J. (1982) **Hollow section joints**, Delft University Press.

Zoetemeijer If. P. (1983) **Sammenvatting van het onderzoek op geboute balkkolom verbindigen**, Technische Hogeschool Delft (in Dutch).

29 ECONOMIC APPRAISAL OF TUBULAR TRUSS DESIGN

W.M.K. TIZANI, G. DAVIES and D.A. NETHERCOT
Department of Civil Engineering, University of Nottingham, UK
T.J. McCARTHY and N.J. SMITH
Department of Civil and Structural Engineering, UMIST, UK

Abstract
The paper reports on a study aimed at linking structural design with economic
appraisal for tubular truss construction using knowledge-based systems' techniques.
It uses accepted methods for design of members and joints in association with a
simulation of the fabrication and construction. Use of this approach enables
designers at the concept stage to accurately assess the relative worth of a number
of structurally adequate though different designs and to select the most appropriate
on objective criteria based on the practicality of actually producing the final
product.
Keywords: Economic Design, Tubular Trusses, Knowledge-based Systems,
Fabrication Led Design, Construction Led Design.

1 Introduction

The design of tubular trusses involves the selection of a framing system, the design
of member sections and the design of connections. The framing systems and the
member sections are normally selected by consultants who often also give an
indication of the types of connections to be used at the joints. The design of these
connections is usually carried out by the fabricators.

Typically, the main measure taken by consultants to reduce cost is minimum
material weight. The fabricators will estimate cost of fabrication based on, among
other factors, weight, complexity of detailing, and estimate of handling time.

It was estimated that it is difficult to alter the weight of a structure by more
than ±10%, Girardier (1993). It was also estimated that it is very easy to revise
the connection work by well in excess of ±50%. Thus, there is a much larger
scope for saving in optimizing for connections, especially since 60% of non-
material costs of steel frames are directly dependent on the connections, Nethercot
(1991). In addition, minimizing the material weight could well lead to connection
stiffening which could increase cost by a larger factor than that saved by reducing
basic steel weight, Girardier (1991) and SCI (1992). Therefore, in order to

Tubular Structures V. Edited by M.G. Coutie and G. Davies.
© 1993 E & FN Spon, 2–6 Boundary Row, London SE1 8HN. ISBN 0 419 18770 7.

effectively reduce cost, the designer should consider the effects of design decisions on fabrication cost. In a wider sense, the designer should also consider the effects of these decisions on other factors with the potential to influence costs, such as ease and speed of construction.

This paper outlines an SERC funded research project for the economic appraisal of structural design. It aims at investigating the feasibility of aiding designers, starting from the early stage of preliminary design, to carry out 'fabrication led' and 'construction led' designs. This investigation is being carried out through the development of a knowledge-based system (KBS), or an expert system, for the design of tubular trusses.

The main points that are touched on in this paper are the following:
- Current practice for the design of tubular trusses.
- The potential savings that could be obtained by carrying out fabrication led and construction led designs.
- The description of an integrated design system for carrying out such designs.
- The structure and operation of this system.
- The structural design and the construction led design procedures used by the system.

2 Current Practice in Design

The design process is currently undertaken in two stages. The first stage is carried out by a consultant and the second by a fabricator. At the first stage, the consultant uses past experience and judgement to determine the most appropriate design. The designer searches for design alternatives within a possible solution space bounded by structural safety and practical constraints. Typically, the consultant produces a detailed design that will include the framing system, and the member sections to be used, but with no in depth attention being paid to the detailing of connections.

At the second stage, the fabricator will design the joints taking into account the constraints given by the consultant, which includes the joints types, the sections to be used and the geometry. Some of these constraints might prove unworkable and modifications may have to be made, e.g. changing sections sizes or introducing stiffeners.

The two stages are usually completely separate and no significant provisions are made in the first for alleviating problems in the second. This is partly due to the specialization involved in each stage and due to lack of effective dialogue between consultants and fabricators, AISC (1990). As a consequence of this current practice economy is, in most cases, compromised or not all the potential savings are made.

3 An Integrated KBS for Structural Design

It is clear from current practice that if fabrication and construction factors were fully taken into account at the design stage, savings could be made. However, in order to do so, the design engineer would need extensive experience in both fabrication and construction techniques in addition to design. Alternatively, access to all the extra knowledge and the necessary backup should be readily available.

It is not feasible that all designers would have had such a wide and extensive experience. However, with the advent of artificial intelligence techniques, it is possible to represent and manipulate expert knowledge within a computer system to give the designer access to this extra knowledge with the necessary backup. This research project aims to acquire this expert knowledge from practising designers, fabricators, and constructors and to formulate it into an integrated design system for producing a fabrication and construction led design approach. It is envisaged that such a system would guide designers through the design procedure into producing economical design starting at the conceptual stage.

The system is composed of a number of modules, as can be seen from Figure 1, the various components being inter-linked via the Control Module. All modules are being developed from scratch except for the Analysis and the Member Design Modules, where an interface from within the Control Module will be developed to link the system with existing analysis and design packages. The 'loose' structure of the system has the advantage of keeping it flexible and to allow future modifications to be carried out relatively easily. However, it is also realised that more effort will be required to allow the system to act as a coherent unit.

The main component of the system is the Economic Appraisal Module which constitutes the knowledge-based system part. It comprises of the Inference Engine, three knowledge bases, i.e. Design, Fabrication and Construction Knowledge Bases, and of the Evaluation Module. This module is being developed using GoldWorks III. GoldWorks III is an expert system shell which is used to represent and manipulate the expert knowledge. It was chosen because it offers all major forms of knowledge representations (rules, frames and object oriented) and both forward- and backward-chaining inference mechanisms, in addition to interfaces to C, LISP and dBase, Gold Hill Inc (1991). It has also been successfully used for the development of other KBSs, McCarthy et al (1991).

The problem solving strategy adopted in the Inference Engine is as follows:
- Selecting adequate structural systems which are most feasible for the specified geometrical and other requirements and comply with site constraints.
- Through the interface with the control module, the structure is analysed and the member sections are selected.
- Design of connections is carried out based on the selected member sections.
- The choice of connection details and member sections are then evaluated taking into account economy of fabrication and construction. A cycle of evaluation and modifications is carried out until an optimised solution is obtained.

- If more than one structural system is feasible, the above three steps are repeated for each of them.
- The alternative systems are evaluated and ranked in the order of the most economical.

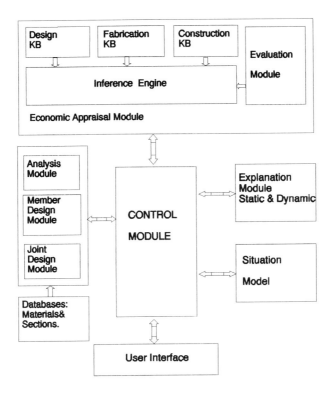

Fig. 1. Schematic Representation of The Integrated Design System

4 Operation of The System

The following stages summarize the operation of the system.

1. The execution starts at the Control Module which requests input from the User Interface.
2. The user will describe the problem at hand via the User Interface. The problem description includes the known structural and constructional constraints. The user could also describe, if known, the fabrication process to be used.

3. The problem description is saved in the Situation Model, and the Design Knowledge Base (included within the Economic Appraisal Module) is consulted for suggestions of structural systems that are suitable for the described problem.
4. The various feasible structural systems are then analysed and the member sections are chosen using the Analysis and Design Modules with input from the material and sections databases. The selection of the member sections is appraised by reconsulting the Design Knowledge Base and if necessary different member sections are chosen based on the criterion of structural efficiency.
5. The Joint Design Module is then used and preliminary joint designs are obtained. The design is then appraised from the fabrication and construction prospectives.
6. The cycle of generating solutions, appraising, and modifying continues until a design is optimized using the Design, Fabrication and Construction Knowledge Bases.
7. The estimated costs of the various feasible structural systems selected in step 3 are then compared, and are ranked starting with the most economical.

In the following three sections, aspects of the operation of the system are expanded on, namely: structural design, fabrication led design, construction led design. This is in addition to the role of the Explanation Model.

5 Procedure of Structural Design

The integrated design system is intended to advise and guide on the structural design of tubular trusses. The system evaluates each design and guides the designer in order to achieve both structural adequacy and efficiency.

Guidance on structural adequacy is given when a proposed solution has failed to fulfil the function of withstanding the loading. In this case advice on remedial actions is given. This could include what aspect(s) or parameter(s) of the design should be modified.

This can be illustrated by an example from the design of tubular joints. If a joint has failed to transmit all forces adequately, the joint could be remedied by either: changing its geometry, e.g. (when applicable) use smaller gap between the bracing members or change the gap to an overlap, or select heavier sections, e.g. chords with larger diameter or using different material, e.g. grade 50 instead of grade 43. The advice given will mainly take into account the cause of the joint failure, e.g. if a K type connection with a gap has failed in punching shear of the chord, the remedial action could be to either increase the ratio of diameter to thickness of the chord and/or increase the ratio of the bracing diameter to the chord diameter, Wardenier (1991).

After achieving structural adequacy, the proposed design solution is evaluated based on the criterion of structural efficiency. In this case advice will be given on whether the geometry and material chosen will result in a system which is structurally efficient, or where necessary the modifications which should be carried out in order to achieve this.

Assuming the joint is structurally adequate, the coefficient of efficiency of the joint is calculated, Wardenier et al (1991), and the choices of the joint geometry, member sections and material are evaluated. Based on this evaluation, advice on modifications that could be made to increase the joint efficiency is given. Examples of such modifications are increasing chord thicknesses and/or decreasing the ratios of brace diameters to chord diameters, etc.

6 Procedure of Fabrication Led Design

As it was mentioned earlier, achieving structural efficiency does not necessarily mean economical design. Therefore, the Inference Engine is used to reason and manipulate the knowledge provided in the Fabrication Knowledge Base in order to select the most economical of the proposed alternative solutions. This applies to the overall structure or to a part of it. The decision on which alternative is the most economical and suited for the job at hand is decided upon with the aid of the Evaluation Module.

The structurally efficient members selected in the structural design part of the procedure described above are then evaluated. If there is a scope for improvements, a more economical arrangement is suggested. For example, the most structurally efficient members would vary in size along the chord length. However, the use of a single section for the whole of the chord may be more economical, i.e. the reduction in fabrication cost is greater than the increase of the cost of the material.

Other examples of possible cost reduction without affecting structural adequacy, are the provision of gap joints instead of overlap ones, British Steel (1991), and providing equal diameters for all internal members while if necessary varying their thicknesses, Tizani et al (1993).

In addition, the system will allow the user to vary the relative importance of parameters used to "judge" the relative worth of different designs. This is necessary in order to be able to model different fabrication practices and to allow for future updates. The importance of this feature was demonstrated by Zayyat et al (1989).

7 Construction Led Design

The Construction led design procedure is to be incorporated within the system at a later stage. It is envisaged that it will follow the same layout and functions of the fabrication led design procedure. The selected structural system is evaluated based on the known site constraints and the preferred detailing for ease and speed of construction.

8 Role of Explanation Module

During a consultation, the Control Module stores the inferences used in the dynamic part of the Explanation Module. The Static part of the Explanation Module helps the user to drive the system.

An explanation facility to provide a rational basis for any advice is an essential part of a KBS and is mainly due to the type of knowledge used in KBSs. The reasoning is dependent on 'shallow knowledge', rules of thumb and heuristics. Some of this knowledge could be subject to changes, dependent on changes to practices in the industry. Also, knowledge bases in KBSs tend to be built incrementally and it is difficult to achieve a complete coverage of all the knowledge in the required area of expertise. It is essential, therefore, for the designer to be aware of the reasoning behind any advice, and to accept it before acting on it. After all, the designer is still responsible for the final design.

This explanation will be provided by the Explanation Module which will retrieve the appropriate knowledge part used for reasoning and display it at the request of the user.

9 Summary and Conclusions

This paper has outlined an integrated design system for the economical design of tubular steel trusses. The system, which is under development, uses artificial intelligence techniques in order to advise and guide designers into using economical structural systems and detailing. The use of these techniques will allow the incorporation of rules of thumb and heuristics within a computerised system.

The aim of the system is to facilitate the practice of carrying out fabrication led and construction led designs. If such a practice is followed, potentially sizeable savings could be realised. This is mainly due to the current practice of separating the selection of the framing systems and member sections, carried out by consultants, and the design of connections and the manufacturing of the frames, carried out by fabricators. The same could be said about the task carried out by consultants and constructors. Also, with the aid of such a system, significant savings may be possible because of the large proportion of the fabrication and construction related costs as compared to material costs.

The system is being developed under the Windows environment using the knowledge-based systems development tool GoldWorks III and the object oriented language C++. It comprises a number of specialised modules (Control, Economic Appraisal, Structural Analysis, Member Design, Joint Design, and Explanation module) and a Situation Model. The Control Module manages the problem-solving strategy and links the different modules together. The Economic Appraisal Module includes an Inference Engine and three knowledge bases, namely: Design, Fabrication and Construction Knowledge Bases, and an Evaluation Module. The function of the Economic Appraisal Module is to generate feasible solutions, evaluate them, and suggest modifications for improving the economy and the structural efficiency of the structure.

The proposed system is still at an early stage of development; therefore the authors would welcome the readers' comments and suggestions.

10 Acknowledgements

The authors would like to acknowledge the financial support of the Science and Engineering Research Council, UK and would also like to thank those who have helped with the interviews used in the knowledge acquisition process.

11 References

AISC (1990) **Economical Structural Steelwork.** Third Edition, Australian Institute of Steel Construction, Australia.

British Steel (1991) **Design of SHS Welded Joints.** British Steel Welded Tubes, UK.

Girardier, E.V. (1991) Customer Led, Construction Led, **Steel Construction,** BCSA, Vol 7, No 1, UK.

Girardier, E.V. (1993) Construction Led - Chapter 4, The Role of Standardised Connections. **New Steel Construction,** February 1993, pp. 16-18.

Gold Hill Inc, (1991) **GoldWorks III,** various manuals, Cambridge, Massachusetts, USA.

McCarthy, T.J., Nouas, Z., and Muhammed, A. (1991) Knowledge-based system for Steel Fabrication, **Structural Engineering Review,** Vol 3, pp. 255-264.

Nethercot, D.A. (1991) Steeling the Mind. **Steel Construction Today,** May 1991, pp. 111-114.

SCI (1992) Economic Design and Shorter Lead Times, **Steel Construction Today,** Vol. 6, No. 5, p. 228.

Tizani, W.M.K., Davies, G., and Nethercot, D.A. (1993), **Knowledge Acquisition About Design and Fabrication Practices of Tubular Trusses,** Report SR93013, Dept of Civil Engineering, University of Nottingham, UK.

Wardenier, J., Kurobane, Y., Packer, J.A., Dutta, D. and Yeomans N. (1991) **Design Guide for Circular Hollow Section (CHS) Joints Under Predominantly Static Loading.** CIDECT, Verlag TÜV Rheinland, Germany.

Zayyat, M.M.M., Nethercot, D.A., and Smith, N.J. (1989), Production Related Design of Heavy Steelwork Box Columns, **Proc. of The 4th Int. Conf. on Civil and Structural Engineering Computing,** Vol 1., pp. 171-176, CIVIL-COMP Press, UK.

PART 9

IMPERFECTIONS

Damage and misalignments
in circular hollow sections

30 RESERVE STRENGTH OF DENTED TUBULAR MEMBERS

A.N. SHERBOURNE and F. LU
Civil Engineering, University of Waterloo, Ontario, Canada

Abstract
Circular tubular members with a dent are studied for their load carrying capacity. Column analysis is performed on the tube using the moment-curvature approach and equations are derived assuming a strain hardening material. However, comparisons with available experimental data are made for a perfectly plastic material due to the lack of strain hardening data in the reported experiments.
Keywords: Columns, Dent, Offshore Structures, Strength, Tubular Members.

1 Introduction

The denting of a bracing tubular member is an unavoidable phenomenon in the engineering of offshore structures. The circular tubular members, used in, for example, offshore jacket platforms, are prone to damage by collision with supply vessels and by the impact of falling objects. In the case of semi-submersibles, it was reported (Ellinas et al. 1987) that, in the North Sea region alone, a total of fifty six collision incidents occurred in the period 1975 to 1986. Thirty four of them caused moderate or severe damage. Most of these incidents involved supply vessels and some occurred in anchor handling. As a result of denting damage, the residual load carrying capacity of the member may be considerably reduced. An appropriate assessment of the remaining strength is warranted to ensure overall structural integrity, or, to make decisions for necessary repair.

Depending on the geometry of the member and the intensity of the collision, the damage is usually classified into three categories (Ellinas 1984): (1) local denting of the tube wall; (2) overall bending; and (3) combined denting and overall bending. In general, a member of larger diameter and shorter length is more vulnerable to denting than a more slender member (Smith 1982). A large amount of experimental and theoretical research on this subject has been reported in the literature. It appears that damaged tubular members were first studied experimentally by Smith et al. (1979). An extensive investigation ensued, notably by Ellinas (1984), Ueda et al. (1985) and Chen et al. (1988). Ellinas (1984) developed simple, design-oriented analytical expressions to estimate the load carrying capacity of damaged tubular members. The analysis was based on

Tubular Structures V. Edited by M.G. Coutie and G. Davies.
© 1993 E & FN Spon, 2–6 Boundary Row, London SE1 8HN. ISBN 0 419 18770 7.

a simplifying assumption that the damaged part is ineffective and the load carrying capacity comes from the reduced section. In comparison, Smith (1982) proposed a similar idea earlier that the strength and stiffness in the dented region be modified by a reduction factor which is a function of dent size, elastic modulus, yield strength and stiffness. Ueda and Rashed (1985) presented an analytical model based on the ultimate strength interaction equation of a dented section. Various other analytical methods were also attempted, e.g. Durkin (1987), Yao et al. (1988), Zhou et al. (1991) and Duan et al. (1993).

The transverse denting behaviour of a tubular member has been studied earlier (Lu 1993). The purpose of the present paper is to investigate the effect of denting damage on the axial load carrying capacity of a tube. The analysis is based on knowledge gained previously for strain hardening beam-columns of rectangular section (Lu and Sherbourne 1992). The dented tubular member is treated as an imperfect column with eccentricity. Assuming an elastic-linear strain hardening material, the moment-curvature-thrust interaction relation is first described. The overall behaviour of the tubular column is examined by considering the equilibrium at the mid-span dented section. The ultimate load carrying capacity for various dent sizes is then obtained.

2 Geometric Consideration

The configuration of a tube with a dent at mid-section is shown in Fig. 1. The axial thrust, P, acts through the centroidal axis of the original undeformed tube. The dent section is assumed to consist of a circular bottom part together with a flat chord across the top. Using the polar coordinate, θ, and assuming that the original radius of the circular section is R_o [Fig. 2(a)], then the radius of the dent section with respect to the center, O, is

$$R = \frac{\pi R_o}{\pi + \sin\theta_o - \theta_o} \quad \text{or} \quad \beta = \frac{R}{R_o} = \frac{\pi}{\pi + \sin\theta_o - \theta_o} \tag{1}$$

The parameter, β, indicates the severity of the dent damage. The location of the dent section centroid, C, can also be represented by θ_c as follows

$$\theta_c = \cos^{-1}\left[\frac{\beta}{\pi}\sin\theta_o(\cos\theta_o - 1)\right] \tag{2}$$

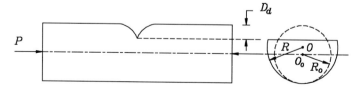

FIG. 1 Configuration of a dented tube

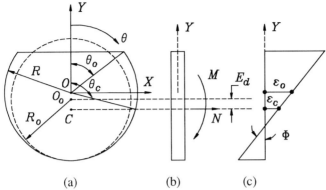

(a)   (b)   (c)

FIG. 2 Dent section under combined loading

Thus, the eccentricity induced in the dent, E_d, or, $e_d(=E_d/R_o)$, is

$$e_d = 1 - \beta(1 + \cos\theta_c) \tag{3}$$

The dent depth, D_d, or, d_d $(=D_d/R_o)$, is expressed as

$$d_d = 2 - \beta(1 + \cos\theta_o) \tag{4}$$

3 Bending-Axial Interaction for the Dent Section

Consider the dent section which is subjected to a tensile axial force, N, at the centroid, C, and a bending moment, M, as shown in Fig. 2(b). Assuming that the strain distribution is linear through the depth of the section [Fig. 2(c)], the normal force and moment can be computed using the following expressions:

$$N = (2Rt\sin\theta_o)\sigma|_{\theta=\theta_o} + \int_{\theta_o}^{\pi}\sigma(2Rt\,d\theta) \tag{5}$$

$$M = M_o - N\cdot(R\cos\theta_c) \tag{6}$$

where $M_o = (R\cos\theta_o)(2Rt\sin\theta_o)\sigma|_{\theta=\theta_o} + \int_{\theta_o}^{\pi}\sigma(R\cos\theta)(2Rt\,d\theta) \tag{7}$

and t is the tube wall thickness.

 If an elastic-strain hardening material is used, the section will experience three stages of deformation: the elastic, primary yield and secondary yield phases. The stress distribution profiles corresponding to the latter two phases are illustrated in Fig. 3. The expressions for N and M_o, or, n and m_o in normalized form, can be obtained as follows. Note that $\varphi = \Phi/\Phi_y = \Phi R_o/\varepsilon_y$ and Φ is the section curvature (Fig. 2).

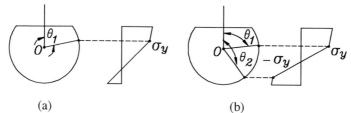

(a)   (b)

FIG. 3   Stress distributions. (a) Primary; (b) Secondary.

Primary Yield

$$
n = \frac{N}{N_p} = \beta\left(1 - \frac{\theta_o}{\pi}\right) + \frac{\varphi\beta^2}{\pi}\sin\theta_o\left[\frac{1}{\beta\varphi} - \alpha(\cos\theta_1 - \cos\theta_o)\right] +
$$
$$
+ \frac{\varphi\beta^2}{\pi}\left\{(1-\alpha)\left[(\theta_1 - \theta_o)\cos\theta_1 - \sin\theta_1 + \sin\theta_o\right] + \right.
$$
$$
\left. + (\theta_o - \pi)\cos\theta_1 - \sin\theta_o\right\} \tag{8}
$$

$$
m_o = \frac{M_o}{M_y} = \frac{\varphi\beta^3}{\pi}\left[\pi - \theta_o - \cos\theta_o\sin\theta_o + 2\sin\theta_o\left(\cos\theta_1 - \frac{1}{\beta\varphi}\right)\right] +
$$
$$
+ \frac{\varphi\beta^3}{\pi}\sin 2\theta_o\left[\frac{1}{\beta\varphi} - \alpha(\cos\theta_1 - \cos\theta_o)\right] +
$$
$$
+ \frac{\varphi\beta^3}{\pi}(1-\alpha)\left[\cos\theta_1(\sin\theta_1 - 2\sin\theta_o) - \theta_1 + \theta_o + \sin\theta_o\right] \tag{9}
$$

where $N_p = 2\pi R_o t\sigma_y$ and $M_y = \pi R_o^3 t\sigma_y$ are used for normalization.

Secondary Yield

$$
n = \frac{\varphi\beta^2}{\pi}\left\{\sin\theta_o\left[\frac{1}{\beta\varphi} - \alpha(\cos\theta_1 - \cos\theta_o)\right] + (\theta_o - \pi)\cos\theta_1 - \sin\theta_o\right\} +
$$
$$
+ \frac{\varphi\beta^2}{\pi}(1-\alpha)\left[(\theta_1 - \theta_o)\cos\theta_1 - \sin\theta_1 + \sin\theta_o + \right.
$$
$$
\left. + (\pi - \theta_2)\cos\theta_2 + \sin\theta_2\right] + \beta\left(1 - \frac{\theta_o}{\pi}\right) \tag{10}
$$

$$
m_o = \frac{\varphi\beta^3}{\pi}\left[\pi - \theta_o - \cos\theta_o\sin\theta_o + 2\sin\theta_o\left(\cos\theta_1 - \frac{1}{\beta\varphi}\right)\right] +
$$
$$
+ \frac{\varphi\beta^3}{\pi}\sin 2\theta_o\left[\frac{1}{\beta\varphi} - \alpha(\cos\theta_1 - \cos\theta_o)\right] +
$$
$$
+ \frac{\varphi\beta^3}{\pi}(1-\alpha)\left[\cos\theta_1(\sin\theta_1 - 2\sin\theta_o) + \right.
$$
$$
\left. + \theta_o - \theta_1 + \sin\theta_o - \cos\theta_2\sin\theta_2 - \pi + \theta_2\right] \tag{11}
$$

Note that Eq (6) should be rewritten in the following form:

$$m = m_o - 2\beta n \cos\theta_c \tag{12}$$

If a constant value of n is specified, the m-φ relationship is obtained through the parameters θ_1 and θ_2 where $\cos\theta_1 - \cos\theta_2 = 2/\beta\varphi$. Equations (8) to (11) can be reduced for a perfect section by letting $\theta_o = 0$. Accordingly, $\beta=1$, $\theta_c=\pi/2$, $e_d=0$ and $d_d=0$ from (1) to (4).

4 Eccentrically Loaded Column

As indicated in Fig. 1, the axial load in service acts on the centroid of the perfect section. This condition induces an eccentricity in the dent section, as given by Eq (3). In order to derive the ultimate load capacity, the dented tube is modelled as a simply-supported column of a dented section, subjected to a load with an eccentricity value of E_c (Fig. 4).

Following the approximate theory used by Chen and Atsuta (1976), it is assumed that the curvature is distributed along the column in the following sinusoidal form

$$\Phi(Z) = \Phi_A + (\Phi_C - \Phi_A)\sin\frac{\pi Z}{L} \tag{13}$$

The deflection, W, can be obtained by integrating (13) and satisfying the boundary conditions that the deflections at the ends are zero.

$$W(Z) = \frac{1}{2}Z(L-Z)\Phi_A + \frac{L^2}{\pi^2}(\Phi_C - \Phi_A)\sin\frac{\pi Z}{L} \tag{14}$$

Considering the equilibrium at the mid-span point, C,

$$M_C = P(E_c + W|_{Z=L/2}) \tag{15}$$

we have, in normalized form

$$m_C = 2pe_c\left[1 + \frac{p}{p_E}\left(\frac{\pi^2}{8} - 1\right)\right] + \frac{p}{p_E}\varphi_C \tag{16}$$

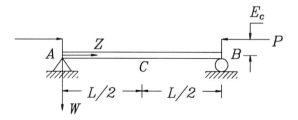

FIG. 4 Eccentrically compressed column

where

$$p_E = \frac{P}{P_E} = \frac{\pi^2 E}{\sigma_y}\left(\frac{r}{L}\right)^2 , \quad P_E = \frac{\pi^2 EI}{L^2} , \quad P_o = 2\pi R_o t \sigma_y \qquad (17a)$$

$$r^2 = \frac{I}{A} = \frac{R_o^2}{2} , \quad I = \pi R_o^3 t , \quad A = 2\pi R_o t \qquad (17b)$$

Essentially, Eq (16) embodies the relationship between P and Φ_C (or, p and φ_C), since m_C is a function of p. The condition for complete collapse is (Lu 1993): $d\varphi_C/dp = \infty$, or, $dp/d\varphi_C = 0$. Thus, (16) reduces to

$$\frac{dm_C}{d\varphi_C} = \frac{p}{p_E} \qquad (18)$$

This equation allows one to determine the deformation phase of the mid-section, if the load P is given. Consequently, the exact value of φ_C and m_C can be obtained from (8) to (12). The allowable eccentricity can then be found from Eq (16). Note that Eqs (8) to (12), derived for a tensile N, are equally valid for the compression P.

5 Results and Discussions

The load(p)-eccentricity(e_c) curves, assuming $\alpha = 0$, are plotted in Fig. 5 for tubes at three denting levels: $d_d = 0.1, 0.5$ and 0.7. Since, for a particular tube, say, the one with $d_d = 0.5$, the inherent eccentricity is e_{tB}, the ultimate load capacity can be located at point B. In the same way, A and C can be found for the other two tubes. The broken line passing through A, B and C represents the ultimate load resistance of a tube with a specified geometry (L/r) and material

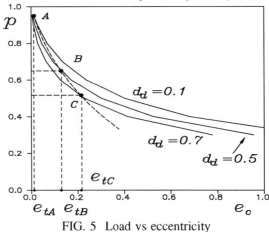

FIG. 5 Load vs eccentricity

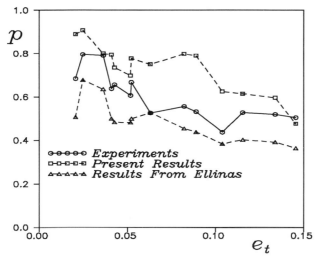

FIG. 6 Comparison with experiments

$(\pi^2 E/\sigma_y)$. In the present case, $L/r = 60$ and the material is that of a structural steel with $E = 200\ GPa$ and $\sigma_y = 300\ MPa$. Note also that e_{tA}, e_{tB} and e_{tC} are 0.01, 0.13 and 0.22 respectively.

The experimental data produced by Taby et al. (1981), as reported by Zhou et al. (1991), are now used to test the predictions made using this scheme. The overall bending imperfection in the original experiments is combined with the dent-induced eccentricity. The results are presented in Fig. 6, where the theoretical data using Ellinas' method (Zhou et al. 1991) have also been included. It is observed that both predictions reflect the feature exhibited in the experiment. However, the present results give an upper bound while the results using Ellinas' method show a lower bound. The difference is, nevertheless, within expectation, since, in the current analysis, only the geometrical degradation in the dent section has been considered. Ellinas, on the other hand, assumes that the damaged part is ineffective and the load is mainly carried by the reduced section. It must be noted that, if the same assumption is adopted, the dent-induced eccentricity becomes appreciably larger, leading to a reduction of the load carrying capacity.

6 Conclusions

The moment-curvature approach has been adopted to obtain the load carrying capacity of a dented tube. Unlike other methods, this approach allows one to incorporate material strain hardening into the analysis, which sometimes may prove necessary as noted by Prion and Birkemoe (1992). The results can be further refined by considering the reduction of material stiffness and strength since an upper bound solution gives an unsafe estimate for the purpose of design.

7 Acknowledgements

The research is sponsored by the Natural Sciences and Engineering Research Council of Canada through grant No. A-1582.

8 References

Chen, W. F. and Atsuta, T. (1976) **Theory of beam-columns: vol 1, In-plane behaviour and design**, McGraw-Hill.

Chen, W. F. and Sohal, I. S. (1988) Cylindrical members in offshore structures. **Thin-Walled Structures**, 6, 153-285.

Duan, L., Chen, W. F. and Loh, J. T. (1993) Analysis of dented tubular members using moment curvature approach. **Thin-Walled Structures**, 15, 15-41.

Durkin, S. (1987) An analytical method for predicting the ultimate capacity of a dented tubular member. **Int. J. Mech. Sci.**, 29(7), 449-467.

Ellinas, C. P. (1984) Ultimate strength of damaged tubular bracing members. **J. of the Structural Division**, ASCE, 110(2), 245-259.

Ellinas, C. P., Williams, K. A. J. and Walker, A. C. (1987) Collision damage effects in floating platforms. in **Mobile Offshore Structures** (eds L. F. Boswell, C. A. D'Mello, and A. J. Edwards), Elsevier Applied Science, London, U. K., 477-498.

Lu, F. (1993) Large plastic deformation of metal tubes as energy absorbing systems. A Ph.D. thesis submitted to University of Waterloo, Canada.

Lu, F. and Sherbourne, A. N. (1993) Load carrying capacity of strain hardening beams under shear, bending and axial forces. **J. Eng. Mech.**, ASCE, Submitted for publication.

Prion, H. G. and Birkemoe, P. C. (1992) Beam-column behaviour of fabricated steel tubular members. **J. Struct. Eng.**, 118(5), 1213-1232.

Smith, C. S. (1982) Strength and stiffness of damaged tubular beam columns. In **Buckling of shells in offshore structures** (eds. J. E. Harding, P. J. Dowling and N. Agelidis), Granada, 1-23.

Smith, C. S., Kirkwood, W. and Swan, J. W. (1979) Buckling strength and post-collapse behaviour of tubular bracing members including damage effects. **Boss'79**, London, England, 303-326.

Taby, J. and Moan, T. (1981) Theoretical and experimental study of the behaviour of damaged tubular members in offshore structures. **Norwegian Maritime Research**, no. 2, 26-33.

Ueda, Y. and Rashed, S. M. H. (1985) Behaviour of damaged tubular structural members. **Trans. ASME, J. Offshore Mech. & Arctic Eng.**, 107, 342-349.

Yao, T., Taby, J. and Moan, T. (1988) Ultimate strength and post-ultimate strength behaviour of damaged tubular members in offshore structures. **Trans. ASME, J. Offshore Mech. & Arctic Eng.**, 110, 254-262.

Zhou, Y. X., Chen, W. M. and Chen, T. Y. (1991) A plastic model for limit analysis of dented tubular members. The 10th Int. OMAE, vol.III-B, 427-432.

31 STRESS CONCENTRATION AT GIRTH WELDS OF TUBULARS WITH AXIAL WALL MISALIGNMENT

L.M. CONNELLY
Stress Engineering Services Inc., Houston, USA
N. ZETTLEMOYER
Exxon Production Research Company, Houston, USA

Abstract

Butt welds of tubular sections have always been common to offshore structures. However, in recent years these connections have become more critical to fatigue life assessments. Potentially critical applications include TLP tendons, caissons, risers of compliant structures, and platform jackets with forged or cast nodes. In such instances, the designer/analyst requires information on the stress concentration factor that is caused by axial wall misalignment of the adjacent tubular sections. This misalignment is usually due to a change in wall thickness, ovalized cross sections, and/or imprecise fabrication controls.

This paper presents the findings of a suite of finite element analyses of both flat plates and tubulars with axial wall misalignment. Flat plate results can be quite sensitive to length and boundary conditions, thus suggesting caution when using finite element or experimental results to represent plate applications. Fortunately, the hoop stress of tubulars within the common diameter-to-thickness range mitigates this sensitivity to boundary conditions and also results in smaller stress concentrations as compared with corresponding flat plate values. However, both the tubular and flat plate stress concentration factors can exceed those values commonly assumed or available in the literature.

One equation for stress concentration has been found to fit all tubular geometries. The recommended stress concentration equation is general in that it accommodates all practical instances of thickness change and eccentricity of midplanes. However, it assumes that a 1:4 transition exists between adjacent surfaces, which in some instances may only be possible if accomplished through welding alone.

Keywords: Misalignment, Eccentricity, Offset, Plate, Tubular, SCF.

1 Introduction

Since the early 1980's, the offshore industry has become increasingly interested in the potential for fatigue problems at tubular girth welds. For conventional steel jackets, welded tubular joints have been the historic weak links. However, in recent times both cast and forged nodes have been used, whereby the critical location for crack growth is more likely to be the girth weld connection between the nodes and the tubular members, or closure girth welds. More importantly, girth welds have primary importance in TLP structures, both for the tendons and the risers.

Tubular Structures V. Edited by M.G. Coutie and G. Davies.
© 1993 E & FN Spon, 2–6 Boundary Row, London SE1 8HN. ISBN 0 419 18770 7.

There are two uncertainties in predicting the fatigue life of girth welds, even if the nominal stress histories are known. One uncertainty is the S/N curve, which is presently being addressed by the girth weld test program headed by the UK Welding Institute (Andrews 1991). Without such data, designers are forced to make use of S/N curves for simple plates given in AWS (ANSI/AWS 1992) and the UK DEn Guidance Notes (UK DEn 1990).

A second uncertainty is the stress concentration factor (SCF) to be used with the S/N curve. The above Guidance Notes recommend that an SCF be established prior to the use of any of the S/N curves. However, the only quantitative SCF guidance is for butt-welded plates with eccentricity, e, where the plates are of equal thickness, T. The formula is :

$$SCF = 1 + 3e/T \tag{1}$$

Equation 1 is simply the result of adding membrane to local bending stress. The local bending stress magnitude assumes that the bending moment caused by the axial load and eccentricity is shared equally by the two plate sections. This assumption holds reasonably well for <u>equal thickness</u> misaligned plates.

The existence of eccentricity is sometimes referred to as thickness mismatch or axial wall misalignment. It is also possible to have angular misalignment. Some amount of both types of misalignment exist in virtually all S/N data, but it is generally discounted in establishing SCFs. Although the various codes give little guidance to either axial or angular misalignment SCFs, Maddox of the UK Welding Institute gave a review of the general literature in a 1985 paper (Maddox 1985). In his paper, Maddox points out that the case of axial misalignment of plates of <u>unequal</u> thickness is sometimes handled as in Eq. 2.

$$SCF = 1 + \frac{6e}{T_1}\left[\frac{T_1^3}{T_1^3 + T_2^3}\right] = 1 + \frac{6e}{T_1}\left[\frac{1}{1 - \left(\frac{T_2}{T_1}\right)^3}\right] \tag{2}$$

where SCF implies the higher stress concentration associated with the thinner plate, T_1. The only difference between the cases of equal and unequal thickness is the assumed distribution of the eccentric moment between the two plate sections according to relative moments of inertia. It can be shown that the largest local stress always exists in the thinner plate, so that is the focus of the calculations in this paper.

Maddox performed tests, where plates had one common surface, to investigate the accuracy of Eq. 2. Generally, he concluded that Eq. 2 is applicable if the thickness exponent is reduced from 3 to 1.5. Equation 2 then becomes:

$$SCF = 1 + \frac{3(T_2 - T_1)}{T_1}\left[\frac{1}{1 + \left(\frac{T_2}{T_1}\right)^{1.5}}\right] = 1 + \frac{6e}{T_1}\left[\frac{1}{1 + \left(\frac{T_2}{T_1}\right)^{1.5}}\right] \tag{3}$$

This reduction of the exponent has the effect of raising the calculated SCF. Maddox recognized that the exponent may be sensitive to plate geometry, boundary conditions, and stress levels (i.e. secondary displacements).

Maddox offered little guidance for tubulars, except that one paper he cited suggested that the case of equal thickness can be handled by introducing Poisson's ratio into Eq. 1 as follows:

$$SCF = 1 + \frac{3e}{T\left(1 - v^2\right)} \qquad (4)$$

For steel, Eqs. 1 and 4 yield similar results.

The FE results presented in this paper emphasize the tubular misalignment condition but have sufficient plate results to evaluate the appropriateness Eq. 3 for plates. This paper does not address angular misalignment, although this condition is believed to have less impact for tubulars than it has for plates.

2 Plate Models

Although the main thrust of this work is tubular girth welds, additional plate analyses were necessary to resolve some remaining plate questions and to provide a basis for comparison between plates and tubulars. Therefore, the first section of this paper addresses issues that are common to both plate and tubular axial misalignment. The effect of taper transitions and mesh density are analyzed from a plate perspective but the results are expected to apply equally well to tubular geometries.

2.1 Common Mid Plane Plate Models

Issues related to mesh refinement, transition taper, and notch effects can be addressed through an evaluation of a simple "thickness transition" plate model with no centerline offset. This "co-planar" geometry is illustrated in Fig. 1a.

Figure 1a depicts the cross section of two types of co-planar geometry–one with and one without a transition section. A 1:4 taper transition between sections of different thickness is recommended by the DEn code to minimize notch effects. (Here, we assume a 1:4 transition could exist, although with unplanned eccentricity, it may have to be effected by welding alone.) The most effective way to evaluate the role of the taper transition on the local stress field in the vicinity of the thickness change is to eliminate eccentric bending moments. Such elimination is accomplished with the coplanar plate model shown in Fig. 1a. The centerlines of the two plate sections in these models are aligned. Consequently, no bending stress will develop under axial loading. The SCF in the thickness transition region will be due solely to notch effects.

The model in Fig. 1a was composed of a thick and thin section. The thicker section (T_2) was 25 mm thick for this case and all cases studied in this paper. This thickness value was chosen to be consistent with earlier Maddox tests, and because it is a reasonable nominal value for both plate and tubular cases. The thin section thickness (T_1) for the co-planar case was chosen to be 12.5 mm. The length of both sections was 200 mm for a total model length of 400 mm (plus transition.) The end of the thick section was fixed and axial tension load was applied to the end of the thin section. The applied load was sized to generate a nominal stress of 100 MPa ($\cong 14.5$ ksi) in the thin section. The length of the model with the transition was longer than the no transition case by an amount equal to the transition length.

The ABAQUS (Hibbitt, Karlsson, and Sorensen 1987) finite element (FE) code was used to analyze this case and all subsequent cases described in this study. The plate geometry in Fig. 1a was modeled with eight-noded, second order, isoparametric, plane strain (solid) elements with reduced integration (CPE8R.) All FE simulations assumed elastic material properties and nonlinear geometry to capture possible second order effects. The non-dimensional parameters that were used to describe the SCF behavior are e/T_1 and T_2/T_1.

2.1.1 Mesh Sensitivity
A mesh sensitivity evaluation was required to assess how well a given mesh density can capture important model results. The model illustrated in Fig. 1a had the "standard" mesh density, which

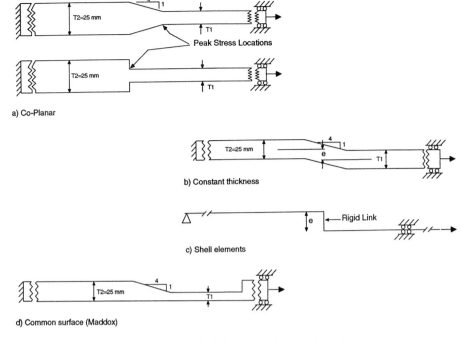

a) Co-Planar

b) Constant thickness

c) Shell elements

d) Common surface (Maddox)

Fig. 1 Model geometries analyzed

consisted of about 1 element for each 6 mm of model length within the transition proper and for a 50 mm length on either side of this transition. There were five elements through the thickness for all but the no-transition case, which resulted in an element dimension of no more than 5 mm in the transverse direction. The elements outside of this critical "transition zone", where the axial stress gradients are small, were about 18.75 mm long.

FE solution accuracy was determined by a comparison of the peak stress results for the standard and refined mesh densities. For this study, peak stresses were located at the reentrant corners of the transition section (Fig. 1a.) To obtain the stress at these points, FE integration point values were extrapolated to the nodes on the surface of the model. The standard mesh was uniformly refined by a factor of two, in the transition zone. The results for both standard and fine mesh density models are provided in Table 1, Cases 1 and 2, respectively. A comparison of peak stress values for Cases 1 and 2 indicates that there was less than a 4% difference between the standard and fine mesh models. Consequently, the standard mesh density was considered to be adequate for the purposes of this study.

2.1.2 Notch Effects

The peak stress for the fine mesh model (Case 2) was about 124.8 MPa compared to the nominal stress of 100 MPa. Comparison of the two stresses suggests an SCF of 1.25 in the transition region adjacent to the thin section. The eccentricity (e) for these models is zero, suggesting that the stress amplification at the transition is due solely to "notch" effects. The intent of this study is to determine SCFs that are due to induced bending stresses only, excluding notch effects, which are normally included in the S/N data. To this end, another characteristic stress was defined for the cases studied. This characteristic stress was determined by a linear extrapolation of the known stress from two points on the model surface. The extrapolation technique is illustrated in Fig. 2a. The extrapolation point closest to the notch is defined as the largest stress value prior to a drop in stress, as one approaches the notch from

Table 1. SCFs for plate models

Case #	Dimensions (mm) Length	T2	T1	e	Tran- stion	e/T1	T2/T1	Stress (MPa) Nominal	Peak	Extrap	SCF	Eq. 3	Eq. 3/ SCF
Common Mid Plane													
1	425	25	12.5	0.00	y	0.00	2.00	100	120.0	98.5	0.99	1.00	1.02
2	425	25	12.5	0.00	y	0.00	2.00	100	124.8	99.0	0.99	1.00	1.01
3	400	25	12.5	0.00	n	0.00	2.00	100	113.4			1.00	
4	400	25	12.5	0.00	n	0.00	2.00	100	125.0	97.8	0.98	1.00	1.02
5	400	25	12.5	0.00	n	0.00	2.00	100	153.9			1.00	
6	400	25	12.5	0.00	n	0.00	2.00	100	176.2			1.00	
Constant Thickness													
7	425	25	25	6.25	y	0.25	1.00	100	193.2	169.3	1.69	1.75	1.03
8	450	25	25	12.50	y	0.50	1.00	100	268.5	231.0	2.31	2.50	1.08
9	500	25	25	25.00	y	1.00	1.00	100	376.7	313.8	3.14	4.00	1.27
10	850	25	25	12.50	y	0.50	1.00	100	281.4	242.0	2.42	2.50	1.03
11	900	25	25	25.00	y	1.00	1.00	100	420.1	353.5	3.53	4.00	1.13
12	1700	25	25	25.00	y	1.00	1.00	100	437.6	367.5	3.67	4.00	1.09
13	450	25	25	12.50	y	0.50	1.00	200	534.8	459.7	2.30	2.50	1.09
14	850	25	25	12.50	y	0.50	1.00	200	558.5	479.9	2.40	2.50	1.04
Thin Shell Model													
15	400	25	25	12.50	n	0.50	1.00	100	251.0	251.0	2.51	2.50	1.00
16	400	25	25	12.50	n	0.50	1.00	100	251.0	251.0	2.51	2.50	1.00
17	400	25	25	25.00	n	1.00	1.00	100	444.0	444.0	4.44	4.00	0.90

Table 2. SCFs for tubular models

Case #	Dimensions (mm) Length	T2	T1	e	Tran- stion	D/T2	e/T1	T2/T1	Stress (MPa) Nominal	Node	Extrap	SCF	Eq. 3	Eq. 3/ SCF
1	439	25	7	0.70	y	25	0.10	3.57	100		114.0	1.14	1.08	0.94
2	444	25	7	2.00	y	25	0.29	3.57	100	441	124.0	1.24	1.22	0.99
3	444	25	7	2.10	y	25	0.30	3.57	100		123.7	1.24	1.23	1.00
4	457	25	7	5.25	y	25	0.75	3.57	100		144.1	1.44	1.58	1.10
5	472	25	7	9.00	y	25	1.29	3.57	100	509	164.5	1.65	2.00	1.21
6	434	25	10	1.00	y	25	0.10	2.50	100	390	113.0	1.13	1.12	0.99
7	442	25	10	3.00	y	25	0.30	2.50	100	441	128.2	1.28	1.36	1.06
8	460	25	10	7.50	y	25	0.75	2.50	100	475	156.9	1.57	1.91	1.22
9	428	25	14	1.40	y	25	0.10	1.79	100		114.5	1.15	1.18	1.03
10	439	25	14	4.20	y	25	0.30	1.79	100		136.8	1.37	1.53	1.12
11	442	25	14	5.00	y	25	0.36	1.79	100	441	142.6	1.43	1.63	1.15
12	444	25	14	5.50	y	25	0.39	1.79	100	441	145.8	1.46	1.70	1.16
13	450	25	14	7.00	y	25	0.50	1.79	100	441	155.2	1.55	1.89	1.22
14	464	25	14	10.50	y	25	0.75	1.79	100		172.4	1.72	2.33	1.35
15	410	25	25	2.50	y	25	0.10	1.00	100		126.6	1.27	1.30	1.03
16	416	25	25	4.00	y	25	0.16	1.00	100	294	143.1	1.43	1.48	1.03
17	430	25	25	7.50	y	25	0.30	1.00	100		168.3	1.68	1.90	1.13
18	475	25	25	18.70	y	25	0.75	1.00	100		208.9	2.09	3.24	1.55
19	480	25	25	21.00	y	25	0.84	1.00	100	294	230.0	2.30	3.52	1.53
Sensitivity Cases														
20	880	25	25	21.00	y	25	0.84	1.00	100	294	229.3	2.29	3.52	1.54
21	480	25	25	21.00	y	25	0.84	1.00	200	294	458.2	2.29	3.52	1.54
22	425	25	12.5	0	y	10	0	2.00	100		100.7	1.01	1.00	0.99
23	425	25	12.5	0	y	25	0	2.00	100		102.5	1.03	1.00	0.98
24	425	25	12.5	0	y	50	0	2.00	100		103.4	1.03	1.00	0.97

the thin section end. This point generally occurs between $0.3T_1$ and $0.4T_1$ away from the notch. The second stress point used for extrapolation purposes is located two element lengths (about 12 mm) away. The stress values at these locations were used to determine the combined membrane and bending stress at the transition, which is provided in Table 1 under the

313

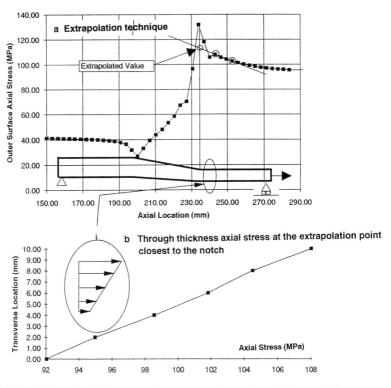

Fig. 2   Through thickness axial stress at the extrapolation point

"Extrapolated Stress" heading. The extrapolated stress for Cases 1 and 2 were 98.5 and 99 MPa, respectively, resulting in SCF values of 0.985 and 0.99. These SCF values compare well with the theoretical SCF value of 1, suggesting that the notch component of the stress riser has been eliminated.

To verify that the notch stress component is not present at the extrapolation point closest to the notch area, a plot of the axial stress through the thickness is provided in Fig. 2b. The linear variation of through-thickness stress illustrated in Fig. 2b suggests that only bending and membrane stresses are acting at this location. Figures 2a and 2b correspond to a model with T_1 of 10 mm and an e of 1 mm, but are representative of all models considered in this study.

Given the information available from the finite element analysis of plate and tubular misalignment, it is possible to separate notch behavior from bending and membrane effects in the notch region, without resorting to extrapolation. However, as illustrated above, the extrapolation approach was very effective, and is consistent with the approach that must be taken when evaluating strain gauge data.

2.1.3 Taper Effect
As mentioned earlier, a 1:4 taper is assumed, even though conditions may require that the taper be achieved through welding alone, rather than by machining or chamfering. To examine the effect of taper, the "no taper" model in Fig. 1a was analyzed and compared with the taper model results discussed above. Four different mesh densities (Cases 3 through 6) were considered to examine notch stress effects. The peak stress from the standard mesh density was 113.4 MPa. The peak stress associated with the transition region mesh densities that were finer than the

314

standard mesh density by factors of 2, 4, and 8 were 125.0, 153.9, and 176.2 MPa, respectively. These results suggest that, as expected, the notch effect associated with this geometry is much greater than that associated with the 1:4 transition. However, the extrapolation technique, discussed above, still removes the notch contribution to the stress riser, as evidenced by the extrapolated SCF value of 0.98 for Case 4 of Table 1.

The only question, then, is how much of a stress raiser is implicit in the S/N data. The notch conditions for the 1:4 taper is reasonably mild and is probably adequately represented by the butt weld database. However, the no taper case is likely unconservatively addressed by the S/N data.

2.2 Constant Thickness Models

An important subset of misalignment geometries is comprised of plate sections of equal thickness with centerlines offset by an eccentricity e. This geometry is especially well-suited to determine the extent to which plate SCF values are affected by eccentricity (e), load magnitude, and specimen length.

2.2.1 Eccentricity Effects

The geometry shown in Fig. 1b was loaded by a longitudinal tension of 100 MPa. The ends of the specimen were constrained to move in the longitudinal direction only, with no end rotation. These conditions were selected to mimic the most likely experimental conditions. Each plate section was 25 mm thick and 200 mm long. A transition section was placed between the plate sections with a length that varied with eccentricity so as to maintain a 1:4 taper. Eccentricity values of 6.25 mm, 12.5 mm, and 25 mm were selected as a representative range of values, with results given in Table 1 as Cases 7, 8, and 9, respectively. The extrapolation technique discussed above was used again to evaluate SCF values of 1.69, 2.31, and 3.14 for Cases 7 through 9, respectively.

Equation 3 was employed for all of the geometries analyzed, with results given in Table 1 under the column labeled "Eq. 3". It can be seen that, for Cases 7 and 8 the calculated values compare well with the predicted values, for e/T_1 values of 0.25 and 0.5, differing by 3% and 8%, respectively. However, for Case 9 ($e/T_1 = 1.0$), the predicted value exceeds the calculated value by 27%. This result suggests that, for this geometry, boundary conditions and/or length effects may be important for e/T_1 values greater than 0.5.

2.2.2 Length Effects

Specimen model length can be an important issue for some plate model geometries. The plate model used by Maddox (Maddox 1985), for example, was found to be very sensitive to length variations as will be discussed in the next section. To assess the effect of model length on calculated SCFs for the constant thickness cases used in this study, a model that was about twice as long as the base case model, with $e/T_1 = 0.5$ and identical loading and boundary conditions, was developed. The results from this model are given as Case 10 in Table 1. The SCF value for this case is 2.42, which is about 5% greater than the corresponding base model result of 2.31 (Case 8.) Therefore, for this eccentricity and misalignment geometry, the base case length of 400 mm (+ transition) is long enough to preclude length effects from affecting the stress field in the transition region. However, the model length necessary to avoid affecting calculated SCF values, is likely to be a function of eccentricity as well as the model's basic geometry. Therefore, two models with an $e/T_1 = 1.0$ and lengths of 900 mm and 1700 mm, were analyzed (Cases 11 and 12.)

The SCFs for Cases 11 and 12 are 3.53 and 3.67, respectively. These values compare with an SCF of 3.14 for the corresponding model with the base case length of 400 mm (+ transition) (Case 9). It can be seen that, as the length increases, the calculated SCFs are converging but at slower rate than that for the models with smaller eccentricities. These results confirm that 1)

315

model length can have important effects on calculated SCF values, and 2) the length required to preclude affects on SCF evaluation can be a strong function of eccentricity.

2.2.3 Load Magnitude

Nonlinear geometry effects can be important for some misalignment geometries (Maddox 1985). All analyses conducted in this work account for potential nonlinear geometry effects. Because these nonlinear effects can increase with load magnitude, both the base case and long constant thickness models were loaded with twice the normal axial tension. The increased load results in a nominal axial stress of 200 MPa ($\approx$29 ksi) and a nominal strain of about 0.1%. Cases 13 and 14 in Table 1 represent the results for the base model length and the long model subjected to twice the normal load. A comparison of SCF values for models with $e/T_1 = 0.5$ and a standard length of 450 mm, result in similar SCFs for both load conditions (2.31 and 2.30 for Cases 8 and 13, respectively.) Similar results were found for the long model (Cases 10 and 14).

The negligible effect of nonlinear geometry on SCF results, for models with $e/T_1 = 0.5$, can be explained by a comparison of the displaced mesh results of the two load cases (Fig. 3a.) Figure 3a depicts the original and displaced mesh associated with Case 11. It can be seen that once the specimen's transition region rotates to align with the specimen's end restraints, little additional rotation or lateral translation will occur with increasing load. For the geometry and boundary conditions under consideration, this rotation/translation is completed in the initial stage of loading.

2.3 Thin Shell/Rigid-Link Model

Other investigators have used various means to evaluate the relationship between plate and tubular misalignment and associated SCFs. One alternative method employs thin shell finite elements as opposed to the current approach, which uses solid finite elements. Shell models are usually appropriate for problems in which membrane and bending stresses are much larger than the associated transverse stress. Shell models are therefore used in problems where "local effects" are not considered important. As stated earlier, in calculating SCFs due to plate and tubular misalignment, an attempt was made to separate notch and bending stress effects. However, the separation of these two effects is not always easily accomplished, and in some cases, the notch effect may alter the evaluation of the "nominal" stress that is desired. Therefore, it is important to know how these different models relate to one another for a range of misalignment geometries.

The shell model shown in Fig. 1c was subjected to loads and boundary conditions that were identical to those associated with the constant thickness cases discussed above. The plate sections were 200 mm long and 25 mm thick. The misalignment was accomplished by means of a "rigid-link" between the two flat plate sections. As the name implies, the rigid-link constrains the displacement and rotation of the adjacent ends of the plate sections to be identical. The length of the rigid-link is equivalent to the model eccentricity.

Three shell cases were analyzed, two with $e/T_1 = 0.5$ and one with $e/T_1 = 1.0$. The first case ($e/T_1 = 0.5$) only restrained the lateral translation of the model ends. The second case constrained lateral translation and rotation of the model ends. The results for these cases are virtually the same and are given in Table 1 as Cases 15 and 16, respectively. This particular misalignment geometry is apparently insensitive to boundary conditions, whether modeled by shell or solid elements. Because these cases were modeled with shell elements, notch effects were not captured and therefore extrapolation was unnecessary. Consequently, the peak stress and extrapolated stress values in Table 1 are the same for all the shell models.

Unlike the solid element models, which simulate plane strain conditions, the shell results are subject to edge effects that give rise to slight differences in stress between corresponding edge and center locations along the length of the model. The values listed in Table 1 represent centerline values.

The shell SCF results compare well with Eq. 1 predictions (2.51 versus 2.50) for $e/T_1 = 0.5$. However, for Case 17, with $e/T_1 = 1.0$, the calculated SCF was 4.44, which compares with the Eq. 1 prediction of 4.0. This represents a difference of 10%, with the predicted value on the unconservative side. Recall that, for the solid element model, a specimen length of 1700 mm (Case 12) was required to obtain an acceptably close comparison with Eq. 1 predictions. Therefore, a plate model represented by shell elements will also require additional length for cases with $e/T_1 > 0.5$.

2.4 Common Surface Model

The work described in Maddox (1985) was based on tests of samples with the geometries depicted in Fig. 1d. This geometry is different from the geometries discussed above in that the thin section is bounded on both ends by 25 mm thick sections. The specimen was constructed such that the thin and thick sections had a common surface. Although this rapid transition from thick to thin to thick is not representative of an actual plate or tubular misalignment geometry, it was likely selected because it simplifies the alignment and testing of a specimen in a test-rig.

An early phase of the current study involved a finite element analysis of this geometry. As part of this work, several parameters were investigated with respect to their affect on the calculated value of SCF. Overall specimen length varied between 600 mm (the length used in Maddox (1985)) and 4600 mm. The effect of specimen boundary conditions and load levels were also investigated. The results of this sensitivity analysis suggested that these parameters can be important, especially for this geometry. For example, it was determined that calculated SCF values can vary by 20% for specimen lengths between 600 mm and 4600 mm with $T_1=14$ mm. The SCF variation with specimen length can be as much as 100% with $T_1=7$ mm. SCF values for the 4600 mm specimens were insensitive to boundary condition changes. However, for the 600 mm specimens, the calculated value of SCF varied by as much as 13%, depending upon rotation constraints applied at the ends of the specimen.

The FE calculated SCF value for $T_1=10$ mm on a 600 mm long specimen with rotation end constraints was 1.97, which may be compared to the measured value of 1.73 from Maddox (1985), a difference of 12%. Possible explanations for some of this difference include the averaging associated with the finite length of the strain gauge and/or test boundary conditions that may only partially restrain specimen translation and rotation. Nonetheless, a difference of only 12% suggests that the finite element analysis captures the important aspects of the common boundary misalignment problem.

Sensitivity to model boundary conditions and specimen length is geometry dependent. The common surface geometry used in Maddox (1985) appears to be more sensitive to these parameters than the base case plate model used in the present study. The increased sensitivity of the Maddox geometry to length and boundary condition parameters can be explained by an examination of the deformed mesh plots of the two geometries (Figs. 3a and 3b.) It can be seen that the rotation and translation associated with the Maddox geometry generally exceeds that of the corresponding base case plate model used in this study.

Axisymmetric models were developed from the plate cross sections shown in Fig. 1d. The sensitivity studies discussed above were carried out with these tubular models along with a study of the sensitivity of SCF to the diameter to thickness ratio (D/T). It was determined that hoop restraint minimizes the sensitivity of SCF to specimen length, boundary conditions, and D/T.

3 Tubular Models

The primary objective of this overall study was to develop a design equation that can be used to estimate the fatigue life of tubular butt-welded connections. To develop such an expression, it was necessary to quantify the misalignment at the connection between two pipe sections.

a) Constant thickness model (e=12.5 mm)   b) Common boundary model (e=9 mm)

Fig. 3 Deformed mesh plots (magnification=30)

This study only addresses mid-surface misalignment that is constant in the circumferential direction. Such misalignment commonly occurs when wall thickness changes at the butt weld and a common boundary OD or ID is maintained. However, mid-surface misalignment can also occur when equal thickness sections are out-of-round or adjacent round sections are not exactly coaxial. Although these types of misalignment are not strictly addressed by the models considered here, conservative approximations for the SCF of such models can be obtained by applying the results of this study to the maximum mid-surface misalignment anywhere around the circumference.

3.1 Geometries

Tubular models were generated from the plane strain plate geometries discussed in the above plate section. Twenty-four individual tubular analyses were performed assuming different combinations of e and T_1.

The models for the tubular analysis are similar to the plate cases in Fig. 1b, except with axial symmetry and $T_1 \leq T_2$. The tubular specimens were modeled with axisymmetric elements (ABAQUS CAX8R) rather than plane strain elements (ABAQUS CPE8R) and had a base case outside diameter of 625 mm. The CAX8R element is an eight-noded bi-quadratic element with reduced integration. As in the preceding plate section, the maximum thickness (T_2) was held constant at 25 mm; therefore, the base case tubular model has a diameter to thickness ratio (D/T_2) of 25. This D/T ratio is considered representative of tubulars used in the offshore industry. As for the plate models, the parameters of primary concern were the nondimensional mid surface offset (e/T_1) and thickness (T_2/T_1.)

The tubular geometries were pinned in the axial direction at the end of the thick section and loaded with an axial force at the end of the thin section. The axial force, in all but a few sensitivity cases, was selected to produce a 100 MPa stress in the thin section. The boundary conditions for the tubular specimens differed from those of the plate model in that no restraint is applied in the radial direction. This is in keeping with our attempt to mimic typical long members. The inclusion of radial restraint would be of little consequence since radial displacement was found to be negligible due to the hoop constraint for axisymmetric models.

Depending upon the particular values of T_1 and e selected, two transitions were sometimes necessary for a particular tubular model. These transitions were generated such that the side of the tubular joint with the largest surface mismatch was given a 1:4 transition that defined the transition section length. The transition on the opposite surface had the same transition length but a smaller thickness discontinuity leading to a more gradual transition. The stress values reported in the following are associated with the side having the 1:4 transition, which in all cases contained the largest stress value.

3.2 Sensitivity Studies

A total of 24 cases with varying e/T_1, T_2/T_1, and D/T were analyzed, with results given in Table 2. The first 19 cases were used to develop the relationship between SCF and the nondimensional parameters e/T_1 and T_2/T_1, whereas cases 20 through 24 were used to determine the sensitivity of this relationship to parameters such as specimen length and D/T.

Before evaluating the relationship between SCF and the primary parameters (e/T_1 and T_2/T_1), it was necessary to ensure that model length and boundary conditions were not impacting the solution for SCFs over the range of parameters used for the tubular study.

Previous plate results suggest that length and boundary conditions will most likely affect models with larger e/T_1 values. Therefore, a model with a base case length of 400 mm (+ transition length) and an e/T_1 of 0.84 (the largest value used in the tubular study) was compared with models with different lengths and load conditions. Cases 20, and 21 indicate that doubling the length and doubling the applied load cause less than a 1% difference in SCF values.

For axisymmetric geometries, there can be a difference between two models with identical e and T_1 parameters. For example, a tubular misaligned butt connection with an e/T_1 of 0.25 with a T_2/T_1 of 0.5 can be modeled with a flush outer diameter or a flush inner diameter at the transition. With the Poisson effect, this can possibly lead to different results. However, Cases 15 through 21 with $e \neq 0$, $T_1 = T_2$, and a D/T = 25 have SCF values that were virtually the same on the inside and outside of the transition region. Therefore, axisymmetric models with equal eccentricities, whether formed near the ID or the OD, yield equivalent SCF values, at least for diameter-to-thickness ratios in the neighborhood of 25.

Early analyses suggested that the effect of D/T on calculated SCF was of little consequence for the range of D/Ts used in the offshore industry. However, the Poisson effect that acts on axially loaded tubulars has not yet been addressed. Cases 22, 23, and 24 estimate the differential amount of radial movement across a thickness transition under the action of axial loads. As axial load is applied to a tubular butt weld, bending will occur at a thickness transition, even though the mid surfaces of both sections are coincident. Cases 22 through 24 suggest that this Poisson effect is on the order of 3% for D/Ts ranging from 10 to 50. Consequently, little error will result by assuming that an SCF of 1 represents all cases with no mid surface misalignment, regardless of thickness variation. This assumption is especially valid when a 1:4 transition is employed at the butt weld.

3.3 SCF Equation

The relationship between SCF and e/T_1 was determined by a suite of cases in which T_2/T_1 was maintained while e/T_1 varied from 0.1 to 0.84. Figure 4 suggests that a linear relation exists between SCF and e/T_1 for a realistic range of T_2/T_1 values. Similarly, it was determined that a nonlinear relationship exists between SCF and T_2/T_1, for a range of e/T_1 values (Fig. 5.) The

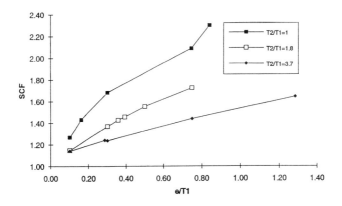

Fig. 4 Tubular SCF versus e/T1 for a range of T2/T1

combined results indicate that the form of the relationship between tubular misalignment SCFs and the nondimensional parameters e/T_1 and T_2/T_1 can be expressed as:

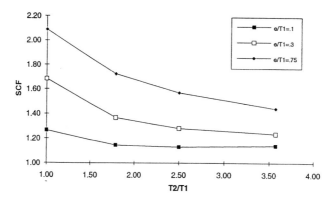

Fig. 5 Tubular SCF versusT2/T1 for a range of e/T1

$$SCF = 1 + a\frac{e}{T_1}\left[\frac{1}{1 + b\left(\dfrac{T_2}{T_1}\right)^n}\right] \tag{4}$$

in which a, b, and n are parameters to be evaluated.

Equation 4 provides the best fit to the tubular SCF data in Table 2 when parameters a, b and n have values of 2.6, 0.7, and 1.4, respectively. With these parameters, Eq. 4 becomes:

$$SCF = 1 + 2.6\frac{e}{T_1}\left[\frac{1}{1 + 0.7\left(\dfrac{T_2}{T_1}\right)^{1.4}}\right] \tag{5}$$

The curve in Fig. 6 plots the SCF values predicted by Eq. 5 versus the FE calculated values for tubulars. The standard error between the predicted and calculated SCF values is less than 6%. Also plotted in Fig. 6 are the predicted SCFs based on the plate equation (Eq. 3). The plate equation predicts SCFs that are, on average, 20% higher (more conservative) than the best fit tubular equation. However, the plate equation becomes more conservative as the SCF increases. For a calculated SCF of 2.3, Eq. 3 predicts an SCF of 3.52 compared with an Eq. 5 prediction of 2.5, a difference of more than 50%.

In view of the difficulty in achieving a 1:4 transition with welding alone, it may be advisable to modify the "best fit" coefficients, represented in Eq. 5 above, so that the predicted values are, for the most part, conservative. With this objective, a second set of coefficients for Eq. 4 were developed. The modified equation is given by:

$$SCF = 1.2 + 6.2\frac{e}{T_1}\left[\frac{1}{1 + 3.1\left(\dfrac{T_2}{T_1}\right)^{1.4}}\right] \tag{6}$$

The leading coefficient of Eq. 6 is 1.2 instead of 1.0 as in Eq. 5. This was done to optimize the upper bound expression given by Eq. 6. A side effect of this optimization is that Eq. 6 predicts an SCF of 1.2 for a geometry with no eccentricity. However, it can be argued that it is prudent for designers to assume some minimum SCF in their designs, even for cases with no eccentricity, to account for unknowns.

Figure 7 replots the predicted versus calculated SCFs with the proposed coefficient modifications as given in Eq. 6. The new coefficients increase the tubular SCF predictions, so

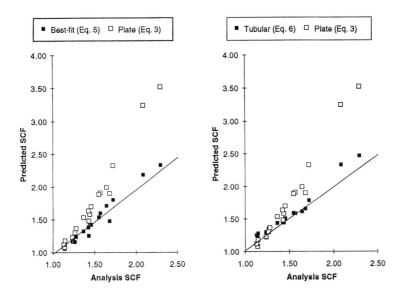

Fig. 6 Best-fit predicted SCFs
versus analysis SCFs for
tubular data

Fig. 7 Comparison of tubular and
plate equations with tubular
data

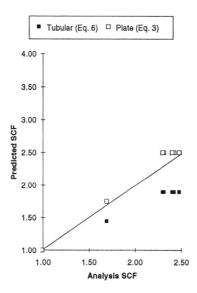

Fig. 8 Comparison of tubular and
plate equations with plate data

that all of the points now fall on the conservative side of the equivalence line. Consequently, Eq. 6 provides an appropriate design equation for prediction of SCFs due to axial misalignment of tubular butt welds.

It is of some interest to compare Eqs. 3 and 6 with the plate data collected and discussed in the first section of this study. Figure 8 plots the SCFs predicted by Eqs. 3 and 6 and compares them with calculated plate SCFs. The calculated SCFs, taken from Table 1, exclude cases in which model length adversely affects SCF values. It can be seen that Eq. 3 provides a reasonable upper bound approximation, as intended by Maddox, to the limited amount of plate data collected. Consequently, Eqs. 2 and 6, which incorporate hoop restraint that is not present in plate data, provide an unconservative estimate to the calculated plate SCFs.

4 Conclusions

The purpose of this study was to develop a design equation for the prediction of SCFs associated with the misalignment of tubular butt welds. The equation developed above provides this result for a realistic range of misalignment parameters. The nondimensional misalignment eccentricity (e/T_1) ranged from 0 to 1.29 and the nondimensional thickness (T_2/T_1) ranged from 1.0 to 3.57. The design equation provides an upper bound value for SCF with a standard deviation on error of less than 6%.

The results of this study suggest that calculated misalignment SCFs for tubular geometries are much less sensitive to length and boundary condition effects than are SCFs for corresponding plate geometries. This insensitivity to boundary conditions is a result of hoop restraint that exists in tubular geometries. SCFs due to misalignment were also relatively insensitive to the range of diameter to thickness ratios that are commonly used in the offshore industry.

Several different misaligned plate geometries were considered in this study for the purpose of comparison with tubular models. It was determined that plate models were particularly susceptible to length and boundary condition effects, and that length effects on misalignment SCFs increase substantially with values of $e/T_1 > 0.5$. For example, in some cases, plate lengths in excess of 70 times the wall thickness may be required to obtain asymptotic SCF solutions. Several plate geometries were also modeled with thin shell elements as opposed to the plane strain or axisymmetric continuum elements used for the majority of cases. The shell element models, which used rigid-links to model the offset, were found to provide adequate SCF results. In fact, the shell models can simplify the misalignment SCF analysis by implicitly eliminating notch effects. Solid element modeling of plate and tubular misalignment geometries requires an extrapolation approach to eliminate notch effects, which are implicitly included in the S-N curve.

The results of this study indicate that, while the Maddox equation provides appropriate SCF values for plate geometries, it was found to be overly conservative for tubular geometries. The design equation developed above for tubular geometries provides appropriate SCF predictions for a wide range of tubular misalignment parameters.

5 References

Andrews, R.M. (1991) Fatigue performance of girth welded tubular connections, The Welding Institute 5611/4/91.

ANSI/AWS (1992) Structural Welding Code/Steel

UK DEn (1990) Offshore installations: Guidance on design and construction, 4th ed.

Maddox, S.J. (1985) Fitness-for-purpose assessment of misalignment in transverse butt welds subject to fatigue loading, in International Institute of Welding (XIII-1180-85).

Hibbitt, Karlsson, and Sorensen (1987) ABAQUS uses manual, Version 4-9.

PART 10

JOINTS

Static strength of planar joints
in circular or rectangular
hollow sections

32 IN-PLANE BENDING STRENGTH OF CIRCULAR TUBULAR JOINTS

B.E. HEALY and N. ZETTLEMOYER
Exxon Production Research Company, Houston, USA

Abstract

Design guidance for strength of circular tubular joints has existed in several design codes for more than a decade. Yet, doubts have long existed about the reliability of these equations. The equations that are in the codes and general literature are not all the same; history has shown they are integrally linked to the database that was used to derive them. Furthermore, codes differ as to whether additional requirements such as a punching shear check need to be part of the design/analysis recipe.

The subject paper takes a new approach to in–plane bending strength. In principle, an identical approach could be taken with respect to other types of loading. The idea is to use calibrated, nonlinear, finite element analysis of a statistically defined set of joint geometries to devise an appropriate capacity equation. This equation is then compared with an updated database that, for in– plane bending, contains over twice the number of data as heretofore considered. The database has undergone an unusual screening process in that data were separated into groups that were considered valid, usable as lower bounds, or unusable. Some data that were historically considered valid are now evaluated as lower bounds or totally unusable.

The work has confirmed that the form of in–plane bending equation now in DnV, CIDECT, and UK guidance is acceptable, although the lead constant should be altered and rearrangement of the equation is advisable. The work has also revealed a totally new need to compare the calculated joint capacity with brace end plastification and select the lowest value for joint utilization assessment. This additional requirement is offset by a suggested elimination of any requirement to check allowable shear stress, as is currently required by DnV, CIDECT, and API. One other recommended relaxation of requirements is that the limiting yield–to–tensile strength ratio used in API and UK guidance be raised 0.80.

Keywords: In–Plane Bending Strength, Finite Element Applications, Simple Tubular Joints, Design Provisions, New Experimental Database

1 Introduction

Since the late 1970s, extensive worldwide effort has been devoted to development of improved strength (i.e. capacity) equations for circular tubular joints. Several key examples of this work are given in (Yura '80), (UEG '85), and (Ochi '84). The efforts eventually led to changes in design rules, as exemplified by (API '91), (UK DEn '90), and (Wardenier '91). These rules are comparable to those adopted by various other codes of practice (DnV '89, Packer '92). Although not all of these rules are identical, they are at least similar to one another and are founded upon an experimental database. Hence, the term "semi–empirical" is often applied to all of the capacity equations.

Tubular Structures V. Edited by M.G. Coutie and G. Davies.
© 1993 E & FN Spon, 2–6 Boundary Row, London SE1 8HN. ISBN 0 419 18770 7.

A primary problem in developing the subject equations has been the establishment of a so–called valid database. Of course, part of the difficulty relates to identification of worldwide data that should be considered for inclusion in the database. Numerous new data have become available in the 1980s, thus making the early data bases of (Yura '80), (UEG '85), and (Ochi '84) somewhat antiquated. However, an even more pressing part of the problem has been screening the data to ensure that those eventually included in the database are valid to an acceptable degree (Yura '80). Early efforts at screening were admirable, but are now known to have deficiencies.

The case of in–plane bending strength is a good example of the suspect nature of the database, as explained in (Zettlemoyer '88). In many cases the test specimen failure was dominated by brace yield-ing, in the strength–of–materials beam bending sense. Hence, a question has arisen as to how to treat such data. It has been US practice to eliminate all cases where the failure was not clearly that of the joint proper (Yura '80, Zettlemoyer '88). This approach has severely limited the databases, especially in the instance of in–plane bending. Furthermore, elimination of data with premature brace yielding is not always consistent with worldwide practice, which has often been to leave such data in the database to represent geometrical conditions that do indeed exist.

The 1980s have also witnessed a growth in the use of calibrated finite element analysis to investi-gate both linear and nonlinear behavior of joints. Although not yet broadly accepted or implemented as a means to develop design guidance, there are some recent examples where FE results have proven quite useful (e.g. van der Vegte '92). FE analysis has the distinct advantage of permitting a comprehen-sive study of geometric and material parameters, the ranges of which are often limited in the existing database. FE investigations also allow study of both the effect of boundary conditions and the applica-tion of simple joint results to more complex joint configurations. Finally, by means of FE one can sort out the question of how failure is defined and ensure that the results used to develop an equation are consistent. This question about failure definition has been found especially important for the case of in–plane bending because failure is typically not exemplified by an instability peak on the moment–rotation curve.

The following paper presents a new assessment of the in–plane bending strength of circular tubular joints. The resulting equation is based upon FE analysis of a statistically defined matrix of T/Y–joint geometries. Special care has been taken to ensure the failure is consistently defined and representative of true joint failure. Comparison of this equation is made to the valid and lower bound data from a new-ly updated and screened database of all known simple joint tests. In terms of design guidance, the paper addresses such issues as the need for a separate shear stress limit, as is present US and CIDECT gui-dance but not UK practice, and how to include premature brace end failure in the utilization check for the joint.

2 Simple Joints

2.1 FE Studies
As outlined in the Introduction, the strategy employed in this paper to develop a new in–plane bending (IPB) design equation is to base this equation on a series of FE results, and then to demonstrate that this equation provides a good lower bound to the experimental data. This section of the paper deals with the numerical development of the new IPB design equation.

2.1.1 FE Validation
Beginning in the mid 1980s, a number of studies have been conducted to demonstrate that, when crack-ing is not an important failure mechanism, nonlinear FE analysis has the ability to predict an exper-imental load–displacement plot with high accuracy (Hayman '86, van der Vegte '91b). Although confi-

326

dence in FE is high within a number of research groups and total reliance on FE is not intended in this paper, it is considered useful to provide an example of FE performance specific to IPB.

The joint chosen to validate FE analysis as applied to the IPB strength of tubular joints is designated TNO–10 (Stol '85) in Tables 3 and 4, which describe the existing experimental database and are presented in Appendix A. This joint was chosen foremost because its capacity is governed by joint failure and is not unduly influenced by brace yielding, a consideration which will be discussed later on. Also, its load–deformation plot does not exhibit a well defined peak and its capacity must be determined through the application of a deformation limit. This limit is established by arbitrarily restricting the brace rotation that corresponds to the maximum achievable IPB capacity before structural collapse intervenes to

$$\theta = \frac{8 F_y L_b}{3 E d} \; ; \qquad L_b = 30d \tag{1}$$

or

$$\theta = \frac{80 F_y}{E} \tag{2}$$

The brace member length, L_b, representative of a typical upper bound found in offshore structures, is assumed to be 30 times the outside diameter of the brace (Yura '80). Equation 2 has been used extensively in the past and is used throughout this research for the sake of consistency.

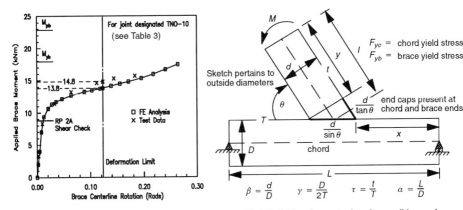

Fig. 1, Moment–rotation plots, TNO joint 10

Fig 2, Definition of geometry, boundary conditions, and applied moment direction in the T/Y–joint numerical database

The moment–rotation plots generated in the test and FE analysis are shown in Fig. 1. The FE analysis was conducted in a manner consistent with those comprising the numerical database. Also plotted are the brace plastic moment, M_{pb}, defined as

$$M_{pb} = F_{yb} \bar{d}^2 t \; ; \qquad \bar{d} = d - t \tag{3}$$

and the brace yield moment, M_{yb}, defined as

$$M_{yb} = \frac{\pi}{4} F_{yb} \bar{d}^2 t \tag{4}$$

In these definitions, the thin shell approximation employing the average shell diameter is used. This approximation is quite accurate for the joints considered in this paper.

In Figure 1, the FE curve is seen to fall quite closely to the experimental data. The capacities determined at the deformation limit differ by less than 7%. This difference is slight given the uncertainties in actual boundary conditions, material properties, nominal dimensions, test rig flexibility and measurements, and other considerations which exist when an actual test is modeled numerically. The good agreement between test and analysis computed for this joint lends confidence that FE analysis provides a reasonably accurate assessment of IPB capacity.

2.1.2 Numerical Database

In order to develop the numerical database, the parameters that have primary influence on the IPB capacity must be identified. The approach taken here is to define the capacity in terms of joint failure alone, so that failure of the brace does not influence the capacity of the joint. Joint failure can be loosely described as collapse of the chord in the vicinity of the brace due to the action of the brace. Certainly, brace deformations in close proximity to the chord will have an impact upon the joint failure, but these effects should be indistinguishable from those of the chord deformations and not be readily identifiable as constituting brace failure.

In this study, the dimensional parameters D, T, d, and t, the brace angle θ, and the chord yield strength F_{yc} are initially considered potentially significant. These parameters, defined in Fig. 2, define the local joint geometry and the material strength of the chord and are considered sufficient to describe true joint failure. The yield strength of the brace, F_{yb}, is indicative of the brace capacity and is assigned a value high enough to ensure that valid joint failure occurs. Valid joint failure is said to be achieved if the capacity of the joint M_u is less than the brace yield moment M_{yb}. Prior to this moment any brace yielding should be localized at the brace–chord intersection and will be an integral part of the joint failure. In a similar fashion, brace failure is said to occur if the joint capacity is greater than the brace plastic moment M_{pb} (Zettlemoyer '88). Brace failure may occur prior to this moment, but should definitely occur beyond. Practically, the point where the influence of brace yielding is acceptable and the capacity represents joint failure is arbitrarily defined as the 95% level; that is, if the capacity computed for the joint is above 95% of the true capacity of the joint, then the computed capacity is accepted.

It can be shown that it is feasible to vary two parameters in order to force valid joint failure: τ and F_{yb}. For various joints in the numerical database, a series of FE analyses were conducted by varying both τ and F_{yb} through the transition from brace to joint failure. These joints were chosen to provide a sampling of β from 0.5 to 0.95 and of γ from 13.3 to 40.0. It was found that it is more reliable to manipulate F_{yb} to ensure joint failure, and that is the choice adopted for the numerical database.

It was also found that the fraction of the difference between M_{yb} and M_{pb} realized when the computed capacity reaches 95% of the true joint capacity ranged from nearly zero to about 0.5. As such, screening against M_{pb} does not eliminate brace failure. Further, there seems to be no specific value of this fraction that generally indicates acceptable joint failure. Consequently, to ensure that brace yielding does not unduly influence the joint capacity, it was decided that experimental data should be screened against M_{yb}.

Length parameters such as the clear distance from the brace footprint to the chord ends, x, and the clear distance from the chord mid–plane to the brace end, y, are assigned values so that the chord and brace end conditions do not unduly affect the joint failure (see Fig. 2). For the numerical database, values of x/D and y/d were maintained at 3.5, so that α and l/d are always greater than 7.0 and 3.5, respectively. FE analyses of the Wimpey/JISSP (Ma '88) test series indicated that short chord lengths, and to a lesser extent short brace lengths, can significantly increase the joint strength when β and γ are high. The same analyses further indicated that the x/D and y/d ratios adopted for the numerical database are sufficient to avoid the undue influence of end conditions caused by short chord and brace lengths.

All key geometric and material parameters are listed for the numerical database in Table 2 of Appendix A. Also listed following the table are the ranges of the dimensional and non–dimensional parameters. It can be seen that a wide range of parameters are covered and that the capacity equation resulting from this database should have an equally wide range of applicability.

The numerical database was originally devised as a two level factorial design based on the 5 geometric parameters D, T, d, t, and θ (Box '78). Such designs indicate well the major trends in the data, are easily augmented for more local explorations, and are simple to interpret. In practice, only the first half factorial was analyzed to get the major geometric trends, requiring 16 runs. The validity of these trends was reinforced by analyzing half of the remaining half factorial, or 8 more runs. The effects of F_{yc} were examined by reanalysis of the latter 8 runs with a higher value of F_{yc}. Finally, the study of the parameter θ was extended with analyses for varying values of θ at the midpoints of the ranges of the other parameters, for a total of 37 runs.

2.1.2.1 FE Modeling
The nonlinear finite element method was used to perform the analyses contained in this study. All analyses were accomplished using the ABAQUS (Hibbit '89) finite element program. As only in–plane loads are considered, in–plane symmetry was exploited in the finite element meshes. Typical meshes contain approximately 500 elements and 1700 nodes, resulting in about 10,000 degrees of freedom. Two typical meshes for both a T– and a Y–joint are shown in Fig. 3. The elements used throughout are eight noded, biquadratic, mid–plane, reduced integration shell elements which include transverse shear effects. Geometric nonlinearities include large displacements and rotations but not large strains.

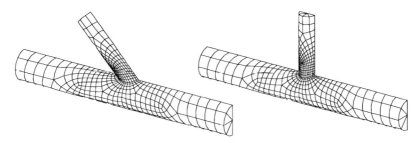

Fig. 3, Typical FE meshes for the numerical T/Y database

In all analyses, explicit modeling of the weld fillets was omitted. It has been found through experience that such modeling usually has a negligible effect on the ultimate strength of T/Y–joints. As accounting for the welds adds an extra level of complexity to the modeling process, it was decided to forego weld modeling given the large number of analyses performed for the T/Y–joint studies.

The simply supported boundary conditions employed at the chord ends are depicted in Fig. 2. These supports are applied to the chord end caps, which are provided both for convenience of load and boundary condition application and to reflect the end conditions commonly utilized in tests. Since the brace and chord lengths are chosen to avoid end effects, the boundary conditions should not influence the capacity and simple support conditions are assumed. If the end conditions do happen to affect the capacity, simple supports are likely to be conservative in that further end restraint will most likely lead to extra strength.

The direction of loading is also indicated in Fig. 2. A concentrated moment causing the brace to rotate in the direction of the acute brace angle is applied to the brace end cap. Based on the previously referenced FE analyses of the Wimpey/JISSP test series, it was found that a moment applied in this direction can lead to significantly less strength than a moment applied in the opposite direction when β

329

and γ are high. In fact, the direction of loading was found to contribute as much to the unusually high capacities recorded for the Wimpey/JISSP test series (see Table 4) as the short chord lengths. As the experimental database includes joints loaded in both directions, the most conservative load direction was chosen to provide lower bound design guidance.

The material behavior of the brace and chord is assumed to be governed by rate independent incremental J_2 flow theory, and isotropic strain hardening is assumed to govern the evolution of the yield surface. The material is initially isotropic with an elastic modulus of 200,000 MPa, and after yielding hardens with a plastic modulus of 2585 MPa. After 8% plastic strain, the material is perfectly plastic. The amount of strain hardening specified is considered representative and conservative in that general materials are more likely to exhibit more, rather than less, hardening. Orthotropic effects in the hoop and longitudinal directions are ignored.

2.1.3 Results

The capacities corresponding to the numerical database are given in Table 2. Table 2 further indicates which joints reach a peak capacity and which joints have capacities determined by the deformation limit. Only 6 joints reached a peak capacity: nos. 3, 4, 10, 19, 26, and 28. Essentially, these joints are characterized by thin chords and/or large brace footprints. The lone exception is joint 28, which combines β= 0.95 with γ= 8. However, the failure mode of this joint is elastic buckling caused by the very high stresses generated in the brace by increasing F_{yb} to ensure joint failure. Joint 7, which has the same geometry but a lower chord yield strength, requires a smaller F_{yb} to guarantee joint failure and encounters the deformation limit well before elastic brace buckling can occur.

Another trend that can be observed from the numerical database is that as θ increases, the joint is less likely to reach a peak. All of the joints experiencing peaks have θ= 45° except joint 19, which has θ= 90°. However, joint 3, which has the same geometry as joint 19 except that θ= 45° (t is different also, but as seen below this should make little difference), also experiences a peak which precedes the deformation limit by a wider margin and thus can be considered more pronounced.

Correlation of failure modes with moment–rotation plots suggests that chord ovalization and local chord plastic buckling are more likely to produce a peak, whereas punching shear behavior is more likely to encounter the deformation limit. Joints with large β and γ, and to a lesser extent θ, are more likely to exhibit the former failure modes, and joints with small β and γ are more likely to experience the latter. Local chord plastic buckling generates the sharpest peak, whereas chord ovalization leads to a gentle peak after which significant post–peak strength is often seen. This post–peak strength appears to result from deformation resembling that of punching shear, which begins to dominate after the chord has ovalized.

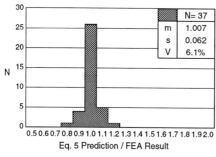

Fig. 4, Histogram of Eq. 5 relative to
computed capacity, numerical database

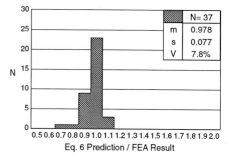

Fig. 5, Histogram of Eq. 6 relative to
computed capacity, numerical database

Regression of the capacities computed by FE analysis and based on the independent variables D, d, T, t, and sin θ yields, after rearrangement,

$$M_u = \frac{17.354 \; \tau^{0.05} \; F_{yc}^{0.91} \; \beta^{2.06} \; \gamma^{1.47} \; T^{3.04}}{\sin \theta^{0.82}} \tag{5}$$

Results of this equation applied to the joints in the numerical database are listed in Table 2, and a histogram relating these results to the computed capacities is shown in Fig. 4. Agreement is seen to be good, with a mean virtually equal to 1.0 and a variation of only 6%.

Examination of Eq. 5 reveals that τ is a very weak parameter and could easily be eliminated. After elimination of τ (actually t in the list of independent variables for regression) and rounding the exponents of other terms, regression to determine the constant yields the following equation

$$M_u = \frac{9.5 \; F_{yc} \; \beta^2 \; \gamma^{1.5} \; T^3}{\sin \theta} \tag{6}$$

Results of Eq. 6 applied to the joints in the numerical database are also listed in Table 2, and a histogram relating these results to the computed capacities is given in Fig. 5. Although the histogram is skewed slightly to the left (more conservative) and the agreement is not as good as it is for Eq. 5, the variation remains only 8% and the form, i.e. the parameters and their exponents, of the equation is simpler. Consequently, Eq. 6 is chosen to represent the mean T/Y capacity of the numerical database. As will be demonstrated, Eq. 6 also serves quite well as a lower bound to the existing experimental data.

Researchers in the past have questioned whether or not the more exact expression, commonly termed K_{bi}, for the section modulus of the brace relative to that of the brace footprint on the chord should be used in lieu of $\sin^{-1} \theta$ (Zettlemoyer '88). In order to answer this question, the mean equation of the numerical database was re–evaluated using K_{bi} in place of sin θ. The resulting equation is nearly identical to Eq. 5, with the exception of the brace angle term. No decrease in the scatter associated with Eq. 5 is obtained by employing K_{bi} in place of sin θ. In fact, more scatter is engendered when τ is discarded and the remaining exponents are rounded. As sin θ is a simpler expression than K_{bi}, and its use leads to an equation that is at least as good and probably better, there is no incentive to use K_{bi} in a design equation.

Table 1, Comparison of Eq. 6 to current design equations

Source	Leading Constant, α	
	$M_u = \dfrac{\alpha \; F_{yc} \; \beta^2 \; \gamma^{1.5} \; T^3}{\sin\theta}$	$M_u = \dfrac{\alpha \; F_{yc} \; d \; T^2 \; \sqrt{\gamma} \; \beta}{\sin\theta}$
Equation 6	9.5	4.75
DnV	12.0	6.0
DEn	10.0	5.0
CIDECT	9.7	4.85
Zettlemoyer '88	9.0	4.5
Yura '85	10.0	5.0

The form of Eq. 6 differs from the form of the design equation currently recommended by DnV, DEn, and CIDECT, which is

$$M_u = \frac{\alpha \; F_{yc} \; d \; T^2 \; \sqrt{\gamma} \; \beta}{\sin \theta} \tag{7}$$

where α denotes a multiplicative constant. Unfortunately, one cannot choose all of the parameters d, T, γ, and β from Eq. 7 independently, so that plots of non–dimensional capacity versus γ or β do not truly isolate the influence of these parameters. By contrast, the parameters T, γ, and β from Eq. 6 can all be chosen independently, so that non–dimensional plots do not include hidden effects from the other parameters.

Through algebraic manipulation, it is easily discovered that the form of Eq. 6 is obtained by a simple transformation of Eq. 7, as other researchers have noticed (Stol '85). In transformation, the lead constant is increased by a factor of 2. Equation 6 can thus be compared to other current or proposed design equations in either form, as shown in Table 1. It is apparent that these current design equations are fundamentally acceptable in form, but that the leading constant should be adjusted.

Figures 6 and 7 contain plots of non–dimensional moment capacity computed by FE analysis against β and γ, respectively. Also plotted is the curve corresponding to the mean Eq. 6. Scatter occurs primarily in the regions of large β and γ. Some of this scatter is due to the rounding of Eq. 5 to Eq. 6, as is apparent by comparison of the histograms. However, most of the scatter is due to the tendency of large β and γ joints to either reach a peak capacity or nearly reach the peak before the deformation limit is crossed. As discussed, these joints exhibit different dominant failure modes than those seen for the joints in the numerical database with lesser values of β and γ.

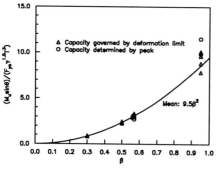

Fig. 6, Non–dimensional capacity
vs. β, numerical T/Y database

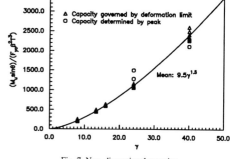

Fig. 7, Non–dimensional capacity
vs. γ, numerical T/Y database

Figures 6 and 7 also show that correlation of the data to the mean equation is outstanding at low β and γ. As discussed previously, the dominant failure mode in this range is shear, and this is precisely the region in which the correlation to Eq. 6 is the best. Further, Fig. 1 shows that the current RP 2A separate shear check versus first yield, computed with the safety factor excluded, is conservative by over 40% for the TNO–10 joint and would needlessly control the design of the joint. The conclusion to be drawn is that the ultimate strength equation derived in this paper incorporates the shear failure mode quite well and an additional shear check is not only unnecessary but often unduly conservative.

2.2 Experimental Database

The new database for in–plane bending capacity of simple joints is given in Tables 3 and 4. Table 3 contains all of the geometric and material information whereas Table 4 lists relevant capacities for each case. These tables include all of the known tests, not just those that have remained after a screening exercise, so that the reader can see how a particular test was evaluated and the results utilized. For each test program a reference is cited, with each reference
being considered the one most often cited for the particular data involved. However, typically not all of the relevant information for a particular test can be found in that reference. In many cases, other source documents had to be obtained in order to complete the tables to the extent shown.

It is of some interest to compare the number of data to that in (Yura '80) and (UEG '85). Ninety–one tests are listed in the tables, whereas only 27 were considered (and 16 validated) in the Yura effort and 37 were evaluated (all validated) in the UEG work. Obviously, the database has increased in size by more than a factor of two in the last decade. Much of the increase has occurred because of tests on load interaction (brace or chord), in which the investigators needed to establish baseline capacities for brace loads acting independently.

Although utilization of the data is discussed in detail below, three concerns of prior validation efforts are worth highlighting at the outset of the discussion. First, the scale of the test specimens is considered adequately large in all cases, even though the diameter of the chords is not always above the cutoffs discussed in (Yura '80) and (UEG '85). The diameters here range from 114 to 508 mm, with a fairly uniform distribution of data across this range. The only portion with poor representation is between 300 and 400 mm.

The second point to keep in mind is that the geometry (diameters and thicknesses, not angles) were typically measured, as opposed to assuming nominal values. (There are some instances where geometry was measured on only one of several nominally identical specimens.) The most critical dimension to have measured is the chord thickness, and almost all data satisfy this need. Hence, all data were considered valid from the measured geometry viewpoint.

The third issue is the quality or type of welding used to connect the brace to the chord. Some of the specimens (e.g. from Stamenkovic '81 and '84) are known to have had fillet welds rather than the full penetration groove welds typically used offshore. Yet, there is scattered evidence that the type of welding rarely alters either the fatigue or strength behavior of a circular tubular joint. Therefore, in evaluating the data a decision was made to disregard the welding type or quality unless the test documentation indicated premature weld failure.

2.2.1 Material Properties

Material yield strengths for the chord and brace members are listed in Table 3. The importance of having the yield value of the brace can not be overstated in the instance of in–plane bending. Often the brace steel is quite different than the chord steel, and it is the brace steel yield that must be used to determine if premature brace failure occurred during the test. There is one instance where the brace yield value was assumed. That was for the Kingston Polytecnic T–joint test series (Stamenkovic '81). The lowest of the measured chord yield strengths was used on the basis that the brace members were known to be of the same type of steel and pipe fabrication process.

Assessment of the data has been based on static yield strength, as measured by common bar or strip specimens extracted from the pipe. The reason for static yield is that the joint tests themselves are conducted in a quasi–static manner; there is interest in accurately representing the yield so that the influence of geometry can be properly deduced. Unfortunately, in most cases the static yield strengths were not determined. Hence, comparative results from several University of Texas programs (Boone '84, Swensson '86, Weinstein '86, Yura '78) and Wimpey's JISSP project (Ma '88) were used to develop an average adjustment factor of 0.93. An approximate static yield value has been estimated using this factor for all data for which a static value was not specifically measured.

The static yield of the brace has been used to assess the yield and plastic moment capacities of the brace. These capacities are designated as M_{yb} and M_{pb} in Table 4. These values are later compared to the moment capacities achieved in the test.

One other bit of material information given in Table 3 is the ultimate tensile strength. API RP2A (API 1991) and and other codes often have a limit on the chord yield value, as a function of the tensile strength. Presumably, this limit accounts for cases where sufficient ductility and/or strain hardening are considered unavailable. API uses a 2/3 value and the UK has a rounded value of 0.7. Yet, Fig. 8 shows that the yield–to–tensile (dynamic) strength ratio for the in–plane data base extends up to about 0.85. Furthermore, Fig. 9 shows the yield values themselves extend up to over 450 MPa. Even more liberal findings are known to exist for the entire axial load data base (671 tests) in (Ochi '84). There is no known evidence of premature failure due to high base metal strength within these databases. Hence, the API restrictions are overly conservative. An allowable ratio of 0.80 can be readily justified, which means that many steels of 345 to 380 MPa nominal yield strength need no adjustment in yield when estimating the joint capacity.

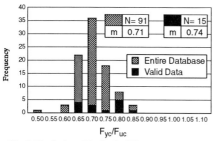

Fig. 8, Distribution of F_{yb}/F_{uc} for IPB experimental database

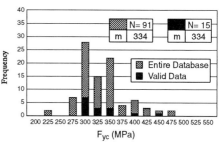

Fig. 9, Distribution of F_{yb} for IPB experimental database

2.2.2 Ultimate Capacity

Table 4 contains several columns relating to ultimate capacity. One labeled simply M_u is generally the maximum value recorded in the test, in all cases measured at the chord surface. Unfortunately, for in–plane bending this maximum is often not a peak on the load–displacement curve. This fact places high importance on reviewing load–displacement curves from each test. Such curves could not be located from one entire program. However, those that were found confirmed that often (especially with γ below 20) the test was stopped before a peak was reached. In a few instances where a peak actually existed, some of the recorded ultimate capacities had to be adjusted to agree with the peak (Stamenkovic '81 and '84).

The general absence of a peak on the load–displacement curve emphasizes the need to consider a displacement limit. (The importance of considering a displacement limit for in–plane bending was not appreciated when Yura assessed the database in the late '70s (Yura '80). Hence, some data from (Gibstein '76) were accepted as–is, even though the load–displacement curves were unavailable.) Although the manner of calculation of the displacement limit is debatable, this paper uses the approach introduced by Yura. Table 4 gives an alternate capacity value, designated M_{ud}, in the instance the load–displacement curve was available and the rotational limit did not exceed either the peak (if there was one) or the end of the test. Only 15 displacement–fixed capacities were thus determined.

For plotting purposes, the controlling capacity from the peak and displacement information has been used. These values have been non–dimensionalized by the chord yield and geometric parameters, as given by Eq. 6.

2.2.3 Disposition Codes

The last column of Table 4 contains one or more disposition codes associated with each test. The word "disposition" is meant to imply how the datum can be treated from the perspective of judging the results of the above FE analyses. There are three general categories of these codes or combinations thereof. First, there are 15 test results that can be considered valid, although the recorded capacity may have to be replaced by the value at the displacement limit. Second, there are 47 data that can be considered lower bounds to joint capacity, whether due to the test being stopped before a peak or displacement limit was reached, due to the capacity being underrated by virtue of premature brace yielding (exceeding M_y, based on the FE results, not just M_p) or, in a few cases, due to known fabrication defects.

The third general category of disposition codes is the one that causes the data to be rejected at this time. One reason is the load–displacement plot is unavailable. There are also other reasons to question some data. For example, the JISSP data for Y joints (Ma '88) for which FE analysis showed that the short chord lengths artificially increased the recorded capacity. However, short chord length or the absence of chord length information was not always cause for data rejection, as seen for some of the cross joints.

Since many data have several disposition codes assigned to them, the approach taken here was to treat the data according to the lowest general category of code shown. For example, a datum with premature brace yielding and without a load–displacement plot was treated as unusable.

2.2.4 Comparisons to FE results

Figures 10 through 13 plot the non–dimensionalized data. Figures 10 and 11 contain plots of all the valid points with respect to β and γ, respectively, while Figs. 12 and 13 plot non–dimensional capacity versus β and γ for all of the known lower bound data. Although the screening of data plotted has been severe, the figures demonstrate that the practical ranges of β and γ are represented by the remaining data.

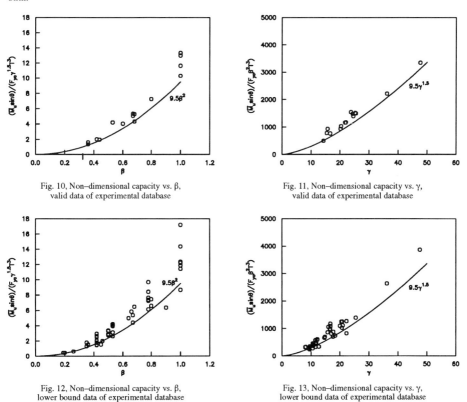

Fig. 10, Non–dimensional capacity vs. β,
valid data of experimental database

Fig. 11, Non–dimensional capacity vs. γ,
valid data of experimental database

Fig. 12, Non–dimensional capacity vs. β,
lower bound data of experimental database

Fig. 13, Non–dimensional capacity vs. γ,
lower bound data of experimental database

The general conclusion drawn from Figs 10–13 is that the lower bound to the data is remarkably well represented by the mean capacity equation derived from FE results. (The few points below the curve in Figs. 12 and 13 are not a concern because they are known to under–represent the actual joint capacity.) There are many possible reasons why the mean FE equation underestimates the test capacities. Some of these include not modeling the welds and less than precise representation of the boundary conditions and/or the material properties (e.g. orthotropic yield conditions or strain hardening modulus). Nevertheless, sorting out the reason(s) for this amount of conservatism is not considered necessary.

One other point to recognize is that the scatter of even the valid data is greater than that experienced in fitting an equation to the FE results. In addition to variations in material properties and boundary conditions, the database has the obvious characteristic of representing all simple joint types, not just the T/Y joints analyzed by FE. Nevertheless, the database scatter is not so large as to suggest that a separate IPB equation is required for each joint type. This finding is consistent with historic development of IPB equations.

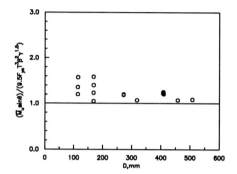

Fig. 14, Non–dimensional capacity vs. chord diameter, valid data of experimental database

Some readers will recall that one objective of the companion T–joint tests at Kingston Polytechnic and Corby was to evaluate the effect of scale (Stamenkovic '84). Unfortunately, a conclusion can not be drawn from the collective data because all but one of the combined results can only be treated as lower bounds. However, it is possible to plot the valid data from the entire database in such a way that some statement about scale can be made. Figure 14 is such a plot. Although there are only 15 points, they do represent the entire range of diameters for the database as a whole. Figure 14 does not suggest there is a need to restrict the database due to scale of the joints included, which is in agreement with the assumption at the outset of the database discussion.

2.3 Design Provisions
To this point the argument has been that joint failure should be evaluated in the pure joint failure mode sense. Mixed mode failure including some degree of brace end yielding would necessitate a capacity expression that includes the brace yield and thickness (or τ) in addition to the usual joint parameters. The fact that present day code provisions with respect to joints do not include these parameters means pure joint failure is all that has been explicitly addressed. Yet, engineers know that platform geometric or material conditions often exist whereby mixed mode or pure brace end failure would be expected. The implicit assumption in most current codes is that premature brace failure, whether or not mixed with joint failure, is considered in the member checks.

Simplified assessment of typical member/joint configurations and relative yield strengths reveals that the safety net assumed to be provided by the member check(s) is by no means guaranteed. For example, even though a brace hypothetically reaches Mp prior to pure joint failure, it is quite possible that the joint will fail first due to lower axial load capacity. In other words, the combination of load–capacity ratios in a utilization check is what ultimately decides which component is perceived to fail first. However, short of much more study of joint capacity when subjected to multiple brace loading, it is necessary to assume that brace end yielding due to bending could reduce the joint capacity due to other brace loads. Therefore, there needs to be a mechanism for including the possibility of premature brace end failure in the calculation of joint utilization values.

The proposal here is that the lower bound joint capacity be represented by either the joint capacity devised above or the plastic moment capacity of the attached brace, whichever is smaller. Use of the plastic moment capacity rather than the brace yield moment is recommended because the joint utilization check refers to a limit state, and because only those joints in the experimental database whose capacities are limited by M_{pb} plot noticeably below Eq. 6. M_{yb} is used to screen the experimental data to ensure joint failure, but generally when the joint capacity is between M_{yb} and M_{pb} it is not far off the true joint capacity.

There is no apparent need to include a chord stress reduction factor along with the plastic moment in the joint unity check because the failure being hypothesized in that instant is on the brace side of the weld. Since large chord load effects are rather common, it is not at all clear how often the plastic moment will govern.

In design it is also necessary to consider resistance or safety factors to apply to the capacities. No attempt is made here to fully rationalize values for the in–plane bending case. However, the 0.95 resistance factor and 1.7 safety factor presently used within API practices appear about right for the pure joint failure mode. (The scatter and bias of the capacity equation relative to the valid data are similar to statistical characteristics of the prior database and capacity equation.) Furthermore, given the uncertainties associated with substituting brace end M_p for M_u in the joint strength utilization check, these same values should probably be used whether pure joint or pure member failure is envisioned.

3 Conclusions

This paper has reported a new investigation of in–plane bending capacity of circular tubular joints. The objective has been to resolve several outstanding concerns about design guidance in existing design codes. Conclusions are based upon extensive FE analyses and both expansion and refined assessment of the historical experimental database. The conclusions are best represented in a list format.

[a] The API equation should be changed. The form of equation currently represented in DnV, CIDECT, and UK guidance is acceptable, although rearrangement is recommended in order to better represent independent variables. Even without rearrangement, the lead constant in these various equations should be somewhat reduced.

[b] The calculated strength of the joint should not exceed the plastic moment capacity of the attached brace. This limit should be imposed when conducting the joint utilization check due to multiple brace loads.

[c] There is no apparent need for a supplemental shear stress limit, as exists in DnV, CIDECT, and API guidance. This limit is particularly onerous for in–plane bending where it controls the capacity at γ values below about 17, depending on the code involved.

[d] For capacity calculations, the API and UK limiting chord yield strength, as a function of the ultimate tensile strength, can be increased to 0.80. Relaxation of this ratio limit means that tensile strength will rarely enter into the calculations for conventional steels.

337

4 Acknowledgements

Development of an updated database and the search for source documents was assisted by MSL Engineering in the UK. Some of the plots and tables were created by James Kahlich, a technician at Exxon Production Research Co.

5 References

Akiyama, N., et al. (1975) Experimental Study on Strength of Joints in Steel Tubular Structures, **JSSC**, Vol. 10, No. 102, June (in Japanese).

American Petroleum Institute (API) (1991) **Recommended Practice for Planning, Designing and Constructing Fixed Offshore Platforms RP 2A**. 19th edition

Boone, T.J., et al. (1984)
Ultimate Strength of Tubular Joints: Chord Stress Effects, in **Offshore Technology Conference Proceedings,** OTC Paper No. 4828, Houston, Texas

Box, G.E.P., Hunter, W.G., Hunter, J.S. (1978) **Statistics for Experimenters**. John Wiley & Sons, Inc.

DnV (1989) **Rules for Classification of Fixed Offshore Installations, Part 3**

Gibstein, M.B. (1976) The Static Strength of T Joints Subjected to In–Plane Bending Moments, **Det norske Veritas**, Rept. No. 76–137, Oslo, Norway

Hayman, B. and Bjornoy, O.H. (1986) Static Strength of Tubular Joints – Phase I Part 2: Establishment of the Approach – Analyses and Tests of T–joints in Compression and Tension, **Veritec Report No. 86–3635**, A.S. VERITEC

Hibbit, D., et al. (1989) **ABAQUS Users Manual**. Version 4.8

Ma, S.Y.A. (1988)
A Test Programme on the Static Ultimate Strength of Welded Fabricated Tubular Joints (JISSP), **OTJ'88 Conference**, Surrey

Mitra, H.S. (1989) Ultimate In–Plane Moment Capacity of T and Y Tubular Joints, **J. Construct. Steel Research**, Vol. 12

Ochi, K., et al. (1984)
Basis for Design of Unstiffened Tubular Joints under Axial Brace Loading, in **IIW Conference on Welding of Tubular Structures**, Doc. XV–561–84, Boston

Packer, J.A., et al. (1992) **Design Guide for Hollow Structural Steel Connections**. Canadian Institute of Steel Construction

Stamenkovic, A. and Sparrow, K.D. (1981)
Experimental Determination of the Ultimate Static Strength of T–Joints in Circular Hollow Steel Sections Subject to Axial Load and Moment, **Conference on Joints in Structural Steelwork**, Teesside Polytechnic

Stamenkovic, A. and Sparrow, K.D. (1984)
In–Plane Bending and Interaction of CHS T– and X–Joints, **IIW Conference on Welding of Tubular Structures**, Boston

Stol, H.G.A., et al. (1985)
Experimental Research on Tubular T–Joints under Proportionally Applied Combined Static Loading, **BOSS'85**, Delft

Swensson, K.D. and Yura, J.A. (1986) Ultimate Strength of Double–Tee Tubular Joints: Interaction Effects, **PMFSEL Report**, U. Texas

Tebbett, I.E., et al. (1979)
The Punching Shear Strength of Tubular Joints Reinforced with a Grouted Pile, in **Offshore Technology Conference Proceedings**, OTC Paper No. 3463, Houston, Texas

UEG (1985) **Static Strength of Simple Welded Joints, Part B, Vol. 2, Design of Tubular Joints for Offshore Structures**. Pub. No. UR33, CIRIA

UK DEn (1990) **Offshore Installations: Guidance on Design, Construction and Certification**. 4th edition

van der Vegte, G.J., et al. (1991a) The Static Strength and Stiffness of Multiplanar Tubular Steel X–Joints, **ISOPE Journal**, Vol. 1, No. 1, March

van der Vegte, G.J., et al. (1991b) Numerical Simulations of Experiments on Multiplanar Tubular Steel X–Joints, **ISOPE Journal**, Vol. 1, No. 3, September

van der Vegte, G.J., et al. (1992) The Static Strength of Tubular Steel X–Joints Reinforced by a Can, **ISOPE Journal**, Vol. 2, No. 1, March

Veritas Sesam Systems A.S (1992) **PRETUBE Users Manual**. Version 5.5–01

Wardenier, J., et al. (1991) **Design Guide for Circular Hollow Section (CHS) Joints under Predominantly Static Loading**. CIDECT

Weinstein, R.M. and Yura, J.A. (1986)
The Effect of Chord Stress on the Static Strength of DT Tubular Connections, in **Offshore Technology Conference Proceedings,** OTC Paper No. 5135, Houston, Texas

Yura, J.A. (1978) Ultimate Load Tests on Tubular Connections, **CESRL Report No. 78–1**, U. Texas

Yura, J.A. (1985)
Connections with Round Tubes, in **Symposium – Hollow Structural Sections in Building Construction**, ASCE Structural Engineering Congress, Chicago, Illinois

Yura J.A., et al. (1980)
Ultimate Capacity Equations For Tubular Joints, in **Offshore Technology Conference Proceedings**, OTC Paper No. 3690, Houston, Texas

Zettlemoyer, N. (1988)
Developments in Ultimate Strength Technology for Simple Tubular Joints, in **Offshore Tubular Joints Conference, OTJ '88**, Paper No. 4, UK

6   Appendix A. Tables Defining The Numerical And Existing Databases

Table 2, Numerical T/Y–Joint Database

joint	D mm	d mm	T mm	t mm	F_{yc} Mpa	θ deg.	β	γ	τ	M_u, FEA kNm	def. lim. governs?	M_u, Eq.5 kNm.	M_u Eq.6 kNm
1	304.8	152.4	6.35	3.81	296.5	45.00	0.50	24.00	0.60	29.861	yes	29.493	29.795
2	508.0	152.4	6.35	3.81	296.5	45.00	0.30	40.00	0.60	21.897	yes	21.820	23.079
3	304.8	289.6	6.35	3.81	296.5	45.00	0.95	24.00	0.60	124.58	no	110.65	107.56
4	508.0	289.6	6.35	3.81	296.5	45.00	0.57	40.00	0.60	80.174	no	81.860	83.316
5	304.8	152.4	19.05	3.81	296.5	45.00	0.50	8.00	0.20	150.42	yes	156.66	154.82
6	508.0	152.4	19.05	3.81	296.5	45.00	0.30	13.33	0.20	123.57	yes	115.89	119.92
7	304.8	289.6	19.05	3.81	296.5	45.00	0.95	8.00	0.20	515.10	yes	587.74	558.90
8	508.0	289.6	19.05	3.81	296.5	45.00	0.57	13.33	0.20	420.25	yes	434.81	432.92
9	304.8	152.4	6.35	6.35	296.5	45.00	0.50	24.00	1.00	29.258	yes	30.257	29.795
10	508.0	289.6	6.35	6.35	296.5	45.00	0.57	40.00	1.00	79.812	no	83.978	83.316
11	508.0	152.4	19.05	6.35	296.5	45.00	0.30	13.33	0.33	128.89	yes	118.89	119.92
12	304.8	289.6	19.05	6.35	296.5	45.00	0.95	8.00	0.33	579.11	yes	602.95	558.90
13	304.8	152.4	6.35	3.81	296.5	90.00	0.50	24.00	0.60	22.060	yes	22.198	21.068
14	508.0	289.6	6.35	3.81	296.5	90.00	0.57	40.00	0.60	61.693	yes	61.609	58.913
15	508.0	152.4	19.05	3.81	296.5	90.00	0.30	13.33	0.20	88.713	yes	87.225	84.798
16	304.8	289.6	19.05	3.81	296.5	90.00	0.95	8.00	0.20	448.34	yes	442.35	395.20
17	304.8	152.4	6.35	6.35	296.5	90.00	0.50	24.00	1.00	21.891	yes	22.772	21.068
18	508.0	152.4	6.35	6.35	296.5	90.00	0.30	40.00	1.00	15.631	yes	16.846	16.319
19	304.8	289.6	6.35	6.35	296.5	90.00	0.95	24.00	1.00	102.75	no	85.434	76.056
20	508.0	289.6	6.35	6.35	296.5	90.00	0.57	40.00	1.00	64.253	yes	63.203	58.912
21	304.8	152.4	19.05	6.35	296.5	90.00	0.50	8.00	0.33	114.51	yes	120.96	109.47
22	508.0	152.4	19.05	6.35	296.5	90.00	0.30	13.33	0.33	90.436	yes	89.481	84.798
23	304.8	289.6	19.05	6.35	296.5	90.00	0.95	8.00	0.33	469.04	yes	453.79	395.20
24	508.0	289.6	19.05	6.35	296.5	90.00	0.57	13.33	0.33	324.40	yes	335.71	306.12
25	304.8	152.4	6.35	3.81	434.4	45.00	0.50	24.00	0.60	41.828	yes	41.752	43.653
26	508.0	289.6	6.35	3.81	434.4	45.00	0.57	40.00	0.60	108.24	no	115.88	122.07
27	508.0	152.4	19.05	3.81	434.4	45.00	0.30	13.33	0.20	178.65	yes	164.06	175.70
28	304.8	289.6	19.05	3.81	434.4	45.00	0.95	8.00	0.20	713.42	no	832.01	818.85
29	304.8	152.4	6.35	6.35	434.4	90.00	0.50	24.00	1.00	30.378	yes	32.237	30.867
30	508.0	289.6	6.35	6.35	434.4	90.00	0.57	40.00	1.00	87.285	yes	89.471	86.314
31	508.0	152.4	19.05	6.35	434.4	90.00	0.30	13.33	0.33	132.51	yes	126.67	124.24
32	304.8	289.6	19.05	6.35	434.4	90.00	0.95	8.00	0.33	654.92	yes	642.39	579.01
33	406.4	228.6	12.7	5.08	365.4	30.00	0.56	16.00	0.40	279.04	yes	268.31	286.24
34	406.4	228.6	12.7	5.08	365.4	45.00	0.56	16.00	0.40	200.30	yes	201.93	202.40
35	406.4	228.6	12.7	5.08	365.4	60.00	0.56	16.00	0.40	167.75	yes	171.01	165.26
36	406.4	228.6	12.7	5.08	365.4	75.00	0.56	16.00	0.40	153.38	yes	156.36	148.17
37	406.4	228.6	12.7	5.08	365.4	90.00	0.56	16.00	0.40	149.69	yes	151.98	143.12

For the definition of parameters, please see Fig. 2

Range of Key Dimensional Parameters   **Range of Key Non–Dimensional Parameters**   **Deformation Limits Applied**

$304.8 \leq D \leq 508.0$ (Discrete: 304.8, 406.4, 508.0)

$152.4 \leq d \leq 289.6$ (Discrete: 152.4, 228.6, 289.6)

$6.35 \leq T \leq 19.05$ (Discrete: 6.35, 12.7, 19.05)

$3.81 \leq t \leq 6.35$ (Discrete: 3.81, 5.08, 6.35)

$296.5 \leq F_{yc} \leq 434.4$ (Discrete: 296.5, 365.4, 434.4)

$0.30 \leq \beta \leq 0.95$

$8.0 \leq \gamma \leq 40.0$

$30.0 \leq \theta \leq 90.0$

$0.20 \leq \tau \leq 1.0$

$F_{yc} = 296.5$: 0.1186

$F_{yc} = 365.4$: 0.1462

$F_{yc} = 434.4$: 0.1738

Table 3. Database: Geometric & Material Properties.

Organization (Author)	ID	D mm	T mm	d mm	t mm	θ deg.	L mm	l mm	γ	β	τ	α	l/d	F_{yc} Dynamic N/mm²	F_{yc} Static N/mm²	F_{uc} N/mm²	F_{yc}/F_{uc} Dynamic	F_{yb} Dynamic N/mm²	F_{yb} Static N/mm²
T-Joints																			
JSSC	B-40-.3	165.5	4.70	42.7	3.30	90	662	728	17.6	0.26	0.70	8.0	17.1	471	[438]	539	0.87	343	[319]
(Akiya-	B-40-.5	165.5	4.50	76.3	2.90	90	662	728	18.4	0.46	0.64	8.0	9.5	471	[438]	539	0.87	373	[347]
ma, '75)	B-70-.2	318.4	4.40	60.5	3.00	90	1274	652	36.2	0.19	0.68	8.0	10.8	441	[410]	539	0.82	333	[310]
	B-70-.4	318.5	4.40	139.8	4.40	90	1274	652	36.2	0.44	1.00	8.0	4.7	441	[410]	539	0.82	343	[319]
	B-100-.2	456.9	4.80	89.1	3.00	90	1828	728	47.6	0.20	0.63	8.0	8.2	402	[374]	539	0.75	363	[338]
	B-100-.4	457.6	4.80	165.2	4.70	90	1830	728	47.7	0.36	0.98	8.0	4.4	402	[374]	539	0.75	471	[438]
Wimpey	W7	508.0	12.70	193.7	6.35	90	1575	508	20.0	0.38	0.50	6.2	2.6	338	[314]	462	0.73	315	[293]
(Tebbett,	W9	508.0	12.70	193.7	6.35	90	1575	508	20.0	0.38	0.50	6.2	2.6	338	[314]	462	0.73	315	[293]
'79)	A17	508.0	7.90	168.3	7.94	90	1575	508	32.2	0.33	1.01	6.2	3.0	335	[312]	463	0.72	312	[290]
	A19	508.0	7.90	168.3	7.94	90	1575	508	32.2	0.33	1.01	6.2	3.0	335	[312]	463	0.72	312	[290]
TNO	1	168.3	5.78	60.6	5.63	90	840	590	14.6	0.36	0.97	10.0	9.7	286	[266]	398	0.72	316	[294]
(Stol,	3	168.7	10.55	59.8	11.10	90	840	590	8.0	0.35	1.05	10.0	9.9	263	[245]	426	0.62	376	[350]
'85)	10	168.3	5.90	114.6	5.95	90	840	590	14.3	0.68	1.01	10.0	5.1	332	[309]	410	0.81	335	[312]
	13	168.5	3.45	60.8	3.81	90	840	590	24.4	0.36	1.10	10.0	9.7	299	[278]	370	0.81	296	[275]
	16	168.5	3.42	114.7	3.90	90	840	590	24.6	0.68	1.14	10.0	5.1	303	[282]	383	0.79	250	[233]
	19	168.8	3.55	168.8	3.55	90	840	590	23.8	1.00	1.00	10.0	3.5	305	[284]	366	0.83	305	[284]
	61	168.1	5.68	168.3	5.78	90	675	590	14.8	1.00	1.02	8.0	3.5	305	[284]	397	0.77	286	[266]
	70	168.4	10.28	114.5	11.31	90	675	590	8.2	0.68	1.10	8.0	5.2	235	[219]	385	0.61	293	[272]
Veritec	1	298.5	10.30	101.2	5.00	90	1865	2500	14.5	0.34	0.49	12.5	24.7	294	[273]	422	0.70	294	[273]
(Gibstein,	2	298.5	10.00	108.2	6.30	90	1865	2500	14.9	0.36	0.63	12.5	23.1	294	[273]	422	0.70	306	[285]
'76)	3	298.5	10.00	108.2	8.00	90	1865	2500	14.9	0.36	0.80	12.5	23.1	294	[273]	422	0.70	309	[287]
	4	219.1	6.30	71.6	18.50	90	1865	2500	17.4	0.33	2.94	17.0	34.9	314	[292]	439	0.72	335	[312]
	5	219.1	8.90	71.6	18.50	90	1865	2500	12.3	0.33	2.08	17.0	34.9	422	[392]	579	0.73	335	[312]
	6	298.5	7.20	101.6	16.00	90	1865	2500	20.7	0.34	2.22	12.5	24.6	294	[273]	422	0.70	417	[388]
	7	219.1	5.50	101.6	16.00	90	1865	2500	19.9	0.46	2.91	17.0	24.6	305	[284]	438	0.70	417	[388]
	8	219.1	8.40	101.6	16.00	90	1865	2500	13.0	0.46	1.90	17.0	24.6	367	[341]	549	0.67	417	[388]
	9	219.1	10.00	101.6	16.00	90	1865	2500	11.0	0.46	1.60	17.0	24.6	368	[342]	579	0.64	417	[388]
	10	219.1	12.30	101.6	16.00	90	1865	2500	8.9	0.46	1.30	17.0	24.6	404	[376]	596	0.68	417	[388]
	11	219.1	6.00	139.7	17.50	90	1865	2500	18.3	0.64	2.92	17.0	17.9	314	[292]	439	0.72	415	[386]
	12	219.1	8.80	139.7	17.50	90	1865	2500	12.4	0.64	1.99	17.0	17.9	422	[392]	579	0.73	415	[386]
	13	219.1	12.30	139.7	17.50	90	1865	2500	8.9	0.64	1.42	17.0	17.9	392	[365]	549	0.71	415	[386]
	14	298.5	7.30	193.7	7.10	90	1865	2500	20.4	0.65	0.97	12.5	12.9	296	[275]	439	0.67	340	[316]
	15	298.5	10.00	193.7	7.10	90	1865	2500	14.9	0.65	0.71	12.5	12.9	294	[273]	422	0.70	340	[316]
	16	298.5	10.00	193.7	7.10	90	1865	2500	14.9	0.65	0.71	12.5	12.9	294	[273]	422	0.70	340	[316]
	17	219.1	5.90	177.8	16.00	90	1865	2500	18.6	0.81	2.71	17.0	14.1	314	[292]	439	0.72	399	[371]
	18	219.1	8.60	177.8	16.00	90	1865	2500	12.7	0.81	1.86	17.0	14.1	422	[392]	579	0.73	399	[371]
	19	219.1	12.50	177.8	16.00	90	1865	2500	8.8	0.81	1.28	17.0	14.1	392	[365]	549	0.71	399	[371]
Kingston	A2	114.2	3.44	48.3	3.99	90	795	310	16.6	0.42	1.16	13.9	6.4	347	[323]	497	0.70	[320]	[298]
Poly.	B2	114.1	4.95	48.4	3.72	90	795	310	11.5	0.42	0.75	13.9	6.4	329	[306]	480	0.69	[320]	[298]
(Stamen-	C2	114.3	5.41	48.4	4.14	90	795	310	10.6	0.42	0.77	13.9	6.4	333	[310]	489	0.68	[320]	[298]
kovic,	D2	113.9	6.01	48.4	4.03	90	795	310	9.5	0.42	0.67	14.0	6.4	352	[327]	512	0.69	[320]	[298]
'81)	E2	114.1	3.45	60.7	4.95	90	795	310	16.5	0.53	1.43	13.9	5.1	388	[361]	507	0.77	[320]	[298]
	F2	114.1	4.92	60.6	4.85	90	795	310	11.6	0.53	0.99	13.9	5.1	320	[298]	471	0.68	[320]	[298]
	G2	114.0	6.05	60.6	4.71	90	795	310	9.4	0.53	0.78	13.9	5.1	349	[325]	515	0.68	[320]	[298]
	H2	114.2	3.45	76.1	4.33	90	795	310	16.6	0.67	1.26	13.9	4.1	356	[331]	492	0.72	[320]	[298]
	J2	114.1	4.92	76.0	4.00	90	795	310	11.6	0.67	0.89	13.9	4.1	330	[307]	453	0.73	[320]	[298]
	K2	114.3	5.41	75.9	4.51	90	795	310	10.6	0.66	0.83	13.9	4.1	342	[318]	499	0.69	[320]	[298]
	L2	114.1	6.03	76.0	5.00	90	795	310	9.5	0.67	0.83	13.9	4.1	362	[337]	498	0.73	[320]	[298]
	M2	114.2	3.42	89.0	4.85	90	795	310	16.7	0.78	1.42	13.9	3.5	347	[323]	489	0.71	[320]	[298]
	N2	114.3	5.41	89.3	4.77	90	795	310	10.6	0.78	0.88	13.9	3.5	345	[321]	498	0.69	[320]	[298]
	P2	114.0	5.96	89.1	4.95	90	795	310	9.6	0.78	0.83	13.9	3.5	361	[336]	504	0.72	[320]	[298]
	Q2	114.2	3.42	114.1	5.02	90	795	310	16.7	1.00	1.47	13.9	2.7	341	[317]	497	0.69	[320]	[298]
	R2	114.1	4.95	114.2	5.01	90	795	310	11.5	1.00	1.01	13.9	2.7	335	[312]	478	0.70	[320]	[298]
	S2	114.1	5.93	114.3	4.87	90	795	310	9.6	1.00	0.82	13.9	2.7	359	[334]	506	0.71	[320]	[298]

Table 3 (Cont.). Database: Geometric & Material Properties.

Organi-zation (Author)	ID	D mm	T mm	d mm	t mm	θ deg.	L mm	l mm	γ	β	τ	α	l/d	F_{yc} Dynamic N/mm²	F_{yc} Static N/mm²	F_{uc} Dynamic N/mm²	F_{yc}/F_{uc} Dynamic	F_{yb} Dynamic N/mm²	F_{yb} Static N/mm²
T-Joints (cont.)																			
Corby	TCC-1	273.4	12.65	219.5	12.40	90	2134	864	10.8	0.80	0.98	15.6	3.9	290	[270]	482	0.60	316	[294]
(Stamen-	TCC-2	272.6	8.00	218.8	8.16	90	2134	864	17.0	0.80	1.02	15.7	3.9	284	[264]	430	0.66	276	[257]
kovic,	TCC-3	273.0	5.95	219.0	6.27	90	2134	864	22.9	0.80	1.05	15.6	3.9	304	[283]	457	0.67	290	[270]
'84)	TCC-4	273.0	12.48	114.3	6.00	90	2134	464	10.9	0.42	0.48	15.6	4.1	233	[217]	485	0.48	367	[341]
	TCC-5	273.0	7.70	114.3	6.00	90	2134	464	17.7	0.42	0.78	15.6	4.1	284	[264]	430	0.66	367	[341]
	TCC-6	273.0	5.98	114.3	6.00	90	2134	464	22.8	0.42	1.00	15.6	4.1	304	[283]	457	0.67	367	[341]
	TCC-7	168.3	6.64	76.1	4.85	90	2134	516	12.7	0.45	0.73	25.4	6.8	353	[328]	439	0.80	346	[322]
Y-Joints																			
Wimpey	1.9	508.0	12.40	203.0	12.40	45	1575	1000	20.5	0.40	1.00	6.2	4.9	322	[299]	483	0.67	475	[442]
/JISSP	1.10	508.0	12.40	406.0	12.40	45	1575	1000	20.5	0.80	1.00	6.2	2.5	317	[295]	448	0.71	387	[360]
(Ma,	1.11	508.0	7.90	406.0	7.90	45	1575	1000	32.2	0.80	1.00	6.2	2.5	278	[259]	424	0.66	378	[352]
'88)	1.12	508.0	8.00	508.0	8.00	45	1575	1000	31.8	1.00	1.00	6.2	2.0	300	[279]	428	0.70	405	[377]
DT-Joints																			
Kingston	XCC-1	114.3	3.70	114.7	5.00	90	795	310	15.4	1.00	1.35	13.9	2.7	292	[272]	449	0.65	311	[289]
Poly.	2	113.9	3.62	88.9	5.04	90	795	310	15.7	0.78	1.39	14.0	3.5	292	[272]	449	0.65	357	[332]
(Stamen-	3	114.9	3.66	60.9	4.85	90	795	310	15.7	0.53	1.33	13.8	5.1	292	[272]	449	0.65	363	[338]
kovic,	4	114.8	3.59	48.4	4.95	90	795	310	16.0	0.42	1.38	13.9	6.4	292	[272]	449	0.65	365	[339]
'84)	5	114.7	4.85	114.3	4.90	90	795	310	11.8	1.00	1.01	13.9	2.7	311	[289]	462	0.67	311	[289]
	6	114.0	4.95	89.0	5.00	90	795	310	11.5	0.78	1.01	13.9	3.5	311	[289]	462	0.67	357	[332]
	7	115.0	4.71	60.7	5.00	90	795	310	12.2	0.53	1.06	13.8	5.1	311	[289]	462	0.67	363	[338]
	8	115.1	4.75	48.3	4.95	90	795	310	12.1	0.42	1.04	13.8	6.4	311	[289]	462	0.67	365	[339]
	9	114.3	6.10	114.7	4.90	90	795	310	9.4	1.00	0.80	13.9	2.7	349	[325]	478	0.73	311	[289]
	10	114.4	6.01	89.0	5.01	90	795	310	9.5	0.78	0.83	13.9	3.5	349	[325]	478	0.73	357	[332]
	11	114.4	6.16	60.6	5.00	90	795	310	9.3	0.53	0.81	13.9	5.1	349	[325]	478	0.73	363	[338]
	12	114.6	6.05	48.3	4.95	90	795	310	9.5	0.42	0.82	13.9	6.4	349	[325]	478	0.73	365	[339]
	13	272.1	6.15	218.6	6.30	90	2134	864	22.1	0.80	1.02	15.7	4.0	304	[283]	457	0.67	290	[270]
	14	272.4	6.25	114.4	6.20	90	2134	464	21.8	0.42	0.99	15.7	4.1	304	[283]	457	0.67	367	[341]
U. Tex. (Boone, '84)	I7	408.0	8.05	274.0	6.60	90	3516	1396	25.3	0.67	0.82	17.2	5.1	334	321	470	0.71	339	331
(Wein-stein, '86)	I24	407.0	7.98	407.0	7.98	90	3516	1397	25.5	1.00	1.00	17.3	3.4	348	336	443	0.79	348	336
(Swens-son, '86)	I43	407.0	7.98	142.0	6.65	90	3516	1397	25.5	0.35	0.83	17.3	9.8	350	337	444	0.79	295	273
U. Delft (Vegte, '91)	X5	409.0	10.00	246.0	10.35	90	2440	1225	20.4	0.60	1.03	11.9	5.0	318	[296]	425	0.75	284	[264]
K-Joints																			
U. Tex. (Yura, '78)	A2-X-90	507.0	11.38	326.0	7.34	90	3591	1080	22.3	0.64	0.64	14.2	3.3	383	353	516	0.74	410	386
	A2-X-30	507.0	11.38	456.0	9.50	30	3591	1905	22.3	0.90	0.83	14.2	4.2	383	353	516	0.74	366	335
Wimpey	3.1	508.0	12.40	254.0	12.40	45	3200	1100	20.5	0.50	1.00	12.6	4.3	280	[260]	436	0.64	377	[351]
/JISSP	3.2	508.0	12.60	508.0	12.60	45	3200	1100	20.2	1.00	1.00	12.6	2.2	349	[325]	504	0.69	390	[363]
(Ma, '88)	3.3	508.0	12.20	254.0	12.20	45	3200	1100	20.8	0.50	1.00	12.6	4.3	274	[255]	426	0.64	373	[347]
	3.4	508.0	12.30	254.0	12.30	45	3200	1100	20.7	0.50	1.00	12.6	4.3	310	257	445	0.70	378	[352]
	3.5	508.0	12.70	508.0	12.70	45	3200	1100	20.0	1.00	1.00	12.6	2.2	350	[326]	494	0.71	377	[351]
	3.6	508.0	12.10	254.0	12.10	45	3200	1100	21.0	0.50	1.00	12.6	4.3	294	[273]	449	0.65	369	[343]

Nomenclature

D	chord diameter	l	brace length to chord surface	F_{yc}	chord steel yield stress
T	chord thickness	γ	chord thinness ratio	F_{uc}	chord steel ultimate tensile stress
d	brace diameter	β	brace-chord diameter ratio	F_{yb}	brace steel yield stress
θ	brace angle wrt chord	τ	brace-chord thickness ratio		
L	chord length, not including end fixtures	α	chord length to radius ratio	**Symbols**	
				[ ]	estimated value

Table 4. Database: Capacities & Disposition Codes.

Organization (Author)	ID	M_{yb} kN-m	M_{pb} kN-m	M_u kN-m	M_u / l kN	$80\,F_{yc}$ / E Rad.	$80\,F_{yc}$ / Deg.	M_{ud} kN-m	$\overline{M}_u \sin(\theta)$ / $F_{yc}\,T^3\,\gamma^{1.5}$	$\overline{M}_u \sin(\theta)$ / $F_{yc}\,T^3\,\beta^2$	Disposition Code(s)
T-Joints											
JSSC	B-40-.3	1.19	1.64	2.11	2.9	0.1752	10.0	**	0.63	696.99	c
(Akiyama,	B-40-.5	4.10	5.42	6.28	8.6	0.1752	10.0	**	2.00	740.23	c
'75)	B-70-.2	2.30	3.07	3.33	5.1	0.1641	9.4	**	0.44	2639.98	c
	B-70-.4	19.59	25.74	14.91	22.9	0.1641	9.4	**	1.96	2215.15	a
	B-100-.2	5.71	7.51	6.08	8.3	0.1495	8.6	**	0.45	3866.85	c
	B-100-.4	40.50	53.05	18.04	24.8	0.1495	8.6	**	1.33	3347.77	a
Wimpey	W7	49.66	65.32	77.20	152.0	0.1257	7.2		1.34	824.66	e, g
(Tebbett,	W9	49.66	65.32	79.10	155.7	0.1257	7.2		1.37	844.96	e, g
'79)	A17	44.44	59.29	36.95	72.7	0.1246	7.1	**	1.32	2191.62	g
	A19	44.44	59.29	35.94	70.7	0.1246	7.1	**	1.28	2131.71	g
TNO	1	3.60	5.02	4.80	8.1	0.1064	6.1	4.4	1.54	660.76	b, c
(Stol, '85)	3	6.19	9.37	13.20	22.4	0.0978	5.6	11.6	1.79	321.43	b, c
	10	16.34	21.90	15.80	26.8	0.1235	7.1	14.8	4.33	503.37	b
	13	2.52	3.41	2.70	4.6	0.1112	6.4	2.2	1.60	1479.80	b
	16	8.46	11.14	7.50	12.7	0.1127	6.5	7.3	5.30	1397.63	b
	19	21.15	27.50	19.60	33.2	0.1135	6.5	**	13.32	1544.50	a
	61	30.84	40.62	37.50	63.6	0.1135	6.5	36.5	12.34	700.54	b, c
	70	23.51	32.95	39.00	66.1	0.0874	5.0	36.0	6.47	327.98	b, c
Veritec	1	9.47	12.66			0.1094	6.3				c, e, g
(Gibstein,	2	13.82	18.64			0.1094	6.3				c, e, g
'76)	3	16.89	23.13			0.1094	6.3				c, e, g
	4	10.61	16.91	8.24	3.3	0.1168	6.7		1.56	1056.70	e
	5	10.61	16.91	17.75	7.1	0.1570	9.0		1.49	600.75	c, e
	6	31.14	46.00	14.32	5.7	0.1094	6.3		1.49	1211.20	e
	7	31.14	46.00	11.67	4.7	0.1135	6.5		2.78	1150.00	e
	8	31.14	46.00	25.79	10.3	0.1365	7.8		2.71	592.87	e
	9	31.14	46.00	34.91	14.0	0.1369	7.8		2.81	474.37	c, e
	10	31.14	46.00	53.94	21.6	0.1503	8.6		2.90	358.78	c, e
	11	70.71	101.55	25.79	10.3	0.1168	6.7		5.24	1005.72	e
	12	70.71	101.55	58.84	23.5	0.1570	9.0		5.01	541.15	e
	13	70.71	101.55	88.26	35.3	0.1458	8.4		4.89	320.02	c, e
	14	59.23	78.21	53.45	21.4	0.1101	6.3		5.40	1185.32	e
	15	59.23	78.21	78.46	31.4	0.1094	6.3		4.98	681.47	c, e, f
	16	59.23	78.21	85.62	34.2	0.1094	6.3		5.43	743.66	c, e
	17	112.17	155.94	40.50	16.2	0.1168	6.7		8.44	1025.43	e
	18	112.17	155.94	98.07	39.2	0.1570	9.0		8.64	596.58	e
	19	112.17	155.94	160.83	64.3	0.1458	8.4		8.71	343.00	c, e
Kingston	A2	1.69	2.34	2.33	7.5	0.1291	7.4	**	2.62	989.96	c
Poly.	B2	1.61	2.22	2.16	7.0	0.1224	7.0	*	1.49	323.49	c, d
(Stamenkovic,	C2	1.75	2.42	2.44	7.9	0.1239	7.1	**	1.45	277.57	c
'81)	D2	1.71	2.37	3.09	10.0	0.1309	7.5	*	1.49	241.18	c, d
	E2	3.33	4.59	4.02	13.0	0.1443	8.3	*	4.03	958.41	c, d
	F2	3.30	4.55	5.88	19.0	0.1190	6.8	*	4.20	582.16	c, d
	G2	3.19	4.39	5.42	17.5	0.1298	7.4	*	2.61	266.92	c, d
	H2	4.93	6.65	4.64	15.0	0.1324	7.6	**	5.07	768.95	a
	J2	4.99	6.72	7.74	25.0	0.1228	7.0	**	5.36	477.12	c
	K2	5.07	6.85	10.07	32.5	0.1272	7.3	**	5.83	453.63	c
	L2	5.53	7.51	9.44	30.5	0.1347	7.7	**	4.39	288.26	c
	M2	7.62	10.23	8.54	27.5	0.1291	7.4	*	9.69	1088.81	c, d
	N2	7.56	10.15	13.30	42.9	0.1283	7.4	**	7.63	429.00	c
	P2	7.76	10.44	15.19	49.0	0.1343	7.7	*	7.23	349.93	c, d
	Q2	13.38	17.79	14.88	48.0	0.1269	7.3	**	17.20	1175.19	c
	R2	13.38	17.79	21.24	68.5	0.1246	7.1	**	14.36	561.06	c
	S2	13.08	17.37	23.72	76.5	0.1335	7.7	**	11.42	339.51	c

Table 4 (Cont.). Database: Capacities & Disposition Codes.

Organization (Author)	ID	M_{yb} kN-m	M_{pb} kN-m	M_u kN-m	$\dfrac{M_u}{l}$ kN	$\dfrac{80 F_{yc}}{E}$ Rad.	Deg.	M_{ud} kN-m	$\dfrac{\overline{M}_u \sin(\theta)}{F_{yc} T^3 \gamma^{1.5}}$	$\dfrac{\overline{M}_u \sin(\theta)}{F_{yc} T^3 \beta^2}$	Disposition Code(s)
T–Joints (cont.)											
Corby	TCC-1	116.24	156.48	127.87	148.0	0.1079	6.2	**	6.59	363.37	c
(Stamen-	TCC-2	70.37	92.98	70.80	82.0	0.1056	6.1	**	7.44	812.68	c
kovic, '84)	TCC-3	58.43	76.55	54.40	63.0	0.1131	6.5		8.31	1419.48	e
	TCC-4	17.93	24.04	32.00	69.0	0.0867	5.0	31.1	2.04	421.22	c, f
	TCC-5	17.93	24.04	18.80	40.5	0.1056	6.1	18.6	2.07	879.98	c
	TCC-6	17.93	24.04	15.50	33.3	0.1131	6.5		2.35	1462.53	e
	TCC-7	5.85	7.93	6.64	12.9	0.1313	7.5	*	1.53	337.91	c, d
Y–Joints											
Wimpey	1.9	147.37	199.28	206.00	206.0	0.1198	6.9	200.0	2.67	1551.12	b, c, g
/JISSP	1.10	526.96	691.62	762.00	762.0	0.1179	6.8	*	10.34	1500.75	c, d, g
(Ma, '88)	1.11	339.09	440.19	376.00	376.0	0.1034	5.9	**	11.44	3265.42	c, g
	1.12	582.47	753.36	712.00	712.0	0.1116	6.4	**	19.70	3524.45	c, g
DT–Joints											
Kingston	XCC-1	13.09	17.40	10.85	35.0	0.1086	6.2	**	12.99	783.57	a
Poly.	2	8.75	11.78	6.76	21.8	0.1086	6.2	*	8.41	861.69	d
(Stamenkovic,	3	3.75	5.16	3.47	11.2	0.1086	6.2	**	4.20	931.88	a
'84)	4	2.26	3.18	2.36	7.6	0.1086	6.2	*	2.94	1056.71	c, d
	5	12.75	16.94	15.92	51.4	0.1157	6.6	**	11.87	486.28	c
	6	8.71	11.73	10.10	32.6	0.1157	6.6	**	7.37	472.38	c, f
	7	3.81	5.26	5.09	16.4	0.1157	6.6	*	3.95	604.55	c, d
	8	2.25	3.17	3.22	10.4	0.1157	6.6	*	2.46	590.32	c, d
	9	12.86	17.08	18.29	59.0	0.1298	7.4	**	8.66	246.54	c, f
	10	8.73	11.75	12.69	40.9	0.1298	7.4	**	6.13	297.58	c
	11	3.80	5.24	6.70	21.6	0.1298	7.4	**	3.12	315.49	c
	12	2.25	3.17	3.70	11.9	0.1298	7.4	*	1.77	289.80	c, d
	13	58.53	76.69	49.70	57.5	0.1131	6.5	**	7.26	1170.93	a
	14	18.43	24.76	14.11	30.4	0.1131	6.5	**	2.01	1160.04	a
U. Tex. (Boone, '84)	17	119.80	156.24	114.60	82.1	0.1284	7.4	114.3	5.35	1513.47	b
(Weinstein, '86)	I24	328.85	426.96	255.90	183.2	0.1344	7.7	**	11.64	1498.73	a
(Swensson, '86)	I43	24.96	33.29	30.10	21.6	0.1348	7.7	29.0	1.31	1391.14	b, c
U. Delft (Vegte, '91)	X5	114.43	151.90	113.05	92.3	0.1183	6.8	110.5	4.04	1032.83	b
K–Joints											
U. Texas	A2-X-90	220.99	287.75	272.50	252.3	0.1412	8.1	**	4.98	1266.91	c
(Yura, '78)	A2-X-30	488.15	634.57	694.20	364.4	0.1412	8.1	**	6.35	824.78	c
Wimpey	3.1	190.08	253.99	219.00	199.1	0.1042	6.0	**	3.36	1247.62	c
/JISSP	3.2	859.59	1121.82	857.00	779.1	0.1298	7.4	**	10.31	933.35	a
(Ma, 88)	3.3	185.47	247.65	208.00	189.1	0.1019	5.8	205.0	3.30	1253.10	b, c
	3.4	189.27	252.82	198.00	180.0	0.1153	6.6	195.0	2.74	1028.06	b, c
	3.5	837.04	1092.60	1030.00	936.3	0.1302	7.5	*	12.21	1092.35	c, d
	3.6	182.20	243.18	191.00	173.6	0.1094	6.3	*	2.90	1115.30	c, d

Nomenclature

M_{yb} yield moment capacity of brace
M_{pb} plastic moment capacity of brace
M_u ultimate capacity recorded in test
M_{ud} capacity at Yura displacement limit
$\overline{M}_u$ minimum of M_u and M_{ud}

Symbols

*   post end of test
**   post peak

Disposition Codes

a)   Recorded M_u totally OK
b) M_u replaced by M_{ud}
c) M_u or M_{ud} preceded by brace yielding
d)   Neither M_{ud} nor peak actually reached
e) $M_u - \theta$ plot unavailable or incomplete
f)   Fabrication defect or premature weld failure
g)   Other reasons to question data

344

33 PLASTIC MECHANISM ANALYSIS OF T-JOINTS IN RHS SUBJECT TO COMBINED BENDING AND CONCENTRATED FORCE

X.L. ZHAO and G.J. HANCOCK
The University of Sydney, Australia

Abstract

A model of the strength of T-joints in Rectangular Hollow Sections (RHS) with $\beta < 1.0$ subject to combined bending and concentrated force is developed based on experimental observations of joint tests performed by the authors. The action in the chord (moment) rather than the chord normal stress is used in the development of the interaction curves. The reduction of the plastic moment capacity of inclined yield lines under axial force is considered in the study. This reduction is based on a new model for inclined yield lines under axial force developed by the authors. The predicted interaction curve for the case where $\beta = 0.5$ is compared with test results of the authors. A design formula is derived based on the plastic mechanism analysis for T-joints with $\beta < 1.0$ under combined bending and concentrated force.
Keywords: Design, Failure, Load Combinations, Models, Tubes

1 Introduction

The load capacity of an RHS connection is reduced by chord end loads (Wardenier (1982)). The effect of the axial compressive forces in the chord on the strength of T and K joints in RHS has been reported by CIDECT (1986). The effect of the bending moment in the chord on the strength of T-joints in RHS has been studied experimentally by Zhao and Hancock (1991a). It was found by Zhao and Hancock (1991a) that for $\beta = 1.0$, the effect of the bending moment is so small that it needs not be taken into account. This is in agreement with the rules described in IIW (1989), Eurocode 3 (1992) and Packer et al (1992). It was also found by Zhao and Hancock (1991a) that for $\beta = 0.5$, the effect of the bending moment becomes significant when the bending moment exceeds half of the fully plastic moment capacity of the section. The theoretical analysis for T-joints in RHS with $\beta < 1.0$ under concentrated force alone has been studied by Zhao and Hancock (1991b). The purpose of this paper is to study theoretically the interaction behaviour for T-joints in RHS with $\beta < 1.0$ under combined bending and concentrated force using plastic mechanism analyses.

Tubular Structures V. Edited by M.G. Coutie and G. Davies.
© 1993 E & FN Spon, 2–6 Boundary Row, London SE1 8HN. ISBN 0 419 18770 7.

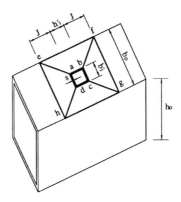

Fig.1. Modified Kato model

2 Model for Combined Actions

It was found by Zhao and Hancock (1991a) that the typical load versus connection deformation curve for the combined action case becomes flat after the failure of a T-joint, which implies that the effect of the membrane force in the chord and the strain hardening of the material are small and might be ignored in the model. For the interaction tests with large moment to concentrated force ratio, no significant web mechanism was found in the tests performed by Zhao and Hancock (1991a).

Based on the above observations, the model for T-joints subject to combined bending moment and concentrated force can be approximated by the "modified Kato model" defined by Zhao and Hancock (1991a). It is shown in Fig. 1, in which $b'_1 = b_1 + 2s$ and $h'_1 = h_1 + 2s$ where b_1 is the width of the branch member, h_1 is the depth of the branch member, s is the size (horizontal leg length) of the fillet weld and J is the model size. The details of hinges are summarised in Zhao and Hancock (1993c).

3 Model Stress and Dimensionless Moment

In the preparation of design formulae in limit state format, it is better to use the actions in the chord rather than the chord normal stress, since use of chord normal stress cannot distinguish between first yield of the section and its fully plastic capacity (Zhao and Hancock (1992)). In this paper, the dimensionless moment $(\frac{M}{M_{pt}})$ is used for the interaction curves, where the value of M_{pt} is the fully plastic moment capacity of an RHS section. However, the value of the stress in the chord is needed in order to calculate the reduction in the plastic moment capacity of yield lines. The model stress is based on the assumption that the normal stress in the chord is only caused by the bending moment. The model stress distribution is assumed such that it gives the same value of bending moment as that produced by the elastic-plastic stress distribution for an RHS beam.

346

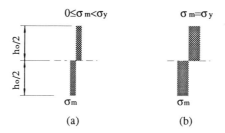

$0 \leq \sigma_m < \sigma_y$ $\sigma_m = \sigma_y$

ho/2   ho/2

σ_m σ_m

(a)   (b)

Fig.2. Model stress distribution

From the model stress distribution (Fig. 2), the dimensionless moment can be expressed as:

$$\frac{M}{M_{pt}} = \frac{Z_p \times \sigma_m}{Z_p \times \sigma_y} = \frac{\sigma_m}{\sigma_y} \tag{1}$$

in which, Z_p is the plastic section modulus and $\sigma_m \in [0, \sigma_y]$ is the model stress. Further details of the assumptions and errors associated with the model stress distribution are given in Zhao and Hancock (1993c).

4 Plastic Moment Capacity of Yield Lines

The plastic moment capacity of each of the hinge types is summarised in Zhao and Hancock (1993c). For hinges parallel to the direction of σ_m, the plastic moment capacity of the hinges is not reduced. For hinges perpendicular or inclined to the direction of σ_m (see Fig. 3), the lower bound solutions for the reduced plastic moment capacity (M_{ph}) have been derived and experimentally verified by Zhao and Hancock (1993a, 1993b). The simple expression based on the the Tresca yield criterion given by Zhao and Hancock (1993b) is used in this paper.

5 Determination of Model Size (J)

The total virtual change of the internal work is:

$$\delta W_{int} = \sum_{i=1}^{5} k_i \delta W_i \tag{2}$$

in which, k_i and δW_i are given in Zhao and Hancock (1993c). The virtual change of the external work done by the load (P) is:

$$\delta W_{ext} = P \delta \Delta \tag{3}$$

From the virtual work principle, the load P can be expressed as:

$$P = \sum_{i=1}^{5} k_i \frac{\delta W_i}{\delta \Delta} = \sum_{i=1}^{5} k_i P_i \tag{4}$$

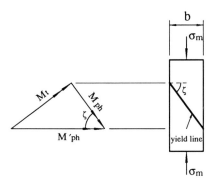

Fig. 3. An inclined yield line

The model size (J) can be determined from the minimum condition:

$$\frac{dP}{dJ} = \sum_{i=1}^{5} k_i \frac{dP_i}{dJ} = 0 \tag{5}$$

in which, the expressions for $\frac{dP_i}{dJ}$ are given in Table 3 of Zhao and Hancock (1993c).

6 Interaction Curves

The load P can be determined from Eq. (4) as a function of β, $\frac{h_1}{b_o}$, $\frac{s}{b_o}$ and $\frac{M}{M_{pt}}$, i.e.,

$$P = f(\beta, \frac{h_1}{b_o}, \frac{s}{b_o}, \frac{M}{M_{pt}}) \tag{6}$$

The dimensionless moment ($\frac{M}{M_{pt}}$) and the dimensionless load ($\frac{P}{P_f}$) are used in the interaction curves. The normalizing value M_{pt} ($= Z_p \times \sigma_y$) is the fully plastic moment capacity of the section. The normalizing value P_f is the load when $\frac{M}{M_{pt}} = 0$, which can be calculated by substituting $\frac{M}{M_{pt}} = 0$ into Eq. (6).

For $\beta = 0.5$, the predicted interaction curve based on Eq. (6) is compared in Fig. 4 with the test results by Zhao and Hancock (1991a), where the measured value of $\frac{s}{b_o} = 0.07$ has been used. The points on the $P/P_f = 0$ axis represent the test results from the pure bending tests. The increase in capacity above M_{pt} for these points is most likely due to strain hardening of the material and the strength enhancing effect of cold-forming especially in the corners. The increase in capacity above M_{pt} for the points in the interaction test results is most likely due to the above reasons plus the finite size of the branch member welded to the chord member. It can be seen from Fig. 4 that the shape of the predicted interaction curve from the plastic mechanism analysis is generally in agreement with that of the test results for each RHS section, especially for section S3B1C23 which had the most slender walls ($\frac{b_o}{t} = 23.4$) and hence smallest corner radii.

Similar to the $\beta = 0.5$ case, the interaction curves for other $\beta < 1.0$ values are

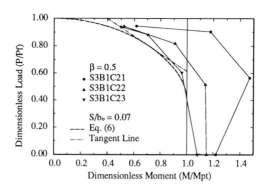

Fig. 4. Dimensionless moment versus dimensionless load

plotted in Fig. 5, where the value of $\frac{s}{b_o} = 0.07$ is used and an SHS branch member is assumed. It can be seen from Fig. 5 that the effect of the bending moment on the strength of T-joints decreases as β increases. This trend is in agreement with the test observations by Zhao and Hancock (1991a). This trend is also similar to that obtained by CIDECT (1986) for the case of axial compressive force acting in the chord and to that given by IIW (1989) and AWS (1992).

7 Proposed Design Formula

The derivation of the proposed design formula is based on the interaction curves shown in Fig. 5. A linear relationship is assumed for the proposed design formula as:

$$K_1(\frac{P}{P_f}) + (\frac{M}{M_{pt}}) \leq K_2 \tag{7}$$

It is assumed that for $\frac{M}{M_{pt}} \leq 0.4$, the effect of the bending moment on the strength $(\frac{P}{P_f})$ is so small that Eq. (7) passes the point $\frac{P}{P_f} = 1.0$ and $\frac{M}{M_{pt}} = 0.4$ below which no interaction is assumed to occur. The predicted interaction curve is approximated by a tangent line to the curve from the above point as shown in Fig. 4 for $\beta = 0.5$. For a certain value of β, the tangent line can be drawn from this point to the predicted interaction curves in Fig. 5. If η is set to be the value of $\frac{P}{P_f}$ at the intersection point of the tangent line and the vertical line with $\frac{M}{M_{pt}} = 1.0$, a pair of coordinates (β, η) is obtained. A linear regression formula can be obtained as:

$$\eta = 0.335 + 0.580\beta \tag{8}$$

From the two points $(\frac{M}{M_{pt}}, \frac{P}{P_f}) = (0.4, 1.0)$ and $(\frac{M}{M_{pt}}, \frac{P}{P_f}) = (1.0, \eta)$, a linear function can be written as:

$$(\frac{P}{P_f} - 1.0) = (\frac{1.0 - \eta}{0.4 - 1.0})(\frac{M}{M_{pt}} - 0.4) \tag{9}$$

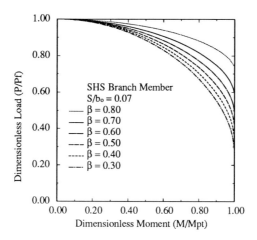

Fig. 5. Dimensionless moment versus dimensionless load

The factors K_1 and K_2 in Eq. (7) are determined from Eq. (9) as:

$$K_1 = \frac{0.6}{1 - \eta} \tag{10}$$

$$K_2 = K_1 + 0.4 \tag{11}$$

in which, η is given in Eq. (8).

An interaction formula has been given by IIW (1989) and AWS (1992), which was expressed as a function of the chord normal stress. However, the existing interaction formula can be rewritten in the following format as:

$$(\frac{P}{P_f}) = 1.3 - \frac{0.4}{\beta}(\frac{M}{M_{pt}} \times S) \tag{12}$$

where S is the shape factor of an RHS section.

For $\beta = 0.4$, 0.6 and 0.8, the proposed interaction formula (Eq. (7)) is compared in Fig. 6 with the existing interaction formula (Eq. (12)) where $S = 1.22$ is used as an example. It can be seen from Fig. 6 that the use of the fully plastic moment rather than the chord normal stress produces higher curves.

8 Conclusions

- A model for T-joints in RHS with $\beta < 1.0$ subject to combined bending and concentrated force has been developed based on experimental observations. The model is a development of the the "modified Kato model" proposed by Zhao and Hancock (1991a). The action in the chord (moment) rather than the chord normal stress is used for the interaction curves. The reduction of the plastic moment capacity of yield lines has been considered in the study.

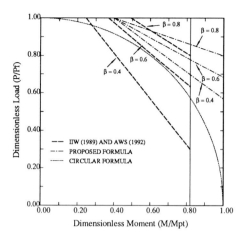

Fig. 6. Dimensionless moment versus dimensionless load

- For T-joints in RHS with $\beta = 0.5$, the predicted interaction curve has been compared with the test results of Zhao and Hancock (1991a). It has been found that the plastic mechanism prediction is a reasonable lower bound to all of the results for the sections tested with the best estimates provided for the thinner sections with smaller corners. For the thicker sections, strain-hardening and the effect of cold-work in the corners increase the capacity.

- For T-joints in RHS with $\beta < 1.0$, the predicted interaction curves have been given. It has been found that the effect of the bending moment on the strength of T-joints decreases as β increases. A linear interaction formula for use in design is derived based on the plastic mechanism analysis for T-joints with $\beta < 1.0$ under combined bending and concentrated force. The formula includes the variation of capacity with β. It produces higher interaction curves than those produced by the existing interaction formula given by IIW (1989) and AWS (1992).

9 Acknowledgments

The authors are grateful to Tubemakers of Australia Limited for supplying test specimens and additional financial support. Thanks are given to the University of Sydney for the UPRA Scholarship and the Centre for Advanced Structural Engineering of the University of Sydney for additional support of the first author.

10 References

AWS (1992), "*Structural Welding Code – Steel*", ANSI/AWS D1.1-92, 13th. ed., American Welding Society, Miami, Fla.

CIDECT (1986), "The Strength and Behaviour of Statically Loaded Welded Connections in Structural Hollow Sections", *CIDECT Monograph*, No.6, Corby.

351

IIW (1989), IIW Subcommission XV-E, "Design Recommendations for Hollow Section Joints – Predominantly Statically Loaded", 2nd. ed., *IIW Doc. XV-701-89*, IIW Annual Assembly, Sept., Helsinki, Finland.

Packer et al (1992), "CIDECT Design Guide for Rectangular Hollow Sections (RHS) Joints Under Predominantly Static Loading", Verlag TÜV Rheinland, Köln, Federal Republic of Germany.

Wardenier, J. (1982), "Hollow Section Joints", Delft Univ. Press, Delft, the Netherlands.

Zhao, X.L, and Hancock, G.J.,(1991a), "T-joints in Rectangular Hollow Sections Subject to Combined Actions", *J. Struct. Engrg.*, ASCE, 117(8), 2258–2277.

Zhao, X. L, and Hancock, G. J., (1991b), "Plastic Mechanism Analysis of T-Joints in RHS Under Concentrated Force", *J. Singapore Struct. Steel Society*, 2(1), 31–44.

Zhao, X. L, and Hancock, G. J., (1992), "T-joints in Rectangular Hollow Sections Subject to Combined Actions", Closure for discussion by J. A. Packer and J. Wardenier, *J. Struct. Engrg.*, ASCE, 118(9), 2639–2640.

Zhao, X.L, and Hancock, G.J., (1993a), "Theoretical Analysis of Plastic Moment Capacity of an Inclined Yield Line Under Axial Force", *Thin-Walled Structures*, 15(3), 185–208.

Zhao, X.L, and Hancock, G.J., (1993b), "Experimental Verification of the Theory of Plastic Moment Capacity of an Inclined Yield Line Under Axial Force", *Thin-Walled Structures*, 15(3), 209–233.

Zhao, X.L, and Hancock, G.J., (1993c), "Plastic Mechanism Analysis of T-Joints in RHS Subject to Combined Bending and Concentrated Force", *Research Report*, No. R673, School of Civil and Mining Engineering, University of Sydney, Sydney, Australia.

34 ON CONVERGENCE OF YIELD LINE THEORY AND EXPERIMENTAL TEST CAPACITY OF RHS K- AND T-JOINTS

T. PARTANEN and T. BJÖRK
Lappeenranta University of Technology, Finland

Abstract
The capacity of 54 RHS K-joints experimentally tested at low temperatures and 28 T-joints tested at room temperature was calculated using yield line theory and Cidect or EC3 formulae. Most of the joints did not meet the joint geometry validity criteria of Cidect. The yield line method was modified to take into account stress biaxiality, reductions due to out-of-plane shear stress, axial membrane stress and the true location of plastic hinges. The results of K-joint tests, where the gap between the braces was too narrow compared to recomendations, show predominantly ductile fracture behaviour and acceptable convergence between the yield line theory, EC3 predictions and test yield capacity. The test capacities of T-joints not meeting the size requirements of Cidect were in agreement with the Cidect recommendations.
Keywords: Capacity, RHS-Joints, Static Loading, Experimental Tests, Yield Line Theory.

1 Introduction

Based on Cidect Monograph No 6 (1986), the present recommendations of EC3 (1991), IIW (1989) and Brittish Steel (1992) for calculating the design capacity of rectangular hollow section joints are similar and determined by either theoretical calculations following the yield line theory or statistical analysis of experimental test results. Several alternative T-joint capacity formulae have been determined theoretically, while a single K-joint capacity formula is based on experimental results. Concerning chord face yielding of T-joints, one single formula is used to define the ultimate, characteristic and design capacities. In the case of K-joints, the characteristic capacity is 10% higher than the design strength value; the mean ultimate capacity is 10% higher than the design strength (constants 6.9/1.1, 6.9 and 7.7 in the capacity formulae). The validity of the design capacities calculated by the Cidect equations is restricted by the choice of joint and section dimensions. The purpose of this study is to test, both experimentally and analytically, specimens not meeting the joint and member size requirements of Cidect (1986). The results of 30 K-joints and 15 T-joints experimentally tested at Lappeenranta University of Technology (abrev. LUT) are presented. The complementary tests to earlier K-joint tests of Niemi et al

Tubular Structures V. Edited by M.G. Coutie and G. Davies.
© 1993 E & FN Spon, 2–6 Boundary Row, London SE1 8HN. ISBN 0 419 18770 7.

(1988) and Niemi (1990), 24 specimens, were also carried out at low temperatures, while the T-joints were tested at room temperature.

In this study, the capacity of the joints tested at LUT, T-joints tested at the University of Sydney by Zhao and Hancock (1991) and T-joints tested at Chiba University by Morita et al (1990) were calculated using a modified yield line theory, Partanen (1991), which takes into account the biaxial stress at the plastic hinges, and the guidelines of Cidect (1986). At the end of the report the results are discussed and some conclusions are made concerning the validity of present recommendations.

2 Materials and methods

2.1 Theory
Classic yield line theory can be refined by taking into account shear stress and membrane stress reductions in the moment capacity. The total capacity can be further separated to the sum of the capacities of individual yield line pairs. Partanen (1991) presented modifications to the classic yield line theory to take into account stress biaxiality in a plastic hinge and shear stress correction in biaxial stress state. The capacity of two heel sides of a K- or T-joint ("Knife Edge Capacity") was determined as being valid in the whole range $0 < \beta < 1$, $\beta =$ brace/chord width ratio. The capacity of a joint using the method presented is the sum of shear capacities of unit width plastic hinge pairs multiplied by their respective yield lengths and the capacity of the heel side (or sides). The modifications to the general yield line theory were verified by materially nonlinear FEM-analysis. Good convergence was generally obtained, accuracy within +-10%. In this paper the theory presented is further refined to take into account the possibility that a plastic hinge forming at the wall of the brace in the fillet welded joint gives the lowest shear capacity.

The shear capacity for a unit length, q_p, of a plastic hinge pair in a wide gap is obtained from eq. 1, where t_0 is the thickness of the plate, g is the general gap between the hinges, and q_{p0} is the pure shear capacity of the plate, f_{y0} being the uniaxial yield strength of the chord wall at the hinges. When taking into account shear stress reduction in the shear capacity, the shear capacity, q_p, is obtained from eq 2, see also Fig. 1.

$$\frac{q_p}{q_{p0}} = \frac{t_0}{g}, \quad q_{p0} = \frac{f_{y0}}{\sqrt{3}} t_0 \tag{1}$$

$$\frac{q_p}{q_{p0}} = \frac{1}{\sqrt{1 + (g/t_0)^2}} \tag{2}$$

Knife Edge Capacity, N_{ke}, of a transverse flat bar, external width b_1, welded on a rectangular hollow section chord, plate thickness t_0, and external width b_0, is obtained from eq. 3, where g_l is the gap on the longitudinal sides, $g_l = \frac{1}{2}(b_0 - t_0 - b_1 - 2\sqrt{2}a)$, and a is the throat thickness of the fillet weld.

354

$$N_{ke} = \pi f_{y0} t_0^2 \frac{1}{\sqrt[4]{(2g_l/t_0)^2 + 3.43}} \sqrt{\frac{b_0}{t_0} - 1} \qquad (3)$$

2.1.1 Refinement of eqs. 1-3

Full width joints ($\beta = 1.0$) can carry the vertical component of the axial yield force, q_N, of the weakest member: minimum of {brace thickness $t_1 \sin \theta$, throat thickness of weld, a, web thickness of chord, t_{0w}}, where θ is the angle between the brace and the chord. When material design strengths, f_{yi}, differ, the thicknesses have to be multiplied by the corresponding design strengths, eq. 4.

$$q_N(\beta = 1.0) = \min\{t_{0w} \cdot f_{y0}, a \cdot f_{yWd}, t_1 \cdot f_{y1} \cdot \sin \theta\} \qquad (4)$$

The capacity of transverse plastic hinges in the joints of RHS sections is corrected by the membrane stress reduction, Φ, due to axial stress, f_0, in the chord member, eq. 5.

$$q_{pm} = [1 - (f_0/f_{y0})^2] q_p = \Phi q_p \qquad (5)$$

Plastic hinges are assumed to be located at the boundary of fillet welds - either one hinge at the outside toe of the fillet weld in the chord wall - or one hinge at the root of the fillet weld in the chord wall, and the other at the toe of the fillet weld in the brace wall. The latter failure mode is called *effective gap reduction*, determining the capacity of narrow gaps, having thin brace walls compared to chord wall thickness. While the shear force flow, q_p, has to be transmitted from the center line of one plate to the center line of the other, it is possible that the reduced bending capacity of the brace wall is less than required to carry the shear flow from the weld toe to the centerline of the brace wall. Therefore the shear flow capacity, q_p, of eq.2 has to be reduced further due to *effective gap reduction*. Figure 1 illustrates the two competing mechanisms of a unit depth plate strip on the longitudinal sides, fig.1a, and in the transverse gap between the brace walls, fig.1b. The discontinuity of the plastic hinge mechanism in the corner areas of the brace has been assumed as a minor effect in the total capacity of the joint.

When taking into account the membrane stress reductions in brace and chord web wall hinges and equating the internal and external works in the plastic hinge mechanism, W_i and W_o, the shear force capacity, q_l, of the mechanism on a longitudinal gap side, g_l, is defined using the earlier definitions as:

$$\frac{q_l}{q_{p0}} = \frac{g_l}{t_0} \frac{3}{2} \left[\sqrt{1 + \frac{4}{3}\left(\frac{t_0}{g_l}\right)^2 \left(1 + \frac{1}{2}\left(\frac{t_1}{t_0}\right)^2\right)} - 1 \right] \approx \frac{t_0}{g_l}\left(1 + \frac{1}{2}\left(\frac{t_1}{t_0}\right)^2\right) \qquad (6)$$

The shear force capacity in the gap of a K-joint between the braces, g_0, can be defined in a corresponding way. Now the membrane stress reduction, Φ in eq.5, in the capacity of the chord wall hinge will be taken into account. Thus the shear capacity, q_t, in the transverse gap between the brace walls is defined as:

$$\frac{q_t}{q_{p0}} = \frac{g_t}{t_0}\sin^2\theta \frac{3}{2}\left[\sqrt{1+\frac{4}{3}\left(\frac{t_0}{g_t}\right)^2\left(\Phi+\left(\frac{t_1}{t_0}\right)^2\right)\frac{1}{\sin^2\theta}}-1\right] \approx \frac{t_0}{g_t}\left(1+\left(\frac{t_1}{t_0}\right)^2\right) \tag{7}$$

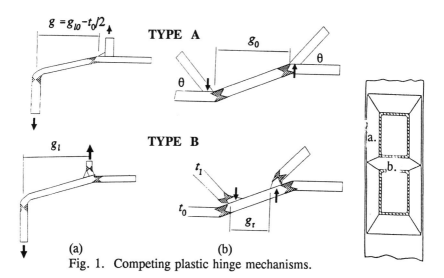

$g = g_{lo} - t_0/2$

TYPE A

g_0

θ

θ

g_1

TYPE B

t_1

t_0

g_t

a.

b.

(a)   (b)

Fig. 1. Competing plastic hinge mechanisms.

If the right side of eqs. 6 or 7 is less than defined by eq. 2, then the *effective gap reduction* defines the capacity of the joint.

Total capacity of a T-joint, N_q, is the sum of the Knife Edge Capacity, N_{ke}, and the minimum of $\{q_p,q_1,q_N\}$ multiplied by the total yield length, $2l_1$ (or twice the brace depth, h_1).

Total vertical shear force capacity, N_q, of a K-joint is the sum of three components: (i) Half of the Knife Edge Capacity of the heel side, $\frac{1}{2}N_{ke}$; (ii) Capacity of the transverse gap side being the minimum of $\{q_p,q_t,q_N\}$ multiplied by the transverse yield length, b_0; (iii) capacity of the longitudinal sides being the minimum of $\{q_p,q_1,q_N\}$ multiplied by the total yield length $2l_1$ $(=2h_1/\sin\theta+g_t)$. The axial force capacity of a brace, N_1, is obtained from the geometry by dividing N_0 by $\sin\theta$.

2.2 Experimental test verification

Yield line theory presented by Partanen (1991), modified according to this paper, was used to predict the capacity of 54 K-joints and 15 T-joints experimentally tested at Lappeenranta University of Technology. These joints of cold formed RHS sections did not meet the normal size or shape requirements of EC3 (1991), IIW (1989) or Cidect (1986). The three test series have been identified by researchers Lehtinen, Soininen and Björk. The testing was supervised by prof. Niemi. Only the test series of Lehtinen have been reported, by Niemi et al (1988) and Niemi (1990).

The theoretical model presented was also used to calculate the capacity of 6 small-sized RHS T-joints experimentally tested and analyzed by Zhao and Hancock (1991), and 7 large transverse flat bar T-joints of Morita et al (1990).

K-joint tests of Lehtinen and Soininen at LUT: The main purpose of the two K-joint test programs was to define the deformation capacity of gapped joints of cold formed steel sections at low temperatures. Experimental test arrangements, test specimens, material and more detailed geometry data have been reported by Niemi et al (1988), Niemi (1990) and in the unpublished paper of of Soininen et al (1991).

In the first test program of Lehtinen (Niemi 1988) 24 joints were tested at different low temperatures. The chord section was a rectangular hollow section 250·150·8 and the braces were square hollow sections 120·120·8. In series A the RHS chord was in the horizontal position ($\beta=0.48$) and in series B the chord member was in the vertical position ($\beta=0.80$). The angle, θ, between the chord and the brace was 45° in the series A and 60° in the series B. The nominal gap between the braces, g, was 65 mm in series A, and 30 mm in series B. The nominal weld leg length was 9 mm (varying between 5 mm and 13 mm). Two different materials, Fe 430 D and RAEX 355 F, having almost equal strength values, were used. The average yield strength, measured from each specimen, varied from 396 MPa to 456 MPa. In the corner area the yield strength was on average 13% higher, being also higher than the average ultimate strength, from 435 MPa to 504 MPa. The value of the pretensioning load in the chord member was kept constant at 1225 kN in all tests.

In the second test program of Soininen 5 series, A,B,C,D,E each containing 6 specimens, were tested at different low temperatures, the temperature of specimens 4-6 in series B,C,D,E was kept -60°, and -40° in A3-A6. The first specimen in each series was tested at room temperature, the second at -20°. Pretensioning load in the chord was kept constant at 760 kN, except 380 kN of specimens C6,D6,E6 and zero of specimens A6,B5,C5,D5,E5. All the sections were square hollow sections. All the chord sections were 150·150·t_0. The braces were of two type of sections: 100·100·t_1 in series A,B,E and 70·70·t_1 in series C and D. The nominal weld leg length was 6 mm (varying between 4 mm and 8 mm). The average yield and ultimate strengths of the material used, Fe 510 D, measured from the flat side of the chord sections, were 465 MPa and 530 MPa. The dimensions are shown in Table 1.

Table 1.   K-joint dimensions of Soininen (1991).

Series	t_0	t_1	β	θ	g	g_0	g_{10}	l_1
A	7.8	6.2	0.67	45°	16	13 - 25	16 - 21	151 - 158
B	7.8	6.2	0.67	45°	4	2.5 - 6.5	17 - 18	156 - 161
C	7.9	4.9	0.47	60°	32	21 - 28	33 - 34	95 - 100
D	7.8	4.9	0.47	60°	4	1.5 - 5	32 - 34	93 - 95
E	7.9	8.1	0.67	45°	4	1.5 - 6	17 - 18	160 - 163

The symbols used in Table 1 are: thickness of chord, t_0; thickness of brace, t_1; width ratio between brace and chord, β; angle between the brace and the chord, θ; nominal gap between the brace walls, g; the true gap between the braces (= weld toe to weld toe length), g_0; transverse distance from the weld toe parallel to the chord to the outer side of the chord (= edge gap), g_{10}; distance between the cross weld toes of the bracing member, l_1 (= depth of brace, h_1, plus two weld legs).

T-joint tests of Björk at LUT: 15 T-joint hot-rolled RHS specimens, having constant chord member size and varying brace member dimensions were tested at room temperature. The brace member was loaded in tension in such a way that the bending of the chord cross section was prevented by continuous support system. The chord member, size 199·100·7.8, was in the horizontal position, member length 500 mm. The thickness of brace, t_1, was 2 mm in specimens 1-4, 3.9 mm in specimens 5-11 and 6 mm in the rest. The average yield and ultimate strengths of the material used, Fe 44, measured from the flat side of the chord sections, were 288 MPa and 422 MPa. The dimensions of specimens are shown in Table 4, where the symbols of Table 1 apply. The nominal weld leg length was 6 mm (varying between 4 mm and 8 mm: see columns g_{10} and l_1 in Table 4).

3 Results

3.1 K-joint test results of Soininen
Although tested at very cold temperatures, up to -60°C, most of the specimens failed in a ductile manner. Deformation capacity was lower at cold temperatures than at room temperature. Ductile failed joints at cold temperatures normally had at least the same capacity as the specimens tested at room temperature. The results are shown in Table 2.

Symbols and remarks of Table 2:

T Type of plastic hinge mechanism in the transverse gap betwen the braces; A = plastic hinge in the chord, only; B = plastic hinge in the brace, too; see Fig.1.

PF Dimensions Passed or Failed the criterias of Cidect; FG = Failed due to Gap, g; Fe=Failed due to brace eccentricity, e. Failure mode was chord face Yielding.

N_{ty} Yield capacity according to tests.

N_{tu} Ultimate test capacity.

N_{cy} Calculated yield capacity according to the yield line model presented.

N_{cu} Calculated ultimate capacity according to the model presented (=Yield capacity multiplyed by the strength ratio).

N_{yCI} Calculated design capacity according to EC3.

N_{uCI} Calculated mean ultimate capacity according Cidect (N_{uCI} = 1.22 N_{yCI} , in Table 2 only).

When only one result is given, it applies to all specimens in the series considered.

Table 2. K-joint test results of Soininen, in (kN).

Spec	β	N_{ty}	N_{tu}	N_{cy}	T	$\dfrac{N_{ty}}{N_{cy}}$	N_{cu}	N_{yCI}	PF	$\dfrac{N_{ty}}{N_{yCI}}$	N_{uCI}
A1		600	745	657		0.91	749			0.82	928
A2		600	738	634		0.95	723			0.82	901
A3	0.67	-	784	660	A	1.18	753	736	FG	1.07	901
A4		660	831	712		0.93	811			0.90	901
A5		800	1081	831		0.96	947			1.09	907
A6		850	1021	705		1.20	853			1.15	906
B1		850	1023	900		0.94	1026			1.15	928
B2		900	1129	910		0.99	1045			1.22	906
B3	0.67	900	1058	897	A	1.00	1028	736	FG	1.22	910
B4		-	807	925		0.87	1029			1.10	923
B5		1000	1234	947		1.05	1100			1.36	906
B6		-	-	897		-	1022			-	906
C1		330	485	294		1.12	335			0.78	525
C2		340	531	300		1.13	342			0.81	525
C3	0.47	340	470	305	A	1.11	348	421	Fe	0.81	525
C4		300	469	291		1.03	332			0.71	525
C5		370	502	348		1.06	397			0.88	525
C6		310	476	321		0.97	366			0.74	525
D1		440	489	423		1.04	484			1.04	515
D2		470	533	423		1.11	484		Fe	1.11	515
D3	0.47	470	548	420	B	1.11	478	421	+	1.11	515
D4		450	565	440		1.02	501		FG	1.07	525
D5		480	586	456		1.05	520			1.14	515
D6		440	533	459		0.96	524			1.04	525
E1		850	965	892	A	0.95	1049			1.15	918
E2		860	1056	968	B	0.89	1122			1.17	918
E3	0.67	900	1053	950	B	0.95	1080	736	FG	1.22	918
E4		-	544	960	B	-	1058			-	918
E5		1120	1243	1000	B	1.12	1133			1.52	918
E6		1020	1156	990	B	1.03	1140			1.38	918

3.2 K-joint test results of Lehtinen

The results are shown in Table 3, where the symbols of Table 2 apply. Design capacities were calculated using the strength of the corner area (13% higher than the average yield stress), remark 1). Mean ultimate capacity was obtained from Niemi et al (1988) and Niemi (1990), where the average strength was used as the yield stress (N_{uCI} 7% higher than N_{yCI} of this table), remark 2). Failure mode is chord face yielding.

3.3 T-joint tests of Björk

The results are shown in Table 4. The symbols of Table 2 apply with the following explanations of column PF: **Fx** = Cidect validity criteria Failed; **FY** = failure mode chord face Yielding; **FI** = Interaction $0.8 < \beta < 1.0$; **FE** = failure due to Effective width criteria, $\beta = 1.0$; **PI** = Cidect validity criteria Passed, failure Interaction $0.8 < \beta < 1.0$. Due to thin wall thickness of brace plastic hinges were assumed to form both at the chord face and at the brace wall, type B mechanism dominant, except the full width joints.

Table 3. K-joint test results of Lehtinen, in (kN).

Spec.	β	N_{ty}	N_{tu}	N_{cy} 1)	$\dfrac{N_{ty}}{N_{cy}}$	N_{cu}	N_{ycI} 1)	PF	$\dfrac{N_{ty}}{N_{ycI}}$	N_{ucI} 2)
1.1A		485	770	378	1.27	416			0.61	829
2.1A		490	835	407	1.20	448			0.62	885
3.1A	.48	500	800	382	1.31	420	791	Fe	0.63	818
4.1A		519	830	327	1.59	360			0.66	861
5.1A		490	830	363	1.35	400			0.62	817
1.3A		431	615	327	1.30	360			0.58	764
2.3A		421	645	363	1.16	400			0.57	828
3.3A	.48	475	755	338	1.40	373	736	Fe	0.65	741
4.3A		473	795	372	1.27	409			0.64	789
5.3A		519	940	327	1.59	360			0.71	745
1.2B		1000	1075	906	1.10	996			1.23	889
2.2B		1029	1165	886	1.16	976			1.27	836
3.2B	.80	1091	1120	892	1.22	981	810	P	1.34	867
4.2B		1079	1150	879	1.22	968			1.33	854
5.2Ba		-	605	912	-	1000			-	873
5.2Bb		1054	1060	1041	1.01	1145			1.42	912
1.4B		891	985	890	1.00	980			1.10	863
2.4B		921	970	890	1.03	980			1.15	854
3.4B	.80	882	1065	831	1.06	914	805	P	1.10	820
4.4B		931	1060	874	1.06	961			1.16	859
5.4Ba		991	1025	822	1.20	904			1.23	808
5.4Bb		-	550	812	-	893			-	810

Table 4 T-joint test results of Björk, in (mm),(kN).

Spec	β	g_{10}	l_1	N_{ty}	N_{tu}	N_{cy}	$\dfrac{N_{ty}}{N_{cy}}$	N_{cu}	N_{ycI}	PF	$\dfrac{N_{ty}}{N_{ycI}}$	N_{ucI}
1	.20	75.	52.	75	138	79	0.95	116	89	FY	0.84	130
2	.30	64.	51.	74	177	86	0.86	126	96	FY	0.77	140
3	.50	44.	93.	125	185	127	0.98	185	131	FY	0.95	192
4	.60	35.	95.	142	260	146	0.97	214	151	FY	0.94	222
5	.60	32.	51.	110	263	128	0.86	186	129	FY	0.86	189
6	.80	14.	94.	242	395	254	0.95	373	226	FY	1.07	332
7	.90	8.	93.	347	475	377	0.92	552	378	FI	0.92	555
8	.80	14.	141.	300	400	322	0.94	472	268	FY	1.11	393
9	.90	3.5	135.	460	625	466	0.99	683	468	FI	0.98	687
10	.80	10.	177.	390	735	377	1.03	552	302	FY	1.30	443
11	1.0	0.	185.	600	946	678	0.89	994	694	FE	0.86	1020
12	.80	13.	96.	275	409	281	0.98	412	226	FY	1.22	332
13	.90	4.	96.	395	611	399	0.99	585	390	FI	1.01	573
14	.90	3.5	135.	525	796	480	1.09	703	465	PI	1.13	683
15	1.0	0.	142.	740	1189	680	1.09	996	608	FE	1.22	893

3.4 T-joint tests of Zhao and Hancock (1991)

The chord and brace members were of rectangular hollow sections, chord $102 \cdot 102 \cdot t_0$ in series S1B1C2i, chord $102 \cdot 51 \cdot t_0$ in series S1B1C1i, brace $51 \cdot 51 \cdot 4.9$ in all tests. Weld leg length varied from 6.5 mm to 7 mm. Thus the yield line length l_1 was a constant length of 64 mm. Results are shown in Table 5. Due to the risk of buckling the two lowest values of N_{yc} for $\beta = 1$ are not valid. Joints passed the joint size requirements of Cidect. N_{yCI} capacities were calculated according to web buckling criteria using effective depth $h_e = h_0/2$.

Table 5. T-joint test results of Zhao and Hancock (1991), as a function of the true strength, in (kN).

Spec	t_0 (mm)	f_y (MPa)	f_u	N_{ty}	N_{tu}	N_{cy}	$\dfrac{N_{ty}}{N_{cy}}$	N_{cu}	N_{yCI}	$\dfrac{N_{ty}}{N_{yCI}}$	N_{uCI}
$\beta=0.5$	Flat:										
S1B1C21	9.5	435	454	422	438	330	1.28	344	301	1.48	314
S1B1C22	6.3	363	455	175	245	125	1.40	157	110	1.59	138
S1B1C23	4.0	378	471	67	67	59	1.13	74	46	1.45	57
$\beta=0.5$	Corner:										
S1B1C21		513	558	422	438	389	1.08	423	354	1.19	389
S1B1C22		539	573	175	245	185	0.95	197	165	1.06	175
S1B1C23		543	560	67	67	84	0.80	87	67	1.0	70
$\beta=1.0$ $h_e=h_0/2$	Flat:										
S1B1C11	4.9	322	418	316	-	-264	1.20	-	238	1.33	
S1B1C12	3.2	325	431	163	-	162	(1.)	-	138	1.18	
S1B1C13	2.0	373	467	76		111	-	-	79	0.96	

3.5 T-joint tests of Morita Yamamoto and Ebato (1990)

The width of the chord face section, b_0, was 250 mm except for specimen FN1, where b_0 was 500 mm. Yield strength of the material used, JIS SM50A, was assumed to be 337 MPa. According to the ultimate capacities calculated by Morita et al, ultimate strength was on average 35% higher than the yield strength. The thickness of the transverse bracing flat bar varied from 32 mm to 40 mm. Other dimensions are shown in Table 6.

Table 6. Flat bar T-joint results of Morita et al (1990).

Spec.	Dimensions (mm)				Test results, in (kN)							
	β	t_0	b_1	l_1	N_{ty}	N_{tu}	N_{cy}	$\dfrac{N_{ty}}{N_{cy}}$	N_{cu}	N_{yCI}	$\dfrac{N_{ty}}{N_{yCI}}$	N_{uCI}
FN1	.81	22	406	54	1293	1764	1470	0.88	2000	1692	0.77	2280
FN2a	1.0	16	250	51	1451	1936	1312	1.10	1770	1428	1.02	1930
FN2b	.46	16	115	46	476	875	447	1.06	603	528	0.91	713
FN2c	.52	16	130	46	607	961	483	1.26	652	564	1.08	761
FN2c-T	.52	16	130	54	697	962	465	1.50	627	576	1.21	777
FN3	.72	25	180	53	1961	2355	1630	1.20	2200	1920	1.02	2590
PN3	.72	25	180	53	1401	1746	1220	1.14	1650	1485	0.95	2000

The capacities of the joints were calculated in a normal manner except for two cases: i) FN2a ($\beta=1.0$), the capacity of the longitudinal sides was calculated as the uniaxial tensile strength of the web; ii) PN3 = 75% of FN3.

4 Discussion

T-joint results: All the T-joint test results of hot-rolled RHS sections of Björk were within 86-109% of the capacity predicted by the yield line method presented in this report, while using the Cidect characteristic capacity equations the convergence was within 77-130%. If the brace heigth, h_1, had been used as the yield line length instead of the weld toe to toe length, l_1, the slighty non-conservative results obtained by the yield line method would become slightly conservative (prediction 90-116%). Good predictions using the Cidect formula were obtained with joint ratios $b_1/t_1=60$, $h_1/b_1=0.4$ and $h_1/b_0=0.2$, see Table 4. If the results of the analysis of this test are applied to predict the capacity of joints made of cold-rolled RHS sections, where the yield strength of corner areas is normally higher than the average strength, even more conservative predictions would have been obtained.

Cidect RHS T-joint formulae predicted well the yield and ultimate capacities of Knife Edge Loaded flat bar T-joints of Morita et al (1990), if nominal strength values were used. The capacity of T-joints of Zhao and Hancock (1991) were strongly dependant on the material strength: the capacity calculated according to the Cidect recommendations, N_{yCI}, using the corner area yield stress, was in agreement with the yield line capacity, N_{cy}, and correlated well with the test results, also. The arguments of Zhao and Hancock (1991) concerning the over-conservativism of the Cidect capacity formula of T-joints, are not confirmed by the results of this study.

K-joint results: The yield capacity of joints in the tests of Soininen (1991) corresponded within 87-120% the capacity calculated according to the modified yield line theory of this study. The design formula of EC3 can be non-conservative (74% in series C) or over-conservative (152% in series E) for predicting the onset of yielding in a K-joint.

In K-joint tests of Lehtinen the non-conservatism of the empiric EC3 capacity formula for predicting yielding was even more pronounced. If an average yield strength, instead of the corner strength, were used, the EC3 formula would predict from 65% to 160% of the capacity obtained from the tests. The modified yield line approach would always give conservative predictions, ranging from 113% to 180%, when the average yield strength is used.

When predicting the ultimate capacity of joints at low temperature and having small β ($\beta \approx 0.5$), or wide gap, g_0, the Cidect mean capacity formula was still slightly non-conservative (Lehtinen's series A and Soininen's series A,C). In series of narrow gap or large β, both the yield line theory and Cidect mean capacity formula gave conservative predictions. The capacity of K-joints subjected to loading at very cold temperatures was not affected by brittle behaviour except in three extreme cases at -60°C, E4, 5.2Ba and 5.4Bb. Thus the capacity of an RHS K-joint at a low temperature can be predicted by the yield line theory or the EC3 design capacity formula.

6 Conclusions

Plastic hinge forming at the wall of the brace can reduce the capacity of an RHS joint from the capacity defined by assuming the plastic hinges are located at the fillet weld toes of the chord wall.

The design capacity formula of EC3 for predicting the yield capacity of K-joints having wide transverse gap and small width ratio, β, is non-conservative, when compared with the results based on yield line theory and experimental test capacity based on moderate plastic deformation.

Yield line theory predictions are normally conservative, when determining the capacity of joints having small β, and when the average yield strength is used. When defining the strength of joints having narrow gap between the plastic hinges, the yield line theory predicts the capacity of a joint as accurately as the strength of the material can be defined.

7 References

EC3 (1991) Design of Steel Structures, Part 1 - general rules for building.

CIDECT (1986) **The Strength and Behaviour of Statically Loaded Welded Connections in Structural Hollow Sections**, Monograph No 6., (eds. T:W: Giddings and J.Wardenier) Comite pour de Developpement et l'etude de la Construction Tubulaire.

British Steel (1992) **Design of SHS Welded Joints**. PC program manual, Welded Tubes The Tubemasters.

IIW (1989) **Design Recommendations for Hollow Section Joints-Predominantly Statically Loaded**. International Institute of Welding Doc. XV-701-89.

Morita K., Yamamoto N. and Ebato K. (1990) Analysis of the Strength of Unstiffened Beam Flange to RHS Column Connections Based on the Combined Yield line Model, in **Tubular Structures, 3th International Symposium, Lappeeenranta 1989** (eds. E. Niemi & P. Mäkeläinen), Elsevier Applied Science, London 1990, pp. 164-171,ISBN 1-85166-474-2.

Niemi E., Lehtinen J. & Sorsa I. (1988) **Behaviour of Rectangular Hollow-section K-joints at Low Temperatures**. CIDECT Report n:o 5AQ-88/13E.

Niemi E. (1990) Behaviour of Rectangular Hollow-section K-joints at Low Temperatures, in **Tubular Structures, 3th International Symposium, Lappeeenranta 1989** (eds. E. Niemi & P. Mäkeläinen), Elsevier Applied Science, London 1990, pp. 19-27,ISBN 1-85166-474-2.

Partanen T. (1991) On Convergence of Yield Line Theory and Nonlinear FEM Results in Plate Structures, in **Tubular Structures, 4th International Symposium, Delft 1991** (eds. J. Wardenier & E.P. Shahi), Delft University Press, Delft 1991, pp. 313-323.

Soininen R., Niemi E. & Zhiliang Z. (1991) **Studies of the Fracture Behaviour of RHS K-joints at Low Temperatures.** Unpublished test report, L:ranta U.T.

Zhao X-L. and Hancock G.J. (1991) T-joints in Rectangular Hollow Sections Subject to Combined Actions. **Journal of Str. Eng.**, Vol 117, No.8, Aug. 1991.

35 STUDIES OF THE BEHAVIOUR OF RHS GAP K-JOINTS BY NON-LINEAR FEM

Z. ZHANG and E. NIEMI
Lappeenranta University of Technology, Finland

Abstract

Results of studies of the behaviour of RHS gap K-joints are described. The main issues addressed include the prediction of load-bearing behaviour, the effect of gap size on joint stiffness, the stress distribution in the tension brace and the maximum strain in the gap region, and the membrane effect in the joint. Four joints with different configurations were tested and analysed using the finite element program ABAQUS. The gap size (g) and the brace/chord ratio (β) are the two main parameters investigated. Special care has been taken to model the gap region, plastic hinge and chord pretension loading correctly. Good agreement between the finite element predictions and test results has been obtained.

Keywords: RHS K-joints, FEM analysis, Membrane effect, Gap size effect.

1 Introduction

Rectangular hollow section (RHS) K-joints are widely used in onshore and offshore structures, and the ultimate strength and stiffness are of dominant interest. The yield-line method has been successfully used to predict the strength of RHS T-, Y- and X-joints. Although some efforts have been made to analyse RHS K-joints using yield-line [6] and modified yield-line [5] methods, due to the complex geometry and the membrane and gap size effects, the strength formula in the latest IIW Recommendations [3] is still based on semi-empirical results for chord face failure. The nonlinear finite element method has been used to study the behaviour of RHS K-joints [4], however, significant variations (about 40%) between FEM and experimental results were found.

Moreover, the effect of gap size on the behaviour of RHS K-joints is not well understood. Recent tests [7] have shown that the gap size has an effect on the ultimate strength of RHS K-joints. It is also known that there is a membrane effect in RHS joints. However, although studies have been reported [8,9] on the membrane effect in X and T joints using simplified models, the effect is difficult to assess using simple

Tubular Structures V. Edited by M.G. Coutie and G. Davies.

models. In order to obtain a deeper understanding of all these factors on the behaviour of RHS K-joints, including joint stiffness and stress distribution in the brace, full nonlinear (both material and geometrical) finite element analyses have been performed. It is considered that correct choice of finite elements, and modelling of plastic hinges are very important in order to obtain good results. Special attention is paid to modelling the gap region.

2 Finite Element Analysis

The non-linear finite element analyses were performed using the general purpose finite element program ABAQUS [1]. The finite element meshes were generated by the I-DEAS Finite Element Modelling program [2]. The results in reference [4] showed that a four-node thin shell element in ABAQUS was not suitable to model the joints, and the authors suggested that it is necessary to use thick shell or solid elements in order to model the wall thickness and weld legs properly. However, preliminary runs have shown that there are only very small differences between the thin shell and thick shell element analysis results. This is perhaps due to the fact that in ABAQUS the transverse shears are always treated elastically, and the chord width/thickness ratio of the RHS joints analysed is about 20. In this paper, more accurate eight-node parabolic thin shell elements are used to model all the joints. Although it is known that modelling the joints, especially the weld legs, using solid elements could be expected to give more accurate predictions, the method was not used in this work because of the consideration of the computational cost and the difficulty of pre- and post-processing.

2.1 Specimen and material

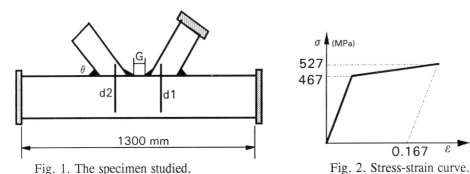

Fig. 1. The specimen studied.   Fig. 2. Stress-strain curve.

Fig. 1 shows the specimen tested and analysed in the studies. The boundary conditions are the same as in reference [4]. d1 and d2 defined in [7] are the dimensions from which the deformations were determined. The three joint variables in the studies are gap size (G), brace angle (θ) and brace/chord width ratio (β). A constant chord pretension load of 760 kN was applied to each joint. Since the real stress-strain relation in the joint was not available, the average stress-strain data of the joint material given in reference [7] was used in all the analyses, and approximated as an elastic-plastic model which is shown in Fig. 2. For the chord, a square hollow section with nominal dimensions 150×150×8 mm was used. Two brace dimensions were used, 100×100×6.1 and 70×70×5 mm, for $\beta=0.67$ and $\beta=0.47$ respectively.

2.2 Analysis program

Four specimens were tested and analysed. Table 1 shows the analysis program, in which the real gap size was measured from weld to weld and the FEM gap size is the length between the centre points of the shells.

Table 1. Analysis program

	β	$\theta(°)$	Real gap (mm) G	FEM gap (mm) g
Joint A	0.67	45	22.0	32.0
Joint B	0.67	45	5.1	20.8
Joint C	0.47	60	26.0	38.0
Joint D	0.47	60	0.0	11.5

2.3 Modelling details

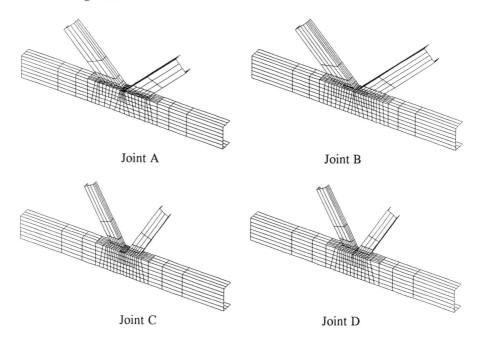

Joint A   Joint B

Joint C   Joint D

Fig. 3. Finite element meshes used in the studies.

Four meshes were generated for the joints, shown in Fig. 3. Due to the symmetry of geometry, boundary conditions and loading of the joints, only one-half of the joint was modelled. The number of elements in the models varied between 438 and 650. In joints B and D, experiments showed that the yield line passed into the braces, due to the very narrow gap. As can be seen in Fig. 3, very fine meshes were used in chord regions near the gap in these cases.

In practice, the direction of chord pretension loading rotated during brace loading. This rotation was not considered in reference [4], but undoubtedly affects the analysis results. In this work, a spring element (ABAQUS element SPRINGA) which can trace the node rotations is used to model the chord pretension loading. Multi-point constraints Nos. 2 and 7 in ABAQUS were used in refining meshes and modelling the rigid links simulating boundary conditions. Seven integration points in the thickness direction of the shell were used. The automatic incrementation modified RIKS method was chosen in the FEM solution. The tolerance for individual forces and moments was set to about 0.5 % of the respective maximum values.

3 Results of finite element analysis and discussion

3.1 Load-Displacement Relationships

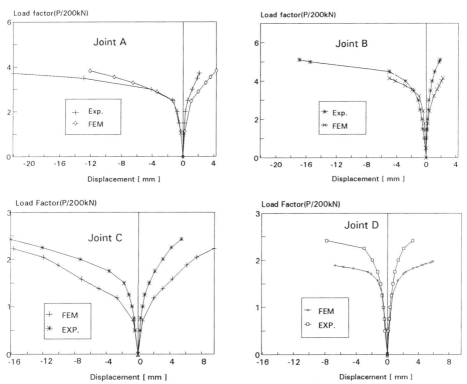

Fig. 4. Load-Displacement relationships of the joints.

Fig. 4 shows computed and experimental load-displacement curves for the joints. It can be seen in Fig. 4 that, in general, the FEM solutions give lower predictions than the experimental results. Good agreement for joints A and B ($\beta=0.67$) was obtained. The agreement is better than that in reference [4], in which approximate 40 % scatter in ultimate strength prediction was found. However, the FEM predictions give more flexible results than experiment for joints C and D ($\beta=0.47$). One reason for the large scatter may be that the effect of weld legs in the small gaps has not been modelled correctly and the mesh is still not fine enough. Another reason may be that the real stress-strain curve is different from the one in the analyses. The RHSs in the tests were cold formed with pre-hardening at the corners of the sections. The real stress-strain curve for the material at the corner is not available, but it is expected that the yield stress is much higher than the average value used in the analysis.

The effect of the gap size on the behaviour of the joints is shown in Fig. 5. It can be seen that, for joints A and B ($\beta=0.67$), a decrease in gap size greatly increases the stiffness and strength of the joints. With a reduction of 12 mm in FEM gap size, the stiffness of joint almost doubles. Due to the inability to predict fracture load, the FEM program cannot predict the numerical values of the ultimate strength increase. For $\beta=0.47$, a similar increase in stiffness as the previous example resulting from a decrease in the gap size can be also seen, Fig. 5. However, very little post-yielding load-bearing capacity is found in Joint D.

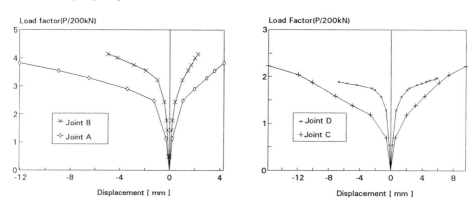

Fig. 5. Comparisons of the gap size effect on load-bearing behaviour.

3.2 Membrane effect in the joints
The membrane effects in the joints are shown in Fig. 6. It can be seen that the membrane effect depends on the joint geometry. Reference [4] gives that for a chord width/thickness ratio of approximate 20, and for $\beta=0.8$, there is very little difference between analyses using MN (material nonlinearity) only and both MN and GN (geometrical nonlinearity), although both MN and GN give a slightly lower prediction than MN only, due to the possibility of chord wall buckling failure. In present studies we find that, for $\beta= 0.47$, the strength will double due to the membrane effect in joint C, whereas for $\beta=0.67$, an approximate 30% increase in strength is found in joint A.

Also, it can be observed that the membrane effect depends on the gap size. A small gap size tends to decrease the membrane effect.

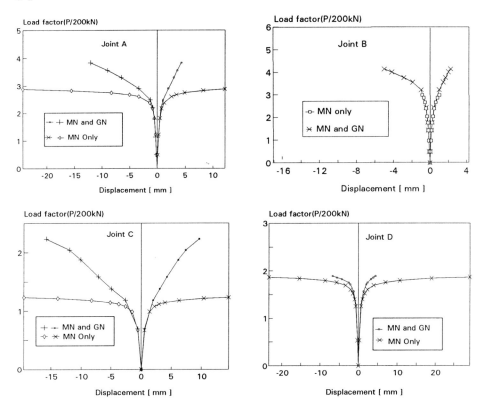

Fig. 6. Membrane effect in the joints.

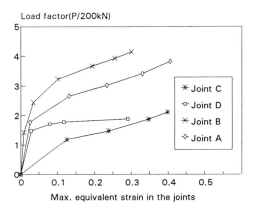

Fig. 7. The effect of joint geometry on maximum strain in the gap region.

3.3 Maximum strain in the gap region

The effect of the joint geometry on the maximum equivalent strain at integration points in the gap region is shown in Fig. 7. It is found that the maximum strain in the gap regions of the four joints appeared at the same positions, near the corner of the tension brace, in outside layers. It can be seen that, for $\beta = 0.67$ (joints A and B), a decrease in the gap size reduces the maximum strain in the gap region. This can effectively explain the increase in the ultimate strength of joint B found in the experiment. However, for $\beta = 0.47$ with zero real gap (joint D), a much smaller maximum strain is observed in the early stages of loading, but at later stages a large increase in the maximum strain is observed.

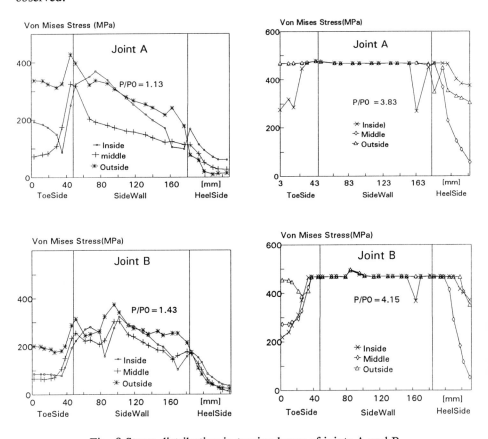

Fig. 8 Stress distribution in tension brace of joints A and B

3.4 Stress distribution in tension braces

Fig. 8 and 9 shows the von Mises stress distribution in tension braces at pre- and post-yielding stages of joints A, B, C, D (P0=200kN). By comparing the distributions in joints A and B, it can be seen that, before yielding, the stress concentrates at the corner of the side wall near the gap, while a more even distribution is observed in joint

370

B with the peak stress appearing at 1/3 of the length of the side wall (near the gap). This finding could explain the internal movement of the yield line in joints with a very narrow gap.

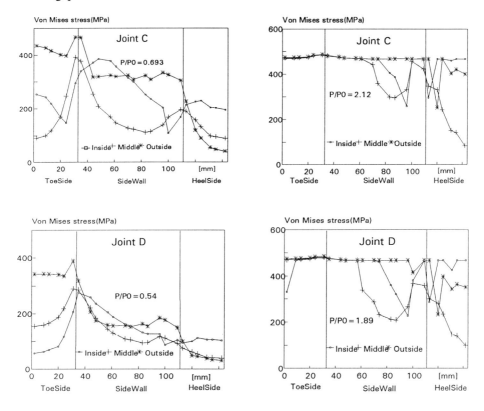

Fig. 9. Stress distribution in tension brace of joints C and D

4 Conclusions

Four specimens were tested and analysed using nonlinear FEM. Improved agreement between the FEM and experimental load-displacement relationships was obtained compared with a previous study [4]. From the studies, the following conclusions can be drawn:

(1) It is important to use high-order thin shell elements in ABAQUS for the analysis of RHS K-joints. When modelling, special care must be taken with the yield-line pattern of the joints. Correct modelling of the chord pretension loading is also important. For joints with a small gap, it is important to use solid elements to model the weld legs in order to obtain accurate predictions.
(2) The membrane effect in the joints depends on joint geometry. With a chord width

/ thickness ratio of approximate 20, for $\beta < 0.67$, the strength will increase by 30-100% due to the membrane effect; while for $\beta > 0.8$, no membrane effect needs to be considered in design, despite the possibility of chord web buckling failure.

(3) The joint geometry, especially the gap size, will affect the maximum strain value in the gap region. For $\beta = 0.67$, the maximum strain in joint B with a very small gap is almost half the maximum strain in joint A with a normal gap. This finding may explain the experimental ultimate strength increase in Joint B [7].

(4) The gap size affects the stress distribution in tension braces. With a very narrow gap, the stress concentration position moves from the corner to the inside of the brace. This confirmed the finding of previous work [7].

5 References

[1] ABAQUS User's Manual, Version 4.8. (1989) Hibbit, Karlsson & Sörensen, Inc.

[2] I-DEAS, Finite Element Modeling User's Guide (1990) Structural Dynamics Research Corporation.

[3] International Institute of welding (1989) Design Recommendations for Hollow Section Joints - Predominantly statically loaded. **Document No. XV**-701-89 (IIW).

[4] Koskimäki, M. and Niemi, E. (1989) Finite element studies on the behaviour of rectangular hollow section K-joints, in **Tubular Structures** (Eds. by E. Niemi and P. Mäkeläinen), Elsevier Applied Science, pp 28-37.

[5] Niemi, E. (1986) On the deformation capacity of rectangular hollow section K-joints-a modified yield-line approach. **Mechanical Engineering Department Publication**, No. 37, Lappeenranta University of Technology.

[6] Packer, J. A. Davies G. and Coutie, M. G. (1980) Yield strength of gapped Joints in rectangular hollow section trusses. **Institution of Civil Engineers**, Part 2, V69, p 995-1013.

[7] Soininen, R. Niemi, E. and Zhang, Z. L. Studies of the fracture behaviour of RHS K-joints at low temperatures. To be published as **Mechanical Engineering Department Publication**, Lappeenranta University of Technology.

[8] Wardenier, J. (1982) **Hollow Section Joints**. Delft University Press.

[9] Zhao, X.-L. and Hancock, G. J. (1991) Plastic mechanism analysis of T-joints in RHS under concentrated force. **Research report** No. R644, School of Civil and Mining Engineering, The University of Sydney.

36 TESTS OF K-JOINTS IN STAINLESS STEEL SQUARE HOLLOW SECTIONS

K.J.R. RASMUSSEN, F. TENG and B. YOUNG
University of Sydney, Australia

Abstract
The paper describes a test program on welded stainless steel K-joints fabricated from square hollow section brace members and chords. Design rules are proposed for stainless steel K-joints by adopting the rules of the CIDECT Recommendations for carbon steel tubular structures and replacing the yield stress in these recommendations by a proof stress. It is shown that both the 0.2 % and 0.5 % proof stresses can be used for the design strength and that the serviceability limit state is not critical if either of these stresses is used. The proof stresses are based on the properties of the finished tube.
Keywords: Tests, Stainless Steel, Square Hollow Section, Welded Joints, Design Recommendations.

1 Introduction

In Australia, cold-formed stainless steel tubular sections have been marketed for structural purposes since 1989. To stimulate the use of these sections in buildings, design guidelines were recently developed at the University of Sydney for the bending and compression strengths of stainless steel square and circular hollow sections (Rasmussen & Hancock 1990, 1992). These guidelines complemented the ASCE Specification for the Design of Cold-formed Stainless Steel Structural Members (ASCE 1990) which is based mainly on tests of light gauge open sections.

The present paper is a further development in the research of stainless steel tubular sections and concerns the strength of welded K-joints of square hollow sections (SHS). It describes a series of tests and provides guidelines for the design of these joints.

Extensive research was performed on welded connections of *carbon* steel SHS during the 1960's, 70's and 80's. The research was described in detail in Section 6 of CIDECT Monograph No. 6 (CIDECT 1986) and was compiled in a recent publication by CIDECT (1992). The design guidelines proposed in this paper adopt the CIDECT Recommedations while incorporating the material properties specific for stainless steel. These include a rounded stress-strain curve with no distinct yield plateau and different properties in tension and compression. As a consequence of the rounded stress-strain curve, deformations of stainless steel joints generally exceed those of carbon steel joints, and hence particular attention needs to be paid to deformations of stainless steel joints under service loads.

The manufacturing process of the stainless steel tubes used in the test program

Tubular Structures V. Edited by M.G. Coutie and G. Davies.
© 1993 E & FN Spon, 2–6 Boundary Row, London SE1 8HN. ISBN 0 419 18770 7.

involved cold-forming into a circular shape, welding and subsequent sizing into a square shape. It is well known that the plastic deformations induced by this process enhance the material properties of the tubes. For carbon steel tubular joints, the CIDECT Recommendations (Section 2 of CIDECT 1986) permit utilisation of this enhancement by allowing the yield and tensile strengths to be obtained from the finished product rather than the coil strip. Furthermore, it was shown in Rasmussen & Hancock (1990, 1992) that the bending and compression strengths of stainless steel tubes can be based on the enhanced properties and that the design strengths become very conservative when based on the properties of the annealed material. Consequently, the design guidelines proposed in this paper are based on the properties of the finished product.

2 Test program

2.1 Material

The tests were performed on SHS of austenitic stainless steel of type 304L, having a Nickel content between 8 and 13 %, a Chromium content between 18 and 20 %, and a maximum Carbon content of 0.035 %. The tubes were cold-rolled from annealed coils of strip.

In all tests, the chord consisted of an 80×80 mm SHS. The chord members were selected from the same batch and so could be expected to have the same material properties. A longitudinal coupon was cut from the centre of a wall which formed a 90° angle with the wall containing the weld. (The stress-strain characteristics vary around the tube, because the plastic deformation of the section during rolling is non-uniform, and are generally lowest in the wall perpendicular to the wall containing the weld). The static 0.2 % and 0.5 % tensile proof stresses and the tensile strength were obtained as $\sigma_{0.2} = 450$ MPa, $\sigma_{0.5} = 520$ MPa and $\sigma_u = 690$ MPa respectively, and the equivalent elongation after fracture was measured as 67 %, indicating a very ductile material. The initial Young's Modulus was obtained as 191 GPa and the proportionality stress was estimated at 150 MPa. Only the tensile properties were determined because the compressive stress-strain curve of the finished SHS is higher than the tensile stress-strain curve (Rasmussen & Hancock 1990, 1992), and hence the tensile properties may conservatively be used in design. The properties of the 51×51 mm and 38×38 mm brace members were not measured because the stress-strain curves of the these sections are nearly the same as those of the 80×80 mm SHS (CASE 1990). The curves are slightly higher and so can conservatively be taken as that for the 80×80 tube. Furthermore, in all tests involving 51×51 mm and 38×38 mm brace members, failure occurred by plastification of the chord or by fracture of the weld rather than failure of the brace members.

2.2 Specimen fabrication

Three brace member widths of 38, 51 and 80 mm were welded to 80 mm wide chords providing ratios (β) of brace width to chord width of 0.47, 0.64 and 1.0 respectively. For each value of β, the brace members were connected at angles (θ) of 30°, 45° and 60° so that nine tests were performed in total. The measured cross-section dimensions are shown in Table 1 using the nomenclature defined in Fig. 1. The values of plate width, thickness and corner radius were averages of measurements of all four sides or corners of the cross-sections.

374

Table 1. Measured specimen dimensions.

Spec.	Chord			Brace			Angle		Ratio	Gap	O'lap	Ecc
	b_0	t_0	r_0	$b_{1,2}$	$t_{1,2}$	$r_{1,2}$	θ_1	θ_2	β	g	q	e
	mm						-	-	-	mm		
S38-30	80.7	3.10	5.1	38.1	3.13	5.3	29°45'	30°05'	0.47	58	–	0
S38-45	80.4	3.10	4.5	38.1	3.14	4.8	45°00'	44°25'	0.47	22	–	0
S38-60	80.7	3.40	5.4	38.2	3.10	5.3	59°35'	59°35'	0.47	20	–	16
S51-30	80.0	3.30	5.3	51.0	3.00	5.3	29°55'	29°55'	0.64	35	–	0
S51-45	80.3	3.10	5.1	51.0	3.00	5.1	44°55'	44°55'	0.64	15	–	0
S51-60	80.3	3.36	5.5	51.0	3.05	5.1	61°10'	61°10'	0.64	–	16	0
S80-30	80.3	3.16	5.3	80.4	3.36	5.3	30°00'	30°00'	1.0	–	41	–9
S80-45	80.5	3.30	5.1	80.3	3.10	5.1	46°05'	44°30'	1.0	–	30	0
S80-60	80.3	3.35	5.3	80.5	3.10	5.5	62°00'	60°05'	1.0	–	45	0

Fig. 1. Definition of symbols.

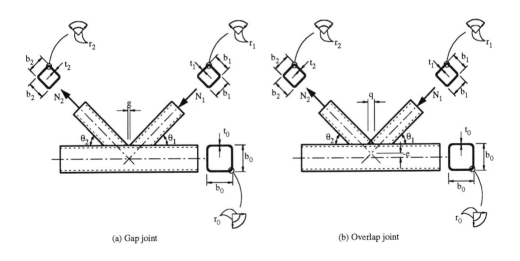

(a) Gap joint   (b) Overlap joint

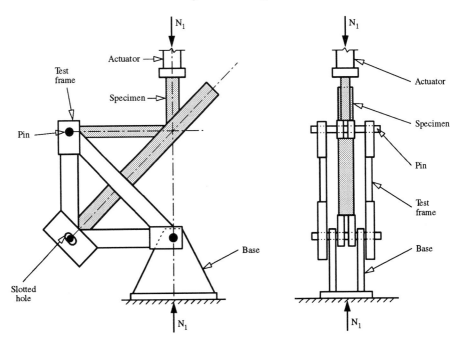

Fig. 2.   Test rig.

The joints were designed such that the centreline of the brace and chord members intersected at the same point. However, specimens S38-60 and S80-30 included an eccentricity (e) in order to comply with the CIDECT Recommendations (1992) for the minimum gap and overlap sizes. The eccentricity was assumed to be negative when the intersection of the brace member centrelines was towards the top face of the chord, as shown in Fig. 1b. The gap (g) and overlap (q) were measured as the distance between the toes of the brace members, as shown in Figs 1a and 1b respectively.

The welds were designed according to the AWS Specification (AWS 1990) and laid using Manual Metal-Arc Welding. A 3.25 mm electrode of type E308L-16 with nominal 0.2 % proof stress, tensile strength and elongation of 400 MPa, 610 MPa and 40 % respectively was used for all welds.

2.3   Test procedure

The specimens were tested in the rig shown in Fig. 2. The pins and the slotted hole of the triangular test frame allowed the concentric member forces to be applied by a single actuator. The applied force, which was also the force in the compressive brace, is denoted by N_1. As also shown in the figure, there was no force applied at the unsupported end of the chord.

The instrumentation consisted of transducers measuring deflections of the top face and the sidewalls of the chord. One set of transducers measured the indentation (u) of the top face of the chord caused by the compressive brace member, as shown in Fig. 3.

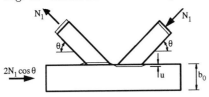

Fig. 3. Indentation of chord top face.

The tests were controlled by incrementing the stroke extension. This allowed the loading to be continued after reaching the ultimate load. Readings of the applied load (N_1) and the transducers were taken approximately one minute after applying an increment of stroke extension, thereby allowing the stress relaxation associated with plastic straining to take place and hence the load to stabilise.

2.4 Test strengths and failure modes

The maximum load applied to the compressive brace member is shown as N_{1u} in Table 2a. The table also describes the observed failure modes, using the same classification as that of the CIDECT Recommendations (1992). In the tests of specimens S38-45, S38-60 and S51-45, the indentation (u) was of the order of 10 % of the chord width at ultimate, while in the test of specimen S51-60, the indentation (u) was approximately 5 % of the chord width. Consequently, failure of these specimens involved large plastic deformations of the chord top face. However, the ultimate load was governed by fracture of the weld rather than tearing of the chord face or brace members. Most likely, tearing occurred in the weld because the weld material had a lower tensile strength and elongation than the finished tube. This indicates that the electrode (E308-16), which is generally recommended for welding 304L stainless steel material, may not be appropriate for structural applications of 304L tubes for which the properties are enhanced by cold-forming.

Failure of the full width specimens was generally associated with small deformations. In the tests of specimens S80-45 and S80-60, the ultimate load was governed by tearing of the weld whereas the ultimate load of specimen S80-30 was governed by local buckling of the chord. The latter mode involved local buckling of the entire cross-section at a load close to the stub column strength of the chord. This mode should be classified as section failure rather than joint failure, and would be considered while designing the chord in engineering practice.

3   Design guidelines

3.1   General

In the absence of a distinct yield stress of stainless steel, the approach adopted in the ASCE Specification (1990) for the Design of Cold-formed Stainless Steel Structural Members is to use strength equations derived from those for carbon steel cold-formed members and to replace the material constants for carbon steel by their equivalent for stainless steel. For example, the strength equation for compression members involves the tangent modulus (E_t) rather than Young's modulus (E) and the 0.2 % proof stress rather than the yield stress. (The 0.2 % proof stress is defined as the stress on the stress-strain curve at which the *plastic* strain equals 0.2 %).

The same approach is extended to joints in the present paper by adopting the CIDECT strength equations and replacing the yield stress by a proof stress in these

Table 2. Limit states loads and failure modes.

Specimen	Failure mode	N_{1u}	$N_{1\sigma_{0.2}}$	$N_{1\sigma_{0.5}}$	$\dfrac{N_{1u}}{N_{1\sigma_{0.2}}}$	$\dfrac{N_{1u}}{N_{1\sigma_{0.5}}}$
		(kN)	(kN)	(kN)		
S38-30	A	156	116	135	1.34	1.16
S38-45	A+C	157	93	107	1.69	1.46
S38-60	A+C	150	88	101	1.70	1.49
S51-30	A	233	167	193	1.40	1.21
S51-45	A+C	187	125	144	1.50	1.30
S51-60	C	211	141	163	1.50	1.30
S80-30	G	(295)	210	243	(1.40)	(1.22)
S80-45	C	341	202	233	1.69	1.46
S80-60	D+C	391	296	342	1.32	1.14
Average					1.52	1.32
Standard deviation					0.16	0.14

A	Plastic failure of chord face
C	Tension failure of bracing member or weld
D	Local buckling of bracing member
G	Local buckling or chord

Key to failure modes

a) Ultimate limit state

Specimen	N_{1s}	$N_{1\sigma_{0.2}}/1.5$	$N_{1\sigma_{0.5}}/1.5$	$\dfrac{N_{1s}}{N_{1\sigma_{0.2}}/1.5}$	$\dfrac{N_{1s}}{N_{1\sigma_{0.5}}/1.5}$
	(kN)	(kN)	(kN)		
S38-30	144	77	90	1.87	1.60
S38-45	78	62	72	1.26	1.09
S38-60	71	59	68	1.21	1.04
S51-30	222	111	129	2.00	1.72
S51-45	132	83	96	1.59	1.38
S51-60	154	94	108	1.65	1.42
S80-30	–	–	–	–	–
S80-45	–	–	–	–	–
S80-60	350	197	228	1.78	1.54
Average				1.62	1.40
Standard deviation				0.30	0.25

b) Serviceability limit state

equations. The proof stress shall be one obtained from the finished tube, as described in Section 2.1.

3.2 CIDECT design recommendations

According to the CIDECT Recommendations (1992), the strength of *gap* K-joints with square chords shall be determined using,

$$N_1 = 8.9 \frac{f_{y0} \, t_0^2}{\sin \theta} \frac{b_1}{b_0} \gamma^{0.5} f(n) \tag{1}$$

where $\gamma = b_0/(2t_0)$, f_{y0} is the yield stress of the chord, $f(n)$ is a function of the compressive force in the chord, and it has been assumed that the brace members have the same width (b_1) and form the same angle (θ) with the chord. The function $f(n)$ is given by,

$$f(n) = 1.3 + \frac{0.4}{\beta} n \leq 1.0 \tag{2}$$

where

$$n = -\frac{2 N_1 \cos \theta}{A_0 \, f_{y0}} \tag{3}$$

For *overlap* K-joints, the strength shall be determined using,

$$N_1 = \begin{cases} f_{y1} t_1 \left[\left(\frac{O_v}{50} \right) (2b_1 - 4t_1) + b_e + b_{e(OV)} \right] & \text{for } \ 25 \leq O_v < 50 \\[2mm] f_{y1} t_1 \left[2b_1 - 4t_1 + b_e + b_{e(OV)} \right] & \text{for } \ 50 \leq O_v < 80 \\[2mm] f_{y1} t_1 \left[2b_1 - 4t_1 + b_1 + b_{e(OV)} \right] & \text{for } \ O_v \geq 80 \end{cases} \tag{4}$$

where f_{y1} is the yield stress of the brace members and it has been assumed that the brace members are square. The following definitions apply:

$$O_v = \frac{q}{b_1/\sin(\theta)} 100 \tag{5}$$

$$b_e = \frac{10}{b_0/t_o} \frac{t_0}{t_1} b_1 \leq b_1 \tag{6}$$

$$b_{e(Ov)} = 10 \, t_1 \leq b_1 \tag{7}$$

These expressions are the same as those included in the CIDECT Recommendations (1992) when using that the brace members have equal dimensions and yield stress.

It is proposed to adopt eqns (1,4) for the design of stainless steel K-joints by substituting appropriate proof stresses for the yield stresses (f_{y0}, f_{y1}). The design strengths resulting from this approach using the 0.2 % and 0.5 % proof stresses for *both* f_{y0} and f_{y1} are shown as $N_{1\sigma 0.2}$ and $N_{1\sigma 0.5}$ respectively in Table 2a. The proof stresses were $\sigma_{0.2} = 450$ MPa and $\sigma_{0.5} = 520$ MPa, as described in Section 2.1.

3.3 Comparison of design strength with test strength

Ultimate limit state

The ratios $(N_{1u}/N_{1\sigma 0.2}, N_{1u}/N_{1\sigma 0.5})$ of the test strength to the design strengths based on the 0.2 % and 0.5 % proof stresses are shown in the 6th and 7th columns of Table 2a. For both proof stresses, these ratios are greater than one for all tests, indicating conservative design strengths.

The mean values of the ratios are 1.52 and 1.32 using the 0.2 % and 0.5 % proof stresses respectively, and so the 0.5 % proof stress provides the most accurate design strengths. However, it may be preferable to adopt the 0.2 % proof stress in order to use a readily available material constant which is also the equivalent yield stress used in the ACSE Specification (1990).

It should be noticed that the CIDECT Recommendations are limited to a yield stress of 355 MPa. This limit is imposed partly because most test data was obtained for joints with yield stresses less than 355 MPa and partly because carbon steel joints with yield stresses greater than 355 MPa may not have adequate ductility. The latter aspect is not of concern in most stainless steel structures, since stainless steels generally have high ratios of tensile strength to the 0.2 % (or 0.5 %) proof stress and high values of elongation after fracture. The present test program demonstrates that the limit of 355 MPa need not be imposed on cold-formed stainless steel joints fabricated from 304L austenitic stainless steel.

The use of E308-16 electrode with a nominal 0.2 % proof stress of 400 MPa which is less than that of the tube (450 Mpa) was found to be satisfactory although the ultimate failure of several specimens involved tension failure of the weld.

Serviceability limit state
It was proposed in CIDECT Monograph No. 6 (CIDECT 1986) that joint deformations under service loads should be limited to 1 % of the chord width (b_0). It was also shown that this limit would not be exceeded if the design strength was based on the CIDECT strength rules.

The design of stainless steel structures is more likely to be governed by the serviceability limit state than carbon steel structures because the rounded stress-strain curve precipitates growth of deformations at loads well below ultimate. The load at which the measured indentation (u) equalled 1 % of the chord width is shown as N_{1s} in Table 2b. (Values of N_{1s} were not obtained for specimens S80-30 and S80-45 because the indentation was less than 1 % of the chord width at ultimate for these specimens).

The load (N_{1s}) may be compared with serviceability design loads determined by dividing the joint strengths $N_{1\sigma_{0.2}}$ and $N_{1\sigma_{0.5}}$ by 1.5. (The value of 1.5 is consistent with the CIDECT Recommendation (1992) of using a load factor of 1.5 on the design strength in allowable stress design).

The ratios $(N_{1s}/(N_{1\sigma_{0.2}}/1.5), N_{1s}/(N_{1\sigma_{0.5}}/1.5))$ of the test to design serviceability load based on the 0.2 % and 0.5 % proof stresses are shown in the 5th and 6th columns of Table 2b. For both proof stresses, these ratios are greater than one for all tests, and hence the serviceability limit state will not be reached if the ultimate strength is calculated as either $N_{1\sigma_{0.2}}$ or $N_{1\sigma_{0.5}}$. However, this is a result of the conservatism of the ultimate strength equations, since specimens S38-45 and S38-60 would reach the serviceability limit state before reaching the ultimate limit state. This follows from the fact that the ratio $N_{1s}/(N_{1\sigma_{0.2}}/1.5)$ (or $N_{1s}/(N_{1\sigma_{0.5}}/1.5)$) is less than $N_{1u}/N_{1\sigma_{0.2}}$ (or $N_{1u}/N_{1\sigma_{0.5}}$) for specimens S38-45 and S38-60.

4 Conclusions

A test program on K-joints of stainless steel SHS has been presented, consisting of square brace members welded to square chords. The angle between the chord and brace members was varied in the tests as were the ratios of brace width to chord width.

It was proposed that stainless steel SHS K-joints may be designed using the CIDECT Recommendations (1992) by replacing the yield stress by a proof stress which could be either the 0.2 % or the 0.5 % proof stress. The mean ratios of test strength to design strength were 1.52 and 1.32 when using the 0.2 % and 0.5 % proof stress respectively, indicating a conservative design approach. However, this conservatism was required to ensure that the serviceability limit state would not be reached before the ultimate limit state.

5 Acknowledgements

The authors are grateful to BHP Coated Products Division - Stainless for its support of the test program described in this paper. The authors also wish to thank Professor Greg Hancock for his comments to this paper.

6 References

ASCE (1992), Specification for the Design of Cold-formed Stainless Steel Structural Members, *American Society of Civil Engineers*, New York.

AWS (1990), Structural Welding Code D1.1, *American Welding Society*, Miami.

CASE (1990), 'Compression Tests of Stainless Steel Tubular Columns', *Investigation Report No. S770*, Centre for Advanced Structural Engineering, School of Civil and Mining Engineering, University of Sydney.

CIDECT (1986), The Strength and Behaviour of Statically Loaded Welded Connections in Structural Hollow Sections, Monograph No. 6, *Comite International pour le Developpement et l'Etude de la Construction Tubulaire, (International Committee for the Development and Study of Tubular Structures)*, British Steel Corporation.

CIDECT (1992), *Design Guide for Rectangular Hollow Sections (RHS) Joints under Predominantly Static Loading*, Auths J.A. Packer, J. Wardenier, Y. Kurobane, D. Dutta & N. Yeomans, Comite International pour le Developpement et l'Etude de la Construction Tubulaire, Verlag TUV Rheinland, Cologne.

Rasmussen, K.J.R. & Hancock, G.J., (1990), 'Stainless Steel Tubular Columns - Tests and Design', *Proc.*, Tenth International Specialty Conference on Cold-formed Steel Structures, St Louis, pp. 471-491.

Rasmussen, K.J.R. & Hancock, G.J., (1992), 'Stainless Steel Tubular Beams - Tests and Design', *Proc.*, Eleventh International Specialty Conference on Cold-formed Steel Structures, St Louis, pp. 587-610.

37 LOAD-BEARING BEHAVIOUR OF HOLLOW SECTIONS WITH INSERTED PLATES

F. MANG, Ö. BUCAK and D. KARCHER
Versuchsanstalt für Stahl, Holz und Steine (Testing Laboratory for Steel, Timber and Stone), University of Karlsruhe, Germany

Abstract
Twenty six static tensile tests have been carried out to failure on circular and rectangular hollow sections at the University of Karlsruhe. These tests form the basis for an appropriate elaboration of the design standards.
   The most important parameter for the load-bearing capacity of such joints is the geometry of the tube and in particular the non-dimensional parameter D/t. This dependence is presented by diagrams, from which a reduction factor for the load-bearing capacity of such tube-plate connections can be determined.
Keywords: Hollow Section, Tube to Plate Connections, SNCF (Strain Concentration Factor), Ultimate Load, Design Recommendation.

1 Introduction

Hollow sections are increasingly applied to the construction of halls, cranes and bridges. The German standard DIN 18 808 [1], presently valid for the design of hollow section joints (K-, N-, KT- and L-joints respectively), gives regulations for joints which are interconnected directly tube to tube. For the application of joints with gusset plates or inserted plates, which can be easily fabricated, general steel construction rules are referred to, which do not consider the strength depending on design. Up to now, few detailed investigations of such tube jointing sleeve connections are available. This research program fills the gap in order to establish the basis for a simple design procedure and provides recommendations for DIN 18 808 [1] and Eurocode 3 [3].
   Normally, at the end of the web members, plates, forks or halved I-sections have been welded for bolted connections. Rules in standards for this type of connection are not available. Since experimental data was lacking, experimental and theoretical investigations have been carried out in Karlsruhe. Through a CIDECT project,

Tubular Structures V. Edited by M.G. Coutie and G. Davies.
© 1993 E & FN Spon, 2–6 Boundary Row, London SE1 8HN. ISBN 0 419 18770 7.

the results of the connection c) of Figure 1 have been
investigated and documented [4]. In the following,
investigations on connection type e) of Figure 1 are
described and the possibilities of design methods are
shown.

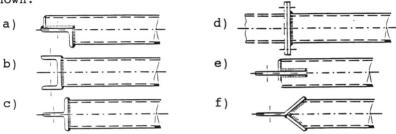

Figure 1. Different types of construction for the ends of
the bracing member

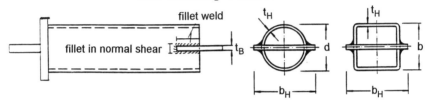

Figure 2. Form of the test specimen

2 Symbols and designations

A	cross sectional area of hollow sections
A_w	area of the weld
t_0	wall thickness of the hollow sections
a	throat thickness of the fillet weld
b_P	width of the plate
t_P	thickness of the plate
l	length of the weld
f_y, f_u	material yield and ultimate stress
$d_0, D, \varnothing$	external diameter of CHS
b_0, B	outer width of RHS
d_0/t_0 or b_0/t_0	wallslenderness ratio for CHS and RHS
P_d, P_k	tensile design resistance and capacity of joint
γ_m, γ_f	partial safety factors for resistance and action
δ [%]	elongation by tensile tests
$\varepsilon, \varepsilon_{el}, \varepsilon_{pl}, \varepsilon_{nom}$	strain, elastic, plastic, nominal respectively
ε_{hs} (σ_{hs})	hot spot strain (stress)
σ_{max}, σ_B	nominal maximum stress of connection

$\sigma_{0.2}$	nominal yield stress of connection
CHS, RHS	Circular and Rectangular Hollow Section
E	modulus of elasticity of steel
	(210 kN/mm^2)
FEM	Finite Element Method
SCF	Stress Concentration Factor
SNCF	Strain Concentration Factor

3 Parameters for the load bearing tests

The following parameters are of importance:
- plate thickness (t_p)
- inserted length of the plate (equal to the weld length)
- ratio d_0/t_0
- type of weld (fillet weld or three-plate weld)

The test specimens have been defined in a way that an answer could be given to the parametric influence mentioned here.

4 Test specimens and testing procedure

20 tests on CHS specimens and 6 tests on RHS specimens have been carried out at Karlsruhe. For the experimental work, a T-joint formed from thick plates (see Figure 1c) was welded on one end of the hollow section, so that the failure had to occur in the plate-to-tube connection. The maximum load-bearing capacity as well as the force spread were the interesting points of the static tensile tests.

It becomes evident from the test results that the load-bearing capacity was always below the calculated load-bearing capacity of the tubular sections.

The static load was increased in small steps up to the maximum load. After each loading step, the load was reduced to the preload (≈ 10 kN). From the measurements (see section 5), the yield and the maximum loads can be determined. The strain gauge measurements have been used for the determination of $\varepsilon_{nom} = (\varepsilon_{el} + \varepsilon_{pl})$ (see Figure 3).

5 Measurements

Measurements have been carried out to determine the elongation of the specimens and the partial deflection of the critical points of the specimens using displacement transducers.

Strain gauges were used to determine the strain (and stress) distribution in the critical parts of the connections. Figure 4 shows the location and numbering of these gauges on the specimen. At the midlength of tube, 4 linear strain gauges were attached in order to

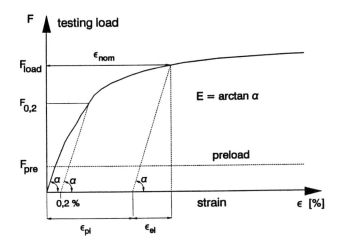

Figure 3. Determination of the nominal strain

determine the possible share of moments. The test
specimens were also provided with strip gauges at the weld
toe at the end of the inserted plate. From these gauges,
hot spot strains at the weld toes were determined. For
three test specimens, 5 rosette gauges were applied
additionally at the end of the inserted plate. The photo
of Figure 5 give an impression about the rosette gauges on
the specimens.

The hot spot strain ε_{hs} at the weld on the critical
zones has been determined by linear extrapolation. In
order to avoid the local effects of the weld toe, the
first extrapolation point is taken at $0.4 \cdot t_0$, but not
smaller than 4 mm.

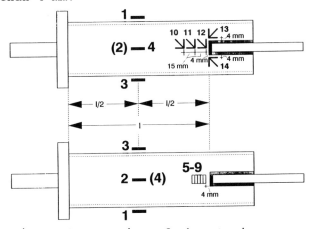

Figure 4. Location of the strain gauges

Figure 5. Test specimen after the test (typical crack)
location of the rosette gauges

6 Test results

Bear axial loading was checked from the four midlength
gauges. Usually, the stresses from the moments were within
5% to 10% of the axial stress and the eccentricities (e =
N/M) were in the range of 1 to 4 mm. These were due to
dimensional inacurracies of the test specimens and
unavoidable constraint upon installation into the testing
machine.

The evaluation of the strain gauge chains supplied the
stress concentration factor SCF. This factor is defined as
the ratio σ_{hs}/σ_{nom}. A linear extrapolation was made from
which the values given in Table 1 were determined.

	Dimension	weld length	material	SCF
	82.5x4.0	75	Fe 360	3.15
	101.6x2.9	135	Fe 360	2.39
	101.6x4.0	90	Fe 360	3.07
CHS	101.6x4.0	180	Fe 360	2.65
	101.6x4.0	165	Fe 510	2.37
	109.7x4.0	125	Fe 360	2.85
	193.7x5.0	255	Fe 360	2.01
	100x100x8	140	Fe 360	2.60
RHS	100x100x8	170	Fe 510	3.04
	140x140x5	190	Fe 360	2.54

Table 1. Determined SCF (stress concentration factor)

The stress concentration is in the range between 2 and 3. For static load, an evaluation against yielding in the total tube area would not be economical and some local yielding is allowed.

When evaluating the rosette gauges, the direction angles of the principal stresses were important. A modification of the direction angles could not be ascertained, e.g. no significant load transposition takes place.

The load bearing respectively the yielding load of the connection have been also evaluated. This is illustrated in the diagrams 1 to 4 (Figure 6 to 9). When determining the yield load of the connection, a straight line is arranged parallel to the initial slope of the curve through the point with an elongation of 0.2% (Figure 3).

The tube geometry expressed by the non-dimensional ratio D/t, proved to be the governing parameter of the load-bearing capacity. The length of the weld does not play such an imporant role as assumed previously.

7 Evaluation of the test data

The ratio $P_{max}/(A \cdot f_{u,k})$ decreased in dependence on the geometrical parameter d_0/t_0 (b_0/t_0) of the tube. No considerable increase of the load could be achieved by a longer weld. This is due to the given load spread from the tube to the plate (stress concentration at the end of the inserted plate). A large part of the load is transferred from the tube to the plate in the area of the fillet in normal shear. The weld only contributes to a small extent to the load transfer on its total length. Only for about one third of the yield load of the tube, the area of the fillet in normal shear comes to yielding. A considerable load transfer through the total weld can only be achieved near the yield load. This has been determined by the application of strain gauges and crackle lacquer as well as by a finite element calculation. The tests were evaluated with regard to the load-bearing capacity, the course of the force lines (load spread) and the stress concentration. In Section 8, the design is made based on this evaluation.

8 Calculation methods (design recommendations)

The tests available have been evaluated with regard to the strength depending on design and have been recorded in diagrams. With this, the ratio of actually achieved load and theoretically achievable load of the total hollow section ($P_{max}/(A \cdot f_{u,k})$ respectively $P_{0.2}/(A \cdot f_{y,k})$) has been recorded via the non-dimensional value d_0/t_0.

In two diagrams (Figures 6 and 7), the achieved yield load of the test specimen (strength depending on design) has been related to the theoretical yield load of the hollow section ($P_{0.2}/(A \cdot f_{y,k})$). In two further diagrams (Figures 8 and 9), the ratio load-bearing capacity/theoretical ultimate load at rupture of the hollow section ($P_{max}/(A \cdot f_{u,k})$) has been recorded. In the diagrams 1 and 3 (Figures 6 and 8) a generating curve of an envelope has been defined above which are all test values. In the diagrams 2 and 4 (Figures 7 and 9) a statistical evaluation of results is presented.

The design is realized by common methods. The inserted plate, the weld and the hollow section are calculated. Only for the allowable load of the hollow section, a reduction is made in dependence on the ratio d_0/t_0. The reduction factor can be taken from the diagrams. It becomes evident from the diagrams that a reduction has to be made since the theoretical yield or load-bearing capacity of the tube cannot be reached. The diagrams represent a first evaluation of such tests. These diagrams are to be extended in their range of validity through further tests.

The following prerequisites have to be kept for the application of the diagram:

1) static tensile load,
2) consideration of welding requirements
3) $A_{cover\ plate} > A_{tube}$,
4) sufficcient lateral excess of the cover plate
5) Thickness of the cover plate:
   $t_L \geq 0.5 \cdot \sqrt{(D_{tube}^2/16 + A_{tube})} - 0.125 \cdot D_{tube}$

In the following, an example for the handling of the proposed calculation methods is given by means of diagrams.

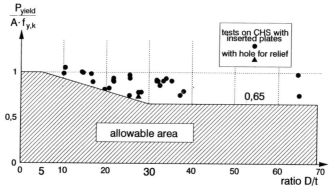

Figure 6. Experimental values and design diagram 1 based on yield

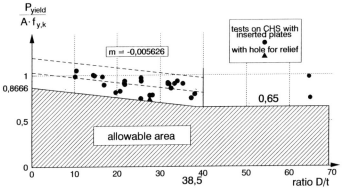

Figure 7. Statistical evaluation of the test results and design diagram 2 based on yield

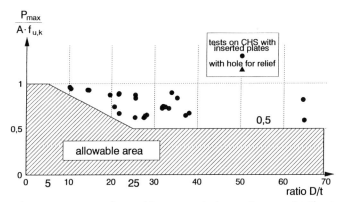

Figure 8. Design diagram 3 based on ultimate

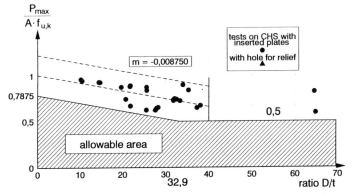

Figure 9. Design diagram 4 based on ultimate

9 Example

Additional calculation of a test specimen. The calculation is performed according to **DIN 18 800**, part 1.

internal force from the calculation:
tensile load $\quad$ P_k = 150 kN
- (CHS) selected 82.5 x 4.0
- real tube dimensions: $\emptyset$ 82.5 x 4.2 $\quad$ A_{tube} = 1033 mm²
- characteristical values of the material of the tube

$$f_{y,k} = 335 \ N/mm^2$$
$$f_{u,k} = 439 \ N/mm^2$$

- plate: $\quad$ b_P x t_P = 123 x 10 mm

characteristical values of the material of the tube are
used

thick. $t_P \quad \geq 0.5 \cdot \sqrt{(D_{tube}^2/16 + A_{tube})} - 0.125 \cdot D_{tube}$

(see point 5 $\quad \geq 0.5 \cdot \sqrt{(82.5^2/16 + 1033)} - 0.125 \cdot 82.5 = 8.8$ mm

on page 7) $\quad t_P = 10$ mm ≥ 8.8 mm

width $b_P \quad \geq D_{tube} + 4 \cdot 8.8 = 117.7$ mm

(see points 3+4 $\quad \geq 82.5 + 4 \cdot 8.8$ mm $= 117.7$ mm

on page 7) $\quad b_L = 123$ mm ≥ 117.7 mm

- weld length: l = 110 mm

throat thickness: a = 4 mm

DIN 18 800 [2], part 1

min a $\quad \geq 2$ mm

min a $\quad \geq \sqrt{(max \ t)} - 0.5 = \sqrt{(10)} - 0.5 = 2.66$ mm

max a $\quad \leq 0.7 \cdot min \ t = 0.7 \cdot 4$ mm $= 2.8$ mm

according to DIN 18 808 [1], the throat thickness should
be equal to the smaller wall thickness t = 4 mm

weld length:
l $\geq 6.0 \cdot a = 24$ mm resp. $l_{min} = 30$ mm
l $\leq 150 \cdot a = 600$ mm

Calculations:
$P_d = P_k \cdot \gamma_f = 150 kN \cdot 1.35 = 202.5$ kN

plate: $\quad \sigma = P_d/A = 202.5/1230 = 164 \ N/mm^2$

$\quad \leq 305 \ N/mm^2 = f_{y,d} = f_{y,k}/\gamma_m = 335/1.1$

weld: $\quad A_w = \Sigma(a \cdot l) = 4 \cdot 4 \cdot 110 = 1760 \ mm^2$

$\quad \tau_{\parallel} = 202.5/1760 = 115 \ N/mm^2 \leq 207 \ N/mm^2$

$\quad = \sigma_{w,R,d} = \alpha_w \cdot f_{y,k}/\gamma_m = 0.95 \cdot 240/1.1$

tube: $\quad D/t = 10.6$

- design diagram 1 $\quad k_{y1} \approx 0.8$
- design diagram 2 $\quad k_{y2} = -0.005626 \cdot 19.6 + 0.8666$

$\quad = 0.756$

$\Rightarrow P_{y,d} = min k_y \cdot A \cdot f_{y,d} = 0.756 \cdot 1033 \cdot 305 = 238.3 kN$

$P_k = 202.5$ kN ≤ 238.3 kN $= P_{y,d}$

- design diagram 3 $\quad k_{u3} \approx 0.64$
- design diagram 4 $\quad k_{u3} = -0.008750 \cdot 19.6 + 0.7875$

$\quad = 0.616$

$\Rightarrow P_{u,d} = min k_u \cdot A \cdot f_{u,d} = 0.616 \cdot 1033 \cdot 439/1.1 = 254$ kN

$P_k = 202.5$ kN ≤ 254 kN $= P_{u,d}$

10 Conclusion and Overview

This paper presents the results and analyses of experimental and numerical investigations on hollow section connections with inserted plates. Experimental investigations include 26 tests.

The most important parameters for the load-bearing capacity of connections with inserted plates are geometric parameters of the hollow sections.

For the numerical work, the test specimens investigated experimentally have been analysed numerically. Comparisons have been done between the results of experiments and those of numerical work.

The objective of the research program was to establish design rules for the special types of connections.

11. References

[1] N.N.,DIN 18 808 "Stahlbauten",Tragwerke aus Hohlprofilen unter vorwiegend ruhender Beanspruchung, (Steel Structures Consisting of Hollow Sections Predominantly Static Loaded), Ausgabe Oktober 1984

[2] N.N., DIN 18 800, Teil 1 "Stahlbauten",Bemessung und Konstruktionen (Steel Structures; Design Construction), Ausgabe November 1990

[3] Eurocode 3, Bemessungen und Konstruktion von Stahlbauten (Design of Steel Structures)

[4] Gräßlin,Rohrer
Versuchstechnische und theoretische Untersuchungen der Tragfähigkeit von Plattenverbindungen runder und rechteckiger Hohlprofile
Seminararbeiten an der Versuchsanstalt für Stahl, Holz und Steine, 1988 und Januar 1989

[5] Mang, F., Tragfähigkeit von Hohlprofilen mit einge-steckten Laschen (Load-Bearing Capacity of Hollow Sections with Inserted Plates). Final report of the research project, No. 585/89 IfBt, Berlin 1992

[6] N.N., Recommended fatigue design procedure for hollow section joints, part 1, hot spot stress method for nodal joints, IIW Doc. XV-582-85/XII-1158-85, International Institute of Welding, IIW-XV-E (1985)

[7] Wardenier, Kurobane, Packer, Dutta, Yeomans, Berechnung und Bemessung von Rundhohlprofilen unter vorwiegend ruhender Beanspruchung (Calculation and Design of Circular Hollow Sections under Predominantly Axial Load), Comité International pour le Développement et l'Etude de la Construction Tubulaire, Verlag TÜV Rheinland GmbH, Köln 1991

PART 11

JOINTS

Static strength of
three dimensional joints

38 NEW ULTIMATE CAPACITY FORMULAE FOR MULTIPLANAR JOINTS

J.C. PAUL
Obayashi Corporation/University of Tokyo, Japan
Y. MAKINO and Y. KUROBANE
Kumamoto University, Japan

Abstract

The lack of data on multiplanar joints is reflected in a different treatment of multiplanar joints in the various design codes. Traditionally the design codes treat multiplanar joints as a series of uniplanar joints. The interaction between the different planes is ignored, while the uncertainty about the real behavior is covered by the safety factor.

In this paper new semi–empirical formulae for the ultimate capacity are proposed for multiplanar TT– and KK–joints, based on a screened data base of all the available experimental data and an analytical ring model of a multiplanar TT–joint.

Keywords: Multiplanar tubular joints, Ultimate capacity, Capacity formulae

1 Introduction

Tubular three dimensional space frames of circular hollow sections are frequently used for off– and on–shore applications. The intersecting braces, which are connected by a weld to the continuous chord, often lie in different planes making multiplanar joints unavoidable features of space frames.

Studies of the behavior of multiplanar joints are scarce when compared to their uniplanar counterparts. As the need for a more realistic assessment of the ultimate behaviour of multiplanar joints becomes imminent increasing research is devoted to this topic in order to establish improved guidance when designing multiplanar joints. Experimental research is limited to multiplanar XX–, TT– and KK–joints. (Akiyama et al. (1974), Makino et al. (1984, 1993), Scola et al. (1990) ,Van der Vegte et al. (1991), Paul et al. (1991, 1992a) and Mouty et al. (1992)). Earlier proposed interaction formulae for multiplanar TT– and KK–joints are shown in Table 1. The TT– and KK–joint configurations are shown in Fig. 1.

Despite these research efforts, the basis for designing multiplanar joints remains insufficient and the commonly accepted design codes do not provide extensive, nor succinct, guidance on multiplanar joints. The general practice is to design multiplanar joints as a series of uniplanar joints ignoring the interaction between the different planes.

The American Welding Society Structural Welding Code AWS–D1.1 (1992) is the only design code which provides general design criteria applicable to any type of non–overlapping multiplanar joints. Comparisons with the data base of Paul (1992b) show that the multiplanar features in this design code hardly increase the accuracy of predictions for multiplanar TT– and KK–joints. The American Petroleum Institute Recommended Practice 2A (1991) provides formulae only for uniplanar joints, of which the K–joint formulae is more accurate in the prediction of multiplanar KK–joints than predictions with the AWS Code (Paul et al., 1993c).

The Cidect Design Guide for Circular Hollow Section Joints (Wardenier et al., 1991) proposes criteria for multiplanar XX– ,TT– and KK–joints based on initial investigations into the ultimate behavior of these joints.

Although it appears impossible to develop formulae which are both accurate and applicable

Tubular Structures V. Edited by M.G. Coutie and G. Davies.
© 1993 E & FN Spon, 2–6 Boundary Row, London SE1 8HN. ISBN 0 419 18770 7.

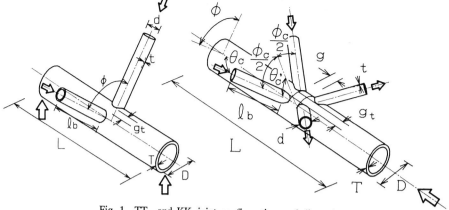

Fig. 1. TT– and KK–joint configuration and dimensions.

to a wide range of geometries and loadings due to the limited test data available, this paper attempts to derive simple interaction formulae incorporating observed multiplanar effects.

2 Need for new interaction formulae

The interaction formulae for multiplanar TT– and KK–joints of Table 1 predict the multiplanar variations in the ultimate capacity fairly accurately. However, these interaction formulae are based on regression models which are only suited for one of the two failure types that occured. When applying these formulae beyond the variation ranges of the experimental data caution becomes necessary. The following presents some important issues when developing new interaction formulae.

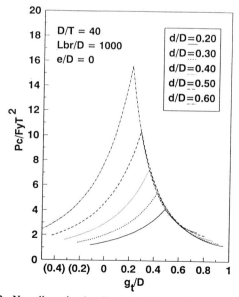

Fig. 2. Non–dimensional collapse load $P_{c,TT}$ as function of ζ_t and β.

Table 1.   Mean and CoV of the test to prediction ratios for TT- and KK-joints.

		Prediction formula mean ultimate capacity		Mean	CoV
T-joints	Kurobane (1984)	$$P_{u,T} = 4.83\ [1+4.94\ (\frac{d}{D})^2](\frac{D}{T})^{0.233}(\frac{L}{D})^{-0.45}\frac{F_y T^2}{\sin\theta} \qquad (1)$$		1.195	0.142
TT-joints	Scola (1990)	$$\frac{P_{u,TT}}{P_{u,T}} = 1.34-0.69\ \arcsin(\frac{g_t}{D}) \qquad (2)$$		1.111	0.108
	Paul (1991)	$$\frac{P'_{u,TT}}{P'_{u,T}} = a(\phi)e^{\,b(\phi)\frac{g_t}{D}} \qquad (3)$$ $$a(\phi) = 1.4018\sin\phi - 0.1975\cos\phi$$ $$b(\phi) = -2.3255\sin\phi + 0.7700\cos\phi$$		0.999	0.065
	This paper			1.026	0.101
K-joints	Kurobane (1984)	$$P_{u,K} = f_0\, f_1\, f_2\, f_3\, f_4\ F_y T^2 \quad with \quad f_0 = 2.11(1+5.66\frac{d}{D}),$$ $$f_1 = [1 + \frac{0.00904(\frac{D}{T})^{1.24}}{\exp(0.508\frac{g-3.04}{T}-1.33)+1}](\frac{D}{T})^{0.209},$$ $$f_2 = \frac{1-0.376\cos^2\theta}{\sin\theta},\quad f_3 = 1+0.305n_0-0.285n_0^2,$$ $$n_0 = \frac{N}{\pi(D-T)TF_y},\quad f_4 = (\frac{F_y}{F_u})^{-0.723} \qquad (4)$$		0.966	0.103
KK-joints	Makino (1984)	$$\frac{P'_{u,KK}}{P'_{u,K}} = 0.664(1+1.66\frac{d}{D})(1-0.675\frac{g_t}{D}) \qquad (5)$$		1.037	0.096
	Makino (1984)	$$\frac{P_{u,KK}}{P_{u,K}} = 0.965(1-0.127\frac{g_t}{D}) \qquad (6)$$		1.014	0.097
	Mouty (1992)	$$\frac{P_{u,KK}}{P_{u,K}} = 1-0.49\frac{g_t}{D} \qquad (7)$$		1.120	0.098
	Paul (1992)	$$\frac{P'_{u,KK}}{P'_{u,K}} = 0.656(1+1.142\frac{d}{D})(1+0.727\frac{g}{D}) \qquad (8)$$ $$\frac{P'_{u,KK}}{P'_{u,K}} = 0.712(1-0.976\frac{g_t}{D})(1+1.606\frac{d}{D})(1+0.500\frac{g}{D}) \qquad (9)$$		1.005	0.064
	Paul (1992)	$$\frac{P'_{u,KK}}{P'_{u,K}} = 0.601\ f_1\, f_2\, f_3 \quad with\ f_1 = 1 \quad for\ \frac{g_t}{D} \le 0.2,$$ $$f_1 = 1.264-1.318\frac{g_t}{D} \quad for\ \frac{g_t}{D} > 0.2,$$ $$f_2 = 1+1.496\frac{d}{D},\quad f_3 = 1+0.557\frac{g}{D} \qquad (10)$$		1.004	0.066
	This paper			1.011	0.070

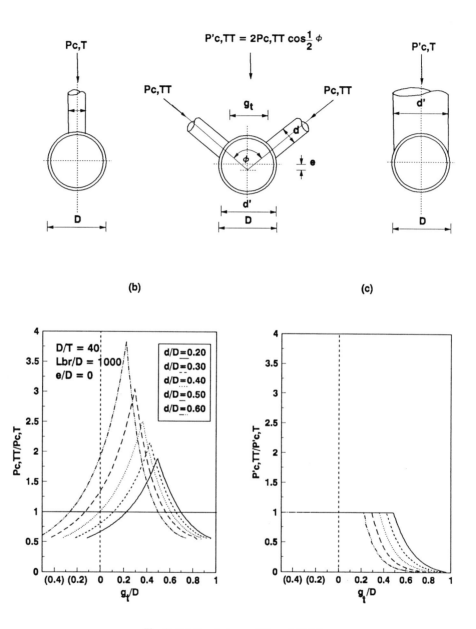

Fig. 3 Relation between TT- and T–joints:
(a) Definition. (b) Collapse load ratio $P_{c,TT}/P_{c,T}$ as functions of ζ_t and β.
(c) Collapse load ratio $P'_{c,TT}/P'_{c,T}$ as function of ζ_t and β.

2.1 Distinction between different failure types

Based on a closed ring analysis of multiplanar TT–joints, the ultimate capacity responded differently as a function of ζ under the two governing types of plastic failure, as shown in Fig. 2 (Paul et al., 1993b). For a constant value of β, the ultimate capacity of the TT–joint increases with ζ for the failure type without local deformation between the compression braces (type 1). In this failure type which will also occur for joints with out-of-plane overlapping braces, the two compression braces act as one brace and penetrate the chord simultaneously. At a certain value of ζ, the failure type changes into a failure type with local deformation between the compression braces (type 2) and the ultimate capacity decreases with ζ.

For the ring model the multiplanar coefficient μ_{TT} decreases to 0.5 for $\phi=0°$ ($\zeta=-\beta$ with e=0), which can be easily understood as the two brace loads coincide in this case. For $\phi=180°$, the multiplanar coefficient also drops below 1, supporting the commonly accepted trend that uniplanar X–joints have capacities lower than similar T–joints. This trend is shown in Fig. 3 (b).

The formulae of Scola et al. (1990) and Paul et al. (1991) for TT–joints, and Mouty et al. (1992) and Makino et al. (1984) for KK–joints do not make a distinction between the different failure types. Formulae of Scola et al. (1990), Mouty et al. (1992) and Makino et al. (1984) predict an increase or decrease of the ultimate capacity with an increase of ζ over the whole range of ζ, as shown in Fig. 4.

The occurrence of a maximum capacity at a certain value of ζ is treated only by the formula of Paul et al. (1991) valid for $60°<\phi<120°$, based on a quite complicated model which does not reflect the expected trends for ϕ approaching $0°$ or $180°$, as demonstrated in Fig. 5.

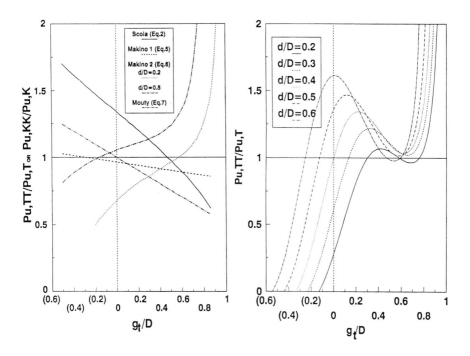

Fig. 4. μ_{TT} or μ_{KK} as functions of ζ and β (Eq. 2, 5, 6, 7).

Fig. 5. μ_{TT} as functions of ζ and β (Eq. 3).

2.2 Distinction between different models

Based on analyses with the ring model (Paul et al., 1993b) it can be concluded that the interaction formulae for multiplanar TT–joints, relating the ultimate capacity of TT–joints to T–joints, are most suitably represented by two different formats for the failure types 1 and 2. For joints with failure type 1, the ultimate capacity of the TT–joint $P'_{u,TT}$ in terms of the resultant of the two brace forces can be related best to the capacity $P'_{u,T}$ of a T–joint with a brace diameter d'. For joints with failure type 2 the ultimate brace load $P_{u,TT}$ can be related best to the ultimate capacity $P_{u,T}$ of a T–joint.

The ring model identified influences of some geometric variables on the TT– to T–joints capacity ratio as shown in Fig. 3 (Paul, 1992b). The capacity ratio $P'_{u,TT}/P'_{u,T}$ for the ring model is equal to 1, while the capacity ratio $P_{u,TT}/P_{u,T}$ is a function of β and ζ_t. For KK–joints capacity ratios are also expected to be a function of ζ_j, while for the capacity ratios $P'_{u,TT}/P'_{u,T}$ and $P'_{u,KK}/P'_{u,K}$, an influence of β, φ or ζ_t is expected to describe the variations for a constant value of β'.

In the KK–joint interaction formulae of Paul et al. (1992a) a distinction was made between the failure types 1 and 2, but the model used relates $P'_{u,KK}$ to $P'_{u,K}$ for both failure types. In sharp contrast with expectations the multiplanar coefficient μ_{KK} becomes either plus or minus infinity when φ approaches 180°, as demonstrated in Fig. 6.

2.3 Assessment of occurring failure mode and type

In the test of Scola et al. (1990), Mouty et al. (1992), Makino et al. (1984), Paul et al. (1991, 1992a) failure modes different from the described failure types 1 and 2 occurred. Altough joints with different failure types were used in the former analysis to derive prediction formulae (Paul et al.; 1991, 1992a) only joints with a clear failure type 1 or 2 are used in this paper.

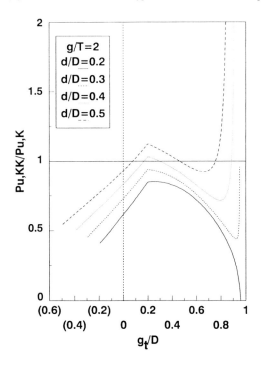

Fig. 6. μ_{KK} as functions of ζ_t and β (ζ_j=2, γ=20, Eq. 10).

400

3 Data base for multiplanar TT– and KK–joints

A screened data base of multiplanar TT– and KK–joints with symmetrical and anti–symmetrical brace loading was compiled based on the research activities of Akiyama et al. (1974), Makino et al. (1984, 1993), Scola et al. (1990), Mouty et al. (1992) and Paul et al. (1991,1992a) by Paul (1992b). Joints which failed by plastic deformation of the chord wall under the compression braces were included in this data base as opposed to those with failure modes such as flexural and local buckling and failure by cracks without sufficient plastic deformation. Joints with uncommon or unclear testing conditions were also omitted. The following represents the variable ranges of the data base of symmetrical loaded joints:

TT–joints (19 joints): $0.222 < \beta < 0.732$, $60° < \phi < 120°$, $e/D = 0$, $34.40 < 2\gamma < 44.71$;

KK–joints (39 joints): $0.224 < \beta < 0.471$, $60° < \phi < 90°$, $-0.17 < e/D < 0.25$, $31.99 < 2\gamma < 49.07$, $0.82 < \zeta_i < 16.72$.

4 New interaction formulae for multiplanar TT– and KK–joints

4.1 Models, regression analysis and resulting formulae
A distinction is made between joints with failure types 1 or 2. Formulae for the joints with other failure types cannot be derived due to a lack of test data. Linear models in Eq. 11 to 14 of Table 2 were selected; they take a multiplicative form and were transformed by taking the natural logarithm, assuming a log–normal distribution of the error ε. Linearization and iterative techniques were used to fit the models by minimizing the sum of squares of lnε. The prediction of the ultimate capacity of the uniplanar T– and K–joints was performed using the formulae of Kurobane et al. (1984) of Eq. 1 and 4. Assesment of the fit was made by comparing the coefficient of variance (CoV) of the test to prediction ratios. Several small modifications of the models in Eq. 11 to 14 were made during the regression analysis to enhance the fit.

The simple formulae given in Eq. 15 to 18 of Table 2 are proposed for the prediction of the mean ultimate capacity of multiplanar TT– and KK–joints. The formulae compare well with the data base (which also includes 4 joints with other failure types), resulting in test to prediction ratios ranging from 0.90 to 1.19 for TT–joints and from 0.85 to 1.16 for KK–joints. The test to prediction ratios showed no variations with ζ_o, β, ϕ or ζ_i (Paul, 1992b).

4.2 Minimum method
The simplest method of using the formulae in Eq. 15 to 18, is to calculate the ultimate capacities for failure types 1 and 2, using the Eq. 15/17 for TT–joints or Eq. 16/18 for KK–joints. The minimum of the two values is taken as the ultimate capacity. By using this minimum method, the failure type does not need to be determined. In the test to prediction ratios for TT–joints, the use of this method results in a mean of 1.026 and a CoV of 0.101. For KK–joints, the minimum method results in a mean of 1.011 and a CoV of 0.070.

4.3 Validity range
Extrapolation of the usage to a wider range of geometric variables appears justified when Fig. 7 is taken into consideration. The formulae for failure type 1 can also be applied to out–of–plane overlapping joints as this type of failure is similar to the failure type of non–overlapping joints with failure type 1. The multiplanar coefficient is approximately 0.5 for joints with $\phi = 0°$ this forming the justification in supporting the use of the formulae for a type 1 failure over the interval $0° < \phi < 60°$. For the formulae for failure type 2, the multiplanar coefficient is inversely proportional to ϕ. A TT–joint with $\phi = 180°$ becomes an X–joint, with a capacity lower than a T–joint. This trend is captured by the interaction formulae. However, the X– to T–joints capacity ratio decreases with β for $0.2 < \beta < 0.6$ (Kurobane et al., 1984), in contrast with the pattern observed in Fig. 7 (d). Consequently, only an extrapolation up to 150° is proposed here. Due to a lack of test data for KK–joints under symmetrical loading and with an out–of–plane angle ϕ greater than 90°, only an extrapolation for joints with an–out–of–plane angle up to 120° is proposed.

401

Table 2.   Models and resulting interaction formulae.

	Failure type	TT–joints	KK–joints
Models	Type 1	$\dfrac{P'_{u,TT}}{P'_{u,T}} = c_1(1+c_2\dfrac{d}{D})\varepsilon$ (11)	$\dfrac{P'_{u,KK}}{P'_{u,K}} = c_1(1+c_2\dfrac{d}{D})(1+c_3\dfrac{g}{T})\varepsilon$ (12)
	Type 2	$\dfrac{P_{u,TT}}{P_{u,T}} = c_1(1+c_2\dfrac{d}{D})(1+c_3\dfrac{g_t}{D})\varepsilon$ (13)	$\dfrac{P_{u,KK}}{P_{u,K}} = c_1(1+c_2\dfrac{d}{D})(1+c_3\dfrac{g_t}{D})(1+c_4\dfrac{g}{T})\varepsilon$ (14)
Formulae	Type 1	$\dfrac{P'_{u,TT}}{P'_{u,T}} = 0.747(1+0.586\dfrac{d}{D})$ (15) $m = 1.001,\ CoV = 0.079$ $n = 3,\ k = 2$	$\dfrac{P'_{u,KK}}{P'_{u,K}} = 0.746(1+0.693\dfrac{d}{D})(1+0.741\dfrac{g}{D})$ (16) $m = 1.001,\ CoV = 0.059$ $n = 11,\ k = 3$
	Type 2	$\dfrac{P_{u,TT}}{P_{u,T}} = 1.329(1-0.336\dfrac{g_t}{D})$ (17) $m = 1.003,\ CoV = 0.081$ $n = 14,\ k = 2$	$\dfrac{P_{u,KK}}{P_{u,K}} = 0.798(1+0.808\dfrac{d}{D})(1-0.410\dfrac{g_t}{D})$ $(1+0.423\dfrac{g}{D})$ (18) $m = 1.003,\ CoV = 0.076$ $n = 26,\ k = 4$

Table 3.   Design strength for CIDECT and multiplication factors for the mean ultimate capacity.

	Prediction formula for the design value of the ultimate capacity		multiplication factor
T–joints	$N^* = (2.8 + 14.2\beta^2)\,\gamma^{0.2}\dfrac{F_y\,T^2}{\sin\theta}\,f(n')$	(19)	1.1/0.85
K–joints	$N^*_c = (1.8 + 10.2\beta)\gamma^{0.2}\left[1+\dfrac{0.024\gamma^{1.2}}{\exp(0.5\dfrac{g}{T}-1.33)+1}\right]f(n')\dfrac{F_y\,T^2}{\sin\theta_c}$ $N^*_t = N^*_c\dfrac{\sin\theta_c}{\sin\theta_t}$ $f(n') = 1\ \text{ for }\ n' = \dfrac{\sigma_{ch}}{F_y} \geq 0$ $f(n') = 1 + 0.3n' - 0.3n'^2\ \text{ for }\ n'<0$	(20)	1.1/0.74

Table 4.   Mean and CoV of the test to CIDECT prediction ratios for multiplanar TT– and KK–joints.

Prediction formula		$F_y=\min[\sigma(0.2\%),0.8F_u]$	
		mean	CoV
TT–joints (mean)	normal (constant=1.0)	1.264	0.154
	minimum method	1.083	0.111
KK–joints (mean)	normal (constant=0.9)	1.328	0.102
	minimum method	1.245	0.085

With respect to TT–joints, an extension for eccentricities e other than zero appears justified as the main describing out–of–plane variables, β' for failure type 1 and ζ_t for failure type 2, are functions of the eccentricity e. The influence of the eccentricity e is captured indirectly. The ring model confirms this proposal indicating that there is no influence on the ultimate capacity of the eccentricity e for a constant value of β' for failure type 1, while the influence of e can be ignored when compared with the influence of ζ_t for failure type 2 (Paul, 1992b). For both TT– and KK–joints, an extension of e/D is proposed from –0.25 to 0.25. This results in the following application range of the Eq. 15 to 18:

TT–joints: $0.2<\beta<0.6$, $0°<\phi<150°$, $-0.25<e/D<0.25$, $30<2\gamma<45$.

KK–joints: $0.2<\beta<0.5$, $0°<\phi<120°$, $-0.25<e/D<0.25$, $30<2\gamma<50$, $2<\zeta_i<14$.

5 Comparison with CIDECT

5.1 Formulae for uniplanar and multiplanar joints

The design strength formula for uniplanar K–joints provided by CIDECT is based on simplifications of the K–joint formula of Kurobane (1984) given in Eq. 4 and a statistical treatment accounting for the variations of the basic parameters as well as for the test results (Wardenier, 1982). The same procedure was been performed for T–joints, based on a T–joint formula preceding that of Eq. 1 (Kurobane et al., 1980), with slightly different coefficients. Based on statistical analysis, factors are calculated with which the mean ultimate capacity has to be multiplied to obtain the characteristic strength. The design strength was then obtained by dividing the characteristic strength by the material and joint partial safety factor γ_m, as given in Table 3. The validity range is: $0.2\leq\beta\leq1.0$, $\gamma\leq25$, $30°\leq\theta_i\leq90°$, $-0.55\leq e_{ip}/D\leq0.25$. The CIDECT Monograph 6 (Giddings et al., 1986) defines the design value of F_y as the lower of either the yield stress or 80% of the ultimate strength F_u. For the F_y/F_u ratio the mean value of 0.66 has been adopted based on the use of hot rolled sections. However, the KK–joints in the data base are manufactured from cold formed sections with a mean of the F_y/F_u ratio of 0.83. Accordingly the factor $(0.83/0.66)^{-0.723}=0.847$ is used to correct for this influence for the K–joint formula in Eq. 20.

For multiplanar joints without out–of–plane overlaps correction factors are given which are to be used with the formulae for uniplanar joints. For multiplanar TT–joints this factor is 1.0, while the factor for multiplanar KK–joints is 0.9. The validity range is restricted for out–of–plane angles ϕ between 60° and 90°. For out–of–plane overlapping TT– and KK–joints, the resultant load in the compression braces $P'=2P\cos(\phi_o/2)$ is taken instead of P and in the Eq. 19 (T–joint) or 20 (K–joints) $(\beta+\beta')/2=(d'+d)/2D$ replaces $\beta=d/D$, while θ'_c replaces θ_c. Note that in concept this is similar to the method used to derive the ultimate capacity formulae for TT– and KK–joints with a failure type 1.

5.2 Comparison of experimental ultimate capacities with code predictions

The test to CIDECT mean prediction ratios for TT–joints vary from 0.96 to 1.73 with a mean of 1.221 and a CoV of 0.154. The test to CIDECT prediction ratios for KK–joints vary from 1.04 to 1.58 with a mean of 1.328 and a CoV of 0.102. The results are summarized in Table 4. The quality of the prediction can be characterized by the mean and CoV of the ratios of test capacities to mean strength predictions. The smaller the CoV and the closer the mean is to 1 the better the prediction.

5.3 Influence of the multiplanar coefficients

The use of the multiplanar interaction formulae in Eq. 15 to 18 instead of the constants is illustrated in Table 4. For TT–joints a reduction for the CoV of 28% and 14% for the mean is obtained using the minimum method. The reduction in the CoV with the Kurobane formulae using the minimum method was 29%, thus indicating that the simplifications for the uniplanar T–joint have only little influence on the effectiveness of the multiplanar interaction formulae for TT–joints.

For KK–joints, a reduction in the CoV of 17% is achieved using the minimum method instead of the constant factor, while the mean is reduced by 6%. A reduction of 32% is achieved by the minimum method using the Kurobane formulae, which shows that the simplifications for the

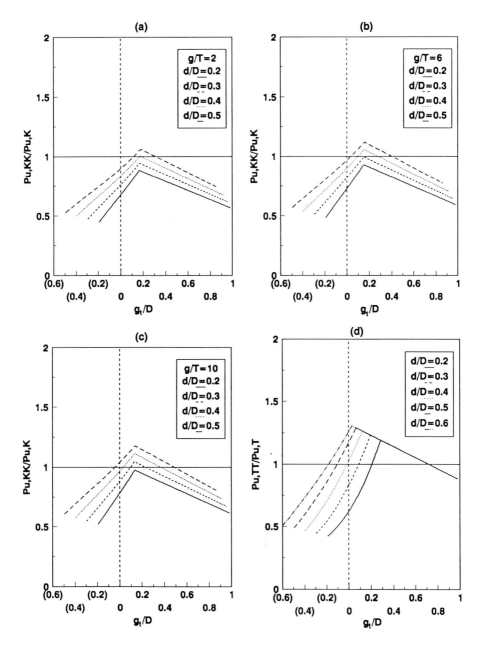

Fig. 7 (a),(b),(c) μ_{KK} as functions of ζ_l, ζ_t and β (γ=20, Eq.16/18).
(d) μ_{TT} as functions of ζ_t and β (γ=20, Eq.12/14).

uniplanar K–joint have a considerable influence on the effectiveness of the multiplanar interaction formulae for KK–joints. A more extensive treatment is presented by Paul (1992b).

6 Conclusions

Existing formulae proposed by Scola et al. (Eq. 2), Makino et al. (Eq. 6) and Mouty et al. (Eq. 7) were based on a failure type 2 model and propose a increase of the $P_{u,TT}/P_{u,T}$ ratio or $P_{u,KK}/P_{u,K}$ ratio with the decrease of ζ. This is in contrast with observations made with the ring model as well as the test on TT– and KK–joints. The formula of Mouty et al. underestimates the capacity over the entire range of geometric variables.

The predictions with multiplanar interaction formulae proposed by Makino et al. (Eq. 5) and Paul et al. (Eq. 3 and Eq.'s 8 to 10) based on a failure type 1 model provide improvements over predictions with uniplanar joint formulae. However, extrapolations beyond the ranges of their data base is not feasible. The newly proposed semi–empirical multiplanar interaction formulae (Eq. 15 to Eq. 18) incorporate the observed multiplanar effects while predicting the ultimate capacity to a reasonable high accuracy. Not only do the formulae show an improvement in accuracy over prediction with formulae for uniplanar joints and existing multiplanar intcraction formulae they also provide extension of the validity ranges beyond the data base ranges. In the proposed validity ranges, the multiplanar coefficient μ, the factor with which the uniplanar joints capacity has to be multiplied to obtain the multiplanar joint capacity, ranges from 0.5 to 1.3, indicating that multiplanar effects are significant.

The use of the newly proposed multiplanar interaction formulae (Eq. 15 to 18), instead of the recommended constant factors, together with the Cidect formulae for T– and K–joints improves the accuracy of the predictions for TT– and KK–joints.

7 References

Akiyama, N., Yajima, M., Akiyama, H., & Othake, A. (1974) Experimental Study on Strength of Joints in Steel Structures. **J. of Japanese Soc. of Steel Construction**, Vol. 10, No.102, pp. 37–68 **(in Japanese)**.

American Petroleum Institute (1991) **Recommended Practice for Planning, Designing and Construction of Fixed Offshore Platforms**. API Recommended Practice 2A (RP–2A), 19th edition, Washington, U.S.A..

American Welding Society (1992) **Structural Welding Code–Steel**, 14th edition, ANSI/AWS D1.1-92, Miami, U.S.A..

Giddings, T.W., & Wardenier, J. (1986) **CIDECT Monograph No. 6: The Strength and Behaviour of Statically Loaded Welded Connections in Structural Hollow Sections**.

Kurobane, Y., Makino, Y., & Mitsui (1980) Re–Analysis of Ultimate Strength Data of Truss Connections of in Circular Hollow Sections, **IIW Doc. XV–461–80**, Lissabon, Portugal.

Kurobane, Y., Makino, Y., & Ochi, K. (1984) Ultimate Resistance of Unstiffened Tubular Joints **Journal of Structural Engineering**, ASCE, Vol.110, No. 2, pp. 385–400.

Makino, Y., Kurobane, Y., & Ochi, K. (1984) Ultimate Capacity of Tubular Double K–joints in **Proc. 2nd. Inter. Conference on Welding of Tubular Structures**, Boston, U.S.A., pp. 451–458.

Makino, Y., Kurobane, Y., Takagi, & M., Hori, A. (1993) Diaphragm Stiffened Multiplanar Joints for Rectractable Roofs in **Proc. 4th. Int. Conf. on Space Structures** Guilford United Kingdom.

Mouty, J., & Rondal J. (1992) Study of the Behavior under Static Loads of Welded Triangular and Rectangular Lattice Girders made with Circular Hollow Sections, **CIDECT Report 5AS–92/1**, Final Report, University de Liege/Valexy.

Paul, J.C., Ueno, T., Makino, Y., & Kurobane, Y. (1991) The Ultimate Behavior of Circular

Multiplanar TT–Joints in **Proc. 4th. Inter. Symposium on Tubular Structures**, Delft, The Netherlands, pp. 448–460.

Paul, J.C., Ueno, T., Makino, Y., & Kurobane, Y. (1992a) Ultimate Behavior of Multiplanar Double K–joints in **Proc. 2nd. Inter. Offshore and Polar Engineering Conference**, San Francisco, U.S.A., pp. 377–383/ (1993a) **J. of Offshore and Polar Engineering**, Vol. 3, No. 1., pp. 43–50.

Paul, J.C. (1992b) **The Ultimate Behavior of Multiplanar TT– and KK–joints made of Circular Hollow Sections**", Doctoral Dissertation Kumamoto University, Kumamoto, Japan.

Paul, J.C., Makino, Y., & Kurobane, Y. (1993b) Ultimate Capacity of Tubular Double T–Joints under Axial Brace Loading, **J. Construct. Steel Research**, Vol. 24, pp. 205–228.

Paul, J.C., Makino, Y., & Kurobane, Y. (1993c) The Ultimate Capacity of Multiplanar TT– and KK–joints –Comparison with AWS and API Design Codes– in **Proc. 3nd. Inter. Offshore and Polar Engineering Conference**, Singapore.

Scola, S., Redwood, R.G., & Mitri, H.S. (1990) Behavior of Axially Loaded Tubular V–joints", **J. Construct. Steel Research**, Vol. 16, pp. 89–109.

Vegte, G.J. van der, Koning, C.H.M., Puthli, R.S., & Wardenier, J. (1991) The Static Strength and Stiffness of Multiplanar Steel X–joints **Int. J. of Offshore and Polar Engineering**, Vol. 1, No. 1.

Wardenier, J. (1982) **Hollow Section Joints**, Delft University Press, Delft, The Netherlands.

Wardenier, J., Kurobane, Y., Packer, J.A., Dutta D., & Yeomans N. (1991) **Design Guide for Circular Hollow Sections (CHS) Joints under Predominantly Static Loading**, CIDECT, Verlag TÜV Rheinland, Köln, Germany.

8 Appendix. Nomenclature

e	Overall eccentricity (if al center lines meet in one point)
e_{op}	Out–of–plane eccentricity $=e\sin(\phi/2)$
F_y	Yield stress
F_u	Ultimate strength
k	Number of fitted constants
m	Mean
N^*_i	Design strength, i=c(ompr.), t=t(ens.) (CIDECT)
n	Number of joints
$P_{c,T}\ (P_{c,TT})$	Collapse brace load T–joint (TT–joint)
$P_{u,T}\ (P_{u,TT})$	Ultimate brace load T–joint (TT–joint)
$P_{u,K}\ (P_{u,KK})$	Ultimate brace load K–joint (KK–joint)
$P'_{c,T}\ (P'_{u,T})$	Collapse (Ultimate) brace load T–joint with brace diameter d'
$P'_{c,TT}\ (P'_{u,TT})$	Resultant collapse (ultimate) load TT–joint
$P'_{u,K}$	Ultimate brace load T–joint with brace diameter d' and angle θ' between the brace and chord
$P'_{u,KK}$	Resultant ultimate load KK–joint
β	Diameter ratio $=d/D$
β'	Diameter ratio $=d'/D$ $=\sin(\phi/2+\arcsin\beta+\arcsin(2e_{op}/D))$
θ_i	In–plane angle between the chord and the brace, i=c(ompr.),t(ens.)
θ'_i	Angle between the planes in which the braces lay and the chord, i=c(ompr.), t(ens.)
γ	Chord thinness $=D/(2T)$
$\mu_{TT}\ (\mu_{KK})$	Multiplanar coefficient TT–joint $=P_{u,TT}/P_{u,T}$ (KK–joint $=P_{u,KK}/P_{u,K}$)
ϕ	Out–of–plane angle between the planes braces in which the braces lay
ϕ_i	Angle between the in– and out–of–plane braces, i=c(ompr.), t(ens.)
ζ_i	In–plane gap factor $=g/T$
ζ_t	Out–of–plane gap factor $=g_t/D$ $=\sin(\phi/2-\arcsin\beta+\arcsin(2e_{op}/D))$

39 THE STATIC STRENGTH OF T-JOINTS UNDER BENDING IN DIFFERENT DIRECTIONS

H. FESSLER, T.H. HYDE and Y. KHALID
Department of Mechanical Engineering,
University of Nottingham, UK

Abstract
36 tin-lead alloy models of T joints without weld fillets have been held at one
end of the chord only and loaded by pure bending moments applied to the brace.
The direction of bending was changed from out-of-plane bending to in-plane
bending in 15° intervals. Results are presented for two brace thicknesses.

Some of the thin-brace models had casting defects at the saddles. These
defects were measured and the models were also tested. The bending strength of
these defective models is compared with the strength of models without defects.

Failures start on the compressed side of the brace but the joints carry
increasing loads till tearing occurs on the tensile side. In in-plane bending, the
joints were significantly stronger when the bending was away from the chord
support than towards it. This conclusion may be important in assessing the tests
of steel models which are usually held at one end of the chord only.
Keywords: Tubular Joints, Static Strength, Bending

1   Introduction

The static strength of tubular joints for offshore structures has been investigated
by testing models in simple loading modes, including in-plane or out-of-plane
bending. This paper deals with combinations of these two loading modes which
must occur in practice but do not appear to have been studied. The resultant of
any combination of in-plane and out-of-plane bending is bending in some
intermediate direction α (see Fig. 1). Pure bending has been chosen because the
effects of shear forces are usually very small in offshore structures and the
combination of bending with axial forces will be dealt with in a subsequent
paper.

The bending moments were applied at the end of the brace and the reaction to
most of them was carried at one end of the chord, ie. loading the chord as a
cantilever in combined bending and torsion. This type of reaction was used
because it is often employed in testing steel models but for some tests the chord
was simply supported at both ends.

T joints have been chosen because their strength is generally smaller than that
of Y joints and T joints avoid any possible complications due to interaction

Tubular Structures V. Edited by M.G. Coutie and G. Davies.
© 1993 E & FN Spon, 2–6 Boundary Row, London SE1 8HN. ISBN 0 419 18770 7.

between the brace inclination θ and the direction of bending α. Tests of single brace joints are reported here; bending of multi-brace joints will be reported later.

Initially, it was intended to make the models with brace/chord thickness ratio τ = ½ but when difficulties arose in casting models with 1 mm thick braces, τ was increased to 1. The models with casting defects were also tested and are reported on in this paper.

All tests of joints of offshore structures are model tests because the structural joints are much too big for laboratory tests. Tin-lead models can be smaller than steel models and require much smaller loads to fail them because their UTS is also much lower than steel's. The models and the calibration specimens were die-cast in our laboratory to their finished shape. This is a very efficient method of making models of identical shape, thereby eliminating experimental scatter due to shape variation. Scatter due to variation of material properties was minimised by casting four calibration specimens with each model.

2 Nomenclature

B_α	maximum bending moment in a test, see Fig. 1
B_α^*	$B_\alpha/\sigma_u T^2 d$
D, d, T, t	see Fig. 2
L, S	see Fig. 1
α, δ	see Fig. 1
β	d/D
γ	D/2T
τ	t/T
σ_y, σ_u	yield and ultimate tensile strength, respectively
ψ_1, ψ_2	extent of blow hole at junction, see Fig. 1

3 Model material

A tin-lead alloy (50% Sn, 47% Pb, 3% Sb) has been shown to be a suitable material (1). The method of die casting this material in mild steel moulds has been described (2). It has an ultimate tensile strength $\sigma_u \simeq 60$ N/mm², a yield/ultimate strength ratio $\sigma_y/\sigma_u \simeq 0.7$ and ductility of 30% ± 10%; the last two properties are within the range of structural steels used in offshore structures, making it a suitable model material.

The castings are homogenous, ie. there is no difference between weld and 'parent' plate (from which the chord and brace are rolled) and there are no heat-affected zones. The models without defects are therefore representative of structural joints which have been heat-treated after making full-penetration welds without defects. The defective models had 'through-thickness' holes or cracks across the brace at the saddle, extending over angles ψ_1 and ψ_2 (see Fig. 1).

The material age-softens, ie. the tensile strength reduces during storing at room temperature. Typically, $\sigma_u = (74.0 - 11.8 \log_{10}t)$ N/mm^2 where t is the time in days from casting to testing. Errors due to this variation are eliminated by testing tensile specimens (which have been cast with the model) within less than a week of the model test and correcting their strength according to the above equation. Further details about this material have been presented in Ref. 1.

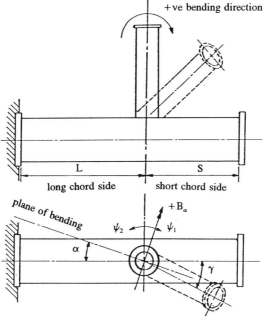

Fig. 1. +L type loading of cantilevered T joints.
Note: Dotted lines show brace after loading.

4 Model shape

The shape and nominal dimensions are shown in Fig. 2. They give frequently found shapes, ie. $\beta = 0.49$, $\gamma = 18$, $\tau = 0.96$ or $\tau = 0.59$. The plain chords were 5.2 diameters long.

The mould in which these integral models were cast was for a YT shape but the Y brace was carefully cut off for the tests reported here. The steel moulds, cores, tensile specimen and method of die-casting have been described (2). No weld fillets were modelled.

For one thin-brace and one thick-brace model, the actual diameters and wall thicknesses were measured in eight section planes, identified in Fig. 2 by numbers 1 to 5 in the chord and 6 to 8 in the braces. The diameters D or d were measured in the four compass directions stated in Tables 1 and 2 and defined in Fig. 2. The wall thicknesses T and t were measured at positions, S, SE, E and NE. Chords and braces were straight, the braces were perpendicular to the chords and their axes intersected, ie. the inaccuracies were too small to be measured. Only the measurements near the brace-chord junction are shown, the others were similar. These results show that the chords and braces were accurately round and that the wall thickness variations were very small.

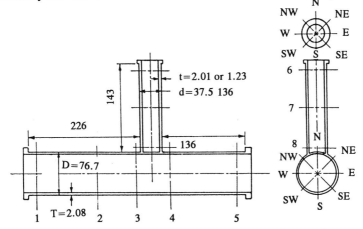

Fig. 2. Dimensions in mm and positions of sections for measurements of D, T, d, and t.

5   Loading Rig

The loading rig shown in Fig. 3, was a modification of an existing rig (2). The axis of the brace before loading is always horizontal and the plane of bending is always vertical. The different directions of loading are achieved by changing the inclination $\propto$ (see Fig. 1) of the chord axis from the vertical.

Rapid loading at constant speeds was carried out in an Instron 1193 testing machine. The base B (Fig. 3) of the loading rig was clamped to the base of the testing machine. Pure bending of the braces was produced by pairs of chains C, fixed to sprocket wheels which were rigidly attached to the ends of the braces. The upper chains were connected to the load cell L at the top of the machine and the lower chains were connected to the moving crossbar M.

It may be seen from the left hand view in Fig. 3 that it is necessary for the chord and supports to move vertically if a pure bending moment is to be exerted on the brace. If this freedom is restricted, shear forces are set up in the model and the brace loading is like that of a propped cantilever with the maximum bending moment in line with the upper chains C. Deflection of the brace and twisting of the chord change the relative positions of brace and chord ends as seen from the plan view. To avoid parasitic loads, the chord end support is free to rotate on a linear bearing as well as sliding up and down on it. The weight of this linear bearing, attachments and model are counterbalanced by weight W.

Table 1. Chord and brace diameters in mm

Direction	Thick-brace model			Thin-brace model		
	chord section		brace section	chord section		brace section
	3	4	8	3	4	8
N - S	76.62	76.80	37.55	76.60	76.35	37.42
NW - SE	76.70	76.75	37.45	76.72	76.45	37.56
W - E	76.65	76.71	37.35	76.82	76.86	37.52
SW - NE	76.68	76.74	37.60	76.75	76.65	37.50
max difference	0.07	0.09	0.25	0.22	0.51	0.14

Table 2. Chord and brace wall thicknesses in mm

Direction	Thick-brace model			Thin-brace model		
	chord section		brace section	chord section		brace section
	3	4	8	3	4	8
N - S	2.08	2.07	2.08	2.09	2.11	1.30
NW - SE	2.10	2.08	2.07	2.05	2.12	1.28
W - E	2.06	2.05	2.12	2.11	2.08	1.26
SW - NE	2.09	2.06	2.11	2.08	2.12	1.29
max difference	0.04	0.02	0.04	0.06	0.04	0.04

To permit seven different combinations of in-plane and out-of-plane bending to be applied, the chord end support can be set accurately at $\propto$ = 0°, 15°, 30°, 45°, 60° 75° and 90° by a dowel in jig-bored holes.

6   Experimental procedure

After the base of the loading rig had been clamped in the required position, the chord end support was set in the required loading direction $\propto$ and the model was clamped to the end attachment. The testing machine calibration was checked and the recording pen was 'zeroed' at the required load scale of the recorder. The loading attachment was clamped to the end of the brace and the pairs of chains C were connected to the load cell L and the moving crossbar M.

The tests were carried out at the maximum speed of the Instron testing machine, 500 mm/min and stopped as soon as possible after the maximum load had been reached.

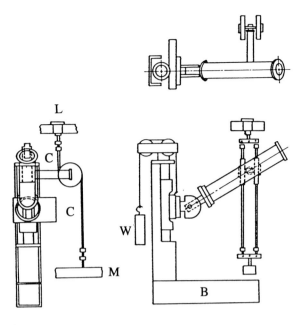

Fig. 3. Loading rig for cantilevered T joints, set for $\alpha = 60°$.

Samples of the testing machine records are reproduced in Fig. 4. Each of these graphs is an accurate record of a test, but detailed quantitative comparisons between the graphs would have to allow for variations in tensile strength of the material. Some of the curves in Fig. 4 show two distinct maxima. The first occurs when collapse occurs at the compressed side of the brace, the second is associated with tearing on the tensile side. The curves in Fig. 4 are typical of all the results. After loading, the models were examined for tears and plastic deformations of the joints. When unloaded, after testing, the braces were straight, as near as could conveniently be measured with a straight edge. For $0 < \alpha < 90$, their inclination δ (see Fig. 1) was greater than α. All models failed at the junction of brace and chord.

7   Results

The failure loads B_α are presented in non-dimensional form as B_α^* to make them directly applicable to offshore structures. The definition $B^*_\alpha = B_\alpha / \sigma_u T^2 d$ has been used by other authors, e.g. (5), and is satisfactory for this paper.

The non-dimensional failure loads of the models without defects are presented in Table 3 and Fig. 5. They show that there is no significant difference between long-chord (L) and short-chord (S) results. They also show that there is no significant difference between thin-brace and thick-brace results; this makes it unlikely that weld fillets would change the static strength. However, there are significant differences between +ve and -ve loading; the -ve loading which causes brace compression on the side nearest to the support (see Fig. 1) gives significantly lower strengths than corresponding +ve loading. Of course, there is no difference in out-of-plane bending ($\propto$ = 90°).

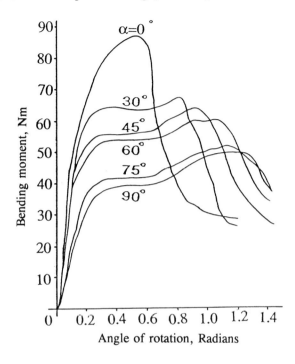

Fig. 4. Some testing machine records for cantilevered T joints.

Fig. 5 shows that, for positive and negative bending, there are sinusoidal variations of strength, i.e.

$$B^*_\alpha = B^*_0 + (B^*_{90} - B^*_0) \sin\alpha \qquad (1)$$

Because the joints are weaker in out-of-plane bending B_{90} than in in-plane bending B_0, the braces also move 'sideways' during the plastic deformation which causes collapse, as indicated by dashed lines in Fig. 1. This sideways movement was measured as δ-α and these results are shown in Fig. 6. There were no measurable movements for $\alpha = 0°$ and $\alpha = 90°$, showing that the models and their alignment in the test rig were accurate. The dashed line in Fig. 6 would be reached if the braces moved 'all the way' to the out-of-plane bending direction. This indicates that the sideways movement is significant but small.

The lower limit of the scatter band in Fig. 6 has a sinusoidal shape and fits

$$(\delta\text{-}\alpha) = 0.09 \sin 2\alpha = (0.18 \cos\alpha) \sin\alpha \qquad (2)$$

where the angles are measured in radians. $(0.18 \cos\alpha)$ is proportional to the out-of-plane component of the load (see Fig. 1); thus Equ. 2 suggests that the sideways movement (δ-α) varies sinusoidally with the out-of-plane component of the load.

Table 3. Non-dimensional strengths B* of cantilevered models

Direction see Fig. 1	Thick-walled brace active chord length		Thin-walled brace active chord length	
$\alpha°$	L	S	L	S
+0	9.39	8.58		7.09
+30	7.13	6.26		
+45	6.01, 6.16, 6.49	6.67, 5.82		
+60	6.49		5.81 (1)	
90	4.64, 5.25			
-75		5.36		
-45		5.42, 4.86		
-30		5.95, 5.57		
-15	5.52, 6.68		5.92	
-0		6.40, 6.56		

<u>Note</u>   (1) small blow hole, $\psi_1 + \psi_2 = 12°$

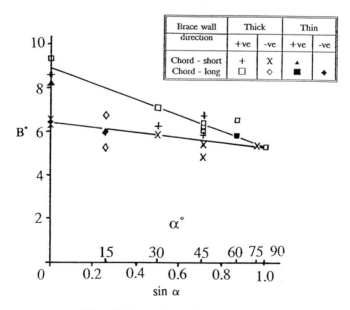

Fig. 5. Strength cantilevered T joints.

414

After ψ_1 and ψ_2 (see Fig. 1), the extent of the blow holes or cracks in the defective thin-brace models, had been measured, these models were also tested. The directions of bending were arranged so that the defects straddled the plane of bending. Of course most values of B^* were lower than the expected mean strengths, as given by the two straight lines in Fig. 5. The differences, divided by the appropriate mean values, give the reductions which are presented in Fig.7. The two straight lines in that diagram suggest that the reduction in strength was between 8% and 36% per radian of hole or crack around the junction of brace and chord.

As part of another programme, nine thick-brace models were also tested in bending with the chord simply supported on ball joints at both ends. The out-of-plane couple was reacted at one end only. These results for $\alpha = 0$, 45° and 90° are shown in Fig. 8 and compared with the mean cantilevered-chord values in Table 4.

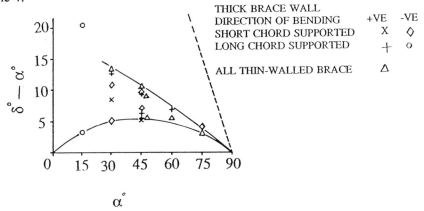

Fig. 6. Sideways movement of braces.

8   Comparisons with other work

Valid comparisons require the models to have the same shape, the same loading and support system and the same loading sequence. Although many static strength tests of T joints have been reported, only three sources of bending tests refer to sufficiently similar shape. None applied pure bending moments, they applied single transverse forces at the end of the brace, thereby loading it in shear as well as in bending.

The Society of Steel Construction of Japan published a study of tubular joints in marine structures (6) in Japanese. This includes in-plane and out-of-plane bending of T joints but the available copy gave insufficient details as shown by the ? in Table 4 which includes their relevant results. Yura, Zettlemoyer and Edwards (4) presented a paper to the 1980 Offshore Technology Conference which included one relevant in-plane bending result.

Makino, Kurobane, Takizawa and Yamamoto (3) tested T joints made of different steels in out-of-plane, cantilever bending. Their chords were very short (see Table 4) and built into rigid, pivoted blocks at both ends. The distance from the chord to the transverse force (causing bending) was 800 mm = 7d.

The Department of Energy published parametric equations as part of its Guidelines (5). For in-plane bending of T joints

$$B^* = (6.20 \ \beta - 0.27) \ \gamma^{0.5} \tag{3}$$

and for out-of-plane bending of T joints

$$B^* = 8.64 \ \beta + 1.88 \tag{4}$$

These equations are also evaluated in Table 4.

Table 4. Comparison of mean values with steel model results and predictions

Source	Material		Shape		Loading				Strength
	type	$\frac{\sigma_y}{\sigma_u}$	β	γ	L/D (2)	$\alpha°$	chord support	direc-tion of bending	$B/\sigma_u \ T^2 d$
authors	tin-lead	0.7	0.49	18	2.9	0	one end	-ve	6.4
authors	tin-lead	0.7	0.49	18	2.9	0	one end	+ve	9.0
authors	tin-lead	0.7	0.49	18	2.9	0	simple	-	9.2
Ref. 6	steel	?	0.46	19	?	0	?	?	8.6(1)
Ref. 4	steel	?	0.46	20	?	0	?	?	12.4(1)
Equ. 3	any	-	0.49	18	-	0	?	?	11.9
authors	tin-lead	0.7	0.49	18	2.9	90	one end	-	5.3
authors	tin-lead	0.7	0.49	18	2.9	90	simple	-	5.8
Ref. 3	steel	0.57	0.53	17	1.5	90	simple	-	4.9
Ref. 3	steel	0.89	0.53	18	1.5	90	simple	-	6.0
Ref. 3	steel	0.95	0.53	18	1.5	90	simple	-	5.7
Ref. 6	steel	?	0.46	19	?	90	?	-	5.5(1)
Equ. 4	any	-	0.49	18	-	90	-	?	6.1

Notes 1 calculated with σ_y because σ_u not known
2 L is defined in Fig. 1
? indicates that no information is available in the Ref.

9   Discussion

Table 3 and Fig. 8 show that for nine loading and support conditions, two or more 'repeat' tests were carried out. Experimental scatter is here described by the maximum differences between B^* values of repeat tests. The mean value of these nine differences is 0.6; $B^* \simeq 6$ (see last column of Table 4). Hence it is assumed that B^* values are reliable to $\pm 5\%$. The small number of results do not warrant more sophisticated statistical analysis.

Direction	$+0^0$	-15^0	$+30^0$	$+45^0$	$+60^0$	-75^0	90^0
Symbol	O	▷	▽	x	▵	◁	+

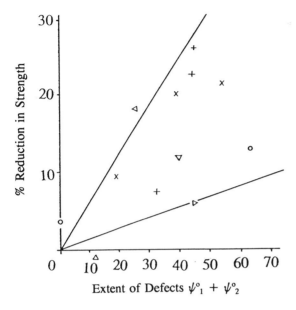

Fig. 7. Reduction of strength due to defects in thin-braced models.

10   Conclusions

Out-of-plane bending has lower strength than in-plane bending; intermediate values of strength and movement of the brace follow sinusoidal relationships (see Equ. 1 and 2).

Cantilevered models (supported at one end only) are weaker in in-plane bending towards the support than models whose chords are simply supported at both ends (see Table 4).

The brace wall thickness does not influence the strength of the joints (see Fig. 5).

Holes or through-thickness cracks at the brace-chord junction reduce the bending strength by 8% to 36% per radian of cracked brace.

In out-of-plane bending there is good agreement with steel model results and a parametric equation. Comparisons of in-plane bending results with those of others suffer from uncertainty of support conditions (see Table 4).

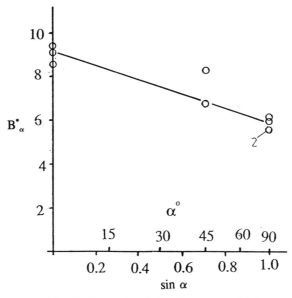

Fig. 8. Strength of simply supported joints.

11   References

1 Fessler, H. and Hassell, W. (1990) **A model technique for static strength of tubular joints** Steel Construction Today, Vol 4, pp. 104-112.
2 Fessler, H. Hassell, W. and Hyde, T.H. (1992) **A model technique for bending strength of tubular joints** J. Strain Analysis, Vol 27, pp. 197-210.
3 Makino, Y. Kurobane, Y. Takizawa, S. and Yamamoto, N. (1986) **Behaviour of tubular T and K joints under combined loads**. 18th OTC Houston, Texas pp. 429-438.
4 Yura, J.A. Zettlemoyer, N. and Edwards, I.F. (1981) **Ultimate capacity of circular tubular joints** ASCE, Vol 107, pp. 1965-1984.
5 **Background to New Static Strength Guidance for Tubular Joints in Steel Offshore Structures** Offshore Technology Report, Department of Energy OTH 89-308.
6 **Study of tubular joints used for marine structures** (1972) Japanese Society of Steel Construction.

12   Acknowledgements

To assist in carrying out this work, some joints were provided from an HSE co-ordinated program. The opinions of the authors are not necessarily those of HSE. The authors also thank Mr R Pickard and other University technicians for their skilled, enthusiastic work.

40 STATIC STRENGTH OF MULTIPLANAR DX-JOINTS IN RECTANGULAR HOLLOW SECTIONS

D.K. LIU and J. WARDENIER
Delft University of Technology, The Netherlands
C.H.M. de KONING and R.S. PUTHLI
TNO-Building and Construction Research, Rijswijk,
The Netherlands

Abstract
In this paper the results are given of an experimental and numerical investigation into the static behaviour of multiplanar welded DX-joints (and uniplanar X-joints for comparison) in rectangular hollow sections with $\beta = 0.6$. Both experimental and numerical results show that multiplanar joints with unloaded out-of-plane braces give higher values for the strength than comparable uniplanar joints. For multiplanar joints with in-plane braces axially loaded or loaded by in-plane bending, there is a clear influence of compression and tension loaded out-of-plane braces on the joint capacity for the joints considered.
Keywords : Static strength, Ultimate load, Maximum load, Joint in RHS

List of symbols

F_v	:	axial load (in in-plane braces);
F_h	:	proportional load (in out-of-plane braces);
M_{ipb}	:	in-plane bending moment; at the intersection of the centrelines of chord and brace;
b_0	:	external width of chord;
b_1	:	external width of the braces;
f_y	:	yield stress (in longitudinal direction);
f_u	:	ultimate stress (in longitudinal direction);
h_0	:	depth of chord (= b_0 for square section);
h_1	:	depth of the braces (= b_1 for square section);
ℓ_0	:	chord length; $\ell_0 = 6b_0$
ℓ_1	:	brace length; $\ell_1 = 5b_1$
t_0	:	wall thickness of chord;
t_1	:	wall thickness of the braces;
β	:	brace to chord width ratio b_1/b_0;
2γ	:	width to wall thickness ratio of chord b_0/t_0;
$\epsilon(\%)$	:	permanent elongation;
E	:	modulus of elasticity of steel, taken as 0.21 x 10^6 N/mm^2;

1 Introduction

It is standard practice in current design philosophy to treat multi-

Tubular Structures V. Edited by M.G. Coutie and G. Davies.
© 1993 E & FN Spon, 2–6 Boundary Row, London SE1 8HN. ISBN 0 419 18770 7.

planar joints as a series of uniplanar connections for which design guidelines are available. Each plane is therefore treated in isolation with no reference to the effect of (and load in) out-of-plane braces. Initial investigations, Vegte et al. (1991) and Paul et al. (1989a), on multiplanar joints have shown that, depending on the geometry and loading conditions, this may result in actual strengths which are significantly lower or higher than the ultimate strength for uniplanar joints.

From the existing codes of practice such as API (1987), AWS (1984), IIW (1989) and Eurocode 3 (1992), only the AWS includes the multiplanar effects for joints in circular hollow sections (CHS). This as well as other codes is based on extensive series of tests on uniplanar joints. Very few test results, especially for multiplanar joints in rectangular hollow sections (RHS) are available. In keeping with the aim of relating the research to practice, and to fill the gap in knowledge, a series of multiplanar DX-, DT- and DK-joint tests, together with numerical analyses, have been defined for the ECSC research programme "The development of design methods for the cost-effective applications of multiplanar joints; Phase 2 - "Static strength of multiplanar connections in rectangular hollow sections", which is carried out jointly by two ECSC countries, namely the United Kingdom and The Netherlands. The participating organizations are: The Steel Construction Institute, British Steel, University of Nottingham, Delft University of Technology and TNO Building and Construction Research, Rijswijk.

The aim of the research programme is to complement recent European research programmes on multiplanar joints in CHS and RHS, coupled with the development of numerical modelling of such joints and to establish European guidelines for multiplanar joints in structural hollow sections.

This paper presents the results of an experimental and numerical investigation on multiplanar welded DX-joints, and uniplanar X-joints

Table 1   Joints considered

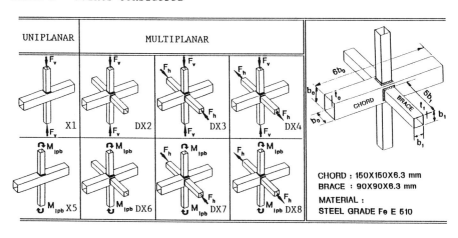

$(\beta = b_1/b_0 = 0.6, \ 2\gamma = b_0/t_0 = 23.8, \ \text{and} \ \tau = t_1/t_0 = 1.0)$

420

for comparison, in rectangular hollow sections. Two series of four joints listed in Table 1, consisting of three multiplanar DX-joints and 1 uniplanar X-joint are numerically and experimentaly investigated. The first series of joints(X1,DX2,DX3 and DX4) are loaded in the in-plane braces by axial load and the second series(X5,DX6,DX7 and DX8) by in-plane bending. The geometrical parameters are all the same.

The influence of axially loaded and unloaded out-of-plane braces on the static strength and deformation capacity has been determined for different types of loading on the in-plane braces. Comparisons are carried out between the results of the experiments and the numerical work. Comparisons have also been made of the experimental and numerical results with European codes (EC 3, 1992), and the CIDECT design guide (Packer et al., 1992b) for rectangular hollow sections.

2   Experimental work

2.1 General
All test specimens are welded with basic electrodes (trade name Conarc 60G) in accordance with the standards ASME SFA-5.5, E9018G, DIN 8529 and EY5565 1NiMoBH5. The braces are welded to the chord with fillet welds, with a throat thickness equal to the wall thickness of the connected brace. All welding is carried out in welding position 2G according to ASME Section IX.

The hollow sections used for the specimens are hot finished, steel grade Fe 510 D, in accordance with Euronorm EN 10025. The actual mechanical properties f_y (yield stress), f_u (ultimate stress), and ϵ (permanent elongation) of the hollow sections have been determined with tensile tests and carried out in accordance with Euronorm 2-80 "Tensile tests for steel". The tensile coupons were taken in the longitudinal direction along one of the flat sides and at the corner. The measured mechanical properties for the chord and braces are summarized in Table 2.

Table 2. Measured material properties

MEMBERS	JOINTS	STOCK No.	f_y (N/mm^2)	f_u (N/mm^2)	$\epsilon_\%$
CHORD	ALL	1	411	543	32
	X1-DX4	1	444	551	32
BRACES	X5-DX7	2	406	520	33
	DX8	3	433	550	30

During the test, the ends of the axially loaded in-plane braces of the joints are pin-ended. The chord is supported in the lateral and longitudinal direction, to prevent displacements in these directions.

421

The axial load F_v in the in-plane braces is applied by means of a jack at one end of the brace.

On the specimens DX3 and DX4 a proportional load $F_h = 0.6F_v$ (compression load for DX3, and a tension load for DX4), is applied in the out-of-plane braces. The load is applied by means of a jack mounted in a frame enclosing the out-of-plane braces.

For the joints loaded in bending a compression force is applied to one end of the chord and the test specimen is supported at the ends of the in-plane braces. The chord is supported in two directions by lateral supports, to prevent lateral displacements in any direction. The load in the chord is applied by means of a jack and measured with a dynamometer. On the specimens DX7 and DX8 a proportional load $F_h = M_{ipb}/100.46$ (compression load for DX7, and a tension load for DX8) is applied in the out-of-plane braces. Under this relationship of $F_h = M_{ipb}/100.46$, the axial stress in the out-of-plane braces is 30% of the maximum bending stress in the in-plane braces.

2.2 Experimental results

The relationships between the axial load F_v and the measured overall deflection (the sum of the identations in the two opposite faces of the chord) of the joints are shown in Figure 1a. The overall deflection is defined as the change in distance between points, close to the chord wall, in the middle of the flat sides parallel to the chord axis, of the two in-plane braces(see Figure 2). The mode of failure for the specimens X1 to DX4 is plastic failure in the chord face. The test load based on the serviceability limit of 1%b_0 and ultimate load based on Yura (Yura et al., 1980) deformation limit, are determined and listed in Tables 3a and 3b. The Yura deformation limit is based on an assumption that a member reaches its practical deformation limit when the strain along its entire length is four times the yield strain. For the joints considered the Yura deformation limit is $4f_{y1}L/E$ for the axially loaded joints and the Yura rotation limit is $80f_{y1}L/E$ for the joints loaded by in-plane bending, where an L of 30 times the brace width is used.

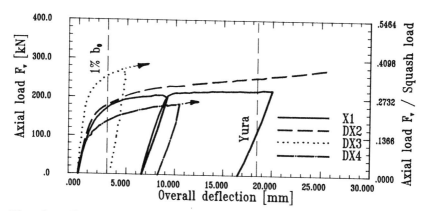

Fig. 1a   The experimental load-deformation curves for axially loaded joints

422

The relationships between the in-plane bending moments M_{ipb} and the rotations are given in Figure 1b. M_{ipb} is taken at the intersection of the centrelines of chord and braces. The rotation is defined as the in-plane rotation of the in-plane braces relative to the chord. The mode of failure for the specimens is the plastic failure in the chord faces in combination with <u>initiation of cracks</u> in the chord faces at the weld toe of the brace corners. The ultimate in-plane moments based on the Yura (Yura et al., 1980) deformation limit are determined for specimens X5, DX6, DX7 and DX8, as shown in Table 4.

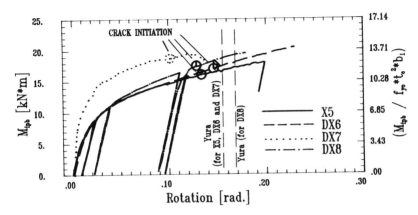

Fig. 1b   The experimental moment-rotation curves for the joints loaded by in-plane bending

3   Numerical analysis

3.1 Finite element strategy for the numerical work

The numerical analyses are carried out using the general purpose finite element program MARC. Based on the preliminary study and the previous numerical work (Vegte et al., 1991), eight noded thick shell elements (MARC type 22) are chosen for the numerical modelling of the joints. Both geometrical and material nonlinearities are considered. The measured actual dimensions of the test specimens, including the fillet welds, are used in the numerical modelling. The engineering stress-strain relationships obtained in the experiments are converted to true stress-strain relationships using the Ramberg-Osgood theory (EC 3, 1992), to fit the requirements of MARC. As an example, the finite element meshes and the boundary conditions for the multiplanar joints DX2 and DX6 are shown in Figure 2.

For the joints which are only loaded in the in-plane braces, the load is applied by displacement control, while for the joints which are loaded simultaneously in both in-plane and out-of-plane braces, load is applied by force control so that the load in the out-of-plane braces is maintained proportional to the load in the in-plane braces.

3.2 Numerical results

The relationships between the axial load and the deflection for the

joints X1 to DX4 are given in Figure 3a. The loads corresponding to
the deformation limits of 1%b_0 and Yura are summarized in Tables 3a
and 3b, respectively.

For the joints loaded by in-plane bending, the numerically determined moment-rotation relationships for the in-plane brace are given
in Figure 3b. The moments corresponding to the Yura rotation limit
are summarized in Table 4.

4 Comparision of the experimental and numerical work

In general there is a reasonable agreement between the experimental
and numerical results. Also, the deformed shapes of the tested
specimens and in the numerical work at ultimate loads agree very well
(Koning et al, 1992a). It is observed that the experimental and
numerical moment-rotation curve for joint DX7 diverge from each other
for a part of the trajectory. It is considered that this divergence
could be because of the boundary conditions for axial compression in
the out-of-plane braces for the experiments, which are neither
ideally fully rigid or rotation free, but depend on the relationship
between the axial forces and stiffness of the testing system. This is
not the case for the tension loaded out-of-plane braces, where the
agreement between experimental and numerical results is good.

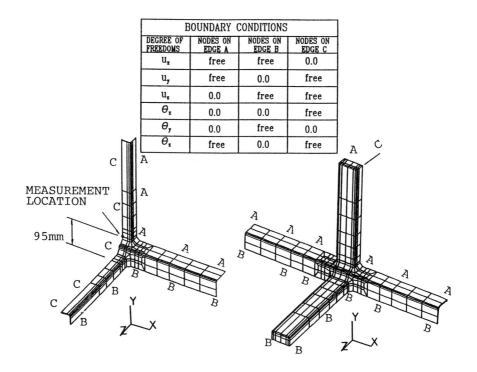

BOUNDARY CONDITIONS			
DEGREE OF FREEDOMS	NODES ON EDGE A	NODES ON EDGE B	NODES ON EDGE C
u_x	free	free	0.0
u_y	free	0.0	free
u_z	0.0	free	free
θ_x	0.0	0.0	free
θ_y	0.0	free	0.0
θ_z	free	0.0	free

MEASUREMENT
LOCATION

95mm

Fig. 2 The finite element meshs for joints DX2 and DX6

424

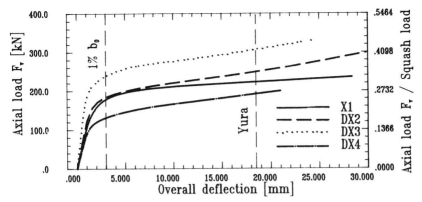

Fig. 3a The numerically determined load-deformation curves for
axially loaded joints

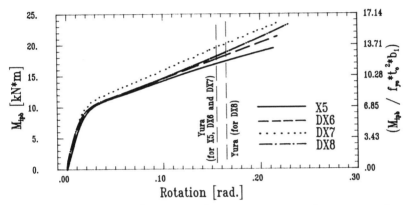

Fig. 3b The numerically determined moment-rotation curves for the
joints loaded by in-plane bending

5 Comparison of the experimental and numerical results with existing design rules

EC 3 (1992) and the CIDECT Design Guide (Packer et al., 1992b) give,
based on the yield line theory (Wadenier, 1982), similar design
guidance for <u>uniplanar</u> RHS joints subjected to axial load or in-plane
bending. For the axially loaded multiplanar joints in RHS, only
CIDECT gives recommendations to account for the effect of out-of-
plane loaded braces, where a correction factor of 0.9 is recommended
to be applied to the uniplanar T- and X-joint resistances, to get the
multiplanar joint resistances for axially loaded cases. For the
multiplanar joints in RHS subjected to in-plane bending, there is no
design guidance available at present. The comparisions are tabulated
in Table 5. It should be noted that in Table 5 the strengths determined from the codes are the design strengths.

Table 3a   Summary of the experimental and numerical results on the basis of 1%b_0 deformation limit for axially loaded joints

JOINTS	TEST LOAD (kN)		MULTIPLANAR EFFECT ON ULTIMATE LOAD		NUM. RESULT
	EXP.	NUM.	EXP.	NUM.	EXP. RESULT
X1	171	195	—	—	1.14
DX2	180	205	1.05	1.05	1.14
DX3	250	240	1.46	1.23	0.96
DX4	132	143	0.77	0.73	1.08

Table 3b   Summary of the experimental and numerical results on the basis of the Yura deformation limit for axially loaded joints

JOINTS	ULTIMATE LOAD (kN)		MULTIPLANAR EFFECT ON ULTIMATE LOAD		NUM. RESULT
	EXP.	NUM.	EXP.	NUM.	EXP. RESULT
X1	222	242	—	—	1.09
DX2	266	300	1.20	1.24	1.13
DX3	>275	323	>1.24	1.35	—
DX4	>184	231	>0.83	0.95	—

Table 4   Summary of the experimental and numerical results on the basis of the Yura deformation limit for the joints loaded by in-plane bending(based on the moment at the intersection of the centrelines of chord and brace. At the chord face the moment is 14% lower)

JOINTS	ULTIMATE MOMENT (kN*m)		MULTIPLANAR EFFECT ON ULTIMATE MOMENT		NUM. RESULT
	EXP.	NUM.	EXP.	NUM.	EXP. RESULT
X5	16.8	17.0	—	—	1.01
DX6	17.5	17.8	1.04	1.05	1.02
DX7	19.3	19.7	1.15	1.16	1.02
DX8	19.2	19.0	1.14	1.12	0.99

Table 5a   The comparison between the results of the codes and the present work on the basis of 1%b_0 deformation limit

JOINTS	TEST LOAD (kN)		ULTIMATE LOAD BASED ON	EXP. RESULT	NUM. RESULT
	EXP.	NUM.	EC 3 AND CIDECT	CODE STRENGTH	CODE STRENGTH
X1	171	195	153	1.12	1.27
DX2	180	205	153	1.18	1.34
DX3	250	240	137	1.82	1.75
DX4	132	143	137	0.96	1.04

Table 5b The comparison between the results of the codes and the
          present work on the basis of the Yura deformation limit

JOINTS	ULTIMATE LOAD (kN)		ULTIMATE LOAD BASED ON	EXP. RESULT	NUM. RESULT
	EXP.	NUM.	EC 3 AND CIDECT	CODE STRENGTH	CODE STRENTH
X1	222	242	153	1.45	1.58
DX2	226	300	153	1.74	1.96
DX3	>275	323	137	>2.00	2.36
DX4	>184	231	137	>1.34	1.67

6   Discussion and conclusions

- The experimental work indicates that the strength of the joints
  loaded in in-plane bending is governed by plastic failure of the
  chord face in combination with initiation of cracks at the tension
  side of the chord faces. Since material cracking is not modelled
  in the numerical work, the numerical analysis indicates that the
  strength of the joints is governed only by plastic failure of the
  chord face.
- The present work shows that the multiplanar joints give higher
  values for the strength than the comparable uniplanar joint, when
  in-plane braces are axially loaded and the out-of-plane braces
  are unloaded.
- For multiplanar joints with in-plane braces axially loaded there
  is a clear influence of compression and tension loaded out-of-
  plane braces on the joint capacity for the present work.
- The present work shows that for multiplanar joints, when in-plane
  braces are loaded by in-plane bending and the out-of-plane braces
  unloaded, the bending capacity is almost the same as for uniplanar
  joints.
- For multiplanar joints with in-plane braces loaded by in-plane
  bending, there is a clear influence of compression and tension
  loaded out-of-plane braces on the joint capacity for the present
  work. The influence of compression and tension loaded out-of-plane
  braces is almost the same, provided that the tension load is the
  same value as the compression load.
- The 1%b_o deformation serviceability criterion (1% for each chord
  face) is not governing for these joints compared to the ultimate
  loads based on the Yura deformation limit and adopting a load
  factor of about 1.4.
- The Yura limit for axially loaded joints gives minimum values of
  1.45 for ratios of experimental to design loads. However, the
  deformation is very large as a deformation criterion, giving about
  16%b_o deformation (8% for each chord face) for all joints.
- To determine the multiplanar effect properly more work has to
  carried out for other geometry parameters and loading conditions.

7 Acknowledgemens

Appreciation is extended to the European Community of Steel and Coal (ECSC) for sponsoring the total research programme. Thanks are due for additional financial contribution to the programme by British Steel (UK) and for the Dutch part of this work to Centrum Staal and the Dutch Ministry of Economic Affairs.

8 References

American Petroleum Institute (1987) Recommended Practice for Planning, Designing and Constructing Fixed Offshore Platforms, API-RP2a, 17th Edition.

American Welding Society (1984) Structural Welding Code, AWS D1.1-84.

Eurocode No. 3 (1992) Design of Steel Structures, Part 1.1: General Rules and Rules for Buildings, European Committee for Standardisation, ENV 1993-1-1.

International Institute of Welding (1989) Design Recommendations for Hollow Section Joints - Predominantly Statically Loaded, IIW Doc. XV-701-89.

Koning, C.H.M. de, Liu, D.K., Puthli, R.S., Wardenier, J. (1992a) The development of design methods for the cost-effective applications of multiplanar connections, Phase 2: Static strength of multiplanar connections in rectangular hollow sections, Experimental and numerical investigation on the static strength of multiplanar welded DX- and X-joints in R.H.S., TNO-Bouw Report: BI-92-0129/21.4.6161, Stevin-TU-Delft Report: 6.92.28/A1/11.08.

Packer,J.A., Wardenier,J., Kurobane, Y., Dutta, D., Yeomans,N. (1992b) Design guide for rectangular hollow section (RHS) joints under predominantly static loading, CIDECT Report 5AZ-17/91, University of Toronto.

Paul, J.C., Valk, van der C.A.C., Wardenier, J., (1989a) The static strength of circular multiplanar X-joints, IIW International Symposium on Tubular Structures, Finland.

Vegte, J. v.d., Koning, C.H.M. de, Puthli, R.S., Wardenier, J. (1991) Static behaviour of multiplanar welded joints in circular hollow sections, TNO-IBBC Report: BI-90-106/63.5.3860, Stevin-TU-Delft Report: 25.6.90.13/A1/11.03.

Wardenier, J. (1982) Hollow section joints. Delft University Press, Delft, The Netherlands.

Yura, J.A., Zettlemoyer, N., Edwards, I.F. (1980) Ultimate Capacity Equations for Tubular Joints, OTC Proceedings, Vol. 1, No. 3690.

41 THE BEHAVIOUR OF THREE DIMENSIONAL RECTANGULAR HOLLOW SECTION TEE JOINTS UNDER AXIAL BRANCH LOADS

G. DAVIES, M.G. COUTIE and M. BETTISON
Department of Civil Engineering,
University of Nottingham, UK

Abstract
The paper describes tests carried out on three dimensional RHS Tee joints as part of a European investigation of the effect of axial loads in out-of-plane bracings on the behaviour of RHS joints. The programme also included cross joints and K joints. Eight Tee joints were fabricated, six of which had out-of-plane bracings, where a branch to chord width ratio of 0.6 was chosen for all branches, with a chord wall slenderness of 22. One of the joints had no out-of-plane bracings attached, so that a comparison of behaviour could be made with that of an identical planar connection for which design recommendations already exist. The force in the in-plane branch was always compressive while out-of-plane branches were subject to axial tension, compression or zero forces. Two different boundary conditions for the ends of the out-of-plane branches were considered, viz free to move vertically, or alternatively restrained by end shear to ensure branches remained parallel as they moved with the chord. The latter method of loading was also chosen since a free end would have been subject to a P/Δ effect with compressive branch load as the branches rotated. The effect of this constraint is discussed.The load-local deformation characteristics are presented, and the various methods of measurement discussed. Comparisons are made with the planar joint, at both serviceability and strength limit states, and interaction curves drawn.
Keywords: Tubular Structures, Welded Joints, Tee Joints, Rectangular Hollow Sections, Strength, Stiffness, Design, Experiments.

1 Introduction

The strength of planar Tee and Cross joints in Rectangular Hollow Sections (RHS) is given in various design recommendations in terms of the yield line solution, where the major parameters are the width ratio (β) and the chord slenderness (h_0/t_0). For Tee joints (as opposed to Cross joints) the loaded chord span is an important parameter, as described in Section 2 where the mode of failure can vary from local deformation failure for short spans to general flexural plastification of the chord for longer spans.

The experimental work described in this paper was designed to examine the effect of welding additional RHS branches to the out-of-plane chord faces of the Tee joints, and the variation of stiffness and axial strength of the joint under various ratios (and sense) of out-of-plane axial force. This formed part of a combined European investigation carried out with Delft (DX joints), and Corby (KK joints). It was also to form the basis of comparisons with Finite Element (FE) simulation of joint behaviour. Care has therefore been taken to record measured material and dimensional

Tubular Structures V. Edited by M.G. Coutie and G. Davies.

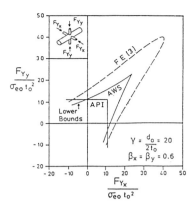

Fig.1. Interaction contours according to API, AWS and FE (Paul (1989)) results.

properties, as well as nominal ones.

Earlier testing and FE work by Paul (1989) has indicated that a significant variation in the strength of a Circular Hollow Section (CHS) joint can take place due to the axial loading of out-of-plane members. However yield line modelling indicates that only modest enhancement occurs for RHS joints when loads are applied in the same sense, Davies et al (1991). The values allowed in codes of practice for CHS joints are indicated in Figure 1.

The limitation on symmetry of the Tee joint (compared with the X), meant that care had to be taken to define the freedom required at the outer end of each branch, due to the effect of the unsymmetric chord sidewall distortion-Figure 2. The in-plane force normally produces a lateral deflection at the ends of the out-of-plane bracings (opbs), which under a compressive load could produce P/Δ instability effects as shown in (b) It was therefore decided to test the specimens in two ways:-

(a)   with complete freedom of opbs - Figure 2(b)
(b)   with shear provided at the ends of 'opbs', such that these branches moved with the overall movement of the chord - Figure 2(c).

A comparison of the strength of the three dimensional joints is presented for both modes of testing in terms of the strength of the planar joint, measured at various local deflection levels.

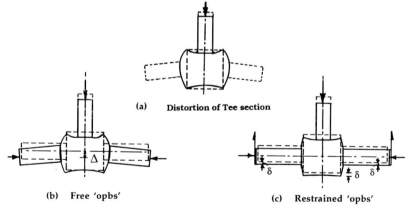

(a)   **Distortion of Tee section**

(b)   **Free 'opbs'**

(c)   **Restrained 'opbs'**

Fig.2. Effect of out of plane branch end restraint

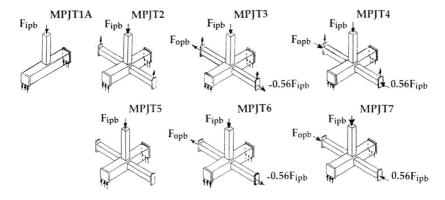

Fig.3. Range of tests carried out on Tee joints.

2 Tests

The total span of the chord was chosen as approximately $5h_0+h_1$ or 875 mm, the largest span which could be accepted without producing flexural plastification. It was agreed to use 150 x 150 RHS for the chord and 90 x 90 RHS for all the branches to give a β ratio of 0.60, using an equal wall thickness to give the chord slenderness of 24. The test specimens are shown in Figure 3. Wall thicknesses, weld sizes, and material properties were measured and are shown in Table 1.

The local deflection of the joint was difficult to measure and was carried out by taking the mean from four calibrated Linear Potentiometers which were clamped to the in-plane-bracing (ipb), and measured the relative deflection to the extreme corners of the chord as shown in Figure 4. The initial arrangement for MPJT', had fin plates spot welded to the corners. However sufficient rotation of these corners occurred to vitiate the readings - the test was then repeated as MPJT1A, and an alternative arrangement was used for all other tests as shown. This worked well. Out-of-plane local deflections were measured in a similar manner, the movement of one 'opb' relative to the other being taken. The local deflection for each side was then assumed to be half this value.

Table 1. Nominal and measured RHS properties.

		DIMENSIONS AND PROPERTIES				NOMINAL
		CHORD (i=0)		BRACE (i=1)		RATIOS
		Nominal	Actual	Nominal	Actual	
b_i	mm	150.0	150.0	90.0	90.5	$\beta = 0.6$
h_i	mm	150.0	149.5	90.0	89.5	
t_i	mm	6.3	6.2	6.3	6.2	$b_0/t_0 = 23.8$
A	mm^2	3600	3505	2090	2062	
f_y N/mm^2		355	420	355	423	$b_1/t_1 = 14.3$
f_u N/mm^2		490	546	490	530	
Weld a mm		-	-	6.3	6.9	
f_s *N/mm^2		-	392	-	422	

* Based on squash tests

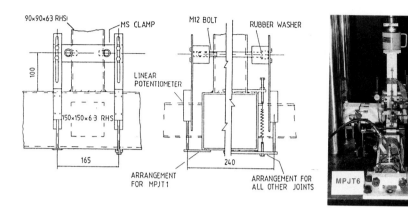

(a) Local deflection measurement.   (b) Testing rig.

Fig.4. Joint testing arrangements.

The testing was carried out within the frame of an Instron testi:.g machine. The desired ratio of axial force in 'opb' to 'ipb' was provided by using interdependent hydraulic jacks with the appropriate ram areas. Deformation control for the loading was provided through the Instron machine. The general view of Joint MPJT6 in the testing rig is shown in Figure 4(b).

Electrical resistance strain gauges were mounted on the branches and the chords to check magnitude and axiality of forces, before the tests to failure were carried out.

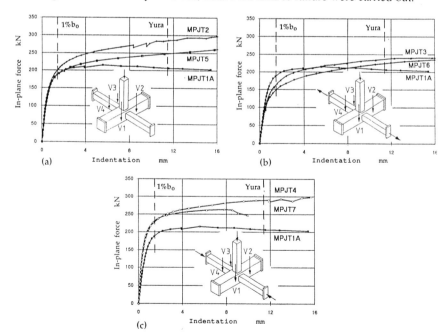

Fig.5. 'ipb' load v 'ipb' displacement for the tests.

432

Table 2. Summary of test results.

Joint No.	MPJT1 *MPJT1A	MPJT2	MPJT3	MPJT4	MPJT5	MPJT6	MPJT7
		- - - - restrained - - - -			- - - - free - - - -		
Elastic Stiffness kN/mm	--- *233	271	208	356	264	207	352
Load F_{ip} for an Indentation of 0.3% b_o (kN)	138 *105	122	94	160	119	93	159
Load F_{ip} for an Indentation of 1% b_o (kN)	192 *189	210	166	232	190	150	234
Load F_{ip} for an Indentation of 5.0% b_o (kN)	207 *212	268	223	280	250	209	264
Load F_{ip} for an Indentation of 11.5mm (YURA) (kN)	--- *207	285	236	290	248	223	---
Maximum Recorded Load (kN)	210 *214 (3.7%)**	310 (14.0%)	246 (10.0%)	300 (10.5)	266 (12.0%)	238 (11.7%)	264 (5.3%)

* Repeated Test   ** Vertical Branch Indentation as % b_o

3 Results

The 'ipb' axial load is plotted against the 'ipb' local deflection in Figure 5(a) where a comparison between three joints is given-the basic planar joint(MPJT1A), the unrestrained joint with unloaded 'opbs' (MPJT5) and the restrained joint with unloaded 'opbs' (MPJT2). A similar comparison but with 'opbs' in tension (MPJT6, MPJT3) is given in (b) and that of 'opbs' in compression (MPJT7, MPJT4) in (c).

The mode of failure was similar in most joints involving some degree of chord wall plastification associated with large local deflection. However in joint MPJT7 the 'opb' loading system became unstable and was finally dominated by P/Δ effects after the maximum load was reached. For joint MPJT3 chord wall rupture adjacent to the weld in the out-of-plane bracing connection was the final cause of failure.

Results are also given in Table 2.

4 Discussion

Figure 5 shows that there is some fall-off of capacity for large deflections, as the side walls of the planar model (MPJT1A) displace outwards. This negative stiffness is

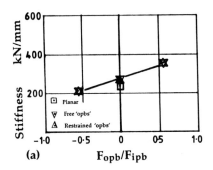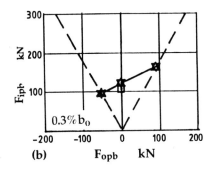

Fig.6. Elastic behaviour of the joints.

not seen in three dimensional joints failing with local chord plastification, but does appear with 'opb' instability in MPJT3, and chord wall rupture in MPJT7.

The addition of welded but free 'opbs' shown in Figure 5(a) clearly stiffens and strengthens the joint, and prevents the fall off of capacity experienced with planar joint MPJT1A. Restraining the ends of the 'opbs' significantly increases the stiffness and strength compared with the planar joint. The increase of strength for MPJT2 at the Yura (1981) deflection limit is about 30% above that of MPJT1A.

The performance of the three dimensional joints under tensile 'opbs' shown in Figure 5(b) indicates the reduced stiffness performance, particularly with free ends (MPJT6), and that MPJT3 with restrained 'opbs' is still below that of the planar joint. From a capacity point of view the three dimensional joints only exceed that of the planar joint for indentations in excess of 6% and 4% respectively. It is worth noting that the unrestrained 'opb' joint does not reach the capacity of the restrained joint.

The effect of the compressively loaded 'opbs' is clearly seen in Figure 5(c). The increased stiffness and strength demonstrated for both MPJT4 and MPJT7 in contrast to the planar joint is also clearly seen.

Non -linearity of the load-indentation curves is seen to occur at indentations less than the commonly accepted 1% b_0 local serviceability deflection limit.

The strength estimate of the Planar Tee Joint (MPJT1A) based on $\beta = 0.6$ in the yield line formulation of IIW (1989) is 152 kN, using the measured values of Table 1. The corresponding value based on the effective $\beta_e = 0.759$, which allows for the weld size and wall thickness is 233 kN. The design value for this joint using the same IIW formula but based on nominal values for both material and geometric properties is 131 kN. The serviceability check would be obtained by dividing the above design value by a factor between 1.4 and 1.5, giving 94-87 kN. It is readily seen that this is well within the 1% b_0 value - even for the worst case shown in Figure 5.

Table 2 shows the values abstracted from the load deflection curves for each joint, including elastic stiffness, and loads at indentations of 0.3% b_0, 1% b_0, 5% b_0 and 11.5mm (Yura). Interaction lines are plotted based on these results in Figure 6, 7and 8. The elastic behaviour is illustrated in Figure 6. The variation of 'ipb' stiffness is shown in Figure 6(a), in terms of the axial load ratio for both branches, while (b) shows the in-plane branch load at a deflection of 0.3%b_0.It is seen that the stiffness is increased by the addition of the 'opbs' and by increasing the compression force ratio, but is not significantly influenced by bracing restraint.

1%b_0, has frequently been used to indicate a sensible limit of local deflection for the seviceability conditions. Figure7(a) shows the variation of axial load sustained in the joint under that deflection condition. Figure7(b) shows the variation of $F_{5\%bo}/F_{1\%bo}$ for the various joints giving a value between 1.22 and 1.4, which ensures that this serviceability criterion is not exceeded under normal conditions.

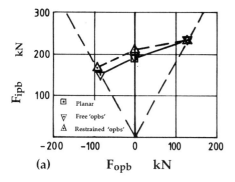
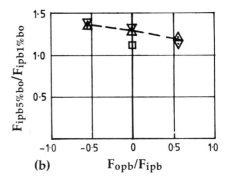

Fig.7. Variation of 'ipb' axial load at assumed serviceability limit of $1\%b_0$

The gain in joint strength for unrestrained and restrained 'opb' ends is clearly shown in Figure 8, at two local deflection levels.

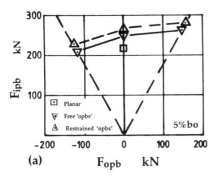
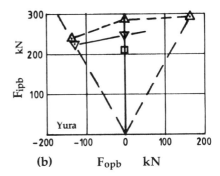

Fig.8. Variation of axial force capacity at $5\%b_0$, and Yura displacement conditions.

5 Conclusions

The experimental tests carried out indicate that the restraint conditions at the ends of the out-of-plane members had a significant effect on the strength of the three dimensional joint in comparison with the planar joint.

Welding in the out-of-plane members enhanced joint stiffness and strength, and ensured there was no fall off in capacity with increasing deformation.

There is a significant increase in joint stiffness depending on whether the out-of-plane branches are in tension or compression, but much less effect is produced by the end restraint.

For both in-plane and out-of-plane branches in compression a significant part of the increase in joint strength at larger deflections derives from the welding in or restraining of the out-of-plane branches rather than on the magnitude of the axial force compression ratio. At a ratio of 0.56 the total enhancement above the planar joint is about 30%.

The rate of fall-off of compressive joint capacity with out-of-plane branches in tension is more noticeable, although at a load ratio of -0.56 the three dimensional joint strengths are still comparable to the planar joint.

The increase in compressive strength of three dimensional RHS joints for positive load ratios is much less than that previously observed for CHS joints.

Care needs to be taken in generalising these results for extremes of branch width ratio and chord slenderness - particularly if the mode of failure changes.

6 Notation

a	fillet weld throat thickness
b_o, b_1	breadth of RHS chord and branch respectively
h_o, h_1	depth of RHS chord and branch respectively
F_{ipb}, F_{opb}	axial force in-plane bracing and out-of-plane bracing respectively
$F_{1\%bo}$, $F_{5\%bo}$	in-plane branch force at 1%, 5% chord width respectively
t_o	Chord wall thickness
β	branch to chord width ratio b_1/b_o
β_e	effective width ratio $(b_1+2\sqrt{2}a)/(b_o-t_o)$

7 Acknowledgements

The partners in the ECSC programme are the Steel Construction Institute, British Steel, University of Nottingham TNO Building and Construction Research and Delft University of Technology. Appreciation is extended to the ECSC for permission to publish this paper.

8 References

API (1987) **Recommended Practice for Planning, Designing and Constructing Fixed Offshore Platforms,** American Petrolium Institute, RP2A, Texas.

AWS (1988) **Structural Welding Code**, American Welding Society, 1-88, USA.

Davies, G. and Morita, K. (1991) Three Dimensional Cross Joints Under Combined Branch Loading, in **Tubular Structures, 4th International Symposium** (ed. Wardenier, J. and Panjeh Shahi, E.), Delft University Press, pp324-333.

IIW (1989) Design Recommendations for Hollow Section Joints - Predominantly Statically Loaded, **International Institute of Welding,** Doc XV-70-89, UK.

Paul, J. C., Van Der Valk, C. A. C. and Wardenier, J. (1989) Static Strength of Circular Multi-planar X Joints, **Tubular Structures, 3rd International Sympoium**, Elsevier Applied Science.

Yura, J. A., Zettlemoyer, N. and Edwards, I. F. (1981) Ultimate Capacity of Circular Tubular Joints, **Journal of the Strucural Division, ASCE**, Vol. 107, No. ST10, pp1965-1984, USA.

42 RECTANGULAR HOLLOW SECTION DOUBLE K-JOINTS – EXPERIMENTAL TESTS AND ANALYSIS

N.F. YEOMANS
British Steel – Tubes and Pipes, Corby, UK

Abstract
This paper describes the results of a series of tests on rectangular hollow section double K-joints (KK-joints) carried out at British Steel's Swinden Laboratories for British Steel, General Steels - Welded Tubes at Corby, UK. The work was carried out on three series of three tests each in which the various important geometric parameters were varied and the results were compared to current IIW/CIDECT and, now, Eurocode No. 3, Annex K recommended design formula for uni-planar RHS K-joints. The work has shown that these uni-planar design formulae can be used for the design of multi-planar joints with only minor modifications to one of the parameter limits and an overall reduction factor to take account of the multi-planar interaction effects.
Keywords: Analysis, Connections, Double K-Joints, Multi-planar, Rectangular Hollow Section, Test.

1 Introduction

The work described in this paper is part of a much larger research project on statically loaded multi-planar rectangular hollow section (RHS) joints, the overall aim of which is to produce design recommendations for such joints, and so exploit the undoubted economic, aesthetic and structural advantages of hollow sections in such constructions. The experimental parts of the project are being carried out by British Steel, England on double K-joints (the subject of this paper), Nottingham University, England on treble T-joints and by Delft University, The Netherlands on double X-joints. In addition a series of numerical analyses and simulations are being undertaken by The Steel Construction Institute, England and Delft University.

Tubular Structures V. Edited by M.G. Coutie and G. Davies.

2 Objectives

Design formulae for tubular joints in plane frame constructions are well known and are already included in
various design standards and recommendations (Eurocode
1992, IIW 1989). However, very little work has been carried out on multi-planar joints, especially with rectangular hollow section (RHS) members by Redwood (1983), Giddings (1985) and Scola (1990); much more work has been done
on joints with circular hollow section (CHS) members by
Makino (1984), Paul (1991) and Mouty (1992) and recommendations for their design have been established.

The objective of the work described here are:- (a) to
determine if the current recommendations for uni-planar
joints can be used for multi-planar joints by using a simple modification, and if not, (b) determine what the
design formulae should be.

3 Test Specimens and Procedure

3.1 Test Specimen Details
A total of nine specimens were tested split into three
series of three specimens each, depending on their bracing
size. All the specimens consisted of two pairs of bracings, each pair being welded to the chord on adjacent
faces with all bracings at 45° to the relevant chord face.
The specimen details are given in Table 1 and Figure 1.

All the material used was to BS 4360:grade 43c (now EN
10210: grade S275J0H) for the steel grade and to BS
4848:Pt2 for dimensions. Generally the yield strength of
this material is well above the nominal value of $275N/mm^2$;
therefore, in order to ensure that there would be no premature weld failures during the tests it was decided that
all welding electrode material would be a grade higher in
strength and equivalent to a grade 50 steel ($355N/mm^2$
yield strength).

3.2 Specimen Instrumentation
All the specimens were instrumented in the same way with
strain gauges to measure axial loads and bending moments
in all the joint members and with linear deflection transducers to measure local joint deformations.

3.3 Test Procedure
The inherent problem with load testing complex joints in a
test rig, as opposed to a full framework, is the specimen
to test rig interface and rigid body movements of the
specimen in the test rig. All of which tend to generate
uncharacteristic moments in the specimen. Due to this problem, each specimen was loaded several times, up to a
maximum of 25% of the predicted failure, and the set-up

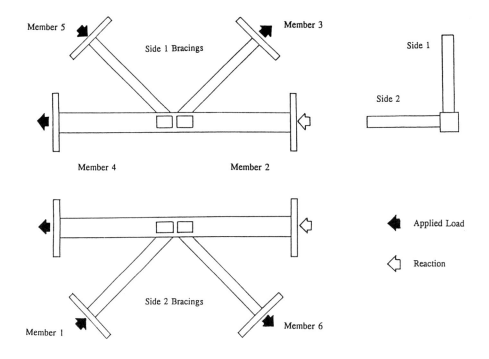

Fig.1. General layout of test specimens

adjusted to reduce these moments to an acceptable level.
An attempt was also made to offset the specimen from a
true alignment so that rigid body movements brought each
specimen into alignment at about 75% of predicted failure.
Even after these precautions, quite large bending strains
were recorded in some specimens during the test to failure.

4   Analysis of Results

4.1 Predicted Joint Capacity Formulae
The joint capacities were calculated using the normal
recommended plane frame design formulae in Eurocode
(1992), IIW (1989), suitably modified to change them from
design capacity to mean experimental formulae. The details
of these modifications are described below.

4.1.1 Chord face deformation criterion, N_{def}
The design formula constant (8.9) must be multiplied by
1.22 to increase it to the mean test value. Additionally,
because of the joint symmetry other simplifications can be
made to the formula:-the bracings are all square and of

the same size, thus the term $(b_1+h_1+b_2+h_2)$ becomes $4b_1$; in the chord end load function the load factor, n, becomes $[2 N_{(brace\ mean)} Cos\theta_1/(f_{y0} A_0)]$. Thus the bracing load to produce failure by chord face yielding N_{def} is given by:-

$$N_{def} = 10.9\ f_{y0}\ t_0^2\ (b_1/b_0)\ (b_0/t_0)^{0.5}\ f(n)\ /\ Sin\theta_1 \qquad (1)$$

$$\text{with } f(n) = 1.3 + 0.4\ n\ (b_0/b_1) \leq 1.0$$
$$\text{and } n \text{ negative for compression.}$$

4.1.2 Chord shear criterion, N_{shr}
The IIW shear formula assumes, depending on the gap between bracings, that part of the chord cross face can resist shear. In the multi-planar case, however, it cannot. The shear formula is a lower bound so no correction can be made for mean failure values. Thus the lower bound formula for bracing load for chord shear failure is:-

$$N_{shr} = f_{y0}\ A_0\ /\ (2\sqrt{3}\ Sin\theta_1) \qquad (2)$$

4.1.3 Chord punching shear criterion, N_{cps}
The effective width constant has been increased from 10.0 to 12.5, and since the bracings are all square and of the same size the h_1 terms can be replaced by b_1. Hence, the mean formula for chord punching shear becomes:-

$$N_{cps} = f_{y0}\ t_0\ b_1\ [2/Sin\theta_1 + 1 + b_{eps}/b_1]\ /\ (\sqrt{3}\ Sin\theta_1) \qquad (3)$$

$$\text{with } b_{eps}/b_1 = 12.5\ t_0/b_0 \leq 1.0$$

4.1.4 Bracing effective width criterion, N_{bew}
The same modifications as for chord punching shear apply, thus the mean failure formula for bracing effective width becomes:-

$$N_{bew} = f_{y1}\ t_1\ b_1\ [3 - 4t_1/b_1 + b_{eff}/b_1] \qquad (4)$$

$$\text{with } b_{eff}/b_1 = 12.5\ (t_0/b_0)\ (f_{y0}/t_0)\ /\ (f_{y1}/t_1) \leq 1.0$$

4.1.5 Chord combined shear and axial load criterion
This relates to the combined action of shear and axial force in the chord gap between the bracings, in these tests the loading was so arranged that there was, theoretically, zero axial force at this point, and hence this criterion will not apply.

4.1.6 General serviceability criterion
Because of the possibility of relatively large deformations in the chord of tubular joints, historically it has been accepted that a deformation limit, at normal applied load levels, of 1% of the chord depth, h_0, be used. Assuming this load to be F_δ with a minimum overall load factor

of 1.4, this will restrict the capacity in these cases to 1.4 F_δ, ie F_{sl}, even if the actual test maximum load, F_{max}, is higher than this.

For this reason two test loads are shown in Table 2, the load due to reaching the deformation limit, F_{sl}, and the maximum load attained in the test, F_{max}.

4.1.7 Predicted joint failure loads

The resulting predicted test failure loads, calculated from the above formulae and using the actual material properties, the actual material geometry and the most optimistic material orientation, are given in Table 1, where the underscored values are the lowest predicted failure loads and have been used for the analysis in Table 2

Table 1. Joint geometry and predicted capacity.

Spec	Chord member	Bracing member	Gap	Ecc.	Predicted capacity			
ref no	b_0 x h_0 x t_0 (mm)	b_i x h_i x t_i (mm)	g (mm)	e (mm)	N_{def} kN	N_{shr} kN	N_{cps} kN	N_{bew} kN
KK01	150x150x5.0	60x60x6.3	65	0	195	383	334	467
KK02	150x150x6.3	60x60x6.3	65	0	245	471	428	489
KK03	150x150x10	60x60x8.0	65	0	365	700	695	580
KK04	150x150x5.0	90x90x6.3	30	3.6	295	383	500	647
KK05	150x150x6.3	90x90x6.3	30	3.6	367	471	639	681
KK06	150x150x10	90x90x8.0	30	3.6	540	700	1026	964
KK07	150x150x5.0	150x150x6.3	13	37.6	495	383	–	1011
KK08	150x150x6.3	150x150x6.3	13	37.6	610	471	–	1055
KK09	150x150x10	150x150x6.3	13	37.6	900	700	–	1217

4.2 Test failure loads and analysis

The test failure loads, based on both the chord face serviceability limit, F_δ, (see section 4.1.6) and the maximum load attained in the test, F_{max}, are given in Table 2; also included is a brief description of the type of failure and the ratios of actual to predicted failure load. It can be seen that all the load factors based on F_{max}, except for specimens KK01 and KK02 are above 1.0, indicating that the predictions are safe if these two specimens are excluded; on the other hand almost half of those based on the deformation limiting load, F_{sl}, are below 1.0. However, the constant of 1.4 used to factor up from the chord face serviceability limit, F_δ, to the deformation limiting load, F_{sl}, is an arbitary figure and any value from 1.4 up to about 1.6 could have been chosen. Indeed if

Table 2. Predicted capacities and test loads.

Spec ref no.	Calc. cap. N_{cal} (kN) fail mode	Test max loads (kN)			Load factor		Comments on test failure mode
		load at $.01b_0$ F_δ	service limit F_{sl}	max test load F_{max}	$\dfrac{F_{sl}}{N_{def}}$	$\dfrac{F_{max}}{N_{cal}}$	
KK01	195 N_{def}	82	115	154	0.59	0.79	Gross chord face deformation
KK02	245 N_{def}	140	196	215	0.80	0.88	Gross chord face deformation
KK03	365 N_{def}	365	511*	393	1.40	1.08	Yielding of the chord in the gap and brace out-of-plane buckling
KK04	295 N_{def}	188	263	400	0.89	1.36	Gross chord face deformation plus signs of chord shear
KK05	367 N_{def}	245	343	399	0.93	1.09	Chord face deformation and excessive bending moments in bracings
KK06	540 N_{def}	615	861*	674	1.59	1.25	Chord face deformation and in-plane buckling of bracing
KK07	383 N_{shr}	540	756*	603	1.53	1.57	Compression buckling of the chord and indications of chord shear in the gap
KK08	471 N_{shr}	652	913*	663	1.50	1.41	Extensive yielding of the chord in the gap between the bracings
KK09	700 N_{shr}	#	#	799	-	1.14	Extensive yielding of the chord in the gap between the bracings

* $1.4F_\delta$ not reached   # deformation of $0.01b_0$ not reached before failure

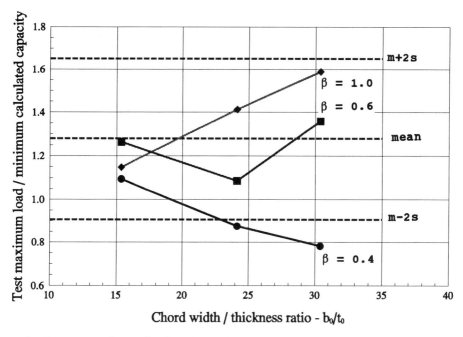

Fig. 2. Test maximum load against minimum predicted load

a value of 1.6 had been used only the first two specimens, KK01 and KK02, would have given values below 1.0. Figure 2 shows the maximum test loads F_{max}, related to the minimum predicted failure load in a graphical manner and although the scatter is larger than than one would like, the results are all safe, with the exception of KK01 and KK02. Also shown in this figure are the mean plus and minus two standard deviation values of the results, when the two low values from specimens KK01 and KK02 are excluded. Prior to these tests it had been suggested that the plane frame joint design formulae should be used for multiplanar joints of this type but with a reduction factor of 0.9. The results here suggest the same value, since the mean minus two standard deviations value is 0.91. The two specimens that failed at loads below the minimum predicted ones can be excluded by modifying the validity range for the minimum bracing to chord width ratio. For uni-planar joints this limit is $(0.1 + 0.01b_0/t_0)$ with an over-riding minimum of 0.35. If the limit is modified to $(0.1 + 0.015b_0/t_0)$ then the two specimens which gave low results are excluded, and the recommendations become acceptable.

5 Conclusions

The results of the nine tests on multi-planar rectangular hollow section joints have shown that the current design recommendations for the calculation of the capacities of similar uni-planar joints can be used provided that the resulting capacities are reduced by a factor of 0.9.

It is also necessary to modify the range of valadity for the bracing to chord with ratio from $(0.1 + 0.01b_0/t_0)$ to $(0.1 + 0.015b_0/t_0)$, to prevent small bracings being used with thin chords.

6 Acknowledgements

This work has been undertaken within the framework of and with the financial support of ECSC Research Agreement No. 7210/SA/830.

7 References

Eurocode (1992) **Eurocode No.3:** Design of Steel Structures: Part 1.1:General Rules and Rules for Buildings:Annex K: Hollow Section Lattice Girder Connections.

International Institute of Welding (1989) Design Recommendations for Hollow Section Joints-Predominantly statically loaded. 2nd Edition, **IIW Doc No. XV-701-89.**

Redwood,R.G. Bauer,D.B. (1983) Behaviour of HSS Triangular Trusses and Design Considerations. **Canadian Symposium on HSS**, April-May 1983, CIDECT.

Giddings,T.W. (1985) The Development of Recommendations for the Design of Welded Joints between Steel Structu ral Hollow Sections and H-sections. **Commission of Euro pean Communities**, Technical Steel Research, Report No. EUR 9462 EN.

Scola,D. Redwood,R.G. and Mitri,H.S. (1990) Behaviour of Axially Loaded Tubular V-joints. **Journal of Constructional Steel Research**, Vol. 16, No.2.

Makino,Y. Kurobane,Y. Ochi,K. (1984) Ultimate Capacity of Tubular Double K-joints. **Welding of Tubular Structures, Proceedings of the 1st International Conference of IIW**, Boston, Pergamon Press.

Paul,J.C. Ueno,T. Makino,Y. Kurobane,Y. (1991) The Ulti mate Behaviour of Circular Multi-planar TT-joints. **Pro ceedings of the 4th International Symposium on Tubular Structures**, Delft, Delft University Press

Mouty,J. Rondal,J. (1992) Study of the Behaviour under Static Loads of Welded Triangular and Rectangular Lattice Girders made with Circular Hollow Sections. **CIDECT** report No.5AS-92/1.

8 List of Symbols

8.1 Joint geometry and material properties

b_i	width of member i - mm
e	joint noding eccentricity - mm
g	gap between bracing members - mm
h_i	depth of member i - mm
t_i	thickness of member i - mm
θ	angle between bracings and chord - °
f_{yi}	yield strength of member i - N/mm^2
i	suffix, $_0$ - chord, $_1$ - compression bracing, $_2$ - tension bracing

8.2 Test loads, and calculated capacities, kN

F_δ	test load at service load deflection limit
F_{sl}	maximum test load for service load, F_δ
F_{max}	maximum load attained in test
n	ratio of actual chord load to chord yield load
N_{bew}	calculated capacity for bracing effective width
$N_{(brace\ mean)}$	mean value of the four bracing loads
N_{cal}	minimum calculated capacity
N_{cps}	calculated capacity for chord punching shear
N_{def}	calculated capacity for chord face deformation
N_{shr}	calculated capacity based on chord shear

43 POST-YIELD AND POST-PEAK BEHAVIOUR OF TUBULAR JOINTS IN OFFSHORE STRUCTURES

M. LALANI
MSL Engineering Limited, Ascot, UK

Abstract
Tubular joints exhibit substantial post-yield reserve strength. At loads two to five times the joint first yield load (depending on joint type, geometry and loadcase), the peak load is attained. In recent times, there has been an increasing recognition that the post-yield and post-peak behaviour of tubular joints (reserve strength) is an important consideration in non-linear frame analysis for pushover where joints represent the weak link. This paper addresses the reserve strength of tubular joints by reference to the present day knowledge in this field.
Keywords: Tubular Joints, Offshore, Reserve Strength, Post-Yield, Post-Peak.

1 Introduction

A number of codes and guidance documents provide recommendations which relate to the design, construction and inspection of tubular joints. These recommendations have been derived from an interpretation of research results and in-service performance experience. The object of establishing tubular joint design criteria is to dimension the joints so that they perform satisfactorily in service and achieve a reasonable balance between economy and risk of failure. Typically, it is assumed by the designer that these requirements are implicitly met by satisfying the tubular joint provisions contained in documents such as API RP2A (1991) and the HSE Guidance Notes (1990). Advances in tubular joint's technology are encompassed in codified documents and recommended practices through amendments or updates as new research findings become available, although there is usually a time lag between the availability of research findings and its inclusion in design practices.

In recent times, there has been an increasing recognition that the post-yield and post-peak behaviour of tubular joints (reserve strength) is an important consideration in non-linear jacket pushover analysis where joints represent the weak link. This paper addresses the reserve strength of tubular joints by

Tubular Structures V. Edited by M.G. Coutie and G. Davies.
© 1993 E & FN Spon, 2–6 Boundary Row, London SE1 8HN. ISBN 0 419 18770 7.

reference to the present day knowledge base in this field. The notation adopted is identical to API RP2A, unless otherwise noted.

2   Failure modes for tubular joints

It is instructive to examine the failure modes of tubular joints, in order to assess the yield and post-yield response. The mode of failure is dependent on the type of joint, loading conditions and the geometrical parameters defining the joint. Tests carried out to date have identified several types of failures, namely plastification of the chord, cracking and gross separation of brace from chord, cracking of the brace, local buckling, shear failure of the chord between adjacent bracings, and lamellar tearing of the chord wall under brace tension loading (often considered as a material-related failure).

Typical load-deformation curves for axially loaded joints are shown in Figure 1. For brace axial tension loading, yielding of the chord around the brace and distortion of the chord cross-section occurs. As the load increases, a crack at the 'hot-spot' which eventually leads to gross separation of the brace from the chord may be initiated. Failure in compression loaded T/Y and DT/X joints is usually associated with buckling and plastic deformation of the chord wall. The stiffness and capacity of DT/X joints can be less than those of T/Y joints, but depend on the sense of the relative brace loading. Although the failure modes are similar with regard to deformations local to the brace/chord intersection, there are clearly differences in the way that brace axial loads are reacted by the chord; global bending of the chord can occur in T/Y joints but not in DT/X joints. In addition, DT/X joints are subjected to double ovalisation. The failure mechanism of balanced axially loaded K joints (tension in one brace, compression in other brace) largely depends on the gap between the two brace members. As the gap reduces, the strength of the joint can increase due to the increased bending stiffness of the chord wall between the braces. Chord plastic deformations and 'punching' failure are the most common modes of failure for these joints. However, for large β ratios, shear failure of the chord section between the two braces member can occur.

For joints subject to in-plane moment, failure occurs due to fracture through the chord wall on the tension side of the brace and plastic bending and buckling of the chord (and/or brace local to the intersection) on the compression side. For out-of-plane moment loaded joints, local buckling of the chord wall in the vicinity of the brace saddle on the compression side occurs, resulting in reduced stiffness. Failure is usually associated with fracture on the tension side of the brace after excessive plastic deformations.

It has been observed that tubular joints have a substantial reserve beyond first yield (Wardinier, 1982). (Note that joint yield is defined from the linear portion of the joint load-deformation curve, as opposed to the chord yield which is determined from tensile coupon tests or stub-column tests of the chord

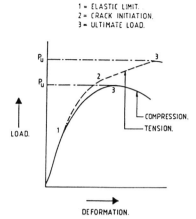

Fig.1. Failure nodes for tubular joints.

material). The joint 'softens' beyond the point of first yield, and joint stiffness reduces, until the maximum load-carrying capability of the joint is attained (Point 3, Figure 1).

3 Reserve strength technology

The analysis of the behaviour of existing offshore steel platforms requires the application of fundamental structural engineering practice. Initially, a linear elastic analysis for storm loading is conducted for the jacket structure, modelling, the actual condition of the platform (as-built plus all subsequent modifications). The joint components are subjected to code checking or specific assessment depending on whether they are undamaged or damaged.

Where component utilisation exceeds unity (i.e. acting load is greater than allowable load), following application of various refinements (e.g. accounting for multiplanar action, Davies (1991)), the integrity of the platform as a whole may be investigated by assessing the capability of the structure to redistribute loads away from these components to other parts of the structure. This 'redundancy' analysis can be undertaken by omitting the affected components from the analysis or, should omission make other components marginal, by replacing them with an appropriate set of reduced stiffnesses in the model.

Advanced structural assessment involves a non-linear analysis of the complete structure to assess and examine its ultimate load-bearing capacity, using reserve strength technology. In this manner, the redundancy aspects can be assessed more accurately than by linear elastic procedures. Further, this non-linear analysis (commonly referred to as pushover analysis) allows the ultimate limit state of the complete structural system to be appraised, vis-a-vis component-based assessments.

The non-linear stiffness characteristics for joints have a significant bearing on frame behaviour for connection-dominated collapse mechanisms, and the reserve beyond first yield is dependent on the configuration, geometry, load type and ductility of the joints. Ductile yielding at a joint in a statically indeterminate structure (such as typical offshore jackets) will redistribute internal forces so that larger loads can be applied. The manner of the redistribution is dictated by the availability of alternative load paths and the interaction between member and joint non-linear behaviour. Loads can be increased until hinges occur at sufficient locations to create a collapse mechanism. Fundamental in the application of reserve strength technology is the need to understand the behaviour of tubular joints, from two specific standpoints:

(a) For jacket structures where a joint or joints represent the weak link, how does the joint behave in the post-yield and post-peak phase and what are the joint redistribution characteristics?

(b) For jacket structures where members are a weak link, how does a joint react to loads redistributed from member load-shedding?

The solution lies in the understanding of the behaviour of tubular joints in the post-yield and post-peak regime.

4 Present day status of joints technology

A number of design guidance documents and recommended practices are available which contain approaches to design based on an interpretation of research data and field experience. API RP2A (1991) and the HSE Guidance Notes (1990) are used in the offshore industry, whilst the IIW (1989) and CIDECT (1991) recommendations are deployed extensively for land-based structures. All present day codes deal with peak load, ie. the maximum load that can be sustained by a tubular joint. From a reserve strength and pushover analysis standpoint, the designer is faced with the following issues for a given joint configuration:

(a) What is the load at which joint 'softening' commences?
(b) What is the deformation at which joint 'softening' commences?
(c) What is the deformation at which joint peak load is reached?

It is important in a pushover analysis to accurately model the linear and non-linear stiffness characteristics of tubular joints which may be deemed to be the potential 'weak' links in the subject jacket structure. This is because the non-linear stiffness characteristics will dictate the extent to which load-shedding will occur as increasing jacket loads are applied. The lack of codified methodologies in this respect is not surprising, given the lack of data and information in this area. In the following section, the development of joint stiffness characteristics is addressed.

5 Development of joint stiffness characteristics

At the present time, the designer can, and does, adopt one of the following options in pushover analysis where the joint forms the weak link:

(a) Assume the joint is rigid up to peak load and fully plastic at peak load. This approach does not require any information concerning the deformation at peak load. Whilst adequate for initial investigations of system strength, this approach has a number of obvious shortcomings, as it does not allow for load-shedding prior to attainment of peak load which will occur in practice.

(b) Undertake a study of available steel model data to assess their applicability to the joint(s) under consideration.

(c) Conduct non-linear analysis or experimentation to develop appropriate stiffness curves for the joints under consideration. This approach is often prohibitively expensive, and still leaves the designer with the task of adjusting the derived stiffness curves to ensure compatibility with the global non-linear system analysis package being used.

It has been recognised that, ideally, a series of closed-form expressions are required to enable the designer to establish stiffness curves for tubular joints to be used in reserve strength calculations. This recognition has stemmed from an observation that tubular joint and system reserve strength calculations are increasingly being conducted as part of offshore structural integrity investigations. It is against this background that a project was commenced at MSL Engineering in 1992 on behalf of a major offshore operator. The project, now a combined industry effort, is aimed at developing closed-form algorithms for stiffness curve definition, and concentrates specifically on:

(a) the load at which joint 'softening' commences (axial, IPB and OPB)
(b) the deformation at which joint 'softening' commences (axial, IPB and OPB)
(c) the deformation at which joint peak load is reached (axial, IPB and OPB)
(d) the load at first crack, for tension loaded joints
(e) the deformation at first crack, for tension loaded joints.

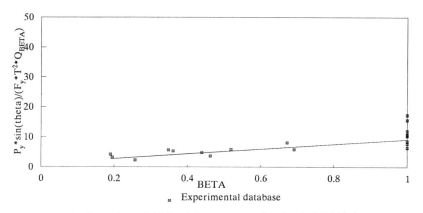

Fig.2.　First yield load for compression loaded DT joints.

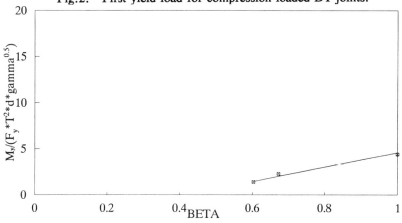

Fig.3.　First yield load for IPB loaded DT joints.

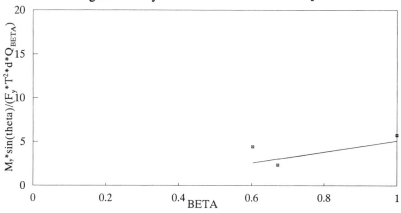

Fig.4.　First yield load for OPB loaded DT joints.

All available steel model test data have been collated, and the load-deformation curves for each specimen have been extensively interrogated to allow extraction of the relevant information to meet the above aims. The databases created have been subjected to rigorous screening, consistent with the approach adopted in the current HSE static strength project (HSE, 1991). The assessments below concentrate on DT joints.

Figures 2, 3 and 4 show the final product from extensive data assessments for DT joints subjected to axial compression, IPB and OPB loads, respectively. These figures relate to the development of algorithms for attainment of first yield load. Figures 5, 6 and 7 reflect comparable plots for deformations at first yield. The following comments are worthy of note in relation to these figures:

(a)   A tight banding of results can be observed from Figure 2 for medium β's. It can also be noted that this figure encompasses a larger dataset than the other plots, which reflect a paucity of data in specific areas. Note that the reduced dataset in Figure 5 compared with Figure 2 is a reflection of the validity range for the predictive algorithm in Figure 5.

(b)   A least squares analysis has been conducted for the dataset noted in Figure 2, and the derived equation is as follows:

$$\frac{P_y \sin\theta}{F_y T^2} = (1.2 + 8\,\beta)\,Q_\beta \tag{1}$$

where P_y is the predicted mean load at joint first yield, and all other terms consistent with API RP2A.

(c)   The prediction models noted in Figures 5, 6 and 7 represent adjusted, amplified and calibrated models, reflecting deformation dependancy on β, γ, τ and θ (API RP2A terminology). It can be observed that the prediction models adequately capture the deformations observed, at least within the bounds of the available data and the accuracy expected from these computations.

6   Closure

The need to assess the post-yield reserve strength of tubular joints within the context of frame system reserve strength calculations for offshore structures is increasing. The present day technology in this field is being expanded through a project currently underway as described in this paper. One finding from the project is the paucity of data in specific areas to allow new yield strength and

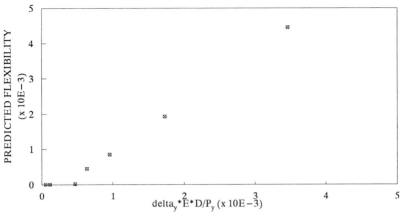

Fig.5.   First yield deformation for compression loaded DT joints.

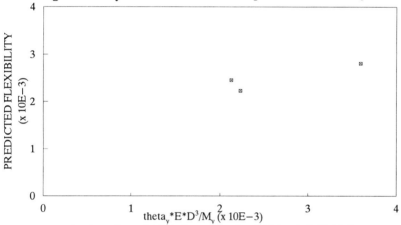

Fig.6.   First yield deformation for IPB loaded DT joints.

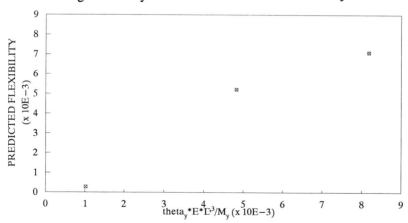

Fig.7.   First yield deformation for OPB loaded DT joints.

deformation expressions to be developed. One medium that is currently being explored to generate the required data relates to the technique developed at Nottingham University. This technique (Fessler H and Hassell W, 1991) uses a lead-tin alloy to create tubular joints of the desired configuration and geometry. Lead-tin alloy tests are attractive from a cost standpoint and, provided the technique is calibrated for the application under consideration, the evidence to date suggests that the results using this method are technically acceptable. It is expected that the combined industry effort, will add substantially to the present day knowledge base in this field.

References

American Petroleum Institute (1991) **Recommended practice for planning, designing and constructing fixed offshore platforms**. API RP2A, Nineteenth Edition.

Davies G. and Morita K. (1991) Three dimensional cross joints under combined axial branch loading. **4th International Symposium on Tubular Structures**, Delft.

Fessler H. and Hassell W. (1991) Static strength of K joints under axial loading. **4th International Symposium on Tubular Structures**, Delft.

HSE (1990) **Offshore installations: guidance on design and construction**. 4th Edition, HMSO.

International Institute of Welding (1989) **Design recommendation for hollow section joints - predominantly statically loaded**. IIW-XV E, IIW Document XV-701-89.

Wardinier J., (1982) **Hollow section joints.** Delft University Press, Delft 1982, ISBN, 90.6275.084.2.

Wardinier J., Kurobane Y., Packer J.A., Dutta D. and Yeomans N. (1991) **CIDECT design guide for circular hollow section (CHS) joints under predominantly static loading.** Verlag TUV Rheinland.

44 THE STRENGTH OF BALL JOINTS IN SPACE TRUSSES – PART II

T. TANAKA, H. KANATANI and M. TABUCHI
Kobe University, Japan

Abstract
The strength of hollow steel spheres used as nodes in tubular space
trusses is investigated. In this paper, the exact collapse load is
obtained for hollow steel spheres subjected to uniaxial force by
limit analysis. It is assumed that the shell material is rigid-plas-
tic and obeys the yield condition which has been derived by Onat and
Prager (1954) based on the Tresca yield criterion. The analytical results are verified through comparison with test results. Further, a
practical formula for predicting the strength of hollow steel spheres
is proposed on the basis of the complete numerical solution.
Keywords: Ball Joint, Space Truss, Limit Analysis, Collapse Load,
Spherical Shell, Axisymmetric Load.

1 Introduction

This paper describes a basic study on the strength of a hollow steel
sphere (called "ball" hereinafter) subjected to the direct force of
circular tubes welded to the ball. Methods of limit analysis are applied to find the strength of balls subjected to uniaxial force. The
analytical model is a spherical shell subjected to axisymmetric line
loads as shown in Fig.1. The authors (1987) reported collapse loads
for this model based on the upper-bound theorem. However, the ratio
of the measured maximum load to the analytical result decreases as
tube-to-ball diameter ratio increases. In this paper, the complete
numerical solution is obtained for this model. The mechanism for this
model consists of two yield hinge circles, one circle is the loading
position denoted by the angle ϕ_α and the other circle is the unknown
position denoted by the angle ϕ_β, and a yield zone bounded by the two
hinge circles, as shown in Fig.1. The complete solution is obtained
by solving a particular boundary value problem of differential equations.

We append a new boundary condition which reduces computation work
for solving a problem of this kind. Moreover, as the discontinuous
transition type at a yield hinge circle on this spherical shell has
not been presented in any previous paper, we present the solution
method for the new transition type.

The analytical results are verified through comparison with test
results. Further, a practical formula for predicting the strength of
balls is proposed on the basis of the complete numerical solution.

Tubular Structures V. Edited by M.G. Coutie and G. Davies.
© 1993 E & FN Spon, 2–6 Boundary Row, London SE1 8HN. ISBN 0 419 18770 7.

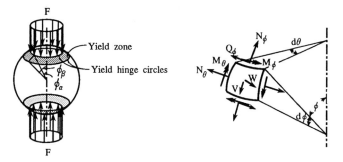

Fig.1. Analytical model   **Fig.2. Element of shell**

2 Basic equations

2.1 Equations of equilibrium and compatibility

The stress resultants on a shell element subjected to axisymmetric loads are three resultant forces N_ϕ, N_θ and Q_ϕ and two resultant moments M_ϕ and M_θ as shown in Fig.2. The well-known equations of equilibrium have been given by Timoshenko(1940), and are specialized here for spherical shells subjected to circular force, e.g. in the range of $\phi_\alpha \leq \phi \leq \pi/2$, as follows:

$$n_\phi' = \{n_\theta - (n_\phi + f) - (n_\phi + f)\tan^2\phi\}\cot\phi, \quad m_\phi' = \{m_\theta - m_\phi - \omega(n_\phi + f)\tan^2\phi - \omega f\}\cot\phi. \qquad (1)$$

The primes indicate differentiation with respect to ϕ, and

$$n_\phi = \frac{N_\phi}{N_0}, \quad n_\theta = \frac{N_\theta}{N_0}, \quad m_\phi = \frac{M_\phi}{M_0}, \quad m_\theta = \frac{M_\theta}{M_0}. \qquad (2)$$

Mo and No are the yield bending moment and the yield normal force per unit length. The dimensionless force f is defined by

$$f = \frac{F}{2\pi R N_0}, \qquad (3)$$

and the dimensionless shell parameter ω is defined by

$$\omega = \frac{R N_0}{M_0} = \frac{4R}{T}, \qquad (4)$$

where R and T denote the radius and the thickness of a spherical shell respectively.

Dimensionless velocities are v=V/R and w=W/R, where V and W are the velocity components shown in Fig.2. The generalized strain rates associated with the stress resultants n_θ, n_ϕ, m_θ and m_ϕ have been given in terms of reduced velocities by Hodge (1959) as follows:

$$\varepsilon_\theta = v\cot\phi - w, \quad \varepsilon_\phi = v' - w, \quad \kappa_\theta = -\frac{\cot\phi}{\omega}(v+w'), \quad \kappa_\phi = -\frac{1}{\omega}(v+w')', \qquad (5)$$

where ε and κ denote generalized rates of extension and generalized changes of curvature respectively.

2.2 Yield condition and flow rule

The yield condition for shells of revolution composed of a material which obeys the Tresca yield criterion has been obtained in parametric representation by Onat and Prager (1954). Later, Lance and Onat (1963) replaced parametric representations by explicit ones as shown in Table 1. The yield condition is composed of twelve hypersurfaces.

Each constituent surface is denoted by G or H with appropriate subscripts as shown in the first column of Table 1. The inequalities in the third column of Table 1 delimit the surface and determine its boundaries.

For all points within one yield surface defined by yield function $\Phi=1$, the flow rule is written as follows:

$$\varepsilon_\theta = \lambda \frac{\partial\Phi}{\partial n_\theta}, \quad \varepsilon_\phi = \lambda \frac{\partial\Phi}{\partial n_\phi}, \quad \kappa_\theta = \lambda \frac{\partial\Phi}{\partial m_\theta}, \quad \kappa_\phi = \lambda \frac{\partial\Phi}{\partial m_\phi}. \tag{6}$$

where λ is a nonnegative deformation parameter.

Table 1. Yield condition

Face	Yield Condition	Restrictions
$G_\theta^\pm$	$\pm(m_\theta-m_\phi)+(n_\theta-n_\phi)^2+(\frac{m_\phi}{2n_\phi}\pm n_\theta)^2-1$ $=0$	$\mp m_\phi-2n_\phi(n_\theta-n_\phi)\geq0, \quad \pm m_\phi+2n_\theta n_\phi\geq0,$ $1-\left|\frac{m_\phi}{2n_\phi}\pm n_\theta\right|\geq0, \quad 1-\left|\frac{m_\phi}{2n_\phi}\mp n^\phi\right|\geq0.$
$G_{\theta\phi}^\pm$	$\pm m_\theta+n_\theta^2+\{\frac{m_\theta-m_\phi}{2(n_\theta-n_\phi)}\pm n_\phi\}^2-1$ $=0$	$\pm(m_\theta-m_\phi)+2n_\theta(n_\theta-n_\phi)\geq0, \quad \mp(m_\theta-m_\phi)-2n_\theta(n_\theta-n_\phi)\geq0,$ $1-\left|\frac{m_\theta-m_\phi}{2(n_\theta-n_\phi)}\mp(n_\theta-n_\phi)\right|\geq0, \quad 1-\left|\frac{m_\theta-m_\phi}{2(n_\theta-n_\phi)}\pm(n_\theta-n_\phi)\right|\geq0.$
$G_\phi^\pm$	$\mp(m_\theta-m_\phi)+(n_\theta-n_\phi)^2+(\frac{m_\theta}{2n_\theta}\pm n_\phi)^2-1$ $=0$	$\pm m_\theta+2n_\theta n_\phi\geq0, \quad \mp m_\theta+2n_\theta(n_\theta-n_\phi)\geq0,$ $1-\left|\frac{m_\theta}{2n_\theta}\pm n_\phi\right|\geq0, 1-\left|\frac{m_\theta}{2n_\theta}\mp n_\theta\right|\geq0.$
$H_\theta^\pm$	$\pm m_\theta+n_\theta-1=0$	$\pm m_\theta\geq0, \quad \pm m_\theta+2n_\theta(n_\theta-n_\phi)\geq0, \quad \pm(m_\theta-m_\phi)+2n_\phi(n_\theta-n_\phi)\geq0.$
$H_\phi^\pm$	$\pm m_\phi+n_\phi^2-1=0$	$\pm m_\phi\geq0, \quad \mp(m_\theta-m_\phi)-2n_\theta(n_\theta-n_\phi)\geq0, \quad \pm m_\theta-2n_\theta(n_\theta-n_\phi)\geq0.$
$H_{\theta\phi}^\pm$	$\pm(m_\theta-m_\phi)+(n_\theta-n_\phi)^2-1=0$	$\pm(m_\theta-m_\phi)\geq0, \quad \mp m_\phi-2n_\theta n_\phi\geq0, \quad \pm m_\theta-2n_\theta n_\phi\geq0.$

2.3 Stress field and velocity field

Lance and Onat (1963) have obtained the equation of compatibility in terms of stress components for conical shells by use of Eqs.(5) and (6). The equation of compatibility for spherical shells is given by

$$(\frac{\partial\Phi}{\partial n_\theta})'\frac{\partial\Phi}{\partial m_\theta}-\frac{\partial\Phi}{\partial n_\theta}(\frac{\partial\Phi}{\partial m_\theta})'-(\frac{\partial\Phi}{\partial n_\phi}\frac{\partial\Phi}{\partial m_\theta}-\frac{\partial\Phi}{\partial n_\theta}\frac{\partial\Phi}{\partial m_\phi})\cot\phi-\omega(\frac{\partial\Phi}{\partial m_\theta})^2\tan\phi-\frac{\partial\Phi}{\partial n_\theta}\frac{\partial\Phi}{\partial m_\theta}\tan\phi=0. \tag{7}$$

The system of equations consists of two differential equations of equilibrium Eq.(1), one differential equation of compatibility Eq.(7) and the yield condition from one line in Table 1. There are four equations for four unknown stresses. Therefore, if initial conditions and a yield condition are given, the profile of the stress points can be determined.

On the other hand, combination of Eqs.(5) and (6) gives

$$\lambda'=[\{(\frac{\partial\Phi}{\partial m_\phi}-\frac{\partial\Phi}{\partial m_\theta}\sec^2\phi)\cot\phi-(\frac{\partial\Phi}{\partial m_\theta})'\}/(\frac{\partial\Phi}{\partial m_\theta})]\lambda, \quad v'=v\cot\phi-\lambda(\frac{\partial\Phi}{\partial n_\theta}-\frac{\partial\Phi}{\partial n_\phi}), \quad w=v\cot\phi-\lambda\frac{\partial\Phi}{\partial n_\theta}. \tag{8}$$

Eq.(8) shows that the profile of λ and the velocity field of the deformations can be determined by giving the initial conditions. When λ is nonnegative and the stress state satisfies the restrictions of their yield condition, the assumed yield condition is appropriate.

General solutions for six hypersurfaces denoted by H have been derived by Flügge and Nakamura (1965). No solutions in closed form could be found for the remaining six hypersurfaces denoted by G so that recourse is made to numerical integration.

3 Analysis

3.1 Statement of the problem
The problem in this paper is described as follows: Problem A. "Find the critical force f and the angle ϕ_β of the outside yield hinge which are statically and kinematically admissible for given ω and ϕ_α." It is convenient to replace this problem by the following equivalent problem: Problem B. "Find the ϕ_α and ϕ_β which are statically and kinematically admissible for given ω and f". Because ϕ_α and ϕ_β appear only in the boundary conditions.

3.2 Boundary condition
Generally, a yield hinge circle where v and w' are discontinuous must be formed at the boundary between a rigid zone and a yield zone for kinematic admissibility. Flügge and Gerdeen (1968) have presented the boundary conditions which determine the existence of yield hinge circles. For our problem, the boundary conditions are given as follows:

$$n_\theta=n_\phi, \quad m_\phi+n_\phi^2-1=0 \quad \text{at } \phi=\phi_\alpha, \tag{9}$$

$$n_\theta=0, \quad -m_\phi+n_\phi^2-1=0 \quad \text{at } \phi=\phi_\beta. \tag{10}$$

The stress state by Eq.(9) stands at the intersection of three sur faces, i.e. $G_{\theta\phi}^+ \cap H_\theta^+ \cap H_\phi^+$, and one by Eq.(10) stands at $H_\phi^- \cap H_{\theta\phi}^- \cap G_\phi^-$. However, even if the ϕ_β value is given, Eq.(10) can not give a unique stress state. So we append the following boundary condition.

$$(h_\phi^-)'=-m_\phi'+2n_\phi n_\phi'=0 \quad \text{at } \phi=\phi_\beta, \tag{11}$$

where h_ϕ^- denotes the yield function of H_ϕ^-. This condition can be proved by the necessary condition of the static admissibility for $\phi \geq \phi_\beta$. Eqs.(10) and (11) give the following stress state at $\phi=\phi_\beta$.

$$n_\phi= \frac{\omega \sin^2\phi -2f -\{(4+\omega^2)\sin^4\phi+4f(\omega+f)-4\}^{1/2}}{2(1+\sin^2\phi)}, \quad m_\phi=n_\phi^2-1, \quad n_\theta=0, \quad m_\theta=0 \quad \text{at } \phi=\phi_\beta. \tag{12}$$

On the other hand, the value of the deformation parameter λ at $\phi=\phi_\beta$ is determined by using of Eqs.(5) and (6) and by giving an arbitrary discontinuous value for $v(\phi_\beta)$. The initial condition for Eq.(8) is

$$v = 1, \quad \lambda=\cot\phi/(\frac{\partial\Phi}{\partial n_\theta}) \quad \text{at } \phi=\phi_\beta. \tag{13}$$

3.3 Transition at the yield hinge circle
When the stress state at $\phi=\phi_\beta-\Delta\phi$ is governed by the yield condition $H_{\theta\phi}^+$, calculations are easy because solutions for $H_{\theta\phi}^+$ can be found in closed form. However, in the case of our problem, the stress state at $\phi=\phi_\beta-\Delta\phi$ is governed by the yield condition G_ϕ^-. No author has presented a transition type of this kind at a yield hinge circle. Special care must be exercised in calculation in the start of the numerical integration by reason that n_θ' which is derived from Eq.(7) is in an indeterminate form as follows:

$$-2(n_\phi'-n_\theta') -2\zeta(n_\phi'+\zeta n_\theta')+(1+\zeta)(2n_\phi-\omega)\tan\phi=0 \quad \text{at } \phi=\phi_\beta, \tag{14}$$

where

$$\zeta=\lim_{\phi\to\phi_\beta} \frac{1}{2n_\theta^2} (m_\theta-2n_\phi n_\theta). \tag{15}$$

n_θ' and ζ at $\phi=\phi_\beta$ can be determined by using

$$(g_\phi^-)''=2(\zeta^2+2\zeta+1)(n_\theta')^2- (h_\phi^-)''=0 \quad \text{at } \phi=\phi_\beta. \tag{16}$$

Combination of Eqs.(14) and (16) gives

$$n_\theta' = \frac{1}{4}\{\eta - 2(-\xi)^{1/2}\}, \quad \zeta = \frac{\eta + 2(-\xi)^{1/2}}{\eta - 2(-\xi)^{1/2}} \quad \text{at } \phi = \phi_\beta, \tag{17}$$

where

$$\xi = \frac{1}{2}[\{(-\sin^2\phi + 3)(n_\phi^2 + 2f\, n_\phi) + 2f^2\}(\cot\phi + \tan\phi)^2 + (1 - \omega\, f)\frac{1}{\sin^2\phi}], \tag{18}$$

$$\eta = 2(n_\phi + f)(\cot\phi + \tan\phi) + (2n_\phi - \omega)\tan\phi \quad \text{at } \phi = \phi_\beta.$$

3.4 Process of analysis
The analysis starts with a choice of ϕ_β. Initial conditions are determined by Eqs.(12) and (17). The Adams-Bashforth stepping method was used to integrate numerically the system of the differential equations.

The stress profile lies on three yield surfaces, G_ϕ^-, G_θ^+ and $G_{\theta\phi}^+$ as follows: $H_\phi^- \cap H_{\theta\phi}^- \cap G_\phi^- \underset{\phi_\beta}{\overset{\cdot}{\to}} G_\phi^- \underset{\phi_1}{\overset{\cdot}{\to}} G_\theta^+ \underset{\phi_2}{\overset{\cdot}{\to}} G_{\theta\phi}^+ \underset{\phi_\alpha}{\overset{\cdot}{\to}} G_{\theta\phi}^+ \cap H_\theta^+ \cap H_\phi^+$,

where ϕ_1 and ϕ_2 are angles corresponding to the intersection of two surfaces. Initial conditions for the surface G_θ^+ and $G_{\theta\phi}^+$ are provided by continuous conditions of n_ϕ, m_ϕ, ε_θ and κ_θ at ϕ_1 and ϕ_2 respectively. The motion of the stress point on the surface $G_{\theta\phi}^+$ continues until it satisfies either condition in Eq.(10). One procedure for the tentative ϕ_β is terminated at this point $\phi = \phi_\alpha$. However, in general, the obtained pair of ϕ_α and ϕ_β is not a kinematically admissible condition. Because it requires that both conditions in Eq.(10) are satisfied.

By repeating the procedure for various choices of ϕ_β, the exact ϕ_α value which fulfills both conditions in Eq.(10) is obtained as shown in Fig.3. The exact ϕ_α attains its minimum value on the ϕ_α-ϕ_β curve. The reason of this phenomenon is considered as follows:

The collapse load may increase as ϕ_α increases. Therefore, the collapse load corresponding to a certain ϕ_α value can not be kinematically admissible for $\phi > \phi_\alpha$. This means that the exact ϕ_α must be its minimum value on the ϕ_α-ϕ_β curve.

3.5 Numerical calculation results
Numerical calculation results are illustrated for the case of $\omega = 50$. The stress profiles projected on the n_ϕ-m_ϕ plane for the various f are shown in Fig.4. The stress fields for f=0.04, 0.1, 0.2 and 0.5 are

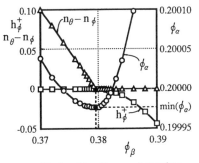

Fig.3. Iterative calculation

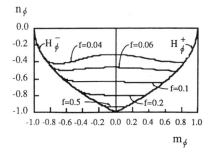

Fig.4. Stress profiles on n_ϕ-m_ϕ plane

shown in Fig.5. The corresponding deformation parameter λ and the deformation velocity fields are shown in Fig.6. The stress state changes to a state of membrane forces being superior from ones of bending moments as ϕ_α increases. When the dimensionless collapse load f is small, i.e. the corresponding ϕ_α is small, the distribution of bending moments is close to that obtained for a clamped circular plate subjected to a concentrated force at its centre. Fig.6 shows that λ is nonnegative for $\phi_\alpha \leq \phi \leq \phi_\beta$; that is to say, λ satisfy the necessary condition for the kinematically admissible state. λ is discontinuous at ϕ_1 and ϕ_2 because the intersection of the two hypersurfaces is not smooth.

Fig.7 shows the f–ϕ_α relations for ω=20~100. Fig.8 shows the relations between a yield zone angle $\phi_y=\phi_\beta-\phi_\alpha$ and ϕ_α. Fig.7 shows that the dimensionless collapse load f increases as ω decreases for a fixed ϕ_α, and this difference in ω decreases as ϕ_α increases. The reason of this tendency is that bending moments are superior as ϕ_α is small and membrane forces are superior as ϕ_α is large. Fig.8 shows that ϕ_y decreases as ϕ_α increases except a neighbourhood of ϕ_α=0. The yield zone angle ϕ_y increases as ω decreases for a fixed ϕ_α.

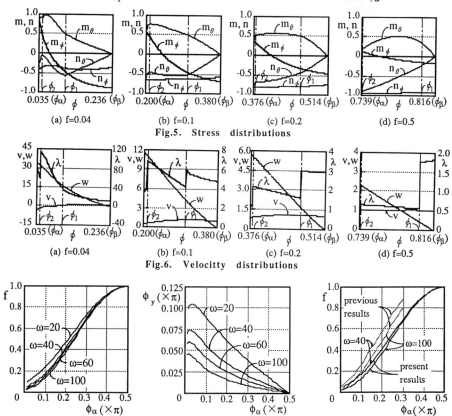

Fig.5. Stress distributions

Fig.6. Velocitty distributions

Fig.7. f–ϕ_α relationship   Fug.8. f–ϕ_y relationship   Fig.9. Comparison with previous analyses

4 Comparison with previous analysis results

The authors (1987) have reported the calculation results based on the upper-bound theorem for this problem, using an approximate yield condition derived for sandwich shells by Hodge (1959). The approximate yield condition is a polyhedron composed of twelve hyperplanes which is inscribed in the exact yield hypersurface. Fig.9 shows the comparison between the two analyses in the case of $\omega=40$ and 100. The collapse load obtained in the previous paper is higher than the one in this paper. However, the tendency of the increase in f with the increase in ϕ_α in the previous paper is nearly equal to that in this paper as ϕ_α values are large.

5 Comparison between tests and analyses

A ball is fabricated from two cold-formed hemispheres. Test specimen and test set up are shown in Fig.10. The dimensions of specimens are listed in Table 2.

The maximum load Fmax and the yield load Fy obtained from the tests are made dimensionless as follows:

$$f_{max}=\frac{F_{max}}{\pi\,(D{-}T)\,T\,\sigma_y}, \qquad f_y=\frac{F_y}{\pi\,(D{-}T)\,T\,\sigma_y}, \tag{19}$$

where Fy is defined as the load when the joint stiffness in the load-deformation curve becomes 1/3 of the initial stiffness. D and T denote a ball diameter and thickness respectively and σ_y is the yield stress of the tensile test specimen taken from a ball.

Table 2. Summary of test specimen

Specimen	Ball $D \times T$(mm)	Tube $d \times t$(mm)
M36-0.12		$\phi\,25.0$
M36-0.23		50.0×9.0
M36-0.35	216.3×6.0	75.0×7.0
M36-0.53		114.3×5.6
M36-0.65		139.8×6.0
M48-0.35	216.3×4.5	75.0×7.0
M48-0.53		114.3×5.6
M24-0.35	216.3×9.0	75.0×12.0
M24-0.53		114.3×12.0
S37-0.36	165.2×4.5	60.0×6.0
L35-0.36	318.5×9.0	114.3×12.0

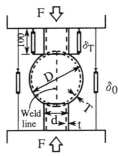

Fig.10. Test specimen

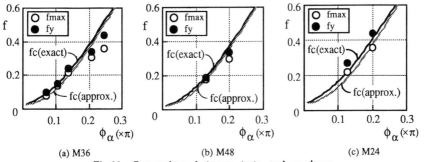

(a) M36   (b) M48   (c) M24

Fig.11. Comparison between tests and analyses

Fig.11 shows the relationship between test results and analytical results for the M36, M48 and M24 series. Here, the loading angle ϕ_α of test specimens is calculated by assuming that the loading position is at the weld toe of the tubes.

The test results sufficiently correspond to the analyses independent of D/T. The mean value of fmax/fc for all specimens is 1.00 and the COV is 0.12, while the mean value of fy/fc is 0.86 and the COV is 0.10.

6 Practical formula for predicting the strength

It is not practical to obtain the strength of the various balls by the numerical calculation method. Therefore, in order to predict the strength of balls, we propose the practical formula which gives a lower bound for the exact collapse load.

The ϕ_β value in Eq.(12) becomes the minimun value as $n_\phi = -1$, i.e.

$$\min(\phi_\beta) = \arcsin\{\frac{(2+\omega)\,f-2}{\omega}\}^{1/2}. \tag{20}$$

For the purpose of obtaining the lower bound, we replace $\min(\phi_\beta)$ by ϕ_α in Eq.(20). The approximate collapse load is

$$f = \frac{2}{2+\omega}\cos^2\phi_\alpha + \sin^2\phi_\alpha. \tag{21}$$

The values by Eq.(21) are shown as fc(approx.) in Fig.11. This formula gives an excellent approximation to the exact solution. The mean value of fmax/fc(approx.) for all specimens is 1.12 and the COV is 0.16, while the mean value of fy/fc(approx.) is 0.97 and the COV is 0.14.

7 Conclusions

1. The exact collapse load is obtained for balls subjected to uniaxial force by the limit analysis.
2. The test results correspond sufficiently to the analytical results.
3. The practical formula for predicting the strength of balls is proposed. The formula gives an excellent approximation to the exact solution.

8 References

Flügge, W. and Gerdeen, J.C. (1968) Axisymmetric plastic collapse of shells of revolution according to the Nakamura yield criterion, in **Proceedings of the twelfth International Congress of Applied Mechanics,** pp.209-220

Flügge, W. and Nakamura, T. (1965) Plastic analysis of shells of revolution under axisymmetric loads, in **Ingenieur-Archiv,** 34. Band, 4.Heft, pp.238-247.

Hodge, P.G,Jr. (1959) **Plastic analysis of structures.** McGraw-Hill Book Co., Inc.,.

Lance, L.C. and Onat, E.T. (1963) Analysis of plastic shallow conical shells, in **Jurnal of Applied Mechanics,** vol.30, pp.199-209.

Onat, E.T. and Prager, W. (1954) Limit analysis of shells of revolution, in **Proceedings of The Royal Netherlands Academy of Science,** vol.B57, pp.535-548.

Tabuchi, M. Kanatani, H. Kamba, T. and Tanaka, T. (1987) The strength of ball joints in space trusses, in **Safety Criteria in Design of Tubular Structures,** Proc. of Int. Meeting,Tokyo, Japan, pp.361-372.

Timoshenko, S.P. and Woinowsky-Krieger, S. (1940) **Theory of plates and shells.** Second ED. McGraw-Hill Book Co., Inc.,.

PART 12

STIFFENED JOINTS

Strength and fatigue behaviour

45 ULTIMATE BEHAVIOR OF DIAPHRAGM-STIFFENED TUBULAR KK-JOINTS

Y. MAKINO, Y. KUROBANE
Kumamoto University, Japan
J.C. PAUL
Obayashi Corporation, Tokyo, Japan

Abstract
This paper discusses the ultimate behavior of two types of diaphragm stiffened double K–joints to be used for long–span trusses. One type of stiffened joint has three stiffening diaphragms, a central diaphragm passing through the chord and two diaphragms welded to inside surfaces of the chord. The other type has only one central diaphragm. Unstiffened double K–joints were also tested. Two cases of loading were considered, the case of gravity load acting on truss frames and that of horizontal loads perpendicular to the spanning direction of truss frames.

Failure modes of joints each with one diaphragm were similar to those of un–stiffened double K–joints. The ultimate capacity of these joints was found to be predictable by utilizing the ultimate capacity equation for unstiffened joints. The joints with three diaphragms failed owing to either tensile failure in welds or shear failure of the chord.
Keywords: Tubular Joint, Multiplanar Joint, Stiffened Joint, Ultimate Behaviour, Failure Mode, Ultimate Capacity.

1 Introduction

The benefits of using circular hollow sections with large diameter to thickness ratios in space trusses have been recognized by designers and architects. However, an increase in the compressive strength of members is earned at the expense of reducing the ultimate resistance of unstiffened tubular joints. The most common joint type used in 3D trusses is a spherical joint composed of a spherical node, bolts and end–cones, which is called ball joint/system joint. However, dimensions of spherical nodes, manufacturing of high strength thick rods for bolts and other problems prevent system joints from being used in space trusses with large diameter members.

This paper proposes methods of stiffening tubular joints using diaphragms. These joints composed of members with large diameter to thickness ratios were found to be advantageous as a design technique. Furthermore, it is found that the ultimate capacities of these stiffened joints could be predicted using the existing ultimate capacity formulae for unstiffened joints.

2 Test

2.1 Specimens
The specimens are unstiffened joints and two types of diaphragm–stiffened joints.

Tubular Structures V. Edited by M.G. Coutie and G. Davies.
© 1993 E & FN Spon, 2–6 Boundary Row, London SE1 8HN. ISBN 0 419 18770 7.

The first type has one diaphragm passing through a chord to which two half lengths of the chord are welded. The second type has two additional diaphragms welded to the inside surfaces of a chord at cross sections where two braces are attached. The second type of specimen is shown in Fig. 1 with definitions of symbols. The measured dimensions are given in Table 1. The direction cosine $\vec{n}$ of a brace axis with respect to the coordinate axes shown in Fig. 1 is:

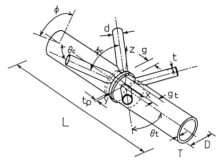

Fig. 1. Specimen and symbols.

$$\vec{n} = (\sqrt{21}/7, \sqrt{7}/7, \sqrt{21}/7)$$

In the specimen designation, the symbol DK means Double K-joints, the symbols S and U denote Loading Case 1 (gravity load : symmetric loads) and Loading Case 2 (horizontal load perpendicular to the spanning direction : unsymmetric loads), respectively. N, S1 and S2 denote unstiffened, stiffened type 1 (single central diaphragm) and stiffened type 2 (three diaphragms) joints, respectively. NP denotes the trial test carried out to check the test set-up performance (pilot test). The data obtained from this test was reliable and hence was used for the unstiffened joint data (Loading Case 1). The NP test for Loading Case 2 was considered unreliable because of errors related to support conditions and was therefore ignored. Due to the thick brace walls, the brace members did not buckle before the joint failed in the test.

Table 1. Dimensions of specimens (measured)

Specimen Designation	D	T	d	t (mm)	t_p (mm)	g	g_t (mm)	θ_c	θ_t	ϕ (deg.)
DK-S-N	215.9	4.4	101.7	5.2	–	52.8	7.9	49.1	49.1	60
DK-S-S1	215.9	4.4	101.7	5.2	5.6	56.1	7.6	49.1	49.1	60
DK-S-S2	215.9	4.4	101.7	5.2	5.6	56.3	7.4	49.1	49.1	60
DK-U-N	215.9	4.4	101.7	5.2	–	52.7	7.9	49.1	49.1	60
DK-U-S1	215.9	4.4	101.7	5.2	5.6	55.8	7.3	49.1	49.1	60
DK-U-S2	215.9	4.4	101.7	5.2	5.6	56.4	7.7	49.1	49.1	60

Note: Dimensions of DK-S-NP specimen were supposed to equal to those of DK-S-N.

Table 2. Results of tensile coupon test

Specimen Nominal Size	σ_y (kN/mm^2)	σ_u (kN/mm^2)	Elongation (%)	Grade of Steel
Chord(216.3ϕ×4.5)	388	470	35.2	STK400
Brace(101.6ϕ×5.7)	426	473	29.8	STK400
Plate(t=6mm)	321	453	38.5	SS400

2.2 Test set-up

Two test devices were used in accordance with the loading cases. The two loading cases are shown in Figs. 2(a) and (b) with the support conditions. The corresponding test set-ups are shown in Figs. 3(a) and (b), respectively. In both cases, the load is applied from a frame through a 100 ton hydraulic jack and a spreader beam to the tension braces. Universal joints at the ends of compression braces allow rotation in all directions. In Loading Case 1, the chord is free at one end and has a pin at the other end allowing in-plane rotation of the truss. In Loading Case 2, the chord is braced at both ends only to prevent the rigid body rotation of the chord in the horizontal plane during testing.

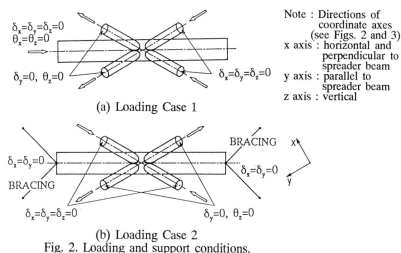

(a) Loading Case 1

Note : Directions of
coordinate axes
(see Figs. 2 and 3)
x axis : horizontal and
perpendicular to
spreader beam
y axis : parallel to
spreader beam
z axis : vertical

(b) Loading Case 2
Fig. 2. Loading and support conditions.

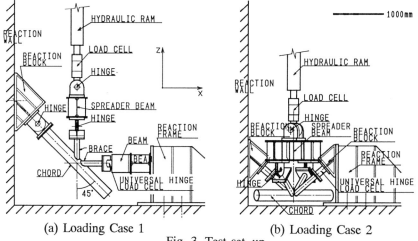

(a) Loading Case 1   (b) Loading Case 2
Fig. 3. Test set-up.

2.3 Measuring system for loads and deformations

Loads were measured by a load cell connected to the hydraulic ram and by load cells attached at the compression brace ends. Strain gauges were placed on the chord and braces to measure the axial and the bending stresses. Stress measure-ments by strain gauges were necessary to monitor the differences between the loads in the two compression braces which arose when a test specimen was not set accu-rately into position. Because of difficulties in measuring local deformations of the joints, displacements of the ram relative to the support points were measured instead.

3 Test results

3.1 Load–deformation relationships and ultimate resistances

The test results are summarized in Table 3. The load–deformation relationships are shown in Figs. 4(a) and (b), where the load is that recorded by the load cell at the hydraulic ram and the deformation direction is that of the ram cylinder. The ultimate capacities are denoted by circles.

Loading Case 1 : The initial stiffness determined from the pilot test is low because of loose connections between the measuring devices and the test set–up. The ultimate capacities of stiffened type 1 and type 2 joints are respectively about 1.5 and 1.9 times those of unstiffened joints. As for the post peak load–deformation relations, the unstiffened and stiffened type 1 joints exhibit a gradual decay in the load, whereas sudden failure due to shear in the chord governed the capacity of the stiffened type 2 joint.

Loading Case 2 : The ultimate capacities of stiffened type 1 and type 2 joints were about 1.3 and 2.2 times those of unstiffened joints. The loads declined owing to cracks extending along the weld toes or in the welds.

Table 3. Test results and predicted ultimate capacities

Specimen Designation	$_t P_u$	$_c P_u$ (kN)	$_{chord} P_u$	δ_u (mm)	K (kN/mm)	Paul et al. (1993) $_c P_u / P_{u,c}$	Kurobane et al.(1984) $_c P_u / P_{u,c}$
DK–S–NP	200	179	498	8.2	101	1.02	–
DK–S–N	201	189	511	5.9	143	1.08	–
DK–S–S1	288	284	749	15.4	135	1.42	–
DK–S–S2	371	361	956	19.5	168	–	–
DK–U–N	207	212	–	26.7	58	–	1.25
DK–U–S1	268	278	–	22.0	66	–	1.32
DK–U–S2	459	445	–	18.7	85	–	–

$_t P_u$: Axial force in tension brace at ultimate load
$_c P_u$: Axial force in compression brace at ultimate load
$_{chord} P_u$: Axial force in chord at ultimate load
δ_u : Deformation at ultimate capacity
K : Initial stiffness
$P_{u,c}$: Predicted ultimate brace load

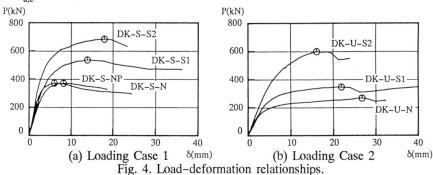

(a) Loading Case 1 δ(mm)   (b) Loading Case 2 δ(mm)

Fig. 4. Load–deformation relationships.

3.2 Failure modes

Loading Case 1: No local deformation of the chord wall in the crotch area between the two compression braces occurred because of the narrow transverse gap. Since the compression braces acted as one member, the chord cross section after collapse profiled as a uniplanar K–joint with a large value of d/D ratio, as shown in Fig. 5(a). The failure mode of DK–S–S1 is similar to that of DK–S–N except that the central diaphragm restrained the deformation of the chord wall. The out–of–plane deformations of the diaphragm occurred as shown in Fig. 5(c). Cracks occurred at the weld toes on the chord crown connecting the tension braces, as was the case for unstiffened joints. For DK–S–S2 specimen, the internal diaphragms restrained the chord wall effectively from local deformations produced by the axial brace forces. The collapse of DK–S–S2 joint was caused by shear stresses in the chord section between the central and the internal diaphragm on the tension brace side.

Loading Case 2: In the unstiffened specimen DK–U–N, local deformations were observed in the area between the compression braces forming a wrinkle running diagonally (the area shown by ① in Fig. 6). The central axis of the chord was

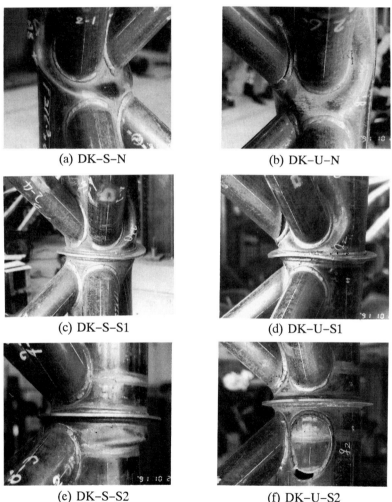

(a) DK–S–N   (b) DK–U–N

(c) DK–S–S1   (d) DK–U–S1

(e) DK–S–S2   (f) DK–U–S2

Fig. 5. Joint deformation after testing.

distorted and curved in the joint area as shown in Fig. 6. Cracks at the weld toes connecting the tension braces appeared in proximity of the maximum load. After reaching the maximum load, cracks extended rapidly and the joint failed. It should be noted that cracks occurred on the tension brace surfaces at the weld toes and not on the chord surface, as shown by ② in Fig. 6. Failure mode of DK–U–S1 specimen was similar to that of DK–U–N except that the deformation of area ① was restrained by the diaphragm. The diaphragm sustained out–of–plane bending with the distortion of the chord axis. In DK–U–S2 specimen, local deformations of the chord were small because of the restraint imposed by the three diaphragms. Final collapse was due to a crack extension along the weld toes on the tension brace surfaces.

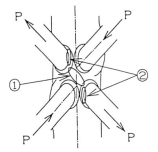

Fig. 6. Joint deformation for Loading Case 2.

4 Estimation of ultimate capacities

4.1 Unstiffened joint in Loading Case 1 (DK–S–N, DK–S–NP)

Ultimate capacities in Loading Case 1 for unstiffened joints were estimated using the existing formula for KK–joints by Paul et al.(1993). As no local deformation of the chord wall in the crotch area between the compression braces in this test took place, the ultimate formulae $P_{u,c}$ was derived as follow:

$$P'_{u,c}=0.746(1+0.693(d/D))(1+0.741(g/D))_K P'_{u,c} \qquad (1)$$

where $_K P'_{u,c}$ is the predicted ultimate capacity of uniplanar K–joints replacing two compression braces by one brace with diameter equal to the transverse distance between the saddles on the outside of the braces.

Herein, $P_{u,c}=P'_{u,c}/(2\cos(\phi_c/2))$, where $P_{u,c}$ is the estimated ultimate brace load. The predicted ultimate capacities are shown in Table 3.

4.2 Unstiffened joint in Loading Case 2 (DK–U–N)

No existing data for Loading Case 2 is found except for this test. The ultimate capacity of DK–U–N joint is greater than that of the uniplanar joint predicted by formula of Kurobane et al.(1984), as compared in Table 3. Furthermore the capacity of DK–U–N is 15% greater than that of the same joints under symmetrical loading (DK–S–N and DK–S–NP). The significant difference in joint capacity between the two loading cases indicates a trend opposite to the chord ovalizing parameter recommended in AWS Code(1992). Namely, AWS Code predicts a smaller resistance for joints under antisymmetrical loads when compared to symmetrically loaded joints with $\phi=60°$.

4.3 Stiffened type 1 joint in Loading Case 1 (DK–S–S1)

In the stiffened joint with one central diaphragm, the ultimate capacity was estimated by the unstiffened KK–joint formula (Paul et al.(1993)) after replacing the longitudinal gap between two braces with the distance between the diaphragm and the compression brace. The predicted ultimate capacity is given in Table 3. Assuming that the gap length is equal to zero millimeter, the predicted ultimate capacity is 286 kN, which approximates the test result. The ultimate capacity for stiffened type 1 joints may be estimated by adjustment of the longitudinal gap length.

4.4 Stiffened type 1 joint in Loading Case 2 (DK–U–S1)

Again, the ultimate capacity was compared with that of the uniplanar joint in which the gap length of K–joint was replaced by the distance between the diaphragm and the compression brace as for the stiffened type 1 joint for Loading Case 1. The predicted ultimate capacity is shown in Table 3.

4.5 Stiffened type 2 joint in Loading Case 1 (DK–S–S2)

Failure of DK–S–S2 specimen was caused by shear in the chord between the central diaphragm and the internal one on the tension brace side as shown in Fig. 5(e). The ultimate shear resistance of the chord according to Wardenier(1982) is represented as follows:

$$Q_p = 2(D-t) \, t \, \tau_y \qquad (2)$$

Herein, τ_y is the shear yield stress which is given below considering the effect of the axial stresses:

$$\tau_y = \frac{\sigma_y}{\sqrt{3}} \sqrt{1-(N/N_y)^2} \qquad (3)$$

The shear resistance is given in Table 4. Additionally, the ultimate capacity of DK–U–S2 represented by a simple model was calculated by using a nonlinear FEM package. The analytical model was a drum with two attached diaphragms loaded by a couple of shear forces. Calculations were carried out by using the program COSMOS/M (Lashkari(1990)), in which four node quadrilateral thick shell elements were employed using the displacement control method. The analysis results corresponded closely with those of the test as is shown in Table 4.

Table 4. Comparison between tested and estimated ultimate capacities

Specimen Designation	Test Result Q_u(kN)	Wardenier (1982) Q_u/Q_p	FE Analysis $Q_u/Q_{u,c}$
DK–U–S2	472	1.15	1.00

Herein, Q represents the shear force on the chord section in the joint, which is the component of the brace axial force perpendicular to the chord axis.

4.6 Ultimate capacity of weld between chord and tension brace

The ultimate capacity of DK–U–S2 joint was governed by cracks which initiated in the weld toes on the tension brace surfaces as shown by ② in Fig. 6. Local deflections that the chord wall sustained before cracks started were smaller than those for the other specimens as observed from the load–deformation curves and also from observation of the joints after failure. According to an examination of weld profiles after testing, the throat thickness was found to be 5.2 mm in crotch area, which is considered as an appropriate size compared with the chord wall thickness and brace wall thickness. The average stress over total throat area at the ultimate joint capacity was calculated on the assumption that the throat thickness is equal to 5.2 mm measured in the crotch area throughout the total weld length surrounding the tensile brace end. Its value was equal to 86% of the maximum shear strength ($\sigma_u/\sqrt{3}$) of the chord/brace material, which shows that the welds were not fully effective in carrying a tensile load in the brace. No doubt stresses concentrated at points of crack initiation because a diaphragm was welded at these locations to the inside of the chord. An effective weld length approach may be a feasible solution to prevent a premature failure of joints due to failures in welds, although this requires further study.

5 Conclusions

In this paper, the improvement in ultimate capacity of tubular KK–joints, which are strengthened by appropriate stiffening diaphragms, is investigated. It is indicated that the ultimate capacities of stiffened tubular joints can be estimated by the utilization of existing formulae. The following results may be concluded:
(1) Stiffening by diaphragms improves the ultimate capacity of joints and is quite appealing in terms of fabrication and cost.
(2) The ultimate capacities of unstiffened KK–joints for Loading Case 1 can be estimated closely by the existing formulae (Paul et al.(1993)).
(3) The joints of stiffened type 2 for Loading Case 1 failed by shear force in the chord between both braces. The ultimate capacities of these joints can be estimated from the shear strength of the chord.
(4) In unstiffened and stiffened type 1 joints for Loading Case 2 and stiffened type 1 joints for Loading Case 1, the ultimate capacities can be estimated approximately by application of the existing formulae for K– or KK–joints. However, no accurate equations for the ultimate capacity are attained at this stage.
(5) According to the recommendations of Architectural Institute of Japan(1990), the strength of welds conforming with the specified size requirements is greater than the joint capacity of unstiffened joints. However, in this study, there were cases where the strength of the welds was lower than the joint capacity.
In order to establish ultimate capacity formulae for Loading Case 2 further research is needed as the influencing parameters have not been investigated in depth. Further, it may be necessary that tests with the other failure mode which introduces local deformation of the chord wall in the crotch area should be conducted.

Acknowledgements

This investigation was accomplished using the test data of a study on retractable roof trusses performed at Kumamoto University in cooperation with Hazama Corporation. The test apparatus was provided by Miike Manufacturing Co. Ltd. Financial assistance was provided also by Nippon Steel Corporation. All support is greatly acknowledged and appreciated.

References

American Welding Society (1992) **Structural Welding Code–Steel**, 14th edition, ANSI/AWS D1.1–92, Miami, USA
Architectural Institute of Japan (1990) **Recommendations for the Design and Fabrication of Tubular Structures in Steel**.
Kurobane,Y. Makino,Y. and Ochi,K. (1984) Ultimate Resistance of Unstiffened Tubular Joints. **Journal of Structural Engineering**, ASCE, Vol.110, NO.2
Lashkari,M. (1990) COSMOS/M User Guide Release Version 1.6.
Makino,Y. Kurobane,Y. Takagi,M and Hori,A (1993) Diaphragm Stiffened Multiplanar Tubular Joints for Retractable Roofs. in **Proc. Int. Conf. Space Structures**, Guildford, UK.
Paul,J.C. Makino,Y. and Kurobane,Y. (1993) New Ultimate Capacity Formulae for Multiplanar Joints. in **Proc. 5th Int. Symposium on Tubular Structures**, Nottingham, UK.
Wardenier,J. (1982) **Hollow Section Joints**. Delft Univ. Press.

46 CORROSION FATIGUE BEHAVIOUR OF STIFFENED TUBULAR JOINTS

P. GANDHI, G. RAGHAVA, D.S. RAMACHANDRA MURTHY
and A.G. MADHAVA RAO
Structural Engineering Research Centre, Madras, India
P.F. ANTO
Institute of Engineering and Ocean Technology, Panvel, India

Abstract
Constant amplitude corrosion fatigue tests were conducted on stiffened
steel tubular T and Y joints to study the behaviour in a synthetic
seawater environment under freely corroding conditions. The corrosion
fatigue tests were conducted at 0.2 Hz frequency which is representative
of sea wave frequency. The test results are presented in this paper.
The results are compared with standard S-N curves recommended by
various codal authorities.
Keywords: Corrosion fatigue, Crack initiation, Crack propagation, Fatigue
life, Offshore structures, Ring stiffeners, Stress concentration, Synthetic
seawater, Tubular joints.

1 Introduction

Exploration and production of oil in deeper waters and more hostile environments have become more and more necessary because of
ever-increasing global demand for oil. In India, there have been
dramatic developments on the oil front during the last two decades with
the discovery of large reserves of crude oil and natural gas, mostly in offshore areas. The majority of offshore production of oil is from the
Bombay High region. At present, there are about 120 offshore platforms
in this region, engaged in the production of oil, all of them being jacket
type structures. Because of the repeated loading due to sea waves,
fatigue is an important factor to be considered in the design of these
structures. Because of the constant exposure of these structures to the
seawater environment, there is a continuous threat of degradation of the
structural material due to corrosion. The combined action of cyclic loading and aggressive environment often results in significant reduction in
the fatigue performance, compared with that obtained under cyclic loading in inert environments. Hence, proper understanding of the corrosion
fatigue behaviour of offshore structures is very important. Fatigue
failure in such structures normally initiates at joints. Hence, knowledge
about the fatigue strength of the joints helps us in evaluating the fatigue
life of these structures.

Tubular Structures V. Edited by M.G. Coutie and G. Davies.
© 1993 E & FN Spon, 2–6 Boundary Row, London SE1 8HN. ISBN 0 419 18770 7.

It has been found that, in a corrosive environment, there is a continuous decrease in fatigue resistance as cyclic life increases. In other words, no knee or endurance limit is noticed for unprotected steel. High stress ratios reduce fatigue resistance at low values of the total hot spot strain range. Slower frequencies cause reduced fatigue resistance for unprotected steel at intermediate values of strain range. Corrosion fatigue tests in a seawater environment are conducted at the natural sea wave frequency of 0.17 to 0.20 Hz (Raghava and Madhava Rao, 1992). It means that the fatigue tests take a much longer time and are more expensive than those in air and this is probably one reason for the limited number of studies reported on corrosion fatigue, particularly with regard to tubular joints.

Optimum cathodic protection restores fatigue resistance to the levels observed in air. The recommended cathodic potential is about -0.85 V Ag/AgCl. Over protection can be detrimental and can increase crack growth rates.

The main seawater characteristics which influence corrosion fatigue are oxygen level, temperature and pH level. The higher the dissolved oxygen content, the lower will be the corrosion fatigue strength. Similarly, the higher the temperature, the lower will be the fatigue strength. Crack growth rate increases at higher temperatures. There is little effect of pH on fatigue resistance over a broad range of values, 4 to 10. Low values of pH (< 4) decrease fatigue resistance, while high values (> 10) improve fatigue resistance to levels similar to those observed in air, particularly with respect to crack initiation.

Considering many variations in seawater from place to place as well as at different depths at the same location, laboratory tests are usually conducted in a synthetic seawater environment prepared as per ASTM specifications (ASTM D1141-1975). This will enable all the test results to be evaluated on a common footing. Typically, seawater will have 5 to 10 ppm dissolved oxygen, 30 to 35 ppt salinity and 7.8 to 8.2 pH level. The seawater characteristics in Bombay High also lie generally within these ranges except the dissolved oxygen content which is lower. Variations within these ranges generally have little influence on corrosion fatigue behaviour. The temperature of seawater in the Bombay High region is higher ($22.1^{\circ}C$ to $29.7^{\circ}C$) when compared to that in the European countries, particularly the North Sea conditions, for which most of the corrosion fatigue results are available. This aspect has to be considered while making recommendations for Indian conditions (Ramachandra Murthy et al, 1992 b).

Extensive analytical and experimental work carried out at the Structural Engineering Research Centre (SERC), Madras, India, have shown that internally ring stiffened tubular joints are structurally more efficient than unstiffened joints. Provision of internal rings helps in reducing stress concentration and improving ultimate and fatigue strengths in air (Gandhi et al, 1991; Ramachandra Murthy et al, 1992 a). In order to evaluate the fatigue performance of these joints in a seawater environment, corrosion fatigue tests have been undertaken at SERC, Madras. This paper describes the results of corrosion fatigue tests.

2 Experimental investigations

2.1 Static tests

Static tests on stiffened T and Y joints were conducted to study stress distribution around the intersection of chord and brace members. These joints were tested under axial loading of the brace. The specimens were extensively instrumented in both chord and brace members with 2 mm rosette strain gauges. The srains were measured at various loading stages. From the measured strain values, principal stresses were calculated, and were extrapolated to the weld toe. The stress concentration factor (SCF) which is the ratio of the maximum stress at the intersection to the nominal stress in the brace away from the intersection was calculated for both the joint types. A typical T joint specimen is shown in Fig. 1. The stress distribution obtained around the intersection of chord and brace tubulars was very useful in locating the hot spots, which are the points of maximum SCF. Figure 2 shows a typical stress distribution plot for a stiffened Y joint.

2.2 Corrosion fatigue tests

Corrosion fatigue tests were conducted by using a sophisticated electro-hydraulic servo-controlled fatigue testing system. The fatigue loads were applied through ± 500 kN and ± 1000 kN capacity actuators and the loading was controlled by control consoles which were interfaced with computer systems. Four stiffened steel tubular T and Y joints were tested under constant amplitude, and at stress ratios of -1 and 0. All the corrosion fatigue tests were conducted under freely corroding conditions. The test frequency was 0.2 Hz. The details of test specimens are given in Table 1.

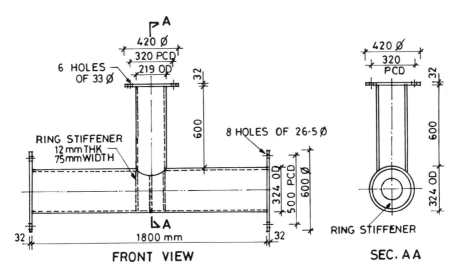

Fig. 1. Typical T joint specimen

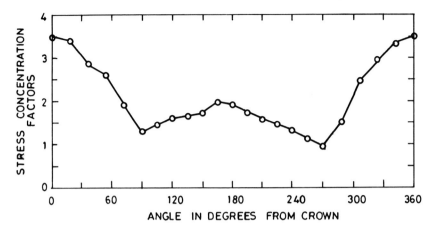

Fig.2. Stress distribution around the intersection for chord member

The intersection portion of the tubular joint was enclosed in a corrosion chamber containing synthetic seawater. The corrosion chamber was fabricated by using transparent perspex sheets. The capacity of the chamber was 500 litres, approximately. Biologically inert synthetic seawater was prepared as per ASTM D1141 specifications without the addition of heavy metals (ASTM D1141-1975). To maintain the level of dissolved oxygen, the seawater was continuously circulated at a rate of about one litre per minute. The temperature of the synthetic seawater was $25^{o}C \pm 5^{o}C$. The pH and dissolved oxygen contents were frequently monitored for the desired levels.

Table 1. Details of test specimens

Sl No	Model No	Material	D	T	d	t	B	Ts	Angle (°)
1	Ts1	IS 226	320	12	220	8	64	10	90
2	Ts2	IS 226	320	12	220	8	64	10	90
3	Ys1	API5L GB	324	12	220	8.18	75	12	60
4	Ys2	IS 226	320	12	220	8	64	10	60

D and T are diameter and thickness of chord member
d and t are diameter and thickness of brace member
B and Ts are width and thickness of ring stiffener
All linear dimensions are in mm

2.2.1 Fatigue crack monitoring

On-line fatigue crack growth measurements were taken at regular intervals during the fatigue tests. To help in crack growth measurement, steel pins/probes of 1 mm diameter were spot welded, at 10 mm spacing along the intersection in both chord and brace members, at hot spot locations. The hot spot locations were based on stress distribution obtained from static tests. The steel probes were connected to a multichannel switching unit, which in turn was connected to a microprocessor based crack microgauge. The crack microgauge works on ACPD (Alternate Current Potential Difference) technique. The whole system was fully automated allowing non-stop crack growth data logging for prolonged periods, typically for several months. The crack depths were monitored at 96 separate sites at the hot spot locations of both chord and brace members. All the tubular joints were tested under axial loading of the brace under freely corroding conditions. The hot spot stress range and the stress ratio were varied for the different specimens. Out of the four specimens tested, two specimens were tested at a stress ratio of -1 and the other two specimens were tested at a stress ratio of 0. In order to evaluate the effect of stress range on fatigue life, the joints were tested at different stress ranges.

The significant stages in the fatigue life of tubular joints are:

Ni : Number of cycles to first visible cracking as detected by visual examination or NDT technique; considered as crack initiation life.

Nc : Number of cycles at which full thickness crack occurs; considered as crack propagation life.

Nf : Number of cycles to complete failure of the specimen.

For design purposes, crack propagation life, Nc, is considered as the fatigue life for tubular members. This is based on the assumption that the joint loses considerable stiffness when the crack reaches the full thickness of the tubular member.

Table 2.   Corrosion fatigue test results

Sl No	Model No	SCF Expt	Hotspot stress range (MPa)	Stress ratio	Fatigue life ($\times 10^5$)		
					Ni	Nc	Nf
					(number of cycles)		
1	Ts1	2.93	233	-1	---	---	0.173
2	Ts2	2.93	102	0	1.814	11.680	12.248
3	Ys1	3.73	204	-1	0.310	1.110	1.290
4	Ys2	3.50	204	0	0.970	1.540	2.030

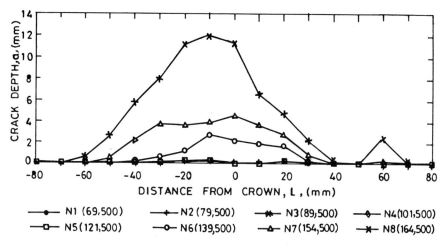

Fig.3. Crack profile development for a stiffened Y joint (YS2)

Results of the four corrosion fatigue tests on T and Y joints are given in Table 2. On-line fatigue crack growth data were obtained for all the joints. A typical crack profile development for a Y joint, YS2, is shown in Fig.3. It can be seen from the figure that multiple cracks initiate at the weld toe at different locations and these multiple cracks grow along both thickness and length directions.

3 Discussion of test results

The crack initiation life, Ni, and the crack propagation life, Nc, were obtained for all the specimens except for the specimen Ts1, which failed prematurely because of weld defect. The locations of fatigue crack initiation for all the specimens were the hot spot locations. The specimens

Table 3. Comparison of experimental results with S-N curves recommended by various codes

Sl No	Model No	Hotspot stress range (MPa)	Fatigue life (x10^5)					% Reduction			
			Expt (Nc)	API	DEn	DnV	NPD	API	DEn	DnV	NPD
1	Ts1	233	0.17	0.49	1.52	0.73	0.34	−65	−89	−76	−49
2	Ts2	102	11.68	18.33	18.20	21.61	9.23	−36	−36	−46	+26
3	Ys1	204	1.11	0.88	2.27	1.26	0.55	+26	−51	−12	+99
4	Ys2	204	1.64	0.88	2.27	1.26	0.55	+87	−28	+31	+195

were tested beyond the crack propagation life till complete failure in order to study their remaining life . The final failure lives are also given in Table 2. It can be seen from the table that the remaining lives are between 5 per cent to 24 per cent. Because of the limited test data available at present, a firm conclusion can not be drawn on this aspect.

The experimental fatigue lives obtained for the tubular T & Y joints are compared with the S-N curves recommended by various codal authorities for design of offshore structures (UEG, 1985), namely, the American Petroleum Institute (API), the UK Department of Energy (DEn), Det norske Veritas (DnV) and the Norwegian Petroleum Directorate (NPD). These S-N curves are basically for unstiffened tubular joints based on air fatigue tests. The experimental fatigue lives and the estimated fatigue lives based on the above S-N curves are given in Table 3. Crack propagation life, Nc, is considered as the experimental fatigue life.

It can be seen from Table 3 that the fatigue lives estimated from API-X curve are found to be unconservative for one joint, omitting the joint Ts1 which failed due to weld defect. The fatigue lives estimated from DEn curve after the imposition of correction factor for thickness are also

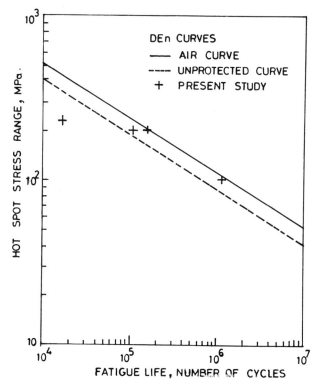

Fig.4. Comparison of experimental fatigue
life with DEn curve

479

given in Table 3. Omitting the specimen Ts1, for all other specimens, the fatigue lives were reduced between 28% to 51% under freely corroding conditions. The DEn guidance note has recommended a penalty factor of 2 on air fatigue lives to be used for unprotected joints exposed to seawater environment. This penalty factor recommended by DEn to account for free corrosion seems to be adequate. The experimental results are compared with DEn curves in Fig. 4. The S-N curves recommended by DnV and NPD assume adequate corrosion protection and no mention is made about the penalty factor for freely corroding conditions. As can be seen from Table 3, omitting the specimen Ts1, the DnV code estimates unconservative lives for two joints to an extent of 46%, and for specimen Ys2, it estimates the fatigue life conservatively to an extent of 31%. The estimated lives based on NPD code are found to be conservative even for freely corroding joints for all the specimens except for the specimen Ts1.

As can be seen from the above observations, the fatigue lives estimated based on the S-N curves recommended by various codal authorities differ a lot.

4 Conclusions

The S-N curve recommended by DEn can be used conservatively for stiffened joints under freely corroding conditions in seawater, after imposing a penalty factor of 2 on air fatigue lives, to account for free corrosion. The same S-N curve can also be used for freely corroding stiffened joints at higher temperatures(around 30°C) after the imposition of a penalty factor of 2 on air fatigue lives.

The S-N curve recommended by NPD can be used conservatively for stiffened joints exposed to freely corroding environment without any penalty factor.

5 Acknowledgement

The authors thank Mr. N.V. Raman, Director, SERC, Madras and Dr. T.V.S.R. Appa Rao, Deputy Director, SERC, Madras for their guidance and encouragement throughout the course of this investigation. This paper is published with the kind permission of the Director, SERC, Madras.

6 References

ASTM D1141-1975 Standard specification for substitute ocean water. American Society of Testing and Materials.

Gandhi, P., Madhava Rao, A.G., Ramachandra Murthy, D.S. and Raghava, G. (1991) Static and fatigue behaviour of internally ring stiffened steel tubular Y joints, in **Tubular Structures** (eds J. Wardenier and E. Panjeh Shahi), Delft University Press, Delft, The Netherlands, pp. 229-238.

Raghava, G. and Madhava Rao, A.G. (1992) Corrosion, Cathodic Protection and Corrosion Fatigue in Steel Offshore Structures, Project Report No. FTL/RR-92-2, Structural Engineering Research Centre, Madras, India.

Ramachandra Murthy, D.S., Madhava Rao, A.G., Gandhi, P. and Pant, P.K. (1992 a) Structural efficiency of internally ring stiffened tubular joints. **Journal of Structural Engineering,** American Society of Civil Engineers, 118, 3016-3035.

Ramachandra Murthy, D.S., Madhava Rao, A.G., Raghava, G. and Gandhi, P. (1992 b) Corrosion fatigue of steel offshore structures - an overview, in the **Third National Congress on Corrosion control**, National Corrosion Council of India, Central Electro Chemical Research Institute, Karaikudi, India, pp. 22-55.

UEG PublicationUR 33 (1985) **Design of Tubular Joints for Offshore Structures**, Volume 2.

PART 13

INNOVATIVE OR UNUSUAL JOINTS

Stiffness, strength and modes of failure

47 LOAD CAPACITY OF INNOVATIVE TUBULAR X-JOINTS

A. EIMANIS and P. GRUNDY
Monash University, Melbourne, Australia

Abstract
A series of static tension and compression tests were performed to determine the load capacity of various innovative tubular X-Joints. The results obtained were compared to the standard joint strength data, encountered in typical tubular cross-braced structures. The testing program adopted a value of $\beta = 1.0$, where β is the ratio of the brace diameter to the chord diameter. The X-Joint specimens were manufactured using Grade 350 steel and 273.1 mm diameter Circular Hollow Sections (CHS). Significant increases in static tensile and compressive strengths were observed in these innovative connections, compared to conventional connections.

Five different joint configurations were tested. These included the standard profile-cut specimen which was compared to the profile-cut grouted sleeve specimen, the flattened-end specimen, the flattened-end grouted sleeve specimen and the flat plate insert specimen. The flattened-end members were the braces, whereas the chord was continuous. In the flat plate insert specimen the flattened brace portion was removed and replaced with a thicker 30 mm plate. The grouted sleeve was positioned on the chord member using expansive grout, which developed a measured radial prestress in the grouted sleeve of up to 13 MPa.
Keywords: Tubular DT-Joints, Flattened-end Connections, Grouted Sleeve Connections, Ultimate Strength.

1 Introduction

Failure in tubular joints in offshore structures are costly events. These financial and human loses have prompted the introduction of various testing programs to study the ultimate and fatigue strengths of various joint configurations. This testing program adopted new and innovative joint design features to increase the static tensile and compressive strengths of the Double-T (DT) joint, which is typically encountered in the offshore jacket structure at the intersection of two cross bracing members of equal diameter.

Because the original tubular frames contain welded contoured connections, the tendency is to use the same repair mode. Unfortunately this method of welded repair proves to be in many situations not only time consuming but expensive. The advantage of the innovative joints are that they can be used either in initial fabrication or in repairs.

Tubular Structures V. Edited by M.G. Coutie and G. Davies.
© 1993 E & FN Spon, 2–6 Boundary Row, London SE1 8HN. ISBN 0 419 18770 7.

1.1 Grouted connections

The main advantages of grouted connections over typical welded connections in offshore jackets are the significant gains in strength because of the increased stiffness in the composite section and the installation time is decreased as the joint has the ability to accommodate high geometrical tolerances. Reduction of underwater repair work also occurs because offshore grouting technology is highly developed. Close fit of the outer sleeve is not required as the damaged section is easily contained within the grout. These grouted joints can be used to strengthen or repair tubular members and connections. New bracing members can also be easily inserted into the jacket structure using grouted sleeves, thus increasing the strength and stiffness of the frame.

The shear bond strength developed at the steel-grout interface greatly influences joint strength. As reported by Foo (1990) chemical and mechanical prestressing increases the shear and ultimate strength capacity of grouted connections. Fatigue life is also enhanced by the elimination of welding from the through member. When failure occurs, it usually happens in one member rather than in both members.

1.2 Flattened-end connections

The flattened-end connection is made first by flattening the end in a press. The flattened-end brace cross-section gradually changes from the flat profile to that of the circular pipe. This tapered section is called the transition region. The flat section has bulbous ends to form a 'dog bone' shape so that rupture does not occur at the side of the flat section during pressing.

This flat brace portion has an optimal length to minimise flexural stresses. This length was calculated using mathematical derivations by Grundy (1989). A suggested solution for this flat length is $F/t > 8$ where F is the flat length and t is the pipe thickness.

If the connection contains flattened-end braces welded to the tubular chord member, the joint behaves as a pinned connection, which essentially eliminates all end moments and ensures rotational stability of the chord. One advantage of the flattened-end connection during fabrication is that it is a simpler connection to form when compared to a profile-cut joint. Welding is less complicated and full welding penetration is achieved throughout the flattened cross-section, because access is available to the section from both sides of the brace. Local external stiffening of the immediate chord area also results in enhanced static strength properties for joints with $\beta = 1.0$.

Fatigue life and static strength are improved in the connection as end moments are eliminated in the chord due to the flattening of the brace ends. This benefit is further increased as the complexity of the joint increases. With a number of brace members terminating at a single node, problems can arise in obtaining a suitable joint geometry. Complex joint geometry, involving overlap joints to reduce excessive brace connection eccentricities results. Using flattened-end braces, the axial loads at the joint can generally be designed to pass through one point along the chord centre-line, thus eliminating such load eccentricities and subsequently induced end moments.

2 Experimental program

The following five different types of innovative tubular DT-Joints with $\beta = 1.0$ were tested under static tensile and compressive loading;

(a) Profile-cut specimen
(b) Profile-cut grouted specimen
(c) Flattened-end specimen
(d) Flattened-end grouted specimen
(e) Flat plate insert specimen

These joints are defined in the Appendix. Vertical brace loading was applied to the joints. The horizontal chord member was unrestrained for the tensile tests, whilst under compressive loading in-plane brace bending and chord rotation was prevented using a lateral support.

2.1 Grout mix

One of the major components of the innovative DT-Joints was the expansive grout. As the grout expanded during curing between the chord and the sleeve, hoop and longitudinal stresses were generated within the steel, resulting in joint prestressing. The expansive agent CSA (calcium-sulpho-aluminate) generates this grout expansion.

The grout mix was developed by Lim (1992). Oil Well Type G cement was used to obtain a slower setting time. The grout mix adopted included Oilwell Cement, Denka CSA and laboratory seawater in the ratio 0.75, 0.25 and 0.36 by weight respectively.

Strain gauges measured the development of radial prestress σ_r for all of the grouted specimens at various outer sleeve surface locations. The radial prestress developed in the grout varied between 6.9 MPa and 13.1 MPa after 28 days of curing.

3 Compression tests

3.1 Ultimate compressive strength

Tables 1 and 2 summarise the compression test results carried out on the various DT-Joints.

Table 1. Comparison of static compressive strength for the DT-Joints tested

Specimen Type	Compressive Load (kN)	% Increase in Compressive Strength
Profile-cut	1067.6	0.0
Profile-cut Grouted	1474.4	38.1
Flat Plate Insert	1565.6	46.6
Flattened-end	1622.8	52.0
Flattened-end Grouted	1733.1	62.3

Maximum compressive load was attained with the flattened-end grouted specimen, where a strength increase of 62.3 % over the conventional profile-cut joint was recorded. The difference in the recorded compressive strengths between the three flattened-end specimens was only 15.7 %.

Of all the flattened-end brace specimens, the flat plate insert specimen had the lowest compressive strength, which was attributed to the brace geometry in the

Table 2. Compression loaded DT-Joint failure modes

Specimen Type	Specimen Failure Mode
Profile-cut	Chord wall bulging and collapse between saddle points.
Profile-cut Grouted	Grouted sleeve wall bulging between saddle points. Less than standard joint due to thicker composite section.
Flattened-end	Buckling and collapse of brace in the transition region.
Flattened-end Grouted	Buckling and collapse of brace in the transition region.
Flat Plate Insert	Brace buckling and collapse in the transition region at the intersection between the brace and flat plate insert.

vicinity of the taper. A bulge developed in the tapered brace region of the plate insert specimen at the intersection between the brace and plate, which effectively increased the included angle and reduced joint capacity.

The compressive strength increase of the flattened-end specimens was attributed to the portion of flat brace that extended past the diameter of the chord member. This flat section increased the effective cross-section through the chord making it act as an external chord stiffener and thus resulting in increased joint strength.

The introduction of the grouted sleeve with the flattened-end resulted in greater compressive strength, but no conclusions could be made from the results as to whether the grouted sleeve would generally contribute to increased joint compressive strength. There was limited chord member deformation because failure of the flattened-end specimen was generally restricted to the brace transition area.

There was an increase in compressive strength of 38.1 % for the profile-cut grouted joint over the standard profile-cut joint. This was the smallest increase of the four innovative joint types tested and was attributed to the grouted specimen having a value of $\beta = 0.84$ with respect to the sleeve, rather than $\beta = 1.0$, as was recorded for the traditional profile-cut specimen. The greater wall thickness of the composite section increased joint strength, although a minor strength decrease also accompanied this reduced β value. The β value is important in determining joint strength because it governs the compressive failure mode which occurs either by plate bending, membrane action or a combination of both.

3.2 Ductility

The load-shortening characteristic of the flattened-end joints in Figure 1 show there is significant deformation capacity. Much of this capacity occurs in the transition zone of the flattened-end members and not in the connection itself, as is observed for the profile-cut specimens.

4 Tension tests

4.1 Ultimate tensile strength

Tables 3 and 4 summarise the results of the tension tests carried out on the various DT-Joints.

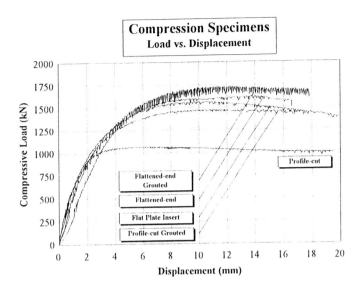

Figure 1. Comparison of the compressive test results for innovative DT-Joints.

Table 3. Comparison of static tensile strength for the DT-Joints tested

Specimen Type	Tensile Load (kN)	% Increase in Tensile Strength
Profile-cut	2205.6	0.0
Profile-cut Grouted	1418.2	−35.7
Flat Plate Insert	#2473.1	10.8
Flattened-end	2761.1	20.1
Flattened-end Grouted	2607.4	15.4

# Denotes that ultimate tensile joint capacity was not obtained due to rig failure.

The maximum static tensile load in the tests of the five innovative DT-Joints was achieved with the flattened-end specimen, where a strength increase of 20.1 % over the traditional profile-cut joint was recorded. The difference in strength of the three flattened-end type connections over the typical profile-cut joint was only 9.3 %.

Of all the flattened-end specimens the lowest tensile strength was noted for the flat plate insert specimen. This was not a true indication of joint ultimate strength because the weld fractured between the base plate and brace before the critical joint region failed. The increased tensile strength of the flattened-end joint was due to the external chord stiffening provided by the portion of the flattened-end brace that extended past the chord diameter.

There was a strength increase of 15.4 % for the flattened-end grouted sleeve joint when it was compared to the profile-cut joint. This slight strength reduction, compared to the flattened-end specimen, was attributed to the region of joint failure.

Table 4. Tension loaded DT-Joint failure modes

Specimen Type	Specimen Failure Mode
Profile-cut	Chord wall straightening. Material rupture at the bottom saddle point weld toe.
Profile-cut Grouted	Grouted sleeve wall straightening reduced between saddle points due to thicker composite section. Material rupture along top and bottom saddle point weld toes.
Flattened-end	Weld rupture at the flattened-end brace intersection outside the chord diameter. Extensive elongation of chord in loading direction.
Flattened-end Grouted	Weld rupture at the flattened-end brace intersection outside the grouted sleeve diameter. Reduced elongation of grouted sleeve in loading direction.
Flat Plate Insert	No region of joint failure recorded. Weld failure in rig at bottom brace and base plate intersection.

As the majority of the load passed through the brace portion that extended around the chord or sleeve, failure occurred at the butt weld between the braces. Grouting resulted in a reduction of the area bypassing the chord, which decreased the ultimate tensile capacity. Additional wall thickness was provided by the grouted sleeve, which reduced joint deformation compared to the chord deformation observed in the ungrouted flattened-end connection.

There was an ultimate tensile strength reduction of 35.7 % for the profile-cut grouted joint when it was compared to the traditional profile-cut joint. This was a major decrease in tensile strength, which was attributed to the joint not having a value of $\beta = 1.0$, but a β value of 0.84 with respect to the sleeve. This β value was critical in determining the ultimate joint tensile strength, as the failure mode was influenced by this variable. The composite section influenced joint behaviour by providing additional wall area bypassing the chord. The composite section did not greatly affect joint strength, as separation of the outer sleeve from the grout interface occurred at the crown.

4.2 Ductility
The flattened-end specimen exhibited the greatest ductility of all the tensile tests, although the majority of the extension occurred within the transition region and flat brace section. Chord ovalling also occurred in the load direction.

The flattened-end grouted sleeve joint was the next most ductile section. The joint still incorporated the increased extension from the flattened-end brace, but the associated chord deformation was reduced due to the strength of the composite section. As the chord side walls were less susceptible to straightening, the overall joint ductility was slightly decreased compared to the ungrouted flattened-end joint.

Final ductility was not recorded for the flat plate insert specimen because weld failure occurred at the base plate. Specimen ductility was marginally reduced, as the degree of chord and flat brace section deformation was limited due to the increased stiffness of the plate insert. The majority of joint deformation for traditional profile-cut specimen was attributed to chord wall straightening between the saddle points. The profile-cut grouted specimen was the least ductile, because side wall deformation was reduced due to the stiffer composite section.

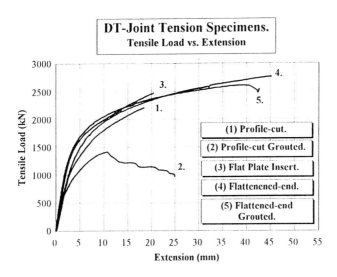

Figure 2. Comparison of the tensile test results for innovative DT-Joints.

5 Conclusion

Significant increases in tensile and compressive strengths by flattened-end specimens can be achieved over the traditional profile-cut joint. The flat brace portion that extends past the chord diameter acts as an external stiffener, thus an increase in chord wall strength is observed.

A significant reduction in tensile strength was observed for the profile-cut grouted specimen, which had a reduced β value as a result of the grouted sleeve, when compared to the ungrouted specimen.

The use of a grouted sleeve or a flattened-end connection has the advantage of separating the axial load paths of the brace and the chord, so that chord capacity is less affected by the joint capacity with respect to the brace. A further advantage of separating this structural action is that damage, when it occurs, is isolated. If cracks do occur, they are not located in the chord meaning that the chord capacity is unaffected. This is an important consideration in structures requiring redundant load paths for reliability.

6 References

Foo J.E.K. (1990) **Prestressed grouted tubular connections.** PhD Thesis, Monash University, Australia.

Foo J.E.K. and Grundy P. (1990) **Behaviour of prestressed grouted connections.** Proc. of Pacific/Asia Offshore Mechanics Symposium, Paper PACOMS-380, Seoul, South Korea.

Grundy P. and Foo J.E.K. (1990) **Performance of flattened tube connections.** International Symposium on Tubular Structures, 25-27 June 1991, Delft, The Netherlands.

Grundy P. (1989) **Squashed tubes in welded frames.** 3rd International Symposium on Tubular Structures 1-2. September 1989, Editors E. Niemi and P. Mäkeläinan, Publisher Elsevier, pp.111-117, Lappeenranta, Finland.

Lim K.S. (1992) **Prestressed grouted pile / sleeve connections for offshore structures.** M.Eng.Sc. Minor Thesis, Monash University, Australia.

7 Appendix

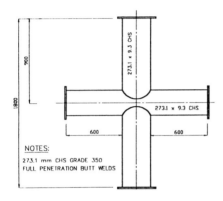

Figure A1. Profile-cut joint.

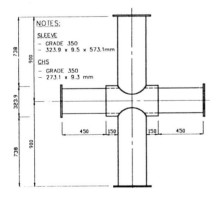

Figure A2. Profile-cut grouted joint.

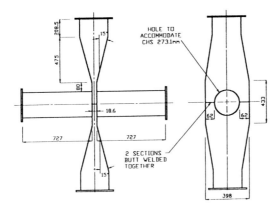

Figure A3. Flattened-end joint.

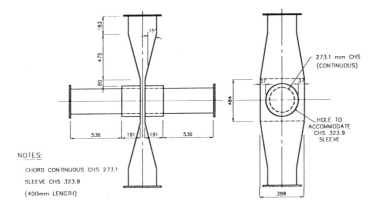

NOTES:
- CHORD CONTINUOUS CHS 273.1
- SLEEVE CHS 323.9
- (400mm LENGTH)

Figure A4. Flattened-end grouted sleeve joint.

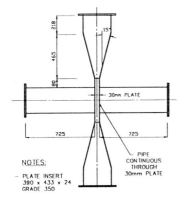

NOTES:
- PLATE INSERT
  390 x 433 x 24
  GRADE 350

Figure A5. Flat plate insert joint.

493

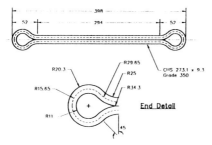

Figure A6. Cross-sectional profile of
the 273.1mm CHS brace.

48 STRENGTH OF T-JOINTS IN BACK-TO-BACK DOUBLE CHORD HSS TRUSSES

A. OSMAN
Cairo University, Giza, Egypt
R.M. KOROL
McMaster University, Hamilton, Ontario, Canada

Abstract
This paper proposes analytical models that are capable of predicting the yield strength of T-joints comprised of hollow square sections (HSS), utilizing the concept of a double-chord under both direct compression and direct tension forces. Plastic failure mechanisms are suggested and the yield line theory is employed to estimate the joints' yield strength. The validity of these models is examined by comparing the analytical predictions with the experimental results.
Keywords: T-Joints, Double Chord Trusses, Yield Line Analysis, Experiments, Strength.

1 Introduction

During the past two decades, several experimental and analytical research programs have been initiated to investigate the behaviour of welded joints in tubular structures (CIDECT 1986, Packer et al 1990). These documents provide valuable information that is relevant to the design and details of various joint types. However, a large amount of this research and development work was aimed mainly towards examining the response of welded joints in single chord triangulated structures (trusses). Other joint configurations such as joints in trusses employing back-to-back or separated double chords has received limited attention.

In this study, the strength of welded joints in double chord HSS trusses is evaluated. Emphasis is being placed on examining the yield strength of T-joints in double chord back-to-back HSS trusses (Fig. 1) under both axial tension and axial compression. Theoretical models that can estimate accurately such joint strength are proposed and their accuracy verified with test results.

2 Theory

2.1 T-joints under direct compression
When T-joints are subjected to direct axial compression, they tend to fail by forming a failure mechanism similar to that illustrated in Fig. 2. As can be seen, this mechanism involves both plastic failure of the chord flanges (dishing effect) and shear deformation of the inner webs.

Examining this simple model shows that the angle α is the only variable needing to be determined. This can be achieved by minimizing the theoretical yield load,

Tubular Structures V. Edited by M.G. Coutie and G. Davies.
© 1993 E & FN Spon, 2–6 Boundary Row, London SE1 8HN. ISBN 0 419 18770 7.

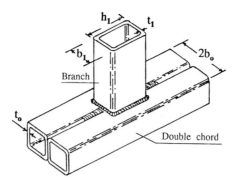

Fig. 1. Typical Double Chord T-joint

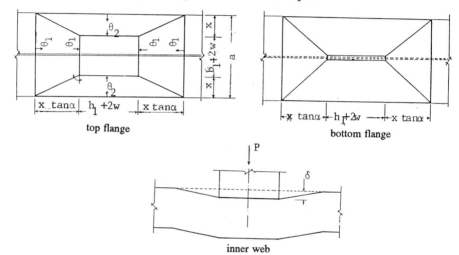

top flange

bottom flange

inner web

Fig. 2. Yield Mechanism Under Compression

P_y^* , which is given by

$$P_y^* = 2t_0^2 \sigma_{yc} \left\{ \left[\frac{(h_1 + 2w)}{2} + x\tan\alpha \right] \left[\frac{1}{x} + \frac{1}{b_0 - t_0} \right] + \frac{a}{x\tan\alpha} \right\} + \frac{4}{\sqrt{3}} \sigma_{yc} t_0 h_0$$

(1)

where
a = $2b_0 - t_0$;
t_0 = the thickness of the chord walls;
b_0 = the width of the chord member;
b_1 = the width of the branch member;
w = the size of the weld between the branch and the chord;
h_0 = the height of the chord member;
σ_{yc} = yield stress of the chord material; and
x = $(a - b_1 - 2w)/2$

Taking $\dfrac{dP_y^*}{d\alpha} = 0$ gives the critical value of α, i.e.

$$\alpha = \tan^{-1}\sqrt{\frac{a(b_0 - t_0)}{x(b_0 - t_0 + x)}} \tag{2}$$

It should be noted that the first term in Eq. (1) represents the flange's contribution to the joint's yield strength, while the second term represents the contribution of the inner webs.

2.2 T-joints under axial tension

For estimating the yield strength of T-joints subjected to axial tension, a failure mechanism that combines both the local yielding of the branch member with the moment yield line mechanism of the chord flanges was proposed. Figure 3 shows such a collapse mechanism.

As can be observed, the proposed failure mechanism is defined by two degrees of freedom — the angle α and the distance y which is half the width of the yielded area in the branch member during formation of the chord flange mechanism. Assuming that hinge lines occur at the weld edges and along the centre lines of the wall thickness, the following expression for a joint's resistance can be obtained as

$$P_y^* = t_0^2\sigma_{yc}\left\{ (h_1 + 2w + 2x\tan\alpha)\left[\frac{1}{x} + \frac{1}{y}\right] + \frac{(2b_0 - t_0)}{x\tan\alpha} + \frac{2(y + t_0)t_1\sigma_{yb}}{t_0^2\sigma_{yc}} \right\} \tag{3}$$

The value of y was assumed to occur simultaneously with yielding of the chord flanges. As such, the minimum value of the load is obtained by setting dP_y^*/dy and $dP_y^*/d\alpha$ equal to zero. The results are expressed as follows.

$$y^5 + xy^4 - \frac{t_0^2}{t_1}(h_1 + 2w)\left[\frac{\sigma_{yc}}{\sigma_{yb}}\right]y^3 - \frac{t_0^2 x}{t_1}(h_1 + 2w)\left[\frac{\sigma_{yc}}{\sigma_{yb}}\right]y^2$$

$$+ \frac{((h_1 + 2w)^2 - 2x(2b_0 - t_0))t_0^4}{2t_1^2}\left[\frac{\sigma_{yc}}{\sigma_{yb}}\right]^2 y + \frac{(h_1 + 2w)^2 x t_0^4}{4t_1^2}\left[\frac{\sigma_{yc}}{\sigma_{yb}}\right]^2 = 0 \tag{4}$$

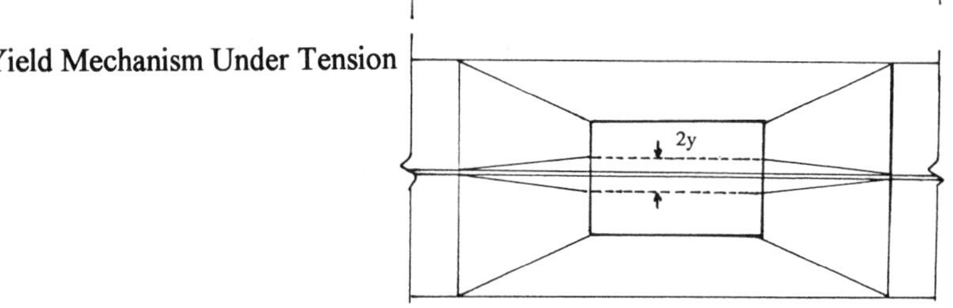

Fig. 3. Yield Mechanism Under Tension

Once y is determined, the corresponding α can be calculated as follows

$$\alpha = \tan^{-1}\sqrt{\frac{y(2b_0 - t_0)}{2x(x + y)}}$$ (5)

Examining Eq. (3) reveals that the joint resistance in tension is comprised of two parts. The first part as represented by the first two terms in the equation is the contribution from the chord flange. The second part, i.e. the third term, is the contribution of the branch member.

3 Experimental Program

To verify the validity of these models, an experimental program was undertaken. In this program, twelve T-joints were tested, six of which were subjected to direct compression while the other six were loaded in tension. A detailed description of the member sizes and weld lengths of the tested specimens are given in Table 1. As can be observed, the specimens were carefully selected to provide joints similar to those used in real practice.

For all the tested specimens, a 350 W grade steel was used. The hollow sections were manufactured by a cold forming, stress relieving process categorized as class H (CSA 1990). Compatible electrodes (E480XX) were used for welding in accordance with the Canadian standard CSA−W59 (1989).

A special set-up was constructed to apply direct compression to the specimens. Figure 4 shows a schematic illustration for the test rig, the instrumentation and the test arrangement. During the test, the load was increased monotonically in equal increments until joint failure. Linear voltage displacement transducers (LVDTs) and dial gauges were used to monitor the specimens' behaviour. In one of these tests (A5), the branch member was filled with concrete to force failure to occur in the joint.

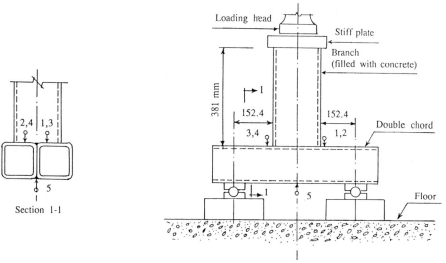

Fig. 4. Setup for the Compression Test.

498

Table 1. Details of Tested Specimens

Test type	Specimen No.	Chord member RHS (h_oxb_oxt_o) mm	Branch member RHS (h_1xb_1xt_1) mm	$\beta = b_1/2b_o$	interchord weld length (mm)
Compression	A1	152.4x152.4x9.53	152.4x152.4x6.35	0.50	76
	A2	152.4x152.4x9.53	152.4x152.4x6.35	0.50	305
	A3	152.4x152.4x6.35	203.2x152.4x6.35	0.50	76
	A4	152.4x152.4x6.35	203.2x152.4x6.35	0.50	305
	A5	152.4x152.4x6.35	254.0x254.0x6.35	0.83	305
	A6	152.4x152.4x6.35	203.2x203.2x6.35	0.67	305
Tension	T1	152.4x152.4x6.35	101.2x203.6x4.78	0.67	500
	T2	152.4x152.4x6.35	203.2x101.6x4.78	0.33	500
	T3	152.4x152.4x6.35	203.2x101.6x4.78	0.33	500
	T4	152.4x152.4x6.35	152.4x152.4x6.35	0.50	500
	T5	152.4x152.4x6.35	101.6x203.6x4.78	0.67	500
	T6	152.4x152.4x6.35	203.2x101.6x4.78	0.33	500

Table 2. Comparison between yield line predictions and experimental results for compression tests

Specimen No.	P_{jy}	P_{yf}^*	P_{yw}^*	P_y^*	P_y^*/P_{jy}
A1	1282	540	1418	1955	1.53[+]
A2	1553	501	1317	1818	1.17
A3	735	225	825	1050	1.42[+]
A4	1051	225	825	1050	1.01
A5	1437	670	907	1577	1.10
A6	1177	298	849	1147	0.97

Where loads are in kN, and

P_{jy} = The joint yield load.
P_{yf}^* = The contribution from the top chord flange to the joint yield load.
P_{yw}^* = The contribution from the chord inner web to the joint yield load.
P_y^* = The predicted joint yield load.
[+] Specimens failed by interchord separation.

Table 3. Comparison between yield line predictions and experimental results for tension tests

Specimen No.	P_{jy}	P_{yc}^*	P_{yb}^*	P_y^*	P_y^*/P_{jy}
T1	433	438	76	514	1.18
T2	485	372	106	478	0.98
T3	422	357	103	460	1.09
T4	419	391	105	496	1.18
T5	527	412	73	484	0.92
T6	561	372	106	478	0.85

Where loads are in kN, and

P_{jy} = The joint yield load.
P_{yc}^* = The contribution from the chord flanges to the joint yield load.
P_{yb}^* = The contribution from the branch to the joint yield load.

For the tension tests, the set-up shown in Fig. 5 was arranged. The tension forces were applied to the branch member through a 2000 kN capacity hydraulic jack. LVDTs and strain gauges were used to record the connection deformations and the stress variation.

4 Experimental Results and Discussion

A comparison between the T-joints' yield loads as predicted from the yield line models, P_y, and the yield loads as recorded from the tests, P_{jy}, are given in Tables 2 and 3 for the compression and tension tests, respectively. As can be seen for specimens loaded in compression reasonable agreement exists between the yield line predictions and the experimental results, with the exception of specimens A1 and A3 which failed primarily due to lack of adequate inter-chord welding.

For the specimens tested under direct tension, the proposed model predicted the yield load reasonably well. It may be of interest to note that in compression, about 75% of the yield resistance is provided by the inner webs and 25% by the flanges. On the other hand, for the tension tests, flanges provided 80% of the specimen's yield resistance and the remaining 20% was provided by branch yielding at the

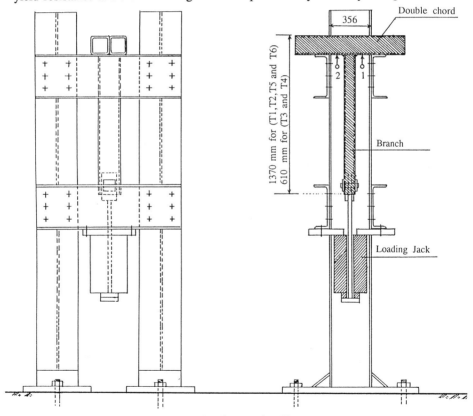

Fig. 5. Setup for the Tension Test

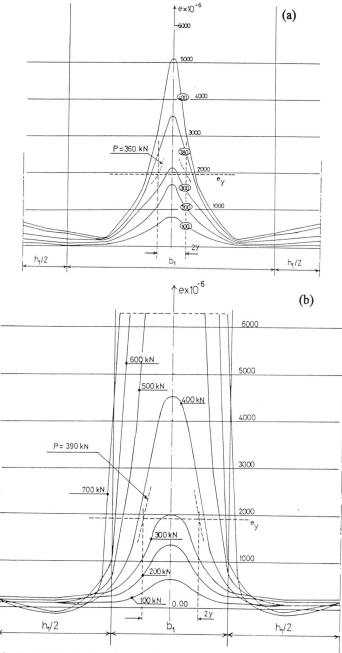

Fig. 6. Strain Distribution Across the Branch Walls at Different Load Levels
a) Specimen T5; b) Specimen T6

junction with the two chords.

Indeed, strain gauges attached to the specimens loaded in tension, indicated that at loads corresponding to the development of a yield area over the branch member with width $2y$ (as calculated from Eq. (4)) marked the beginning of a joint's significant non-linear behaviour. The strain distributions for T5 and T6 shown in Fig. 6 are indicative of the very high strains that develop in the branches at the intersection with the chord inner webs, and provide the basis for validating the proposed analytical model for tension loading.

5 Conclusions

Two analytical models that are capable of predicting the yield strength of double chord T-joints under both compression and tension are proposed. For the compression case, a yield line mechanism that combines chord inner web plastic shear deformation with yield line mechanisms in the top and bottom chord flanges is suggested. For the tension case, a collapse mechanism that combines the tensile yielding of the branch member with the formation of a yield line in the chord's upper flange was developed. The accuracy of these models in predicting the joints' yield strength was judged by comparing predictions with the experimental results. Good agreement was generally found between the predictions and the experimental findings.

6 References

CIDECT monograph No. 6. (1986) The strength and behaviour of statically loaded welded connections in structural hollow sections. Comité International pour de Développement et l'Étude de la Construction Tubulaire, Corby, England.

Packer, J.A. and Cran, J.A. (1990) Hollow sections in Canada. International Institute of Welding Annual Assembly, Montreal, Canada.

Canadian Standards Association. (1990) Steel structures for buildings (limit states design), CAN3−S16.1−M89, Rexdale, Ontario, Canada.

Canadian Standards Association. (1989) Welded steel association (Metal-arc welding), CSA Standard W59−1989, Rexdale, Ontario, Canada.

49 LOCAL FAILURE OF JOINTS OF NEW TRUSS SYSTEM USING RECTANGULAR HOLLOW SECTIONS SUBJECTED TO IN-PLANE BENDING MOMENT

T. ONO
Nagoya Institute of Technology, Japan
M. IWATA
Nippon Steel Corporation, Tokyo, Japan
K. ISHIDA
Aichi Institute of Technology, Toyota, Japan

Abstract
Experiments on local failure of joints subjected to in-plane bending moment were conducted. At the joints of the new truss system, the degree of fixing of the leg of the bracing for in-plane bending moment from the bracing is large compared to that of the conventional type joint. Two types of failure were observed : one is the collapsing of the leg of the bracing by bending and the other is the local failure of the chord near the joint. In the former case, ultimate strength can be estimated by use of a plastic section modulus of the bracing, while in the latter case, ultimate strength can be estimated by assuming the region sharing the stress of chord.
Keyword: Experiment, Rectangular hollow section, In-plane bending moment, Y-shaped joint, Ultimate strength, Failure mode, Discriminant

1 Introduction

The conventional truss system using rectangular hollow sections is formed by butt-welding the bracing members to the plate elements of the chord member. In the Y-shaped joint using rectangular hollow sections proposed by the authors, the chord and bracing are welded together, each being rotated 45 degrees to their axis. (Fig.1) Thus, efficiency at the joint is enhanced.

On statically indeterminate structures or vierendeel trusses, the joints are predominantly subjected to in-plane bending moments or out-of-plane bending moments. This study is an attempt to investigate experimentally, ultimate strength and deformation properties when a T-joint of this new truss system is subjected to in-plane bending moment from the bracing.

2 Design of experiment

In the basic experiment on ultimate strength of joints of the new joint system subjected to in-plane bending moment from the bracing, a T-joint, the simplest form of joint, was studied.

The test specimens used in the experiment were planned, using as parameters the width-thickness ratio of chord and bracing member (D/T, d/t), and the diameter ratio between bracing and chord (d/D).

Tubular Structures V. Edited by M.G. Coutie and G. Davies.
© 1993 E & FN Spon, 2–6 Boundary Row, London SE1 8HN. ISBN 0 419 18770 7.

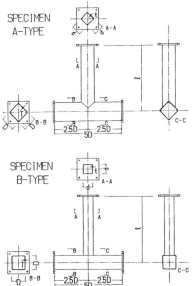

SPECIMEN
A-TYPE

SPECIMEN
B-TYPE

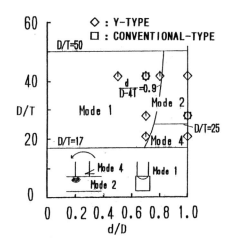

Fig.1. Test specimen

Fig.2. Combination of experimental parameters

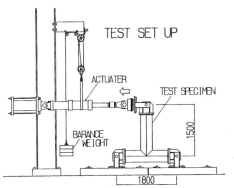

TEST SET UP

ACTUATER

TEST SPECIMEN

BARANCE WEIGHT

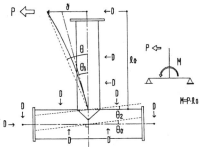

$M = P \cdot \ell_0$

Fig.3. Test set-up

Fig.4. Deformation of specimen

The number of test specimens included 9 for T-joints of the rectangular hollow section Y-truss (A-TYPE) and 2 for T-joints of the conventional type joint (B-TYPE) prepared for comparison purposes. The geometry and dimensions of the specimen are shown in Fig.1 and Table 1. The combination of parameters of specimens are shown in Fig.2. In this figure, the region of failure for conventional type joints of rectangular hollow section given by TABUCHI is shown.

The equipment used for experimentation is outlined in Fig.3. Mechanical properties of the materials used are shown in Table 2.

By measuring the displacements, the deformations of the specimen shown in Fig.4 were obtained. The strains near the connection were evaluated using wire strain gauges.

3 Experimental results

3.1 Failure types
Two failure types were observed (Photo.1 (a),(b)): 1) Collapsing of the leg of the bracing by the bending (Mode B). 2) Cracking of the chord in which the tensioned side of the bracing is jointed and sinking of the compressed side of the bracing into the chord, these phenomena occurring at almost the same time (Mode C).

Bending deformation of the chord side was not noticeable.

3.2 Load-deformation relationships
Fig.5 and 6 show the load-deformation curves. The vertical axis represents the bending moment M due to the bracing, and the horizontal axis represents the angle θ_1 of rotation of the bracing at the joint. In the early stage, there is a linear relationship in elasticity. After the maximum strength is reached, the load drops quickly.

Fig.5 shows the load-deformation curves of the Y-type joint (A-TYPE) and the conventional type joint (B-TYPE). The diameter ratio of specimen B-1 is equal to that of specimen A-1, and the diameter ratio of specimen B-2 is equal to that of specimen A-7. Initial stiffness and ultimate strength of a Y-type joint is greater than that of a conventional type joint. The initial stiffness and yield strength of specimen B-1 whose diameter ratio d/D is 0.7 are remarkably low.

3.3 Shearing strain distribution
Fig.7 shows the shearing strain distribution in the specimen A-8 whose failure mode is Mode C. At the chord, large strain is produced in the chord sides to which the bracing is jointed. By considering the result of other additional specimens, it is shown that the position where the shearing strain does not exist moves toward the top of the chord as the bracing diameter becomes small.

3.4 Ultimate strength of joint
Table 3 shows the value of ultimate strength M_u of joints obtained by the experiment. In Table 3, M_r is the calculated value of ultimate strength of conventional rectangular hollow section joints by using the formula given by TABUCHI, and M_c is the calculated value of

Table 1. List of test specimens

Spec. -No.	Chord			Bracing		d/t	d/D
	D mm	T mm	D/T	d mm	t mm		
A-1	250	6	41.7	175	6	29.2	0.7
A-2	250	9	27.8	175	6	29.2	0.7
A-3	250	12	20.8	175	6	29.2	0.7
A-4	250	6	41.7	125	9	13.9	0.5
A-5	250	6	41.7	200	12	16.7	0.8
A-6	250	6	41.7	250	6	41.7	1.0
A-7	250	9	27.8	250	9	27.8	1.0
A-8	250	12	20.8	250	12	20.8	1.0
A-9	250	12	20.8	250	6	41.7	1.0
B-1	250	6	41.7	175	6	29.2	0.7
B-2	250	9	27.8	250	9	27.8	1.0

Table 2. Mechanical properties of material

t mm	σy tf/cm^2	σu tf/cm^2	E tf/cm^2	Elong. %	note
5.58	3.56	4.47	2150	39.6	A1, A2, A3<B>
5.59	3.68	4.65	2120	39.6	A4, A5, A6<C> A6, A9<B>
5.65	3.39	4.38	2200	38.2	A1, B1<C>
5.66	3.67	4.56	2130	38.2	B1<B>
8.51	3.76	4.77	2150	39.0	A2<C>
8.53	3.87	5.00	1910	37.0	A7, B2<C> A7, B2<B>
8.83	4.62	5.12	2040	33.6	A4<B>
11.38	3.54	4.61	1830	43.5	A8, A9<C> A8<B>
11.39	3.79	4.67	2200	41.6	A3<C>
11.70	3.95	4.75	1990	38.6	A5<B>

Table 3. Comparison between test results and estimated strength

Spec. -No.	Mu tf·cm	Mu/Mr	Mu/Mc	Mo tf·cm	Mpb tf·cm	Mu/Mio	Failure C1, C2, B
A-1	632	—	1.47	677	784	0.93	C1
A-2	877	—	1.06	1130	784	1.12	B
A-3	944	—	0.85	1530	784	1.20	B
A-4	516	—	2.21	470	737	1.10	C1
A-5	909	—	1.51	879	2190	1.03	C2
A-6	1210	1.38	1.29	1270	1700	0.96	C1
A-7	2210	1.18	1.19	2050	2650	1.08	C1
A-8	2920	1.07	1.11	2490	3100	1.17	C1
A-9	1770	0.99	0.68	2490	1700	1.04	B
B-1	129	—	0.30	677	860	0.19	C1
B-2	2170	1.16	1.17	2050	2790	1.06	C1

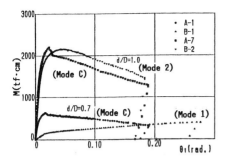

Fig.5. Load-deformation curves

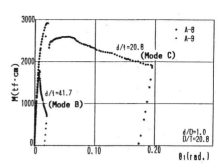

Fig.6. Load-deformation curves

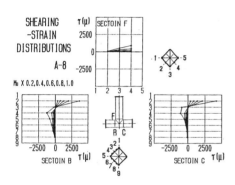

Fig.7. Shearing strain
distributions

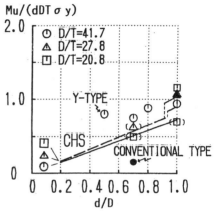

Fig.8. Experimental
ultimate strength

(a) Failure Mode B

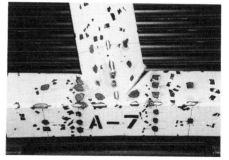

(b) Failure Mode C

Photo.1. Specimen after failure

ultimate strength of circular hollow section joints by using the formula given by CIDECT.

Fig.8 shows the relationships between the experimental ultimate strength Mu and the diameter ratio d/D. The mark having [ ] is the result of the specimen which for the failure mode is Mode B. In this figure, the ultimate strength is nondimensionalized by "dDTσy", considering that the ultimate strength Mu depends on the bracing diameter and the sectional area of the chord. In this figure, the ultimate strength formula for circular hollow section joints given by CIDECT is also shown. The ultimate strength of T-joints made by the new method is superior to that of conventional T-joints. This tendency increases with decreasing diameter ratio d/D and increasing width thickness ratio of the chord D/T.

4 Estimating the ultimate strength

In the case when the failure occurs in the chord, shown in Fig.9, it is assumed that when the shearing stress, uniformly distributed in the effective region Ao, reaches the yield stress, the strength of the region Ao reaches the limit and governs the ultimate strength Mo of the joint. In this condition, following equation is given.

$$M_0 = \sqrt{2/3} \cdot dA_0 \cdot \sigma_{yc} \tag{1}$$

Assuming that the effective region Ao is equal to the effective area obtained from the study of T-joints of the new truss system whose bracing is subjected to axial compression force, Ao is as follows:

$$A_0 = DT \cdot f(d/D) \tag{2}$$

$$f(d/D) = \frac{\sqrt{0.5(3 - 1.41 \cdot d/D)^2 + 3}}{2(1.794 - 0.942 \cdot d/D)} \tag{3}$$

Using eq.1,2 and 3, the ultimate strength formula for T-joints subjected to in-plane bending moment is as follows:

$$M_0 = 0.816 \cdot f(d/D) \cdot dDT \cdot \sigma_{yc} \tag{4}$$

In the case when the failure occurs in bracing, ultimate strength Mpb is given by the next formula.

$$M_{pb} = Z_{pb} \cdot \sigma_{yb} = 1.41 \cdot d^2 t \cdot \sigma_{yb} \tag{5}$$

Fig.10 shows the relationships between Mu and Mo. In this case, the failure occurs in the chord. Fig.11 shows the relationships between Mu and Mpb. In this case, the failure occurs in the bracing. It is shown that ultimate strength of the joint estimated by these formulae is in agreement with the experimental results.

In conclusion, the ultimate strength Mio is given by following:

$$M_{i0} = \min. \{M_0, M_{pb}\} \tag{6}$$

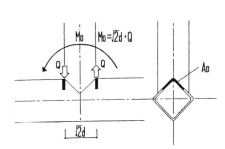

Fig.9. Effective region in chord

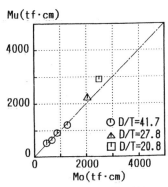

Fig.10. Data compared with eq.4

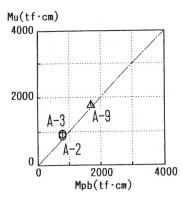

Fig.11. Data compared with eq.5

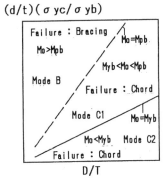

Fig.12. Discrimination of
failure mode

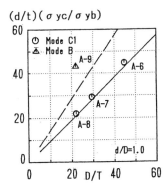

Fig.13. Discriminant and
test results

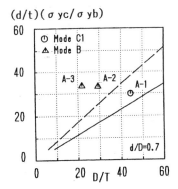

Fig.14. Discriminant and
test results

5 Discrimination of failure mode

The Mode C1 and C2 are established all over again. For the Mode C1, when the failure occurs in the chord, the bracing yields by bending, and for the Mode C2, the bracing is unyielded. The yield moment M_{yb} of the bracing is obtained by the equation.

$$M_{yb}=Z \cdot \sigma_{yb}=0.943 \cdot d^2 t \cdot \sigma_{yb} \tag{7}$$

Using Mo, Mpb and Myb, the discriminants for failure modes are as follows:

$$\frac{d}{t} \cdot \frac{\sigma_{yc}}{\sigma_{yb}} > \frac{1.73}{f(d/D)} \cdot (\frac{d}{D})^2 \cdot \frac{D}{T} \qquad : \quad \text{Mode B} \tag{8}$$

$$\frac{1.73}{f(d/D)} \cdot (\frac{d}{D})^2 \cdot \frac{D}{T} > \frac{d}{t} \cdot \frac{\sigma_{yc}}{\sigma_{yb}} > \frac{1.15}{f(d/D)} \cdot (\frac{d}{D})^2 \cdot \frac{D}{T} \qquad : \quad \text{Mode C1} \tag{9}$$

$$\frac{d}{t} \cdot \frac{\sigma_{yc}}{\sigma_{yb}} < \frac{1.15}{f(d/D)} \cdot (\frac{d}{D})^2 \cdot \frac{D}{T} \qquad : \quad \text{Mode C2} \tag{10}$$

Fig.12 shows these discriminants graphically. Fig.13 and Fig.14 show the relationships between these discriminants and experimental results, and it is apparent that these discriminants are sufficiently effective to discriminate the failure mode.

6 Conclusion

Experiments were conducted on ultimate strength of T-joints in rectangular hollow sections in a new truss system subjected to in-plane bending moment from the bracing. The deformation and stress conditions of the joints were determined as a result. The strength of the joints studied is superior to that of conventional joints. The formula to estimate the ultimate strength with considerable accuracy and the condition to discriminate the failure mode were offered.

7 References

AIJ (1990) Recommendations for the Design and Fabrication of Tubular structures in Steel, AIJ
Ono,T., Iwata,M. and Ishida,K. (1992,1993) An experimental study on ultimate strength of joints of new truss system using rectangular hollow sections Part 1,2, Journal of Struct. Constr. Engng, AIJ, No.436,445
Tabuchi,M. and Kanatani,H. (1985,1986,1986) On the Local Failure of Square Hollow Section Joints Subjected to Bending Moment Part 1,2, 3, Journal of Struct. Constr. Engng, AIJ, No.357,360,362
Wardenier,J., Kurobane,Y., Packer,J.A., Dutta,D. and Yeomans,N. (1991) Design guide for circular hollow section (CHS) joints under predominantly static loading, Verlag TUV Rheinland

50 ULTIMATE STRENGTH FORMULA FOR JOINTS OF NEW TRUSS SYSTEM USING RECTANGULAR HOLLOW SECTIONS

K. ISHIDA
Aichi Institute of Technology, Toyota, Japan
T. ONO
Nagoya Institute of Technology, Japan
M. IWATA
Nippon Steel Corporation, Tokyo, Japan

Abstract
Using a simple model for analysis, the formula for the local ultimate strength of the new joint system can be introduced as follows : $P_u = f(d/D, D/T, \theta, g/T)$. In this formula, the coefficients for the parameters are determined from experimental results. Local ultimate strength of the joint calculated by this formula is in agreement with experimental results. Fatigue strength formulas of this joint system are based on the S-N curves obtained from the fatigue tests, comparing with formulas for the circular hollow section. This joint system possesses fatigue strength equal to or greater than that of the circular hollow section T-joint.
Keywords: Rectangular hollow section, Truss joint, Y-shape jointing method, Design formula, T,K-joint, Ultimate strength, Fatigue strength

1 Introduction

In order to increase the joint efficiency of the rectangular hollow section truss joint, the authors proposed the Y-shape jointing method (a jointing method whereby chord and bracing member are rotated 45 degrees around the member axis, compared to the usual mode of use). However, no design formula for such a joint can be found in the studies made to date. In commercializing this Y-shape jointed truss structure, a formula for obtaining the local ultimate strength of the joint is indispensable. Also, when the possibilities of further applying this joint system for artificial grounds (megastructure) and offshore structures are taken into consideration, data on fatigue strength of the joint will additionally be needed.

The purpose of this paper is to obtain a formula for the local ultimate strength of this joint, as well as a formula for fatigue strength of this newly proposed joint system.

2 Ultimate strength formula

2.1 Ultimate strength formula for T-joints
Fig.1 shows the geometry and dimensions of one T-joint. From the experiments on T-joints, it is observed that local deformation of the

Tubular Structures V. Edited by M.G. Coutie and G. Davies.
© 1993 E & FN Spon, 2–6 Boundary Row, London SE1 8HN. ISBN 0 419 18770 7.

chord is large near the leg of the bracing and it is noted that the mode of deformation of the chord depends only on the diameter ratio d/D. And, the results of experiments show that large shearing and normal strain is produced in the chord sides to which the bracing is jointed.

Using the failure model of the T-joint shown in Fig.2, the ultimate strength Po of T-joint is evaluated. This model is based on the deformation and strain distribution properties of the chord obtained by experiments. In this model, at the region 1 the yield line due to plate bending is formed and at the region 2 the combination of the normal stress σ_n and shearing stress τ reaches the yield stress σ_y. This model puts together the ring model and the shearing model used conventionally. Supporting this model and assuming that the mode of deformation of chord depends only on d/D, following equations are obtained.

$$\tau = (P_0/2 - P_1)/A_0 \tag{1}$$

$$\sigma_n = \sqrt{2} P_0 l / (2A_0 D) \tag{2}$$

$$\sigma_n^2 + 3\tau^2 = \sigma_y^2 \tag{3}$$

$$P_1 = f_1(d/D) \cdot T^2 \sigma_y \tag{4}$$

$$A_0 = f_2(d/D) \cdot 4DT \tag{5}$$

In the above equations, P_1 is the force by which the yield line is formed. Using eqns. 1,2,3,4 and 5, ultimate strength Po is obtained by the following equation. In this equation, function fa(d/D) and fb(d/D) are determined from d/D only.

$$\frac{P_0}{T^2 \sigma_y} = f_a(d/D) + f_b(d/D) \cdot \frac{D}{T} \tag{6}$$

And using Fig.3 and 4, function fa(d/D) and fb(d/D) are as follows:

$$\frac{1}{f_a(d/D)} = 0.211 - 0.147 \cdot \frac{d}{D} \tag{7}$$

$$\frac{1}{f_b(d/D)} = 1.794 - 0.942 \cdot \frac{d}{D} \tag{8}$$

And then, the ultimate strength formula for T-joints is as follows:

$$\frac{P_0}{T^2 \sigma_y} = \frac{1}{0.211 - 0.147 \cdot d/D} + \frac{1}{1.794 - 0.942 \cdot d/D} \cdot \frac{D}{T} \tag{9}$$

Fig.5 shows the influence of prestressing of the chord on the ultimate strength. The magnitude of the effect with the new joint system is shown to be the same as observed in existing joints. This effect is:

$$f_n = 1 + 0.3 \cdot (N/N_y) - 0.3 (N/N_y)^2 \tag{10}$$

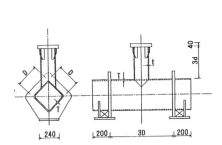

Fig.1. Dimensions of T-joint

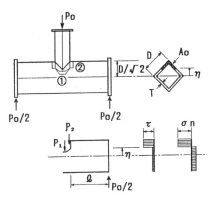

Fig.2. Failure model of T-joint

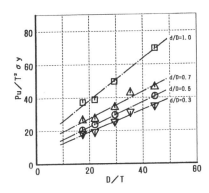

Fig.3. $Pu/(T^2\sigma_y)$-D/T relationships

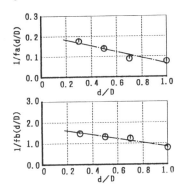

Fig.4. $1/fa(d/D)$, $1/fb(d/d)$-d/D relationships

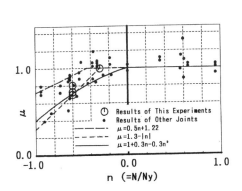

Fig.5. Influence of prestress in chord

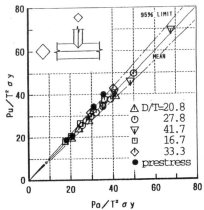

Fig.6. Data for T-joint compared with eq.9

Fig.6 shows the relationships between the experimental values of Pu and the values of Po estimated by eq.9. When the chord is prestressed, the ultimate strength is given by multiplying Po(eq.9) by fn(eq.10). It is apparent that the ultimate strength of the T-joint in the new joint system can be represented by Po in eq.9 with very high accuracy.

2.2 Ultimate strength formula for K-joints

Fig.7 shows the geometry and dimensions of one K-joint. From the experiments on K-joints, it is observed that local deformation of the chord is large near the leg of the compression bracing, and is observed that when dc/D for a K-joint is equal to d/D for a T-joint the deformation mode of the chord of the K-joint is nearly equal to that of the T-joint.

Using the failure model for a K-joint shown in Fig.8, the ultimate strength Pk of the K-joint is evaluated. This model is the same as the failure model for a T-joint. Similar to T-joint case, following equations are obtained.

$$\tau_1 = (Q-Q_1)/(\alpha \cdot A) \tag{11}$$

$$\sigma_1 = N/A + M/(\alpha \cdot A \cdot D/\sqrt{2}) \tag{12}$$

$$\sigma_1^2 + 3\tau_1^2 = \sigma_y^2 \tag{13}$$

$$q_1 = Q_1/(T^2 \sigma_y) \tag{14}$$

In these equations, N, M and Q are as follows:

$$N = P_k \cdot \cos\theta \tag{15}$$

$$M = P_k \cdot D/\sqrt{2} \cdot (\cos\theta - d_c/D) \tag{16}$$

$$Q = P_k \cdot \sin\theta \tag{17}$$

Using eqns. 11,12,13,14,15,16 and 17, ultimate strength Pk is obtained by following equation.

$$\frac{P_k}{T^2 \sigma_y} = \frac{3q_1 \sin\theta}{\{(\alpha+1)\cos\theta - d_c/D\}^2 + 3\sin^2\theta} + \frac{4\alpha}{\sqrt{\{(\alpha+1)\cos\theta - d_c/D\}^2 + 3\sin^2\theta}} \cdot \frac{D}{T}$$

$$\tag{18}$$

In the above equation, α and q_1 are determined at the process deducing the ultimate strength formula for T-joints and they are as follows:

$$\alpha = \frac{\sqrt{0.5(3-\sqrt{2} \cdot d_c/D)^2 + 3}}{8(1.794 - 0.942 \cdot d_c/D)} \tag{19}$$

$$q_1 = \frac{0.5(3-\sqrt{2} \cdot d_c/D)^2 + 3}{3(0.211 - 0.147 \cdot d_c/D)} \tag{20}$$

Fig.9 shows the relationships between the experimental values Pu and the values Pk estimated by eq.18. Fig.10 shows the relationships between the ultimate strength ratio Pu/Pk and the gap g/T. A drop in

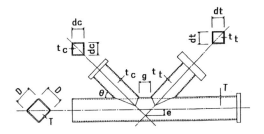

Fig.7. Dimensions of K-joint

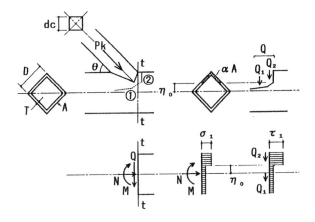

Fig.8. Failure model of K-joint

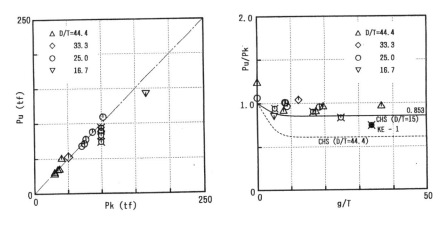

Fig.9. Data for K-joint compared with eq.18

Fig.10. Influence of gap

Pu/Pk as g/T increases is evident. In this figure, the curved lines show the magnitude of the effect in circular hollow sections. It is considered that the magnitude fg of a drop in the ultimate strength of K-joints in the new joint system is approximately presented by the heavy curved line in Fig.10. fg is as follows:

$$f_g = 0.853 + \frac{0.201}{\exp(0.5 \cdot g/T - 1) + 1} \quad (21)$$

In the eq.18, the parameters are intricate. The equation is simplified after going through many trials and processes, and the following equations are obtained.

$$\frac{P_k}{T^2 \sigma_y} = f \cdot f_\theta \quad (22)$$

$$f = \frac{1.33}{1.23 - d_c/D} + 10.8 + \left(\frac{0.374}{2.11 - d_c/D} + 0.266\right) \cdot \frac{D}{T} \quad (23)$$

$$f_\theta = \left(\frac{1 + 0.561\cos\theta - 0.354\cos^2\theta}{\sin\theta} - 1\right)\left(0.95\frac{d_c}{D} + 0.05\right) + 1 \quad (24)$$

And then, the ultimate strength formula for K-joints considering the influence of the gap is as follows.

$$\frac{P_{k0}}{T^2 \sigma_y} = f \cdot f_\theta \cdot f_g \quad (25)$$

Fig.11 shows the relationships between the experimental values of Pu and the values of Pko estimated by eq.25. It is evident that the ultimate strength of the K-joint in the new joint system estimated by eq.25 reproduces the observed ultimate strength with considerable accuracy and on the safe side.

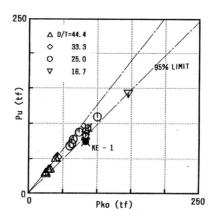

Fig.11. Data for K-joint compared with eq.25

3 Fatigue strength formula

10 T-joint specimens were used for fatigue tests. The specimen
details are shown in Fig.1. Two types of failure were observed : one
is the collapsing of the leg of the bracing (failure:b), and the
other is the collapsing of the chord (failure:c).
  Fatigue characteristics of T-joints are estimated by using the
P_r/P_u ratio (P_r : range of loading, P_u : ultimate strength of the
joint) and the virtual elastic stress range $2\sigma_a$. Fig.12 is an S-N
diagram using P_r/P_u and N_f (N_f : the number of cycles to complete
failure of the specimen), and Fig.13 is an S-N diagram using $2\sigma_a$ and
N_f. The results are plotted in the same diagram without
distinguishing between failure b and c to obtain the fatigue strength
comprehensively. These results are presented approximately as
follows :

$$\log\left(\frac{N_f}{10^6}\right) = -3.02 - 5.57 \cdot \log\left(\frac{P_r}{P_u}\right) \qquad (26)$$

$$\log\left(\frac{N_f}{10^6}\right) = 3.08 - 5.22 \cdot \log(2\sigma_a) \qquad (27)$$

  It is judged from the results that this joint system possesses
fatigue strength equal to or greater than that of the circular hollow
section T-joint.

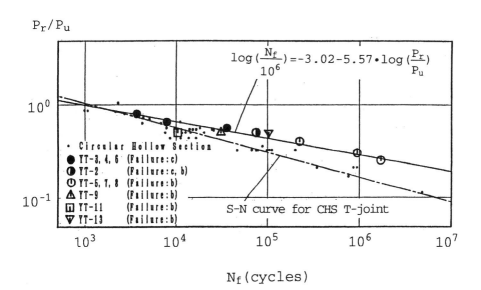

Fig.12. S-N curves (P_r/P_u - N_f relationships)

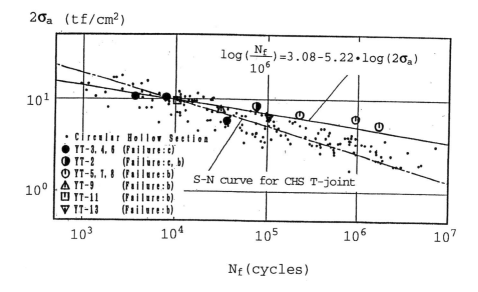

$2\sigma_a$ (tf/cm²)

$$\log\left(\frac{N_f}{10^6}\right) = 3.08 - 5.22 \cdot \log(2\sigma_a)$$

Circular Hollow Section
● YT-3, 4, 6 (Failure: c)
◐ YT-2   (Failure: c, b)
◔ YT-5, 7, 8 (Failure: b)
△ YT-9   (Failure: b)
⊡ YT-11   (Failure: b)
▽ YT-13   (Failure: b)

S-N curve for CHS T-joint

N_f (cycles)

Fig.13. S-N curves ($2\sigma_a$ - N_f relationships)

4 Conclusions

Using a simple model based on the experimental behaviour, the local
ultimate strength formulas for the new joint system were obtained.
It was shown that ultimate strength of the joint estimated by this
formula is in agreement with experimental results. A fatigue
strength formula was also obtained. The formulas in this paper are
of real practical significance.

5 References

AIJ (1990) Recommendations for the Design and Fabrication of Tubular
    structures in Steel, AIJ
Ono,T., Ishida,K., Iwata,M. and Idota,H. (1992) EXPERIMENTAL STUDY OF
    FATIGUE CHARACTERISTIC OF T-JOINT ON Y-TRUSS SYSTEM WITH
    RECTANGULAR HOLLOW SECTION, Journal of Struct. Engng, Vol.38B
Ono,T., Iwata,M. and Ishida,K. (1992,1993) An experimental study on
    ultimate strength of joints of new truss system using rectangular
    hollow sections Part 1,2, Journal of Struct. Constr. Engng, AIJ,
    No.436,445
Wardenier,J. (1982) Hollow Section Joints 1st ed., Delft University
    Press, Delft, the Netherlands

51 FLEXIBILITY COEFFICIENTS OF TUBULAR CONNECTIONS

A. URE, P. GRUNDY, I. EADIE
Monash University, Melbourne, Australia

Abstract
This paper presents a method of determining local joint flexibility (LJF) of
tubular connections using finite element modelling. LJF coefficients are
determined by modelling the chord with braces removed and displacing the
brace-chord intersection line. Results for in-plane tubular connections are
compared with existing formulae for LJF. It is shown that a similar method
may be used to determine the LJF of innovative flattened-end connections. The
practical and structural advantages of flattened-end connections are discussed.
Finite-element models are used to form parametric equations of LJF for a variety
of joint configurations. Parametric equations of LJF are presented and are
compared to equations of flexibility of traditional joints.
Keywords: Local Joint Flexibility, Tubular Joints, Circular Hollow Sections,
Flattened-End Connections.

Notation

D	chord outer diameter
E	modulus of steel
f	flexibility of connection due to local deformation
M	applied moment
P	applied axial load
R	radius of end bulbs of flattened section.
W_{flat}	width of flattened-end section excluding end bulbs
W_{total}	total width of flattened section (outer bulb to outer bulb)
β	brace outer diameter / chord outer diameter
γ	chord outer diameter / 2 x chord thickness
δ	prescribed displacement
θ	prescribed rotation
τ	brace thickness / chord thickness

1 Introduction

Recent research has indicated that the inclusion of joint flexibility in a structural
analysis yields markedly different member actions to those calculated in a
standard structural analysis. In analytical tests performed by Holmas (1987)
certain member bending-moments in a jacket structure changed by approximately
100 percent, while other structural actions were affected to a smaller but still
significant extent. Tests by other authors have led to similar conclusions.

Tubular Structures V. Edited by M.G. Coutie and G. Davies.
© 1993 E & FN Spon, 2–6 Boundary Row, London SE1 8HN. ISBN 0 419 18770 7.

Fessler et al (1986) derived equations predicting the LJF of single and multi-brace tubular joints. The equations for multi-brace joints incorporate the coupling effect of the LJF of braces on other braces. These equations allow a complete LJF matrix to be created for a variety of connection configurations.

While extensive research has been performed deriving parametric equations of LJF of profile tubular joints, there has been virtually no work examining LJF of alternative joint types. Grundy (1989) presented a case for the use of flattened-end connections, in which these connections were claimed to offer several advantages over traditional profile-type connections. Flattened-end connections require less forming in the fabrication process, and are therefore easier to contour form than traditional connections. Structurally, the use of these connections eliminates joint eccentricity which results in member end moments. Flattened-end connections also exhibit considerable out of plane strength, due to the increased moment of inertia of the flattened section compared to a fully circular section. Research performed by Foo (1990) has also shown that the use of composite joints incorporating flattened-end braces significantly reduces stress concentration factors around the brace-chord intersecting region.

Flattened-end connections are shown in this paper to exhibit greater in-plane flexibility than traditional profile-joints. There is, therefore, a greater need to incorporate LJF coefficients of this more flexible connection type in a structural analysis. Finite-element modelling of these connections is unfortunately made considerably more difficult than traditional connections due to the complex shape of the tube in the intersecting region.

LJFs of tubular connections using simplified models were first studied by Holmas et al (1985). In this approach, loads were placed along brace-chord intersection lines and using equations of elasticity LJFs were determined. A similar method of modelling is presented, with several modifications, including the addition of a global boundary restraint. Accuracy of the simplified models is demonstrated for profile joints by comparing the LJF of a model to parametric equations presented by previous authors. This concept is then extended to flattened-end connections. By using this approach, a group of flattened-end connections is modelled and analysed to yield parametric equations for the LJF for a variety of flattened-end connection types. The equations presented allow for the coupling effect of brace actions on other braces.

2 The determination of local joint flexibility coefficients

Columns of the local joint stiffness matrix are created by the individual application of unit displacements (defined as 1 millimetre) or rotations (defined as 10^{-3} radians) on nodes along the brace-chord intersection lines. Such displacements are applied, one brace at a time, normal to the axis of the chord while suppressing all displacements on other braces. Rotation is effected by the application of a linearly varying displacement on nodes along the brace-chord intersection line. This is shown in figure 1. The stiffness terms are calculated by summing the reactions on all nodes along each of the brace-chord intersection lines.

In order to cause local joint deformation, all displacements due to bending need to be eliminated. This has usually been performed in two steps. Firstly, the model is analysed featuring both local and bending displacements. A "beam-element" model is then analysed to determine the bending deformation. The difference in deformation experienced by the two models is assumed to be the deformation due to local distortion of the section.

The problem with this method is the requirement that two models be analysed to obtain one result. It would be more desirable to develop a means of calculating local joint deformations directly. This has been achieved by the application of a global boundary condition eliminating all bending deformation. All nodes are prevented from displacing along the length of the chord. As a result, the only deformations are those due to shear action and wall bending.

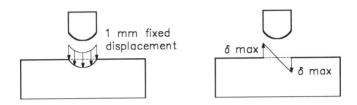

Fig 1. Axial and rotational loading of simplified connection model.

The local joint flexibility matrix (LJF) is calculated by inverting the local joint stiffness matrix. Since flexibility terms are very small (usually of the magnitude of 10^{-6} or less), it is more convenient to express these values in a dimensionless form. This may be done by multiplying terms due to axial loading by E.D and terms due to moment loading by $E.D^3$. The following equations show the relationships between local joint displacement, load and flexibility coefficients:

$$\frac{[\delta]}{D} = \left(ED[f]\right)\frac{[P]}{D} \tag{1}$$

$$[\theta] = \left(ED^3[f]\right)\frac{[M]}{ED^3} \tag{2}$$

The following assumptions are made in the determination of elastic constants of standard profile connections using finite element analysis:

Axial loading from the braces is assumed to cause a uniform displacement over the intersecting region.
Moments are assumed to be transferred from the braces to the chord by a linear displacement distribution.
The chord is infinitely stiff for axial loading, thus only displacements normal to the chord will produce local deformation in the joint.
Local stresses near end supports have no significant influence on stresses near the loaded area.

3 Preliminary testing of simplified finite element model

A finite element model of profile T-joint was created on STRAND6, a finite-element analysis package developed in Sydney, Australia. The joint consisted of a 101.6 x 4.0 CHS brace and a 139.7 x 4.9 CHS chord. The model of the joint was created without the brace, substituting a line of nodes along the brace-chord intersection line. In this case the joint is symmetrical about two axes, so only a quarter of the connection was modelled. Appropriate boundary conditions were applied along lines of symmetry. The finite-element model is shown in figure 2.

The model was displaced by 1 mm along the brace-chord intersection line. The LJF value was calculated by dividing the displacement by the total reaction normal to the chord. The LJF values for axial loading and in-plane bending have been calculated, and these values have been compared to values predicted by several authors in table 1.

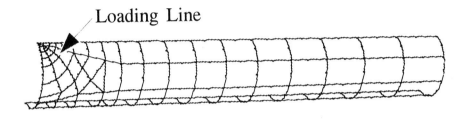

Fig.2. Finite-element model of simplified profile joint.

For axial loading the simplified model exhibits greater LJF than predicted in formulae presented by Fessler and Ueda, although it is only 14 percent greater than the LJF value predicted by Ueda. The LJF coefficient for in-plane bending is within the range of published theoretical results, being closest to the results predicted by Fessler. It can therefore be concluded that the simplified model works reasonably well for the determination of LJF values for profile connections.

Table 1. Comparison of LJF values presented by different authors.

Source	LJF_{Ax} x ED	LJF_{IPB} x ED^3
Simplified Model	236.4	541.3
Fessler	93.3	477.5
Ueda	206.1	772.7
DNV	N/A	633.6

4 Description of detailed and simplified models of flattened-end joint

A detailed model of a typical flattened-end T-joint was analysed to determine whether uniform local displacement is also achieved over the entire brace-chord intersection region for brace axial loading. The model consisted of a flattened-end T-joint, shown in figure 3, which was subjected to a 1 millimetre vertical compressive displacement across the brace cross-section, while each end of the chord provided simple support. Longitudinal translation was suppressed for all nodes in the chord to allow only local deformation of the chord. This is the technique used in the profile joint model.

The T-joint, was modelled as a quarter structure with appropriate boundary conditions applied along the lines of symmetry. A β value of 0.73 was adopted, which is close to the maximum β value possible without the flattened brace width exceeding the chord diameter. If a constant displacement occurs over the brace-chord intersection for this value of β, then it can be concluded that this is applicable for smaller β values, as the brace width is much smaller than the chord diameter. The resulting displacement distribution is therefore more likely to be uniform.

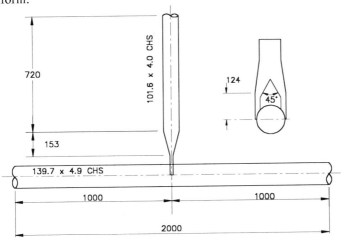

Fig.3. Dimensions of preliminary flattened-end T-joint model.

The flattened-end section, shown in figure 4, is circular at the ends to prevent rupture in the steel, while the transition along the brace from a flattened to a fully circular section is assumed to be linear. The flat section is assumed to have been created using a diamond shaped die with apex angle of 45 degrees. Welds

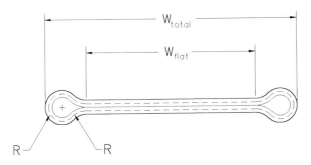

Fig.4. End profile of flattened-end tube.

were modelled using rigid beams linking brace and chord mid-surface nodes. The model was created by drawing the section using AutoCad Release 11 and transferring it to STRAND6 using an interface file.

A simplified finite-element model of this connection, shown in figure 5, was also created for comparison of LJF values with those obtained for the detailed model. This model differed from the detailed model in that the welds and brace were removed, and the brace-chord intersecting region of the flattened section was assumed to be a line. The bulb was not considered to substantially affect the LJF result, and, accordingly, was not included in the simplified model.

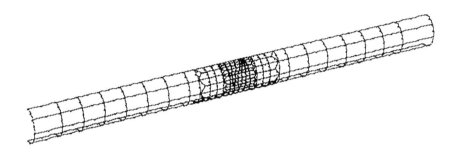

Fig.5. Simplified model of flattened-end Y joint.

The finite element analysis of the models provided significant evidence that flattened end tubular joints may be modelled in a similar manner to profile tubular joints for the determination of the LJF. For the detailed model, the application of a 1 millimetre fixed vertical displacement at the top of the brace produced displacements over the chord-brace intersecting region of between 0.61 millimetres at the crown position and 0.55 millimetres at the saddle position, a variation of less than 10 percent. At the saddle position, load is transferred from the chord to the brace both in bending and in torsion of the chord, and so the chord is stiffer. In addition, the brace is quite stiff at the ends because of the bulbs, and this helps to produce a constant vertical displacement over the brace-chord intersection.

From the tests it can be concluded that the detailed flattened-end joint may be modelled with the application of a constant prescribed displacement over the intersecting region. If the average displacement of 0.58 mm is adopted, then the LJF may be evaluated by dividing this value by the total axial force. On the other hand, the LJF of the simplified model, which is subjected to a fixed displacement across the chord of 1 mm, is given by the inverse of the total normal reaction. A comparison of LJF of the two models is shown in table 2.

Table 2. Summary of results from preliminary tests.

Parameter	Detailed model	Simplified model
LJF x E.D	202.0	204.5

5 Formulation of parametric equations

The use of a statistically efficient selection of a 16 model group enabled the derivation of parametric equations predicting the LJF of flattened-end connections. SYSTAT, a statistical package developed in Illinois, was used for the non-linear regression of finite element results. The following function types were considered in the regression phase of this study: linear models, exponential, logarithm, polynomial (up to fourth order), and trigonometric functions.

Equations based on power functions were successfully fitted to all LJF data in terms of β, γ and θ with coefficient of regression values greater than 0.98. Almost all individual data points were within 10 % of calculated values. The obtained coefficients were close to those shown in table 3, and were adjusted to give simple numbers. This produced only a marginal loss of accuracy. The equations shown are similar to the form of equations of LJF presented by Fessler and Ueda with the exception that Fessler's equation contains an exponential coefficient for the β term.

6 Conclusions

The LJF of flattened-end connections can be satisfactorily determined by the use of simplified models. These models feature the omission of braces, with loads applied along lines approximating the brace-chord intersection. Tests have shown that the LJF of the simplified models produce results within 5% of those obtained from a fully detailed model of the flattened-end connection. The use of simplified models greatly reduces both the pre-processing and processing time, thereby enabling more models to be considered in a statistical analysis.

Parametric equations of LJF for flattened-end connections have been determined from 16 models for each connection group. These models are selected to ensure that the parameters affecting LJF are included without being statistically biased. Results from the analyses have been used in a statistical regression to determine parametric equations of LJF, summarised in table 3.

Using these equations and arbitrary β, γ and θ values, the LJF of 64 independent single-brace flattened-end joints were calculated. The LJF of the series of standard connections of the same dimensions were also calculated using Fessler's equation. The results showed that compared to profile joints, flattened-end joints are between 30 and 400 percent more flexible than a profile joint of the same joint dimensions.

7 References

Fessler, H. Mockford, P. and Webster, J. (1986) **Parametric equations for the flexibility matrices of multi-brace tubular joints in offshore structures.** Proc. Institution of Civil Engineers, Part 2, pages 675-696.

Foo, J. (1990) **Prestressed Grouted Tubular Connections.** Ph.D. Thesis.

Grundy, P. (1989) **Squashed tubes in welded frames.** Third International Symposium of Tubular Structures, Lappeenranta, Finland.

Holmas, T. (1987) **Implementation of tubular joint flexibility in global frame analysis.** Report No. 87-1 Division of structural mechanics, The Norwegian Institute of Technology.

Holmas, T. Remseth, S. and Hals, T. (1985) **Approximate flexibility modelling of tubular joints in marine structures.** SINTEF report 4.12.

Table 3. Summary of LJF coefficients for flattened-end joints.

Type	Geometry	LJF Matrix
Y		$\delta = A \times P$
YT		$\begin{bmatrix} \delta_1 \\ \delta_2 \end{bmatrix} = \begin{bmatrix} A & B \\ B & A \end{bmatrix} \times \begin{bmatrix} P_1 \\ P_2 \end{bmatrix}$
K		$\begin{bmatrix} \delta_1 \\ \delta_2 \end{bmatrix} = \begin{bmatrix} A & C \\ C & A \end{bmatrix} \times \begin{bmatrix} P_1 \\ P_2 \end{bmatrix}$
X		$\begin{bmatrix} \delta_1 \\ \delta_2 \end{bmatrix} = \begin{bmatrix} A & D \\ D & A \end{bmatrix} \times \begin{bmatrix} P_1 \\ P_2 \end{bmatrix}$
KT		$\begin{bmatrix} \delta_1 \\ \delta_2 \\ \delta_3 \end{bmatrix} = \begin{bmatrix} A & B & C \\ B & A & B \\ C & B & A \end{bmatrix} \times \begin{bmatrix} P_1 \\ P_2 \\ P_3 \end{bmatrix}$

LJF Matrix Parameters

A	B	C	D
$\dfrac{2.1\gamma^{\frac{3}{2}}\beta^{-1}}{E.D}$	$\dfrac{2.3\gamma^{\frac{3}{2}}\beta^{\frac{-3}{4}}\theta^{\frac{1}{5}}}{E.D}$	$\dfrac{2.1\gamma^{\frac{3}{2}}\beta^{\frac{-3}{4}}\theta^{\frac{1}{4}}}{E.D}$	$\dfrac{78\gamma^{\frac{2}{3}}\beta^{\frac{4}{3}}}{E.D}$

Parameter Ranges

β	γ	θ
0.3−0.6	22.5−15	30−90° X, Y joints 30−60° K,YT,KT joints

PART 14

TRUSSES AND SPACE FRAMES

Lattice girders and
roof structures

52 STRESS CONCENTRATION FACTORS IN KK-MULTIPLANAR JOINTS MADE OF SQUARE HOLLOW SECTIONS INCLUDING WELDS

E. PANJEH SHAHI
Delft University of Technology, The Netherlands
R.S. PUTHLI
Delft University of Technology, The Netherlands, and
TNO Building and Construction Research, Rijswijk,
The Netherlands
J. WARDENIER
Delft University of Technology, The Netherlands

Abstract
This paper gives the results of numerical investigations
for determination of stress and strain concentration
factors at predefined locations in multiplanar KK-gap and
KK-overlap joints in triangular girders made of rectangular
hollow sections(RHS).
The finite element models of the joints are created
using 8 noded linear solid elements. The weld shape is also
modelled. The strains and stresses are quadratically extrapolated to the weld toes. Comparisons are made between the
numerical results and the experimentally determined strain
concentration factors (SNCFs) of an ECSC experimental
investigation for similar joints in triangular girders.
Keywords: Fatigue, Girder, Welded Multiplanar Joints, SCF,
SNCF, Stress, Strain, Rectangular Hollow Sections, Concentration Factors.

Nomenclature

CHS   Circular Hollow Section
RHS   Rectangular Hollow Sections
SCF   Stress Concentration Factor
SNCF   Strain Concentration Factor
ipb   In-Plane-Bending
opb   Out-of-Plane-Bending
β Brace to chord width ratio
ε Strain
2γ Width to thickness ratio
τ Brace to chord thickness ratio
σ Stress

1 Introduction

In recent years, a considerable amount of work has been
devoted to the understanding of the static and fatigue
behaviour of multiplanar joints in both rectangular and
circular hollow section joints.

Tubular Structures V. Edited by M.G. Coutie and G. Davies.
© 1993 E & FN Spon, 2–6 Boundary Row, London SE1 8HN. ISBN 0 419 18770 7.

Although experimental investigations on fatigue have concentrated mainly on determination of stress concentration factors near the weld toes, numerical research has usually been conducted excluding modelling of the weld dimensions. This numerical research programme sponsored by STW(The Netherlands Technical Foundation) is being carried out at Delft university of technology for determination of stress/strain concentration factors of multiplanar KK-gap and KK-overlap joints in rectangular and circular hollow sections. The finite element modelling includes the weld shape.

The first part of this research programme concernes the calibration of the numerical results with some of the results of a recently completed ECSC experimental research programme conducted in Germany and the Netherlands. The participating organisations in the ECSC research programme are Mannesmannröhren Werke A.G, Düsseldorf and Universität Karlsruhe in Germany and TNO Building and Construction research and Delft University of Technology in the Netherlands.

One of the topics of investigation in the experimental programme has been the determination of SNCFs for isolated KK-joints using CHS and RHS at Universität Karlsruhe and Complete triangular girders with KK-joints using CHS at TNO Building and Construction Research and RHS at Delft University of Technology.

This paper gives the comparison between numerical and experimental results for joints made of RHS in complete girders tested at Delft University of Technology. A total of 4 girders were investigated, each containing 2 KK-gap and 2 KK-overlap joints(50 or 100%). Fig. 1 shows one girder with gap and 100% overlap joints. Joints in the middle of girders experience loads on all members where-as for joints at the ends of girders, chord member in the last bay is unloaded. In this way, the influence of chord loading for joints is also studied.

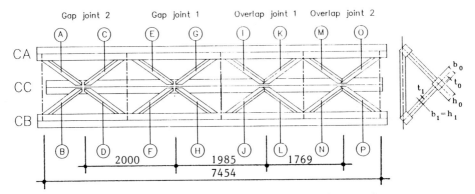

Fig.1. Configuration and dimensions of the girders in RHS.

Table 1. Review of the test series – girders in RHS

Girder	Nominal dimension		β	2γ	τ	Weld Type	Gap Joints e	Overlap Joints %
	Chord b0 x t0	Brace b1 x t1						
1	200x8	80x4	0.4	25	0.5	Fillet	0	100
2	200x16	80x8	0.4	12.5	0.5	Butt	0	100
3	200x8	120x4	0.6	25	0.5	Fillet	0	50
4	200x16	120x8	0.6	12.5	0.5	Butt	0	50

The girder properties and parameters are summarised in Table 1. Width ratios β of 0.4 and 0.6 are considered with 2γ ratio of 12.5 and 25. The thickness ratio τ is 0.5 and the angles between braces in uniplanar and multiplanar directions are 90 degrees.

2 Determination of hot spot stress/strain

Due to the geometry of RHS, the locations where highest stresses(hot spot stress) can occur are at the chord and brace weld toes around the corners of the connecting braces. The magnitude and location of the hot spot stress depends on the type of load being applied to the joint. Therefore, it is necessary to establish fixed positions where the stresses/strains can be measured and the hot spot stress /strain can be calculated.
For a multiplanar KK–joint, stresses are calculated along 5 lines(lines A to E) for each corner of the brace (corners 1 to 4) of the tension brace. See Fig. 2 for locations of measurement lines.

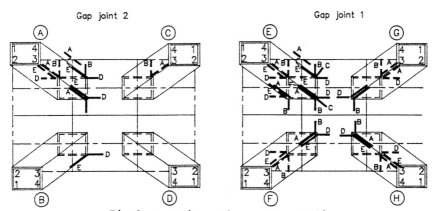

Fig.2. Locations of measurement lines.

All the lines are perpendicular to the weld toe and Primary stress/strains perpendicular to the weld toe are used for extrapolation along these lines. Quadratic extra-

polation to the weld toe was adopted for the study on
uniplanar X and T joints in RHS by van Wingerde et al
(1991) and the experimental part which this paper refers to
Verheul et al, (1993).

This numerical investigation also adopted the quadratic
extrapolation for determination of hot spot stress. Fig. 3
shows the limits for this kind of extrapolation. The region
of influence of notch stress is taken as 0.4t with a minimum of 4 mm.

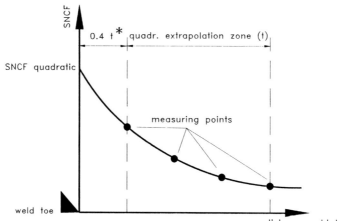

Fig.3. Method of quadratic extrapolation.

The strain or stress concentration factor is then obtained by normalising the hot spot strain or stress with the
nominal strain or stress defined as the sum of axial, in-
plane and out-of-plane strains or stress in the brace
concerned at the intersection of the brace centre line and
the chord top surface, see Fig. 4 for detail. SNCF and SCF
are then defined by the following formulea:

$$SCNF = \frac{\text{Hot Spot Strain}}{\text{Nominal Strain}} \quad \text{and } SCF = \frac{\text{Hot Spot Stress}}{\text{Nominal Stress}}$$

where
nominal strain $= |\varepsilon_{axial}| + |\varepsilon_{ipb}| + |\varepsilon_{opb}|$ and
nominal stress $= |\sigma_{axial}| + |\sigma_{ipb}| + |\sigma_{opb}|$
at the intersection of chord and brace.

3 Finite element modelling and analysis

The I-DEAS software of Structural Dynamic Research Corpora-
tion(SDRC) is used for modelling,analysis and post processing. All numerical work is conducted on a SUN Sparc station. I-DEAS solid modelling is utilised for creation of the
multiplanar joint geometry and the welds. The mesh areas
and mesh volumes are then created using the surfaces of

these geometries in the Finite Element Modelling. After applying suitable load and boundary conditions, I-DEAS Model Solution is used for the numerical analyses.

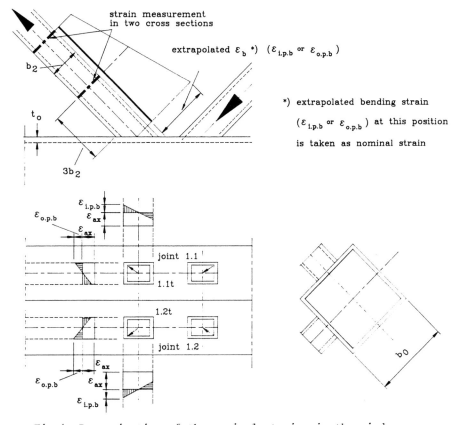

Fig.4. Determination of the nominal strains in the girders.

In modelling the girders investigated experimentally for the numerical investigations, the following approaches have been used:

a 8 noded solid elements used for modelling the welds and other parts of the joint under consideration. Use of solid elements instead of combination of solid and shell elements, eliminates the need for using transition elements or tyings which are not easy to model and to use and are not provided in some finite element packages. It also saves computational time since transition elements consume large amount of CPU, Romeijn et al, (1991). Comparison between 8 noded and 20 noded solid elements have shown insignificant differences in the stress

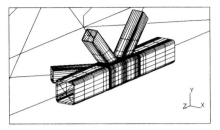

Fig.5. KK-gap joint 1 in girder 1 modelled with solid elements.

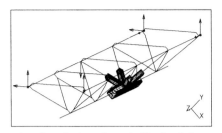

Fig.6. Girder 1 modelled with beam and solid elements

distribution and joint flexibility of KK-joints in RHS, Panjeh Shahi et al, (1993). Therefore, only 8 noded solid elements are used for modelling the RHS joints. (Fig. 5).

b   All other parts of the girders joints and members are modelled with linear beam elements rigidly connected to each other.

c   Welds are modelled according to the actual measured dimensions. This is found to be extremely important. A comparison between a joint modelled with nominal weld size and a joint with actual weld dimensions showed up to 30% difference in the calculated SCF and SNCF's.

d   Each joint under consideration is combined with a beam model of the girder created in item b above to make up the model of the girder as shown in Fig. 6. This method has proved to be suitable, since the results of comparisons for member forces and girder deflections are in close agreement with experimental results. Also, the experimental investigation concludes that when one cracked joint in the girder is repaired or reinforced, very little influence is noted on the strains in the other joints of the girders.

e   Geometrical and material properties of the numerical model are identical to those measured for the experimental specimens.

4 Results of numerical work

The finite element analyses give stresses and strains at nodes on the lines A to E for each corner of the braces. Fig. 7 shows stress contour for corner 4 of one KK-gap joint and Fig. 8 shows the stress contours at the cross section of the weld in the gap region. These stresses/ strains are extrapolated according to Fig. 3 and the extrapolated stress/strains are tabulated in Table 2 for the KK-gap joint 2 in girder 1.

The complete numerical results of finite element analyses for all the joints in all the girders are given by Romeijn et al (1992).

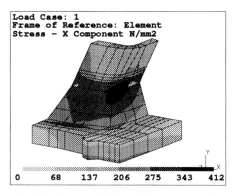

```
Load Case: 1
Frame of Reference: Element
Stress - X Component N/mm2
```

| 0 | 68 | 137 | 206 | 275 | 343 | 412 |

Fig.7. Stress lines for Corner 4
of KK-gap 1 in Girder 3

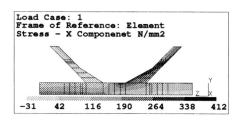

```
Load Case: 1
Frame of Reference: Element
Stress - X Componenet N/mm2
```

| -31 | 42 | 116 | 190 | 264 | 338 | 412 |

Fig.8. Stress lines for the gap
region of KK-gap Joint No.1

Table 2 also gives the numerically and experimentally determined SNCFs and numerically determined SCFs. The SCF/SNCF ratio from the numerical work are also shown in this table.

Table 2 SNCF and SCF values at the weld toe at various
location of the KK-Gap joint no.2 in girder 1

lines	$\epsilon_r {}_{nom}$ (10^{-6})	$S_r {}_{nom}$ (N/mm^2)	$\epsilon_r {}_{hs}$ (10^{-6})	$S_r {}_{hs}$ (N/mm^2)	SNCF	SCF	$\dfrac{SCF}{SNCF}$	Exp. SNCF	$\dfrac{SNCF\ Num.}{Exp.}$
A1A	596	123	1609	436	2,70	3,54	1,31	2,40	1,13
A1B			1734	413	2,91	3,35	1,15	2,18	1,33
A3A			1687	433	2,83	3,51	1,24	2,72	1,04
A3B			1597	327	2,68	2,65	0,99	2,97	0,90
A3D			1811	424	3,04	3,44	1,13	3,33	0,91
A3E			1371	362	2,30	2,94	1,28	2,44	0,94
A4A			3707	821	6,22	6,66	1,07	1,83	3,40*
A4B			1967	476	3,30	3,30	1,00	3,24	1,02
A4D			2127	431	3,57	3,50	0,98	3,91	0,91
A4E			3051	738	5,12	5,99	1,17	2,26	2,27*
B1A			1710	492	2,87	3,99	1,39	3,40	0,84
B3A			1591	404	2,67	3,28	1,23	2,32	1,15
B3B			1824	388	3,06	3,15	1,03	4,12	0,74
B4A			3063	811	5,14	6,58	1,28	4,81	1,07
B4B			1853	432	3,11	3,51	1,13	3,08	1,01
C1D			-473	-128	-0,79	-1,04	1,31	-0,64	1,24
C1E			-1579	-549	-2,65	-2,87	1,07	-2,96	0,91
Girder no. 1 : Joint KKgap no. 2									

*   Undercut of weld toe

The maximum SNCFs in the braces and chords of joints in girders 1 to 4 are presented in Table 3 together with their line No..

Table 3. Maximum numerical SNCFs in girders 1-4

Girder	Joint	SNCF and Line numbers *			
		Brace		Chord	
1	KK-gap1	4.37	E4A	3.75	E4D
	KK-gap2	6.22	A4A	5.12	A4D
	KK-overlap1	2.01	L3A	2.15	K4D
	KK-overlap2	3.01	O1A	2.45	M4D
2	KK-gap1	3.05	H1E	2.53	E4D
	KK-gap2	1.65	A4E	1.70	A4D
	KK-overlap1	2.71	J4E	2.71	K4D
	KK-overlap2	1.92	O1A	1.43	O4D
3	KK-gap1	4.39	F3E	3.55	E4D
	KK-gap2	4.64	B1A	2.90	A4D
	KK-overlap1	1.06	K2E	2.43	J4D
	KK-overlap2	2.44	O1A	1.32	M4D
4	KK-gap1	2.21	F4E	3.27	E4D
	KK-gap2	2.73	B1A	2.19	A4D
	KK-overlap1	1.35	L2E	2.53	J4D
	KK-overlap2	2.25	O1A	1.66	M4D

* Line No. E4A is for SNCF on Brace E, corner 4 and line A
  See Fig. 2 for line locations

Table 4. Comparison of Numerical and Experimental girder deflection.

Girder	Exp. Measured Deflection δ_{exp} (mm)	Numerical Calculated Deflection and the ratio between Numerical and Experimental deflections ($\delta_{num}/\delta_{exp}$)			
		KK-gap2 δ_{num} (mm) ($\delta_{num}/\delta_{exp}$)	KK-gap1 δ_{num} (mm) ($\delta_{num}/\delta_{exp}$)	KK-overlap1 δ_{num} (mm) ($\delta_{num}/\delta_{exp}$)	KK-overlap2 δ_{num} (mm) ($\delta_{num}/\delta_{exp}$)
1	7.97	8.28 (1.04)	8.08 (1.01)	8.02 (1.01)	7.90 (1.01)
2	8.8	8.76 (1.00)	8.68 (1.01)	8.96 (1.02)	8.80 (1.00)
3	11.34	11.57 (1.02)	11.70 (1.03)	11.78 (1.04)	11.63 (1.03)
4	9.0	9.21 (1.02)	9.32 (1.04)	9.03 (1.00)	9.11 (1.01)

5 Calibration of numerical work

The following steps are taken to calibrate the results of the numerical analyses with those found from experimental investigation.
a Comparison between the overall girder deflection measured experimentally and calculated numerically.

A good agreement is achieved with a maximum difference of 4%, See Table 4.

Table 5. Comparison between Numerical and Experimental nominal strains in the tension brace of RHS girders1 to 4.

Girder No.	Joiont No.	Extrapolated Nominal Strains						Axial Num. Exp.	Total Extrapolated Nominal Strains		
		Numerical			Experimental				Num.	Exp.	Num Exp
		Axial	i.p.b	o.p.b	Axial	i.p.b	o.p.b				
1	KK-gap1	288	137	50	270	125	56	1.07	476	451	1.06
	KK-gap2	386	99	109	424	151	20	0.91	596	595	1.00
	KK-overlap1	280	18	15	262	25	38	1.07	314	325	0.97
	KK-overlap2	440	25	164	462	11	177	0.95	630	650	0.97
2	KK-gap1	396	69	43	380	93	69	1.04	508	542	0.94
	KK-gap2	501	44	106	503	67	80	1.00	652	650	1.00
	KK-overlap1	354	22	31	347	52	20	1.02	408	419	0.98
	KK-overlap2	544	133	199	570	107	164	0.95	877	841	1.04
3	KK-gap1	330	132	56	307	180	21	1.07	518	508	1.02
	KK-gap2	370	102	87	365	146	50	1.02	560	561	1.00
	KK-overlap1	303	145	73	290	204	70	1.05	522	564	0.93
	KK-overlap2	420	124	179	394	165	134	1.07	725	693	1.05
4	KK-gap1	296	78	30	281	105	3	1.05	404	389	1.04
	KK-gap2	315	29	79	328	40	47	0.96	423	415	1.02
	KK-overlap1	263	77	13	271	69	36	0.97	354	376	0.94
	KK-overlap2	347	49	122	351	13	128	0.99	518	492	1.05

b   Comparisons are made between member strains in the tension braces of each joint. Although individual components of the element strains show variations, the total nominal strains are in good agreement (See Table 5). For the numerical investigation, stress concentration factors are also calculated. The SCF/SNCF ratios are calculated and plotted in Fig. 9. The average ratio between SCF and SNCF is found to be 1.1, which confirms the work on uniplanar joints using RHS by van Wingerde (1993).

c   Comparisons between extrapolated SNCFs determined numerically and calculated from experimental measurements show good agreement for all the joints. (Fig.10). The average ratio is found to be 1.04.

The finite element strategy applied in this numerical investigation is the basis of further parametric study which is currently under investigation. This study is being conducted for a wide range of KK multiplanar joints and parameters with β = 0.2 to 1.0, 2γ = 12.5 to 25 and τ = 0.5 to 1.0. The angle between the braces and the chord member is 30, 45 and 60 degrees. All joints include weld modelling. 17 load cases are applied to each KK-gap joint, 3 to each brace (axial, in-plane and out-of-plane bending) and a total of 5 on the two ends of the chord member (axial at one end and in-plane or out-of-plane bending at both

SCF − SNCF GIRDERS 1 − 4

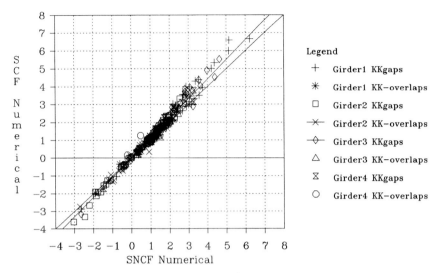

Fig.9. Comparison between Numerical SCF and SNCF.

SNCF for joints of GIRDERS 1,2,3 and 4

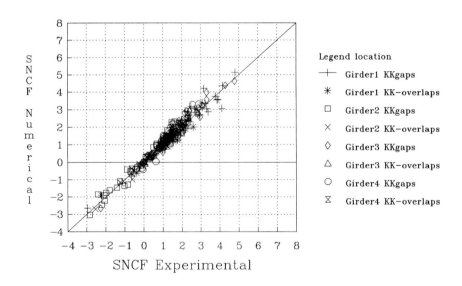

Fig.10. Comparison between Experimental and Numerical
SNCF values.

Table 6. SNCFs for lines A to E at corners 1 and 4
of brace 1 from a KK-gap joint.

Beta=0.2	Gamma=12		Tau=0.5	Teta=45	

SNCF			Lines		

Corners	Case	A	B	C	D	E
Corner1	1	2.280	1.858	0.693	0.064	2.525
	2	0.107	0.118	-0.053	-0.119	-0.057
	3	-0.090	-0.141	-0.017	-0.063	-0.059
	4	-0.019	-0.061	0.190	0.088	0.056
	5	-0.862	-0.174	-0.527	-0.416	-2.160
	6	0.023	0.024	-0.016	-0.024	-0.011
	7	0.029	0.009	0.062	0.034	0.050
	8	-0.004	-0.008	0.009	0.008	0.009
	9	-1.632	-0.584	-0.196	0.085	-1.997
	10	-0.009	-0.021	-0.010	-0.002	-0.016
	11	-0.011	-0.023	-0.025	-0.018	-0.017
	12	-0.007	-0.019	-0.005	0.002	-0.008
	13	-0.103	-0.382	0.748	1.359	-0.161
	14	-0.078	-0.215	0.396	0.770	-0.051
	15	-0.026	-0.056	0.230	0.150	-0.081
	16	-0.062	-0.152	0.324	0.528	-0.038
	17	-0.048	-0.040	-0.037	0.072	-0.056
Corner4	1	2.384	2.231	0.269	0.337	2.672
	2	-0.009	0.301	0.175	-0.235	-0.184
	3	-0.157	-0.190	0.034	-0.053	-0.180
	4	-0.079	-0.152	-0.186	-0.007	-0.023
	5	0.608	0.315	0.114	0.144	0.763
	6	0.002	0.060	0.047	-0.033	-0.031
	7	-0.041	-0.011	-0.033	-0.033	-0.044
	8	-0.014	-0.018	-0.010	-0.014	-0.009
	9	-0.715	-0.369	-0.066	0.004	-0.614
	10	0.013	-0.019	-0.024	0.003	0.015
	11	-0.010	-0.022	0.005	0.001	-0.013
	12	0.004	-0.019	-0.025	0.001	0.008
	13	0.227	-0.419	0.534	1.127	0.358
	14	0.145	-0.222	0.308	0.559	0.238
	15	0.064	-0.058	0.035	0.092	0.066
	16	0.102	-0.176	0.203	0.511	0.178
	17	0.084	-0.027	0.203	0.086	0.087

chord ands). In this way, the influence of different loadings on braces and chord ends on the SCFs and SNCFs at desired locations can be calculated, together with the influence of the unloaded braces. Table 6 shows one typical set of SNCF results for a joint with $\beta=0.25$, $2\gamma=12.5$ and $\tau=0.5$. The angle between the braces and the chord in the longitudinal (uniplanar) direction is 45°. In this table cases 1 to 4 are axial load, cases 5 to 8 are in-plane bending moments and cases 9 to 12 are out-of-plane bending moments on the braces No. 1 to 4. Case 13 is axial load on

the chord while case 14 and 16 are in-plane bending moments
and case 15 and 17 are out of plane bending moments on both
ends of chord. Boundary conditions for each load case on
braces are selected such that, no influence due to secondary bending moments in the chord is present in the SCFs and
SNCFs of each line. Fig. 11 demonstrates all the loading
cases of this parameter study.

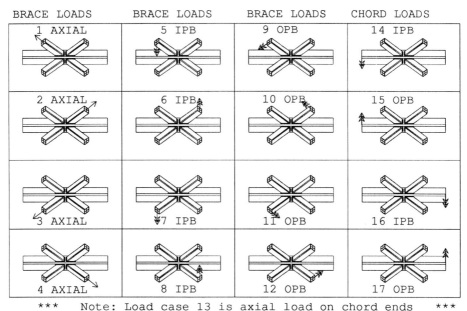

BRACE LOADS	BRACE LOADS	BRACE LOADS	CHORD LOADS
1 AXIAL	5 IPB	9 OPB	14 IPB
2 AXIAL	6 IPB	10 OPB	15 OPB
3 AXIAL	7 IPB	11 OPB	16 IPB
4 AXIAL	8 IPB	12 OPB	17 OPB

*** Note: Load case 13 is axial load on chord ends ***

Fig.11. Load cases for parameter study of isolated joints.
Boundary conditions not shown.

6 Conclusions

a Good agreement is obtained between numerically and
   experimentally determined total nominal strains in the
   tension braces of the joints. The maximum difference is
   between:

   0.94 <Nominal strain ratio(numerical/experimental)< 1.06

b Secondary bending moments are significant in multiplanar
   joints. The ratio between the total nominal strain and
   the nominal axial strain in the braces determined
   numerically varies between 1.10 and 1.72.
   The average is found to be 1.45. It is therefore
   important that for fatigue strength, the secondary
   bending moment are taken into account. In the absence of
   data for bending moments, it is recommended that axial
   strain be multiplied by a factor of 1.5 to allow for

bending moment contribution.

c Good agreement is found between numerically and experimentally extrapolated SNCFs. The average ratio is 1.04.

d The ratio between SCFs and SNCFs calculated numerically has an average of 1.10. Taking this into consideration together with findings from other investigation (van Wingerde et al (1991)), the ratio of SCF/SNCF = 1.1 is proposed for converting SNCF to SCF in multiplanar joints in RHS.

e SNCFs for gap joints are generally larger than those for overlap joints and the largest concentration of strains occur in the chord member around the gap region.

f The maximum SNCFs in the braces vary from 4.39 to 6.22 for multiplanar KK-gap joints with $2\gamma = 25$ (girders 1 & 3) and are slightly lower for the chords where the SNCFs vary from 2.9 to 5.12.

g For multiplanar KK-gap joints with $2\gamma = 12.5$ (girders 2 & 4) the maximum SNCFs for the braces vary from 1.65 to 3.05 and in the chord from 1.70 to 3.27.

h For the 100% overlap multiplanar joints with $2\gamma = 25$, the maximum SNCFs vary from 1.06 to 3.01 in the brace and 1.32 to 2.45 in the chord.

i The 50% overlap multiplanar joints of girders 2 and 4 have maximum SNCFs ranging from 1.35 to 2.71 in the brace and from 1.43 to 2.71 in the chord.

j The girder idealisation shown in Fig. 6 is a very suitable method of analysis for simulating joint behaviour in girders. The use of 8 noded solid elements for modelling rectangular hollow section joints is shown to be a good approach in modelling the weld and the joint members.

7 Acknowledgments

Authors would like to thank the STW (The Netherlands Technical foundation) for allowing the presentation of the numerical results related to this investigation. Appreciation is also extended to ECSC for permission to present the results of the experimental investigation used for the numerical work. The authors would also like to thank the Dutch Ministry of Economic Affairs for their financial contribution. The donation of hollow sections by van Leeuwen Buizen and the financial support by Evers Staalconstructies BV for the Dutch part of the programme is gratefully acknowledged. Finally, thanks are expressed to Mr. D. Dutta of Mannesmannröhren Werke A.G and members of Universität Karlsruhe (Prof. F. Mang, Dr-Ing Ö. Bucak and Dipl-Ing S. Herion) and Mr. N.F. Yeomans of British Steel for their cooperation on the ECSC project

8 References

Eurocode No. 3. (1992) Design of Steel Structures:
   Part 1 -General Rules and Rules for Buildings. CEN 1
   February 1992.
Panjeh Shahi, E Romeijn, A. Puthli, R.S. Wardenier, J.
   (1989) Fatigue strength of multiplanar welded,
   unstiffened hollow section joints and reinforcement
   measures for in-plane and multiplanar joints in repair.
   Literature study, Stevin report No. 25.6.89.40/A1 July
   1989. Delft University of Technology, Delft,
   The Netherlands.
Panjeh Shahi, E. Puthli, R.S. Wardenier, J.(1993)
   Comparison between linear and parabolic solid elemets
   used in modelling a RHS multiplanar KK-gap joint. Report
   to be published.
Puthli, R.S. Wardenier, J. Mang, F. Dutta D.(1992b)
   Fatigue behaviour of multiplanar welded hollow section
   joints and reinforcement measures for repair. Report-
   Part V. TNO-Bouw report: BI-92-0079/21.4.6394 : Stevin
   report: 6.92.17/A1/12.06. (Cnfidential)
Puthli, R.S. Koning, C.H.M. Wingerde, van A.M. Wardenier,
   J. Dutta, D.(1992b)
   Fatigue strength of welded unstiffened RHS joints in
   latticed structures and vierendeel girders. Final
   report, Part IV. Stevin report 25-6-89-37/A1,
   TNO-IBBC report BI-89-102/63.5.3820. (Confidential)
Romeijn, A. Puthli, R.S. Koning, C.H.M. de Wardenier, J.
   (1992b)
   Stress and strain concentration factors of multiplanar
   XX-joints made of circular hollow sections. Second
   International Offshore and Polar Engineering Conference
   (ISOPE-92), San Francisco-USA, June 1992.
Romeijn, A. Panjeh Shahi, E. Wardenier, J. Puthli, R.S.
   Dutta D.(1992a)
   Fatigue behaviour of multiplanar welded hollow section
   joints and reinforcement measures for repair. Report-
   Part IV. TNO-Bouw report: BI-92-0064/21.4.6394 : Stevin
   report:6.92.09/A1/12.06.
Verheul, A. Puthli, R.S. Wardenier, J. Dutta, D.(1992)
   Fatigue behaviour of multiplanar welded hollow section
   joints and reinforcement measures for repair. Report-
   Part II. TNO-Bouw report: BI-92-0036/21.4.6394 : Stevin
   report: 6.92.06/A1/12.06. (Confidential)
Wingerde, van A.M.(1992)
   The fatigue behaviour of T- and X-Joints made of Square
   Hollow Section. PhD Thesis, Delft University of
   Technology, June 1992

53 THEORETICAL AND EXPERIMENTAL STUDY ON THE LOCAL FLEXIBILITY OF TUBULAR JOINTS AND ITS EFFECT ON THE STRUCTURAL ANALYSIS OF OFFSHORE PLATFORMS

B. CHEN, Y. HU and H. XU
Department of Naval Architecture and Ocean Engineering, Shanghai
Jiao Tong University, China

Abstract
Theoretical and experimental studies made by the authors in recent
years on the local flexibilty of tubular joints of offshore platforms
are presented in this paper. Parametric formulae for evaluating the
local joint flexibility are proposed, which are based on the
calculated results by using the Semi-Analyical Method. Results from
steel model and PVC model tests are reported. An equivalent element
modelling the local joint flexibility in the global structural
analysis of offshore platforms is also presented.
Keywords: Local Joint Flexibility, Tubular Joint, Offshore Platform,
Structural Analysis, Model Test of Tubular Joints

1 Introduction

Welded tubular joints are key parts of offshore platforms. In recent
years, much attention has been devoted to the local flexibility of
tubular joints. Conventionally, tubular joints are considered to be
rigid in the global structural analysis of offshore platforms. But
research works revealed that the local joint flexibility will
redistribute the nominal stresses, increase the deformations, change
the natural frequencies and mode shapes of the structure. Therefore,
it is important to take the local joint flexibility into account in
the global structural analysis of offshore platforms.
   Researches on the local joint flexibilty was started at the
beginning of the 1980s (Bouwkamp et al. 1980, Rodabaugh 1980).
Since then a number of parametric formulae for evaluating the local
joint flexibilty has been proposed (Fessler et al. 1986a 1986b,
Ueda et al. 1987, Chen et al. 1990 1992). In recent years,
theoretical and experimental studies on the local joint flexibilty
were made by the authors. The works will be presented in this paper.
First, the definition of the local joint flexibility is given. The
Semi- Analytical Method is applied to the theoretical calculation of
the local joint flexibility. Parametric formulae are proposed on the
basis of the calculated results. Then, results from steel model and
PVC model tests are reported. Finally, an equivalent element
modelling the local joint flexibility in the global structural
analysis of offshore platforms is presented. The stiffness matrix of
the equivalent element is given.

Tubular Structures V. Edited by M.G. Coutie and G. Davies.
© 1993 E & FN Spon, 2–6 Boundary Row, London SE1 8HN. ISBN 0 419 18770 7.

2 Definition and Theoretical Calculation

When a tubular joint is loaded, the chord wall will deform locally and the local flexibility will be caused. The local joint flexibility (LJF) of tubular joints is defined as the local deformation caused by unit external load. For single-brace T and Y type tubular joints (see Fig.1), we have

$$LJF_{AX} = \frac{\delta}{P}, \qquad LJF_{IPB} = \frac{\phi_I}{M_I}, \qquad LJF_{OPB} = \frac{\phi_0}{M_0} \qquad (1)$$

where subscripts AX, IPB and OPB denote the loading cases of axial force, inplane and out-of-plane bending moments, respectively. δ is the local translational displacement at the intersection between the brace and the chord wall, which is caused by axial force P and is in the direction of the brace axis. ϕ_I and ϕ_0 are the local inplane and out-of-plane rotations at the same point, which are caused by inplane bending moment M_I and out-of-plane bending moment M_0, respectively.

In practical calculation, the intersection point between the brace axis and the chord wall (point o in Fig.1) is chosen as the reference point. The total displacement at point o in the direction perpendicular to the chord axis is considered to be the average of the total displacements at crowns and saddles of the intersection curve (points a, b, c and d in Fig.1), which are denoted by Δ_a, Δ_b, Δ_c and Δ_d, respectively. The total inplane and out-of-plane rotations at point o are considered to be the average of the rotation between points a and b, and the average of the rotation between points c and d, respectively. Then we have

$$LJF_{AX} = \frac{\delta}{P} = \frac{1}{P}[\frac{1}{4}(\Delta_a + \Delta_b + \Delta_c + \Delta_d) - v_0]\sin\theta \qquad (2)$$

$$LJF_{IPB} = \frac{\phi_I}{M_I} = \frac{1}{M_I}[\frac{1}{d}(\Delta_a - \Delta_b)\sin\theta - \varepsilon_{I0}] \qquad (3)$$

$$LJF_{OPB} = \frac{\phi_0}{M_0} = \frac{1}{M_0}[\frac{1}{d}(\Delta_c - \Delta_d) - \varepsilon_{0o}]\sin\theta \qquad (4)$$

where θ is the angle between the brace and the chord and d is the brace diameter. v_0, ε_{I0} and ε_{0o} are the deflection, inplane rotation and rotation about the chord axis at point o caused by bending and twisting of the chord as a beam, which can be obtained from the

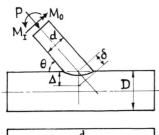

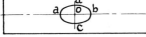

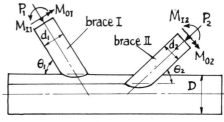

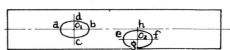

Fig.1. Y Type Joint   Fig.2. Two-brace Multiplanar Joint

544

simple beam theory and must be subtracted from the total deformation.

For multi-brace tubular joints, to account for the interaction between any two braces, it is necessary to define a local flexibility matrix (LJF matrix). In general, for a two-brace multiplanar tubular joint (see Fig.2), the relations between the local displacements and the external loads can be written as

$$\{\delta\} = [F]\{P\} \tag{5}$$

where, $\{\delta\}=[\delta_1, \phi_{o1}, \phi_{I1}, \delta_2, \phi_{o2}, \phi_{I2}]^T$, $\{P\}=[P_1, M_{o1}, M_{I1}, P_2, M_{o2}, M_{I2}]^T$. Subscripts 1 and 2 denote the brace I and II, respectively.

The square matrix $[F]$ in eqn(5) is the LJF matrix. It is symmetric from the reciprocal theorem. Since out-of-plane bending is not coupled with axial tension and compression and inplane bending, elements f_{21}, f_{32}, f_{42}, f_{62}, f_{53}, f_{54} and f_{65} in $[F]$ are equal to zero obviously. In addition, since f_{31} is the local inplane rotation at the intersection between the brace I and the chord wall, which is caused by unit axial force acting on the brace I, it is small in value and can be considered to be equal to zero approximately. Similarly, f_{64} can also be considered to be equal to zero approximately. Therefore, there are only 11 independent elements that need to be determined in $[F]$. $[F]$ can be finally written as

$$[F] = \begin{bmatrix} f_{11} & & & & & \\ 0 & f_{22} & & & \text{sym.} & \\ 0 & 0 & f_{33} & & & \\ f_{41} & 0 & f_{43} & f_{44} & & \\ 0 & f_{52} & 0 & 0 & f_{55} & \\ f_{61} & 0 & f_{63} & 0 & 0 & f_{66} \end{bmatrix} \tag{6}$$

To calculate the elements in the LJF matrix, let the total displacements at points o_1 and o_2 be the average of Δ_a, Δ_b, Δ_c, Δ_d and the average of Δ_e, Δ_f, Δ_g, Δ_h, respectively, let the total inplane rotations at points o_1 and o_2 be the average of the rotations between points a and b and between points e and f, respectively, and let the total rotations about the chord axis at points o_1 and o_2 be the average of the rotations between points c and d and between points g and h, respectively (see Fig.2). Consider in turn the cases in which brace I is loaded by an axial force P_1, an inplane bending moment M_{I1} and an out-of-plane bending moment M_{o1}, and brace II is loaded by an axial force P_2, an inplane bending moment M_{I2} and an out-of-plane bending moment M_{o2}, the elements in the LJF matrix can finally be obtained (Zhu 1992).

The total displacements at crowns and saddles of the intersection curves are calculated by using the Semi-Analytical Method. In this method, the classical theory of thin shells and the techniques of the Finite Element Method are combined. The chord and the braces of a tubular joint are treated as substructures of thin shells. The intersection curve between any two substructures are discretized into finite elements. The method was decribed in detail by Chen et al. (1983 1985). To verify the Semi-Analytical Method, a number of tubular joints was calculated, and the results were compared with those measured from model tests or evaluated from existing parametric formulae (Chen et al. 1990).

3 Parametric Formulae

For the convenience of practical use in designing offshore platforms, a set of parametric formulae for evaluating the local joint flexibility of the planar T, Y, symmetric K and TY type tubular joints is proposed by the authors, which is formulated through regression of the data from the calculation by using the Semi-Analytical Method (Chen et al. 1992). They are listed as follows.

3.1 Formulae for T and Y Type Tubular Joints

$$LJF_{AX} = 3.27\gamma^{2.26}e^{-3.05\beta}\sin^{1.89}\theta/ED \tag{7}$$
$$LJF_{IPB} = 132.49\gamma^{1.70}e^{-4.32\beta}\sin^{1.22}\theta/ED^3 \tag{8}$$
$$LJF_{OPB} = 52.19\gamma^{2.48}e^{-3.95\beta}\sin^{2.01}\theta/ED^3 \tag{9}$$

3.2 Formulae for Symmetric K Type Tubular Joints

$$f_{11} = f_{44} = 3.66\gamma^{2.19}e^{-2.74\beta}\sin^{2.11}\theta/ED \tag{10}$$
$$f_{41} = 1.37\gamma^{2.43}e^{-2.22\beta}\sin^{3.00}\theta/ED \tag{11}$$
$$f_{61} = -f_{43} = 4.76\gamma^{1.84}e^{-2.23\beta}\sin^{1.61}\theta/ED^2 \tag{12}$$
$$f_{22} = f_{55} = 49.59\gamma^{2.48}e^{-3.80\beta}\sin^{2.10}\theta/ED^3 \tag{13}$$
$$f_{52} = 5.61\gamma^{2.81}e^{-1.84\beta}\sin^{3.43}\theta/ED^3 \tag{14}$$
$$f_{33} = f_{66} = 111.49\gamma^{1.73}e^{-3.98\beta}\sin^{1.42}\theta/ED^3 \tag{15}$$
$$f_{63} = -26.33\gamma^{1.33}e^{-3.44\beta}\sin^{0.29}\theta/ED^3 \tag{16}$$

3.3 Formulae for TY Type Tubular Joints

$$f_{11} = 4.54\gamma^{2.16}e^{-2.85\beta}\sin^{2.48}\theta/ED \tag{17}$$
$$f_{41} = 1.30\gamma^{2.55}e^{-2.68\beta}\sin^{2.44}\theta/ED \tag{18}$$
$$f_{61} = 3.96\gamma^{1.91}e^{-2.10\beta}\sin^{1.97}\theta/ED^2 \tag{19}$$
$$f_{22} = 55.37\gamma^{2.45}e^{-3.82\beta}\sin^{2.22}\theta/ED^3 \tag{20}$$
$$f_{52} = 5.56\gamma^{2.86}e^{-2.10\beta}\sin^{2.64}\theta/ED^3 \tag{21}$$
$$f_{33} = 115.57\gamma^{1.74}e^{-4.08\beta}\sin^{1.52}\theta/ED^3 \tag{22}$$
$$f_{43} = -2.78\gamma^{1.99}e^{-1.97\beta}\sin^{0.43}\theta/ED^2 \tag{23}$$
$$f_{63} = -20.90\gamma^{1.44}e^{-3.21\beta}\sin^{0.59}\theta/ED^3 \tag{24}$$
$$f_{44} = 3.55\gamma^{2.21}e^{-2.85\beta}\sin^{0.03}\theta/ED \tag{25}$$
$$f_{55} = 58.07\gamma^{2.43}e^{-3.85\beta}\sin^{0.03}\theta/ED^3 \tag{26}$$
$$f_{66} = 170.02\gamma^{1.62}e^{-4.43\beta}\sin^{-0.01}\theta/ED^3 \tag{27}$$

The above parametric formulae are valid for tubular joints with geometric parameters $\gamma=10-30$ and $\beta=0.4-0.8$.

The results calculated from the above parametric formulae were discussed and verified by the authors (Chen et al. 1992). It can be concluded that these formulae are reliable on the whole in evaluating the local flexibility of tubular joints and can be recommended in use in the global structural analysis of offshore platforms.

4 Model Tests

4.1 Test on Steel Models of T Type Tubular Joints
From March to June of 1990, during a fatigue test on steel models of T type tubular joints, which was carried out by the authors at

Shanghai Jiao Tong University, four models of large scale were
arranged to have their local joint flexibility measured under axial
loading (Chen et al. 1991). Dimensions of the chord and the brace
are D=500mm, T=16mm, L=2400mm, d=250mm (Model No.24 and No.26) and
400mm (Model No.12 and No.13), t=14mm, l=760mm. The Young's Modulus
of the material is $E=218.76kN/mm^2$.

Displacement transducers were used to measure the total
displacements at crowns and saddles of the intersection curve (points
a, b, c and d) of each model. Measured value of LJF_{AX} of each model
was calculated from eqn(2). Dimensionless local flexibility
coefficients of the models measured in the test are listed in Table 1
and compared with those evaluated from parametric formulae proposed
by different authors.

Table 1. Local Flexibility of Steel Models of T Type Tubular Joints

Model No.	$F_{AX}=LJF_{AX}ED$ (1)	(2)	(3)	(4)	Model No.	$F_{AX}=LJF_{AX}ED$ (1)	(2)	(3)	(4)
24	419/522	291	399	355	12	298/225	88	226	142
26	551/638	291	399	355	13	319/216	88	226	142

Note: (1) measured, tension/compression
       (2) Fessler (3) Ueda (4) eqn(7)

Table 2. Local Flexibility of PVC Models of TY Type Tubular Joints

θ_1	$f_{11}(10^{-3}mm/N)$ (1)	(2)	$f_{41}(10^{-3}mm/N)$ (1)	(2)	$f_{61}(10^{-5}1/N)$ (1)	(2)	$f_{22}(10^{-7}1/N\text{-}mm)$ (1)	(2)
45°	0.423	0.375	0.337	0.364	0.154	0.175	1.632	1.819
45°	0.373	0.375	0.263	0.364	0.135	0.175	1.493	1.819
60°	0.864	0.620	0.804	0.597	0.227	0.260	2.474	2.853

θ_1	$f_{52}(10^{-7}1/N\text{-}mm)$ (1)	(2)	$f_{33}(10^{-7}1/N\text{-}mm)$ (1)	(2)	$f_{43}(10^{-5}1/N)$ (1)	(2)	$f_{63}(10^{-7}1/N\text{-}mm)$ (1)	(2)
45°	1.318	1.302	0.676	0.754	-0.134	-0.282	-0.227	-0.128
45°	1.158	1.302	0.771	0.754	-0.141	-0.282	-0.124	-0.128
60°	1.712	2.224	0.894	1.026	-0.163	-0.308	-0.309	-0.145

θ_1	$f_{44}(10^{-3}mm/N)$ (1)	(2)	$f_{55}(10^{-7}1/N\text{-}mm)$ (1)	(2)	$f_{66}(10^{-7}1/N\text{-}mm)$ (1)	(2)
45°	0.662	0.791	3.211	3.786	1.120	1.106
45°	0.633	0.791	3.627	3.786	0.871	1.106
60°	1.069	0.796	3.713	3.809	0.955	1.103

Note: (1) measured (2) calculated

4.2 Test on PVC Models of TY Type Tubular Joints

The test was carried out by the authors at Shanghai Jiao Tong University (Chen et al. 1992 1993). The PVC models include two TY type tubular joints with the angle between the inclined brace and the chord $\theta_1=45°$, and one with $\theta_1=60°$. Dimensions of the chord and the brace are D=164.9mm, T=4.74mm, L=890.0mm, d=90.0mm, t=4.00mm, l=300.0mm. The Young's Modulus of the material is E=3134.3N/mm² for models under axial force and inplane bending moment, and E=4278.1N/mm² for models under out-of-plane bending moment.

After measuring the total displacements at crowns and saddles of the intersection curves of each model when it was loaded by different loads, the measured values of the elements in the LJF matrix of the model were calculated. The measured values of elements in the LJF matrix of the models are listed in Table 2 and compared with those evaluated from the parametric formulae presented in this paper.

5 Equivalent Element Modelling the Local Joint Flexibility

An equivalent element modelling the local joint flexibility in the global structural analysis of offshore platforms is developed by the authors (Hu et al. 1992, Zhu 1992). At the tubular joint, the brace and the chord are considered to be connected through an equivalent element (see Fig.3). For a general multiplanar tubular joint with more than two braces, equivalent elements are introduced to connect any two braces and the chord.

By considering the relation between external loads acting on the tubular joint and local deformations of the joint, stiffness matrix of the equivalent element is derived. The stiffness matrix can be expressed by the following equation.

$$[K] = [B][A]^{-1}[B]^T \qquad (28)$$

For general two-brace multiplanar tubular joints, the equivalent element is a triangle connecting points i, j and k (see Fig.3). The matrices [B] and [A] in eqn(28) are

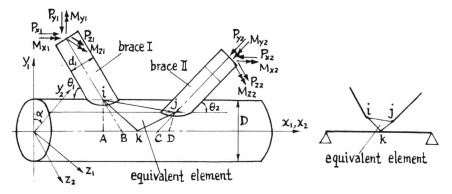

Fig.3. Two-brace Multiplanar Tubular Joint and Equivalent Element

$$[B] = \begin{bmatrix}
1 & 0 & 0 & 0 & 0 & 0 & 0 & 0 & 0 & 0 & 0 & 0 \\
0 & -1 & 0 & 0 & 0 & 0 & 0 & 0 & 0 & 0 & 0 & 0 \\
0 & 0 & 1 & 0 & 0 & 0 & 0 & 0 & 0 & 0 & 0 & 0 \\
0 & 0 & 0 & 1 & 0 & 0 & 0 & 0 & 0 & 0 & 0 & 0 \\
0 & 0 & 0 & 0 & 1 & 0 & 0 & 0 & 0 & 0 & 0 & 0 \\
0 & 0 & 0 & 0 & 0 & 1 & 0 & 0 & 0 & 0 & 0 & 0 \\
0 & 0 & 0 & 0 & 0 & 0 & -1 & 0 & 0 & 0 & 0 & 0 \\
0 & 0 & 0 & 0 & 0 & 0 & 0 & -\cos\alpha & -\sin\alpha & 0 & 0 & 0 \\
0 & 0 & 0 & 0 & 0 & 0 & 0 & -\sin\alpha & \cos\alpha & 0 & 0 & 0 \\
0 & 0 & 0 & 0 & 0 & 0 & 0 & 0 & 0 & 1 & 0 & 0 \\
0 & 0 & 0 & 0 & 0 & 0 & 0 & 0 & 0 & 0 & -\cos\alpha & -\sin\alpha \\
0 & 0 & 0 & 0 & 0 & 0 & 0 & 0 & 0 & 0 & -\sin\alpha & \cos\alpha \\
-1 & 0 & 0 & 0 & 0 & 0 & 1 & 0 & 0 & 0 & 0 & 0 \\
0 & 1 & 0 & 0 & 0 & 0 & 0 & \cos\alpha & \sin\alpha & 0 & 0 & 0 \\
0 & 0 & -1 & 0 & 0 & 0 & 0 & \sin\alpha & -\cos\alpha & 0 & 0 & 0 \\
0 & 0 & -R & -1 & 0 & 0 & 0 & 0 & -R & -1 & 0 & 0 \\
0 & 0 & -Ak & 0 & -1 & 0 & R\sin\alpha & -kD\sin\alpha & kD\cos\alpha & 0 & \cos\alpha & \sin\alpha \\
R & -Ak & 0 & 0 & 0 & -1 & -R\cos\alpha & kD\cos\alpha & kD\sin\alpha & 0 & \sin\alpha & -\cos\alpha
\end{bmatrix} \tag{29}$$

$$[A] = \begin{bmatrix}
\dfrac{1}{k_{x1}} \\
0 & \dfrac{f_{11}}{s_1 s_1} \\
0 & 0 & \dfrac{1}{k_{z1}} \\
0 & 0 & 0 & \dfrac{f_{22}}{s_1 s_1} \\
0 & 0 & 0 & 0 & \dfrac{1}{k_{o1}} & & & & \text{sym.} \\
0 & 0 & 0 & 0 & 0 & f_{33} \\
0 & 0 & 0 & 0 & 0 & 0 & \dfrac{1}{k_{x2}} \\
0 & \dfrac{f_{41}}{s_1 s_2} & 0 & 0 & 0 & \dfrac{f_{43}}{s_2} & 0 & \dfrac{f_{44}}{s_2 s_2} \\
0 & 0 & 0 & 0 & 0 & 0 & 0 & 0 & \dfrac{1}{k_{z2}} \\
0 & 0 & 0 & \dfrac{f_{52}}{s_1 s_2} & 0 & 0 & 0 & 0 & 0 & \dfrac{f_{55}}{s_2 s_2} \\
0 & 0 & 0 & 0 & 0 & 0 & 0 & 0 & 0 & 0 & \dfrac{1}{k_{o2}} \\
0 & \dfrac{f_{61}}{s_1} & 0 & 0 & 0 & f_{63} & 0 & 0 & 0 & 0 & 0 & f_{66}
\end{bmatrix} \tag{30}$$

In eqn(29), $R=D/2$ and α denotes the angle between the y_1 axis and the y_2 axis (see Fig.3). In eqn(30), $s_1=\sin\theta_1$ and $s_2=\sin\theta_2$. k_{x1}, k_{x2}, k_{z1} and k_{z2} are the stiffnesses of the chord wall in the x direction and the z direction, respectively, while k_{o1} and k_{o2} are the rotational stiffnesses of the chord wall about the y axis. Since they are much greater than the stiffnesses in the y direction, they can be considered to approach to infinity in the matrix [A].

With the equivalent element, the local joint flexibility can be taken into account in the global structural analysis of offshore platforms. It can easily be used together with general FEM program such as SAP5. A preprocessing program for automatically generating data, which coincide with the input format of SAP5, has been developed by the authors.

Some numerical examples were calculated (Zhu 1992). The results show that, when the local joint flexibility is considered, the static and dynamic behaviour of the structure is different from that from a conventional method. Therefore, it is necessary to take the LJF into account in the global structural analysis of offshore platforms.

6 References

Bouwkamp,J.G., Hollings,J.P., Malson,B.F. and Row,D.G. (1980) Effect of joint flexibility on the response of offshore structures, in **Proc. Offshore Technol. Conf.**, Houston, TX, Paper OTC 3901

Chen,B.Z., Hu,Y.R. and Tan,M.J. (1990) Local joint flexibility of tubular joints of offshore structures, **Marine Structures**, 3

Chen,B.Z., Xu,H.T., Hu,Y.R. and Pan,H. (1991) Steel model test on local flexibility of tubular joints of offshore platforms, in **Proc. Int. Symp. Marine Structures**, Shanghai, China

Chen,B.Z., Gu,J.M. and Hu,Y.R. (1992) Local joint flexibility of offshore platforms and its parametric analysis, Research Report, Department of Naval Architecture and Ocean Engineering, Shanghai Jiao Tong University

Chen,B.Z., Hu,Y.R., Xu,H.T. and Pan,H. (1993) Parametric analysis and experimental study on the local flexibility of TY-type tubular joints, **The Ocean Engineering** (in press)

Chen,T.Y., Chen,B.Z. and Wang,Y.Q. (1983) The stress analysis and experimental research of tubular joints of offshore drilling platforms, in **Proc. 2nd Int. Symp. Offshore Mechanics and Arctic Engineering**, Houston, TX

Chen,T.Y. and Wang,Y.Q. (1985) The stress analysis of tubular joints of offshore drilling platforms by variational method, in **Proc. 4th Int. Symp. Offshore Mechanics and Arctic Engineering**, Dollas, TX, 1985

Fessler,H., Mockford,P.B. and Webster,J.J. (1986a) Parametric equations for the flexibility matrices of single brace tubular joint in offshore structures, in **Proc. Instn. Civ. Engrs.**, Part 2

Fessler,H., Mockford,P.B. and Webster,J.J. (1986b) Parametric equations for the flexibility matrices of multi-brace tubular joint in offshore structures, in **Proc. Instn. Civ. Engrs.**, Part 2

Hu,Y.R., Chen,B.Z. and Ma,J.P. (1992) Equivalent element representing local flexibility of tubular joints in structural analysis of jacket structure of offshore platforms, **Computational Structural Mechanics and Applications**, 3

Rodabaugh,E.C. (1980) Review of data relevant to the design of tubular joints for use in fixed offshore platforms, **WRC** 256

Tebbett,I.E. (1982) The reappraisal of steel jacket structures allowing for the composite action of grouted piles, in **Proc. Offshore Technol. Conf.**, Houston, TX, Paper OTC 4194

Ueda,Y., Rashed,S.M.H. and Nakacho,K. (1987) An improved joint model and equations for flexibility of tubular joints, in **Proc. 6th Int. Symp. Offshore Mechanics and Arctic Engineering**, Houston, TX

Zhu,J. (1992) Structural analysis of offshore platforms considering local joint flexibility, Master's Thesis, Department of Naval Architecture and Ocean Engineering, Shanghai Jiao Tong University

54 SECONDARY MOMENTS IN RHS LATTICE GIRDERS

M. SAIDANI and M.G. COUTIE
University of Nottingham, UK

Abstract
The paper describes the application of the design rules of British
Standard BS 5950:part 1 to Rectangular Hollow Section (RHS) lattice
girders. Techniques of analysis including joint flexibility have
previously been described. The significance of the forces in the bracing
members, in comparison with those in the chord is discussed in
accordance with BS 5950:Part 1 in which provision is made for the
design of members subjected to combined axial force and bending
moment.
Keywords: Finite Element Analysis, Rectangular Hollow Sections
(RHS), British Standard BS 5950, Design, Lattice Girders, Secondary
Stresses.

1 Introduction

In truss design, Philiastides (1988) concluded that axial forces can be
accurately predicted by simple pin-joint analysis. For chord moments a
rigid-joint analysis gives a good prediction if the joints are overlapped,
and allowance is made for the consequent eccentricity. For gapped joints
the prediction is poorer. Bracing moments are not easily predicted. In a
previous publication, Coutie and Saidani (1989), a technique of analysing
lattice girders made from RHS has been discussed. The technique, based
on the Finite Element (F.E) method, makes allowance for the flexibility of
the joints to be included in the analysis of trusses. However, there is a
considerable amount of computation required. There is therefore a
temptation to ignore the moments particularly in the bracings if this can
be justified. Fig. 1 shows the general layout of half of the truss , designated
T1, on which this technique was demonstrated, with the member sizes (in
mm) , loading and boundary conditions. The truss had noding gap joints.

Tubular Structures V. Edited by M.G. Coutie and G. Davies.
© 1993 E & FN Spon, 2–6 Boundary Row, London SE1 8HN. ISBN 0 419 18770 7.

The aim of the present study is to investigate the significance (in carrying out the design checks) of secondary bending stresses in the members (due to member continuity and fixity at joints) compared to those due to axial forces. In Tables 1 and 2 (see Appendix) are given the axial force and bending moment respectively in the members for an applied ram load of 50kN (these agree well with values obtained experimentally) . The tables are to be read in conjunction with Fig.1. It can be seen that, at first sight, the moment values in the bracings are very small compared to those in the chords (upper and lower) and it is very tempting to ignore them in the design of the members. Similarly, the axial forces in the bracings are also small compared with those in the chords. It therefore becomes relevant to talk in terms of ratios bending effect to axial effect for a better assessment of the components in each member.

It is acknowledged in BS 5950 (1990), Clause 4.10, that secondary moments do occur, but may be safely ignored if the slenderness ratios of bracing and chord members are greater than 100 and 50 respectively. The values for truss T1 are less than 69 and 36 respectively, and therefore secondary moments ought to be considered. However, the corresponding limits in the European code EC 3 (1993) (Annex K, Clause 4) are expressed in terms of a length to depth ratio, and by this code no consideration of secondary moments would be required.

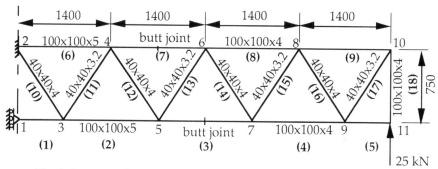

Fig.1. Layout of truss T1 with loading and boundary conditions.

2 Design of members to BS 5950

Clause 4.8.2 of BS 5950 (1990) states that "Tension members should be checked for capacity at the points of greatest bending moments and axial loads, usually at the ends". On the other hand, Clause 4.8.3.1 states that "Compression members should be checked for local capacity at the points of greatest bending moment and axial load (usually at the ends). This capacity may be limited either by yielding or local buckling depending on

552

the section properties. The member should be checked for overall buckling by either the simplified or more exact approach".

Before any check is made the members are first classified as being either plastic, compact, semi-compact or slender. The classification depends on the cross-section dimensions of the member and the material properties. However, the simplified approach given below can be applied to all classes.

3 The design approach

The notations used in the design equations below are explained in the BS 5950:Part 1:1990. For the specific aim of this study, of investigating separately, the effects of B.M and axial force in the members, the simplified approach was adopted. (Use of a more exact approach is allowed by the standard, but with this, the separate influence of axial force and moment is less clearly seen.)

3.1 Tension members with moments
BS 5950 states that tension members should be checked for capacity at points of greatest B.M's and axial forces. The following relationship should be satisfied:

$$\frac{F}{A_e \cdot p_y} + \frac{M_x}{M_{cx}} + \frac{M_y}{M_{cy}} \leq 1 \tag{1}$$

where

F	is the applied axial load in the member
A_e	is the effective area
p_y	is the design strength
M_x and M_y	are the moment capacities about the major and minor axis (in the critical region) respectively
M_{cx} and M_{cy}	are the moment capacities about the major and minor axis in absence of the axial load respectively

3.2 Compression members with moments
For compression members with moments, BS 5950 recommends that these should be checked for local capacity at the points of greatest B.M's

and axial forces. This capacity can be limited by yielding or local buckling depending on the section properties. The member should also be checked for overall buckling.

(i) Local capacity check

The appropriate relationship is similar to that of Eq.(1), with A_g replacing A_e (A_g is the gross cross-sectional area):

$$\frac{F}{A_g \cdot p_y} + \frac{M_x}{M_{cx}} + \frac{M_y}{M_{cy}} \leq 1 \tag{2}$$

(ii) Overall buckling check
In the simplified approach proposed in BS 5950, the following relationship should be satisfied:

$$\frac{F}{A_g \cdot p_c} + m \cdot \left(\frac{M_x}{M_b} + \frac{M_y}{p_y Z_y} \right) \leq 1 \tag{3}$$

where

p_c is the compressive strength

m is the equivalent uniform moment factor

M_b is the buckling resistance moment capacity

Z_y is the elastic section modulus about the minor axis

For a plane truss M_y is normally zero.

4 Design example of members

4.1 General
Some members selected from truss T1 (see Fig.1), were checked using the equations (1), (2) and (3) above. Table 3 (see Appendix) gives a summary of the design of the most critical members (struts, ties, and chords). Also given in the table are the respective ratios of the effects of the axial force and the bending moment in the member. It is worth noting that the

advantage of using the simplified approach (rather than the exact one) is that it shows clearly, and separately, the effects of axial force and bending and their relative significance.

The results show that the bending moment terms are as high as 76% of the axial force term for the local capacity check (Eq.(2)) and 42% for the overall buckling check (Eq.(3)) respectively for the strut 14. This shows that the bending moment effect is almost as significant as the axial force effect and cannot be ignored.

5 Discussion of the results

The significance of the small bending moments in the bracings (compared to those of the chord) is seen above. Traditional design in engineering offices is frequently based on pin-jointed analysis where members are designed on the basis of axial forces, moments arising from joint rigidity being ignored (though any moments arising from joint eccentricity are included later). The secondary moments arising in the bracings due to member continuity often appear small (if calculated at all) and the designer may be tempted to ignore them. The results obtained here highlight the fact that the effect of moments in the bracings may be as important as the axial force and should not be ignored.

It can be seen from Table 3 that in the chord members, the bending moment term is up to 42% of the axial force term (member 6). However, for one of the bracing members (member 14) the corresponding figure is 76%. The influence of bending is therefore more significant in the bracings than in the chord. Although BS 5950 is a limit state design code it allows the use of equations (1), (2) and (3) to be applied to forces obtained from a linear elastic analysis. At collapse the sum of the left hand side terms of equations (1), (2) and (3) should approach unity, and it is of interest to note that this was observed in the tests of truss T1 (Philiastides, 1988).

It would be interesting to apply the checks above, for trusses with different joint configurations and for different load levels, but the authors have not so far been able to carry this out. Such results would be valuable for setting up some general design rules for the bracings.

Similar Clauses occur in EC 3 (Clause 5.4.8), and their application would lead to bending term/axial term ratios of 43% and 57% for the chord and bracing respectively (compare with the above values of 42% and 76% obtained from BS 5950).

6 Conclusions

The importance of recognising and taking into account the bracing moments has been shown through the example of Table 3, the basis for the design being BS 5950. The bending moment terms (expressed as a percentage of the corresponding axial force term) were seen to vary from 76% (in the bracings) to 42% (in the chord). Further checks should be undertaken for other configurations of trusses (and with different joint details) in order to draw general design rules.

7 References

European Prestandard ENV 1993-1-1 **Eurocode 3 - Design of steel structures. Part 1.1: General rules and rules for buildings.**

British Standards Institution (1990) **BS 5950 - Structural use of steelwork in building; Part 1.**

Coutie, M.G. and Saidani, M. (1989) The use of Finite Element techniques for the analysis of RHS structures with flexible joints. **Tubular structures, the 3rd. International Symposium.** Ed. E. Niemi and P. Mäkeläinen. Elsevier Applied Science, Finland 1-2, pp. 224-231.

Philiastides, A. (1988) Fully overlapped rolled hollow section welded joints in trusses. **Ph.D. thesis**, Department of Civil Engineering, The University of Nottingham.

8 Appendix

Table 1. Axial force distribution in the members

Member No.	Axial Force (kN)		
	Pin joint	Rigid joint assumption	Flexible joint (F.E. analysis)
1	186.5	183.1	181.9
2	139.9	138.6	138.1
3	93.2	92.4	91.9
4	46.6	45.9	46.2
5	0.0	2.1	3.6
6	-163.2	-160.7	-159.8
7	-116.6	-115.5	-115.1
8	-70.0	-69.0	-69.0
9	-23.3	-23.1	-24.1
10	-34.2	-33.8	-32.2
11	34.2	32.3	31.3
12	-34.2	-33.6	-33.2
13	34.2	33.9	33.5
14	-34.2	-33.8	-32.9
15	34.2	34.1	32.8
16	-34.2	-33.3	-31.8
17	34.2	30.7	29.8
18	-25.0	-23.1	-23.0

Table 2. Bending moment distribution in truss T1 members

Member No.	End	Bending Moments (Nm) Rigid joint assumption	Flexible joint (F.E.)
1	1	-999.5	-1182.9
	3	-999.5	-1182.9
2	3	-1020.3	-1469.7
	5	-396.6	-277.2
3	5	-506.5	-906.9
	7	-19.2	-40.4
4	7	-275.5	-432.5
	9	259.8	170.4
5	9	-369.2	-113.1
	11	929.7	1265.6
6	2	-1648.6	2300.0
	4	-308.7	150.5
7	4	419.4	666.8
	6	331.3	396.4
8	6	408.3	884.2
	8	-257.2	-99.8
9	8	371.7	517.9
	10	-625.1	-1467.3
10	3	14.5	150.8
	2	45.4	-20.9
11	3	6.2	205.5
	4	-55.0	-244.1
12	5	83.8	289.1
	4	-56.2	-271.7
13	5	26.1	270.7
	6	-44.8	-287.8
14	7	55.7	314.8
	6	-32.3	-307.2
15	7	27.8	302.9
	8	-49.1	-304.6
16	9	79.3	291.1
	8	-65.4	-312.5
17	9	30.2	212.8
	10	-2.7	-18.5
18	10	929.7	1263.8
	11	-622.5	1445.1

Table 3. Design checks of T1 truss members to BS5950

Member No.	Member size	Shear force Fv (N)	Axial force F (kN)	Moment Mx (Nm)	BS 5950* checks				% bending term / axial term	
					Local capacity		Overall buckling		Local	Overall
					F/Ag.py	Mx/Mcx	F/Ag.pc	mMx/Mb	capacity	buckling
14	40x40x4	-606.0	-32.9	314.8	0.21	0.16	0.24	0.15	76%	42%
15	40x40x3.2	-592.0	32.8	304.6	0.26	0.18	-	-	69%	-
6	100x100x5	1535.3	-159.8	2300.0	0.31	0.13	0.32	0.13	42%	22%
2	100x100x5	852.2	138.1	1469.7	0.27	0.08	-	-	30%	-
1	100x100x5	0.0	181.9	-1182.9	0.35	0.06	-	-	17%	-

Notes:

(i) Values shown are the computed values for a ram load of 50 kN

(ii) Table to be read in conjunction of Fig.1

(iii) (*) See Equations (1), (2), and (3)

55 BEHAVIOUR OF RECTANGULAR HOLLOW SECTION TRUSSES WITH STOCKY MEMBERS

E.J. NIEMI
Lappeenranta University of Technology, Finland

Abstract

A simple truss with stocky members is designed from rectangular hollow sections. Its strength is analysed as a pin-jointed truss, as a rigidly-jointed frame, and as a shell structure with nonlinear Finite Element Analysis (FEA). A three criterion analysis method is suggested for taking the secondary stresses into account. The rationale of the proposal is discussed in the light of the nonlinear FEA results.

Keywords: Trusses, Secondary stresses, Finite Elements, Tubular Construction, Rectangular Hollow Sections.

1 Introduction

Trusses with stocky members experience high secondary bending stresses due to rigid joints. According to most design codes, e.g. ENV 1993 (1992), trusses with predominantly static loading may be analysed as pin-jointed trusses, thus ignoring the secondary bending moments. This practice is acceptable because the secondary bending moments vanish at the limit state for the compact cross sections and ductile material specified by the design code. In order to prevent extremely high secondary stresses and strains occurring in trusses with stocky members, the inclusion of a restriction of the ratio member length/profile height in Annex K of ENV 1993 has been proposed. A minimum value of 6 is suggested if the secondary stresses are to be ignored.

If, for some reason, more stocky members are designed, should the analysis in that case be based on the sum of primary and secondary stresses resolved from a rigid jointed model? And what kind of strength criterion should then be specified? One interpretation is proposed by Anon. (1992): If the secondary stresses are excessive (higher than N/mm²), the "truss" should be regarded as a "frame", and the members should be designed for the combined effects. However, such an abrupt jump in the design criterion is not justified, as would be the case if the same design strength were applied as for primary stresses only.

Tubular Structures V. Edited by M.G. Coutie and G. Davies.
© 1993 E & FN Spon, 2–6 Boundary Row, London SE1 8HN. ISBN 0 419 18770 7.

In the design of pressure vessels according to ASME Code, the stress analysis contains several parallel criteria which all must be satisfied. If the same principles were applied to truss structures, the following parallel analyses could be reasonable:

(i)   Primary stress <= design strength
(ii) Range of primary + secondary stress <= 2 times yield strength
(iii) Primary + secondary stress <= 2 times yield strength.

All stresses are calculated according to elastic behaviour. Pinned joints are assumed in the first case, and rigid joints in the latter two cases.

The first criterion is the same as for usual truss structures. It must always be analysed. The second criterion is known as the shakedown criterion. This should guarantee that after some yielding at first loading, the subsequent stress fluctuations remain elastic. The stress range is calculated using characteristic non-factored loads. The second criterion will not become critical for usual trusses but may in the case where the members are stocky and the load fluctuations are large. The third criterion restricts the plastic strains at first loading to an acceptable level.

In fatigue design recommendations given by the International Institute of Welding, IIW (1981) the second criterion is more stringent. In order to guarantee that elastic stress analysis will be sufficient for use in fatigue analysis, the nominal stress range is restricted to 1.5 times yield instead of twice the yield.

The advantage of this three-criterion method is that no abrupt jumps occur. When the members become more stocky, the second or third criterion becomes effective below a certain length/depth ratio. The designer would soon learn when these criteria need to be checked.

In this paper, the rationale of the suggested three-criterion method is studied. A rectangular hollow section truss is designed with stocky members. The nominal primary and secondary stresses are calculated using linear truss and frame analysis methods. The load values are then determined such that all the three criteria are satisfied. Moreover, the truss is modelled with thin shell elements, and a nonlinear analysis is made, assuming both nonlinear material and geometrical nonlinearity. This analysis is used to investigate whether the proposed criteria lead to real shakedown behaviour of the member walls taking into account the local phenomena at the joints.

2 Materials and methods

2.1 Material properties
A high strength steel was chosen with a yield strength of 460 N/mm². A bilinear stress-strain behaviour was assumed according to Figure 1. This is supposed to simulate the behaviour of the cold-formed material of the rectangular hollow sections. The elastic modulus and the Poisson's ratio were assumed as 206,000 N/mm² and 0.3, respectively.

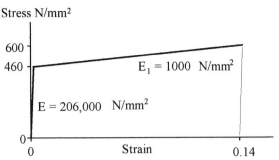

Fig. 1. Assumed stress/strain relationship for the steel material

2.2 Truss design

After initial trials, the design shown in Figures 2 and 3 was chosen. The dimensions satisfy very closely the limiting value of length/depth ratio, equal to 6, for all members. The centre lines of all members intersect at the nodal points, and the loads are applied at nodal points. Thus, no primary bending stresses will occur. The design is not aimed to represent any particularly good or commendable design. As a matter of fact, an attempt was made to find a worst case design.

The rectangular hollow sections do not correspond exactly to any standard profiles. The important values of the cross-sections are given in Figure 2. The diagonal member, D2, was intentionally underdimensioned, because preliminary frame analyses had shown that the actual force in that member is much lower than predicted by the analysis made for a pinned-joint truss.

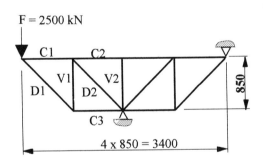

Member	Axial force kN	Size mm	Area mm^2
C1	2500	255x155x15	10870
C2	5000	255x155x15	10870
C3	-2500	255x155x15	10870
D1	-3536	202x202x12.5	9148
D2	-3536	199x199x9	6590
V1	2500	199x149x9	5690
V2	0	199x149x9	5690

Fig. 2. Schematic drawing of the truss showing the predicted ultimate load with corresponding primary member forces.

2.3 Linear frame analysis

A symmetric half of the structure was modelled with rigidly-joined beam elements. The sum of primary and secondary stresses was resolved for both ends of each member. The highest magnitude of stress was used as the criterion for determining the allowable load fluctuation according to criterion (ii).

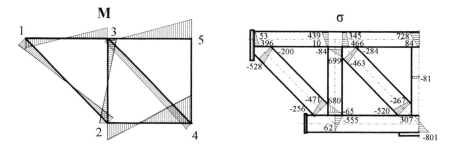

Fig. 3. Frame analysis results: bending moment diagram with node numbers and an overview of theoretical distributions of the primary + secondary stresses.

2.4 Joint strength analysis

The static strengths of the joints were checked according to IIW Recommendations using the program SHS, Issue 2.0, (1992) published by the British Steel. The nominal member forces, corresponding to the ultimate load, given in Figure 2, and a yield strength of 460 N/mm² were used.

2.5 Nonlinear analysis

Due to symmetry, one quarter of the structure was modelled with 4-noded thin shell elements of ABAQUS (1989) program. The pre- and post-processing was done using the I-DEAS VI (1990) Supertab program. Figure 4 shows an overview of the finite element mesh.

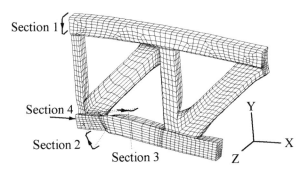

Fig. 4. Thin shell element model of the truss with exaggerated deformations.

The stiffness of each brace-chord intersection, including weld fillet, was modelled by specifying double thickness for those chord elements located between weld toe and root, as well as for the first row of elements in the brace members. The sides of the vertical members which are overlapped by the diagonal members were not fixed to the chord. The round corners of the hollow sections were modelled with inclined plane shell elements for simplicity. Care was taken to get equal cross-sectional areas for the members both in the frame analysis and in the shell analysis.

A thick plate, 500 mm long, was assumed to be welded on the chord at the location of the support. This was modelled by fixing all displacements of the nodes at and inside the periphery of this imaginary plate, including the upper nodes of the inclined corner elements of the chord. The point load was applied on the end plate of the chord.

On the planes of symmetry, displacements in these planes were allowed, but the normal displacement and rotations around the two axes in the corresponding plane were fixed. These boundary conditions prevent antisymmetric local buckling in the compressed chord, but this was not considered to have any effect on the resulting behaviour.

No initial deformations were modelled in order to initiate buckling phenomena. The deformations caused by the variable stiffnesses in the joints were considered large enough to activate realistically any buckling mode.

The ABAQUS version used automatically takes care of nodal force and moment tolerances in the iteration procedure. The number of elements in the model was 2280 and the number of degrees of freedom was 14550. The model was solved on the mainframe computer, CONVEX 3420, of Lappeenranta University of Technology. One nonlinear run from zero-to-maximum load and back to the minimum load required several hours CPU time.

2.6 Loads

The ultimate value for the point load was easily determined according to criterion (i), such that yield load was achieved in the tension chord C2, see Figure 2. The service load of 1562 kN was determined by dividing the ultimate load by a total safety factor of 1.6.

The load range of 2872 kN was chosen such that the highest range of primary + secondary stress from frame analysis reached a value of 920 N/mm^2, or twice the yield. Two fluctuating loading cases were chosen. In the first case, the maximum load was equal to the service load, and the minimum load had a nearly fully reversed negative value. In the second case, the truss was first loaded near the ultimate load determined above. The loads are shown in Table 1.

Table 1. Loads F used in the analyses (kN)

	Truss and frame	Nonlinear analyses	
		Case 1	Case 2
Max F	2500	1562	2300
Min F	-	-1310	-572

3 Results

3.1 Primary plus secondary stresses

Table 2 shows the member forces and stresses from pinned joint analysis, and the forces, bending moments and combined stresses from the linear frame analysis. Figure 3 shows the bending moment diagram and the theoretical stress distributions at nodal points. All values correspond to the ultimate point load of 2500 kN.

Table 2. Primary and secondary stresses

Member	Primary		Primary + secondary		
	N kN	σ N/mm^2	N kN	M kNm	σ N/mm^2
C1 left	2500	230	2440	87.3	396
right	2500	230	2440	-109	439
C2 left	5000	460	4410	30.9	466
right	5000	460	4410	-164	728
C3 left	-2500	-230	-2680	157	-555
right	-2500	-230	-2680	-282	-801
D1 left	-3536	-386	-3330	-87.3	-528
right	-3536	-386	-3330	57.4	-472
D2 left	-3536	-537	-2460	35.2	-463
right	-3536	-537	-2460	57.7	-520
V1 top	2500	439	1750	-105	699
bottom	2500	439	1750	100	680
V2 top	0	0	-230	0	-81
bottom	0	0	-230	0	-81

3.2 Joint strengths

Two joints were analysed: Node 3 as an N-joint and Node 4 as an X-joint. The capacity of Node 3 is 3174 kN, which falls below the primary compression force 3536 kN in D2. The joint was not redesigned due to the reasons explained in Section 2.2. The X-joint in node 4 does not fulfil the validity range $h_1 <= 2b_1$. Actually, this requirement does not apply here because no local buckling should occur due to the compactness of the three individual members joined together. Nevertheless, the capacity was calculated using a reduced joint length of $h = 2b_1 = 398$ mm. The calculated capacity 2355 kN falls below the value ultimate value of the reaction force at the support of 5000 kN. In a real design situation, strengthening, for example gusset plates 20 mm thick, would be needed in this joint supporting the reaction force. It is interesting to compare this result with the nonlinear behaviour of the joint as it is.

3.3 Nonlinear analyses

Figure 4 shows the finite element mesh. Four highly stressed details are also shown in the figure:

(1)   Cross section of the chord C2 at the plane of symmetry;
(2)   Section of the chord C3 along the weld toes at the support plate and the diagonal D2;
(3)   Section of the lower end of D2 along the weld toe ;
(4)   Web of the chord C3 above the support.

Figure 5 shows von Mises stress distributions on the top and bottom surfaces along the four paths shown in Figure 4. The left hand figures show the stresses at service load (Case 1) or ultimate load (Case 2). The right hand figures show the von Mises stress ranges between maximum and minimum load, calculated from ranges of each of the stress components. The stress range distributions were found to be almost equal in both load cases. Therefore, only those from Case 1 are shown.

4 Discussion

The pinned joint analysis suggests that the diagonal member D2 was not adequately designed. In fact, it was intentionally designed so, because the frame analysis had shown that the chord members are bearing fairly large shear forces decreasing the axial load in this member as can be seen from Table 2.

In Case 2 of the nonlinear analysis, the ultimate load predicted by the primary stress analysis could not be reached. The computed run was stopped at a value of 0.92 times that predicted by the primary stress analysis of the members. The reason was the inadequate design of joints. However, the result did not decrease significantly below that predicted.

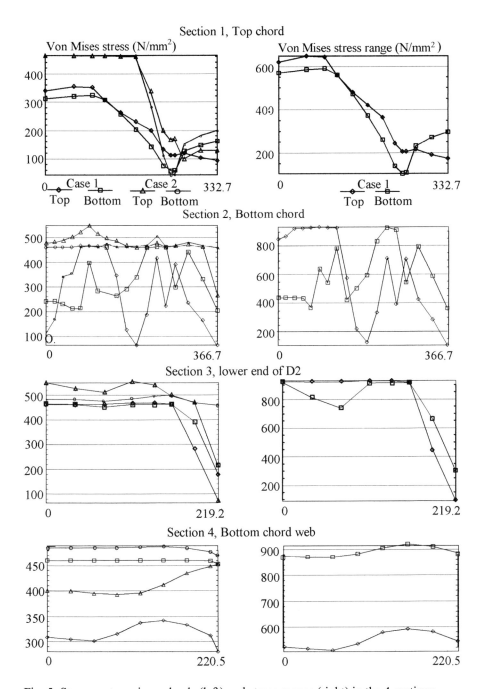

Fig. 5. Stresses at maximum loads (left) and stress ranges (right) in the 4 sections.

567

The maximum von Mises stress, 554 N/mm², is found at the weld toe in D2 at the maximum load, 2300 kN, corresponding to an effective plastic strain of 9.4 %. The reason for this is the bending of sidewalls associated with a yield line mechanism in the chord face due to inadequate joint design, see Figure 4.

In the top chord, subject to a tension force and secondary bending, the maximum von Mises stress equals 464 N/mm², corresponding to an effective plastic strain of 0.4 %. The maximum von Mises stress range found in the top chord equals 649 N/mm² which is less than the range of axial stress of 836 N/mm² predicted by the frame analysis.

In the bottom chord, subject to compression force and secondary bending, the maximum von Mises stress at the maximum load equals 548 N/mm² at the lower corner where the transverse dilatation is restricted by the support plate, see the deformations in Figure 4.

The surface stress ranges exceed slightly the shakedown limit, or twice the yield, in the joints which were inadequately designed. The maximum value of 930 N/mm² corresponds to a plastic strain range of 1 %. If the criterion (ii) is judged to apply to such maximum surface stresses, then the allowable load range should be reduced. According to this study, a safer criterion would be 1.5 times yield stress, instead of 2 times yield. This would be in line with the existing fatigue design recommen-dations of IIW [7] which prescribe a maximum stress range of 1.5 times yield stress.

At the service load, Case 1, the side wall of D2 was already yielding at both surfaces due to the deformation of the chord face. Figure 6 shows the von Mises stress distribution through the thickness in the most stressed element. It is found that about 30 % of the thickness is yielding at service load but there is some capacity remaining in the inner parts. At the ultimate load, plastic strain occurs through the wall. The load range of 2872 kN causes a slight overrun of the shakedown limit of 920 N/mm² only in a thin surface layer, the remaining section behaves elastically.

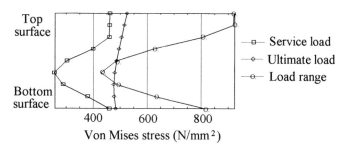

Fig. 6. Stress distributions across the wall thickness in the most stressed element.

5 Conclusions

In this paper a three-criterion approach was proposed for the design of trusses made of hollow sections. The rationale of the approach was studied by analysing a simple truss with stocky members of rectangular hollow sections. The length over depth ratio of the members was equal to 6, a limit value for pinned-joint analysis as proposed by Eurocode 3. In this case, the shakedown criterion was active, putting a limit for the allowable load range. Fully nonlinear FE analyses were made leading to the following conclusions:

1. The ultimate capacity of the welded joint details appeared to be almost adequate, despite their being underdimensioned.
2. Excessive yielding at the ultimate load was found mainly in the joint which was clearly underdimensioned.
3. The allowable load range resolved from the suggested shakedown criterion was fairly high, suggesting that in usual, predominantly statically loaded structures the shakedown criterion can be easily satisfied.
4. The local stress ranges, determined by nonlinear FE analysis, exceeded the shakedown limit only slightly in spite of the underdimensioned joints.
5. The three criterion approach seems to be promising, although its verification needs more studies.

6 Acknowledgment

The element mesh was prepared by Mr. David Thieman from USA, and the computer runs were carried out by Mr. Matti Koskimäki. Their assistance is gratefully acknowledged.

7 References

ABAQUS User's Manual, Version 4.8. (1989) Hibbit, Karlsson & Sörensen, Inc.

Anon. (1992) Compendium of Design Office Problems. **Journal of Structural Engineering**, Vol. 118, No. 12, December 1992.

ASME Boiler and Pressure Vessel Code. Section III, **Rules for Construction of Nuclear Vessels.**

ENV 1993 (1992) Eurocode 3: **Design of Steel Structures**. Annex K, Hollow section lattice girder connections. European Committee for Standardization.

I-DEAS, Finite Element Modeling User's Guide (1990). Structural Dynamics Research Corporation.

IIW (1981) **Design Recommendations for Cyclic Loaded Welded Steel Structures.** IIW Document XIII-998-81.

IIW (1989) **Design Recommendations for Hollow Section Joints - Predominantly statically loaded**. IIW Document XV-701-89.

SHS, Issue 2.00 (1992). **Design of SHS Welded Joints**. British Steel Welded Tubes.

56 NEW CRITERIA FOR DUCTILITY DESIGN OF JOINTS BASED ON COMPLETE CHS TRUSS TESTS

Y. KUROBANE and K. OGAWA
Kumamoto University, Japan

Abstract
Results of a series of tests of 15 complete trusses with CHS members are summarized.
Two of the trusses are space trusses with triangular cross sections. All the trusses were
tested to failure by applying reversals of deflection for 1 to 3 cycles. The trusses showed
diverse failure modes, which included buckling of members, failure of joints and combined member buckling and joint failure. Ductile cracks frequently developed and grew
rapidly during a few cycles of reversed loading, which sometimes led to tensile rupture of
joints or members. When failure of joints occurred before buckling of members, existing
joint capacity formulae based on isolated joint tests were found to predict observed
resistances of joints with good accuracy. However, after buckling of members, joints
came under combined bending and axial loads as a result of stress redistribution and
failed at a load lower than the capacity of joints under axial loads only. New design
criteria for tubular joints to avoid occurrences of failure with insufficient ductility in
trussed structures under the influence of strong earthquakes are proposed.
Keywords: Truss, Circular tube, Tubular joint, Earthquake, Buckling, Ductile fracture,
Shell bending.

1 Introduction

Although extensive studies on unstiffened tubular joints have been performed using
isolated joint specimens in the last 30 years, little is known about interactions between
joint behaviour and frame behaviour. A series of tests on complete trusses was planned
with greater emphasis on the ultimate behaviour of trusses under cyclic loads. The
results of the tests disclosed new evidences that the resistance and deformation of joints
interrelated with buckling and post-buckling behaviour of members. These results raised
new questions regarding the ultimate strength design of trusses. This paper focuses the
discussion on the behavior and design of joints.

2 Summary of tests

2.1 Specimens and test procedures

Tests on 15 different complete trusses were performed. The 6 types of trusses reported
herein, designated as A, B, C, T, F and S, are illustrated in Fig. 1. Each of them has
Warren type K-joints that have the same in-plane angle between the chord and the brace
except for Truss C. Each type of truss includes a few replicate specimens, distinguished
by the number like A1 and A2, with varied dimensions, which are shown in Table 1,

Tubular Structures V. Edited by M.G. Coutie and G. Davies.
© 1993 E & FN Spon, 2–6 Boundary Row, London SE1 8HN. ISBN 0 419 18770 7.

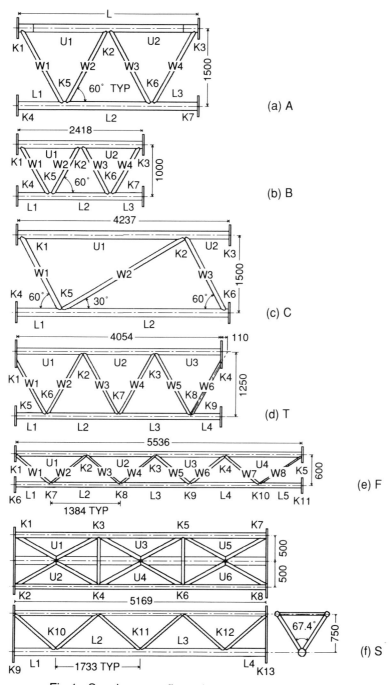

Fig.1. Specimen configuration and designations.

Table 1. Measured dimensions of specimens (mm)

Truss Type	Cross-sectional dimensions				Length L	Height H	K-joint details	
	Upper chord	Lower chord	Brace 2,3	Brace 1,4			gap	eccentricity
A1	165.57 φ x4.27	165.47 φ x4.29	60.62 φ x2.12	60.53 φ x3.82	3572	1500	25.5	0.0
A2	165.53 φ x5.73	165.52 φ x5.72	60.57 φ x2.18	60.52 φ x3.82	3572	1500	25.5	0.0
A3	165.55 φ x4.28	165.58 φ x4.30	60.58 φ x2.20	60.52 φ x3.83	3429	1500	−22.2	−41.3
B1	165.57 φ x4.28	165.53 φ x4.28	60.52 φ x2.19	60.52 φ x3.82	2418	1000	25.5	0.0
B2	165.42 φ x5.73	165.45 φ x5.74	60.57 φ x2.19	60.42 φ x3.78	2418	1000	25.5	0.0
C	165.52 φ x3.80	165.52 φ x3.80	76.42 φ x2.50		4247	1500	−25.0	−41.3
T1	139.83 φ x3.08	139.87 φ x3.15	60.48 φ x2.03		4054	1250	15.0	4.0
T2	139.80 φ x3.19	139.83 φ x3.07	60.48 φ x2.04		4054	1250	−15.0	−23.9
T3	139.93 φ x4.10	139.95 φ x4.04	60.47 φ x2.13		4054	1250	15.0	4.0
T4	139.82 φ x3.11	139.80 φ x3.13	76.50 φ x2.54		4054	1250	−24.0	−15.5
F1	101.75 φ x5.45	101.68 φ x5.44	48.43 φ x2.14		5536	600	−15.0	−25.4
F2	101.65 φ x2.94	101.65 φ x2.93	88.75 φ x2.58		5536	600	−18.0	0.0
F3	101.65 φ x2.90	101.65 φ x2.92	60.60 φ x2.22		5536	600	−33.0	−25.4
S1	101.55 φ x2.95	139.90 φ x3.27	76.33 φ x2.73		5169	750	$g = -38$ $g_t = 1.2$	−35.0 $e_t = 0$
S2	88.96 φ x2.93	139.90 φ x3.25	76.38 φ x2.76		5169	750	$g = -38$ $g_t = 1.2$	−35.0 $e_t = 0$

where L signifies the total chord length measured between the two inside surfaces of flange plates welded at both ends of each chord, and H signifies the distance between the centres of the upper and lower chords. Definitions of the longitudinal gap g, the transverse gap g_t and the eccentricity e are identical to those in the previous papers by the authors (e.g. Kurobane et al. (1989)). In Fig. 1, all the members and joints are numbered following the rules: U1, U2..designate the upper chords; L1, L2.., the lower chords; W1, W2.., the braces (the web members); and K1, K2.., the joints.

Trusses A, B and C have chords relatively strong as compared with braces. The test set-up for these specimens are shown in Fig. 2(a). The chord ends on one side of each truss were fixed via flange joints to a reaction wall. A double-acting ram was used to apply alternating deflections to the cantilevered trusses. Two Lehigh-type lateral bracing systems were mounted at the loading ends of the trusses. Trusses T1 through T4 have chords more slender than those in Trusses A, B, and C in order to investigate the lateral buckling behavior of the chords. The test set-up is similar to that for Truss A as shown in Fig. 2(b) except for using 4 universal joints allowing both in-plane and out-of-plane rotations at the chord ends. The T-shaped loading frame at the loading end of trusses was prevented from any out-of-plane movement by using 4 lateral bracing systems. Trusses F1 through F3 are small-sized lattice

Table 2. Mechanical properties of tube materials

Truss type	Cross-sectional dimensions(mm)	Stub Column Test (N/mm²)			Tensile coupon test (N/mm²)		
		σ_y	σ_m	$E (\times 10^5)$	σ_y	σ_u	Elongation
A,B	165.6 φ x4.3	329	380	2.06	350	441	40%
	165.5 φ x5.7	355	431	2.08	374	458	37%
	60.6 φ x2.2	364	407	2.03	388	446	39%
	60.5 φ x3.8	397	451	2.05	404	435	64%
C	165.5 φ x3.8	469	519	2.07	469	583	29%
	76.4 φ x2.5	364	421	2.07	371	432	37%
T	139.8 φ x3.1	361	393	2.05	384	436	32%
	139.9 φ x4.1	350	407	2.05	369	439	32%
	60.5 φ x2.1	361	408	2.06	411	467	46%
	76.5 φ x2.5	349	405	2.04	380	439	35%
F	101.7 φ x5.4	323	425	2.02	362	439	66%
	101.7 φ x2.9	315	373	2.09	347	424	60%
	88.8 φ x2.6	320	377	2.10	349	428	−
	60.6 φ x2.2	360	407	2.07	376	469	54%
	48.4 φ x2.3	381	449	2.09	407	462	30%
S	139.9 φ x3.3	312	368	2.01	359	451	−
	101.6 φ x2.9	312	374	2.09	349	426	67%
	89.0 φ x2.9	339	422	2.06	398	479	−
	76.4 φ x2.7	332	405	1.97	388	457	64%

girders intended for use in multistory building frames. Each of them was framed as a beam into a portal frame with extended columns as shown in Fig. 2(c). These columns were strong enough to support loads within an elastic region. Cyclic horizontal loads were applied at the tops of the extended columns using two double action rams so that the portal frame swayed between two deflection limits. Trusses S1 and S2 are space trusses with triangular cross sections. The same test set-up as that for Truss A was used.

Strains in all the members were measured at 1 to 5 cross sections for each member using 4 strain gauges at each cross section. Axial forces and bending moments in the members were calculated from these strains and results of stub-column tests, unless once yielded materials sustained yielding in a reversed direction.

Both tensile coupon tests and stub-column tests were performed for all the tube materials from different rolls. The material test results are shown in Table 2.

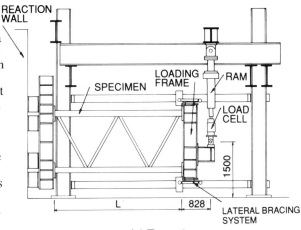

(a) Truss A

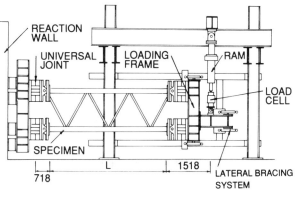

(b) Truss T

2.2 Test results
Load vs. rotation curves are shown in Fig. 3, in which the load refers to the ram load (the sum of two ram loads for Truss F) and the rotation to the deflection of a truss divided by the length of the truss or the rotation of columns at the both ends of a truss for Truss F. Deflections of trusses were determined as vertical displacements of flange plates at the loading end relative to the flange plates at the fixed end

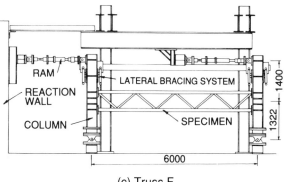

(c) Truss F

Fig.2. Test set-up.

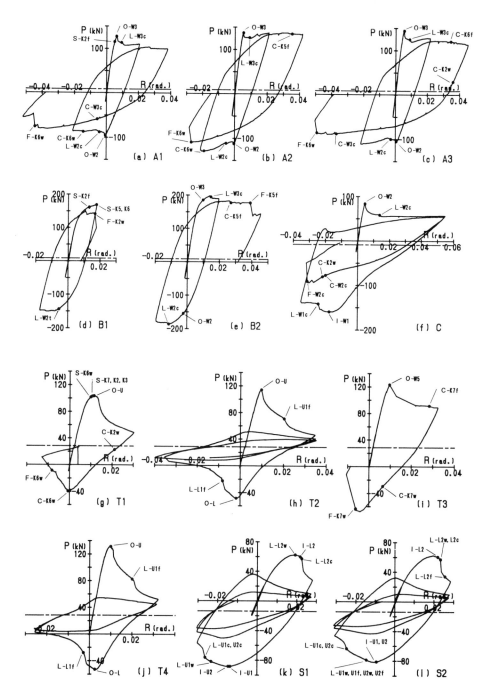

Fig.3. Load vs. rotation relationships for trusses(Continued overleaf).

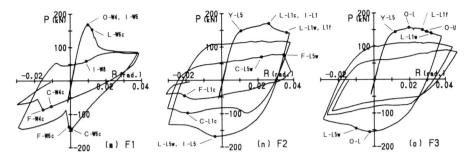

Fig.3. Load vs. rotation relationships for trusses(continued).

(relative to the center of in-plane hinge for Truss T). The horizontal chained lines in Fig. 3 indicate loads at which dead weight of loading apparatus is cancelled and the shear loads in the trusses vanish.

Important events observed during tests are shown by the symbols defined as follows. The first letters denote failure modes: O=out-of-plane buckling, I=in-plane buckling, Y=tensile yielding, S=shell bending deflection in a joint, C=crack detection, L=local buckling either in a member or at a joint, F=ductile fracture due to extended cracks. The following two letters after a hyphen indicate locations of these failure events using the symbols shown in Fig. 2 (e.g. W3 indicates Brace 3). The following lower case letters specify, if necessary, the locations in more detail: t=top, c=center, b=bottom, w=reaction wall side, f=far side from reaction wall. For example, looking at Truss S1, L-L2w shows that the lower chord 2 sustained local buckling at the end close to the reaction wall and then I-L2 shows that the same chord sustained in-plane buckling.

Summarizing Fig. 3, the trusses in which braces buckled (A1, A2, A3, B2, C, T3, F1) show a sudden drop in load due to buckling followed by increase in deflection at a nearly constant load. In these trusses, however, cracks developed in the buckled braces as well as in K-joints as the cycling progressed, leading finally to a complete rupture in each of the trusses. The trusses that failed by buckling of chords (T1, T2, T4, F2, F3, A1, S2) showed continued degradation with the cycle in load-carrying capacity. Truss B1 is the special type that failed only by failures in joints.

3 Ultimate behavior of joints

3.1 Resistance of joints before buckling of members
The ultimate resistance of the K-joint is governed, unless tensile fracture occurs, either by localized shell bending deflection of the chord wall under the compression brace or by local buckling of the compression brace in the region adjacent to the joint. The ultimate resistance $_KN_U$ can be represented by the equation:

$$_KN_U = \min \left(_KN_S , _KN_L \right) \tag{1}$$

where $_KN_S$ and $_KN_L$ denote the resistances of the joint determined by shell bending deflection and local buckling, respectively, which are most accurately predicted by the formulae of Kurobane et al.(1984, 1986).

For all the K-joints in Trusses A, B, C, T and F, the values of the maximum axial force N_m in the compression brace sustained during the first half cycle of loading are shown in Table 3 and compared with the joint resistances $_KN_U$ predicted by Eq. 1. In the same table significant joint deformation or failure modes observed both before buckling of members and after termination of testing are shown, in which definitions of symbols are identical to those used for showing failure modes in Fig. 3. The X symbol signifies that shell bending deflection was unidentified owing to large local buckles in the chord wall near the joint. Test results for Trusses S1 and S2 are not included in this table because the joints were strong enough and showed no visible deformation during testing.

Table 3 shows that, during the first half cycle, shell bending joint failure was invariably observed in Trusses A1, B1 and T1 in which the recorded N_m values exceeded the predicted joint resistances, while no joint failure was found in the rest of trusses in which N_m's were less than the K-joint resistances. One exception in the above statement is Truss F3 that failed by lateral buckling of the lower chord, in which deformations in K-joints were masked by local buckles in the chord wall although N_m was about equal to $_KN_U$. These test results suggest that the ultimate resistance equations based on isolated joint tests predict very well the ultimate resistances of joints in complete trusses.

3.2 Resistance of joints after buckling of members

Table 3 also shows that failures in joints in various modes occurred after the first half cycle of loading even if applied brace loads appeared much smaller than predicted joint resistances. This is due to redistribution of stresses in the members following buckling of members or failure in joints.

One significant example of such sequential failures is shell bending failure of joints after flexural buckling of braces. Axial forces in members framing into K-joints are balanced before braces buckle (this was confirmed by strain gauge measurements). Following buckling of a brace, however, the axial force in the brace is quickly lost. Then, the K-joint comes under combined in-plane bending and axial loads, because the chord carries a part of the shear load that was carried by the brace until it buckled.

The load path of the axial forces in the two braces framing into the K-joint is shown in Fig. 4 for each of 4 different K-joints. As seen, the axial forces in the two braces increase equally with the load following a 45 degree line (the ram

Table 3. Ultimate resistance of K-joints

	Joint	N_m (kN)	$_KN_s$ (kN)	$_KN_s$ (kN)	$\frac{N_m}{_KN_u}$	a	b
A1	K2	119.6	117.8	118.8	1.02	S	S
	K5	133.7	135.5	200.2	0.99		SLC
	K6	119.6	133.7	125.7	0.95		S CF
A2	K2	133.9	217.2	159.3	0.84		L
	K5	202.0	243.5	263.9	0.83		SLC
	K6	133.9	239.8	166.5	0.80		SLCF
A3	K2	131.0	188.1	151.7	0.86		S C
	K5	171.0	218.1	250.1	0.78		SLC
	K6	131.0	216.0	161.4	0.81		SLCF
B1	K2	114.1	108.9	118.2	1.05	S	SLCF
	K5	122.1	136.1	203.8	0.90	S	S
	K6	114.1	136.2	130.6	0.87	S	S
B2	K2	142.3	207.3	156.9	0.91		SL
	K5		245.6	262.0			S CF
	K6	142.3	243.9	168.7	0.84		SLC
C	K2	182.4	400.9	249.5	0.73		SLC
	K5	182.4	359.6	237.7	0.77		SLC
T1	K2	80.5	69.2	93.3	1.16	S	S C
	K3	83.7	76.8	97.7	1.09	S	S
	K6	91.1	95.6	107.7	0.95	S	S CF
	K7	80.5	96.1	107.9	0.84	S	S
	K8	83.7	95.3	107.5	0.88		S
T2	K2	88.5	110.0	115.2	0.80		
	K3	98.5	124.8	121.9	0.81		
	K6	97.0	142.4	129.3	0.75		S
	K7	88.5	144.8	130.2	0.68		S
	K8	98.5	144.7	130.2	0.76		
T3	K2	107.7	115.8	118.8	0.93		
	K3	109.8	128.6	124.5	0.88		S
	K6	112.9	149.0	133.0	0.85		S C
	K7	107.7	150.1	133.4	0.81		SLCF
	K8	109.8	148.8	132.9	0.83		SL
T4	K2	101.0	119.7	148.1	0.84		SL
	K3	113.8	142.2	159.9	0.80		
	K6	101.0	173.6	174.8	0.58		S
	K7	101.0	178.4	177.0	0.57		
	K8	113.8	178.7	177.1	0.64		
F1	K2	108.7	290.1	170.1	0.64		
	K3	113.6	275.1	166.2	0.68		
	K4	112.6	252.5	159.9	0.70		L
	K7	108.7	254.8	160.6	0.68		
	K8	113.6	280.9	167.7	0.68		
	K9	112.6	289.5	170.0	0.66		L
	K10	111.3	287.5	169.4	0.66		
F2	K2	120.4	190.2	189.8	0.63		
	K3	109.3	192.3	190.8	0.57		
	K4	113.7	153.9	172.7	0.74		
	K7	120.4	135.4	163.2	0.89		
	K8	109.3	191.1	190.2	0.57		
	K9	113.7	194.3	191.7	0.59		
	K10	110.9	176.5	183.6	0.63		
F3	K2	105.7	140.4	132.5	0.80		
	K3	100.4	140.3	132.5	0.76		
	K4	101.8	113.3	120.4	0.90		
	K7	105.7	105.2	116.5	1.00	X	
	K8	100.4	137.1	131.1	0.77		
	K9	101.8	145.6	134.7	0.76		
	K10	113.1	132.7	129.2	0.87	X	

Note: joint failure modes observed
  a before member buckling
  b after completion of test

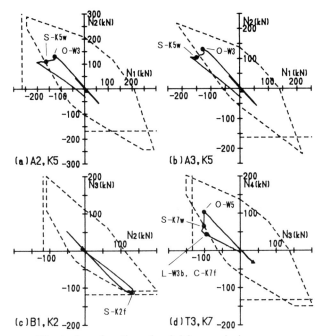

Fig.4. Load paths of axial forces in braces
and ultimate capacity polygons for K-joints

load was started from the point denoted by the triangle, which is not corresponding to the
origin of the graph owing to effects of dead weight of apparatus). In each of Figs.
4(a),(b),(d), it is seen that the axial force in the compression brace, which sustained no
buckling, suddenly increases while that in the tension brace decreases as soon as the other
compression brace immediately adjacent to the tension brace buckles. As the load path
reaches the ultimate capacity polygon shown by dashed lines, shell bending failure of the
joint occurs. In Fig. 4(c), however, the axial load in the compression brace remains
constant while that in the tension brace increases further after the load path reaches the
ultimate capacity polygon, because Truss B1 sustained failures in joints but no buckling
of the braces.

The ultimate capacity polygon for K-joints has already been discussed by the authors
(see Ogawa et al.(1987), Makino et al.(1987), Kurobane et al.(1989)). When the axial
force in the tension brace decreases, the ultimate capacity of the K-joint decreases along
the line liking the two points representing the resistance of the K-joint under balanced
loads and that of the Y-joint under compression. After the K-joint reaches its ultimate
resistance under balanced loads, on the other hand, the K-joint sustains further increase in
load in the tension brace until a tensile (frequently called punching shear) failure takes
place. These rules prescribed by the ultimate capacity polygon interpret the load paths
for the K-joints observed in Fig. 4.

Another example of sequential failures is shell bending failure of K-joints following
lateral buckling of chords, which occurred in Joints K2 and K6 of Truss T4. For this
particular specimen, the axial and bending stresses in the braces measured by strain
gauges at 5 cross sections for each brace are considered reliable throughout the cyclic
loading test because stresses in the braces remained nearly in the elastic region. The
loads applied to these joints are worth examining in detail.

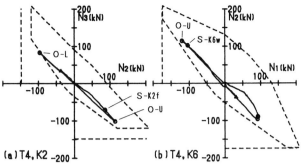

Fig.5. Load paths of axial forces in braces and
ultimate capacity polygons for K-joints.

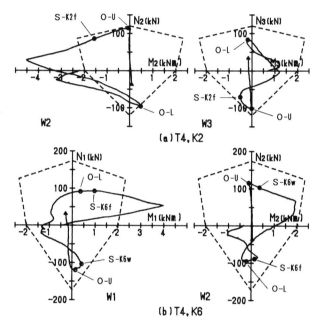

Fig.6. Load paths of OPB and axial loads compared with ultimate
capacity polygons for K-joint

First, the load path of axial loads in the two braces is compared with the ultimate
capacity polygon for each of K2 and K6 in Fig. 5. Both load paths stay on the 45 degree
lines even after loads start to decline as a result of lateral buckling of the chords. Thus
the K-joints are still under balanced loads after lateral buckling of the chords. In contrast
with the previous descriptions regarding Fig. 4, however, shell bending failures appear
after load decay at points inside the ultimate capacity polygons. Next, the load paths of
OPB (out-of-plane bending) and axial loads transmitted to the joints from each of the
braces framing into the same joints are shown in Fig. 6. In this figure new ultimate
capacity polygons drawn by dashed lines for the K-joints under combined OPB and axial
loads are shown. Although no data exists for K-joints under combined OPB and axial

loads, these new polygons were conceived on the basis of the past study on T-joints under combined OPB and axial loads by Kurobane et al.(1991) as shown below.

The ultimate resistance of the K-joint when one of the braces is under an OPB load is given by

$$_K M_{u,OPB} = \frac{0.0991\,d}{1 - 0.785\,d/D}\left(\frac{D}{T}\right)^{0.289} T^2 \frac{1}{\sin\theta}\left(\frac{\sigma_y}{\sigma_u}\right)^{-1.123}\sigma_y \qquad (2)$$

This equation was derived by multiplying $1/\sin\theta$ to the ultimate resistance of the T-joint under an OPB load. The moment resistance of the T-joint decreases linearly when combined with axial compression, while the resistance increases linearly with the combined axial tensile load up to 116% of the pure bending capacity at 72.2% of the tensile capacity and then the resistance decreases as the tensile load is increased further. This behavior of T-joints was copied to formulate ultimate capacity polygons for K-joints under combined OPB and axial loads as shown

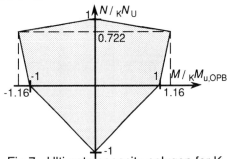

Fig.7. Ultimate capacity polygon for K-joints under OPB and axial loads.

in Fig. 7. It should be noted here that the maximum axial load in the compression brace in the K-joint is given by Eq. 1, while that in the tension brace is determined by the equilibrium of loads.

As seen in Fig. 6, the OPB moments increase and the axial loads decrease in the compression braces immediately after lateral buckling of the chords. When the load paths for the compression braces (subjected to negative axial loads) approach the ultimate capacity polygons, shell bending failures appear and then the OPB moments are unloaded. Similar events are repeated in the tension braces after they change into the compression braces because the loads are reversed. Although the load paths for the tension braces move outside the ultimate capacity polygons, this incurs no great contradiction because both the K-joints having overlapping braces can transmit OPB moments from the tension to the compression braces within the joints.

The above observations show us that in these K-joints shell bending failures occurred at loads lower than the K-joint capacities under balanced loads because combined OPB moments were applied as a result of lateral buckling of the chords.

Most of local buckling failures in K-joints were confirmed after tests were completed. This was due to difficulties involved in observing local buckling failures during testing. However, local buckling was frequently observed in K-joints in which N_m values were close to predicted local buckling resistances.

3.3 Ductile fracture at weld toes
Cracks were frequently observed along the weld toes in K-joints that sustained shell bending or local buckling failures. These cracks grew rapidly as the cycling progresses frequently leading to a complete separation of a member from a joint (See Fig 8). All the cracks extended showing shear slip planes which made an angle of about 45 degrees with the tube surface. Clearly these crack extensions are due to ductile tensile fracture. No flat plane-strain tensile fracture was observed. Similar cracks were observed also in past isolated joint tests just before or after reaching the maximum capacities of the joints due to shell bending failures (See Kurobane et al.(1990)). However, quick extensions of these cracks during a few cycles of repeated loading were first observed in this series of tests.

The material at hot-spots along the weld toes sustains large plastic strains well into a strain-hardening range. The material's toughness deteriorates owing to repeated cold-working. Further, microscopic defects exist due to non-metallic intrusion at the weld toes. These should be the reasons for quick developments of cracks, although no criterion has been identified to predict initiation and extension of ductile cracks at weld toes. Tentative conclu-

Fig.8. Failure of K6 joint in truss T1.

sions drawn from this series of tests can be summarized as:
1. Cracks were found in joints that sustained shell bending, local buckling, or both types of failure.
2. One of strategies to prevent these cracks extending is to keep a reserve of strength for joints so that shell bending or local buckling failure is avoided even after redistribution of stresses. Within a range of these tests, all the joints in which the maximum axial loads during the first half cycle of loading (N_m) were less than 80 % of the predicted joint resistances ($_K N_U$) evaded crack extensions.

4 Proposed design criteria

The results of 15 truss tests revealed that the resistance of joints interacted with frame behavior. Based on these results new criteria for the design of joints and members are proposed and shown below.

When trusses are under static loads like gravity or snow loads, the existing resistance equations based on isolated joint tests are effective to predict the ultimate behavior of the joints in trusses. Namely, no significant effects due to different boundary conditions between the isolated and actual joints (e.g. secondary bending moments and end restraints) have been found. No brittle joint failure under monotonically increasing loads was observed, although K-joints that may sustain premature tensile failure due to an excessively small gap (See Kurobane et al. (1990)) were not used in the trusses in these tests and should be avoided. Trusses may be designed either to have stronger joints than members or vice versa, so far as appropriate values for the resistance factor are assumed with due considerations on failure modes. It should be noted that trusses may fail suddenly when failures of trusses is governed by buckling of slender chords like Trusses T1 through T4.

Two approaches are applicable to the design of trusses under dynamic loads like earthquake loads. The first one is to design the trusses to have sufficient strength so that both the joints and members resist the maximum possible load effects. When shell bending deflection or local buckling occurs in the joints, however, cracks may grow rapidly in a few cycles of repeated loading along the weld toes leading to tensile fractures with insufficient ductility. Although no definite criterion has been established to prevent these cracks, it is tentatively proposed, from the 95 % confidence limit in very low-cycle fatigue test results for T-joints (See e.g. Kurobane (1989)), to design the joints to be 25 % stronger than the maximum load effects.

The other approach is to design the trusses to have sufficient ductility so that they will not collapse under the most unusual external excitations. A similar approach was proposed in API codes (API (1986)). In this latter case joints should be designed against sequential failures including tensile fractures. The present test results indicate that such sequential failures could be avoided when the joints were 25 % stronger than the buckling loads of members. In order to perform the latter design approach, however, the energy absorbing capacity of trusses have to be evaluated based on buckling and post-buckling hysteretic behavior of trusses. Since effective length factors for truss members in an inelastic region are much smaller than those recommended in most design codes, it is possible to design trusses that show good ductility after buckling of members (See Ogawa et al. (1992, 1993)). The latter design approach is frequently preferred to the first one in the earthquake resistant design of building and offshore structures, because to design structures for great earthquakes by requiring that structures remain nearly elastic would be grossly uneconomical and unjustifiable from the probability of such an occurrence.

5 References

American Petroleum Institute (1986) Recommended practice for planning, designing and constructing fixed offshore structures, API RP 2A

Kurobane, Y., Makino, Y. and Ochi, K. (1984) Ultimate resistance of unstiffened tubular joints, **J. Struct. Engrg.**, ASCE, 110, 385-400

Kurobane, Y., Ogawa, K., Ochi, K. and Makino, Y. (1986) Local buckling of braces in tubular K-joints, **Thin-Walled Structures**, 4, 23-40

Kurobane, Y., Ogawa, K. and Ochi, K. (1989) Recent research developments in the design of tubular structures, **J. Construct. Steel Research**, 13, 169-188

Kurobane, Y. (1989) Recent developments in the fatigue design rules in Japan, in **Fatigue Aspects in Structural Design** (eds J. Wardenier and J.H. Reusink), Delft University Press, Delft, pp. 173-187

Kurobane, Y., Makino, Y. and Ogawa, K. (1990) Further ultimate limit state criteria for design of tubular K-joints, in **Tubular Structures** (eds E. Niemi and P. Makelainen), Elsevier Applied Science, London, pp. 65-72

Kurobane, Y., Makino, Y., Ogawa, K. and Maruyama, T. (1991) Capacity of CHS T-joints under combined OPB and axial loads and its interactions with frame behavior, in **Tubular Structures** (eds J. Wardenier and E. Panjeh Shahi), Delft University Press, Delft, pp. 412-423

Makino, Y., Kurobane, Y. and Haraguchi, H. (1987) Tubular K-joints under combined OPB and axial loads (part 4)-Proposed ultimate capacity polygon, Summary Papers, Annual Conf. AIJ, 1017-1018, in Japanese

Ogawa, K., Yamanari, M., Makino, Y., Kurobane, Y., Yamashita, Y. and Sakamoto, S. (1987) Buckling and post-buckling behavior of complete tubular trusses under cyclic loading, **Proc. Offshore Technology Conf.**, OTC 5439, Houston, 161-169

Ogawa, K., Kurobane, Y. and Yamanari, M. (1992) Effective column length and buckling strength of members in complete truss test (Ultimate behavior of circular tubular trusses under cyclic loading-part 2), **J. Struct. and Construct. Engrg.**, AIJ, 438, 157-164, in Japanese

Ogawa, K., Kurobane, Y. and Yamanari, M. (1993) Design effective length factors for chord lateral buckling (Ultimate behavior of circular tubular trusses under cyclic loading-part 3), **J. Struct. and Construct. Engrg.**, AIJ, 443, 117-126, in Japanese

57 ERECTION LOADS IN DOUBLE LAYERED SPACEFRAME ROOF STRUCTURES

P.W. KNEEN
University of NSW, Sydney, Australia

Abstract
This paper describes aspects of erection techniques on the
temporary behaviour of single span, double layered, simply
supported spaceframe roof structures. Many spaceframe geometries are essentially unstable in the kinematic sense
until in their final supported position. However, they are
often constructed at or near ground level and lifted into
their final position with mobile cranes. The number and
location of the lifting points is usually significantly
different from the final supports and hence the distribution of member forces needs to be considered carefully.
Large second order displacements and distortions of the
frame may exist during this operation. A comparison is
carried out with member actions and required sizes when
these effects are taken into account. A large displacement, non-linear stiffness analysis procedure is used for
the work.
Keywords: Spaceframes, Erection, Roofs, Nonlinear analysis.

1 Introduction

Many flat double layered spaceframe structures are assembled at or near ground level and later lifted into
position using mobile cranes.

Whilst being assembled, it is common to use many support points, often with built in height adjustments. In
these situations, the stability of the spaceframe is assured.

In the final position, the effects of the disposition
of support points is analyzed as a matter of course. A

Tubular Structures V. Edited by M.G. Coutie and G. Davies.
© 1993 E & FN Spon, 2–6 Boundary Row, London SE1 8HN. ISBN 0 419 18770 7.

structural analysis using the stiffness method requires the support conditions to be defined and members are designed to resist the actions generated by the assumed loads. For some spaceframe geometries, it is essential that the supports are present as potential mechanisms can be formed.

The traditional square on square offset type lacks torsional rigidity when the nodes are considered as pinned joints. This deficiency is overcome in a single span structure by having adequate perimeter support points.

During erection however, only a limited number of lifting points are used. These are governed partly by the characteristics of the cranes used. When lifting to a significant height, a mobile crane needs to be placed close to the frame and often be extended to its maximum reach. A spreader beam or truss if used will take some of the clearance, particularly if its length is to spread the lifting points over many bays.

The nature of the imposed loads may be different to the final supports. For example, to leave the bottom anchorage nodes free, the slings are attached to the top nodes. Web members thus could carry forces of the opposite sign to the in-service condition. Sling loads could be inclined, and at different angles throughout the lift. See Figure 1. Note that if the lift points were to be located away from the perimeter, the reach of the cranes must be larger.

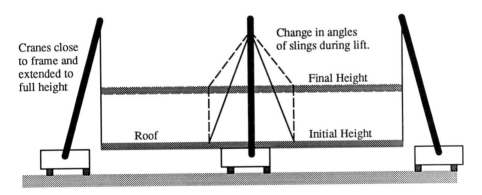

Figure 1 General Lift Arrangements for a Spaceframe

2 Lifting a Two Way Grid by the Edges

For the traditional two way spaceframe, the lifting positions at the mid edge as in Figure 1 would cause the corners to droop significantly. To study the case of a roof, simply supported around the perimeter under a final UDL of 1kPa vertically downwards, we shall consider a rectangular frame of 18x24 bays with chord lengths of 2400mm and a structural depth of 1800mm. It will be designed for the in-service condition first of 3kN per node plus self weight of the frame. This translates to just over 1kPa.

A range of ten suitable tube sizes as follows were chosen for the members and the 3456 member sizes were to be selected by the computer during various stages of the large displacement non linear stiffness analysis.

Section	Comp-capacity	Tens-capacity	Section
1	15000.000	15000.000	60x2.3-M12
2	30300.000	30300.000	76x2.3-M12
3	53300.000	53300.000	88x2.6-M16
4	94000.000	94000.000	102x32-M20
5	143000.000	143000.000	114x36-M20
6	225000.000	225000.000	114x6-1in
7	205000.000	205000.000	127x4-1in
8	328000.000	328000.000	140x5-M30
9	431000.000	431000.000	168x48-M36
10	570000.000	570000.000	168x64-M42

An estimate of volume of material of members was 7.952m^3 or a self weight of 0.246 kPa. The maximum displacement was 207mm which might be considered excessive (Short span/208) and the resultant distribution of tube sizes and ranges of member forces were:

material	#cables	ave.force	max.force	min.force
1	804	-41.8	14129.8	-14250.7
2	866	320.0	28495.3	-30038.4
3	520	2343.1	52487.0	-50508.9
4	286	4112.3	91580.1	-82174.6
5	324	-17094.4	136920.9	-139673.9
6	334	9996.4	221366.9	-221198.0
8	166	-12138.0	316306.1	-316348.3
9	124	35600.4	426865.5	-419377.2
10	32	-52848.9	543758.9	-518927.1

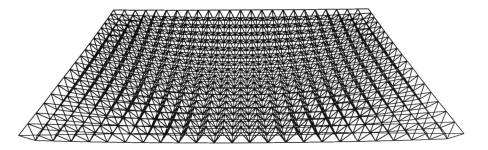

Figure 2 Deflected Form of Simply Supported Spaceframe

After all members have been selected, the effects of the erection will be examined by replacing the supports by cables to the crane hook. A nonlinear large displacement analysis procedure was used to show up any unsatisfactory behaviour due to the lack of torsional stiffness that may be characteristic of this geometry. The slings will be spread over four bays and, for study purposes only, the case at the maximum height will be assumed.

Under the lift operation, the cladding may or may not be present, or it might be stacked in bundles to reduce wind loads, but the assumed design live load will be absent. Thus a UDL of 0.1 kPa will be taken in addition to the self weight of the tubular members to represent the nodes, purlins and cleats. Wind loads are absent.

For this relatively small load, no members increased in size and the structure behaved sensibly as indicated in Figure 3 where the deformations have been scaled by 50.

For an increased UDL of 1kN per node (0.35kPa) to the framing members, to model the weight of cladding, cleats, ducting and purlins, it was found that 88 members had to be increased in size for the lift operation compared to the selected members for the design in the *as built* condition. The increase was just less than one percent of the frame weight. The final distribution of tubes being:

material	#cables	ave.force	max.force	min.force
1	796	136.0	12625.4	−7945.2
2	810	258.6	39915.4	−41124.4
3	536	−171.8	57175.1	−79796.0
4	322	2871.2	92671.5	−93841.1
5	336	−6964.2	100797.4	−140298.2
6	334	402.9	82554.7	−102040.3
8	166	−2001.2	99592.8	−114689.4
9	124	8312.4	145170.3	−147884.9
10	32	−20493.9	189484.3	−186042.0

585

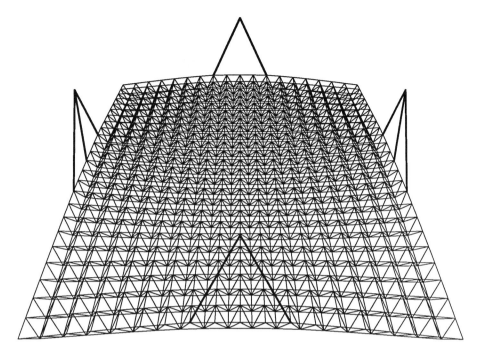

Figure 3 Deflected Form of Spaceframe During Lifting

Over 50% of these changes were changes (increases) by
two section sizes and 11% were by three sections.

3 Lifting a Two Way Grid by Internal Poles

Rather than lift from the edges with mobile cranes, another option is with internally located gin poles. In the
following example, the same 24x18 bay structure had holes
left in the frame 14.4m from each end and 9.6m from each
side. A single top chord node at these locations was
omitted together with all incident members to permit the
gin poles to be erected as shown in Figure 4.

Once again, the original member sizing was done automatically assuming all perimeter nodes were simply supported and a load of 3kN per node in addition to member
self weight was applied. Subsequently, the new guys and
poles were used as the supports and the additional load
was reduced to 1kN per node for the lift. Two non linear
runs were done - one with the frame at the start of the
lift, the other at the top of the lift. In the latter
case, the guys are inducing some larger horizontal loads

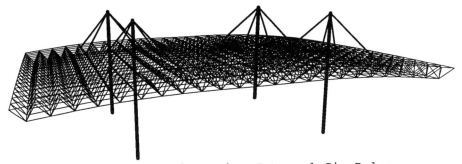

Figure 4 Erection using Internal Gin Poles

into the roof. In both runs, there were 28 members which had to be increased in capacity because of the erection procedure.

This lifting arrangement had much smaller deflections compared to the first, the maximum being 28.4mm. It is likely that final attachment to the permanent support points would be an easy task. Figure 5 indicates the deformed shape in the long span direction. The final distribution of member sizes and forces being:

material	#cables	ave.force	max.force	min.force
1	768	-208.4	9428.0	-11738.1
2	886	478.6	28687.1	-22357.6
3	520	64.3	31663.9	-30789.5
4	286	-1254.6	30881.6	-31632.8
5	316	-3255.8	31327.6	-32391.0
6	330	-2184.7	41534.9	-38431.8
8	162	-4692.3	12310.3	-30850.6
9	124	874.3	16941.1	-15763.3
10	32	-1011.2	9064.4	-8936.9

Of the 28 changes, all were increases from the smallest size to the next available size.

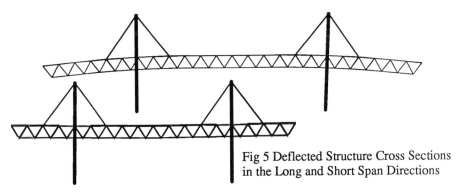

Fig 5 Deflected Structure Cross Sections in the Long and Short Span Directions

587

4 Lifting a Two Way Diagonal Grid by the Edges

Figure 6 shows a similar spaceframe layout with the square on offset square grid orientated with the chords at 45° to the edge perimeter. The different behaviour to the standard two way grid is immediately apparent when compared to Figure 3. For the diagonal orientation, there are strong bands in a diamond layout corresponding to pseudo supporting bands between the middle of adjacent sides, or in this case, the lifting points.

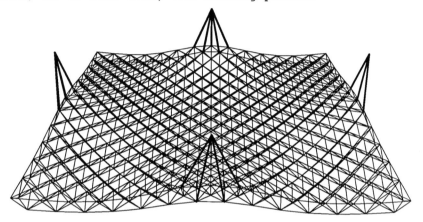

Figure 6 Perspective View of Deflected Diagonal Grid

Figure 7 Cross Sections at an End and Across Mid Span of the Diagonal Grid

5 Conclusion

This paper has examined a few lifting methods for double layered spaceframes. It indicates that there can be significant differences in the required member sizing in cases of perimeter lifting points. It should not be assumed that because the expected design loads are not present during the lift that the member sizing will be adequate, even allowing for increased stress levels.

PART 15

JOINTS

Behaviour under
repeated loading

58 THE INFLUENCE OF SEMI-RIGID JOINTS ON THE CYCLIC BEHAVIOUR OF STEEL FRAMES

K. MORITA
Chiba University, Japan

Abstract
A two-storeyed and two-bayed unbraced frame, in which a composite
steel beam was directly welded to an RHS column without any inner
diaphragms, was tested together with support tests on subassemblages.
The hysteresis loops for the specimens are of the stable spindle type
high in energy dissipation capacity. The effect of flexibility of
the composite joints is almost negligible on the lateral rigidity of
the frame, and a plastic hinge analysis, using the analytical plastic
moment of the composite joint as the plastic hinge moment of the
beam, gives a conservative estimate of the strength.
Keywords: Unbraced Frame, Subassemblage, Semi-Rigid Joint, Strength.

1 Introduction

Unbraced frames of multi-storeyed steel buildings in Japan often
employ a welded rectangular hollow section (RHS) column for its
biaxial moment capacity. The composite joint between an RHS column
and composite beam is usually designed as a rigid, full-strength
joint to provide high lateral rigidity for the serviceability limit
and full frame strength for the ultimate limit demanded of seismic
design. Therefore, this composite joint is usually reinforced with
an inner diaphragm.

From the viewpoint of fabrication efficiency, it is advantageous
to weld a steel beam directly to an RHS column without any inner
diaphragms. As the rigidity and ultimate strength of the joint
without an inner diaphragm depend on the width-to-thickness ratio of
the RHS column and on the width ratio of the beam flange to column
flange, the classification of the joint ranges from rigid to semi-
rigid. For the semi-rigid composite joint between a composite beam
and wide-flange steel column, efforts to establish design principles
for the connection are under way, but the experimental data are quite
limited for semi-rigid composite joints between composite beams and
RHS columns.

In this study, one two-storeyed and two-bayed unbraced frame, in
which a composite steel beam was directly welded to an RHS column
without inner diaphragms, was tested by horizontal loading. And the
test results for the structural performance of the frame are
described in relation to the composite joint behaviour.

Tubular Structures V. Edited by M.G. Coutie and G. Davies.

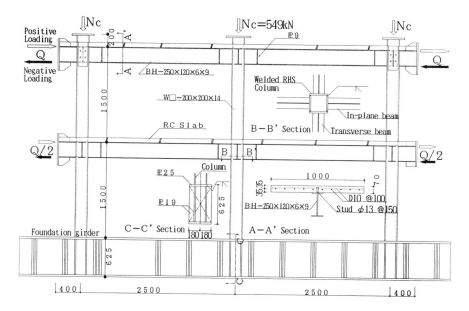

Fig.1. Unbraced frame specimen

2 Test specimens and loading conditions

2.1 Unbraced frame test

The two-storeyed and two-bayed frame specimen, with a storey height of 1500mm and span length of 2500mm, is shown in Fig.1. The column is a welded RHS (□-200 x 200 x 14) and the fully composite beam consists of a welded wide flange section (BH-250 x 120 x 6 x 9) and reinforced concrete (RC) slab of 1000mm width and 70mm thickness. The RC slab is connected to in-plane, transverse steel beams with headed stud-connectors. The beam flange is welded to the column flange with full-penetration weld, and the beam web is welded with fillet weld without an inner diaphragm.

Incremental cyclic reversal horizontal loading was applied to both the first and second floor levels in a ratio of 1 : 2 as shown in Fig.1.

2.2 Subassemblage test

Two subassemblage specimens, one being an interior column joint and the other an exterior column joint, were tested in support. As shown in Fig.2, the size of each member and details of the composite joints are the same as those of the frame specimen.

Asymmetrical incremental cyclic reversal loading was applied at both beam ends for the interior column joint as shown in Fig.2.

The mechanical properties of the steel and concrete materials used are shown in Table 1.

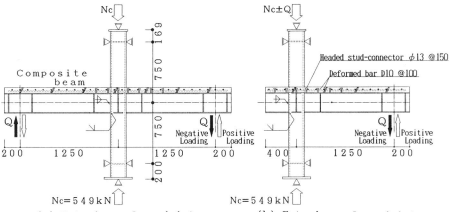

(a) Interior column joint   (b) Exterior column joint

Fig.2. Subassemblage specimens

Table 1. Mechanical properties of materials

Steel		Actual thickness (mm)	Yield point (N/mm²)	Tensile strength (N/mm²)	Young's modulus (kN/mm²)	Elongation (%)
Plate	14mm	13.78	345	516	218	27
	9mm	8.98	368	526	216	24
	6mm	5.90	384	518	213	22
Deformed bar(D10)		—	379	551	196	20

Concrete; Compressive strength=28.6N/mm², Modular ratio=8.69

3 Test results

3.1 Unbraced frame test

The relationship between the applied horizontal load and drift angle of each floor is shown in Fig.3. At first, hair-line cracks were initiated at the RHS column-to-beam flange connection of the interior column joint at a drift angle of 1/100 rad. However, even after the initiation of cracks in the connections, each load-drift angle relationship remained stable, showing a hysteresis loop of the spindle type, and cracks finally extended through the column walls as shown in Fig.4. At the final stage, crushing of the RC slab was also initiated at the corner of the RHS column.

The shear strength and drift angle at general yield are 515kN, and 1/85 and -1/120 rad. for the first storey, and 774kN, and 1/110 and -1/130 rad. for the ground storey.

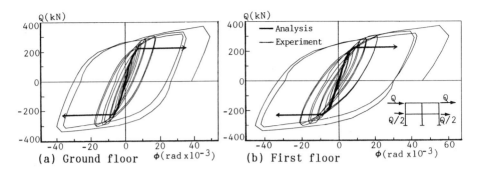

(a) Ground floor $\phi(\text{rad} \times 10^{-3})$ (b) First floor $\phi(\text{rad} \times 10^{-3})$

Fig.3. Load-drift angle relationships for frame

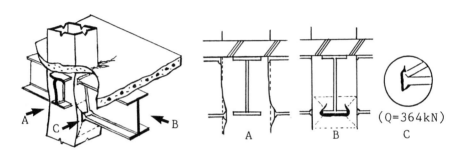

(Q=364kN)

Fig.4. Crack pattern at connection

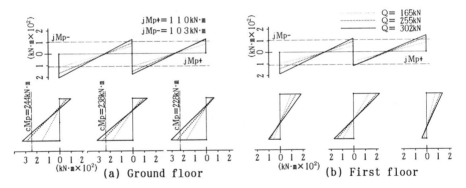

(a) Ground floor   (b) First floor

Fig.5. Moment distributions in frame

The moment distributions of the frame obtained from strain measurements are shown in Fig.5. The moment in each column base at the ground floor has reached its full plastic moment around the general yielding of the frame. As shown in Fig.6, which is the relationship between the applied horizontal load and local deformation of the beam flange against the column wall, all the composite joints have also yielded.

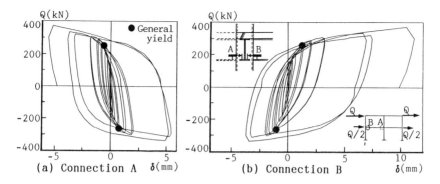

Fig.6. Load-local deformation relationships for frame

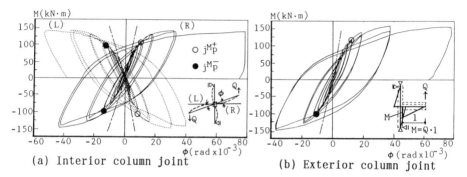

Fig.7. Load-drift angle relationships for subassemblages

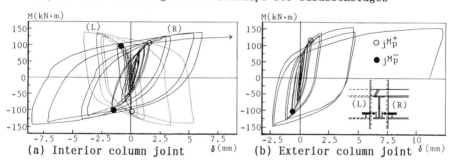

Fig.8. Load-local deformation relationships for subassemblages

3.2 Subassemblage test

The relationships between the applied load and drift angle for the two subassemblages are shown in Fig.7. The positive and negative general yield moments at the beam-to-column connection are jMp⁺ = 118kN·m and jMp⁻ =102kN·m for the exterior column joint, and jMp⁺ = 103, 108kN·m and jMp⁻ =98.7, 100kN·m for the interior column joint.

The relationships between the beam moment at the connection and local deformation of the beam flange against the column wall are

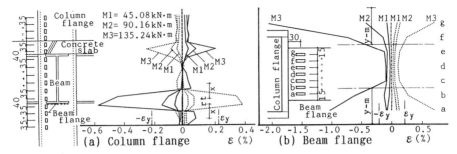

Fig.9. Strain distribution at connection

shown in Fig.8. It can be seen that the general yield strengths of the subassemblages correspond to those of the composite joints.

The failure mode for the composite joints was the same as that of the frame specimen. The maximum moments in the composite joints are jMu$^+$ =156kN·m and jMu$^-$ =147kN·m for the exterior column joint, and jMu$^+$ =144, 138kN·m and jMu$^-$ =138, 143kN·m for the interior column joint.

4 Discussion

4.1 Strength of composite joint

The strain distribution at the connection measured for the interior column joint is shown in Fig.9. By taking this strain distribution into consideration, the analytical model shown in Fig.10 can be assumed for estimating the general yield moment (jMp). The axial strength nPy of the beam flange can be obtained from the yield line theory by assuming moment yield lines at the column walls and partial yield zones on both sides of the beam flange. The maximum moment in the composite joints (jMu) can also be estimated by replacing the yield point with the tensile strength of the steel column and beam materials.

The estimated general yield moments are jMp$^+$ =110kN·m for the positive value and jMp$^-$ =103kN·m for the negative value, and the estimated maximum moments are jMu$^+$ =157kN·m and jMu$^-$ =145kN·m respectively. These correspond relatively well to the experimental values.

4.2 Plastic hinge analysis of unbraced frame

The result of a plastic hinge analysis of the frame specimen is shown in Fig.11. In this analysis, the general yield moments for the composite joint are used as the plastic hinge moments of the composite beam, and the full plastic moment taking account of axial force is used for that of the column. The storey shear strength for the mechanism of the frame is 453kN for the first storey and 680 kN for the ground storey, which are conservative estimates compared with the experimental values. The experimental drift angles for the analytical mechanism load are 1/120 and -1/130 rad. for the first storey, and 1/160 and -1/180 rad. for the ground storey, these values showing high elastic lateral rigidity for the frame.

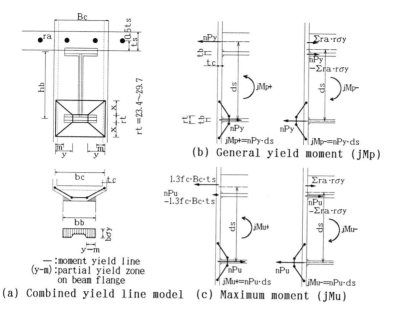

Fig.10. Analytical Model of connection

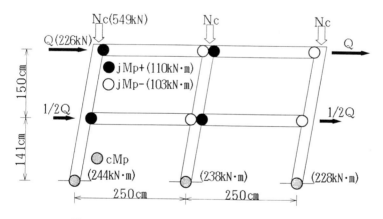

Fig.11. Result of plastic hinge analysis

4.3 Rigidity of unbraced frame

All slab concrete/slab steel is deemed to be effective under a positive/negative moment when calculating the moment of inertia of a composite beam. For the interior spans of the frame, approximately 60% of the span is deemed to be under a positive moment.

The elastic rigidity of the specimens is calculated by neglecting the semi-rigid action of the composite joints.

As the calculated rigidity gives a steeper curve than the

experimental value from the early stage of loading for the
subassemblages as shown in Fig.7, the effect of the semi-rigid action
of a composite joint should be incorporated. For the frame, however,
there is little difference in the elastic lateral rigidity between
the calculated and experimental values shown in Fig.3, so the effect
of the semi-rigid action is small in this case.

4.4 Energy dissipation capacity of unbraced frame
The maximum storey drift angle is 1/17.9 rad. for the first storey,
and 1/22.0 rad. for the ground storey. The hysteresis loop for the
frame is of the stable spindle type up to the maximum strength, and
is high in energy dissipation capacity.

5 Conclusions

From the test results for the unbraced frame with the semi-rigid
composite joints which the steel beams are welded directly to RHS
columns without any inner diaphragms, the following conclusions can
be made:

   (a) The hysteresis loop obtained is of the stable spindle type and
high in energy dissipation capacity.
   (b) The analytical plastic strength of the frame by a plastic
hinge analysis, using the analysed general yield moment of the
composite joint as the plastic hinge moment, gives conservative
estimates for the general yield strengths.
   (c) The effect of the semi-rigid action of the composite joints
is not negligible for the subassemblage specimens, but is small for
the frame specimen.

6 Acknowledgements

The author expresses his thanks to Messrs. K. Ebato, S. Sekine and
T. Sugiyama for their research efforts. This work was financed by
a grant-in-aid for scientific research(B) from the Japanese Ministry
of Education and Culture.

7 References

Eurocode No.4 (1990) Common unified rules for composite steel and
   concrete structures.
Leon, R.T. and Zandonini, R. (1992) Composite joints. **Proc. First
   World Conference on Constructional Steel Design**, Acapulco.
Morita, K., Yamamoto, N. and Ebato, K. (1989) Analysis on the
   strength of unstiffened beam flange to RHS column connections
   based on the combined yield line model. **Tubular Structures**
   (eds E. Niemi and P. Mäkeläinen), Elsevier Applied Science,
   pp. 164-171.

59 BEHAVIOR OF FULL SCALE EXPOSED TYPE STEEL SQUARE TUBULAR COLUMN BASES SUBJECTED TO ALTERNATING LATERAL LOADING UNDER CONSTANT AXIAL FORCE

S. IGARASHI
Osaka Sangyo University, Japan
S. NAKASHIMA and T. IMOTO
Osaka Institute of Technology, Japan
N. NODA
Okabe Engineering Co. Ltd, Chiba, Japan
M. SUZUKI
Asahi Chemical Industry Co. Ltd, Tokyo, Japan

Abstract
The Authors have reported previously on the mechanical behavior of small models of exposed type square tubular column bases subjected to cyclic bending moment and have presented some analytical methods available for structural design. In this study, we examined the same problem on full scale models in which the column bases are connected to concrete foundations using special anchors. Under constant axial force, reversed cyclic lateral loads were applied to the tops of the cantilevered box columns. From this study, the elastic rotational rigidities and ultimate bending strengths of full scale models were confirmed and it was shown that the analytical methods used have good correspondence with experimental results.
Keywords: Exposed Type Column Base, Full Scale Model, Square Tubular Column, Hysteresis Curve, Elastic Rotational Rigidity, Ultimate Bending Strength.

1 Introduction

Column bases are the most important connections in steel structures because their mechanical behavior controls decisively the strengths and rigidities of frames. Especially in seismic zones, the question of how to ensure the necessary rotational rigidities and ultimate strengths of column footings is a very important problem for structural designers.

For design of the footings of the tubular columns, careful consideration of the design problem may be required. One easy solution of the problem is to wrap steel footings with cast in-place reinforced concrete. But sometimes, this method is inadvisable because of the inconvenience of construction of the structure. The Authors have been studying experimentally the behavior of exposed type column bases of square tubular columns and have obtained instructive results.

In this paper, we wish to present some results of our recent research on similar but full-scale models under lateral and axial forces.

2 Experimental procedure

The models used in this study are shown in Fig. 1. The square tubular columns are fixed rigidly to raised column bases cast as part of concrete foundations using anchor bolts having 25~38mm diameter. To ensure high installation precision and rotational rigidities, a special grouting method is also used in this study.

Tubular Structures V. Edited by M.G. Coutie and G. Davies.
© 1993 E & FN Spon, 2–6 Boundary Row, London SE1 8HN. ISBN 0 419 18770 7.

The variable parameters in the tests were the column sizes, thickness of the base plates, number of anchor bolts and axial force ratios for the columns as shown in Table 1.

The sizes of the square tubular columns were 250 × 250 × 12 and 350 × 350 × 16. These are the most widely used column sections in Japan. The thickness of the base plates was 36, 40, 45 and 50 mm. The diameter of the anchor bolts was 25, 29 and 38 mm.

Ten specimens were fabricated and fixed rigidly to the reinforced concrete foundations. To achieve close contact with the concrete surfaces, high strength, non-shrinkage mortar, as listed in Table 3, was grouted through narrow gaps around the anchor bolt holes of the base plates.

The loading conditions were as follows. Constant axial forces applied on the columns by oil jack were in the range N/Ny=0.083~0.333. That is, forces applied to the □-250 series specimens were N/Ny= −0.16, 0.20, 0.333 and 0.083, and 0.20 for the □-350 series. Reversed cyclic loadings were applied to the tops of the cantilevered columns.

Mechanical properties of the materials used here are listed in Table 2 and Table 3.

Table 1 Experimental plan

Specimen No.	Steel column (mm)	Base plate thickness (mm)	Anchor bolt size (mm)	Constant axial force N (kN)	(N/Ny)	Remark
25-3625- 0			D25	0	0.000	
25-3625-523			"	523	0.200	
25-3625-850	□-250x250x12	36	"	850	0.333	
25-3629- 0			D29	0	0.000	
25-3629-850			"	850	0.333	
25-3629-△412			"	-412	-0.160	Tensile force
35-4038-405		40		405	0.083	
35-4538-405	□-350x350x16	45	D38	"	"	
35-5038-405		50		"	"	
35-4538-981		"		981	0.200	

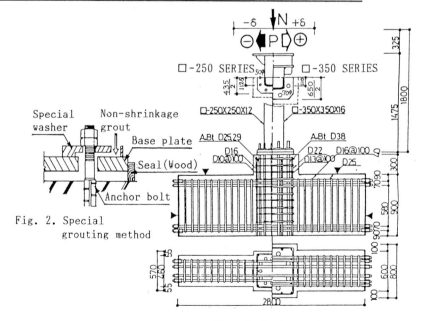

□-250 SERIES   □-350 SERIES

Special   Non-shrinkage
washer   grout
Base plate
Seal(Wood)
Anchor bolt

Fig. 2. Special
grouting method

Fig. 1. Test specimen

600

Table 2 Mechanical properties of steel

Element	Size (mm)	Material	σy (N/mm²)	σu (N/mm²)	Es (×10⁵N/mm²)	ε (%)	Remark
Steel column	t=12	STK400	427	496	2.05	33	□-250 series
	t=16	"	465	536	1.97	28	□-350 series
Base plate	t=36	SM490A	250	417	2.12	30	□-250 series
	t=40	TMCP33	335	515	2.05	32	□-350 series
	t=45	"	426	566	2.12	29	"
	t=50	"	405	553	2.07	30	"
Anchor bolt	D25	SD390	440	624	1.90	16	□-250 series
	D29	"	454	629	1.95	17	"
	D38	"	425	625	1.92	24	□-350 series
Vertical chord reinforcement	D16	SD295	331	498	1.95	21	□-250 series
	D22	"	314	494	1.89	20	□-350 series
Horizontal chord reinforcement	D25	SD345	374	529	1.88	23	□-250 series
	"	"	374	529	1.88	23	□-350 series
Hoop	D10	SD295	355	492	1.93	20	□-250 series
	D13	"	346	492	1.93	19	□-350 series

Table 3 Mechanical properties of concrete and grout

Element	W/C (%)	σc (N/mm²)	Ec (× 10⁴N/mm²)	Remark
Concrete	60	25.1	2.26	□-250 series
	"	22.7	1.89	□-350 series
Grout		55.1		□-250 series
		63.7		□-350 series (t=40, 50)
		46.8		" (t=45)

3 Results of the loading tests

In Fig. 3, the relations between moment and rotational angles of the steel column are shown. The inclined chain lines in these figures are the calculated values of N- δ effects under axial forces.
Noteworthy features of these curves are summarized as follows.

□-250 series:
The shapes of the hysteresis curves of specimens 25-3625-0 and 25-3629-0 are reversed S type loops with slight slips.
The curves of 25-3625-523, 25-3625-850 and 25-3629-850 are stable spindle shapes. In these specimens, no yielding and no plastic elongations of the anchor bolts were observed, and the features of these loops might be due to the elasto-plastic deformations of the columns.
The specimen 25-3629-△412 under tensile axial force also showed a reversed S type loop.

□-350 series:
Unloading curves of all specimens in this series presented, after arriving at 0.015 rad. rotation, initiations of slip phenomena. Following this, reversed S type, loops were drawn. After yielding of the columns, slopes of the hysteresis loops gradually inclined and, after the rotational deformations increased over 0.02 rad., perfect plastic flows were clearly observed. No negative slopes appeared in these loops.
Effects of the thickness of the base plates on the shapes of both curves were unremarkable.

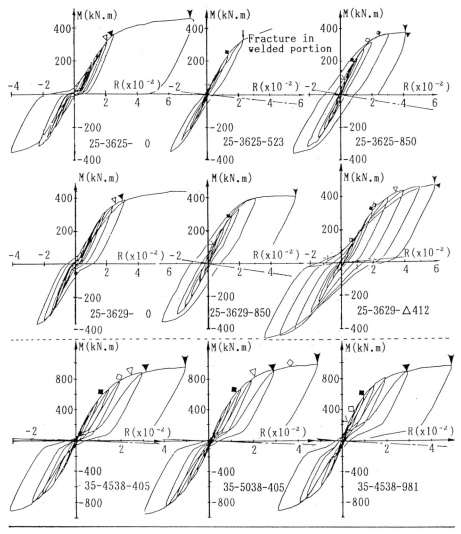

▼: Ultimate strength
▼: Yielding of a group of outermost tensile anchor bolts
∇: Yielding of a tension side anchor bolt
◆: Yielding of compression side steel column flange
◇: Yielding of base plate (tension side)
◇: Yielding of tension side steel column flange
◇: Yielding of base plate (compression side)
∇: Restrained bending moment (Experimental value)

Fig. 3. M-R relations

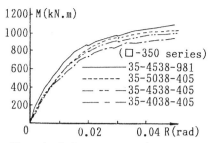

Fig. 4. M-R relations (Envelope curves)

4 Elastic rotational rigidity

Some representative envelope curves of M-R hysteresis loops mentioned above are shown in Fig. 4. The experimental and calculated values of the elastic rotational rigidity K_θ and the restrained bending moments Mn are listed in Table 4.

Fig. 5 shows the relations between elastic rotational rigidities K_θ and axial force N. These experimental eK_θ values relate to the points on the M-R envelope curves having 1/3 of ultimate bending moments Mmax.

Through our previous research on small size models, we obtained the result that the elastic rotational rigidities under zero axial force can be calculated approximately by considerations of deformations of anchor bolts and base plates, and that when some compressive axial forces are applied, rotations of column footings become smaller but, after the moments exceed the restrained moment Mn, rotational behavior is similar to the situation with zero axial force.

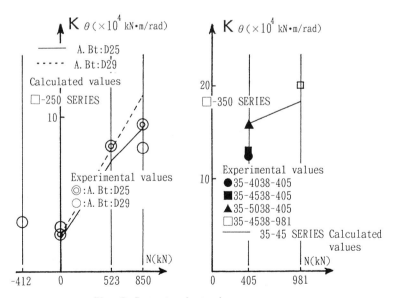

Fig. 5. Rotational rigidity

Table 4 Experimental results

Specimen No.	Rotational rigidity Experimental eKθ(kN.m/rad)	Calculated Kθ(kN.m/rad)	eKθ/Kθ	Restrained moment Experimental eMn(kN.m)	Calculated Mn(kN.m)	eMn/Mn	Ultimate strength Experimental Mmax(kN.m)	Calculated Mu(kN.m)	Mmax/Mu	Failure mode ※
25-3625-0	23200	20900	1.11	–	–	–	426	427	1.00	A, C
25-3625-523	85300	74600	1.14	57.9	45.1	1.28	343	467	0.74	C, E
25-3625-850	92600	89200	1.04	83.4	73.5	1.13	421	478	0.88	C, D, E
25-3629-0	28000	22800	1.23	–	–	–	414	516	0.80	A, C, D
25-3629-850	78600	114000	0.69	90.2	73.5	1.23	452	529	0.86	C, D
25-3629-△412	32800	–	–	–	–	–	450	485	0.93	C, D
35-4038-405	124000	116000	1.07	62.8	54.0	1.16	927	998	0.93	A, B, C, D
35-4538-405	132000	161000	0.82	61.8	54.0	1.14	987	998	0.99	A, B, C, D
35-5038-405	158000	161000	0.98	61.8	54.0	1.14	1020	998	1.02	A, B, C, D
35-4538-981	200000	179000	1.11	152.0	131.0	1.16	1050	1060	0.99	A, B, C, D

※ A: Yielding of anchor bolts B: Yielding of base plate C: Collapse in raiser concrete footing D: Local buckling in steel column E: Fracture in welded portion between steel column and base plate

The analytical values K_θ of elastic rotational rigidity in Table 4 were calculated using the equation (1).

$$K_\theta = (Mo + Mn) / \theta o \qquad (1)$$

K_θ: Elastic rotational rigidity under constant axial compression
Mo: Yield moment under zero axial compression
Mn: Restrained moment under constant axial compression
θo: Rotational angle of steel column base measured at moment Mo under zero axial compression

Here, we assumed the equivalent lengths of the anchor bolts to be 20d (d: diameter of anchor bolt). From these data, we may draw the following conclusions.

Elastic rotational rigidities increase correspondingly with the magnitudes of applied axial force.
Analytical values of elastic rotational rigidity are nearly equal to the experimental results and the assumptions used in equation (1) are appropriate.

5 Strain distribution of grout mortar beneath the base plates

Fig. 7 shows the grout strains measured beneath the base plates under constant axial force and increasing cyclic lateral loading. Maximum strains were obtained near the portion under the base plate along the compression flange of the steel square tubular column under compressive axial force. But, when constant tensile force was applied to the columns, maximum strains appeared in the grout located at the edge of the base plate.

6 Ultimate strength

Experimental and analytical values of ultimate strengths of the specimens are listed in Table 4. These analytical values Mu were introduced from using ultimate strengths of locally compressed concrete beneath the base plates and tensile strengths of the anchor bolts. This method is very similar to that used in our former paper. The local

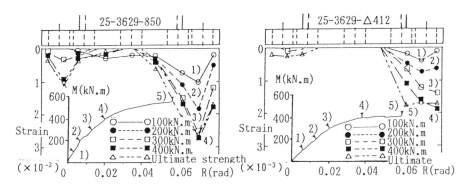

Fig. 7. Strain distribution under base plate

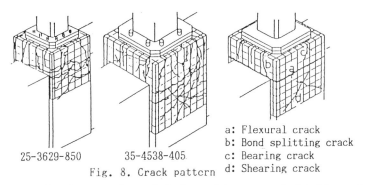

25-3629-850   35-4538-405.
Fig. 8. Crack pattern

a: Flexural crack
b: Bond splitting crack
c: Bearing crack
d: Shearing crack

bearing strengths of concrete beneath the base plates were calculated by applying a coefficient λ to the compressive strengths of the concrete cylinders. (See Fig. 9)

$$\lambda = 0.86(Ae / Ab)^{0.441} \tag{2}$$

Ae: Effective bearing area
Ab: Compressed area

We obtained the above equation from a separate experimental study on cylindrical and block type concrete specimens.

Fig. 10 shows the experimental ultimate strengths Mmax plotted on the M-N interaction curves. From these figures, we may conclude that the analytical and experimental ultimate values have good correspondence with each other.

7 Conclusions

(1) Under constant compressive force, the elastic rotational rigidity and ultimate bending strength of a square tubular column footing increases by increasing the thickness of the base plate and the diameters of the anchor bolts. These values may be calculated approximately by using the equations proposed in this paper.
(2) Through this experimental research, we have clarified the mechanical behavior of exposed tubular column footings under axial and laterally reversed cyclic loading.

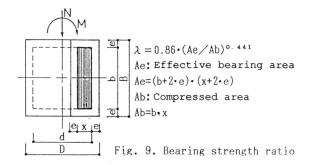

$\lambda = 0.86 \cdot (Ae/Ab)^{0.441}$

Ae: Effective bearing area

Ae = $(b+2 \cdot e) \cdot (x+2 \cdot e)$

Ab: Compressed area

Ab = $b \cdot x$

Fig. 9. Bearing strength ratio

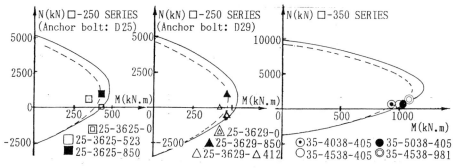

Fig. 10. M-N Interaction curves

(3) The results on the full scale models used in this study can be easily and accurately inferred from experiments on smaller size models. Appropriately designed small size model tests may be substituted for those using full-scale models.

8 References

Igarashi, S., Nakashima, S., Kadoya, H. and Suzuki, M. (1992) Behavior of exposed-type 'fixed' column base connected to riser foundation concrete, Proceedings of the tenth world conference on earthquake engineering, A. A. Balkema, Rotterdam, Brookfield PP. 3197-3200.

Nakashima, S. and Igarashi, S.(1986) Experimental study on square steel tubular column bases embedded in foundation beams, Proceedings of the international meeting, Safety criteria in design of tubular structures, Maruzen, Tokyo. pp. 93-102

Nakashima, S. and Igarashi, S., Kadoya, H., Suzuki, M. and Noda, N. (1991) Experimental study on exposed steel square tubular column bases using a special grout method, Tubular Structures, Delft University press, Delft, pp. 109-118.

60 ABSORBED ENERGY OF CHS AND SHS BRACES CYCLICALLY LOADED IN TENSION-COMPRESSION

J. FARKAS
University of Miskolc, Hungary

Abstract
The absorbed energy is characterized by the area of
hysteresis loop. A brief survey of selected literature
summarizes the experimental results for braces of various
cross-sectional shapes. The post-buckling axial force -
axial shortening relationship is calculated using the
plastic hinge method. The hysteresis loop for the first
load cycle is modified on the basis of experimental results
of several authors considering the degradation due to the
Bauschinger-effect and the effect of residual camber.
Simple closed formulae are given for the area of the
stable hysteresis loop. Based on these formulae, using
numerical examples, the effect of yield stress of steel
and the effect of end restraint is shown, and the CHS and
SHS braces are compared to each other.
Keywords: Cyclic Loaded Struts, Tubular Braces, Post-
Buckling Behaviour, Earthquake-Resistant Design.

1 Introduction

Tubular members of circular, square or rectangular (CHS,
SHS or RHS) sections are widely used in many structures
such as building and offshore structures. Braces play an
important role in the earthquake-resistant design of such
structures. The efficiency of bracing is characterized by
the absorbed energy which can be obtained as the area of
hysteresis loop.

Studies have shown that the first critical overall
buckling strength decreases during the second and third
cycle, but after a few cycles the hysteresis loop becomes
stable. This degradation is caused by the Bauschinger-
effect and by the effect of residual camber as explained
by Popov and Black (1981). Unfortunately, these effects
cannot be considered by analytical derivations, thus, the
characteristics of the stable hysteresis loop will be
taken from the experimental data published in the
literature.

Tubular Structures V. Edited by M.G. Coutie and G. Davies.
© 1993 E & FN Spon, 2–6 Boundary Row, London SE1 8HN. ISBN 0 419 18770 7.

The aim of the present paper is to derive simple closed formulae for the calculation of the area of the stable hysteresis loop. The derived formulae enable designers to analyze the effect of some important parameters such as the yield stress of steel, end restraint and cross-sectional shape, and can serve as a basis of optimization, i.e. the increasing of the energy-absorbing capacity of braces.

2 A brief survey of selected literature

The aim of this survey is to summarize the experimental results relating to the stable hysteresis loop shown schematically in Fig.1.

It can be seen from Table 1 that, for η and z_4 the approximate value of 0.5 is predominantly obtained. The sum of relative axial shortenings $x + |x_4|$ varies in range 5 - 14. On the basis of these data we will consider values $\eta = z_4 = 0.5$ and $x_o = 1$, $x_4 = -10$.

Another important problem is the local buckling. According to many authors, e.g. Lee and Goel(1987), it is recommended to avoid local buckling. Unfortunately, for the limiting D/t or b/t slenderness ratios, in the case of cyclic plastic loading one can find very few proposed values. For CHS Zayas, Mahin and Popov (1982) proposed $\delta_{cL} = (D/t)_L = 6820/f_y$ where f_y is the yield stress in MPa, thus, for $f_y = 235$ $\delta_{cL} = 29$ and for $f_y = 355$ MPa $\delta_{cL}=20$. For SHS or RHS Liu and Goel(1988) proposed $\delta_{sL} = (b/t)_L = 14$ for $f_y = 372$ MPa, thus, we take for $f_y = 355$ MPa $\delta_{sL} = 15$ and for $f_y = 235$ MPa $\delta_{sL} = 15(355/235)^{1/2} = 19$.

The limitation of the strut slenderness plays also an important role. API (1989) proposed $KL/r \leqslant 80$, where K is the end restraint factor, L is the strut length, r is the radius of gyration.

3 Post-buckling behaviour of CHS and SHS braces

The axial force - axial shortening $(P - \Delta)$ relationship (Fig.2) has been derived for CHS struts by Supple and Collins (1980) using the simple plastic hinge method. This method gives a good approximation as it has been verified by the author (Farkas 1992) comparing it with other methods in a numerical example.

$$\Delta = \Delta_{el} + \Delta_{pl} = \frac{PL}{AE} + \frac{\alpha D^2}{4L}\left[\frac{P_y}{P}\cos\left(\frac{\pi}{2}\frac{P}{P_y}\right)\right]^2 \qquad (1)$$

where α is the end restraint factor, for pinned ends $\alpha = 1$, for fixed ends $\alpha = 4$, $P_y = Af_y$ is the squash load.

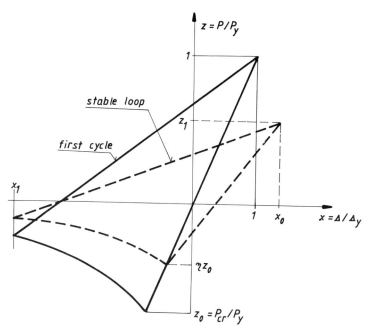

Fig.1. Characteristics of a stable hysteresis loop

Table 1. Characteristics of the stable hysteresis loop
    according to Fig.1

Reference	x_0	x_1	η	z_1	cross-section
Jain (1980)	2	-12	0.5	1	SHS
Liu (1988)	2	-10	0.5	0.5	RHS
Matsumoto (1986)	2	-10	0.5	0.5	CHS
Nonaka (1977)	4	-4	0.5	0.8	▨
Ochi (1990)	1	-10	0.5	0.5	CHS
Papadrakakis (1987)	1	-4	0.5	0.5	CHS
Prathuangsit (1978)	1	-12	0.5	0.5	I
Shibata (1982)	5	-5	0.5	0.5	I

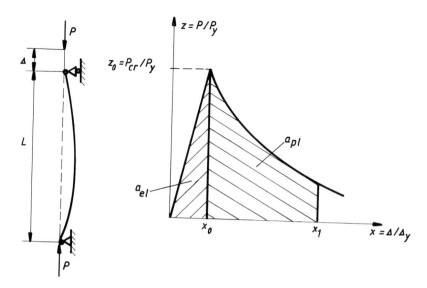

Fig.2. Post-buckling behaviour and the related
specific areas

Using notations $x = \Delta/\Delta_y$, $z = P/P_y$, $z_o = P_{cr}/P_y$
$x_o = \Delta_{el}/\Delta_y = P_{cr}/P_y = z_o$ Eq.(1) can be written in the form

$$x - x_o = C_1\left(\frac{\cos^2(\pi z/2)}{z^2} - C_2\right); \quad C_1 = \frac{\alpha D^2 E}{4L^2 f_y}; \quad C_2 = \frac{\cos^2(\pi z_o/2)}{z_o^2} \tag{2}$$

For $z < 0.4$ the following approximation is acceptable
(Farkas 1992)

$$\cos(\pi z/2) \approx 1 - \pi^2 z^2/8 \quad \text{and} \quad \cos^2(\pi z/2) \approx 1 - \pi^2 z^2/4 \tag{3}$$

and Eq.(2) takes the form

$$x - x_o = C_1\left(\frac{1}{z^2} - \frac{\pi^2}{4} - C_2\right) \tag{4}$$

from which one obtains

$$z = C_1^{1/2}\left(x - x_o + \frac{\pi^2}{4}C_1 + C_1 C_2\right)^{-1/2} \tag{5}$$

The areas shown in Fig.2 can be calculated as follows

$$a_{el} = z_o^2/2 \tag{6}$$

and

$$a_{pl} = \int_{x_o}^{x_1} z\,dx = 2C_1^{1/2}\left[\left(x_1 - x_o + \frac{\pi^2}{4}C_1 + C_1 C_2\right)^{1/2} - \left(\frac{\pi^2}{4}C_1 + C_1 C_2\right)^{1/2}\right] \tag{7}$$

610

According to the author's derivation, for SHS it is

$$\Delta_{pl} = \frac{\alpha \pi^2}{4L}\left(\frac{M}{P}\right)^2 = \frac{\alpha \pi^2 b^2}{16L}\left(\frac{3}{4z} - z\right)^2 \tag{8}$$

and

$$x-x_0 = \Delta_{pl}/\Delta_y = C_3\left(\frac{3}{4z} - z\right)^2 + C_4; \quad C_3 = \frac{\alpha \pi^2 b^2}{16L^2 f_y}; C_4 = C_3\left(\frac{z}{4z_0} - z_0\right)^2 \tag{9}$$

Expressing z from Eq. (9) we obtain

$$z = \frac{1}{2C_3^{1/2}}\left[\left(x-x_0+3C_3+C_4\right)^{1/2} - \left(x-x_0+C_4\right)^{1/2}\right] \tag{10}$$

and the area in the post-buckling range

$$a_{pl} = \int_{x_0}^{x_1} z\,dx = \frac{1}{3C_3^{1/2}}\left[\left(x_1-x_0+3C_3+C_4\right)^{3/2} - \left(x_1-x_0+C_4\right)^{3/2} - \right.$$
$$\left. - \left(3C_3+C_4\right)^{3/2} + C_4^{3/2}\right] \tag{11}$$

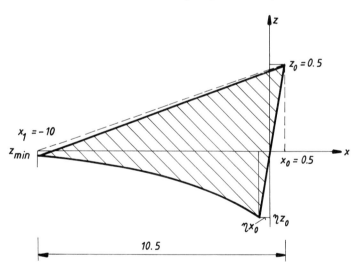

Fig.3. Area of the stable hysteresis loop

4 Formulae for the area of the stable hysteresis loop

As described in Section 2, we consider the stable hysteresis loop according to Fig.3. The whole specific absorbed energy as the area shown in Fig. 3 is

$$\sum a_i = a_{el} + a_{pl} + 10 \times 0.5/2 - 10.5 z_{min}/2 \tag{12}$$

611

For a_{el} and a_{pl} we use Eqs (6) , (7) or (11) , but instead of $x_0 = z_0$ we calculate with $\eta x_0 = \eta z_0$ where $\eta = 0.5$. z_{min} is calculated using Eq.(5) or (10) taking x = 10 and instead of x_0 taking $0.5x_0$.

5 Numerical examples

If the design load in braces is not known, the sizes of cross-sections can be calculated on the basis of slenderness limitations mentioned in Section 2. For CHS braces the mean diameter may be obtained from the constraint on strut slenderness

$$\lambda = KL/r = 8^{1/2}KL/D \leq \lambda_0 = 80; \quad D \geq 8^{1/2}KL/\lambda_0 \tag{13}$$

The thickness can be calculated using the local buckling constraint

$$\delta_c = D/t \leq \delta_{cL} = 6820/f_y \tag{14}$$

Similarly, for SHS braces

$$b \geq 6^{1/2}KL/\lambda_0 \tag{15}$$

and

$$\delta_s = b/t \leq \delta_{sL} = 15\left(355/f_y\right)^{1/2} \tag{16}$$

In Table 2 the calculation of the total area of the stable hysteresis loop is summarized in the case of pinned ends, for CHS and SHS braces made of steels with yield stress of 235 and 355 MPa, respectively. The cross-sections are taken according to ISO/DIS 4019.2 (1979) or ISO 4200 (1980). The first overall buckling force P_{cr} is calculated using the formulae of Eurocode 3 (1992) for curve "b" as follows

$$P_{cr} = \chi P_y; \quad \chi = \left[\phi + \sqrt{\phi^2 - \bar{\lambda}^2}\right]^{-1} \tag{17}$$

$$\phi = 0.5\left[1 + 0.34\left(\bar{\lambda} - 0.2\right) + \bar{\lambda}^2\right]; \quad \bar{\lambda} = \lambda/\lambda_1; \quad \lambda_1 = 93.9\varepsilon, \quad \varepsilon = \left(235/f_y\right)^{1/2}$$

The total absorbed energy is calculated as

$$W_{absorb} = \left(\Sigma a_i\right)f_y^2 AL/E \tag{18}$$

For another calculation details see Section 4.

To complement the numerical examples treated in Table 2, another end restraint case has also been calculated. In the case of SHS and $f_y = 355$ MPa the end restraint is changed from pinned ends to partially restrained ends with K=0.8 and $\alpha = 2$. The ISO cross-section will be 160x10, A=5660mm^2, $z_0 = 0.5675$, $C_3 = 0.4561$, $C_4 = 2.5379$, $a_{pl} = 1.8291$, $\Sigma a_i = 3.6299$ and $W_{absorb} = 73.978$ kNm.

Table 2. Numerical examples. $L = 6$ m, pinned ends, $K = 1$, $\alpha = 1$, $\lambda_0 = 80$; sizes in mm

cross-section f_y (MPa)	CHS 235	355	SHS 235	355
required D, b	$\varepsilon^{1/2}L/\lambda_0 = 212$		$6^{1/2}L/\lambda_0 = 183.7$	
limiting slenderness	29	20	19	15
ISO cross-section	219.1/8	219.1/10	200x10	200x12.5
D, b	211.1	209.1	190	187.5
A (mm^2)	5310	6570	7260	8840
r	74.7	74.0	76.5	75.2
a_{el}	0.0595	0.0390	0.0617	0.0405
a_{pl}	2.0241	1.5808	2.1145	1.7148
$10.5z_{min}/2$	0.7954	0.6288	0.8373	0.6725
$\sum a_i$	3.7882	3.4910	3.8339	3.5828
W_{absorb} (kNm)	31.739	82.586	43.975	114.042

6 Conclusions

The numerical examples in Table 2 show that, if we design the braces considering the strut slenderness limitation and the limitations of local slendernesses D/t or b/t, there will be no significant difference between the specific absorbed energies. In the total absorbed energies are large differencies. W is 38% larger for SHS than for CHS, and it is 160% larger for yield stress of 355 MPa than for 235 MPa.

If the end restraint is changed from pinned ends to partially restrained ends with $K = 0.8$ and $\alpha = 2$, in the case of SHS and 355 MPa, W decreases by 35%. These large changes are caused by the changes in cross-sectional area and in yield stress.

The main conclusion is that, to increase the total absorbed energy, the cross-sectional area and the yield stress should be increased.

In the optimum design of structures the weight or cross-sectional area minimization is predominantly the main aspect (Farkas 1984). In the contrary, the present study shows that, for efficient earthquake-resistant design of braces the A-max concept should be considered.

7 References

American Petroleum Institute (1989) Draft Recommended
  Practice 2A-LRFD, first ed.
Eurocode 3. (1992) Design of steel structures. Part 1.1.
  Brussels, CEN-European Committee for Standardization.
Farkas,J. (1984) Optimum design of metal structures.
  Akadémiai Kiadó Budapest and Ellis Horwood Chichester.
Farkas,J. (1992) Optimum design of circular tubes loaded
  in compression in pre- and post-buckling ranges.
  Struct. Optimization, 5, 123-127.
Jain,A.K. Goel,S.C. and Hanson,R.D. (1980) Hysteretic cycles
  of axially loaded steel members. J.Struct.Div.Proc.ASCE
  106, 1777-1795.
Lee,S. and Goel,S.C. (1987) Seismic behavior of hollow and
  concrete filled square tubular bracing members. Res.
  Rep. UMCE 87-11. Dept of Civil Engng, University of
  Michigan, Ann Arbor.
Liu,Zh. and Goel,S.C. (1988) Cyclic load behavior of
  concrete-filled tubular braces. J.Struct.Engng ASCE
  114, 1488-1506.
Matsumoto,T. Yamashita,M. et al. (1987) Post-buckling
  behavior of circular tube brace under cyclic loadings,
  in Safety Criteria in Design of Tubular Structures,
  Proc. Int. Meeting, Tokyo, 1986.Architectural Institute
  of Japan, pp.15-25.
Nonaka,T. (1977) Approximation of yield condition for the
  hysteretic behavior of a bar under repeated axial
  loading. Int.J.Solids Struct., 13, 637-643.
Ochi,K. Yamashita,M. et al. (1990) Local buckling and
  hysteretic behavior of circular tubular members under
  axial loads. J.Struct.Constr.Engng AIJ No.417. 53-61
  (in Japanese).
Papadrakakis,M. and Loukakis,K. (1987) Elastic-plastic
  hysteretic behaviour of struts with imperfections.
  Eng. Struct. 9, 162-170.
Popov,E.P. and Black,R.G. (1981) Steel struts under severe
  cyclic loadings.J.Struct.Div.Proc.ASCE 107, 1857-1881.
Prathuangsit,D. Goel,S.C. and Hanson,R.D. (1978) Axial
  hysteresis behavior with end restraints. J.Struct.Div.
  Proc.ASCE 104, 883-896.
Shibata,M. (1982) Analysis of elastic-plastic behavior of
  a steel brace subjected to repeated axial force.
  Int.J.Solids Struct. 18, 217-228.
Supple,W.J. and Collins,I. (1980) Post-critical behaviour
  of tubular struts. Eng. Struct. 2, 225-229.
Zayas,V.A. Mahin,S.A. and Popov,E.P. (1982) Ultimate
  strength of steel offshore structures, in Behaviour of
  Offshore Structures, Proc. 3rd Int. Conference, Boston.
  Hemisphere Publ. Co. Washington, Vol.2.pp.39-58.

61 STATIC LOADING TESTS ON CIRCULAR TUBULAR TRUSS STRUCTURE WITH TUBE-TO-TUBE X-JOINTS

K. HIRAMATSU, A. MITSUNASHI, T. MIYATANI
and N. HARA
NTT Power and Building Facilities Inc., Tokyo, Japan

Abstract
Many experiments have been carried out on the behavior of
tube-to-tube x-joints, and design strength formulas have
been presented. However, it is not completely clear how the
behavior after the ultimate strength of tube-to-tube x-
joint in the trusses is reached affects the ultimate behavior of the truss structure.

This paper describes the results of static loading tests
on circular tubular truss structure with tube-to-tube x-
joints subjected to repeated loads. The ultimate behavior
of the joints in the trusses subjected to repeated loads
was discussed. The ultimate behavior of the truss structure
after the ultimate strength of the joints was also
discussed.
Keywords: Tube-to-tube X-joint, Circular hollow section,
Tubular truss structure, Static loading test.

1 Introduction

Many experiments have been carried out on the behavior of
tube-to-tube x-joints, and design strength formulas have
been presented by J. Wardenier et al.(1991), Architectural
Institute of Japan(1990), which we called AIJ in the
following, and other organizations.

However, it is not completely clear what effect on the
ultimate strength of the tube-to-tube x-joint in the truss
framework is affected by repeated loads and the deformation
properties of the joint part. Nor is it clear how the behavior after the ultimate strength of the joint is reached
affects the ultimate behavior of the truss framework.

In this paper, static load tests on a tubular steel
truss framework whose tube-to-tube x-joints are subjected
to repeated loads were carried out. The ultimate behavior
of tube-to-tube x-joints in the trusses subjected to
repeated loads is discussed. The ultimate behavior of the
trusses and the relationship the bracing axial force between and the chord axial force, after the ultimate
strength of the joint is reached, are also discussed.

Tubular Structures V. Edited by M.G. Coutie and G. Davies.
© 1993 E & FN Spon, 2–6 Boundary Row, London SE1 8HN. ISBN 0 419 18770 7.

2 Test Method

2.1 Specimens

Two types of plane truss framework specimens were tested (Fig. 1). They are referred to as specimen A and specimen B in the following. Specimen A has tube-to-tube x-joints in the diagonal members ② of 1st unit and 2nd unit. Specimen B has an tube-to-tube x-joint in the diagonal members ② of the 1st unit. However, the joints between the diagonal members ② and the main members ① are bolted in specimen A and welded in specimen B. The details of the members and joints of the specimen are listed in Table 1.

Specimens A and B both use the same size tube (half diameter to thickness ratio of the chord, D/2T = 16) for the tube-to-tube x-joints. The angle between the bracing member and the chord is 90°. Also, for both specimens, a ultimate strength based on Recommendation for the Design and Fabrication of Tubular Structures in Steel by AIJ(1990) that is lower than the allowable compressive strength for temporary loading based on Design Standard for Steel Structure by AIJ(1973) was chosen so that the ultimate strength of the tube-to-tube x-joint is reached before buckling of the diagonal members ② occurs. The physical properties of the materials used are listed in Table 2.

2.2 Loading apparatus

An overview of the testing setup is shown in Fig. 2. The test specimen is positioned laterally and attached to a reaction framework with fixed equipments and high tensile strength bolts. Two 30-ton actuators fixed to the reaction bed are used to repeatedly apply positive and negative force to loading beam attached to the top of the test specimen by high tensile strength bolts. Roller bearings are used so as not to restrict the plumb line displacement of the point of force application, and the accompanying rotation of the specimen. The filled boxes in Fig. 2 indicate the locations restricting out-of-structure-plane deformation.

To compensate the initial deflection from the weight of the specimen itself and the loading beam for applying force, the loading beam is suspended by means of pulleys and a heavy weight. This weight balancing mechanism is capable of following the horizontal movements of the point at which the loading beam is suspended.

2.3 Measurement method

The load was taken to be the sum of the values detected by shear load cells attached to the front of the two actuators. The displacements of various parts of the specimen were measured by displacement transducers with a capacity 200mm, 100mm, 50mm or 25mm set on a measurement framework that surrounds the specimen and is fixed to the reaction bed. The deformations of tube-to-tube x-joint part

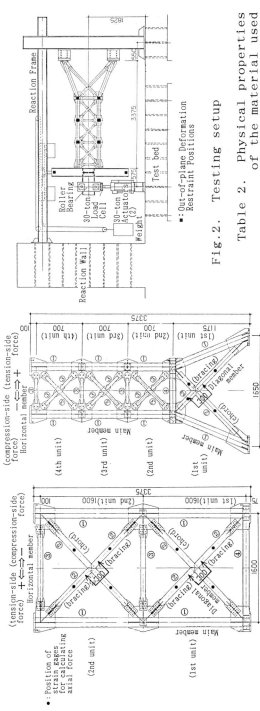

(1) Specimen A

(2) Specimen B

Fig.1. Test specimens

Fig.2. Testing setup

Table 2. Physical properties of the material used

sections specification		σ_y	q_{max}	ε_u
φ139.8x4.0	STK400	4.04	5.04	30
φ114.3x6.0	STK400	4.21	5.24	29
φ89.1x2.8	STK400	4.19	5.20	30
φ76.3x2.8	STK400	3.78	4.66	32
φ48.6x3.2	STK400	4.30	5.16	24
H100x100x6x8	SS400	3.41	4.69	30
PL-9	SM490A	3.58	5.11	27
PL-6	SM490A	3.79	5.54	25
PL-4.5	SS400	2.77	4.38	30
welding rod	D4303	4.18	5.19	29

σ_y : yield point(t/cm²)
q_{max} : tensile strength(t/cm²)
ε_u : elongation(%)

Table 1. Details of members and joints of the specimen

		main member ①	diagonal member ② ③	horizontal member ④
Specimen A	sections of the members	φ139.8x4.0	②φ89.1x2.8 ③φ76.3x2.8	H-100x100x6x8
	the ends of the members	—	tube-plate connection	rigid joints (welded)
	the joints between ① and the members	—	bolted joint(②:8-M12, ③:4-M12) with gusset plates(PL-6) and splice plates (PL-4.5)	rigid joints (welded)
Specimen B	sections of the members	φ114.3x6.0	②φ89.1x2.8 ③φ48.6x3.2	H-100x100x6x8
	the ends of the members	—	tube-plate connection	—
	the joints between ① and the members	welded joint between 1st unit and 2nd unit	welded joint with gusset plate(PL-9)	welded joint with gusset plate(PL-6)	rigid joints (welded)

were measured by displacement transducers with a capacity 50mm or 25mm set on the bracing member. Slippage deformation in the bolted joints was measured with a π -type displacement gage. Strain of the members was measured with from two to four strain gages attached to each measurement point.

3 Results and Discussion

3.1 Behavior of the tube-to-tube x-joint

The solid lines in the Fig. 3 show the relationship between the deformation of the tube-to-tube x-joint and the bracing axial force. The vertical axis values for axial force are calculated from the strain gage. The horizontal axis tube-to-tube x-joint deformation values are the amount of deformation between the two points indicated by the arrows in Fig. 1. The coarse dotted line represents the results of a monotonic load test (zero chord axial force) on a tube-to-tube x-joint removed from the framework, which we call a joint specimen in the following.

For both specimens, there is out-of-plane collapse of the chord when the applied force is in the direction that results in a compressive force on the bracings (called compression-side force in the following); when the direction of applied force results in tension on the bracings (called tension-side force in the following), the chord returns to the original condition. After repetition of this force application, during the tension-side force application immediately following the out-of-plane collapse during the compression-side force application, the bracing axial force does not increase, cracking appears in the chord, and the bracing axial force slowly decreases.

In specimen A, cracking appeared in the chord of the 1st unit in the ⑦ th cycle of tension-side force application, and the bracing axial force of 1st unit decreased. Without cracks appearing in the chord of the 2nd unit, the bracing axial force deformation curve was quite consistent with the load deformation curve of the joint specimen. Cracking appeared in the chords of the 2nd unit in the ⑧ th cycle of tension-side force application, and the bracing axial force decreased.

The fine dotted lines in the Fig. 3 show the relationship between the tube-to-tube x-joint deformation and the load. By comparing the solid lines and fine dashed lines, we can see that although the bracing axial force slowly decreases after reaching the maximum axial force during compression-side force application, the load increases.

The dashed line shows the rigidity obtained from an elastic stress analysis based on the finite element method with shell element modeling. The analytical values agree reasonably well with the measured initial rigidity values.

3.2 Behavior of the truss framework

The dotted lines in the Fig. 4(1) represent the relationship between the load and the displacement of the top of the specimen A(in the direction of force application). That of the specimen B is shown in Fig. 4(2-1).

The rigidity of the truss framework degraded from a very early stage due to slippage in the bolted connections, and with repeated loads a reverse S-shaped loop is described.

Here, we consider the displacement of the top of the specimen, deducting the effect of bolt slippage and locking, using data from the π -type displacement gages and the other displacement transducers. The solid lines in the Fig. 4(1) represent the relationship between the load and the displacement of the top of the specimen A, deducting bolt slippage and locking. That of the specimen B is shown in Fig. 4(2-2).

For both specimens, while out-of-plane deformation of the chord occurs in the ⑦ th of compression-side force application, and the axial force of the bracing slowly decreases (Fig. 3), the truss framework strength increases after that.In the tension-side force application, cracking appears in the chord of 1st unit of specimen A in the ⑦ th cycle. The truss framework strength increases in this cycle. In the ⑧ th cycle, cracking appears in the chords of the 1st and 2nd units, and the truss framework strength gradually decreases. For specimen B, the truss framework strength decreases after cracking appears in the chords.

The displacement of the top part of 1st unit of specimen B (see the fine dotted lines in the Fig. 4(2-2)) is in the direction opposite that of the applied force.

The dashed lines in the Fig. 4(1) and 4(2-2) represent the rigidity from an elastic stress analysis of a truss and beam element model. (The rigidity of the tube-to-tube x-joint of the joint specimen is taken to be the rigidity of the end of the beam element.) The analytical results agree reasonably well with the measured initial rigidity values.

3.3 Axial force ratio of chord and bracing

The solid lines in Fig. 5 show the relationship between the chord axial force and the bracing axial force. The dashed lines and dotted lines represent the allowable strength for temporary loading and the ultimate strength of tube-to-tube x-joint according to AIJ(1990). The solid circles and solid triangles represent the maximum load and elastic limit load of the joint specimen. The dot-dash line and the double-dot dash line show the allowable tensile strength and allowable compressive strength for temporary loading of the chord based on AIJ(1973).

For both specimens, under compression-side force, the bracing axial force decreases after reaching the maximum axial strength, and the chord axial force increases until reaching the yield axial strength.

Under tension-side force, the ratio between the bracing

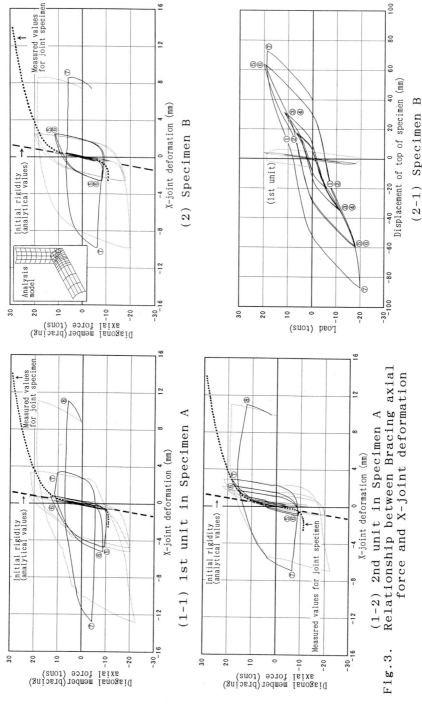

(1-1) 1st unit in Specimen A

(1-2) 2nd unit in Specimen A

Fig.3. Relationship between Bracing axial
force and X-joint deformation

(2-1) Specimen B

(2-2) Specimen B

Fig.4. Relationship between Load and
Displacement of top of specimen

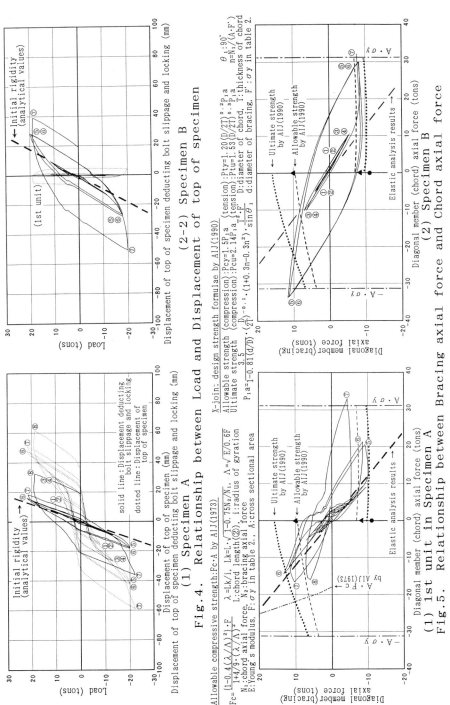

$$F_c = \frac{(1-0.4(\lambda/\Lambda)^2)\cdot F}{1+4/9\cdot(\lambda/\Lambda)^2}$$

Allowable compressive strength:$F_c\cdot A$ by AIJ(1973)

$\lambda = l_k/i$, $L_k = L\cdot\sqrt{1-0.75N_2/N_1}$, $\Lambda = \sqrt{E/0.6F}$
L:chord length(2), i:radius of gyration
N_1:chord axial force, N_2:bracing axial force, A:cross sectional area
E:Young's modulus, F:σy in table 2.

X-join: design strength formulae by AIJ(1990)
Allowable strength (compression):$P_{cy}=1.5P_{1a}$ (tension):$P_{ty}=1.5P_{1a}$
Ultimate strength (compression):$P_{cu}=2.14P_{1a}$ (tension):$P_{tu}=1.53(D/2T)^{0.2}P_{1a}$

$$P_{1a}=1-0.81(d/D)^{2}\big\}^{-0.1}\cdot(1+0.3n-0.3n^2)\cdot\frac{T^2\cdot F'}{\sin\theta_1}\cdot\left(\frac{D}{2T}\right)^{3.5}$$

$\theta_1:90°$
$n=N_1/(A\cdot F')$
D:diameter of chord, T:thickness of chord
d:diameter of bracing, F':σy in table 2.

(1) Specimen A   (2-2) Specimen B

Fig.4. Relationship between Load and Displacement of top of specimen

(1) 1st unit in Specimen A   (2) Specimen B

Fig.5. Relationship between Bracing axial force and Chord axial force

axial force and the chord axial force gradually decreases, but both the bracing axial force and the chord axial force increase. For specimen B, the chord axial force increases until reaching the yield axial strength. However, for specimen A, while the chord axial force increases, cracking appears in the chord before the yield axial strength is reached.

The dashed line in the Fig. 5 shows the analytical results for section 3.2 of this paper. It agrees quite well with the test result ratio for ① st and ② nd cycles of the tension-side force application.

4 Conclusion

Static loading tests were performed in which a circular tubular truss framework having tube-to-tube x-joints is subjected to repeated loads with the purpose of understanding how the behavior of the joint after reaching its ultimate strength affects the ultimate behavior of the truss framework. The results of these experiments revealed the following points.

(1) When a compressive force acts on the bracings, the bracing axial force gradually decreases after the ultimate strength is reached. However, the plane truss framework strength does not decrease. The chord axial force increases to reach the yield axial strength.

(2) When a tensile force acts on the bracings, the ratio between the bracing axial force and the chord axial force gradually decreases, but both the bracing axial force and the chord axial force increase. At the time that cracking appears in the chord, the bracing axial force decreases. At that time, the plane truss framework strength also decreases.

(3) If tensile force acts on the bracings after the out-of-plane deformation of the chord has become large from compressive force applied to the bracings, the bracing axial force does not increase, and cracking appears in the chord with relatively low axial force. At this point, the plane truss framework strength decreases.

5 Reference

Architectural Institute of Japan (1990) **Recommendation for the Design and Fabrication of Tubulars Structures in Steel**

J.Wardenier, Y.Kurobane, J.A.Packer, D.Dutta and N.Yeomans (1991) **Design Guide for Circular Hollow Section (CHS) Joints under Predominantly Static Loading**, Verlag TUV Rheinland

Architectural Institute of Japan (1973) **Design Standard for Steel Structures**

PART 16

FATIGUE

Experimental, numerical
and design approaches

62 GUIDELINES ON THE NUMERICAL DETERMINATION OF STRESS CONCENTRATION FACTORS OF TUBULAR JOINTS

A. ROMEIJN
Delft University of Technology, The Netherlands
R.S. PUTHLI
Delft University of Technology, The Netherlands, and
TNO Building and Construction Research, Delft,
The Netherlands
J. WARDENIER
Delft University of Technology, The Netherlands

Abstract
Tubular joints are one of the most common type of joints in offshore structures. However, regarding the fatigue behaviour, there appears to be no standard method of determining stress concentration factors for such joints. This has led to a divergence in the methods being used. In this paper the results of numerical work on several aspects regarding SCF determination are given, from which guidelines are derived.
Keywords: Tubular Hollow Sections, Stress Concentration Factors, Joints, FE-modelling.

Nomenclature:

F Force.	γ Radius to wall thickness of chord.
L Length of the weld footprint.	β Brace to chord diameter ratio.
M Moment.	τ Brace to chord wall thickness ratio.
x Distance.	θ Angle between brace members.
	α Chord length to chord diameter ratio.

Indices:

		Notation:	
ch	Chord member.	API	American Petroleum Institute.
br	Brace member.	AWS	American Welding Society.
ax	Axial.	CHS	Circular hollow section.
ip(b)	In-plane (bending).	FE	Finite Element.
op(b)	Out-of-plane (bending).	SCF	Stress concentration factor.
t	Torsion.		
a-d	Brace member.		

1 Introduction

In the offshore industry, uniplanar and multiplanar joints made of circular hollow sections are frequently used. The fatigue design of such joints requires knowledge of stress concentration factors (SCFs), which can be obtained experimentally by the use of test specimens.

However, because of high costs using such a method, determination of SCFs based on numerical work together with experimental calibration is widely accepted. From literature studies, it has been found that guidelines for the SCF determination on the basis of numerical finite

Tubular Structures V. Edited by M.G. Coutie and G. Davies.
© 1993 E & FN Spon, 2–6 Boundary Row, London SE1 8HN. ISBN 0 419 18770 7.

element analyses is limited.

This paper deals with results of numerical work on several aspects regarding the SCF determination. Some guidelines for SCF determination are given and the influence of the methods of FE-modelling are discussed. Aspects considered are:

- FE-modelling for determination of SCFs.
- Extrapolation of stresses for determination of SCFs.
- The effect of modelling the weld shape on SCFs.
- The influence of the presence of unloaded brace(s) on the SCFs at the reference brace.
- The influence of boundary conditions on SCFs.
- The influence of torsional brace moments on SCFs.

From publications in recent years it can be concluded that, for the determination of SCFs, different methods are applied which cause problems in the interpretation of the test results as well as the numerical FE results. In the authors' opinion, this problem has occured due to the absence of recommendations on the SCF determination. This paper deals with some important aspects regarding the numerical determination of SCFs, on the basis of which conclusions are made.

2 FE-modelling for determination of SCFs

It has been found that for the determination of SCFs various methods of FE-modelling are applied. The most basic distinction in FE-modelling is whether or not to include the weld shape. It is also noted that the type of finite element used, also has an influence on the results. For one type of multiplanar K-joint considered (γ=24, β=0.40, τ=1.00, θ_{ip}=60° and θ_{op}=180°) the effect of various types of FE models on SCFs has been studied. The SCFs have been determined for both chord and brace members, along the entire intersection of a brace to the chord at regular intervals of 45° (see Figure 1). Fifteen unbalanced load cases have been considered, namely axial load, in-plane bending and out-of-plane bending on the chord and four brace members, indicated as:

$$F_{ax_{ch}} \; ; \; F_{ax_{br;a-d}} \; ; \; M_{ipb_{ch}} \; ; \; M_{ipb_{br;a-d}} \; ; \; M_{opb_{ch}} \; ; \; M_{opb_{br;a-d}} \; .$$

The following four types of FE-models have been used:

a. FE-model using 4-noded thin shell elements; weld shape not included and SCFs defined at the intersection of the midplanes of the connecting walls;

b_1. FE-model using 8-noded thin shell elements; weld shape not included and SCFs defined at the intersection of the midplanes of the connecting walls;

b_2. FE-model using 8-noded thin shell elements; weld shape not included and SCFs defined at the fictitious weld toe location;

c. FE-model using 8-noded solid elements; weld shape included and SCFs defined at weld toe position;
d. FE-model using 20-noded solid elements; weld shape included and SCF defined at weld toe position.
The fictitious weld toe location and the modelled weld shape complies with the AWS specifications.

Because FE-model *d* can be regarded as the most accurate FE-model (as the weld shape is included and the element type has a high degree of accuracy), the results of FE-models *a*, *b* and *c* have been compared to those of FE-model *d*.

Some results of this comparison for the fifteen load cases considered are given in Figures 2-7. In Figures 2 and 3, the SCFs are given for the chord and one brace member (so-called reference brace as explained in Figure 1). It can be concluded that ignoring the weld shape leads to entirely different results (especially for the brace member). The Figures 4 and 5 clarify these differences. In these figures, the stress distribution is shown for the cc-45° and bc-180° positions (see Figure 1) pertaining to load case $F_{ax_{br;a}}$ for FE-models *b* and *d*. The main reason for the differences in SCFs when comparing FE-models b_1 and *d* is the difference in distance between the intersection of the midplane of the connecting walls and the weld toe position (x_{cw}). If this distance is taken into account in the determination of SCFs (in other words, if the SCF is calculated at the fictitious weld toe location), a considerable improvement, especially for the brace members, arises between the results of FE-models b_2 and *d* (see Figure 6 compared to Figure 3). Despite this improvement, it is recommended not to determine SCFs at fictitious weld toe locations without including the weld shape.

When comparing the effect of an element type on the SCFs, it appears that when using identical FE-meshes, an FE-model (*a* and *c*) using an element type without midside-nodes leads to lower SCFs than an FE-model (*b* and *d*) using an element type with midside-nodes. As an example Figure 7 gives the results of the comparison between FE-model *c* and *d* (solid elements).

Based on these results as well as mentioned references, the authors' are of the opinion that when determining SCFs, FE-model *d* (using 20-noded solid elements and including the weld shape) is preferred over FE models *a*, *b* and *c*.

3 Extrapolation of stresses for determination of SCFs

In the authors' opinion, the determination of SCFs should be based on extrapolation of primary stresses perpendicular to the weld toe. This is particularly true for multiplanar joints having significant carry-over effects, where for instance the loading on one brace causes SCFs on other braces.

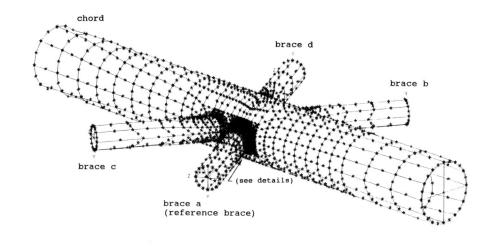

chord

brace d

brace b

brace c

(see details)

brace a
(reference brace)

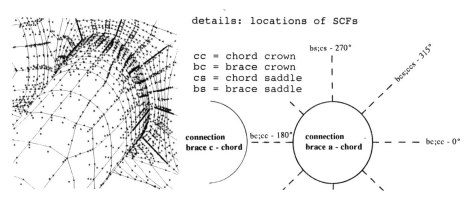

details: locations of SCFs

cc = chord crown
bc = brace crown
cs = chord saddle
bs = brace saddle

bs;cs - 270°

bcs;ccs - 315°

connection
brace c - chord

bc;cc - 180°

connection
brace a - chord

bc;cc - 0°

Fig.1. Basics regarding the SCF determination.

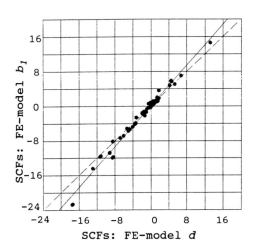

SCFs: FE-model b_1

SCFs: FE-model d

FE-model b_1:
8-noded thin shell elements
and weld shape not included.
SCFs defined at the intersection of the midplanes of
the connecting walls.

FE-model d:
20-noded solid elements and
weld shape included. SCFs
defined at weld toe position.

Fig. 2.
FE-modelling regarding
SCFs chord member.

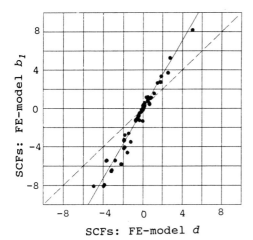

Fig. 3.
FE-modelling regarding
SCFs brace member.

●   FE-model b

○   FE-model d

— —   weld toe position

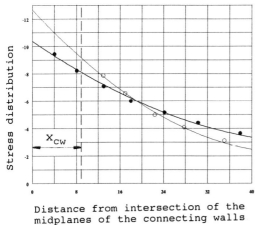

Distance from intersection of the
midplanes of the connecting walls

Fig. 4.
Stress distribution
perpendicular to the
weld toe at chord
cc;45° position.

●   FE-model b

○   FE-model d

— —   weld toe position

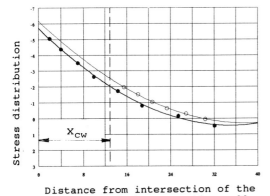

Distance from intersection of the
midplanes of the connecting walls

Fig. 5.
Stress distribution
perpendicular to the
weld toe at brace
bc;180° position.

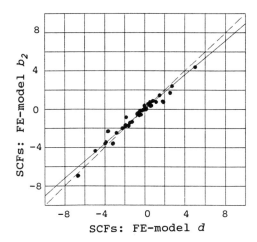

FE-model b_2:
8-noded thin shell elements and weld shape not included. SCFs defined at fictitious weld toe location.

FE-model d:
20-noded solid elements and weld shape included. SCFs defined at weld toe position.

Fig. 6.
FE-modelling regarding SCFs brace member.

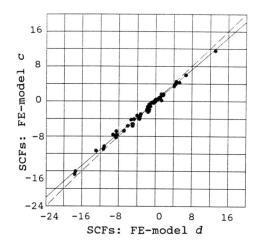

FE-model c:
8-n solid elements and weld shape included. SCFs defined at weld toe position.

FE-model d:
20-n solid elements and weld shape included. SCFs defined at weld toe position.

Fig. 7.
FE-modelling regarding SCFs chord member.

Because of the existence of various extrapolation methods, a study has been carried out using FE-model d, on the influence of extrapolation methods, and the region of extrapolation on SCFs for multiplanar K-joints. The methods of extrapolation considered are linear and parabolic curves obtained by curve fitting through all the points in and around the region of extrapolation considered, and a combination of both. Figure 8 illustrates the use of these three methods. In the combined method a parabolic equation is first determined for the stresses situated in and around the extrapolation region. The stresses from this equation are then analysed at extrapolation points (see Figure 9) from which a linear extrapolation to the weld toe takes place. Another method considered is obtaining SCFs based on the stresses at node locations placed at the weld toes.

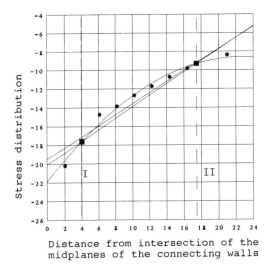

Distance from intersection of the
midplanes of the connecting walls

extrapolation   SCF
method:

- parabolic   : -21.89
- linear   : -19.51
- parabolic/linear: -20.10

Line I : minimum distance[*]
Line II: maximum distance[*]

* see figure 9

Fig. 8.
Influence of extrapo-
lation method on the
determination of SCFs.

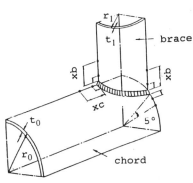

minimum distance:
$a = 0.4t_1$ or $0.4t_0$
    however $a \geq 4$ mm
maximum distance:
chord member:
saddle: 5°
crown : $xc = 0.4(r_1 \cdot t_1 \cdot r_0 \cdot t_0)^{1/4}$
brace member:
saddle and
crown : $xb = 0.65(r_1 \cdot t_1)^{1/2}$

Fig. 9.
Extrapolation region accepted
by the Working Group III,
Tubular Joints, of the ECCS.

The influence of alternative locations of the points of
extrapolation mentioned in Figure 9 on SCFs is also carried
out. The following observations can be made on the basis of
the investigations of these influences.
-    When no extrapolation method is used at all, where the
     SCFs are determined on the basis of the nodal stresses
     at the weld toe location only, substantially smaller
     SCFs (especially for the chord member) are to be
     expected, with respect to SCFs obtained by using an
     extrapolation method. Figure 10 shows the results for
     the chord member for the fifteen load cases analysed.
     The reason for the different results is that at the
     weld toe node, an average stress of the elements
     common to this node is determined, so that an increase
     of wall thickness because of the presence of the weld
     material is clearly noticeable.

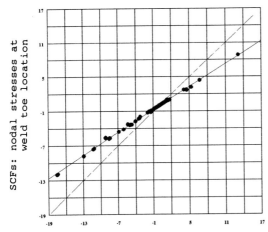

SCFs: nodal stresses at weld toe location

SCFs: extrapolation method parabolic/linear

Fig. 10.
Influence of extrapo-
lation method on the
determination of SCFs
for the chord member.

- With regard to the gradient of primary stresses, except for the immediate vicinity of the weld toe, a slightly parabolic behaviour has been found, which runs over a longer stretch than the region defined according to Figure 9.
- Because of the slightly parabolic stress gradient, different SCFs are obtained by using linear, parabolic or a combined extrapolation method.

On the basis of the results, it would appear that the extrapolation region as shown in Figure 9 is acceptable, and in cases of non-linear stress gradients it is recommended to use a combined extrapolation method as described above. In case of SCF determination at locations between crown and saddle positions, a linear interpolation between the positions of the extrapolation region as described in Figure 9 is proposed.

4 The effect of modelling the weld shape on SCFs

If a weld is included in the FE-modelling for the numerical determination of SCFs, the weld shape to be used should be considered. Also, the fact that in practice the weld shape can show considerable deviations from the desired form should be taken into account.

For one type of multiplanar K-joint with joint parameters $\beta=0.40$, $\gamma=24$, $\tau=1.00$, $\theta_{ip}=60°$ and $\theta_{op}=180°$, the effect of various types of weld shapes on SCFs has been considered. The SCFs have been determined using FE-model d and the combined extrapolation method as described in sections 2 and 3. The weld shapes considered are shown in Figure 11. Weld shape No. 1 is modelled according to the AWS specification for a weld with full penetration accessible from one side. For weld shapes 2 and 3, only increases in the length of the weld footprint (L) have been made, so that for weld shape 2 $L_2 = 1,5.L_1$ and for weld shape 3 $L_3 = 2,0.L_1$. Results about analysed SCFs for the fifteen load cases as described in section 2 are given in Figures 12 and 13. These figures show that an increase of the weld footprint leads to a substantial increase of the SCFs for the brace member and a decrease of the SCFs for the chord member.

This supports previous experimental work and the American codes of practice, where the advantages of improved weld shapes (due to a longer footprint on the chord member) are taken into account.

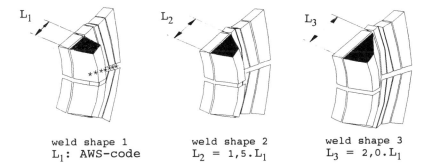

weld shape 1
L_1: AWS-code

weld shape 2
$L_2 = 1,5.L_1$

weld shape 3
$L_3 = 2,0.L_1$

Fig. 11. Investigated weld shapes regarding SCFs.
Weld shape drawn for the location 0°.

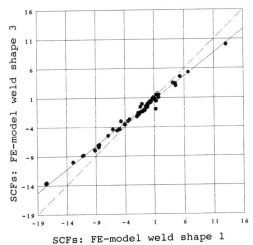

Fig. 12.
Influence weld shape on
SCFs for the chord member.

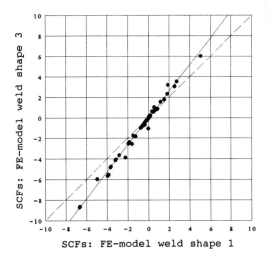

Fig. 13.
Influence weld shape on
SCFs for the brace member.

SCFs: FE-model weld shape 3

SCFs: FE-model weld shape 1

5 The influence of the presence of unloaded brace(s) on the SCFs at the reference brace

To the authors' knowledge, limited information is available on the influence of the presence of unloaded braces on the SCFs of the loaded reference brace (the so-called reference effects) as well on the SCFs of other unloaded braces (the so-called carry-over effects). Mostly when determining the SCFs, an unloaded member is disregarded for both the reference effects as well as the carry-over effects. In order to verify the significance of this omission, research was carried out on the effect of the presence of such members (in-plane as well as out-of-plane). For this purpose, multiplanar K-joints with a variation of number of brace members (see Figure 14) have been analysed several times, namely as a Y-joint (deleting braces b,c and d), as a V-joint (deleting braces c and d), as a K-joint (deleting braces b and d) and as an X-joint (deleting members b and c), and the mutual differences of SCFs have been studied.

5.1 Reference effects
It has been found that substantial differences in SCFs at the loaded reference brace can occur, due to the presence or absence of other unloaded braces.

As an example Table 1 gives for the load cases $F_{ax_{br;a}}$, $M_{ipb_{br;a}}$ and $M_{opb_{br;a}}$ the ratio $\alpha_{y\text{-kk}}$ $(= \dfrac{SCF:\ Y\text{-}joint}{SCF:\ KK\text{-}joint})$ for constant values of γ and τ, while β, θ_{ip} and θ_{op} vary. This table shows that the ratio $\alpha_{y\text{-kk}}$ depends on the variation of the above mentioned parameters, as well as on the type of load case and the location (saddle or crown position) of

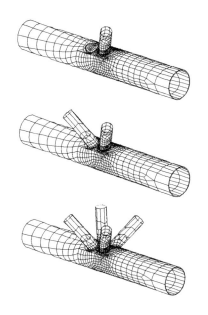
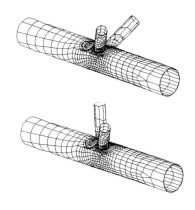

Fig. 14.
Investigated joints regarding reference effects and carry-over effects on SCFs.

the reference brace considered. The ratio α_{y-kk} for constant values of β, θ_{ip} and θ_{op}, while γ and τ vary leads to the additional conclusion that increase of γ and τ, especially for the crown positions, also results in a larger change of α_{y-kk}.

5.2 Carry-over effects
Similar findings as mentioned in section 5.1 for the reference effects are also applicable to the carry-over effects.

Depending on the desired accuracy, a joint parameter range can be indicated (for instance joint types with low values of β, γ, τ and θ_{ip}), in which the influence of the presence of unloaded braces on the SCFs for the reference effects as well as the carry-over effects can be neglected.

6 The influence of boundary conditions on SCFs

For the numerical as well as experimental determination of SCFs, support of the chord member end is necessary for equilibrium. This is particularly true for unbalanced load cases. At these supports, boundary conditions arise, which for instance in case of external loads on a brace member cause moments in the chord member, which especially for geometries with a large value of β, γ, τ and θ_{ip} at certain locations on the intersection area of brace to chord dominate the values of SCFs.

As an illustration, results of SCFs using FE-model d are given in Table 2 for a multiplanar K-joint with joint parameters $\beta=0.40$, $\gamma=24$, $\tau=1.00$, $\theta_{ip}=60°$, $\theta_{op}=180°$ and $\alpha=8.5$.

Table 1. Relation of SCFs (reference effects for constant value of γ and τ) between Y-joint and KK-joint. ** = $|SCF_{Y+KK}| < 0.50$

joint parameters Y-joint ; KK-joint $\gamma = 24$, $\tau = 1.00$ and			load case : F_{ax} brace a ; ratio SCF Y-joint/KK-joint ($\alpha_{y\text{-}kk}$)							
			location intersection area brace a - chord							
β	θ_{ip}	θ_{op}	cc; 0°	cs; 90°	cc; 180°	cs; 270°	bc; 0°	bs; 90°	bc; 180°	bs; 270°
0.25	30	60	1.04	1.00	1.01	1.01	1.00	1.00	1.00	1.02
0.25	30	90	1.00	1.00	1.00	1.00	1.00	1.00	1.00	1.00
0.25	30	180	0.99	1.01	1.00	1.01	1.00	1.00	1.00	1.00
0.40	60	60	0.84	1.11	0.60	1.27	2.71	1.10	-0.10	1.19
0.40	60	90	0.84	1.09	0.57	1.09	**	1.08	0.51	1.08
0.40	60	180	0.82	1.10	0.55	1.10	**	1.09	0.48	1.09
0.60	45	60	0.85	1.10	0.70	1.26	**	1.12	0.68	1.17
0.60	45	90	0.84	1.13	0.43	1.37	0.38	1.15	0.49	1.20
0.60	45	180	0.78	1.19	0.54	1.19	**	1.20	0.68	1.20
			load case : M_{ipb} brace a ; ratio SCF Y-joint/KK-joint ($\alpha_{y\text{-}kk}$)							
0.25	30	60	1.02		1.02		1.00		1.02	
0.25	30	90	1.00		1.00		1.00		1.00	
0.25	30	180	1.00		1.00		1.00		1.00	
0.40	60	60	1.06		0.86		1.05		0.75	
0.40	60	90	1.04		0.85		1.04		0.81	
0.40	60	180	1.03		0.85		1.04		0.81	
0.60	45	60	1.08		0.87		1.05		0.87	
0.60	45	90	1.06		0.85		1.05		0.87	
0.60	45	180	1.04		0.82		1.04		0.81	
			load case : M_{opb} brace a ; ratio SCF Y-joint/KK-joint ($\alpha_{y\text{-}kk}$)							
0.25	30	60		1.00		1.00		1.00		1.01
0.25	30	90		1.00		1.00		1.00		1.00
0.25	30	180		1.00		1.00		1.00		1.00
0.40	60	60		1.08		1.05		1.09		1.03
0.40	60	90		1.09		1.06		1.09		1.06
0.40	60	180		1.07		1.07		1.07		1.07
0.60	45	60		1.18		1.14		1.18		1.10
0.60	45	90		1.13		1.13		1.13		1.09
0.60	45	180		1.10		1.10		1.10		1.10

Table 2. Influence of boundary conditions on SCFs.

load case	boundary conditions of the chord member					
	1. pin-ended		2. fully clamped		3. pin-ended with additional loads	
on brace member	SCFs: location intersection area brace a - chord					
	cc;0°	cc;180°	cc;0°	cc;180°	cc;0°	cc;180°
$F_{ax;a}$	-7.77	-12.53	-5.26	-10.33	-3.51	-8.27
$F_{ax;b}$	3.82	3.08	1.67	1.13	-0.36	-1.18
$F_{ax;c}$	-2.69	-2.76	-0.56	-0.82	1.49	1.50
$F_{ax;d}$	4.28	3.34	1.85	1.13	0.10	-0.92
$M_{ipb;a}$	4.70	-5.97	4.69	-5.97	4.58	-6.07

Three alternatives are used, namely:
1. SCFs determined with chord member ends pin-ended.
2. SCFs determined with chord member ends fully clamped.
3. SCFs determined with chord member ends pin-ended and a correction applied to the SCFs to account for the moments introduced in the chord member.
Especially for the carry-over effects, Table 2 shows large differences for all three boundary conditions.
In the authors' opinion, correct values of SCFs are found only when no influence is present from boundary conditions. This means that for the numerical determination of SCFs, for chord as well as braces, compensating moment(s) should be applied on the chord member ends or reanalysing of the SCFs should take place on the basis of load cases $M_{ipb_{ch}}$ and $M_{opb_{ch}}$ on the chord member ends.

Such an approach is very difficult to simulate in experimental work, so that SCFs (particularly at crown positions) are influenced by the boundary conditions.

7 The influence of torsional brace moments on SCFs

As a rule, in the determination of SCFs and fatigue life, torsion moments $M_{t_{br;a-d}}$ are disregarded because of their low contribution compared to other types of loads such as $F_{ax_{ch}}$; $F_{ax_{br;a-d}}$; $M_{ipb_{ch}}$; $M_{ipb_{br;a-d}}$; $M_{opb_{ch}}$ and $M_{opb_{br;a-d}}$.
However, for multiplanar tubular structures, load situations might occur in which the contribution of SCFs due to torsional moments should be taken into account.
Investigations have been carried out on SCFs for the reference effects as well as carry-over effects for multiplanar K-joints with torsional moment on braces. From comparisons of SCFs due to $M_{t_{br;a-d}}$ with SCFs due to $M_{opb_{br;a-d}}$,

a strong similarity has been found. It appears that as an approximation, SCFs due to $M_{t_{br;a-d}}$ can be determined by resolving $M_{t_{br;a-d}}$ into $M_{opb_{br;a-d}}$ and $M_{w_{br;a-d}}$ according to the principle as explained in Figure 15, in which the contribution of $M_{w_{br;a-d}}$ can be neglected. As an illustration of this, Figure 15 shows results of SCFs using FE-model d for a multiplanar K-joint with $\gamma=24$, $\beta=0.60$, $\tau=1.00$, $\theta_{ip}=30°$ and $\theta_{op}=90°$ for the load cases $M_{t_{br;a-d}}$.

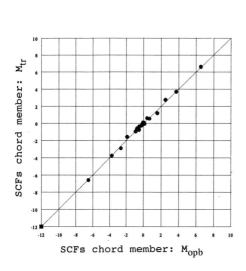

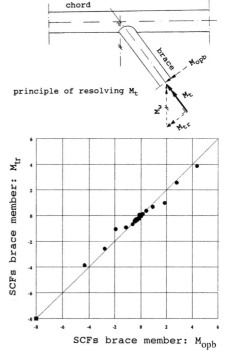

principle of resolving M_t

Fig. 15.
Relation between SCFs for a KK-joint with brace members loaded by M_{opb} and M_t.

8 Final Remarks

It has been found that results regarding the numerical determination of stress concentration factors (SCFs) of tubular joints are greatly influenced by the method used, which causes problems in the interpretation of numerical FE-results as well as experimental results.
    In this paper the influence of main aspects like:
• *FE-modelling (element type and weld shape)*
• *Extrapolation method*
• *Effect of weld shape*

- *Presence of unloaded brace(s) on the SCFs at reference brace*
- *Boundary conditions*
- *Torsional brace moments*

on the SCF values is given, from which several guidelines are given in the seperate sections. These approaches are intended to contribute towards a more uniform method of determining stress concentration factors.

9 Acknowledgements

Appreciation is extended to the Netherlands Technology Foundation (STW) for allowing the presentation of the numerical results carried out in the framework of the project DCT 80.1457.

10 References

American Petroleum Institute. "Recommended Practice for Planning, Designing and Construction Fixed Offshore Platforms" API Recommended Practice 2A (RP 2A), 19th Edition Washington, U.S.A., August 1991.

American Welding Society "Structural Welding Code - Steel" 13th edition ANSI/AWS D1.1-92, Miami, Florida U.S.A., 1992.

Back, de J. Size effect and weld profile effect on fatigue of tubular joints. Proc. Safety Criteria in design of tubular structures. February 1987.

Efthymiou. M. Development SCF formulae and generalised influence functions for use in fatigue analysis. Conference on recent developments in tubular joints technology, organized by UK UEG Offshore Research, October 1988.

Kailash C. Gulati and Wen J. Wang. An Analytical Study of Stress Concentration Effects in Multibrace Joints Under Combined Loading. Proc. OTC'82. OTC 4407.

Lalani, M. Improved Fatigue Life Estimation of Tubular Joints. Proc. OTC'86. OTC 5306.

Lloyd's Register of Shipping (1986). Stress Concentration Factors of Tubular Complex Joints. Final report N° 4: Unstiffened Multiplanar Joints.

Romeijn, A., Puthli, R.S., Wardenier, J. Finite element modelling of multiplanar joint flexibility in tubular structures. Proc. ISOPE'92.

Romeijn, A., Puthli, R.S., Wardenier, J., de Koning C.H.M. Stress and strain concentration factors of Multiplanar joints made of circular hollow sections. Proc. ISOPE'92.

Romeijn, A., Puthli, R.S., Wardenier, J., de Koning C.H.M., Dutta, D. Fatigue behaviour and influence of repair on multiplanar K-joints made of circular hollow sections. Proc. ISOPE'93.

Zienkiewicz, W. FE method: Mcgraw-Hill book company: fourth edition.

63 EXPERIMENTAL AND NUMERICAL DETERMINATION OF STRESS CONCENTRATION FACTORS OF FILLET WELDED RHS K-JOINTS WITH GAP

S. AL LAHAM and F.M. BURDEKIN
Structural Assessment Group, Department of
Civil & Structural Engineering, UMIST, UK

Abstract
The work presented in this paper investigates the stress concentration behaviour of fillet welded K-joints with gap involving rectangular hollow sections (RHS) members with 45° angle between members when subjected to balanced axial loads in the bracings. The linear-elastic finite element (FE) technique was employed, using the ABAQUS program, to study the influence of some geometrical parameters on the fatigue performance of these joints. Parametric formulae which allow the determination of the hot spot stress concentration factors (SCF's) at critical positions in the joints, based on the FE results, have been developed. The effect of different angles between members was also investigated. The theoretical results was compared with results obtained using existing formulae and verified by experimental measurements of SCF's at the critical locations of tested specimens.
Keywords: Fatigue, Fillet Weld, Stress Concentration Factor, RHS, Brick Elements.

1 Introduction and Background

Designers are often required to assess the fatigue performance of hollow section joints by considering the stress levels at potential crack locations. In these types of joints, the flexibility of the walls of rectangular members leads to additional bending stresses developing which depend upon the ratios of member sizes connected. In considering the structural behaviour of RHS connections, also, it is necessary to consider all relevant modes of failure. Burdekin et al (1989) carried out an investigation into the fatigue behaviour of square hollow section double T-joints, they found that for the fillet weld sizes used, the mode of failure changed from failure by fatigue through the fillet weld throat for high β ratios of $\beta = 0.65$, to failure by fatigue cracks at the fillet weld toe for lower values of $\beta = 0.39$. The best fatigue design methods available require knowledge of the stress concentration factors due to bending of the walls at the weld toe position, and knowledge of the stresses in the welds themselves.

An extensive experimental programme was launched in 1975, sponsored by the European Coal and Steel Community (ECSC), and CIDECT. The research project yielded valuable data from which design methods for fatigue loaded RHS joints could be derived. The research was concluded with recommended S_r-N curves and completed in 1980. The design methods and the recommended curves were reported by Dutta et al (1982).

Tubular Structures V. Edited by M.G. Coutie and G. Davies.
© 1993 E & FN Spon, 2–6 Boundary Row, London SE1 8HN. ISBN 0 419 18770 7.

The most practical and efficient method to represent the local behaviour of a hollow section joint is by the use of parametric formulae, which give SCF's at the "hot spot" locations. Mang et al (1989) employed the finite element technique, using shell elements without modelling the welds at the intersection area, to develop parametric equations for the calculations of the strain concentration factors (SNCF's) of square hollow section joints with gap and overlap under axial loading in bracing members.

In the scope of a recent research programme sponsored by the ECSC, fundamental investigations were carried out by Puthli et al (1988a) and van Wingerde et al (1988), they also used a finite element program to investigate the fatigue behaviour of T- and X-joints made of square hollow sections. They modelled the joint using a combination of shell elements for the section walls, while 3-dimensional brick elements were used around the intersection area. They presented their results in the form of parametric formulae. They also concluded from experiments that fatigue strength decreases with increasing joint size as a consequence of the thickness effect. Puthli et al (1988b) used the FE method for the determination of SCF's of welded K-joints between square hollow sections by modelling the joints using shell elements with no account taken of the weld material modelling.

In the present work three-dimensional brick elements have been used throughout the model, taking into account modelling the fillet weld material. A parametric study on the FE results was conducted in order to develop formulae which allow the determination of the hot spot SCF's at critical locations of the joints. The effect of different angle between members was investigated. An experimental programme was conducted to verify the theoretical results.

2 Modelling and Analysis of RHS Fillet Welded K-Joints with Gap

Stress levels in RHS connections need to be determined in order to assess the fatigue behaviour of structures containing these types of joints. RHS joints experience significant stress gradients and concentrations, and very often local yielding can be developed at quite low loading levels.

The analyses were conducted using the finite element package ABAQUS (1988). The type of elements used were the 3-D 20-noded brick element (C3D20R) with reduced integration. Meshes were generated using the general purpose mesh generating program PATRAN. The wall thickness of the chord member under the intersection area, where the stress gradient is expected to be sharp, was modelled by two layers of brick elements to enable the bending of the wall to be modelled accurately. In addition, to make the analyses even more accurate, the member's corner radii was modelled. The fillet welds were also modelled using 3-D brick elements, leaving the unpenetrated land between the bracing members and the chord member to be modelled as a gap (inherent to the use of fillet welds). In regions of the chord member away from the intersection area and in the bracing members a single layer of brick elements was used in a relatively coarser mesh than the intersection area, because stresses are expected to be more evenly distributed. The length of the chord member was chosen in such a manner that the end conditions do not influence the stress field around the intersection area, whereas the lengths of the bracing members were taken to be almost equal to the distance between the intersection area and the point of inflection of the brace. The mesh was designed in a way that it maintained aspect ratios and element-inside angles within acceptable limits as far as possible. Each finite element mesh contained about 1200 brick elements with around 24000 degrees of freedom.

Availability of the optimization facility in PATRAN allowed the process of mesh generation to be easier. A typical mesh used for SCF's calculation is shown in Fig. 1a.

As far as boundary and loading conditions are concerned, all nodes at the chord ends were restrained from moving in all the three directions X, Y and Z (i.e. fully fixed ends), while the ends of the braces were free to move, which corresponds to the test rig conditions used in this research. Because of the advantage of symmetry only half the connection was modelled, and therefore all nodes in the X-Y plane (i.e. Z = 0 plane) were restrained in the Z-direction. Uniformly distributed pressure loads were applied on the surfaces of elements at the ends of the bracing members, one tension and the other compression, in order to simulate the axial loading case.

3 Parametric Study of the Results of the Finite Element Analyses

Finite element analyses were conducted on twenty five different geometrical conditions to study the influence of the geometrical parameters β, τ and 2γ on the stress concentration factors. These geometrical conditions are given in table 1. The angle of bracing to chord members was kept at 45° and the gap g', between the weld toes, was taken as 20 mm in all cases. Nominal dimensions were used throughout the analyses, and the bracing members dimensions kept always at 100 x 50 x 5 mm, allowing the chord member sizes to vary in order to check the effect of the variation of joint parameters under the same loading level.

A zoomed-in deformed shape of a typical mesh of RHS K-joints with gap is shown in Fig. 1b. The maximum principal stresses were evaluated at different positions in both the chord and tension brace members, these positions for SCF's calculations are shown in Fig. 2. The direct stresses, normal to the weld toe at these positions, were also calculated. The difference between the principal and direct stresses was found to be small, and typically it was found that direct stresses normal to the weld toe are (20 - 25)% lower than the maximum principal stresses. The adopted method of extrapolation was the quadratic extrapolation rule, in which stresses were evaluated at nodes which are away from the weld toe position and then a quadratic extrapolation for maximum principal stresses to the weld toe for each position around the intersection was conducted. In fact, the IIW (1985) did not specify any method of extrapolation, it is believed, however, that the quadratic extrapolation method results in a more accurate prediction of the stress field at the weld toe position. Results of the finite element analyses at the weld toe position and up to a distance from the weld toe, which is equal to 0.4 of the member thickness, were ignored in order to eliminate the influence of the weld notch (i.e. similar to the procedure used by Puthli et al (1988b)).

The finite element results are tabulated in table 1. With those numerically established SCF values for the twenty five fillet welded joints, regression analyses were conducted and SCF formulae were established for the highest stressed positions, these are given in table 2 together with the range of validity. These positions are line 1C, line 2C and line 5C in the chord as shown in Fig. 2, other positions were found to be less critical. For all positions in the brace, stresses were about (25 - 30)% lower than that in the corresponding positions in the chord. It is believed that this is due to the fact that hot spot positions in the chord member experience higher additional bending stresses than those in the bracing members. Table 1 also shows the stress concentration factors which are calculated from the SCF formulae given in table 2. It can be seen that the calculated SCF's from the formulae are within 6% of those found from the finite element analyses.

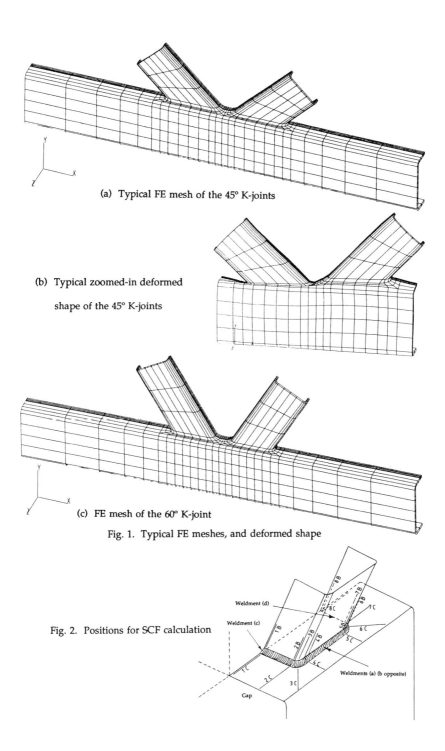

(a) Typical FE mesh of the 45° K-joints

(b) Typical zoomed-in deformed

shape of the 45° K-joints

(c) FE mesh of the 60° K-joint

Fig. 1. Typical FE meshes, and deformed shape

Fig. 2. Positions for SCF calculation

Table 1. Geometries, SCF's from FE and formulae

Model Ref	Brace $h_1 \times b_1 \times t_1$ (mm)	Chord $h_0 \times b_0 \times t_0$ (mm)	β	τ	2γ	line 1C: FE results	formula	formula/FE	line 2C: FE results	formula	formula/FE	line 5C: FE results	formula	formula/FE
SCF631	100x50x5	120x80x5	0.63	1.0	16	5.41	5.43	1.004	5.33	5.44	1.021	5.5	5.59	1.016
SCF632	100x50x5	120x80x6.3	0.63	0.79	12.7	3.53	3.52	0.997	3.33	3.42	1.027	3.18	3.18	1.00
SCF633	100x50x5	120x80x8	0.63	0.63	10	2.28	2.23	0.978	2.06	2.12	1.029	1.84	1.81	0.984
SCF634	100x50x5	120x80x10	0.63	0.5	8	1.43	1.39	0.972	1.27	1.31	1.031	1.09	1.04	0.954
SCF635	100x50x5	120x80x12.5	0.63	0.4	6.4	0.86	0.82	0.953	0.75	0.79	1.053	0.67	0.64	0.955
SCF501	100x50x5	150x100x5	0.50	1.0	20	7.11	7.49	1.053	6.9	6.81	0.987	8.1	8.47	1.046
SCF502	100x50x5	150x100x6.3	0.50	0.79	15.87	4.52	4.76	1.053	4.22	4.1	0.972	4.81	5.1	1.06
SCF503	100x50x5	150x100x8	0.50	0.63	12.5	2.86	3.03	1.059	2.50	2.46	0.984	2.94	3.09	1.051
SCF504	100x50x5	150x100x10	0.50	0.5	10	1.84	1.95	1.06	1.5	1.48	0.987	1.85	1.92	1.038
SCF505	100x50x5	150x100x12.5	0.50	0.4	8	1.18	1.26	1.068	0.94	0.9	0.957	1.18	1.24	1.051
SCF401	100x50x5	200x125x5	0.40	1.0	25	8.57	8.58	1.001	9.1	9.15	1.005	11.67	11.7	1.003
SCF402	100x50x5	200x125x6.3	0.40	0.79	19.84	5.52	5.58	1.011	5.8	5.91	1.019	7.41	7.48	1.009
SCF403	100x50x5	200x125x8	0.40	0.63	15.63	3.73	3.63	0.973	3.61	3.81	1.055	4.6	4.74	1.03
SCF404	100x50x5	200x125x10	0.40	0.5	12.5	2.42	2.38	0.983	2.34	2.48	1.06	2.92	3.00	1.027
SCF405	100x50x5	200x125x12.5	0.40	0.4	10	1.58	1.55	0.981	1.53	1.62	1.059	1.91	1.85	0.969
SCF331	100x50x5	250x150x5	0.33	1.0	30	10.25	10.35	1.01	12.0	12.19	1.016	15.6	15.65	1.003
SCF332	100x50x5	250x150x6.3	0.33	0.79	23.8	6.77	6.79	1.003	7.82	7.63	0.976	10.1	9.98	0.988
SCF333	100x50x5	250x150x8	0.33	0.63	18.75	4.51	4.42	0.98	4.79	4.71	0.983	6.26	6.24	0.997
SCF334	100x50x5	250x150x10	0.33	0.5	15	3.02	2.88	0.954	2.92	2.87	0.983	3.87	3.82	0.987
SCF335	100x50x5	250x150x12.5	0.33	0.4	12	1.96	1.84	0.968	1.75	1.69	0.966	2.25	2.20	0.978
SCF251	100x50x5	300x200x5	0.25	1.0	40	14.8	14.92	1.008	15.35	15.47	1.008	17.6	17.79	1.011
SCF252	100x50x5	300x200x6.3	0.25	0.79	31.75	10.13	9.76	0.963	9.94	9.82	0.988	11.8	11.66	0.988
SCF253	100x50x5	300x200x8	0.25	0.63	25	6.27	6.28	1.002	6.00	6.14	1.023	7.45	7.59	1.019
SCF254	100x50x5	300x200x10	0.25	0.5	20	3.94	4.00	1.015	3.72	3.80	1.022	4.77	4.95	1.038
SCF255	100x50x5	300x200x12.5	0.25	0.4	16	2.4	2.43	1.013	2.2	2.25	1.023	3.22	3.18	0.988

Table 2 Parametric formulae

Line No	SCF formulae (Fillet welded k-joints) with the following limits $0.25 \leq \beta \leq 0.63$ $6.4 \leq 2\gamma \leq 40$ $0.4 \leq \tau \leq 1.0$ $\theta = 45°$	Range
1C	$SCF = -7.202 + 28.941\,\beta - 29.54\,\beta^2 + \beta\,\gamma\,(3.032 - 5.266\,\beta - 12.212\,\beta^2 + 22.283\,\beta^3)$ $+ \beta\,\gamma\,\tau\,(9.947 - 58.692\,\beta + 133.567\,\beta^2 - 100.811\,\beta^3)$	$0.25 \leq \beta \leq 0.63$
2C	$SCF = -19.754\,\beta + 76.797\,\beta^2 - 73.29\,\beta^3 + \beta\,\gamma\,(4.573 - 23.587\,\beta + 32.595\,\beta^2)$ $+ \beta\,\gamma\,\tau\,(0.55 + 18.031\,\beta - 38.724\,\beta^2)$ $SCF = -19.754\,\beta + 76.797\,\beta^2 - 73.29\,\beta^3 + \beta\,\gamma\,(24.268 - 123.891\,\beta + 202.487\,\beta^2 - 105.764\,\beta^3)$ $+ \beta\,\gamma\,\tau\,(-3.231 + 21.065\,\beta - 22.78\,\beta^2)$	$0.25 \leq \beta < 0.4$ $0.4 \leq \beta \leq 0.63$
5C	$SCF = 12.304 - 90.589\,\beta + 144.826\,\beta^2 + \beta\,\gamma\,(-1.344 + 19.163\,\beta - 36.294\,\beta^2)$ $+ \beta\,\gamma\,\tau\,(-0.12 + 20.678\,\beta - 38.588\,\beta^2)$ $SCF = -18.184 + 67.291\,\beta - 59.258\,\beta^2 + \beta\,\gamma\,(12.049 - 44.346\,\beta + 38.291\,\beta^2)$ $+ \beta\,\gamma\,\tau\,(-3.356 + 22.396\,\beta - 22.904\,\beta^2)$	$0.25 \leq \beta \leq 0.4$ $0.4 < \beta \leq 0.63$

Fig. 3 shows the calculated stress concentration factors from the derived equations for different geometrical parameters at the critical locations, namely lines 1C, 2C and 5C. It can be seen from the figure that the stress concentration factor increases with increasing τ ratio for given β and γ ratios. Fig. 4 shows a comparison between the SCF's from the formulae at different locations for the same geometrical parameters. As can be seen, line 5C always has a higher stress concentration factor than lines 1C and 2C. It is also clear from this figure and table 1 that the resulting SCF's for lines 1C and 2C from both the FE analyses and the formulae, are very close to each other.

4 Proportion of bending and membrane Stresses in the Chord Wall

The parametric equations for the determination of stress concentration factors give only the hot spot stress on the outer surface at the position under investigation. Another important factor is the variation of stresses through the wall thickness. Finite element studies have been carried out at UMIST by Chan (1987) to determine the degree of bending in the chord wall for a number of different T- and K- joints made of CHS members.

The definition of degree of bending adopted in this work is the ratio of the bending stresses to the total peak stress at the surface (i.e. hot spot stress). This enables the hot spot stress concentration to be converted directly into membrane and bending components. The results of the finite element analyses for the degree of bending for the analysed geometries are shown in Fig. 5, which are for lines 1C, 2C and 5C.

With regard to defect assessment, it is important to differentiate between bending and membrane stresses as the latter produce a stress field tending to pull the crack apart to a much greater degree than bending stresses. The revised BSI document PD 6493 (1991) approach for defect assessment includes this bending effect and as such the methods employed contain less conservatism than the old document.

5 Comparison with DELFT Results

In their results, Puthli et al (1988b) suggested that for moment percentage higher than 20%, the governing strain concentration factors lie in the gap region (i.e. line 2C) and there is also no large change in the strain concentration factors with variation in moment. It should be noted that the validity range of their formula is not identical to the validity range proposed in this piece of work. However, because of the fact that this is the only available information concerning K-joints, an attempt was made to compare the stress concentration factors calculated from the formula for line 2C (see table 2) with that obtained using their equation. Fig. 6 shows graphically this comparison for different τ ratios and a given 2γ ratio as (20.0). It can be seen from this figure that the resulting SCF calculated from the parametric formula is generally higher than that proposed by Puthli et al (1988b). It is believed that this is due to the fact that RHS has a deeper section than that of SHS (i.e. the total applied load is higher, although it is distributed over a larger area), and due to the modelling of welds which provide a sharp notch and lead to higher stresses at the weld toe. This is more realistic than the use of shell elements to model the joint without the modelling of welds.

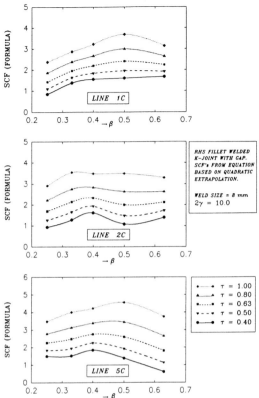

Fig. 3. Variation of Stress Concentration Factor with Geometrical
Parameters β , 2γ , and τ for Axially Loaded K-Joints with Gap

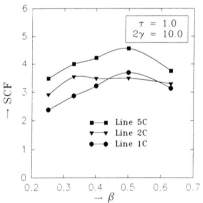

Fig. 4. Comparison of SCF's from formulae
at different locations

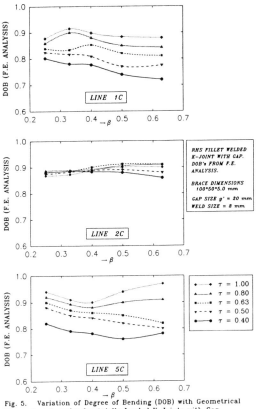

Fig. 5. Variation of Degree of Bending (DOB) with Geometrical Parameters β and τ for Axially Loaded K-Joints with Gap

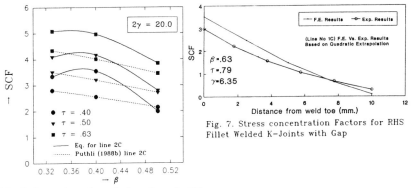

Fig. 6. Comparison of results from formula with results from Puthli (1988b), line 2C

Fig. 7. Stress concentration Factors for RHS Fillet Welded K-Joints with Gap

6 Effect of Different Angle

In practice, generally speaking, Warren trusses are designed in such a manner that the angle between members is kept at 60° in order to reduce the effective length of the members. However, the use of high angles increases the number of joints in the truss, hence increasing the cost of material, welding and cutting. There is a strong economic case for using angles in the range 40° - 50°, provided buckling resistance is maintained. Packer and Frater (1986 and 1987) studied the performance of fillet weldments in hollow section connections in order to determine the weld size requirements for satisfactory joint behaviour. They carried out a total of 26 tests on different joint arrangements by varying the angle θ from 30° to 90° and the weld size. They measured the strain along weldments c and d, see Fig. 2, by placing strain gauges as close as possible to the fillet weld toe. They concluded that if the bracing to chord member angle is 30° weldment d can be considered fully effective, but for angles between 60° and 90° weldment d is totally ineffective. It should be noted, however, that the joint assembly does not reflect the actual conditions for K-joints. Because of this and as a sensitivity study a finite element analysis was carried out in the present work for K-joint having a bracing to chord member angle of 60°.

The general mesh configuration is shown in Fig. 1c, which is similar to the mesh used earlier but having an angle of 60° between members. Loading and boundary conditions are also similar to those used earlier. Stress concentration factors were evaluated at different locations around the tension brace to chord intersection area. The members dimensions used are identical to those used for model SCF502 (see table 1). Table 3 shows the calculated stress concentration factors at different locations in the cases of θ = 45° and 60°. It can be seen that the stress concentration increases when increasing θ from 45° to 60°. It is believed that this is due to the fact that the overall length of the intersection decreases when increasing θ, this in turn reduces the overall fillet weld throat area.

The stress concentration factors for lines 7C and 8C (i.e. weldment d) are much lower than those of lines 1C and 2C (i.e. weldment C), and lines 4C and 5C (i.e. weldments a and b). It should be noted, however, that the SCF's for lines 7C and 8C are slightly higher when θ = 60° than when θ = 45° indicating that weldment d is carrying higher loads when increasing the angle θ to 60°. In fact, when comparing the stress concentration factor for line 7C with that of line 2C and the SCF for line 8C with that of line 1C, in both cases, an increase of 10% for line 7C and an increase of 15% for line 8C were observed (see table 3).

7 Experimental Work

As well as the finite element technique which was used to determine the linear elastic SCF's at critical locations, a test programme which consists of 5 full scale connections with different β and τ ratios as given in table 4 was conducted. The main objective of the test programme was to correlate the SCF test results with the theoretically estimated SCF using the finite element technique. All specimens were fabricated so that the angle of bracing to chord members was kept at 45°. The gap size between bracing members was also restricted to a constant value throughout the test programme, consistent with the FE analysis work. The results obtained from experimental measurements were compared with those from the finite element analysis.

The strain measurements from the strain gauges near to the weld toes were converted to stresses and quadratic extrapolation of stresses to the weld toes was

Table 3. Effect of different angle, Model SCF502.

	Line No.	(1) $\theta = 45°$	(2) $\theta = 60°$	Ratio $\frac{(2)}{(1)}$
Weldment (c)	1C	4.52	5.644	1.25
	2C	4.22	5.406	1.281
Weldments (a,b)	4C	4.88	6.266	1.284
	5C	4.81	6.662	1.385
Weldment (d)	7C	2.07	2.92	1.41
	8C	1.16	1.71	1.47

Table 4. SCF's from FE, Parametric equations and Experimental measurements

Joint Ref	β	τ	2γ	Line Number	SCF Test	SCF Parametric Equation	Difference (%)	SCF Finite Element	Difference (%)
EK126R	0.63	0.79	12.7	1C	2.99	3.52	+17.7	3.53	+18.0
				2C	2.96	3.42	+15.5	3.33	+12.5
				5C	2.88	3.18	+10.4	3.18	+10.4
EK128R	0.63	0.63	10	1C	2.00	2.23	+11.5	2.28	+14.0
				2C	1.79	2.12	+18.4	2.06	+15.0
				5C	1.49	1.81	+21.4	1.84	+23.4
EK156R	0.5	0.79	15.87	1C	3.89	4.76	+22.3	4.52	+16.1
				2C	3.57	4.1	+14.8	4.22	+18.2
				5C	3.9	5.1	+30.7	4.81	+23.3
EK256R	0.33	0.79	23.8	1C	6.26	6.79	+8.4	6.77	+8.6
				2C	7.48	7.63	+2.0	7.82	+4.5
				5C	7.71	9.98	+29.4	10.1	+30.9
EK258R	0.33	0.63	18.75	1C	4.35	4.42	+1.6	4.51	+3.6
				2C	3.98	4.71	+18.3	4.79	+20.3
				3C	5.39	6.24	+15.7	6.26	+16.1

carried out. Fig 7 shows a typical extrapolation of stresses to the weld toe from the measured stresses away from the weld toe in comparison with the corresponding results from the finite element analysis. In table 4, the resulting stress concentration factor values obtained from the finite element analyses, the derived parametric formulae and the experiments are given and the differences are provided in percentages. Clearly, it can be seen that the resulting stress concentration factor values from both numerical analyses and the derived formulae are (3.6 - 30.9)% and (1.6 - 30.7)% higher than the experimental results respectively, with most of the results lying in the range of (10 - 20)% higher than the experimental results. It is believed that this may be caused by the facts that the principal stresses in the analyses are at different angle from those assumed in the experiments, by the finite length of the strain gauges (2mm), and that the actual section thicknesses were (1.5 - 2.0)% larger than the nominal thicknesses used in the finite element analyses. However, the results of the finite element analyses and the derived parametric formulae are in reasonable agreement with the experimental results. Also, by examining Fig. 7 it can be seen that there is a very good agreement between the stress patterns, in the vicinity of the weld toe, for the finite element and experimental cases.

The measured stress concentration factors increased when decreasing the width ratio β for the same thickness ratio τ, but decreased when decreasing the thickness ratio τ for the same width ratio β. This is consistent with the finite element results discussed earlier. For higher β ratios of 0.5 0.63, lines 1C and 2C (which are the lines on the gap side of the K-joint) give the largest stress concentration values. On the other hand, for lower β ratio of 0.33 line 5C is dominant.

8 Conclusions

The finite element studies, using 3-dimensional solid modelling, carried out in this piece of work for the determination of stress concentration factors at critical locations in the joint have dealt with various geometrical parameters. The establishment of reliability of the finite elements results has been accomplished by comparison with existing parametric solutions, namely Puthli et al (1988b).

Parametric formulae which allow the determination of the hot spot stress concentration factors at critical positions in the joint have been developed from the linear elastic finite element analyses. The reliability of the presented parametric formulae has been confirmed by comparison with experimental measurements obtained from the test programme. It was shown that the use of the parametric formulae derived results in a slightly higher SCF's values than the actual measured values, and hence leads to a safe design.

The sensitivity study on the effect of different angles between members has shown that the joint shows lower stress concentrations when the angle between members is reduced from 60° to 45°.

9 References

British Standard Institution, PD6493 revised (1991) **Guidance on methods for assessing the acceptability of flaws in welded structures.**
Burdekin, F.M. Saket, H.K. Thurlbeck, S.D. and Frodin, J.G. (1989) Aspects of assessment of defects in welded joints and related reliability analysis treatments, **European Symp. on Elastic Plastic Fracture Mechanics**, Freiburg.

Chan, W.T. (1987) **The fatigue behaviour of offshore structural tubular nodes using fracture mechanics analysis.** PhD Thesis, UMIST.

Dutta, D. Mang, F. and Wardenier, J. (1982) **Fatigue behaviour of welded hollow section joints.** CIDECT Monograph No 7, Constrado.

Hibbit, Karlsson and Sorensen (1988) **ABAQUS Users Manual,** version 4.7.

International Institute of welding (1985) Recommended fatigue design procedure for hollow section joints. **IIW Doc No XV-582-85,** IIW Annual assembly, Strasbourg.

Mang, F. Herion, S. Bucak, O. and Dutta, D. (1989) Fatigue behaviour of K-joints with gap and with overlap made of RHS. in **Third Int. Symposium on Tubular Structures,** Lappeenranta Univ. of Technology, Lappeenranta, Finland.

Packer, J.A. and Frater, G.S. (1986) Performance of fillet weldments in hollow section connections. in **Int. meeting on safety criteria in design of tubular structures,** Tokyo, Japan.

Packer, J.A. and Frater, G.S. (1987) Weldment design for hollow section joints. **IIW Doc. No. XV-644-87, CIDECT report No 5AN-87/1-E.**

Puthli, R.S. de Koning, C.H.M. Wardenier, J. and Dutta, D. (1988a) The fatigue behaviour of X-joints made from square hollow sections. in **Int. Conference on Weld Failures,** London, UK, paper No. 3.

Puthli, R.S. Wardenier, J. de Koning, C.H.M. Van Wingerde, A.M. and van Dooren, F.J. (1988b) Numerical and experimental determination of strain (stress) concentration factors of welded joints between square hollow sections. **Heron,** Vol. 33, No 2.

van Wingerde, A.M. Verheul, A. Wardenier, J. Puthli, R.S. and Dutta, D. (1988) The fatigue behaviour of T-joints made of square hollow sections. in **Int. conference on weld failure,** London, UK, Paper No. 4.

64 COMPARATIVE INVESTIGATIONS ON THE FATIGUE BEHAVIOUR OF UNIPLANAR AND MULTIPLANAR K-JOINTS WITH GAP

F. MANG, S. HERION and Ö. BUCAK
Versuchsanstalt für Stahl, Holz und Steine,
University of Karlsruhe, Germany

Abstract
Nearly all tests on the fatigue behaviour of hollow section joints published in the last few years were carried out on uniplanar test specimens. Regarding multiplanar K-joints, few experiments have been made in different countries.

In the last years extensive theoretical evaluation and practical tests were performed to investigate the fatigue behaviour of uniplanar K-type square hollow section joints leading to the design formulae for this type of joint at the University of Karlsruhe. A major part of these research results have been incorporated in the Eurocode 3 "Design of Steel Structures" initiated by the European Community.

The paper includes the results of additional finite element investigations carried out in Karlsruhe for the German research association DFG [Mang, 1991]. First results of parameter studies for uniplanar and multiplanar joints, which have been made in Karlsruhe for a Ph.D. thesis [Herion, 1993] are given. The complete work will be available in 1993/1994.
Keywords: Fatigue, Hollow Sections, Uniplanar and Multiplanar K-joints

Symbols and definitions

N_f	:	number of cycles to failure
$S_{r,h.s.}$	:	hot spot stress range
$S_{r,ax,nom}$	:	nominal stress range due to axial force
$S_{r,ipb,nom}$	:	nominal stress range due to in-plane bending moment
$S_{r,opb,nom}$	:	nominal stress range due to out-of-plane bending moment
ϵ_{ax}	:	strain due to axial force
ϵ_{ipb}	:	strain due to in-plane bending moment
ϵ_{opb}	:	strain due to out-of-plane bending moment
SCF	:	stress concentration factor
SNCF	:	strain concentration factor
t	:	wall thickness of a member hollow section
b	:	width of a member hollow section
g	:	gap between the toes of the bracings
Θ	:	angle of inclination between the axes of chord and bracing
β	:	b_1/b_0
2γ	:	b_0/t_0
τ	:	t_1/t_0
ξ	:	g/b_1
g'	:	g/t_0

Tubular Structures V. Edited by M.G. Coutie and G. Davies.
© 1993 E & FN Spon, 2–6 Boundary Row, London SE1 8HN. ISBN 0 419 18770 7.

1 SNCF and SCF

Well-known is the fact that stress or strain concentration occurs in a loaded construction at the places of discontinuity of form and cross section causing localized peak stresses, which can be many times higher than the nominal stresses calculated according to the simple elastic theory. This phenomenon is also observed at the welds, where the stress or strain concentration takes place due to the changeover from basic to weld material at the weld toe depending on the shape of the weld toe, the weld angle and the weld defects (undercut, excess weld metal, overlap etc.). In short, the localized peak stresses appear in welded structures caused by 1) geometrical stress concentration, 2) stress concentration due to weld.

While the role played by the local stress or strain concentration is of minor importance for statically loaded joints due to redistribution of stresses by local yielding of the material, it is vital for joints under fatigue loading. The initiation of fatigue cracks occurs at the location of largest, so-called "hot spot" stresses, which, therefore, must be taken into account for fatigue design.

The phenomenon of stress concentration is especially significant for welded hollow section joints in lattice structures, where bracings and chords of different stiffness intersect and the loading direction changes depending on the joint configuration. In order to incorporate stress concentration into the hollow section fatigue design, it was necessary to carry out fatigue tests and plot curves with the measured hot spot stress ($S_{r,h.s.}$) or strain ($\epsilon_{r,h.s.}$) ranges against the loading cycles of failure (N_f). For practical design of hollow section joints, only the effect of the geometrical stress concentration, which depends on the various geometrical parameters, such as $\beta, \tau, 2\gamma, \xi$ and g' has to be known, as the basic $S_{r,h.s.} - N_f$ curves are determined for specific details. The measurement of hot spot strains is made in a manner that the local weld effects are not included.

In order to simplify the calculation of the hot spot stresses of lattice girder joints of hollow sections, it is useful to relate the hot spot stress in the joint at the weld toe to the nominal stress in the bracing. The factor of relationship, known as the stress concentration factor SCF, is defined as

$$SNCF = \frac{\epsilon_{max,weld\ toe}}{(\epsilon_{ax} + \epsilon_{ipb} + \epsilon_{opb})\ brace\ in\ tension} \tag{1}$$

Further,

$$SCF = 1.1 \cdot SNCF$$

The calculated SNCF values depend on the type of joint (such as T, X, K), type of loading (axial force, in-plane and out-of-plane bending moment) and geometrical parameters. The hot spot stress range at a particular location is calculated according to the following equation:

$$S_{r,h.s.} = SCF_{ax} \cdot S_{r,ax,nom} + SCF_{ipb} \cdot S_{r,ipb,nom} + SCF_{opb} \cdot S_{r,opb,nom} \tag{2}$$

The determination of hot spot stresses at the weld toe of the intersection of bracing and chord is made either experimentally by means of strain gauge measurements or theoretically using the finite element method.

The geometrical strains at the weld toe were determined with strain gauges using the square extrapolation method for rectangular hollow section joints because of highly non-linear stress gradients.

For modelling the joint using finite elements, shell elements and solid elements are most suitable. The following two models should be used:

- model I using solid elements only,
- model II with shell elements and solids in the area of the welds (the connection of both element types was done by constraint equations).

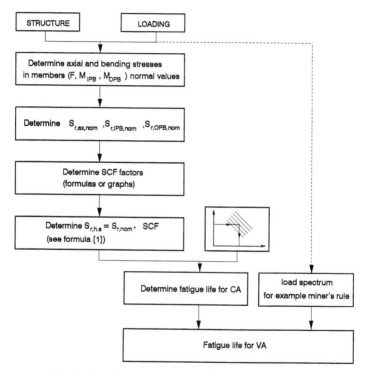

Fig. 1. Design procedure for RHS joints (sheme)

For both models, special work is caused by the simulation of the welds respectively the connection of the shells to the solids. For further information see [Mang 1991] and [Mang 1992].

In order to find out a simple means to calculate the hot spot stresses, parametric formulae and design curves, respectively, for SCF were developed using regression analyses based on an extensive parameter study with a computer program.

The ultimate aim is to give to the designers in practice a calculation method for RHS joints under fatigue load, as described in Fig. 1.

2 Uniplanar K-type joints made of square hollow sections

Sponsored by the European Community, Brussels, and CIDECT (Comité International pour le Développement et l'Etude de la Construction Tubulaire), Düsseldorf, an extensive research program was conducted to investigate into the fatigue behaviour of uniplanar square hollow section joints of T-, X- and K-types. This work was performed jointly by the Delft University of Technology and TNO-IBBC, Delft (T- and X-joints) and the University of Karlsruhe (K-type joints with gap and overlap).

The experimental part of the investigations of K-joints with overlap consisted of four series with a total of 12 joints, and of K-joints with gap of five series with a total of 24 joints. Joints with a stress ratio of R = + 0.1 have been checked at various stress levels. The test rig and manufacturing procedure are the same as for the previous investigations of K- and N-joints also sponsered by ECSC.

The objective of the experimental investigations was to measure the strain concentration in the critical areas of the joint, in order to be able to make comparisons with the results of the numerical investigations, using the finite element method. In addition, fatigue tests were necessary for the determination of S-N curves, in order to be able to check the design recommendations given by Eurocode 3 and others.

The maximum strains were determined in six and fourteen extrapolation lines, respectively, placed near the corners of a joint. Since the influence of the strains directly at the toe of the welds could not be determined due to local weld notch, extrapolation methods had to be used following the agreement made by the experts involved in the European Offshore Program.

As the axial force and in-plane bending components were not determined separately, the bending components were indirectly included, which means that the SCF were calculated as follows:

$$SCF = \frac{S_{r,h.s.}}{S_{r,ax,nom}}$$
and not as shown in formula (1).

In order to take account of the secondary bending moments in the joints, caused by the actual bending stiffness of the joint and joint eccentricities, the recommendation of the fatigue rules of the IIW-Subcommission XV E [IIW,1985] has to be applied. This recommends to multiply the stress range due to axial loading by the factors in table 1, when in-plane bending stress is not known, implying that the nominal bending stress due to secondary moments is generally about 50 % of the nominal axial stress (or 33 % of the nominal total stress).

Table 1. Factors to be applied to K-joints in square hollow sections, which take secondary bending moments into account

Type of K-joints	chords	bracings
gap joints	1.5	1.5
overlap joints	1.5	1.3

Based on a statistical analysis of all relevant experimental data (T-, X-, K-joints) design curves $S_{r,h.s.}$ - N_f for square hollow section joints (also applicable to rectangular hollow section joints) were derived as shown in Fig. 2.

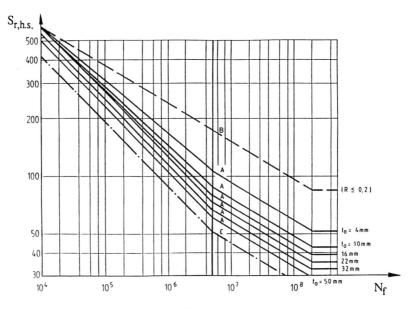

Fig. 2. $S_{r,h.s.}$ - N_f design curves

Parallel to the tests, parametric studies were carried out by means of the finite element method. A total of 103 analyses have been carried out using the finite element method, 48 for K-joints with overlap and 55 for K-joints with gap. With this, the decisive parameters have been varied in wide ranges. The angle between the brace and the chord axis has been kept constant at 45°. But the range of validity can be increased from 35° to 60°, according to the indications in literature [Dutta, 1981]. These studies resulted in formulae as given in tables 2 and 3.

3 Multiplanar double K-joints of square hollow sections

A further research program dealing with the fatigue behaviour of multiplanar double K-joints (Fig. 3) as is found in triangular girders has just been finished (Puthli, 1991). Recently available results of the strain measurements by means of strain gauges under static loading as well as fatigue tests are presented in the following. The aim of the investigation was to determine a design method for multiplanar double K-joints following a similar procedure to that for uniplanar joints described in the first part of this paper. The layout of the strain gauges can be seen in Fig. 3.

Table 2. SCF formulae (maximum value) for K-joints with gap in square hollow sections subjected to balanced axial forces and secondary moments

Member	
	Parametric range: $0.35 \leq \beta \leq 1.0, \quad 12.5 \leq 2\gamma \leq 25.0, \quad 0.25 \leq \tau \leq 1.0,$
	$0.25 \leq \xi \leq 0.75, \quad 1.5 \leq g' \leq 7.0, \quad 35° \leq \theta \leq 55°$
	where: $\xi = g' / (2\gamma\beta) \quad (= g / b_1)$
Brace	$SCF = 3.63\,\tau\,(2\text{-}\tau) + 0.336\,\xi\,\gamma^2\,(0.3 - 0.01\,\xi\,\gamma) - 4.81\,(\frac{\gamma g'}{100})^2$
	$- 2.2 + 0.044\,\gamma\,\beta\,(6.38 - \gamma\,\beta^2)$
chord	$SCF = 1.1\,\tau\,(0.00288\,\gamma^3 + g') + 5.73\,\xi\,(1 - 0.178\,\xi^2\,g') - 0.166\,\xi^3\,g'^2 - 1.73$

Table 3. SCF formulae (maximum value) for K-joints with overlap in square hollow sections subjected to balanced axial forces and secondary moments

Member	
	Parametric range: $0.35 \leq \beta \leq 0.7, \quad 12.5 \leq 2\gamma \leq 25.0, \quad 0.4 \leq \tau \leq 1.0,$
	$-0.40 \leq \xi \leq -1.0, \quad -17.0 \leq g' \leq -2.5, \quad 35° \leq \theta \leq 55°$
	where: $\xi = g' / (2\gamma\beta) \quad (= g / b_1)$
Brace	$SCF = 0.144\,\beta\,\gamma^2\,(1 - 0.813\,\beta^2) + 3.23\,\xi^2\,(1.94\,\tau^2 - 1.9\,\tau^3 - (\frac{\gamma}{10})^3)$
	$- 0.26 + 1.84\,\beta\,\tau$
chord	$SCF = -40.22\beta\,\gamma^2\,(1 - 0.59\,\xi^2) + 0.028\,\gamma^2\,(8.9\,\beta + \tau) - 5.41\,\gamma\,\beta^3$
	$- 0.008\,\xi^2\,\gamma^3 + 2.109\,\xi^6 - 4.24$

Fig. 4 and 5 show that in multiplanar joints maximum stresses occur at the external edges and, as can be seen in Fig. 4, at the edge of the chord member. Therefore, intensive investigations into the behaviour of thes e areas were necessary. One result of these finite-element-investigations is given in Fig. 5 showing the situation inside the chord looking on to the edge between the braces. The experimental data were confirmed by these results.

657

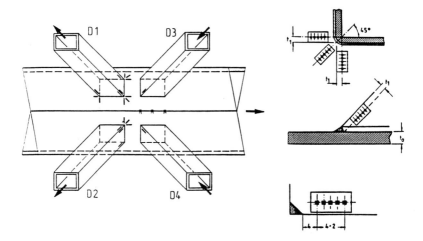

Fig. 3.   Position of strain gauges

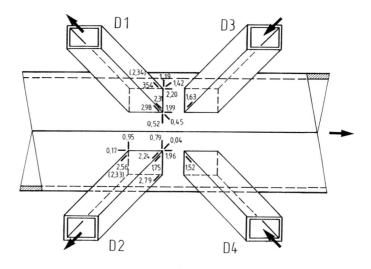

Fig. 4. Results of strain gauge measurements

Fig. 6 shows some typical modes of failure as they are found for uniplanar joints as well as for multiplanars.

Finally, all test results of the fatigue tests obtained for multiplanar K-joints were plotted in one diagram (Fig. 7) together with some data for uniplanar joints using the SCF formula mentioned above. In addition to this, the scatter band for uniplanar K-joints has also been plotted considering the wall thickness effect.

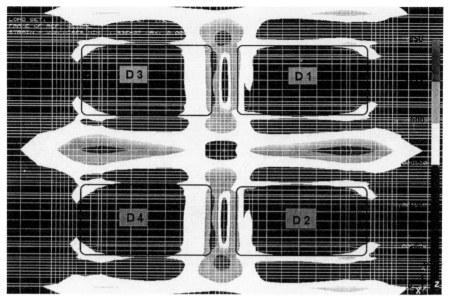

Fig. 5. Von-Mises-strains inside the chord

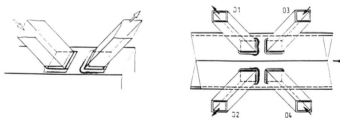

Type I: **Failure of the joint**
Starting from the toe of the weld, material together with the brace is broken out from the chord.

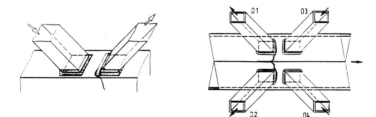

Type II: **Failure of the chord**
Starting from the weld in the chord, the cracks runs diagonally to the longitudinal axis through the chord.

In some cases, a crack can be observed in the chord along the edge between both braces in tension.

Fig. 6. Types of failure   a)

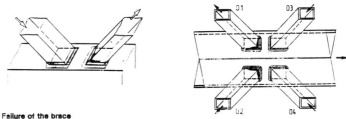

Type III: **Failure of the brace**
The brace breaks at the weld and runs diagonally through the brace
in case of K-joints with gap, the crack starts in the brace in tension.

Fig. 6. Types of failure   b)

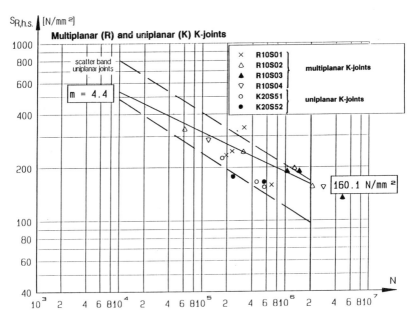

Fig. 7. $S_{r,h.s.}$ - N_f lines (series R10S01 to R10S04 without thickness correction)

It can be seen that the uniplanar joints K20S51 and K20S52 are well within the scatter band, if the wall thickness effect has been taken into account. It also becomes evident that the test results for the multiplanar joints, without considering the wall thickness effect, are well within the scatter band of the uniplanar joints. The two series K20S51 and K20S52 (uniplanar K-joints with thickness correction), however, do also fit.

660

In parallel to the procedure for uniplanar joints, parametric studies for the determination of SNCF for multiplanar K-joints are being carried out at present [Herion, 1993]. Fig. 8 is representative of the first results of these studies, which also include the influence of axial force and in-plane and out-of-plane bending (IPB and OPB). They will be published early in 1994.

$$\beta = 0,5 \quad 2\gamma = 25$$

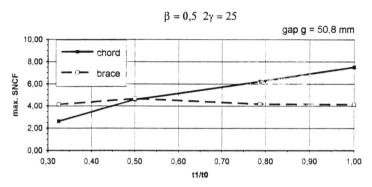

Fig. 8.   SNCF versus for β = 0.5 and g = 50.8 mm

4 Conclusions

Fatigue tests were carried out on uniplanar and multiplanar joints of hollow sections. On the basis of these experimental data and first results of a finite element analysis, the authors make the following recommendations for the investigated parameter ranges:

-   The design line $S_{R,h.s.}$ - N_f for uniplanar K-joints (RHS) can be used.
-   Regarding multiplanar joints no thickness correction is necessary.
-   Using the finite element method diagrams can be found to determinate SNCF (β, τ, 2γ, ξ, g'). These diagrams will be given available in the end of 1993.

Providing convenient design recommendations and formulae for stress concentration factors these investigations should be extended to other parameter ranges.

5 References

International Institute of Welding IIW-XVE "Recommended Fatigue Design Procedure for Hollow Section Joints", Part 1 "Hot Spot Stress Method for Nodal Joints" Doc XV-582-85/XIII-1158-85.
Dutta, D., Mang, F., Wardenier, J. (1981) Fatigue Behaviour of Welded Hollow Section Joints. **Cidect Monograph No. 7**
Herion , S. (1993/94) Untersuchungen über SNCF Faktoren an Konstruktionen aus Hohlprofilen (Investigations on SNCF factors on Constructions made of Hollow Sections) in preparation, PhD thesis, University of Karlsruhe

Mang, F., Herion, S. (1991) Räumliche Hohlprofilverbindungen (Multiplanar Hollow Section Joints). **Deutsche Forschungsgemeinschaft DFG Research Programme MA 761-6-1,** TU Karlsruhe, Versuchsanstalt für Stahl, Holz und Steine

Mang, F., Herion, S., Bucak, Ö. (1992) Comparative Investigation on the Fatigue Behaviour of Uniplanar and Multiplanar K-Joints with Gap. **Second International Offshore and Polar Engineering Conference (ISOPE 92),** Elsevier, London

Puthli, R. Wardenier, J. van Wingerde A.M. Verheul, A. de Koning, C.H.M. (1988, 1991) Fatigue Behaviour of Multiplanar Welded Hollow Section Joints and Reinforcement Measures for Repair. **ECSC research programme 7210/-SA 114.** Interim Reports, TNO Delft

65 PROPOSED REVISIONS FOR FATIGUE DESIGN OF PLANAR WELDED CONNECTIONS MADE OF HOLLOW STRUCTURAL SECTIONS

A.M. VAN WINGERDE
University of Toronto, Canada
Delft University of Technology, The Netherlands
J.A. PACKER
University of Toronto, Canada
J. WARDENIER
Delft University of Technology, The Netherlands
D. DUTTA
Mannesmannröhren-Werke A.G., Düsseldorf, Germany
P.W. MARSHALL
University of Newcastle-upon-Tyne, UK

Abstract
This paper aims to present a proposed update for the fatigue design of structures made of hollow structural sections, which are welded together without the use of additional stiffeners. By presenting a design example, a comparison between the current AWS and proposed design guidelines is made.
Keywords: Fatigue, Hollow Structural Sections, Hot Spot Stress, SCFs

Symbols and Notation

A	Cross sectional area of member considered.
N_f	Number of cycles to failure.
$S_{r,h.s.}$	Hot spot stress range = $SCF \cdot \sigma_r$
W	Elastic section modulus of member considered.
a	Weld throat thickness.
b	External width of member considered.
h	External height of member considered (for square sections: h=b).
l	Length of member considered between points of contra flexure or simple supports.
r	Outside corner radius of member considered (rectangular hollow sections only)
t	Wall thickness of member considered.
w	Weld dimension parallel to member considered.
β	Brace to chord width ratio b_1/b_0.
2γ	Width to wall thickness ratio of the chord b_0/t_0.
σ_r	Nominal stress range (stress range according to beam theory).
τ	Brace to chord wall thickness ratio t_1/t_0.

Subscripts
Member   0=chord, 1=brace
Loading   a=axial stress, m=in-plane bending stress

1 Introduction

The existing fatigue design rules given by the IIW (International Institute of Welding, 1985), or in the EC3 (Eurocode No. 3, 1992) and in the AWS design code (American Welding Society, Structural Welding Code D1.1, 1992) are all based upon research results for circular hollow sections only. Nowadays, these rules are fairly crude compared to the overall level of current codes: often based on the nominal stress approach, with inconsistent hot spot stress definitions, insufficient thickness correction and no (or primitive) SCF formulae. As such, they no longer reflect the current knowledge on the subject, which has been extended due to past and ongoing research programmes, especially within the European Community, e.g. van Delft et al. (1987), Puthli et al. (1989) and Thorpe and Sharp (1989). As a result, a more precise approach based on the hot spot stress method can now be established to replace the previous nominal stress and hot spot stress approach for the fatigue analysis of connections between hollow structural sections. The proposal presented in this paper includes $S_{r.h.s.}$ - N_f lines, parametric formulae for stress concentration factors (SCFs) of connections made of rectangular or circular hollow sections, comments on the thickness effect and the effect of fatigue improvement measures. The hot spot stress definitions are taken from van Wingerde (1992), and these utilize a geometric stress which includes bending in the wall near the weld, but exclude the notch effect, and can be determined by applying a quadratic extrapolation. The thickness effect for a wall thickness up to 16 mm is described by van Wingerde (1992), whereas for larger wall thicknesses the proposed correction of the British Department of Energy guidelines by Thorpe and Sharp (1989) is used. The parametric formulae for connections between circular hollow sections are taken from Efthymiou (1988), for use in conjunction with the new DEn $S_{r.h.s.}$ - N_f line, van Delft et al. (1987) and Thorpe and Sharp (1989). The parametric formulae for T- and X-connections between square hollow sections are taken from van Wingerde (1992).

The work on square hollow sections has been carried out in the framework of CIDECT (Comité International pour le Développement et l'Étude de la Construction Tubulaire) and ECSC (European Coal and Steel Community) research programmes.

The aim of these research programmes has been to establish a better design method for the fatigue strength of square hollow section connections, based on the hot spot stress method. The results are to be proposed for inclusion in the AWS D1.1 and Eurocode 3, as well as in the CIDECT and IIW design guidelines. In this way, a uniform approach to this specialized topic can be realized. In the experimental investigations, the strain concentration factors are measured at various locations around the connection for comparison with results of the numerical investigations and $S_{r.h.s.}$ -N_f lines are determined. The numerical work provides SCF values at weld toes for a range of parametric variations in the connection dimensions. These results form the basis for a set of parametric formulae. These formulae allow the determination of the SCF values at the weld toes of the brace and chord, depending on the non-dimensional parameters (β, 2γ and τ). The resulting parametric formulae can be checked by plotting:

- The hot spot stress range determined from the nominal stress range of the tests multiplied by the SCF of the proposed design formulae versus the number of cycles to failure of the tests.
- The hot spot stress ranges that were measured by means of strain gauges on the test specimens versus the number of cycles to failure of the tests.

A comparison of the scatter bands of the two sets of data points serves to show the accuracy of the formulae, as compared to the inherent scatter of the fatigue test results (van Wingerde, 1992).

2 Hot spot stress definition to be used for structural hollow sections

2.1 Type of stress to be considered

Only Stresses perpendicular to the weld toe are considered rather than principal stresses since:

- Stresses perpendicular to the weld toe can easily be measured, using simple strain gauges, rather than rosettes. Even when rosettes are applied, the stress components in the thickness direction are still ignored.
- Even when principal stresses are determined, it still requires all stress components to be extrapolated separately in order to extrapolate the principal stress (see section 2.2).
- Also, hot spot principal stresses caused by various load cases cannot be superimposed (see section 2.3).
- Only the stress component perpendicular to the weld is enlarged by the presence of the global weld shape and the notch. This view is supported by the direction of the crack growth, which is typically along the toe of the weld during the crack initiation stage.

2.2 Type of extrapolation to be carried out

In order to exclude local stress concentrations due to weld geometry and irregularities at the weld toe, which are heavily dependent upon fabrication and hard to determine, an extrapolation is to be carried out for the hot spot stress approach. Two methods of extrapolation are described below: the quadratic and the linear extrapolation. In principle, the quadratic extrapolation has to be used: this method requires more data to be reliable and is slightly more elaborate but can describe a non-linear stress increase near the weld toe more accurately. However, for most simple circular hollow section connections (except overlapped K connections), the stress increase is fairly linear, so that a linear extrapolation will be sufficiently accurate to be allowed as an alternative for these sections.

Both methods start by fitting a curve through all data points (the thick line in Fig. 1).

Linear extrapolation (simple circular hollow section connections only)

Two points on the curve determined from all data points are used: the first is 0.4 t from the weld toe, with a minimum of 4 mm. The second point is taken to be 0.6 t further from the weld toe, but this value is less critical.

Quadratic extrapolation

The first point is again 0.4 t from the weld toe, with a minimum of 4 mm. The second point on the curve used for the quadratic extrapolation is taken 1.0 t further from the weld toe. The quadratic extrapolation is carried out based on:

-    The first and second point on the curve based on all data points.
-    All data points between the first and second points on the curve (for t>10 mm, this means from 0.4 t to 1.4 t from the weld toe)

In this case, the curve passing through all data points supplies two additional data points as a basis for the extrapolation. By means of the least squares method, a quadratic curve is fitted through all these points, obtaining the quadratic SCF.

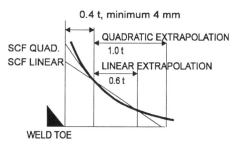

Fig. 1. Extrapolation procedure

2.3 Positions where the hot spot stress is determined

In order to be able to determine the total hot spot stress caused by axial forces and in-plane bending moments on chord and brace (see eqn. 1 in section 2.4), it is necessary to establish fixed positions where the SCFs are determined. For circular hollow sections these are the crown and saddle of chord and brace, whereas for rectangular hollow sections the stresses are considered along five lines A to E on the chord and brace (see Fig. 2).

2.4 Resulting hot spot stress definition

The hot spot stress is thus defined as the **extrapolated** stress at the toe of the weld, **along the lines of measurement** considered. The stress concentration factor SCF is defined as the hot spot stress divided by the nominal stress which causes this hot spot stress. The total hot spot stress is then a function of all nominal stresses in all members of the connection multiplied by their stress concentration factors. In the case where only axial forces and in-plane bending moments are considered, the total hot spot stress can be determined by:

$$S_{\text{rh.s.}} = \sigma_{m1} \cdot SCF_{m1} + \sigma_{a1} \cdot SCF_{a1} + \sigma_{m0} \cdot SCF_{m0} + \sigma_{a0} \cdot SCF_{a0} \tag{1}$$

As a consequence, the hot spot stresses found may underestimate the 'true' hot spot stress if the direction of the principal stresses deviates from these lines, especially if the stress concentration is less pronounced. Here, the stresses at other positions or in other directions or at the inside of the members may be higher. Therefore, a minimum value of 2.0 is specified for SCF_{a1} and SCF_{m1} in the proposed design recommendations.

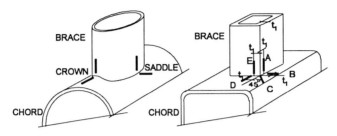

Fig. 2. Position of lines A to E.

3 Proposed design rules

3.1 Basic $S_{\text{rh.s.}}$ -N_f line to be used for circular hollow sections

An extensive investigation by the U.K. Department of Energy on fatigue has been carried out on the basis of 400 test results, see Thorpe and Sharp (1989). The resulting $S_{\text{rh.s.}}$ - N_f line is proposed for inclusion in the new DEn design guidelines and is also the basis for the proposal in this paper. As the DEn line runs at a 1:3 slope until 10 million cycles and then at a 1:5 slope until 100 million cycles, the general shape of the line is very similar to the EC3 $S_{\text{rh.s.}}$ - N_f lines. To enable future inclusion in EC3, this line is translated into an EC3 classification of 114. Note that this revised $S_{\text{rh.s.}}$ - N_f line, which only differs from the proposed DEn line in the high cycle region (> 5 million cycles) is also suggested for the AWS.

3.2 Basic $S_{\text{rh.s.}}$ -N_f line to be used for rectangular hollow sections

In order to maintain a close correlation with existing design guidelines and yet obtain a good agreement with the test results, a statistical evaluation of test data and the proposed design guidelines according to EC3 has been performed. After carrying out a statistical analysis of test data based on square hollow sections, a class of 90, together with a thickness correction, was found to be optimal (van Wingerde, 1992). This line allows a hot spot stress range of 90 N/mm² (13 ksi) for N_f=2 million cycles. The slope of the line is 1:3 until 5 million cycles. For higher numbers of cycles the line becomes horizontal for constant amplitude loading (no fatigue damage), or runs at a slope of 1:5 until 100 million cycles as adopted in EC3 and becomes horizontal thereafter for variable amplitude loading. The lines for circular and rectangular hollow sections are shown in Fig. 3, together with a number of other major $S_{\text{rh.s}}$ - N_f lines. The lines are shown for a groove (butt) welded connection with t=16 mm to show the basic $S_{\text{rh.s.}}$ - N_f lines of the proposal. The IIW line A is taken from the International Institute of Welding (1985).

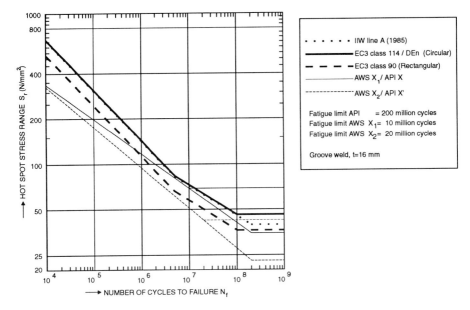

Fig. 3. Major $S_{r_{h.s.}}$ - N_f lines for hollow section connections.

3.3 Correction factors for wall thickness

The basic $S_{r_{h.s.}}$ - N_f lines are used for the basic wall thickness of 16 mm.

1. For wall thicknesses below 16 mm, down to 4 mm, a positive correction factor is applied between $N_f=1,000$ and $N_f=5,000,000$. For larger N_f all lines are parallel to the basic line again (for wall thicknesses below 4 mm, the influence of the root might be governing, leading to a much lower fatigue strength).
2. For wall thicknesses larger than 16 mm, the correction factor of the new DEn guidelines is followed, since the tests of the research programme on square hollow sections did not include specimens with t>16mm.

The equations of the various cases are given in Table 1. However, a designer would typically use the resulting $S_{r_{h.s.}}$ - N_f lines for various wall thicknesses which are shown in Fig. 4 directly.

3.4 Correction factors for weld type

The $S_{r_{h.s.}}$ - N_f lines and parametric formulae are valid for groove (butt) welds.

*   For fillet welds with a throat thickness a≥ t, the same $S_{r_{h.s.}}$ - N_f line and formulae can be applied. A study (van Wingerde, 1992) has shown that in this case the SCFs in the brace are about 40% higher, so that the SCFs in the brace from the parametric formulae of Table 2 are to be multiplied by 1.4. The same correction is suggested for circular hollow sections.
*   The potentially higher fatigue strength of an improved weld shape or an improved weld toe is not (yet) accounted for in the design guide proposal. However, a wealth of data seems to suggest that considerable gains on the fatigue strength can be made.

3.5 Parametric formulae to be used

For connections made of rectangular hollow sections, formulae are available for T- and X-connections only. Formulae for K-connections are being developed at the University of Toronto, in cooperation with Delft University of Technology. No data is available for non-square hollow sections. However, it is believed that chords with 0.5≤ h/b≤ 2 would not show considerable differences in their SCFs, so that the formulae can be used here as well. Formulae for circular hollow sections can be found in Efthymiou (1988). The combination of the Efthymiou formulae and the DEn $S_{r_{h.s.}}$ - N_f line has been verified by van Delft et al. (1987), although on the basis of the old DEn line.

Table 1. Equations for the $S_{r_{h.s.}}$ - N_f lines for hollow structural sections.

t	N_f	$10^3 < N_f < 5 \cdot 10^6$
t < 16mm Circular	$10^3 < N_f < 5 \cdot 10^6$	$\log(S_{r_{h.s.}}) = \frac{1}{3} \cdot (12.476 - \log(N_f)) + 0.11 \cdot \log(N_f) \cdot \log\left(\frac{16}{t}\right)$
	$5 \cdot 10^6 < N_f < 10^8$ (variable amplitude only)	$\log(S_{r_{h.s.}}) = \frac{1}{5} \cdot (16.327 - \log(N_f)) + 0.737 \cdot \log\left(\frac{16}{t}\right)$
t ≥ 16mm Circular	$10^3 < N_f < 5 \cdot 10^6$	$\log(S_{r_{h.s.}}) = \frac{1}{3} \cdot (12.476 - \log(N_f)) + 0.30 \cdot \log\left(\frac{16}{t}\right)$
	$5 \cdot 10^6 < N_f < 10^8$ (variable amplitude only)	$\log(S_{r_{h.s.}}) = \frac{1}{5} \cdot (16.327 - \log(N_f)) + 0.30 \cdot \log\left(\frac{16}{t}\right)$
t < 16mm Rectangular	$10^3 < N_f < 5 \cdot 10^6$	$\log(S_{r_{h.s.}}) = \frac{1}{3} \cdot (12.151 - \log(N_f)) + 0.11 \cdot \log(N_f) \cdot \log\left(\frac{16}{t}\right)$
	$5 \cdot 10^6 < N_f < 10^8$ (variable amplitude only)	$\log(S_{r_{h.s.}}) = \frac{1}{5} \cdot (15.786 - \log(N_f)) + 0.737 \cdot \log\left(\frac{16}{t}\right)$
t ≥ 16mm Rectangular	$10^3 < N_f < 5 \cdot 10^6$	$\log(S_{r_{h.s.}}) = \frac{1}{3} \cdot (12.151 - \log(N_f)) + 0.30 \cdot \log\left(\frac{16}{t}\right)$
	$5 \cdot 10^6 < N_f < 10^8$ (variable amplitude only)	$\log(S_{r_{h.s.}}) = \frac{1}{5} \cdot (15.786 - \log(N_f)) + 0.30 \cdot \log\left(\frac{16}{t}\right)$

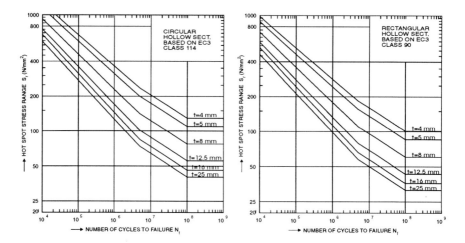

Fig. 4. $S_{r_{h.s.}}$ - N_f lines according to EC3 class 90/114, with thickness correction.

Table 2. SCFs for T- and X-connections made with rectangular hollow sections.

SCFs for connections loaded by a bending moment on the brace (SCF_{m1})

LINE B	$SCF=(-0.011+0.085 \cdot \beta-0.073 \cdot \beta^2) \cdot 2\gamma^{(1.722+1.151 \cdot \beta-0.697 \cdot \beta^2)} \cdot \tau^{0.75}$
LINE C	$SCF=(0.952-3062 \cdot \beta+2.382 \cdot \beta^2+0.0228 \cdot 2\gamma) \cdot 2\gamma^{(-0.690+5.817 \cdot \beta-4.685 \cdot \beta^2)} \cdot \tau^{0.75}$
LINE D	$SCF=(-0.054+0.332 \cdot \beta-0.258 \cdot \beta^2) \cdot 2\gamma^{(2.084-1.062 \cdot \beta+0.527 \cdot \beta^2)} \cdot \tau^{0.75}$
LINES A,E	$SCF=(0.390-1.054 \cdot \beta+1.115 \cdot \beta^2) \cdot 2\gamma^{(-0.154+4.555 \cdot \beta-3.809 \cdot \beta^2)}$
Minimum SCF:	$SCF_{m1} \geq 2.0$
Fillet welds:	Lines A,E: $SCF_{m1}=1.40 \cdot SCF_{formula}$ (if β is close to 1.0, line A cannot have a fillet weld)

SCFs for connections, loaded by an axial force on the brace (SCF_{a1})

LINE B	$SCF=(0.143-0.204 \cdot \beta+0.064 \cdot \beta^2) \cdot 2\gamma^{(1.377+1.715 \cdot \beta-1.103 \cdot \beta^2)} \cdot \tau^{0.75}$
LINE C	$SCF=(0.077-0.129 \cdot \beta+0.061 \cdot \beta^2-0.0003 \cdot 2\gamma) \cdot 2\gamma^{(1.565+1.874 \cdot \beta-1.028 \cdot \beta^2)} \cdot \tau^{0.75}$
LINE D	$SCF=(0.208-0.387 \cdot \beta+0.209 \cdot \beta^2) \cdot 2\gamma^{(0.925+2.398 \cdot \beta-1.881 \cdot \beta^2)} \cdot \tau^{0.75}$
LINES A,E	$SCF=(0.013+0.693 \cdot \beta-0.278 \cdot \beta^2) \cdot 2\gamma^{(0.790+1.898 \cdot \beta-2.109 \cdot \beta^2)}$
Minimum SCF:	$SCF_{a1} \geq 2.0$
X-conn. ,β=1.0:	Line C: $SCF=0.65 \cdot SCF_{formula}$ and Line D: $SCF=0.50 \cdot SCF_{formula}$
Fillet welds:	Lines A,E: $SCF_{a1}=1.40 \cdot SCF_{formula}$ (if β is close to 1.0, line A cannot have a fillet weld)

SCFs for connections, with loads on the chord (SCF_{m0}, SCF_{a0})

LINE C	$SCF=0.725 \cdot 2\gamma^{0.248 \cdot \beta} \cdot \tau^{0.19}$
LINE D	$SCF=1.373 \cdot 2\gamma^{0.205 \cdot \beta} \cdot \tau^{0.24}$
LINES B,A,E	negligible i.e.: $SCF=0$

Range of validity:

$0.35 \leq \beta \leq 1.0$	$12.5 \leq 2\gamma \leq 25.0$	$0.25 \leq \tau \leq 1.0$
$1.0 \leq r/t \leq 4.0$	$0.5 \leq h_0/b_0 \leq 2.0$	$h_1/b_1 = 1.0$

Note: The number of decimal places in the formulae is chosen in order to have the same number of decimals for all terms (except the 2γ terms for line C) and are no reflection of the accuracy or sensitivity of the formulae.

4 Design Example

To show the differences between the proposed and an existing design recommendation (AWS D1.1, Chapter 10, see also Marshall (1992)), a design example will be performed for a square hollow section T-connection.

Chord:	200 x 200 x 12.5 mm, $A_0 = 8973$ mm^2, $W_0 = 513444$ mm^3, $l_0 = 1500$ mm
Brace:	140 x 140 x 5.0 mm, $A_1 = 2661$ mm^2, $W_1 = 114711$ mm^3
	$\beta = 140/200 = 0.7$; $2\gamma = 200/12.5 = 16$; $\tau = 5/12.5 = 0.40$.
Weld:	The brace is 5 mm thick and therefore a 45° fillet weld with a throat thickness of 5 mm is assumed (legs are $5 \cdot \sqrt{2} = 7.1$ mm)
Problem:	To determine the nominal axial force range on the brace for the design of a T-connection at 2 million cycles with constant amplitude loading.

669

Table 3. SCFs in the T-connection considered

Line	Due to axial force in the brace SCF_{a1}	Due to bending moment in the chord SCF_{m0}	Resulting stress concentration $S_{r.h.s.}/\sigma_{a1}$
A, E	10.26^{1}	0	10.26
B	4.50	0	4.50
C	4.27	0.99	6.01
D	2.11	1.64	5.00

[1] The SCF in the brace (lines A,E) is multiplied by 1.40, to correct for the weld type, see Table 2.

4.1 Existing design guidelines (AWS D1.1, part 10.7), analysis based on nominal stress approach
- Different lines are to be used for the chord, brace and weld: the chord is to be checked by line K_1, whereas for the branch line ET is used and for the weld line FT. In this case the weld is such that the throat thickness is the same as the branch member wall thickness. The weld is therefore not critical, since ET is far lower than FT.
- An axial stress σ_{a1} in the brace causes a bending moment in the chord and hence a stress σ_{m0} of:
$$\sigma_{m0} = \frac{A1 \cdot \sigma_{a1} \cdot (l_0 - b_1)}{4W_0}$$ (for a simply supported chord). In this case: $\sigma_{m0} = 1.76 \cdot \sigma_{a1}$.
- For the brace, a nominal stress range of about 20 N/mm^2 (2.9 ksi) is allowed, based on line ET. Based on brace checking: $\sigma_{a1} \leq$ 20 N/mm^2 (2.9 ksi).
- For the chord, line K_1 determines the allowable nominal punching shear stress range, which is about 15 N/mm^2 (2.2 ksi). Applying an ovalisation factor (!), taken from circular tubes of 1.7, and with $\tau = 0.4$, this corresponds to a nominal stress range in the brace of $15/(1.7*0.4) = 22$ N/mm^2 (3.2 ksi). Note that this method ignores the influence of the bending moment on the chord.
- In this case brace failure governs the fatigue strength of the connection. The nominal stress range in the brace of 20 N/mm^2 (2.9 ksi) corresponds to an allowable force range in the brace of **53 kN**.

4.2 Existing design guidelines (AWS D1.1, part 10.7), analysis based on hot spot stress approach
- Using the hot spot stress approach, line X_1 can be used regardless of the weld type, since the brace wall thickness is very small.
- Apply the SCF formulae of the proposal as given in Table 2. (The AWS does not include parametric formulae for the determination of the SCF anyway). This involves checking lines A, B, C, D and E for the given non-dimensional connection parameters β, 2γ and τ. The SCFs and the ratio between the total hot spot stress and the nominal axial stress in the brace are given in Table 3 for lines B, C and D in the chord and lines A and E in the brace. Note that the hot spot stress definition on which the parametric formulae are based is not consistent with the current AWS.
- As there exists both a nominal stress in the brace and a nominal stress in the chord due to the induced bending moment, the total hot spot stress (to be checked for all lines A to E) becomes:
$$S_{r.h.s.} = \sigma_{a1} \cdot SCF_{a1} + \sigma_{m0} \cdot SCF_{m0} = \sigma_{a1} \cdot (SCF_{a1} + 1.76 \cdot SCF_{m0})$$
- The governing total stress concentration factors are 6.01 for the chord (line C) and 10.26 in the brace (lines A,E).
- Use of line X_1 results in an allowable hot spot stress range of 100 N/mm^2 (14.4 ksi). Since the brace has the highest SCF, the brace governs the fatigue design.
The allowable nominal stress range is 100/10.26=10 N/mm^2 (1.4 ksi), which corresponds to an allowable force range of **26 kN** in the brace, less than half the value found from the nominal stress approach, due to the hot spot stress definition of the SCF formulae being inconsistent with the definition on which the $S_{r.h.s.} - N_f$ line is based.

4.3 Proposal, based on hot spot stress approach

- Only the hot spot stress method is used.
- As can be seen in Fig. 3, the basic $S_{r_{h.s.}}$ - N_f lines of the new proposal and line X_1 are very close for N_f=2 million cycles. As the new proposal results in an allowable hot spot stress range of 90 N/mm^2 (13.0 ksi) rather than the 100 N/mm^2 (14.4 ksi) of line X_1, the new proposal initially seems more conservative.
- However, the line given for the proposal is valid for t=16 mm, rather than for 5.0 and 12.5 mm for the members to be checked here. As can be calculated from the equations in Table 1, or seen in Fig. 4, the favourable thickness correction factor of $0.11 \cdot \log(N_f) \cdot \log(16/t)$ produces an increased allowable hot spot stress. For N_f=2,000,000 the chord (t=12.5 mm) has an allowable hot spot stress of 106 N/mm^2 (15.3ksi) and for the brace (t=5 mm) the allowable hot spot stress is even 200 N/mm^2 (28.9 ksi).
- The allowable nominal stress range in the brace is then found by dividing the allowable hot spot stress range for brace and chord by the highest total stress concentration factor for the member considered. This results in an allowable nominal stress range in the brace of:
  Based on chord failure criterion: 106/ 6.01 = 18 N/mm^2 (2.6 ksi)
  Based on brace failure criterion: 200/10.26 = 19 N/mm^2 (2.8 ksi)
  In this case, chord failure governs the allowable force range in the brace which will be
  18 · 2661=**47 kN**, or about the same as presently allowed by the AWS D1.1 nominal stress approach. Note however, that the SCFs given by the parametric formulae for τ=1.0 vary between 2 and 30 within the range of validity of the AWS. For the punching shear method in case τ=1.0, punching shear stress equal to nominal axial stress in the brace, curve K_1 is about 6.7 times lower than curve X_1, which together with the ovalisation factor of 1.7 corresponds to a SCF of about 11.
- It should be noted that the position of the $S_{r_{h.s.}}$ - N_f line is dependent on the definition of the hot spot stress, so that it is important to use a specified combination of $S_{r_{h.s.}}$ - N_f line and parametric formulae rather than picking them from different sources! The use of a $S_{r_{h.s.}}$ - N_f line without matching parametric formulae, as is currently the case in AWS D1.1, is therefore not recommended. The new $S_{r_{h.s.}}$ - N_f lines as presented in Table 1 and Fig. 4 were determined in conjunction with the parametric formulae.

5 Conclusions

- At present, neither the EC3 nor the AWS D1.1 contain very sophisticated rules and parametric formulae for the design of structures made of hollow sections loaded in fatigue. This represents a good opportunity for introducing consistent design rules, thereby at least uniting these two design recommendations in this field. The proposed design procedures are backed up by extensive tests as well as numerical analyses and are expected to avoid the current excessive over- or underestimation of the fatigue capacity. In addition, fabrication costs can be lowered for smaller wall thicknesses yet still utilize their inherently higher fatigue strength.
- The hot spot stress method replaces a large number of $S_{r_{h.s.}}$ - N_f lines in the AWS and includes important influences, such as the influence of the connection geometry and the bending moment in the chord. The example of combining the parametric formulae with the AWS X_1 line serves to illustrate the dangers of combining an $S_{r_{h.s.}}$ - N_f line with parametric formulae which are based on another hot spot stress definition.
- The comparison between hot spot and punching shear methods in section 4.3 shows that the connections which are currently treated as having the same SCF, in reality show a wide variation in SCFs. The hot spot stress method therefore represents an important increase in accuracy over the punching shear stress method. Still, the difference in $S_{r_{h.s.}}$ - N_f lines for circular and rectangular hollow sections of about 20% serves to remind the researcher that the hot spot stress approach is not the ultimate answer to the fatigue analysis either.

6 Remaining issues

- Parametric formulae for the SCFs of K-connections made of rectangular hollow sections should be derived. Then, a comprehensive design method for connections of hollow structural sections can be established. Work on the K-connections is being carried out at the University of Toronto.
- Future extensions might include some research on the behaviour of non-square rectangular hollow structural sections, other types of connections, multiplanar connections and combinations of rectangular and circular hollow structural sections. Work on multiplanar connections is being done at Delft University of Technology, TNO Building and Construction research and the Universität Karlsruhe. The results can be incorporated when they become available in 1994/1995.

7 Acknowledgements

Thanks to ECSC, CIDECT, IPSCO Inc. and Mannesmannröhren-Werke A.G. for their financial support for the research programmes.

8 References

American Welding Society (1992) **Structural Welding Code /Steel**, ANSI/AWS D1.1-92, 13th edition, Miami, Florida, U.S.A.

Delft, D.R.V. van, Noordhoek, C. and Da Re, M. L. (1987) The results of the European fatigue tests on welded tubular joints compared with SCF formulas and design lines. **Steel In Marine Structures (SIMS '87),** Elsevier applied science publishers ltd., Amsterdam/London/New York/Tokyo, Delft, the Netherlands, pp.565-577.

Efthymiou, M. (1988) Development of SCF formulae and generalised influence functions for use in fatigue analysis, 2nd revision. **Offshore Tubular Joints Conference (OTJ '88)**, USG Offshore Research, Englefield Green Near Egham, United Kingdom.

EC3 (1990) **Eurocode no. 3, Design of Steel Structures**, Part 1 - General Rules and Rules for Buildings, Final Draft (november 1990), Issued to Liaison Engineers, February 1989, Report prepared for the Commission of the European Communities, Directorate General, Internal Market and Industrial Affairs.

International Institute of Welding, Subcommission XV-E (1985) Recommended fatigue design procedure for hollow section joints. **IIW Annual Assembly 1985**, Strasbourg, France, IIW doc. XV-582-85.

Marshall, P.W. (1992) Design of welded tubular connections: Basis and use of AWS code provisions. **Developments in Civil Engineering, Volume 37**, Elsevier applied science publishers ltd., Amsterdam/London/New York/Tokyo

Niemi, E. (1992) Recommendations concerning stress calculation for fatigue analysis of welded components. **IIW Annual Assembly 1992**, Madrid, Spain, IIW doc. XV-797-92.

Puthli, R.S., Koning, C.H.M. de, Wingerde, A.M. van, Wardenier, J. and Dutta, D. (1989) Fatigue strength of welded unstiffened R.H.S. joints in latticed structures and Vierendeel girders - Final Report, Part 3: Evaluation for design rules, **TNO-IBBC report BI-89-097/63.5.3820, Stevin report 25.6-89-36/A1**, Delft University of Technology, Delft, the Netherlands.

Thorpe, T.W. and Sharp, J.V. (1989) The fatigue performance of tubular joints in air and seawater. **MaTSU** Harwell Laboratory, Oxfordshire, United Kingdom. (presented during the OMAE '89, but not included in the proceedings)

Wingerde, A.M. van (1992) The fatigue behaviour of T- and X-joints made of square hollow sections. **Heron**, Delft, the Netherlands, Volume 37 No. 2.

PART 17

FINITE ELEMENT VALIDATION

Two and three dimensional joints
in circular hollow sections

66 A COMPARATIVE STUDY ON OUT-OF-PLANE MOMENT CAPACITY OF TUBULAR T/Y JOINTS

E.M. DEXTER
University College of Swansea, UK
J.V. HASWELL
British Gas plc, Newcastle-upon-Tyne, UK
M.M.K. LEE
University College of Swansea, UK

Abstract
This paper reports on a combined numerical and experimental study of the out-of-plane moment capacity of tubular T/Y joints. The aim of the research was to establish a valid and accurate Finite Element model for further studies into the ultimate behaviour of more complex joints. The experimental work consisted of static strength tests on two identical full-scale T/Y joints carried out by British Gas, as part of their offshore research programme. The numerical work was carried out by University College of Swansea, and involved the development and numerical analyses of Finite Element models simulating the joints tested. It was found that, when the test conditions were simulated as closely as possible, the Finite Element results fell centrally between the two test results, the ultimate loads of which differed by only 3%. However it was noted that the ultimate capacities fell 15-19% below the API code allowable capacity.
Keywords: Ultimate strength, Tubular T/Y joints, Out-of-plane bending, Full-scale testing, Finite Element analyses.

1 Introduction

British Gas operates several offshore structures on two offshore gas fields, Morecambe Bay and Rough. The structures are of the fixed steel jacket type, comprising topside modules supported on a piled foundation braced by the jacket. The jacket is a stiff, three-dimensional space frame constructed from tubular members, and it's function is to transfer topside and environmental loads to the piled foundation. The integrity of the jacket must be maintained to ensure the safe operation of the structure in the worst forecast storm.

In order to ensure structural integrity, regular in-service inspections are required by the regulatory authorities. It is the responsibility of the operator to identify critical areas of the structure for inspection. To provide a safe, cost effective inspection policy, British Gas is developing methodologies to identify and assess critical regions of the jacket under both fatigue and ultimate loading. The fatigue assessment methodology is now complete (Haswell(1992)) and work is progressing to complete the ultimate load assessment methodology. A programme of theoretical and experimental research has been carried out at the British Gas Engineering Research Station (ERS) to support the development and validation of the methodologies.

Current ultimate strength design guidance is based on a limited database of reliable test results. For instance, the API(1991) ultimate strength equations were formulated from the database screened by Yura et al.(1980). Much research effort has been expended on expensive and complex testing of

Tubular Structures V. Edited by M.G. Coutie and G. Davies.
© 1993 E & FN Spon, 2–6 Boundary Row, London SE1 8HN. ISBN 0 419 18770 7.

joints of simple geometries and loadings. In spite of such effort, the behaviour of some simple joints is still not fully understood. This is particularly true for T/Y joints subjected to out-of-plane bending (OPB), as the reliable database is relatively small. It is now generally accepted that the way to further understanding the behaviour of complex joints, subjected to the multi-directional loadings occurring in reality, is via non-linear numerical analysis using techniques such as the Finite Element Method (FEM). However, for such a technique to be used with confidence and in an efficient and cost effective manner, the strategy adopted must be well founded and calibrated with experimental data.

The British Gas research programme included full-scale ultimate strength tests on OPB moment loaded T/Y joints. As part of a collaborative research programme, University College of Swansea have carried out detailed non-linear FE analyses of the joints tested by British Gas. This paper summarises the results of the investigation into the ultimate behaviour of OPB moment loaded T/Y joints, which included;

(a) an assessment of the existing static strength database,
(b) static strength tests on two identical T/Y joints with geometric parameters, $\alpha=17$, $\beta=0.375$, $\gamma=18$, $\tau=0.42$.
(c) FE modelling of the joint tests,
(d) comparison of results from tests and FE models with the ultimate strength predictions of the API code.

2 Assessment of existing database and API capacity equation

The API(1991) formula for the allowable non-dimensional capacity of an OPB moment loaded T/Y joint is:

$$\frac{M_a \sin\theta}{F_y T^2 d} = 0.8 (3.4 + 7\beta) Q_\beta \tag{1}$$

where:

$$Q_\beta = 1.0 \; for \; \beta \le 0.6 \;, \qquad Q_\beta = \frac{0.3}{\beta (1 - 0.833\,\beta)} \; for \; \beta > 0.6 \tag{2}$$

This formula was proposed by Yura et al.(1980). The most recent database screened by DEn(1990) is almost identical to that used by Yura et al. A full review of the current static strength database for out-of-plane bending is given in Table 1. The results published by Stol et al.(1985) (TNO) and Ma(1988) (JISSP) have only become available since the API equation was formulated. The data is scattered, due to the effects of parameters not included in the API formula, viz:

θ effect   Test results of only four Y joints (all with large β ratios) are available and all have been severely under-predicted by the API formula. This appears to suggest that the increase in strength of a Y joint, relative to that of a T joint, is greater than $1/\sin\theta$.

α effect α values for the tests reviewed by Yura et al. are not known. The remaining joints tested all had α ratios of 8 or 10. The measured capacities of the JISSP joints, for which $\alpha=8$ and $\beta\geq0.8$, were all severely under-predicted by the API formula. This value of α may not be sufficiently large to prevent end effects

Table 1. Existing database

| Test prg. & specimen numbers | Joint parameters | | | | | | Test Result | measured strength |
	θ	β	α	γ	τ	F_y kNmm^{-2}	$\dfrac{M_u \sin\theta}{F_y T^2 d}$	API
JISC(1972)								
BL-40-0.3	90	0.260	-	17.5	-	.471	4.09	0.98
BL-40-0.5	90	0.462	-	18.4	-	.471	5.46	1.03 *
BL-70-0.2	90	0.190	-	36.2	-	.441	4.10	1.08 *
BL-70-0.4	90	0.439	-	36.2	-	.441	5.53	1.07 *
BL-100-0.2	90	0.195	-	47.6	-	.402	4.12	1.08 *
BL-100-0.4	90	0.361	-	47.6	-	.402	4.36	0.92 *
YURA(1978)								
G1	90	0.338	-	22.8	-	.352	4.39	0.95 *
G2	90	0.338	-	22.8	-	.352	4.83	1.05 *
H1	90	0.644	-	22.8	-	.352	7.29	1.15 *
H2	90	0.644	-	22.8	-	.352	7.86	1.24 *
I1	90	0.899	-	22.8	-	.352	13.40	1.30 *
I2	90	0.899	-	22.8	-	.352	13.45	1.31 *
E1	30	0.899	-	22.8	-	.352	14.96	1.45 *
E2	30	0.899	-	22.8	-	.352	14.90	1.45 *
TNO(1985)								
4	90	0.35	10	8.0	1.1	.254	4.97	1.06
71	90	0.68	8	8.0	1.1	.234	10.07	1.51
2	90	0.36	10	15.0	1.0	.295	5.19	1.10
11	90	0.68	10	15.0	1.0	.321	6.25	0.94
62	90	1.00	8	15.0	1.0	.291	14.92	1.00
14	90	0.36	10	24.0	1.1	.299	5.92	1.25
17	90	0.68	10	24.0	1.0	.308	8.95	1.34
20	90	1.00	10	24.0	1.0	.304	18.61	1.24
JISSP(1988)								
1.2	45	1.00	8	31.8	1.0	.276	42.2	2.82
1.13	90	0.80	8	20.5	1.0	.300	12.8	1.59
1.14	45	1.00	8	20.2	1.0	.295	16.1	2.00

* API database

enhancing the joint capacity, especially for joints with large β ratios.

γ effect   The API database contained no joints possessing both large β and large γ ratios, and the API formula shows no dependence on γ. However it seems that γ has an effect, particularly for joints with high β ratios. The under-prediction of the JISSP results is attributed in part to the high values of γ, given the high β ratios of the specimens.

τ effect τ ratios are not quoted in the existing API database. As both the brace load carrying capacity and the chord local bending reaction depend upon member thickness, τ ratio effects would be expected.

Quantification of the above proposed effects would obviously require further selective testing or numerical analysis. However, although the test results are scattered, the API formula only over-predicts the capacity of four test results, the largest being by 8%. It is therefore unlikely that this formula will severely over-predict results of future tests.

3 Joint testing

Static strength tests on two identical full-scale T/Y joints were carried out by British Gas at ERS. Figure 1. shows the self-reacting joint assembly, joint dimensions and geometric parameter values. The OPB load was applied as a shear load at the top of the braces; the experimental joints being referred to in terms of their saddle locations, 'AB' and 'CD' as shown in Figure 1.

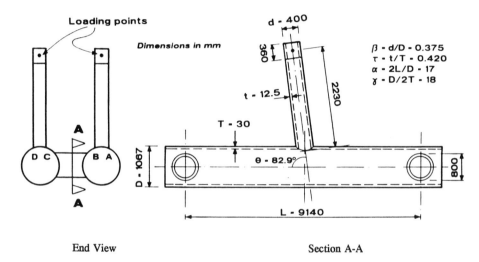

End View   Section A-A

Figure 1. Joint test rig

Displacement transducers were fixed to the test joints in order to measure the displacements of the chords and braces in the out-of-plane direction. Strains in both circumferential and longitudinal directions were measured with strain gauges fitted to the inside and outside of the chord at each of the four saddle positions.

Since the hydraulic rams had a limited amount of travel, the joints could not be tested to failure in one run. Instead, the joints were loaded nearly to the extent of the rams' travel and then unloaded, allowing the insertion of spacers before reloading. A maximum of four runs was possible. Smaller increments of force were applied as the gradient of the force-displacement curve decreased. During testing, the braces did not deform appreciably, but the chord walls adjacent to the saddles in compression buckled inwards noticeably.

Of the two braces, brace 'AB' displaced more quickly than brace 'CD' although the loadings applied to each were equal. The load-displacement curve for joint 'AB' is shown in Figure 2, in the form of non-dimensional moment plotted against brace rotation. The non-dimensional moment, $M\sin\theta/F_y T^2 d$, was calculated using a yield stress value $F_y=0.55$ kNmm^{-2}. This was obtained from tensile tests carried out on material specimens taken from the same batch of material as that used to fabricate the joints tested and reported here.

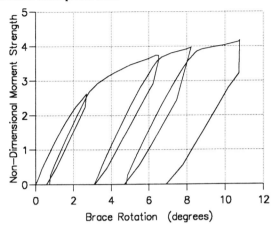

Figure 2. Load-displacement curves for the four runs for joint 'AB'

4 Numerical modelling

The entire joint was modelled since there are no planes of symmetry for a Y joint subjected to OPB loading. The FE model was dimensionally identical to the test joints. The self-reacting cross braces were not modelled. Instead, the chord was modelled up to the centre of the cross bracing, and thick end plates were inserted and suitably restrained to simulate the end condition. Brace end stiffening in the test joints was also modelled by including thick end plates.

Analyses were performed using the general purpose FE package ABAQUS(1989). A convergence study was performed and the most cost-effective mesh, shown in Figure 3, was then selected for further studies. Full details are given by Dexter(1992). The test conditions were simulated as faithfully as possible; the material stress-strain curve specified being a piecewise linear approximation of the true stress-strain curve obtained from tensile testing, the loading being pure out-of-plane bending applied to the brace end. FE results were compared with the results obtained from experimental tests.

5 Results and comparisons

Load-displacement results for the two test joints and the FE model are given in Figure 4. For the test results, only the portions of the curve rejoining the loading path are plotted. It is seen that the FE result falls between the two test results. Since there is no peak load, the ultimate load is taken at a deformation limit similar to that defined by Yura et al.(1980). The loads at the brace deflection angle $\phi=10°$ are shown in Figure 4.

The strains and displacements measured at specific points on the chords of the test joints were compared with those calculated and the correlations were generally poor. This was considered to be

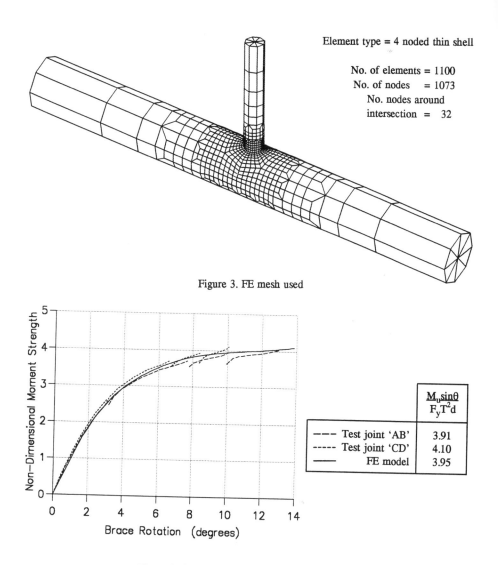

Figure 3. FE mesh used

Figure 4. Comparison of load-displacement curves

due to differences between the local stiffness and global boundary conditions of the experimental joints and the FE model. The FE model was constructed using shell elements and did not include the weld. The predicted strain gradients at the intersection would therefore differ from those of the experimental joint. In addition, residual stress effects were not included in model predictions (Dexter(1992)). The global boundary conditions of the experimental joints differed from those applied to the FE model, which would influence the predicted chord torsional response to OPB load. Finally, the effects of rigid body movement and self-weight displacements on the experimental results were not considered.

The failure modes of the test joints and FE model were also compared. Figure 5. shows that there is very good agreement between the radial deformations of the test joints (at the end of the third run) and the corresponding numerical deformations. The slightly higher calculated radial deformations are due to the omission of the intersection weld in the shell model.

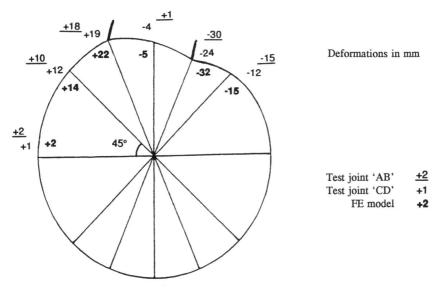

Figure 5. Radial deformation of chord at intersection

The ultimate load results are compared in Table 2. with the predictions of the API allowable capacity formula. This formula had been seen as unlikely to severely over-predict joint capacity. However, all the results presented have been over-predicted by at least 15%. A possible reason for this over-prediction is the high α ratio ($\alpha=17$), of the joints tested. The joints in the limited data-base, from which the API formula was derived, probably had much smaller α ratios (typically, $\alpha=8$), and therefore end effects may have enhanced the joint capacities. It would be expected, however, that this effect be weaker for joints with lower β ratios.

Table 2. Comparison of results with API capacity formula

	$\dfrac{M_u \sin\theta}{F_y T^2 d}$	measured strength API
FE model	3.954	0.82
Test 'AB'	3.912	0.81
Test 'CD'	4.102	0.85

6 Conclusions

The current work has successfully correlated experimental and numerical results, of the ultimate capacity of tubular T/Y joints subjected to OPB loading. It should be noted that such close numerical simulation of the tests was only possible due to the availability of all test information and result data. An accurate FE strategy has been established and validated. Further ultimate capacity investigations into the behaviour of joints of more complex geometries and/or loading conditions may now be confidently pursued.

However, in spite of the good correlation, both test and FE results have been severely over-predicted by the API allowable capacity formula. The 'α effect' has been suggested as one reason for this, and investigations to quantify this and other possible effects are currently underway.

7 Acknowledgements

The first author acknowledges the financial support of the Science and Engineering Research Council and British Gas, whilst carrying out the work reported in this paper. The authors wish to thank British Gas for permission to publish this paper, and acknowledge technical contributions from colleagues at the Engineering Research Station.

8 Nomenclature

d	joint brace diameter	T	chord wall thickness
D	joint chord diameter	α	chord length parameter = 2L/D
F_y	material yield stress	β	diameter ratio = d/D
L	joint chord length	γ	chord thinness ratio = D/2T
M	moment load	θ	in-plane brace angle
M_u	ultimate moment load	τ	wall thickness ratio = t/T
t	brace wall thickness	φ	out-of-plane brace deflection angle

9 References

ABAQUS (1989) **ABAQUS Users' Manual**, Hibbitt, Karlsson and Sorensen, Inc.

API, American Petroleum Institute (1991) **Recommended practice for planning, designing and constructing fixed offshore platforms**, API RP2A, 19th edition.

DEn, Department of Energy (1990) **Background to new static strength guidance for tubular joints in steel offshore structures**, HMSO, London.

Dexter, E.M. (1992) **Ultimate strength of tubular T/Y joints in offshore platforms**, M.Sc. Thesis, University College of Swansea.

Haswell, J.V. (1992) The safe operation of offshore jacket structures containing fatigue cracks. **Int. Conf. on Offshore Mechanics and Arctic Engineering.**

Ma, S.Y.A. (1988) A test programme on the static ultimate strength of weld fabricated tubular joints. **Offshore Technology Journal.**

Stol, H.G.A., Bijlaard, F.S.K., Puthli, R.S. (1985) **Strength and stiffness of tubular joints. Static strength of welded tubular T joints under combined loading.** Sections I,II and III. The Institute TNO for building materials and building structures, Delft.

Yura, J.A., Zettlemoyer, N., Edwards, I.F. (1980) Ultimate capacity equations for tubular joints. **Offshore Technology Conference. pp. 113-126**

67 THE INFLUENCE OF BRACE ANGLE AND INTERSECTION LENGTH ON TUBULAR JOINT CAPACITY

H.M. BOLT
Billington Osborne-Moss Engineering Ltd, Maidenhead, UK
P. CROCKETT
The University of Nottingham, UK

Abstract
Static capacity equations for tubular joints under axial loading include a $\sin\theta$ term to account for the inclination of the brace to the chord. In the UK Health and Safety Executive Guidance Notes, an additional factor K_a allows for the increase in intersection length with brace angle. For a 45° Y joint, this results in a further 20% increase in the characteristic capacity over the otherwise identical T joint.

However, recent ABAQUS finite element analyses by Billington Osborne-Moss Engineering Limited, conducted over a wide parameter range, have indicated that intersection length does not have a significant additional influence on capacity. Furthermore, a re-evaluation of experimental results has not provided support for the use of the K_a factor. The paper presents the background to the original adoption of K_a and the reasons for its rejection based on the results of the numerical analyses for a range of T, Y, X and DT joints and a review of the experimental database. The paper closes with a review of the implications for current practice and illustrates the findings with a recent reassessment of a critical node for a North Sea Operator.
Keywords: Tubular Joints, Static Strength, Design, Finite Element Analysis, Offshore Structures

1 Introduction

The capacity of simple tubular joints (P) is generally expressed in non-dimensionalised form as:

$$\frac{P \sin\theta}{F_y T^2} = f \text{ (geometric parameters, eg. } \beta \text{ etc)} \qquad (1)$$

where F_y = chord yield stress
$\quad\quad\quad T$ = chord wall thickness
$\quad\quad\quad \theta$ = angle between brace and chord
$\quad\quad\quad \beta$ = brace to chord diameter ratio.

Tubular Structures V. Edited by M.G. Coutie and G. Davies.
© 1993 E & FN Spon, 2–6 Boundary Row, London SE1 8HN. ISBN 0 419 18770 7.

Different parametric expressions are given in design codes depending on the availability, screening and interpretation of data. Codes and guidance used for the design of offshore structures co-exist with significant differences. For example, API RP2A (1991) gives a lower bound expression for T and Y joints as

$$3.4 + 19\beta \tag{2}$$

whereas the Health and Safety Executive (HSE) Guidance Notes (1990a) use characteristic formulae which, for compression loaded T or Y joints, take the form:

$$(2 + 20\beta) \sqrt{Q_\beta} \, K_a \tag{3}$$

where Q_β is a function of β to allow for enhanced capacities at high β (>0.6) and K_a is a function of θ allowing for the greater intersection length associated with an inclined brace. The expressions are evaluated for a single brace joint with β ratio of 0.6 ($Q_\beta=1.0$) in Table 1. Notwithstanding differences between the API and HSE guidance in relation to β for the base T joint case, it can be concluded that for certain joint geometries, K_a can be an important factor.

Table 1. Influence of K_a on allowable capacities

Brace angle θ (degrees)	API Lower bound Eqn (2)	HSE Lower characteristic Eqn (3)
90	14.8	14.0
45	14.8	16.9
30	14.8	21.0

The expression K_a allows for the increase in the length of the brace chord intersection as θ reduces. The exact expression K_a' is a complex function of β and θ and is given by:

$$K_a' = x + y + 3(x^2 + y^2)^{0.5} \tag{4}$$

where $x = \dfrac{1}{2\pi\sin\theta}$ and $y = \dfrac{(3-\beta^2)}{3\pi(2-\beta^2)}$

The expression is given in AWS D1.1 (1990) together with a slightly conservative approximation, K_a:

$$K_a = \frac{1 + 1/\sin\theta}{2} \tag{5}$$

It is this form which is adopted in the HSE Guidance Notes (1990a).

The basis for the introduction of K_a is given in the background to the HSE Guidance Notes (1990b). The justification for this influence of brace angle was developed through the following steps:

(a) K_a assumed to be valid.
(b) Mean equations to fit T/Y and K/YT data derived incorporating K_a.
(c) Applicability of formulae specific to Y and K joints investigated.
(d) Having justified K_a for Y and K joints, and in the absence of X joint data (only DT), K_a assumed to be equally valid for X joints.
(e) Recent X joint data from JISSP (Joint Industry Static Strength Project) (1989a & b) used to show that K_a improves their correlation with original DT joint database.

The problem at the key step (c) was that only three of the 42 single brace compression joints had inclined braces and these (and all the K joints considered) had 45° brace angles. Ideally the validation of K_a would have been based on results over a range of θ with all other parameters remaining constant. However, the available data were a constraint to the scope of the investigation and it could only be concluded that K_a 'appeared to be justified'.

The use of experimental results in this way is inevitably problematic and incomplete and the investigators had to interpret the available data. However, five years or more on, parallel analyses using the finite element method can make a significant and cost-effective contribution in isolating and quantifying the influence of individual parameters. Such an analytical investigation into the validity of K_a is presented below.

2 Finite element analyses

A series of compression Y joint analyses were conducted with intersection angles of 46.1° and 90° for various brace sizes giving β values of 0.4, 0.667 and 0.9. The parameters may at first sight appear to be a curious selection. However, as will be shown in Section 5, the analyses were performed as part of an investigation into the ultimate response of a complex multiplanar node in an offshore jacket structure. Mesh convergence studies were conducted and the Y joint in Figure 1 is representative of the modelling finally adopted.

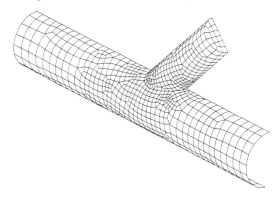

Fig.1. Typical FE mesh adopted - shown for the 46.1° Y joint

The joint models were established using the SDRC-IDEAS modelling package (1990) and analysed on a Sun Sparc II workstation using ABAQUS (Hibbit et al, 1991). Eight noded thin shell elements with 7 layers and a 2x2 integration scheme were utilised throughout. In all analyses modelling of the welds was neglected, the chord length was six times the diameter ($\alpha = 12$) and the ends of the chord were fixed (again to relate to the principal investigation). Although the end fixity might influence the absolute capacities calculated, the relative T and Y capacities considered here should be independent. Table 2 compares the various results and it is clear that the inclusion of K_a causes the results to diverge.

Table 2. Comparison of joint capacities (P_Y, P_T) with and without account of K_a

Geometry	Ratio of capacities $P_Y Sin\theta$ to P_T	Ratio of capacities $P_Y Sin\theta / K_a$ to P_T
$\beta = 0.4$	1.044	0.874
$\beta = 0.667$	0.983	0.823
$\beta = 0.9$	0.903	0.756
Average	0.977	0.818

The results are presented in Figure 2 superimposed on the database used in developing the HSE guidance. The mean line fit to the data is also shown and it can be seen that the T joint results are in good agreement. However, inclusion of K_a in the nondimensionalisation, displaces the Y joint results downwards.

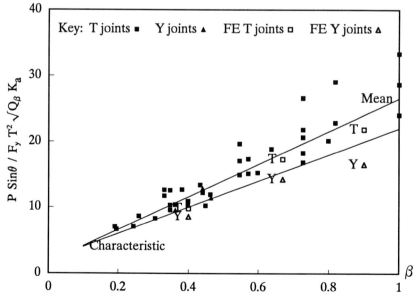

Fig.2. FE results for T and Y joints in comparison with HSE dataset (1990b)

Further analysis of X and DT joints corresponding to the middle case with $\beta=0.667$, showed that the normal component of the peak X joint load and the DT joint capacity were in good agreement with $P_X Sin\theta/P_{DT} = 1.067$, whereas inclusion of the K_a term reduced the ratio to 0.893. Figure 3 shows these results graphically, suggesting that without K_a better correlation of the X joint result with the DT joint dataset would be achieved.

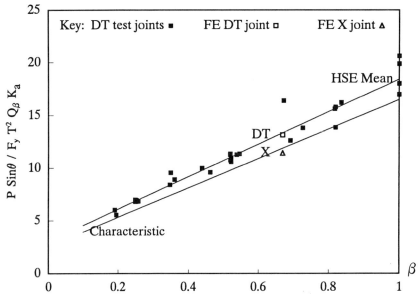

Fig.3. FE results for DT and X joints in comparison with HSE dataset (1990b)

3 Re-review of experimental data

Having cast doubt on the applicability of K_a through these analyses, it is necessary to re-review the experimental data. For the Y joint data in compression in the HSE Guidance background document (1990b), the measured and predicted capacities are in the ratio 0.86, 1.02 and 0.88, and for the two data points in the tension dataset the ratios are both 0.82. The trend is for these to be low values, below unity; without K_a they would increase. In themselves these results cannot give conclusive support for the analytical findings reported above, because of the number of parameters which vary between specimens.

The more recent JISSP programme (1989a & b) was reported after the HSE database was compiled and offers additional data, particulary since nominally identical joints were tested with different brace angles. Reference has already been made to the JISSP X joint data in Section 1, for which $\beta=1.0$ joints with brace angles of 60° and 75° were tested but no base 90° data were generated. The non-dimensionalised results were all above the scatter of the main database and the few percent reduction due to K_a was therefore beneficial. However, it is possible that other factors were at play (perhaps associated with the $\beta=1.0$ geometries) so it is perhaps unwise to draw conclusions with respect to K_a particularly in the absence of reference DT joint data. Zettlemoyer (1988) has also made this point.

However, within the JISSP programme, a series of compression T and Y joints were also tested at $\beta=0.8$ and 1.0. The companion results are taken from Reference 6 and compared in Table 3 and Figure 4. As for the analytical results, it would appear that K_a causes the T and Y results to diverge, whereas including only $Sin\theta$ they are well correlated.

Table 3. Comparison of JISSP Y and T joint capacities with and without K_a

Geometry	Ratio of capacities $P_Y Sin\theta$ to P_T	Ratio of capacities $P_Y Sin\theta/K_a$ to P_T
$\beta=0.8,\ \gamma\approx20$	1.043	0.864
$\beta=0.8,\ \gamma=31.75$	1.017	0.843
$\beta=1.0,\ \gamma=31.75$	1.028	0.852
Average	1.029	0.853

Since the JISSP work was implemented, it has been recognised that the absolute capacity of the T and Y joints was enhanced by the short chord in the specimens. However, there is no reason to conclude that the use of the data in a relative sense is not valid.

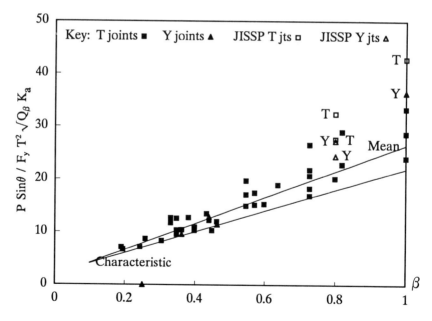

Fig.4. T & Y joint comparisons from HSE dataset (1990b) and JISSP (1989a & b)

4 Implications for practice

Together the numerical results and the review of experimental data presented here, indicate that K_a is not valid for Y and X joints, providing further support to the views presented by Zettlemoyer in 1988. It should be emphasised however that full validation of design equations is dependent on the content of the accepted database and it should be noted that the HSE are funding an expansion and rescreening of the underlying database.

However, this work is ongoing and if K_a is not valid, the implications for design and assessment using the current HSE Guidance Notes need to be considered. The term appears in the numerator of the expression for allowable capacity and takes values greater than unity for angles less than 90°. For Y and X joints its inclusion is therefore unconservative.

5 Relevance to benchmarking

Although the analyses are in themselves instructive, it is relevant to record their motivation. BOMEL had been commissioned by Amoco (UK) Limited to perform a nonlinear collapse analysis of a complex multiplanar node in the NW Hutton jacket structure (Ward et al, 1992). As part of that work, the basic modelling was to be verified against established data and, typically for contractors in the UK, reference was made to the the background data and mean equations underlying the HSE Guidance (1990a and b).

The first analysis performed was for the critical $\beta = 0.667$ intersection with the chord as a Y joint. The poor correlation with the database was shown in Figure 2. Despite extensive checking, convergence studies etc, the results for the X component of the full joint was also some 20% below the mean prediction. Comparison was made with earlier T joint analyses undertaken by van der Valk (1988) which correlated well with the database. To resolve the discrepancy the T joint was modelled and it was as a progression from that analysis that the influence of K_a was identified.

This experience demonstrates the importance of accurate and comprehensive data for validation, if analyses of more complex geometries are to be performed with confidence.

6 Conclusions

The principal conclusion from both the analysis and the review of the database presented in this paper is that the adoption of K_a in design equations for the static strength of Y and X joints under axial loading, is not valid and furthermore is unconservative.

The use of K_a was implicit in the development of T/Y and K/YT equations in the HSE Guidance Notes and it is therefore not appropriate simply to replace K_a with unity. If K_a is not included, other coefficients will change (particularly influencing the final K joint equations). A complete review is therefore required.

Good empirical data are vital for validating complex joint analyses but caution is required to ensure that incomplete knowledge is not misleading.

Acknowledgement

The permission of Amoco (UK) Exploration Company to publish this work arising from an investigation undertaken directly for the Company is gratefully acknowledged. Work was undertaken by Peter Crockett at Billington Osborne-Moss Engineering Limited (BOMEL) under an SERC-CASE award for which BOMEL is the collaborating authority. The support of the Science and Engineering Research Council is acknowledged.

References

American Petroleum Institute. (1991) Recommended practice for planning, designing and constructing fixed offshore platforms. API RP2A, 19th Edition.

American Welding Society. (1990) Structural welding code. AWS D1.1-90.

Health and Safety Executive (formerly Department of Energy). (1989a) Static strength of large scale tubular joints - Engineering Assessment. Offshore Technology Report OTH 89 297, HMSO.

Health and Safety Executive (formerly Department of Energy). (1989b) Static strength of large scale tubular joints - Test programme and results. Offshore Technology Information OTI 89 543, HMSO.

Health and Safety Executive (formerly Department of Energy). (1990a) Offshore installations: Guidance on design and construction. 4th Edition, HMSO.

Health and Safety Executive (formerly Department of Energy). (1990b) Background to guidance on static strength of tubular joints in steel offshore structures. Offshore Technology Report, OTH 89 308, HMSO.

Hibbit, Karlsson and Sorenson. (1991) ABAQUS user manual. Version 4.9.

SDRC. (1990) IDEAS-Finite element modelling user's guide.

Van der Valk, C.A.C. (1988) Factors controlling the static strength of tubular T joints. Conference on Behaviour of Offshore Structures, BOSS'88.

Ward, J.K., Billington, C.J. and Smith, J.K. (1992) Strength assessment of a multiplanar joint in a North Sea jacket structure. Conference of the International Society of Offshore and Polar Engineering, San Francisco.

Zettlemoyer, N. (1988) Developments in ultimate strength technology for simple tubular joints. Offshore Tubular Joints Conference, CIRIA-UEG, Surrey.

68 EXPERIMENTAL AND FINITE ELEMENT STUDIES INTO THE ULTIMATE BEHAVIOUR OF CRACKED AND UNCRACKED RECTANGULAR HOLLOW SECTION FILLET WELDED K-JOINTS WITH GAP

F.M. BURDEKIN and S. AL LAHAM
Structural Assessment Group, Department of
Civil & Structural Engineering, UMIST, UK

Abstract
The work described in this paper investigates the ultimate behaviour of uncracked and cracked rectangular hollow section (RHS) fillet welded K-joints with gap. The non-linear elastic-plastic finite element (FE) technique has been successfully employed to calculate the plastic collapse loads of uncracked and cracked K-joints with gap, under balanced axial loads in the bracings. An experimental programme, conducted at room and low temperatures has confirmed and agreed well with the FE analyses findings.
Keywords: Ultimate Strength, Cracked, Uncracked, RHS, Fracture.

1 Introduction and Background

In offshore tubular joint construction, the normal practice is for tubular joints to be specified as full penetration butt welds. However, fillet welded construction of circular or RHS sections is generally used for thin walled members in roof trusses and cranes. Attention has been focused in recent years on the use of finite element analysis to estimate collapse loads of tubular connections. A programme of work has been in progress at UMIST to determine ultimate strength behaviour of both CHS and RHS members, both in the uncracked and cracked conditions. Burdekin and Frodin (1987) reported three dimensional elastic plastic finite element analyses of three circular tubular double T-joint geometries, without and with cracks, loaded in axial tension. In each case, ABAQUS shell elements were used in a relatively coarse mesh. They found that through thickness cracks of up to 25% of the chord intersection length had relatively little effect on the load deflection behaviour and joint stiffness. Cheaitani (1991) has investigated ultimate strength of circular member 45 K-joints under axial loading for three different β ratios, for uncracked and cracked conditions and has suggested that the effect of cracks on the ultimate plastic collapse strength can be estimated by reduction factors depending on the cracked area and geometry applied to lower bound estimates of uncracked strength. This has been used as a basis for defect assessment in tubular joints using BSI Document PD6493, Burdekin et al (1992).

In rectangular hollow section joints, the design load capacities are usually calculated by semi-empirical formulae. These formulae are based on ultimate load results from tests carried out on isolated joints, Wardenier (1982). Very little information is available regarding the use of the finite element technique as a tool to determine ultimate strength behaviour of RHS joints. In fact, research into this area seems to have started only about five years ago, Roodbaraky (1988) and Koskimaki et al (1989).

In this paper, non-linear elastic-plastic finite element analysis has been

Tubular Structures V. Edited by M.G. Coutie and G. Davies.
© 1993 E & FN Spon, 2–6 Boundary Row, London SE1 8HN. ISBN 0 419 18770 7.

employed to study the ultimate behaviour of RHS fillet welded K-joints with gap for the cases of uncracked and cracked geometries. The effect of the important β and τ ratios on the strength of the joint has also been investigated. A sensitivity study of the effect of weld size on the ultimate behaviour of the joint has been carried out. The theoretical results were compared with results obtained using existing formulae and verified by an experimental programme.

2 FE Modelling, Material, Loading and Boundary Conditions

In the present work, higher order 20-noded brick elements were used for elastic-plastic analysis of the K-joints to model the stiffness of the joint accurately. FE meshes were designed in such a manner that the smallest elements were used in regions of high stress gradients, with gradually increasing element sizes further away from the high stress gradient regions.

Fig. 1 shows a typical basic mesh used in the present analyses, which was modified depending upon the joint geometries. An average of 780 20-noded brick elements was used with an average number of degrees of freedom of 13600. Collapsed 20-noded brick elements with mid-side nodes kept at their original positions were used to model the weld elements. The unpenetrated part between chord surface and brace straight-cut ends, which is inherent in the use of fillet welds, was also modelled. All meshes were generated using PATRAN. Efficient analysis was achieved by means of mesh optimization using PATRAN, prior to each finite element analysis using ABAQUS. The elastic-plastic analysis for some typical meshes was preceded by a linear-elastic one in order to check the performance of the mesh, loading and boundary conditions. A typical complete run required memory size of 10 Mbytes. The average CPU time used for a complete elastic-plastic analysis was 4400 seconds. The material model used here gives the yield and post yield stresses as a function of plastic strain. At the time of analysis test specimens from the experimental programme were not ready and the stress-strain curve which was used by Burdekin and Frodin (1987) has been adopted in this piece of work, see Fig. 2. The yield stress of the material used in their analysis is 8% lower than the actual yield stress of the material tested later in this programme. Therefore it was decided to scale up the FE results for ultimate strength comparisons with experiments. For RHS joints large displacements can occur which are usually associated with membrane forces in the chord face. Since equilibrium conditions between internal and external forces have to be satisfied, geometrical non-linearity had to be modelled, using the NLGEOM parameter in ABAQUS.

As far as the boundary conditions are concerned, all nodes of both ends of the chord member were restrained from movement in the X-, Y- and Z-directions (i.e. fully fixed ends). Because of the advantage of symmetry only half the connection was modelled, and therefore all nodes in the X-Y plane (i.e. Z=0 plane) were restrained in the Z-direction. Nodes at the free ends of the bracing members have not been restrained (i.e. free ends). These boundary conditions correspond to the test rig conditions used later in the laboratory tests in this research. Both braces were loaded at the free ends. Uniformly distributed axial pressure was applied stepwise up to the target of the expected load capacity of the joint.

In order to investigate the effect of cracks on the ultimate behaviour of RHS joints, surface cracks with a semi-elliptical shape were introduced in the gap region at the weld toe of the tension brace. Two layers of brick elements were used under the intersection area between chord and bracing members. The first layer from the chord surface was given the shape of a semi-elliptical crack. The surface crack was introduced at the weld toe position by releasing some of the nodes of neighbouring elements at the crack position, and allowing these elements to share the same nodes along one edge only which formed the crack front.

3 Results of the FE Analysis (Load-Displacement Curves)

A total of 16 different joint geometries, as given in table 1, was analysed. These joint dimensions were chosen in such a manner that the effect of both β and τ ratios on the ultimate behaviour of the joints could be studied. The angle between braces and chord was taken as $45°$ for all the geometries analysed. The gap (g') between the weld toes in the gap region was also fixed at 20.0 mm. These 16 different joint geometries were later analysed by introducing two different semi-elliptical surface cracks in the gap region at the weld toe of the tension brace. The dimensions of those semi-elliptical cracks were 3.2x18 and 4x24 mm, referred to as C1 and C2. This resulted in a reduction in both punching shear area and the cracked side chord wall area of the range of 3-10% and 6-37.7% respectively over the range of geometries considered. The displacements were defined as the average axial displacements of the braces with respect to the chord original axis (i.e. before loading). Fig. 3 shows the load-displacement curves of one particular joint geometry for all the cases analysed, ie. uncracked and two cracked geometries. For this particular case, there is a reduction of about 6% in the predicted ultimate strength of the joint when introducing a surface crack having a crack depth to chord thickness ratio of $a/t_o = 0.51$ and a crack length to brace width ratio of $2c/b_2 = 0.36$. This reduction, however, becomes about 9% when those ratios are increased to $a/t_o = 0.63$ and $2c/b_2 = 0.48$. This means a reduction in strength of 3% is observed as a result of an increase in those two ratios of 24% and 33% leading to an increase in area reduction for both punching shear and the cracked side chord wall of 3.1% and 9.58% respectively.

The finite element predictions of load-displacement curves for the joints analysed have been plotted. Fig. 4a shows typical load displacement curves of the tension brace, while Fig. 4b is of the compression brace. Each figure represents the results of a particular β ratio analysed, for different τ ratios, in the cases of uncracked and cracked joints. In all the cases analysed the load-deflection curves have reached a plateau prior to failure, indicating that there is sufficient strain capacity to allow for yielding to take place, which made the identification of the failure load relatively easy. It was generally found that in most cases the deformation of the compression brace is higher than that of the tension brace. It can be seen from these curves that the cracked and uncracked cases remain similar particularly in the linear elastic part of the curves, indicating very little effect of the introduction of surface cracks in the gap region. The ultimate loads reached in the analyses are summarised in table 2 where it can be seen that the finite element analysis for all the models predicted very little effect of cracks on the load deflection behaviour and joint stiffness. In fact, a maximum reduction in the joint collapse strength of 20% was predicted for the worst case analysed (eg. model FK255 with the larger crack size). For a fixed brace size increasing β ratio increases the collapse strength, whilst increasing τ ratio, keeping β constant leads to a decrease in the collapse strength. Alternatively, increasing the chord thickness strengthens, whilst increasing the chord width weakens as would be expected from bending flexibility of the chord face.

A non-dimensional failure load was used to check the efficiency of the joints. The joint efficiency here is defined by the parameter ($N_{1,2} / b_o^{0.5} t_o^{1.5} \sigma_{eo}$). It should be noted here that Wardenier and Stark (1978) used this parameter and concluded that it was in best agreement with the test results and gave the lowest

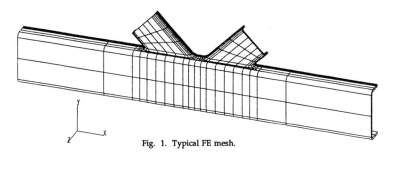

Fig. 1. Typical FE mesh.

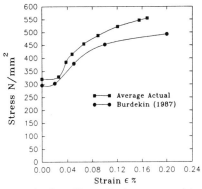

Fig. 2. True Stress–Strain curve used in ABAQUS

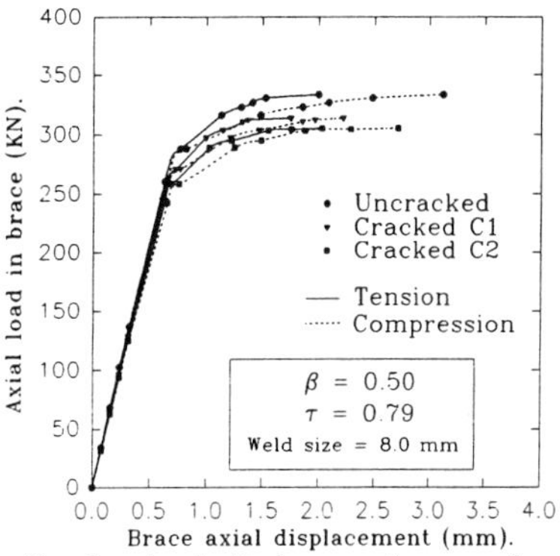

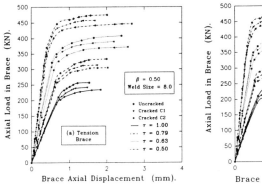

Fig. 3. Load–displacement curves for a particular geometry.

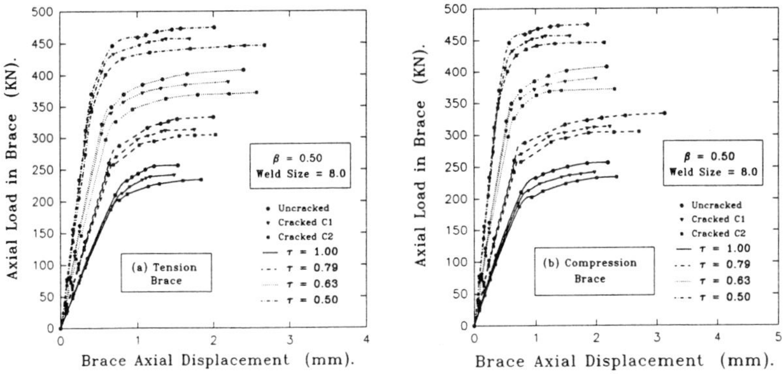

Fig. 4. Load–displacement curves, FE results.

Table 1. Joints dimensions.

Weld size = 8 (mm)		θ = 45°	g' = 20 (mm)	
Joint Ref	Brace $h_{1,2}xb_{1,2}xt_{1,2}$ (mm)	Chord $h_oxb_oxt_o$ (mm)	β	τ
FK125	100x50x5	120x80x5.0	0.63	1.00
FK126	100x50x5	120x80x6.3	0.63	0.79
FK128	100x50x5	120x80x8.0	0.63	0.63
FK121	100x50x5	120x80x10.0	0.63	0.50
FK155	100x50x5	150x100x5.0	0.50	1.00
FK156	100x50x5	150x100x6.3	0.50	0.79
FK158	100x50x5	150x100x8.0	0.50	0.63
FK151	100x50x5	150x100x10.0	0.50	0.50
FK205	100x50x5	200x125x5.0	0.40	1.00
FK206	100x50x5	200x125x6.3	0.40	0.79
FK208	100x50x5	200x125x8.0	0.40	0.63
FK201	100x50x5	200x125x10.0	0.40	0.50
FK255	100x50x5	250x150x5.0	0.33	1.00
FK256	100x50x5	250x150x6.3	0.33	0.79
FK258	100x50x5	250x150x8.0	0.33	0.63
FK251	100x50x5	250x150x10.0	0.33	0.50

Table 2. FE results.

Weld size = 8 (mm)		θ = 45°	g' = 20 (mm)		
Joint Ref	Non Cracked (kN)	Cracked C1 3.2*18 (mm) (kN)	Ratio $\frac{Cracked\ C1}{Uncracked}$	Cracked C2 4*24 (mm) (kN)	Ratio $\frac{Cracked\ C2}{Uncracked}$
FK125	301.79	282.70	0.937	273.75	0.907
FK126	351.52	335.25	0.954	328.13	0.933
FK128	413.53	389.80	0.943	380.43	0.920
FK121	496.73	480.67	0.968	470.79	0.948
FK155	257.01	242.21	0.942	234.34	0.912
FK156	333.26	313.42	0.940	305.07	0.915
FK158	400.00	381.41	0.954	364.69	0.912
FK151	474.83	457.10	0.963	446.25	0.940
FK205	204.96	187.84	0.916	168.82	0.824
FK206	292.17	275.31	0.942	259.94	0.890
FK208	378.15	359.70	0.951	337.33	0.892
FK201	456.52	442.09	0.968	427.00	0.935
FK255	139.23	123.29	0.885	111.08	0.798
FK256	225.21	209.88	0.932	194.19	0.862
FK258	301.60	288.25	0.956	276.05	0.915
FK251	410.86	396.45	0.965	383.30	0.933

Reduction of both Punching shear area and chord wall area due to the presence of cracks

JOINT REF	Punching shear area reduction % (Cracked C1)	Chord wall area reduction % (Cracked C1)	Punching shear area reduction % (Cracked C2)	Chord wall area reduction % (Cracked C2)
FK125	6.03	22.62	10.05	37.70
FK126	4.79	17.95	7.980	29.92
FK128	3.77	14.14	6.280	23.56
FK121	3.02	11.31	5.030	18.85
FK155	6.03	18.10	10.05	30.16
FK156	4.79	14.36	7.980	23.94
FK158	3.77	11.31	6.280	18.85
FK151	3.02	9.050	5.030	15.08
FK205	6.03	14.48	10.05	24.13
FK206	4.79	11.49	7.980	19.15
FK208	3.77	9.050	6.280	15.08
FK201	3.02	7.240	5.030	12.06
FK255	6.03	12.06	10.05	20.11
FK256	4.79	9.570	7.980	15.96
FK258	3.77	7.540	6.280	12.57
FK251	3.02	6.030	5.030	10.05

scatter. Fig. 5 shows the calculated efficiencies of the finite element analyses. Clearly, the finite element results show an increase in efficiency with an increasing β ratio for the same τ ratio. Also, for each β ratio analysed increasing τ ratio increases the efficiency. However, for the lower β ratios of (0.33 - 0.4) the finite element results show very little increase in efficiency with an increasing τ ratio. This increase becomes more pronounced for the higher β ratios of (0.5 - 0.63). It should be noted that the value of σ_{eo} used here is the same yield stress of the chord used in the finite element analyses.

4 Effect of Weld Size

In the previous analyses all the joints were modelled having a weld leg length of 8 mm (i.e. a throat thickness equal to 1.13 the wall thickness of the connected bracings) therefore the welds were not expected to be critical for failure, although high stress gradients in the welds were expected to occur due to the modelling of the unpenetrated land between members. As a sensitivity study on the effect of lower weld size, joint FK156 (see table 1) was modelled with a weld size of 5 mm. Very little work was required to modify the mesh, which was achieved by modifying the session file of PATRAN. The load deformation curve has been plotted and compared with that of the case of 8 mm fillet welded model as shown in Fig. 6. It can be seen from this figure that the joint with the lower fillet weld size has failed at much lower ultimate strength. In fact, the reduction in weld throat thickness in this case of 37%, led to about 29% reduction in strength. Fig. 7 shows the average Von Mises stress distribution in the weld throat around the intersection of the tension brace, for both cases, under the joint failure load of the 5 mm weld model. It is clearly shown that the model with the higher weld size experiences about 21% lower stresses. It should be noted that the weld metal was given the same material properties as the parent material in these analyses whereas in practice weld metal for steels often overmatches the parent properties significantly. Smaller weld sizes may be acceptable for guaranteed overmatching situations when proven by analysis or testing.

5 Experimental Programme

The test programme consisted of 16 full scale connections with different β and τ ratios. Two different low temperatures were considered namely - 20°C and - 50°C. The same rectangular brace section was used in all tests allowing only the chord dimensions to vary. All specimens were fabricated so that the angle of bracing to chord members was kept at 45°. The gap size between bracing members was also restricted to a constant value throughout the test programme, consistent with the FE analysis work. Tests were grouped depending upon the test and specimen conditions (e.g. the letter R was added to the main reference of the joints to indicate room temperature test, L20 and L50 for the low temperature tests of -20°C and -50°C respectively, H for heat-treatment, C1 and C2 for the small and large cracks respectively). All RHS members were hot finished grade 43C steel. A 600 kN, tension-compression, test rig was used to carry out the experimental programme. In order to cool the RHS specimens, thick polystyrene mats of 75 mm thickness were cut to form a cooling cabinet surrounding the specimen. Liquid Nitrogen was used in order to cool the test specimens.

To obtain mechanical properties of material used, tensile tests have been carried out at the end of the programme on specimens cut from the RHS used for the braces and chords. A total of 12 tensile coupons was made, six specimens were tested at room temperature, three were tested at -20°C and the remaining three specimens were tested at -50°C. The average engineering stress-strain curves for the material used at different temperatures are shown in Fig. 8. The calculated average yield stresses for the - 20°C and -50°C temperatures were 9% and 18% higher than the room temperature ones respectively.

6 Comparison with Theoretical Results

Figs. 9 and 10 show the load-displacement curves of the tension and compression braces, respectively, for a typical tested geometry in comparison with the FE results. Figures 11 and 12 show the load-displacement curves, of the tension and compression braces respectively, for the cracked specimens in comparison with the FE results. In the case of specimens having a high β ratio of (0.63), low overall deformations were observed with clear yielding of the compression brace and the mid-section of the chord, but no crack initiation was detected. Reasonably good agreement between the finite element results and the experimental results was observed. It is clearly shown, however, that the finite element results predict higher stiffnesses of the joints than those found experimentally. When testing joint geometries having such a high β ratio at low temperature of (-20°C) the joint failed in a similar manner to the room temperature one, but lower deformation occurred. No significant change in the ultimate strength of the joint was observed and the joint failed at a slightly higher load than that reached in the room temperature test. Testing the same geometry at (-50°C) in the as-welded condition led to brittle fracture, which is believed to have started in the weld root, but not until a load above the room temperature yield load had been reached. The most interesting observation from Figs. 9 and 10 is that when testing the same geometry at (-50°C) after heat-treating the specimen, no brittle failure occurred and the load deformation curve reached a plateau with deformation reaching almost that of the room temperature test. Again no significant change in the ultimate strength of the joint was detected.

In the case of test specimens with medium β ratio of (0.5), more than one failure mode for these joints was observed. Specimens initially failed by extensive plastic deformation of the chord face, i.e. the push-pull failure mechanism, followed by yielding of bracing members and local buckling of the chord side walls under the compression brace. In the late stages of loading, cracks have initiated at the weld toe in the chord face. When testing this geometry at (-20°C), the joint failed by a combination of failure modes but no crack initiation was detected, and there was slightly lower deformation than in the room temperature test. When testing the cracked joints having this β ratio the joints behaved in a similar manner to the uncracked one, but no cracking in the chord wall was detected. Tearing and propagation of the pre-introduced cracks did not occur. The measured ultimate strength of these joints was very slightly lower than the measured strength of the uncracked joint. This, unlike the finite element predictions, indicate that the presence of surface cracks with depth of around 64% of the chord thickness and a surface length of approximately 50% of the brace width (i.e. 30% reduction in the cracked side chord wall area) do not have a significant effect on the overall strength of the joint at room temperature, for this particular tested geometry.

For test specimens with β ratio of (0.33), the room and -20°C temperature tests showed initial failure by an extensive plastic deformation of the chord face, with lower deformation for specimens having low τ ratio, followed by yielding of the

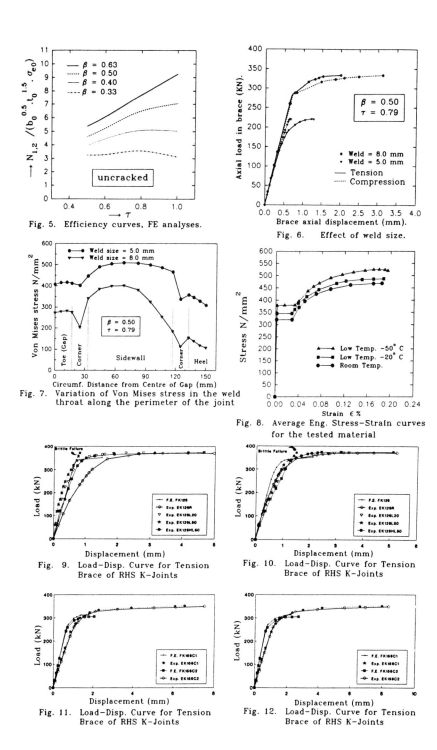

Fig. 5. Efficiency curves, FE analyses.

Fig. 6. Effect of weld size.

Fig. 7. Variation of Von Mises stress in the weld throat along the perimeter of the joint

Fig. 8. Average Eng. Stress-Strain curves for the tested material

Fig. 9. Load-Disp. Curve for Tension Brace of RHS K-Joints

Fig. 10. Load-Disp. Curve for Tension Brace of RHS K-Joints

Fig. 11. Load-Disp. Curve for Tension Brace of RHS K-Joints

Fig. 12. Load-Disp. Curve for Tension Brace of RHS K-Joints

compression brace and initiation of cracks at the weld toe in the tension brace. Slight buckling of the chord side walls under the compression brace was also observed. The deformation capacity of the joint decreased with decreasing temperature. It should be noted that at (-20°C) and after extensive deformation and initiation of cracks, joints EK256L20 and EK258L20 failed in a ductile manner, while joints EK256L50 and EK256HL50, tested at (-50°C) in the as-welded and heat-treated conditions respectively, failed in a brittle manner with very low deformation. However, no significant change in the joint strength was observed.

It is clear from table 3 that the FE results are in the range 1% higher to 12% lower than the experimental results. Also, the finite element results show stiffer overall behaviour of the joints, and give lower strengths than those measured experimentally. This is probably due to the fact that the stress-strain curve which was used in the F.E. analyses was not the actual stress-strain curve of the tested material, and due to the use of nominal thicknesses of the members in the analyses, which were generally lower than the actual measured.

7 Comparison with British Steel's design guide and CIDECT's formula

Table 4 shows the finite element results together with the experimental results, for the uncracked cases at room temperature, in comparison with results obtained using the CIDECT formula and the British Steel Guide (1989).

The British Steel guide gives allowable joint loads in the form of tables depending on the joint geometries, in accordance with BS 449. However it also provides guidance on how to modify the results according to BS 5950. The development of this guide is based on the CIDECT Monograph 6 (British Steel (1986)), and the IIW design recommendations (1981). It can be seen from table 4, that the guide does not provide results for all the geometries under investigation, whilst the CIDECT ultimate strength formula is valid for all the geometries. It is also clear that when comparing results obtained from the F.E.analyses, the CIDECT formula and the British Steel guide with the experimental results almost all the results under estimate the strength of the joints. One exception when using the CIDECT formula for joint FK258 was detected, where the formula predicted a (9%) higher ultimate load than the experimental finding. When comparing the finite element results with those obtained from the CIDECT formula and British Steel guide, it was found that the FE results tend to predict higher strengths for joints with $\tau \geq 0.79$ with an exception in cases where $\beta = 0.33$, but tend to predict lower strengths for joints with $\tau \leq 0.63$. However, reasonable agreement between results obtained from the formula, the guide and the finite element analysis was found.

It should be noted that the CIDECT formula is based on the experimental data reported by Wardenier and Stark (1978), in which the members sizes are smaller than those under investigation particularly the member thicknesses, and the type of weld used was a butt weld (i.e. lower effective width ratio β than that when using fillet welds). It should be noted, also, that the boundary conditions for these tests were different from those used in the analysis and experimental work described here.

8 Effect of Cracks on Fracture and Fatigue

The ultimate strength of RHS joints may be determined by plastic collapse, buckling or fracture. The effects of cracks, or of the unpenetrated land of fillet

Table 3. Comparison between experimental and FE results

Weld size = 8 (mm)			θ = 45°		g' = 20 (mm)		
Joint Ref.	Joint Condition	F.E. (KN)	Exp Room temp (KN)	Ratio F.E./Exp	Exp (-20°c) (KN)	Exp (-50°c) (KN)	Exp (-50°c) Heat-treated (KN)
FK126	Uncracked	351.52	370.6	0.95	371.79	371.2	374.09
FK128	Uncracked	413.53	467.3	0.88	468.3		
FK156	Uncracked	333.26	354.15	0.94	350.4		
FK156	Cracked (C1)	313.42	351	0.89			
FK156	Cracked (C2)	305.07	348	0.88			
FK256	Uncracked	225.21	257.21	0.88	258.5	262	252
FK258	Uncracked	301.60	297.3	1.01	298		

Table 4. Comparison with British Steel's design guide and CIDECT's formula

Joint Ref	(a) FE failure load (kN)	(b) Exp (kN)	FE/Exp (a)/(b)	(c) CIDECT chord failure eq. (kN)	Ratio (a)/(c)	Ratio (c)/(b)	(d) British Steel Guide (1989) (kN)	Ratio (a)/(d)	Ratio (d)/(b)
FK125	301.79			222.96	1.35		226.5	1.33	
FK126	351.52	370.6	0.95	315.34	1.11	0.85	328.32	1.07	0.89
FK128	413.53	467.3	0.88	451.24	0.92	0.97	467.64	0.88	1.0
FK121	496.73			630.63	0.79		654.48	0.76	
FK155	257.01			197.84	1.3		237.8	1.08	
FK156	333.26	354.15	0.94	279.81	1.19	0.79	310.08	1.07	0.88
FK158	400.00			440.4	0.91		416.04	0.96	
FK151	474.83			559.58	0.85		617.88	0.77	
FK205	204.96			176.95	1.16				
FK206	292.17			250.27	1.17				
FK208	378.15			358.13	1.06				
FK201	456.52			500.5	0.91				
FK255	139.23			159.92	0.87				
FK256	225.21	257.21	0.88	226.18	0.996	0.88			
FK258	301.60	297.3	1.01	323.66	0.93	1.09			
FK251	410.86			452.32	0.91				

welds, on fracture may be determined using the assessment diagram methods of BSI Document PD 6493:1991. This requires calculation of the fracture ratio parameter K_r (ratio of elastic stress intensity factor to fracture toughness) and the collapse load ratio parameter S_r or L_r (ratio of applied to collapse or yield loads). The elastic stress intensity factor can be determined from the stress concentration factor and the degree of bending for the particular joint geometry as given by Al Laham and Burdekin (1993).

For the crack geometry C2 in the present work, (4 mm deep x 24 mm long) in a 150x100x6.3 chord RHS K-joint connection under axial loading this approach can be used to show the toughness requirement to allow the joint to develop its full plastic collapse load strength. The resulting required toughness is estimated to be about 160MPa√m (or a CTOD value of about 0.4 mm) which is consistent with the observed experimental behaviour.

As far as fatigue is concerned the SCF data given by Al Laham and Burdekin (1993) can be used directly with standard design methods for fillet welded joints, taking account of thickness and weld profile effects, or alternatively a fracture mechanics approach can be adopted using stress intensity factors and the Paris Law.

9 Conclusions

The elastic-plastic finite element technique has been used in this paper in order to investigate the effect of some influencing parameters on the ultimate strength of RHS fillet welded K-joints with gap in the case of uncracked and cracked geometries. The use of three-dimensional solid elements is necessary in order to model the welds and stiffnesses of the joint accurately. Experimental tests have been carried out in order to verify the F.E. results.

The finite element results, for the uncracked cases, were in the range 1% higher to 12% lower than the experimental results. In other words the finite element results are adequately on the safe side. It was also found, in general, that the CIDECT formula and British Steel guide predict lower strength than the experimental results. The ultimate strength of the joint decreased with increasing τ ratio for the same β ratio. The finite element results showed an increase in efficiency with an increasing β ratio for the same τ ratio.

A finite element sensitivity study, (with both weld metal and parent material having the same properties), on the effect of using lower weld leg length has shown a considerable reduction in the strength of about 29% when the weld leg length is reduced from 8 mm to 5 mm for a 5 mm thick brace wall (i.e. about 37% reduction in the weld throat thickness), which led to higher stresses taking place in the welds..

When the effects of cracks on plastic collapse strength are considered, the FE results investigated cases where the punching shear area was reduced by between 3% and 10%, or the individual chord wall area was reduced by between 6% and 37.7%. The resulting reduction in strength was closer to the punching shear area reduction for the higher β ratios but tended to be closer to the chord wall area reduction for the lower β ratios. A methodology has been outlined for assessing fatigue and fracture behaviour. However, it was shown experimentally that there was an almost negligible effect of the surface crack at room temperature on the ultimate strength of the joints. More experimental work is required in order to check the effect of surface cracks on the ultimate strength of the whole range of geometries at room and low temperature.

In the experimental results, no reduction in strength was observed when testing the specimens at low temperature. In the case of low temperature testing at (-20°C) all specimens failed in a ductile manner, whilst, in the case of low temperature testing at (-50°C) most specimens failed in a brittle manner without reduction in the strength, except in one case where the heat-treated specimen failed in a ductile manner. Because of the low number of heat-treated specimens tested, no clear conclusion can be drawn, as far as the effect of heat-treatment on the mode of failure is concerned. More work is required in order to investigate this effect.

The small discrepancies between FE and experimental results are thought to be mainly due to differences in the actual material stress-strain curves for the two cases, together with small differences in the actual measured section sizes.

10 References

Al Laham, S. and Burdekin, F.M. (1993) Experimental and numerical determination of stress concentration factors of fillet welded RHS K-joints with gap. in **Proc. 5th Int. Symposium on Tubular Structures,** Nottingham.

British Steel Welded Tubes (1986) **The strength and behaviour of statically loaded connections in structural hollow sections.** CIDECT Monograph 6.

British Steel (1989) **Design of SHS welded joints.,** Corby, Northants, TD297/3E/1989.

Burdekin, F.M. and Frodin, J.G. (1987) Ultimate failure of tubular connections. **Marinetech Northwest, Cohesive programme on defect assessment DEF/4,** UMIST, Interim report.

Burdekin, F.M. Thurlbeck, S.D. and Cowling, M.J. (1992) Defect assessment in offshore structures application of BSI Document PD 6493:1991, in **OMAE Vol III part B,** pp 411-419.

Cheaitani, M.J. (1991) Ultimate failure of tubular connections. **Defect assessment in offshore structures, MWG project DA709,** UMIST, final report.

International Institute of Welding Sc.XV-E (1981) Design recommendations for hollow section joints-predominantly statically loaded. IIW Annual assembly, Oporto, Portugal, **Document XV-491-81 revised.**

Koskimaki, M. and Niemi, E. (1989) Finite element studies on the behaviour of rectangular hollow section k-joints. in **Third international symposium on tubular structures,** Lappeenranta, Finland, Pages 28-37.

Roodbaraky, K. (1988) **Finite element modelling of tubular cross joints in RHS.** PhD Thesis, Nottingham University.

Wardenier, J. (1982) **Hollow section joints.** Delft University press, Delft University of Technology, Delft, Netherlands.

Wardenier, J. and Stark, J.W.B. (1978) The static strength of welded lattice girder joints in structural hollow sections. Parts 1-10, **CIDECT final report 5Q/78/4,** Delft University of Technology, Netherlands.

69 THE BEHAVIOUR AND STATIC STRENGTH OF PLATE TO CIRCULAR COLUMN CONNECTIONS UNDER MULTIPLANAR AXIAL LOADINGS

G.D. DE WINKEL and H.D. RINK
Delft University of Technology, The Netherlands
R.S. PUTHLI
Delft University of Technology, The Netherlands and
TNO Building and Construction Research, Rijswijk,
The Netherlands
J. WARDENIER
Delft University of Technology, The Netherlands

Abstract
This paper forms part of a large research programme on "Semi-rigid connections between I-beams and tubular columns" and describes the behaviour and static strength of welded multiplanar connections between plates and circular hollow section columns under various axial loadings, based on experimental tests and numerical simulations. The influence of concrete filling of the column is also taken into account. The numerical results give acceptable agreement to the experimental work.
Keywords: Circular Hollow Sections, Welded Connection, Finite Element Analyses, Plate, Multiplanar, Semi-Rigid.

List of symbols

β	-	Flange width/column diameter ratio
ϵ_i	-	Maximum strain in the "i" material
2γ	-	Wall thickness/column diameter ratio
d_0	-	Column diameter
e	-	Eccentricity
F_j	-	Axial load on plate j (j =1, 2)
$f_{u,i}$	-	Ultimate stress in member i
$f_{y,i}$	-	Yield stress in member i
h_1	-	Beam height
l_0	-	Column length
l_1	-	Plate length
F_1	-	Axial load in the in-plane plate
F_2	-	Axial load in the out-of-plane plate
F_{max}	-	Maximum axial load in the plate
t_0	-	Wall thickness of the column
t_1	-	Wall thickness of the plate
w_1	-	Plate width

indices:

i	-	i = 0 for column, 1 for plate
CHS	-	Circular hollow section
ECSC	-	European Coal and Steel Community
FE	-	Finite element
STW	-	Stichting Technische Wetenschappen (Technology Foundation)

Tubular Structures V. Edited by M.G. Coutie and G. Davies.
© 1993 E & FN Spon, 2–6 Boundary Row, London SE1 8HN. ISBN 0 419 18770 7.

1 Introduction

This paper forms part of a large research programme on "Semi-rigid connections between I-beams and tubular columns" and describes the behaviour and static strength of welded multiplanar connections between plates and circular hollow section columns under various axial loadings, based on experimental tests and numerical simulations.
  Semi-rigid connections between I-section beams and tubular columns can be used economically for buildings and offshore structures. The lack of stiffening plates allows the fabrication of these connections in a cost effective way. The aim of the research is to investigate the behaviour of isolated parts of a connection and their influence on the strength and stiffness behaviour of the complete connections. The topic of the paper is the investigation into the flange-column connection behaviour under various axial loadings. In the numerical work, plates are therefore welded at mid-height of the column by one sided single V butt welds, to simulate the welding of an I-beam flange to the column, as used in offshore deck structures. The influence of a concrete infill in the column, as used in buildings is also considered.
  Eight experimental tests relevant to this paper are carried out, which include different combinations of loading, flange width to column diameter ratio, and columns with or without a concrete infill.
  The experimental tests are simulated by means of finite element analyses including geometrical and material non-linearity. For these analyses, actual measured dimensions and weld sizes are used. Also, the actual material properties obtained from tensile coupon tests are used.
  The results of the experimental tests, used in this paper, are carried out in the framework of ECSC research programme 7210/SA/611. The numerical work is being carried out in the framework of research programme DCT91.1904 of the Dutch Technology Foundation STW.

2 Experimental work

Eight experimental tests on plate to CHS column connections have been carried out. In these tests the ratio β between the width of the plate and the column diameter and the load ratio F2/F1 between the axial forces on the plates in the two orthogonal planes are varied. Another variation is use of columns with and without a concrete filling. The load cases and the nominal dimensions of the test specimens are summarized in Table 1. All the columns are CHS 323.9x9.5. The diameter/wall thickness ratio $2\gamma = 34$, the plate width/diameter ratio β = 0.37 and 0.52. The steel grade used for all members is FE 510D, according to EN 10025. The quality of the concrete in the reinforced concrete infill is C35/C45 and the steel grade of the reinforcement is S 500, both according to Eurocode 2. The design of the composite column is based on a fire resistance (ECCS, 1988) of 60 minutes for a 5 storey building, where the column carries a floor area of 7.2 x 7.2 metres. In Table 2 all measurements and the results of the tensile coupon tests are listed.

Table 1. Cases of the test programme considered

	Nº	F_2/F_1	concrete filling	β	d_0/t_0
	1C1	0	no	0.37	34
	1C2	0	yes	0.37	34
	1C3	0	no	0.52	34
	1C4	0	yes	0.52	34
	1C5	-1	no	0.37	34
	1C6	+1	no	0.37	34
	1C7	-1	no	0.52	34
	1C8	+1	no	0.52	34

Table 2. Measured properties of the test specimens

Nº	d_0 [mm]	t_0 [mm]	w_1 [mm]	t_1 [mm]	$f_{y,0}$ [N/mm^2]	$f_{u,0}$ [N/mm^2]	$f_{y,1}$ [N/mm^2]	$f_{u,1}$ [N/mm^2]
1C1	324.4	9.48	119.9	9.90	392	512	396	521
1C2	324.4	9.48	119.9	9.90	392	512	396	521
1C3	324.4	9.48	170.0	11.53	392	512	392	516
1C4	324.4	9.48	170.0	11.53	392	512	392	516
1C5	324.4	9.48	119.7	9.97	392	512	398	514
1C6	324.4	9.48	119.7	9.97	392	512	398	514
1C7	324.4	9.48	170.0	11.53	392	512	396	521
1C8	324.3	9.44	170.0	11.53	392	512	396	521

2.1 Test setup

The test specimens are placed in the test rig with the column in a horizontal position. For the uniplanar load cases, the load is applied vertically on the lower plate, using a hydraulic jack. The upper plate is pin supported to the reaction frame. The jack load is measured using a load cell. During the test, the column is maintained in a horizontal position by using a hydraulic jack at one end of the column. The vertical displacement is measured at both ends of the column, using displacement transducers. If a difference is measured, the column is automatically balanced into a horizontal position.

For the multiplanar load cases, the same test setup is used as for the uniplanar load case, when applying the vertical load. For the horizontal (multiplanar) load, an additional horizontal reaction frame is used, independent of the vertical reaction frame. In the horizontal frame, a hydraulic jack is used for applying the horizontal axial force. A load cell is used to measure the horizontal load. The horizontal frame is maintained in a horizontal position during testing.

The overall ovalization of the connection is measured at three locations, in the middle and at the two edges of the plates, at a distance of 25 mm from the column wall. These measurements are taken for both the horizontal and vertical plates. Additional strain gauge

measurements have been carried out on the plates.
The out-of-plane displacement of the plates is prevented at two
locations on the plates by the use of lateral supports. The locations
are at approximately 1/3 and 2/3 of the plate length. This avoids
global buckling of the plates, in case of compression.

Initially, the load is applied by force control. After first
occurrence of plastification, displacement control is applied for the
uniplanar load situations. For the multiplanar load cases, the ratio
between the load of the vertical jack and the load of the horizontal
jack is always maintained.

2.2 Results

The connection behaviour is presented in axial load - ovalization
diagrams. The axial load is measured with load cells. The ovalization
is obtained from the displacement transducer measurements. After
reaching the maximum load, the testing has been continued to obtain
information about the deformation capacity and possible failure modes.

The load-ovalization diagrams are shown in Figures 1 to 8. The test
results are summarized in Table 3.

All the connections without a concrete infill failed by yielding of
the column wall. The connection 1C2, with a concrete infill and a
tensile uniplanar load, failed by punching shear of the column wall. In
this case cracks are observed in the column wall, at the welds toes of
the tension plates in the corners of the plate. Connection 1C4 failed
by yielding of the plates under compression. The maximum load for this
test is 15% lower than the squash load of the plates. This could be due
a combination of bending and axial force. Theoretically, the
relationship between normal force and bending moment for a rectangular
crossection is $(M/Mp) + (N/Np)^2 = 1$. With substituting $Mp =$
$.25*f_y*w_1*t_1^2$, $Np = f_y*w_1*t_1$, and $M = e*N$, where e is the eccentricity of
the axial force in thickness direction, it can be derived that $e =$
$(t_1/4)*(Np/N)$. Substituting $t_1=11.53$ mm and $Np/N = 1/.85$ gives $e= 3.4$
mm. Therefore, an eccentricity of 3.4 mm gives a 15% lower ultimate
load than the squash load. The strain gauge measurements also show a
considerable amount of bending in the plates, which is unavoidable
because of the one sided single V butt welds between plate and column.

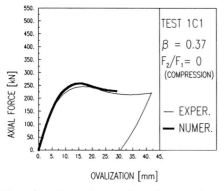

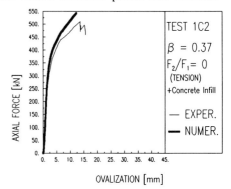

Fig. 1.   Experimental and numeri-   Fig. 2.   Experimental and numeri-
          cal load-ovalization   cal load-ovalization
          curves for 1C1   curves for 1C2

706

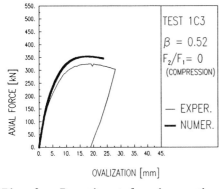

Fig. 3. Experimental and numerical load-ovalization curves for 1C3

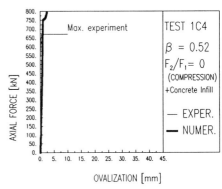

Fig. 4. Experimental and numerical load-ovalization curves for 1C4

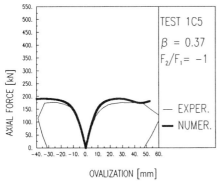

Fig. 5. Experimental and numerical load-ovalization curves for 1C5

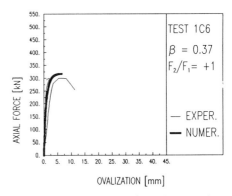

Fig. 6. Experimental and numerical load-ovalization curves for 1C6

3 Numerical work

The experimental tests are simulated by means of finite element analyses, which include both geometrical and material non-linearity. The finite element models are calibrated against the experimental results.

3.1 Method of analyses

For the FE models, eight noded thick shell elements are used, with four integration points at Gauss locations in seven layers across the thickness (a total of 28 integration points), using Simpson integration. It is shown by the authors and others, Wardenier and Puthli, (1991) and Vegte et al., (1992), that using these elements with a proper mesh refinement can give good agreement with experimental results.

The experimentally determined engineering stress-strain curves, obtained from tensile coupon tests, are translated to the true-stress - true-strain relationships, using the Ramberg-Osgood relationship as described in Background Documentation Eurocode 3, (1989).

The load is applied using displacement control for the uniplanar load cases and force control for the multiplanar load cases.

For the solution process, the updated Lagrange procedure is used, allowing large curvatures during deformations, MARC Manual, (1990).

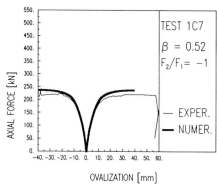

Fig. 7. Experimental and numerical load-ovalization
curves for 1C7

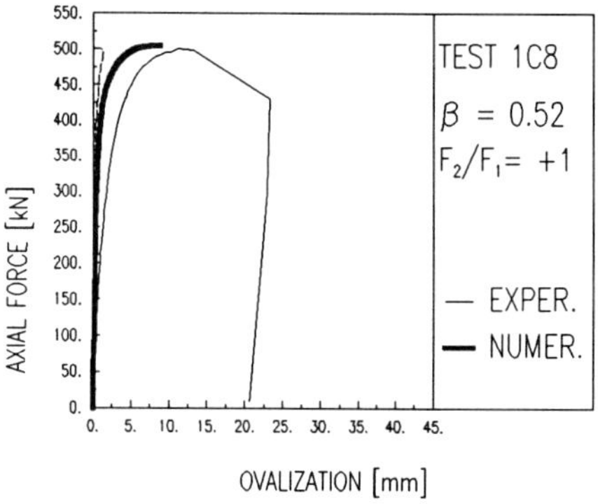

Fig. 8. Experimental and numerical load-ovalization
curves for 1C8

During testing the column wall around the connection of the test with the plate under tension load is observed to pull away from the concrete infill. The concrete infill is therefore modelled as a rigid contact surface for the tensile load cases (see Figure 9). The characteristics of the rigid contact surface are full resistance against compression and no resistance against tension.

The linear elastic deformations of the concrete infill and the adhesive bond between the concrete infill and the column are neglected with this method. However, in reality, these influences are small in comparison with the total ovalization for the tensile load case (1C2). For the compression load case (1C4) this is different. The contribution of the linear elastic deformations of the concrete infill in the total ovalization is relatively large. However, the total ovalization of the column at maximum load is very small. As can be seen in Figure 4, the stiffness is almost infinite in comparison with the stiffness of the other connections.

An advantage of this method of approach is the simple modelling of the concrete infill. Only a few commands are needed to generate the rigid surface. An other advantage is the memory requirements for the finite element analyses. A comparable finite element model where the concrete infill is modelled with 20 noded brick elements needs about four times the computer memory for calculation. Although the model with the rigid contact surface uses about three times more cpu time, than the model without a contact surface, this is still less than needed for a FE model, with the concrete infill modelled with brick elements.

The "punching shear" failure mode with cracking of the column wall is not yet implemented in the numerical models.

For the modelling, the pre- and post processor SDRC-IDEAS is used. The finite element analyses are carried out with the MARC (releases K4.2 and K5) finite element general purpose computer program. For interfacing between IDEAS and MARC the MENTAT 5.4.3 computer program is used.

3.2 Modelling
The finite element models include actual measured dimensions. The measured dimensions are averaged for each component of a test specimen. The influence of the welds is also simulated, by using shell elements. In a preliminary study, Wardenier and Puthli, (1991) the influence of weld modelling on strength and stiffness is investigated. It is shown that the method of the weld modelling has a considerable influence on both strength and stiffness of the connections.

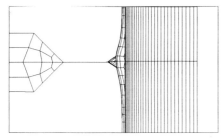

Fig. 9.   Deformed FE mesh of 1C2,
          viewed from the inside of
          the column

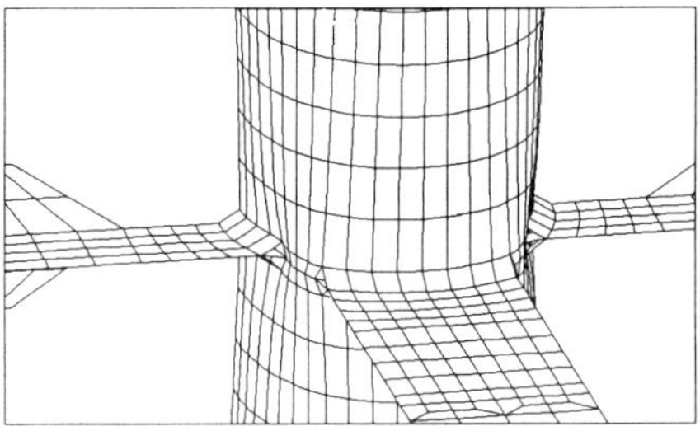

Fig. 10. Deformed FE mesh of 1C3

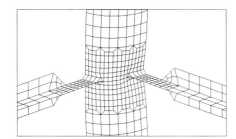
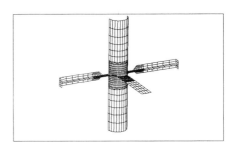

Fig. 11. Deformed FE mesh of 1C5

Fig. 12. FE Mesh of 1C7

For the modelling, symmetry planes have been used where possible. However, in some cases there is an obvious anti-symmetric failure found in the experimental tests. In such cases a half FE model is used instead of a quarter. Each side of the column is given a different wall thickness, corresponding to averaged values for each side. Two types of material properties are used, one for the column and one for the plates.

Typical FE meshes are shown in Figures 9 to 12. In Figures 1 to 8 the numerically obtained load-ovalization diagrams are shown. The numerical results are summarized in Table 3.

4 Comparison of experimental and numerical results

The agreement between the experimental and numerical results is
accepable (see Figures 1 to 8). The deformed shapes of the test
specimens and the finite element models agree well. The differences
between the results of the numerical models and the experimental tests
are quantified in Table 3. The largest difference between the
experimental and numerical results is found for test 1C4. The ultimate
load of the test specimen is .85*Np, while the FE model even exceeds
Np, with continuing loading on the FE model, due to the work hardening
behaviour of steel. In general, the numerical results are up to 8%
higher than the experimental results. The differences are acceptable.

Table 3. Comparison of experimental and numerical results

N$^{\circ}$	F_2/F_1	Concrete filling of column	β	Fmax Expt. [kN]	Fmax Num. [kN]	Num. ――― Expt.
1C1	0	no	0.37	245.3	257.9	1.05
1C2	0	yes	0.37	510.8	542.6*	1.06
1C3	0	no	0.52	325.0	354.0	1.08
1C4	0	yes	0.52	670.8	756.0*	1.12
1C5	-1	no	0.37	175.6	191.0	1.08
1C6	+1	no	0.37	300.8	317.0	1.05
1C7	-1	no	0.52	220.1	236.0	1.07
1C8	+1	no	0.52	499.9	504.2	1.00

* No maximum for the FE model, numerical maximum taken at the same
displacement as at the maximum of the experiment.

5 Concluding remarks

The static behaviour of multiplanar connections between plates and CHS
columns under combinations of axial loadings on the plates can be
simulated by means of geometrical and material non-linear finite
element analyses. The agreement between experimental and numerical
results is acceptable, however the numerical results are generally
higher than the experimental results.
   The behaviour of this type of connection when the column is filled
with concrete can be simulated with the use of a rigid contact surface,
although the elastic deformations of the concrete is neglected.
   For small wall thicknesses of the column, the punching shear failure
mode will also be critical for the compression load cases. This is due
to crushing of the concrete infill.
   At this moment, no design formulae are available for the static
strength of multiplanar plate to CHS column connections. Design
formulae for uniplanar connections are available, but cannot directly
be used for multiplanar connections, because the failure modes are

different for uniplanar and multiplanar connections. Van den Broek et al. (1990) investigated the multiplanar load effect, but insufficient data was available for design formulae. Initial proposals for design formulae for multiplanar X-joints with CHS members, based on a ring model approach have been made by Paul, (1988). This ring model approach can also be used for plate to CHS connections.

In the framework of a Ph.D. research, further parametric studies are carried out, on the basis of finite element analyses, using experimental verification.

6 References

Background Documentation Eurocode 3 (1989) **Design of Steel Structures, Part 1: General Rules and Rules for Buildings, Chapter 3: Design Against Brittle Fracture**, Commission of the European Communities, Brussels, Belgium.

Broek, T.J. van den, Puthli, R.S. and Wardenier, J. (1990) The influence of multiplanar loading on the strength and stiffness of plate to tubular column connections. **Proceedings International Conference Welded Structures '90**, The Welding Institute, Cambridge, UK.

ECCS (1988)- Technical Committee 3 - **Fire Safety of Steel Structures**, Technical Note "Calculation of the Fire Resistance of Centrally Loaded Composite Steel-Concrete Columns Exposed to the Standard Fire, ECCS, Brussels, Belgium.

MARC Manual (1990), **Volume A**, Revision K.4, Marc Analysis Research Corporation, Palo Alto, California, USA.

Paul J.C. (1988), **The Static Strength of Tubular Multiplanar Double T-Joints** (MPhil. thesis), RKER.88.076, KSEPL, Rijswijk, The Netherlands.

Vegte, G.J. van der, Puthli, R.S. and Wardenier, J. (1992) Static Strength of Uniplanar Tubular Steel X-Joints Reinforce by a Can. **International Journal of Offshore and Polar Engineering**, Vol.2, No. 1, March 1992, page 1-6, The International Society of Offshore and Polar Engineers, Golden, Colorado, USA.

Wardenier, J. and Puthli, R.S. (1991) **Semi-rigid Connections Between I-beams and High Strength Steel Tubular Columns, Technical Report 2.** Delft University of Technology, Delft.

70 ULTIMATE CAPACITY OF AXIALLY LOADED MULTIPLANAR DOUBLE K-JOINTS IN CIRCULAR HOLLOW SECTIONS

S.R. WILMSHURST and M.M.K. LEE
University College of Swansea, UK

Abstract

This paper presents data on multiplanar double K-joints subjected to axial brace loading generated using the finite element method. A parametric study has been carried out on the ultimate strength of the joint type under consideration, the main parameters investigated being: $\gamma=12$, $\beta=0.24$, 0.32 & 0.4, $\zeta_t=0.11$, 0.20 & 0.28, $\zeta_l=2$, 4, 6, 8, 10 & 12. In addition, further analyses were performed to examine to influence of the γ and τ ratios on static strength. The numerical results, together with other existing experimental data, were assessed in the light of the predictions obtained from the multiplanar formulation of the AWS code.

Keywords: Double K-joints, Finite Element Analyses, Multiplanar, Ultimate Strength, C.H.S.

1 Introduction

Tubular joints are frequently used in structures such as offshore drilling platforms, towers, masts and long span roofs. These joints are generally of multiplanar configurations: they are composed of members lying in different planes. Traditionally, the design of tubular connections follows guidelines derived from the interpretation of test results of uniplanar joints. The normal practice adopted is a plane-by-plane assessment, with no due consideration given to any possible multiplanar effects on strength.

Study of multiplanar effects on static strength of tubular joints was started in the early 1980s when a series of multiplanar double K-joints was tested in Japan (Makino et al., 1984). This was then followed by the first attempt to incorporate multiplanar effects in design guidance. The AWS Structural Welding Code D1.1(1985) proposed an ovalisation parameter, α_o, to account for the

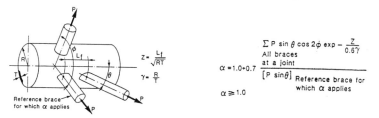

Fig.1. The α_o formulation in AWS D1.1

Tubular Structures V. Edited by M.G. Coutie and G. Davies.
© 1993 E & FN Spon, 2–6 Boundary Row, London SE1 8HN. ISBN 0 419 18770 7.

ovalising effect of loading in out-of-plane braces, Fig. 1. The formulation of the α_o parameter was based on elastic considerations (Marshall & Luyties, 1982), and it had not been calibrated against test results due to a lack of data at that time. Nevertheless, it represented a significant advance in tubular joint technology: it was a pioneering attempt in quantifying capacities of multiplanar joints and its all encompassing approach eliminated the design requirement of joint classification.

Further research on multiplanar joints worldwide thus followed, including the work of Mitri et al. (1987), Nakacho et al. (1989), Scola et al. (1990) and Paul et al. (1993) on V- (or TT) joints, the work of Paul et al. (1989) and van der Vegte et al. (1991) on multiplanar X- (or DT) joints and the work of Mouty & Rondal (1990) and Paul et al. (1992) on multiplanar double K-joints.

An assessment of the AWS formulation was carried out by Lalani and Bolt (1990), who compared AWS predicted joint capacities with test data on multiplanar V- and double K-joints. They concluded that AWS tended to over-predict strength for joints with large out-of-plane gaps and under-predict capacity for small out-of-plane gaps. They also proposed the following in-plane and out-of-plane modifiers to enhance the reliability of the AWS formulation:

in-plane $\quad Q_{gi} = 1.4 - 2\ g_i/D \geq 1.0$

out-of-plane $\quad Q_{go} = 1.4 - 2\ g_o/D \geq 1.0$

Further assessments by other researchers (Scola et al., 1990 and Paul et al., 1992a, 1993) also indicate that the AWS approach does not capture the multiplanar effects adequately.

Although much attention has been given to multiplanar double K-joints, previous work has focused mainly on geometries with high γ ratios and almost exclusively on axially loaded cases. A numerical study on this joint type was therefore initiated at University College of Swansea aimed at filling the gaps of the existing database on axially loaded joints and producing a database for moment loaded joints.

This paper presents preliminary findings of the work on axially loaded joints, and attention is directed towards examining the effect of γ and τ ratios. The results are interpreted with special reference to the AWS formulation.

2 Review of previous work

Previous research efforts on multiplanar double K-joints were exclusively experimental, and mainly concentrated on specimens with high γ ratios (higher than 17). An exception to this was the work by Mouty and Rondal (1990) in which several specimens had low γ and τ and values. A brief review of all previous work is given below.

2.1 Makino et al (1984)

A total of twenty joints was tested. The specimens had high γ and τ ($17.0<\gamma<24.7$ and $0.75<\tau<0.95$) and low β ($0.22<\beta<0.37$ for $\phi=60°$). All joints failed by plastification of the chord wall at the intersection of the compression braces. One joint failed prematurely and was omitted from the analysis of the results. Makino et al. identified two types of failure mode : type 1 in which the two compression braces acted together with no local deformation of the chord wall in the gap area between them, and type 2 in which there was significant local deformation of the chord wall in the gap area in the form of a radial fold. Type 1 failure mode occurred in joints with small transverse gaps while joints with larger transverse gaps would yield type 2 failure mode. Although the

713

transition from one failure type to another must also in some way be governed by the chord wall thickness i.e. the γ effect, this was not quantified. Using regression analysis, Makino et al. developed two equations for predicting ultimate strength, one for each failure type.

2.2 Mouty and Rondal (1990)

The report by Mouty and Rondal has been cited and assessed in some detail by Paul (1992a). The test programme of 96 specimens was divided into three groups, two of which consisted of specimens with special brace end preparations and the third contained 34 joints where the braces had normal profile cuttings. The parameter ranges for the programme were: $60°\leq\phi\leq90°$, $6.9<\gamma<18.0$, $0.2<\beta<0.5$ and $0.25<\tau<1.00$.

The test set-up was different from other work in that both chord ends were free to move and the brace ends were fully restrained. This induced high secondary moments. When the ultimate loads were non-dimensionalised using the uniplanar K-joint prediction of Kurobane (1984), it was found that the capacity ratios were lower than those determined using the test data of Makino et al. (1984 and 1992) and Paul (1992a). The capacity ratios for Mouty and Rondal varied from 0.51 to 0.81, in comparison with 0.84 to 1.09 for Makino et al. (1984, 1992) and 0.79 to 1.23 for Paul (1992a). Consequently there was concern about the validity of some of the test results (Paul, 1992b).

2.3 Makino et al. (1992)

As part of an investigation into the ultimate behaviour of axially loaded multiplanar double K-joints with diaphragm stiffeners, two symmetrically loaded non-stiffened joints were tested as a reference. The parameters used for the specimens were : $\phi=60°$, $\beta=0.45$, $\tau=1.18$, $\zeta_t=0.04$, $\theta_c=\theta_t=45°$, $\gamma=24.5$ and $\zeta_l=12.0$. The failure mode of both specimens was plastic failure of the chord at the intersection of the compression braces. Both joints had a type 1 failure which would be expected as the transverse gaps were extremely small.

2.4 Paul (1992a)

The test programme consisted of 18 multiplanar double K-joints and their geometric parameters were $\phi=60°$ & $90°$, $0.22\leq\beta\leq0.47$, $1.05\leq\tau\leq1.09$, $0.04\leq\zeta_t\leq0.52$, $\theta_c=\theta_t=45°$, $\gamma=18.3$ and $1.8\leq\zeta_l\leq14.1$. All specimens failed by plastification of the chord wall under the compression braces. Cracks formed in the chord wall or the tension brace weld when the in-plane gaps were small. Three types of chord wall failure occurred. Failure type 1 occurred when $\zeta_t\leq0.21$; failure type 2 occurred when $\zeta_t\geq0.28$ for $\phi=60°$ and in all but two specimens with $\phi=90°$; the remaining two failed in an unsymmetrical mode where plastic local deformation occurred under one of the compression braces and the two K-joints acted independently. Paul stated that this failure type occurred when a large transverse gap was combined with a small longitudinal gap, although this type was probably more likely to occur due to a combination of the above and an eccentricity in the test set-up.

3 Numerical programme at Swansea

The numerical work on axially loaded joints at Swansea was aimed at complementing research already undertaken and the emphasis was on geometries with low γ ratios. A series of fifteen joints having the following parameters was analyzed: $\gamma=12$, $\beta=0.24$, 0.32 & 0.4, $\zeta_t=0.11$, 0.20 & 0.28,

$\zeta_1=2, 4, 6, 8, 10$ & 12. The ultimate capacities of these joints were found to be very close to the AWS predictions. When non-dimensionalised by the AWS formula incorporating the small in-plane gap modifier, the range of the results was 0.96 to 1.05 with a mean of 1.01. The same ratio for Paul's results was 1.26 to 1.45 with a mean of 1.35. It is evident that the AWS formulation is very conservative at high γ ratios and requires modification to enhance its reliability. A further series of analyses was therefore carried out to investigate the effect of γ on the ultimate strength. Details of the finite element analysis matrix are summarized in Table 1.

Table 1 Details of the finite element analysis matrix

JOINT	β	g_i/D	γ	τ
DK-03	0.24	0.32	9.0	1.0
			12.0	0.8
			15.0	1.0
			21.0	1.0
DK-09	0.32	0.32	9.0	0.6
			9.0	1.0
			12.0	0.8
			12.0	1.2
			15.0	0.6
			15.0	1.0
			18.0	1.2
			21.0	1.4
DK-07	0.32	0.15	9.0	1.0
			12.0	0.8
			15.0	1.0
			21.0	1.0

4 Numerical procedure

The numerical analyses were performed using the general purpose F.E. package ABAQUS (1989). An in-depth convergence and calibration study was carried out to verify the accuracy and validity of the numerical model. As there was one plane of symmetry, the model consists of only half of a joint. A typical mesh, shown in Fig. 2(a), had about 1500 four noded quadrilateral shell elements. The calibration study indicated that, in order to obtain good correlation with test data, the effect of the presence of the welds had to be modelled. This was achieved by introducing an additional ring of shell elements at the brace-chord intersection, as shown in Fig. 2(b). The weld profile used follows AWS guidelines for fillet welds. Full details of the numerical modelling are given in Wilmshurst and Lee (1993). The boundary and loading conditions are shown in Fig. 2(a).

5 Results and discussion

5.1 γ and τ interaction
As pointed out in section 2, there has been concern about the relatively low strength of the joints tested

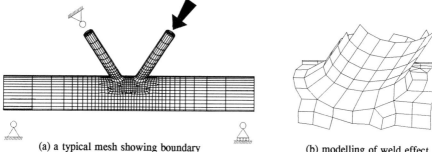

<table>
<tr><td>(a) a typical mesh showing boundary conditions and loading.</td><td>(b) modelling of weld effect</td></tr>
</table>

Fig.2. Details of numerical model

by Mouty and Rondal. Many of their joints failed by brace buckling, but even those that were reported to have failed by chord yielding had strength lower than expected. This may partly be attributed to the low γ and τ ratios. The failure modes of the joints analyzed in this study were examined in detail and are shown in Fig. 3. Also included in the figure are joints tested by Mouty and Rondal which failed by brace buckling. The joint analyzed and located in zone 1 failed prematurely due to extensive plasticity in the compression brace, Fig. 4(a), whereas those joints in zone 3 failed due to chord plastification alone, Fig. 4(c). The joints in zone 2 had plasticity in both the braces and the chord, Fig. 4(b). Although there is not enough data to clearly delineate the exact boundaries of these zones, joints having γ and τ ratios in zone 2 have to be treated with care as, to be demonstrated later, their strengths are overestimated by AWS.

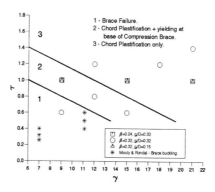

Fig.3. Interaction between γ and τ

5.2 Effects of γ

All Swansea F.E. results were non-dimensionalised with respect to the AWS predictions and are plotted versus γ in Fig. 5. Also included in the figure are test data from Makino et al. (1984, 1992) and Paul (1992a). The data are presented separately for the two types of failure mode, and for both with and without the small gap modifiers. In general, joints with non-dimensional out-of-plane gap $\zeta_t \leq 0.2$ failed in mode type 1 whereas joints with $\zeta_t \geq 0.2$ failed in mode type 2.

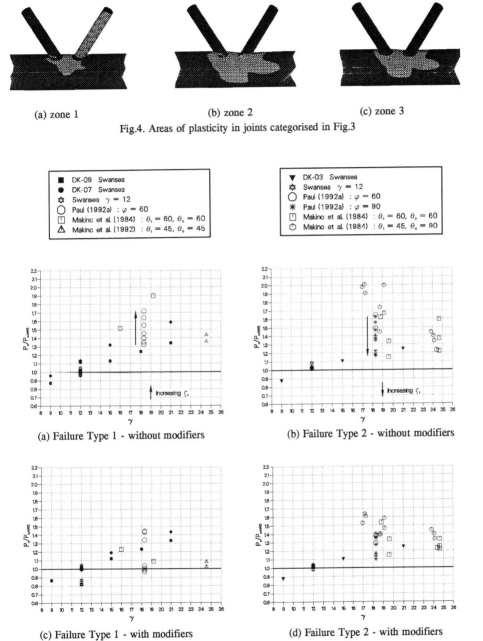

(a) zone 1   (b) zone 2   (c) zone 3

Fig.4. Areas of plasticity in joints categorised in Fig.3

(a) Failure Type 1 - without modifiers

(b) Failure Type 2 - without modifiers

(c) Failure Type 1 - with modifiers

(d) Failure Type 2 - with modifiers

Fig.5. Test to AWS prediction ratios versus γ

With reference to Fig. 5, the following observations can be made:

a. For both types of failure mode, the unmodified AWS formulation gives good predictions to all Swansea joints with $\gamma = 12$.

b. For joints with γ ratios less than 12, AWS over-predicts for both failure modes. However, it has been mentioned that these joints had plasticity in both the braces and the chord (they lie in zone 2 of Fig. 3). There is thus the possibility that the capacity of the joints might have been reduced due to yielding in the braces. Confirmation of this, or otherwise, is required perhaps by analysing an extra joint with the same γ ratio but a higher τ ratio, say about 1.4.

c. For both failure modes, there is the trend that the unmodified AWS becomes more conservative with increasing γ. Taking into account of the scatter of the results, the degree of under-prediction shows no distinction between the modes. For failure mode type 1, the unmodified AWS is less conservative for joints with small ζ_t (small transverse gaps), for a particular value of γ. However, the trend is reversed for joints that failed in type 2 mode. This difference indicates that the two failure mechanism perhaps should be treated separately as far as the ultimate load prediction is concerned.

d. The modifiers proposed by Lalani and Bolt were meant to reduce the amount of under-prediction for joints with small gap terms. Applying them to the AWS formulation has certainly enhanced its reliability, as indicated in Figs. 5(c) & 5(d). However, the under-predictions still amount to over 40% for type 1 and over 60% for type 2. In addition, some joints with $\gamma = 12$ are now over-predicted by over 20%. The modifiers thus require further examination and the trend suggests that they might be γ dependent.

The expression for α_o, as shown in Fig. 1, contains an exponential decay term which is a function of γ, hence the effect of γ can be accounted for by possibly modifying α_o. However, the limitations that α_o has to be at least equal to one, and that for certain geometries, some of the terms within the summation may actually disappear, render this proposition unviable. A separate γ correction factor will thus have to be proposed which, ideally, should be included in the new in-plane and out-of-plane modifiers.

6 Conclusions

Some numerical data on axially brace loaded multiplanar double K-joints has been presented. This data was evaluated together with existing experimental data with special reference to the AWS multiplanar joint formulation. The interaction between γ and τ was highlighted. γ was found to have a significant influence on static strength and such influence has not been accounted for in the AWS formulation. The small gap modifiers proposed by other researchers to enhance the AWS formulation in predicting joint strengths have also been shown to require further examination.

7 Acknowledgement

The first author acknowledges the financial support from the Science and Engineering Research Council whilst carrying out the work reported in this paper.

8 Nomenclature

g_i in-plane gap
g_o out-of-plane gap τ wall thickness ratio $=t/T$
α_o ovalisation parameter θ_c in-plane angle between comp. braces and chord
β diameter ratio $= d/D$ θ_t in-plane angle between tension braces and chord
γ chord thinness ratio $= D/2T$ ζ_l non-dimensional in-plane gap $= g_i/T$
ϕ out-of-plane brace angle ζ_t non-dimensional out-of-plane gap $= g_o/D$

9 References

ABAQUS (1989) **ABAQUS Users' Manual**, Hibbitt, Karlsson and Sorensen Inc.

AWS D1.1 (1985) **American Welding Society**, Structural Welding Code.

Kurobane, Y., Makino, Y. and Ochi, K. (1984) Ultimate resistance of unstiffened tubular joints, **J. Structural Engineering**, ASCE.

Lalani, M. and Bolt, H.M. (1990) Strength of multiplanar joints on offshore platforms, **Int. Symposium on Tubular Structures**.

Makino, Y, Kurobane, Y. and Ochi, K (1984) Ultimate capacity of tubular double K-joints, **IIW Conference**.

Makino, Y, Kurobane, Y., Ueno, T. and Paul, J.C. (1992) Ultimate behaviour of diaphragm stiffened tubular double K-joints, **Report to Hazama Corporation**, Japan (in Japanese).

Marshall, P.W. and Luyties, W.H. (1982) Allowable stresses for fatigue design, **Behaviour of Offshore Structures Conference**.

Mitri, H.S., Scola, S. and Redwood, R.G. (1987) Experimental investigation into the behaviour of axially loaded tubular V-joints, **CSCE Centennial Conference**.

Mouty, J. and Rondal, J (1990) Etude du comportement sobs charge statique des assemblages soudes de profils creux circulaires dans les poutres de sections triangulaires et quadrangulaires (in French).

Nakacho, K., Okada, M. and Ueda, Y (1989) Stiffness and yield strength of simple V-joints of offshore structures, **J. Japanese Naval Institute** (in Japanese).

Paul, J.C., Valk, C.A.C. van der and Wardenier, J (1989) The static strength of multiplanar X-joints, **Inter. Symposium on Tubular Structures**.

Paul, J.C. (1992a) The ultimate behaviour of multiplanar TT- and KK-joints made of circular hollow sections, **Ph.D. Thesis**, Kumamoto University, Japan.

Paul, J.C. (1992b) Private communication.

Paul, J.C., Makino, Y. and Kurobane, Y (1992) Ultimate behaviour of multiplanar double K-joints, **Int. Offshore and Polar Engineering Conference**.

Paul, J.C., Makino, Y and Kurobane, Y (1993) Ultimate resistance of tubular double T-joints under axial brace loading, **J. Constructional Research**, 24.

Scola, S, Redwood, R.G. and Mitri, H.S. (1990) Behaviour of axially loaded tubular V-joints, **J. Constructional Research**, 16.

Vegte, G.J. van der, Koning, C.H.M., Puthli, R.S. and Wardenier, J (1991) Numerical simulations of experiments of multiplanar steel X-joints, **J. Offshore and Polar Engineering**.

Wilmshurst, S.R. and Lee, M.M.K. (1993) Nonlinear FEM study of ultimate strength of tubular multiplanar double K-joints, **Int. Conf. on Offshore Mechanics and Arctic Engineering**.

PART 18

FINITE ELEMENT VALIDATION

Two and three dimensional joints and members in rectangular hollow section

71 SEMI-RIGID CONNECTIONS BETWEEN PLATES AND RECTANGULAR HOLLOW SECTION COLUMNS

L.H. LU
Delft University of Technology, The Netherlands
R.S. PUTHLI
Delft University of Technology, The Netherlands and
TNO Building and Construction Research, Rijswijk,
The Netherlands
J. WARDENIER
Delft University of Technology, The Netherlands

Abstract
Welded multiplanar connections between I-beams and rectangular hollow
section columns are attractive for use in offshore deck structures and
in buildings. In order to investigate the influence of individual
components, such as flanges of I-beams on the connection behaviour
(interaction effects), numerical research has been carried out on
connections between plates at one or two levels and rectangular hollow
section columns.

Calibrations have been done by comparing the numerical results with
the results from experiments carried out in the framework of ECSC
project 7210-SA/611 at Delft University of Technology and TNO Building
and Construction Research. Good agreement has been found between
numerical and experimental work.

Finally, comparisons between numerically determined static strength
and available formulae from literature are shown.
Keywords: Plate to RHS Column Connections, Numerical Simulations,
Static Behaviour.

List of symbols
a_i : throat thickness of the welds i=1,2.....
b_0 : outer width of the RHS column
b_1 : plate width or flange width of I-beams
h_1 : height of I-beams
l_0 : length of RHS column
l_1 : length of plates or I-beam flanges
t_0 : wall thickness in the middle of RHS column walls
t_c : corner thickness of RHS column
t_1 : thickness of plates or I-beam flanges
$t_{1,w}$: thickness of I-beam web
w_i : the weld leg between flanges and RHS column, i = 1, 2,...
$f_{y0,c}$: yield stress in the corner of RHS column
$f_{y0,m}$: yield stress in the middle of RHS column
$f_{y1,f}$: yield stress in plates or I-beam flanges
$f_{y1,w}$: yield stress in I-beam web
$F_{d,eff}$: design connection resistance for effective width failure
$F_{s,num}$: axial load at serviceability deformation limit of $1\%b_0$
$F_{u,exp}$: experimental maximum load at deformation limit
$F_{u,num}$: numerical maximum load at deformation limit

Tubular Structures V. Edited by M.G. Coutie and G. Davies.
© 1993 E & FN Spon, 2–6 Boundary Row, London SE1 8HN. ISBN 0 419 18770 7.

F_i :   axial load on plates or flanges of I-beams, i=1,2
β :   width ratio b_1/b_0
2γ :   width to thickness ratio of RHS column b_0/t_0
RHS   :   rectangular hollow section
CIDECT: Comité International pour le Développement et l'Étude de la
            Construction Tubulaire
ECSC   :   European Coal and Steel Community

1 Introduction

Until now, very little research work has been carried out on connections between I-beams and rectangular hollow section columns. No standard design formulae are available to determine the static ultimate strength of such connections. With increased applications of I-beam to tubular connections, the static behaviour of such connections needs to be investigated urgently and the design formulae need to be established.

     With this aim, experimental research on I-beam to RHS column connections is carried out in the framework of ECSC project 7210-SA/611, and numerical research will be carried out in the framework of a Delft University "BEEK" project. In this paper, the interaction effect of the flanges of the I-beams will be shown.

     For calibration with experiments, the tested specimens of ECSC project 7210-SA/611 are modelled numerically. Average values of measured dimensions and material properties from the test specimens have been used. Eight noded thick shell elements are used to model the connections. Non-linear material behaviour has been taken into account.

     The finite element analysis results are compared with the experimental results. Comparisons are also made between the numerically determined static strength and design value suggested by Wardenier(1982) and Packer, et al. (1992).

2 Research programme

The numerical work is carried out on 10 plate to RHS column connections as summarized in table 1, 2 uniplanar connections and 8 multiplanar connections, which are divided into series I and II. Two uniplanar connections up1R1 and up1R3 (identical connections to 1R1 and 1R3) are not in the ECSC experimental programme, but are analysed under compression to determine the influence of the unloaded out-of-plane plates on the connection behaviour (multiplanar effect). For series I, three different loading cases are considered, namely F_2/F_1 = 0, 1, -1 to investigate the influence of the loads on the out-of-plane plates on the static behaviour of the connections. Two values of parameters β (0.4 and 0.6) and one of 2γ (30) are selected. The average values of the measured dimensions and material properties for the members of the connections used in numerical models are given in tables 2 to 4.

Table 1 Research Programme for Plate to RHS Column Connections

Connections	Configurations of Connections and Loading Conditions			
Uniplanar	up1R1 F_1 F_1 RHS#1 120*10#3	up1R3 F_1 F_1 RHS#1 170*12#1		
	Series I			Series II
Multiplanar	1R1 F_1 F_1 RHS#1 120*10#3	1R3 F_1 F_1 RHS#1 170*12#1	1R5 F_2 F_1 F_1 F_2 RHS#1 120*10#4	2R1 F_1 F_1 RHS#2 IPE240#3
	1R6 F_2 F_1 F_1 F_2 RHS#1 120*10#4	1R7 F_2 F_1 F_1 F_2 RHS#1 170*12#1	1R8 F_2 F_1 F_1 F_2 RHS#1 170*12#1	2R3 F_1 F_1 RHS#2 IPE360#2

Table 2 Average Values of Measured Properties for RHS Columns

Columns with Stock nos.	l_0 mm	h_0 mm	b_0 mm	t_0 mm	t_c mm	$f_{y0,m}$ N/mm^2	$f_{y0,c}$ N/mm^2
RHS #1	1800	299.9	299.9	9.82	13.0	435	392
RHS #2	1800	300.0	300.0	9.74	12.5	449	455

Table 3 Average Values of Measured Properties for Plates and I-beams

Plates and I-beams with Stock nos.	l_1 mm	h_1 mm	b_1 mm	t_1 mm	$t_{1,w}$ mm	$f_{y1,f}$ N/mm^2	$f_{y1,w}$ N/mm^2
120*10 #3	600		119.6	9.89		398	
120*10 #4	600		120.5	10.0		388	
170*12 #1	850		170.0	11.5		380	
IPE240 #3	600	242.1	120.0	9.74	6.42	420	478
IPE360 #1	850	363.8	169.7	12.8	8.15	335	435
IPE360 #2	850	363.4	169.2	12.0	8.34	403	442

Table 4 Average Values of the Measured Welds Sizes in mm

Connections	W_1	W_2	W_3	W_4	W_5	W_6
1R1	12.3	10.7	6.5	6.2	9.6	11.4
1R3	13.9	9.4	7.0	6.4	9.7	12.9
1R5	13.0	11.0	6.5	6.2	10.8	11.1
1R6	12.4	10.1	6.8	6.0	10.2	10.1
1R7	14.9	9.4	6.9	5.7	10.3	9.5
1R8	14.6	7.8	6.4	5.3	11.1	11.4
2R1	13.6	9.3	6.9	5.9	9.5	12.1
2R3	16.8	10.4	7.6	5.7	14.5	15.5

3 Numerical modelling

Pre- and post processing have been performed by using the program
IDEAS. The finite element analyses are carried out with program MARC.
Eight noded thick shell elements (MARC element type 22) are used for
modelling of the connections, which give more reasonable results than
four noded thick shell elements (MARC 75) (Lu, et al., 1992).
   Due to symmetry in geometry and loading, only a quarter or an eighth
of the connections has been modelled as appropriate. In figure 1,
typical finite element meshes are shown for multiplanar connections for
series I and II.
   The welds between the plates or I-beam flanges and RHS columns are
also modelled with thick shell elements. The throat thickness a_1, a_2,
and a_3 are taken as the weld element thicknesses.(see table 4).
   In order to investigate the influence of the material stress-strain
behaviour, the FE analyses are performed on 1R1 and 1R3 twice using two
different material models, material 1 and material 3 in figure 2.
Material 1 has an ideal elastic-plastic stress-strain relationship,
while for material 3, strain hardening is taken into account, which
means that the engineering stress-strain relationship (material 2)
obtained from the material test has been translated into a true
stress - strain curve using the Ramberg-Osgood relationship (Eurocode
3, 1992). The figure 3 shows the numerical results for 1R1 and 1R3
obtained by using these two different material models. It can be seen

that the difference between the numerical results for the static behaviour by using two material models is small. However, throughout this study, the true stress-true strain relationship (material 3) is used, because it gives a more representative material behaviour.

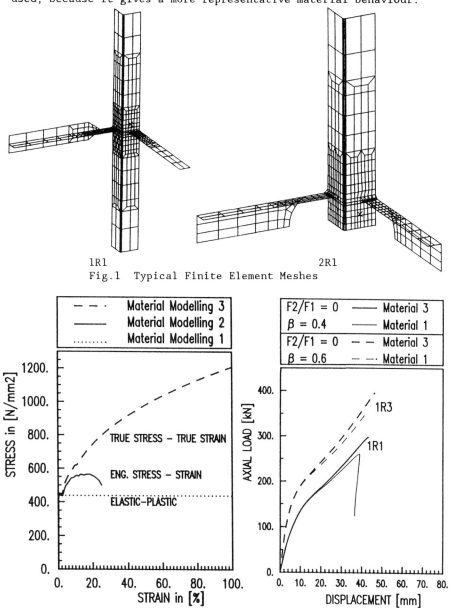

1R1   2R1
Fig.1   Typical Finite Element Meshes

Fig.2 Material Models

Fig.3 The Influence of
the Material Models

During the numerical analyses, displacement-control has been chosen for the connections subjected to $F_2/F_1 = 0$, 1, while load-control is used for the connections loaded by $F_2/F_1 = -1$. This method of loading is similar to that applied in the experimental work.

4 Numerical and experimental results

The load-displacement curves obtained from the numerical analyses are shown in figures 4 to 6. For calibration of the numerical work, the experimentally determined load-displacement curves for the ECSC research project 7210-SA/611 are also plotted in these figures. The influence of the multiplanar loading and the influence of the β values on the static behaviour of the connections can be seen. The influence of the unloaded out-of-plane plates or I-beams flanges on the connection behaviour (multiplanar effect) can be seen from figure 7.

5 Calibration of the numerical work

Figures 4 to 6 show the comparisons between numerical and experimental results for all connections. As no maximum peak load is obtained in the numerical and experimental work, various deformation limit criteria for static loading have been investigated. The most appropriate criterion for the present work has been found by Korol and Mirza (1982), where a value of $1.2t_0$ is used as the failure criterion of the connections loaded by axial forces. In table 5, the values $F_{u,num}$ and $F_{u,exp}$ at this deformation limit are given as the maximal loads of the connections, also the ratio of $F_{u,num}/F_{u,exp}$. These values of $F_{u,num}/F_{u,exp}$ show that the numerical values at the deformation limit lie about 2% to 9% higher than the tests results, so that the numerical modelling is in good agreement with the experiments.

6 Observations of the numerical results

From the limited FE results for the axially loaded multiplanar connections between plates and RHS columns with $\beta = 0.4$ and 0.6, the following observations are made :

- The numerical modelling of these connections, according to the above mentioned method, gives good agreement with the experiments. (see figures 4 to 6).
- For connections 1R6 and 1R8, when out-of-plane plates are proportionally loaded under compression at the same time as in-plane plates ($F_2/F_1=1$), the stiffness and static strength increase, as shown in figure 4 and 5, with respect to identical connections with out-of-plane plates unloaded ($F_2/F_1=0$).
- For connections 1R5 and 1R7, when out-of-plane plates are proportionally loaded under tension at the same time as in-plane plates ($F_2/F_1=-1$), the stiffness and static strength decrease, as shown in figure 4 and 5, with respect to identical connections with out-of-plane

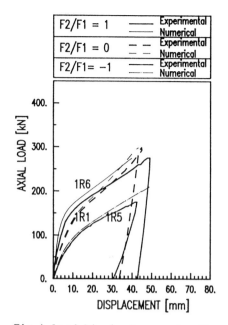

Fig.4 Load-Displ. Curves for Plate to RHS Connections with $\beta = 0.4$

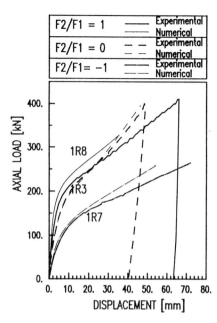

Fig.5 Load-Displ. Curves for Plate to RHS Connections with $\beta = 0.6$

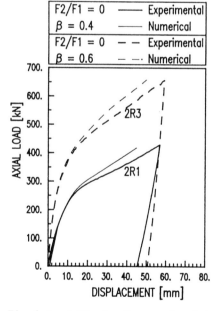

Fig.6 Load-Displ. Curves for I-beam Flanges to RHS Column Connections

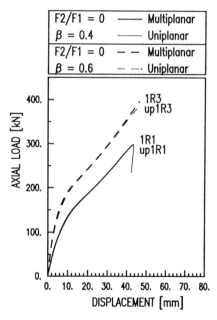

Fig.7 Multiplanar Effect for Plate to RHS Column Connections

plates unloaded $(F_2/F_1=0)$.

- For all connections considered, an increase of the β value results in an increase of the stiffness and static strength of the connections when they are loaded in the same way. (see figures 4 and 5)
- The ultimate strength of the connections with I-beam flanges (two levels of plates) is about 1.63 (for $\beta = 0.4$) to 1.97 (for $\beta = 0.6$) times that for connections with one level of plates. (see figures 4,6).
- The stiffness and strength of multiplanar plate connections are almost identical to those of uniplanar connections (see figure 7). It is evident that the difference is more perceptible for larger β values.

7 Comparison with several design formulae

The $F_{u,num}$ have been also compared with the design values $F_{d.eff}$ according to the CIDECT formulae for effective width failure, which governs for the plate to RHS column connection for $\beta \leq 0.85$. In table 5, the ratio of the $F_{u,num}$ /$F_{d,eff}$ is shown. In addition, for axially loaded hollow section joints, an indication of 1% of the chord width b_0 is generally used as the serviceability deformation limit [Eurocode 3]. For the present work, for indication only, $F_{s,num}$ at a serviceability criterion and the ratios of $F_{u,num}$ / $F_{s,num}$ are also given in table 5.

Table 5 shows that except for connections loaded with $F_2/F_1 = 1$ (1 1R6 and 1R8), the ratios of $F_{u,num}$ / $F_{s,num}$ and $F_{d,eff}/F_{s,num}$ are greater than 1.5. Therefore, it is considered that for these connections the serviceability limit is too conservative. Furthermore, the ratios of $F_{u,num}$ /$F_{d,eff}$ show, for instance, that for connections loaded with F_2/F_1 = -1 (1R5 and 1R7), the design failure load is higher than $F_{u,num}$. This may be due to the reduction of the membrane action.

Table 5 Comparisons of the Numerical Results with Design Codes

Connections	$F_{u,num}$ [kN]	$F_{u,exp}$ [kN]	$\dfrac{F_{u,num}}{F_{u,exp}}$	$F_{s,num}$ [kN]	$F_{d,eff}$ [kN]	$\dfrac{F_{u,num}}{F_{s,num}}$	$\dfrac{F_{d,eff}}{F_{s,num}}$	$\dfrac{F_{u,num}}{F_{d,eff}}$
up1R1	200.49	-	-	95.04	167.29	2.11	1.760	1.198
up1R3	258.64	-	-	144.78	237.29	1.79	1.639	1.090
1R1	202.10	191.15	1.06	96.43	167.29	2.10	1.735	1.201
1R3	260.85	254.75	1.02	153.33	237.78	1.70	1.551	1.097
1R5	141.00	135.58	1.04	67.86	168.55	2.08	2.484	0.837
1R6	217.45	199.50	1.09	146.43	168.55	1.49	1.151	1.290
1R7	175.12	165.21	1.06	100.71	237.78	1.74	2.361	0.736
1R8	283.22	259.83	1.09	208.93	241.38	1.36	1.155	1.173
2R1	323.66	308.25	1.05	162.86	340.77	1.99	2.092	1.341
2R3	501.50	477.62	1.05	311.43	480.48	1.61	1.542	1.472

8 Conclusions

Based on the results of the FE analyses and the comparison with the experimental results and the design formulae for the failure modes, the following conclusions can be drawn :

1). The numerical work gives good agreement with experiments.

2). For in-plane compression loaded connections, compression-loaded out-of-plane plates increase the stiffness and static strength compared to unloaded out-of-plane plates connections.

3). For in-plane compression loaded connections, tension-loaded out-of-plane plates decrease the stiffness and static strength compared to unloaded out-of-plane plates connections.

4). An increase of β results in an increase of the stiffness and static strength of the connections.

5). The strength of the connections with $h_1/b_0=2$ and 2.12 of I-beam flanges (with two levels of plates) is about 1.63 to 1.97 times the strength of the connections with one level of plates.

6). For all connections with β values of 0.4 and 0.6, except for $F_2/F_1=-1$, the maximum loads from the numerical analyses are in good agreement with the governing formula (Wardenier 1982).

7). Multiplanar plates to RHS column connections with $F_2/F_1=0$ give almost the same stiffness and strength as identical uniplanar connections.

8). The serviceability deformation limit is somewhat too small. A further study of deformation criterion is needed.

9 Acknowledgements

Appreciation is extended to ECSC for the permission to publish the results of the experimental tests. The tests are further sponsored by Rijkswaterstaat and Van Leeuwen Buizen. The parameter study will be carried out in the framework of a Delft University "Beek" project.

10 References

Eurocode 3, Design of Steel Structures, (1992) Part 1: General Rules and Rules for Buildings.

Korol, R.M., Mirza, F.A. (1982) Finite Element Analysis of RHS T-Joints. **ASCE**, pp2081-2098.

Lu, L.H., Puthli, R.S, Wardenier, J. (1992) The Static Strength of Multiplanar Connections between I-beams and Rectangular Hollow Section Columns. **Stevin Report 6.92.24/A1/11.10**, Delft.

Packer, J.A., Wardenier, J., Kurobane, Y., Dutta, D., Yeomans, N. (1992) Design Guide for Rectangular Hollow Section (RHS) Joints under Predominantly Static Loading. Publications **Verlag TÜV Rheinland GmbH, Köln.**

Wardenier, J. (1982) Hollow Section Joints. Delft University Press, Delft.

72 NUMERICAL INVESTIGATION INTO THE STATIC BEHAVIOUR OF MULTI-PLANAR WELDED T-JOINTS IN RHS

Y. YU and D.K. LIU
Delft University of Technology, The Netherlands
R.S. PUTHLI
Delft University of Technology, The Netherlands and
TNO Building and Construction Research, Rijswijk,
The Netherlands
J. WARDENIER
Delft University of Technology, The Netherlands

Abstract
This paper deals with a numerical investigation into the static
behaviour of multiplanar tubular T-joints in rectangular hollow
sections. Eight node thick shell elements are used in the FE
modelling. A so-called conservative equivalent thickness is taken as
the thickness of the shell element in modelling the fillet weld.
   The multiplanar effect, the influence of the load applied on the
out-of-plane braces and the influence of the constraints applied at
the ends of the out-of-plane braces in order to keep the out-of-plane
braces horizontal, are determined with regard to the load capacity at
a deformation limit of 1%b_0 as well as the deformation limit defined
by Yura. Comparisons between the numerical and experimental results
are carried out. Furthermore, the numerical results are also compared
with the results using the CIDECT design guide formula.
Keywords: Load Capacity, Static Strength, Rectangular Hollow
Section, Multiplanar T-joint, FE Model.

Symbols

F_v	:Axial load applied to the in-plane brace;
F_{vu}	:Ultimate axial load applied to the in-plane brace;
F_h	:Proportional axial load applied to the out-of-plane braces;
t_w	:Thickness of the weld in the FE model (see fig.1.)
FE	:Finite element;
RHS	:Rectangular hollow sections;
β	:Brace to chord width ratio;
2γ	:Chord width to wall thickness ratio;
τ	:Wall thickness ratio between brace and chord member.

1 Introduction

Considerable research in the field of the ultimate behaviour of
simple uniplanar joints has been undertaken in the past three
decades. However for multiplanar joints in rectangular hollow
sections (RHS), experimental or numerical results are very scant. In
order to get an insight into the static behaviour of such joints, six
axially loaded multiplanar T-joints and one axially loaded uniplanar

Tubular Structures V. Edited by M.G. Coutie and G. Davies.
© 1993 E & FN Spon, 2–6 Boundary Row, London SE1 8HN. ISBN 0 419 18770 7.

T-joint (for comparison) which are simply supported at the ends of the chord have been studied in this work, which forms part of an ECSC research programme. The in-plane-brace is loaded in compression for all the seven joints. The out-of-plane braces are subjected to different loading and constraint conditions.

2 The research programme

As part of the ECSC research programme, the experimental work is carried out at the University of Nottingham. The geometrical parameters are not varied and are fixed at $\beta=0.6$, $2\gamma=23.8$ and $\tau=1$. The numerical simulations of the experiments (this present work) are also included in this ECSC research programme. The test series (table 1) consist of one uniplanar (MPJT1A) and six multiplanar T-joints, which are simply supported at the ends of the chord. The in-plane brace for each of the 7 joints is loaded in compression. The out-of-plane braces are unloaded (MPJT2, MPJT5), loaded in tension (MPJT3, MPJT6) or loaded in compression (MPJT4, MPJT7). For joints MPJT2 to MPJT4, the out-of-plane braces are maintained horizontal. However, for joints MPJT5 to MPJT7, the out-of-plane braces are allowed to displace freely.

Measured dimensions and material properties of the joints are taken from the work at the University of Nottingham (see Davies et al 1992a, 1992b).

Table 1.   Research programme

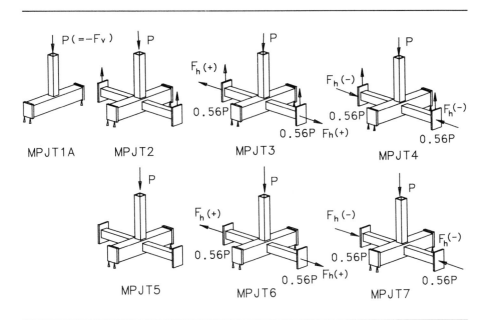

3 Numerical analyses

The main characteristics of the finite element analyses are as follows:
- The FE models are generated using a pre- and post processing package I-DEAS.
- A so-called conservative equivalent weld thickness is used as the thickness of the 8 node thick shell (degenerated solid) element in order to model the fillet weld. It is found that when weld modelling is excluded, much lower strength and stiffness values are obtained than when the weld modelling is included. As the weld dimensions vary along the length, the mechanically equivalent magnitude of the weld has to be used when the weld area is modelled with shell elements. According to Koning et al (1992) and Yu et al (1993), it is concluded that either good agreement or conservative results are obtained in comparison to the experimental results, by choosing the conservative equivalent weld thickness as shown in fig.1, i.e. $t_w=0.5(a_v*a_h)/l$.
- The numerical analyses are carried out using the general purpose FE program MARC.
- Large displacement and large strain options in the program are chosen. The updated Lagrange procedure is used.
- The Von Mises yield criterion and the isotropic strain hardening rule are used.

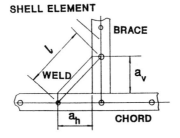

Fig.1. Numerical model of the weld with thickness t_w

3.1 Finite element meshes and load simulations

Using geometrical and load symmetry, a quarter of the joint has been modelled. Figs.2, 3 and 4 show the finite element meshes used for uniplanar joint MPJT1A, multiplanar joints (MPJT2, MPJT3, MPJT4) with the out-of-plane braces kept horizontal, and multiplanar joints (MPJT5, MPJT6, MPJT7) with the out-of-plane braces free, respectively.

In fig.3, points A1 and A2 are tied together in the vertical direction in order to keep the out-of-plane braces horizontal, ie. points A1 and A2 have the same vertical displacements. In the finite element analysis, this is called tying or constraint.

The plates attached to the ends of the chord and the ends of the out-of-plane braces are also modelled as thick shell elements, using the material properties of the chord.

For the uniplanar joint, the compressive load at the end of the brace is applied by displacement control.

For the multiplanar joints, the loads are applied by load control, because proportional loads are applied to the out-of-plane braces.

For both the displacement control and the load control, the 'independent' nodes at the end of the in-plane brace are tied to one 'dependent node' (node A in fig.3.) in the direction of the axis of the in-plane brace, so that all the tied nodes have the same axial displacement as the 'dependent' node, see fig.2,3 and 4.

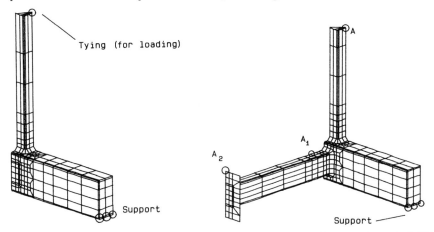

Fig.2. Joint MPJT1A   Fig.3. Joints MPJT2, MPJT3 and MPJT4

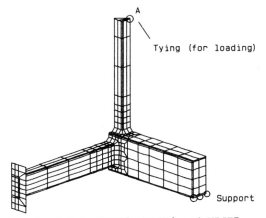

Fig.4. Joints MPJT5, MPJT6 and MPJT7

4 Numerical results

4.1 General remarks
The static behaviour of the joints is shown in the curves of the in-plane brace load versus chord top face indentation (see fig.5. and fig.6). When the in-plane brace is loaded in compression, both the

top and the bottom of the chord displace downwards globally. The
local indentation of the brace into the chord top face is measured as
the in-plane brace displacement (at a position of 100 mm above the
chord top face) minus the bottom face displacement of the chord, as
in the experiments. The load capacity F_{vu} at the 1%b_0 deformation
limit and at the deformation limit defined by Yura (1981) are
determined from the in-plane brace load versus indentation curves
(see fig.5. and fig.6.). The 1%b_0 deformation limit is considered as
the serviceability limit of the joints investigated, which lies at
the vicinity of the kink of the applied in-plane brace load versus
indentation curves. The Yura deformation limit is in the plastic part
of the load versus indentation curves.

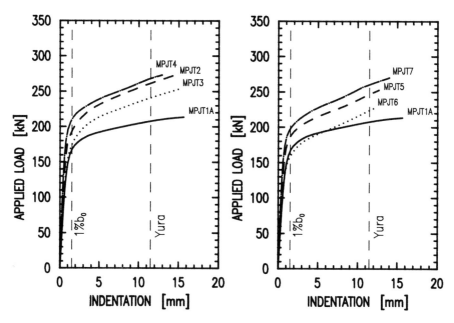

Fig.5. Multiplanar effect
(with constraint)

Fig.6. Multiplanar effect
(without constraint)

4.2 Calibration of the numerical models
In the report of Yu et al (1993), comparisons between the numerical
and experimental results are given. The largest difference in load
capacity at the 1%b_0 deformation limit is 12%, if MPJT7 is not taken
into account where horizontal rotation of the out-of-plane braces
occurred during the experiment. The largest difference in static
strength at the Yura deformation limit is 9%. It is concluded that
either good agreement or conservative results are obtained with
respect to the experimental results.

4.3 The multiplanar effect
Figs.5 and 6 show the load versus indentation relationship for the
uniplanar and multiplanar joints under different load conditions. The

multiplanar effect can be clearly seen. The load capacity for all the multiplanar joints except joint MPJT6 is higher than that of the uniplanar joint. For joint MPJT6, where the out-of-plane braces are in tension and free to displace laterally at the ends, the load-indentation curve is slightly lower than for the uniplanar joint at the load stages of about 130 kN to 195 kN. The load capacity for multiplanar joints shows no maximum value and increases due to membrane action with an increase in the chord indentations, whereas the load-indentation curve for the uniplanar joint is very flat.

Tables 2 shows quantitatively the influence of the multiplanar effect of the joints at the $1\%b_0$ and the Yura deformation limits. It is concluded that the multiplanar effect is more pronounced when the in-plane and out-of-plane braces are loaded in an identical manner (both in compression).

Table 2. Multiplanar effect

Def. crit.	Load capacity	MPJT1A	MPJT2	MPJT3	MPJT4	MPJT5	MPJT6	MPJT7
$1\%b_0$	F_{vu} [kN]	166	192	174	209	187	159	197
$1\%b_0$	$F_{vu}/F_{vu,MPJT1A}$	1.0	1.16	1.05	1.26	1.13	0.96	1.19
Yura	F_{vu} [kN]	208	261	241	269	247	224	260
Yura	$F_{vu}/F_{vu,MPJT1A}$	1.0	1.25	1.16	1.29	1.19	1.08	1.25

4.4 Influence of the load F_h applied to the out-of-plane braces
There are three load ratios for the multiplanar joints in this work: the out-of-plane braces loaded in tension (F_h/F_v=-0.56), unloaded (F_h/F_v=0) and loaded in compression (F_h/F_v=0.56). Table 3 indicates that the load capacity of the joints at the $1\%b_0$ deformation limit decreases by up to 15% for a load ratio of -0.56 and increases by up to 9% for a load ratio of 0.56, with respect to a load ratio of 0.0. The load capacity at the Yura deformation limit decreases by up to 9% for a load ratio of -0.56 and increases by up to 5% for a load ratio of 0.56, with respect to a load ratio of 0.0.

Table 3. Influence of the load applied to the out-of-plane braces

Deformation limit	Load ratio F_h/F_v			
	-0.56	0.56	-0.56	0.56
	$\dfrac{F_{vu,MPJT3}}{F_{vu,MPJT2}}$	$\dfrac{F_{vu,MPJT4}}{F_{vu,MPJT2}}$	$\dfrac{F_{vu,MPJT6}}{F_{vu,MPJT5}}$	$\dfrac{F_{vu,MPJT7}}{F_{vu,MPJT5}}$
$1\%b_0$	0.91	1.09	0.85	1.05
Yura	0.92	1.03	0.91	1.05

4.5 Influence of the constraints applied to the out-of-plane braces

Table 4 shows that, owing to the constraints applied to the out-of-plane braces (Point A2 is constrained to A1 in the vertical direction for joints MPJT2, MPJT3 and MPJT4, see fig.3.), there is an increase in load capacity at the $1\%b_0$ deformation limit by up to 9%, and an increase in strength at the Yura deformation limit by up to 8%.

Table 4.   Influence of constraints applied to the out-of-plane braces

Deformation limit	$F_{vu,MPJT2}$	$F_{vu,MPJT3}$	$F_{vu,MPJT4}$
	$F_{vu,MPJT5}$	$F_{vu,MPJT6}$	$F_{vu,MPJT7}$
$1\%b_0$	1.03	1.09	1.06
Yura	1.06	1.08	1.03

5   Comparisons with the CIDECT design guide formula

Table 5 shows that the load capacity of the multiplanar joints at the $1\%b_0$ deformation limit is much higher than that according to the CIDECT design guide formula (see Packer et al (1992)) for some cases (maximum 54% higher). For the uniplanar joint, the load capacity at the $1\%b_0$ deformation limit is in line with the CIDECT design guide formula. The strength at the Yura deformation limit is significantly higher than that for the CIDECT design guide formula (maximum 98%). This is because the formula for the CIDECT design guide is based upon chord face yielding and limited deformations. However, the Yura deformation limit is at a load stage associated with large deformations.

Table 5.   Numerical comparison with the CIDECT design guide formula

	MPJT1A	MPJT2	MPJT3	MPJT4	MPJT5	MPJT6	MPJT7
F_{vu} ($1\%b_0$ [kN])	166	192	174	209	187	159	197
F_{vu} (Yura [kN])	208	261	241	269	247	224	260
F_{vu} (CIDECT [kN])	161	135	135	135	145	142	143
$1\%b_0$/CIDECT	1.03	1.42	1.29	__1.54__	1.27	1.12	1.38
Yura/CIDECT	1.29	1.93	1.79	__1.98__	1.70	1.58	1.82

6   Conclusions

Based upon the numerical investigation for the present work, where fixed geometrical parameters of $\beta=0.6$ $2\gamma=23.8$ $\tau=1$ are used, the following conclusions can be drawn:

Using eight node thick shell elements for both the parent material and the fillet weld, where a so-called conservative equivalent thickness is used, either good agreement or conservative values with respect to the experimental results are obtained.

The most pronounced multiplanar effect of the joints investigated is when all the braces are loaded axially in compression. The load capacity at the $1\%b_0$ deformation limit is increased by up to 26% more than that of the uniplanar joint. The strength of the joints at the Yura deformation limit is increased by up to 29% higher than that of the uniplanar joint. When the in-plane and out-of-plane braces are loaded in opposite directions, the multiplanar effect is least pronounced among the three load cases.

The constraints used to keep the out-of-plane braces horizontal increase the load capacity at the $1\%b_0$ deformation limit by up to 9%, and the load capacity at the Yura deformation limit by up to 8%.

The load ratio between the axial loads on the out-of-plane and the in-plane braces of 0.56 increases the load capacity at the $1\%b_0$ deformation limit by up to 9% and the strength at the Yura deformation limit by up to 5% with respect to the multiplanar joints with a load ratio of 0.0. On the other hand, for a load ratio of -0.56, the load capacity at the $1\%b_0$ deformation limit decreases by up to 15% and the load capacity at the Yura deformation limit by up to 9%.

The load capacity at the $1\%b_0$ serviceability deformation limit nearly agrees with that according to the design strength formula for uniplanar RHS joints in the CIDECT design guide. However, the CIDECT design guide formula for multiplanar RHS T-joints under axial load is in most of the cases too conservative when compared with the strength at the deformation limits considered (serviceability and Yura deformation limits).

Further parameter studies are required in order to quantify the multiplanar effect of the joints in rectangular hollow sections in the whole range of geometry parameters.

7 Acknowledgements

The partners in the ECSC programme are the Steel Construction Institute, British Steel, the University of Nottingham, TNO Building and Construction Research and Delft University of Technology. Appreciation is extended to the ECSC for sponsoring the research programme and permission to publish this paper.

8 References

Davies, G., Coutie, M.G., Bettison, M. (1992a)
    The static strength of multiplanar hollow section joints-Tee joints, research agreement 7210/SA/830, Technical report No. 2, Number SR91051, the University of Nottingham.
Davies, G., Coutie, M.G., Bettison, M. (1992b)
    The static strength of multiplanar hollow section joints-Tee joints, research agreement 7210/SA/830, Technical report No. 3, Number SR92035, the University of Nottingham.

Koning, C.H.M. de, Liu, D.K., Puthli, R.S., Wardenier, J. (1992)
The development of design methods for the cost-effective
applications of multiplanar joints, experimental and numerical
investigation on the static strength of multiplanar welded DX- and
X-joints in R.H.S., TNO-Bouw report No. BI-92-0129/21.4.6161
Stevin report No. 6.92.28/A1/11.08, the Netherlands.

Packer, J., Wardenier, J. Kurobane, Y., Dutta, D., Yeomans, N. (1992)
Design guide for rectangular hollow section (RHS) joints under
predominantly static loading, ISBN 3-8249-0089-0, Verlag TUV
Rheinland GmbH, Köln 1992.

Yu, Y., Liu, D.K., Puthli, R.S., Wardenier, J. (1993)
The development of design methods for the cost-effective
applications of multiplanar joints, numerical investigation into
the static strength of multiplanar welded T-joints in R.H.S.
TNO-Bouw report No. BI-92-180 ,Stevin report No. 6.92.37/11.08,
the Netherlands.

Yura, J.A. (1981)
Ultimate capacity of circular tubular joints, Journal of the
Structural Division, Proceedings of the American Society of Civil
Engineers. ASCE, Vol. 107, No. ST10. ISSN 0044-8001/81/0010-
1965/$01.00.

73 AN INTERACTION DIAGRAM FOR THREE-DIMENSIONAL T-JOINTS IN RECTANGULAR HOLLOW SECTIONS UNDER BOTH IN-PLANE AND OUT-OF-PLANE AXIAL LOADS

G. DAVIES and P. CROCKETT
University of Nottingham, UK

Abstract
The three dimensional behaviour of rectangular hollow section Tee joints is examined in the presence of largely axially loaded in-plane and out-of-plane brace members using a finite element approach. Interaction resistance capacity diagrams are presented for such joints with unrestrained and restrained out-of-plane braces, which confirm the prediction of the yield line approach that there is no significant gain in strength beyond that provided by the physical presence of the out-of-plane braces for these joints, in contrast to those formed from circular hollow sections.
Keywords: Welded joints, Rectangular Hollow Section, Multiplanar, Strength, Interaction Diagrams, Finite Element

1 Introduction

Investigations by Paul et al(1989) using the finite element(FE) approach have indicated the possibility of significant gains of stiffness and strength for three dimensional(3D) axial loading of cross joints in Circular Hollow Sections(CHS). This has not been found to be so for the resistance capacity when using Rectangular Hollow Sections(RHS), either from the yield line analysis approach described by Davies et al (1992) or indeed in the experimental work described by them (1993), as part of a larger European programme. The advantage of the FE approach is that once validated, it can be easily adapted to examine the various effects of different loading regimes, boundary conditions and variation in material and geometric properties. Comparisons are made in this paper between an FE model and the limited experimental results available for the brace to chord width ratio $\beta = 0.6$, in order to calibrate the results, after which a range of out-of-plane (OPB) axial load actions are considered.

Tubular Structures V. Edited by M.G. Coutie and G. Davies.

Table 1. Material properties and dimensions.

		DIMENSIONS AND PROPERTIES				NOMINAL
		CHORD (i=0)		BRACE (i=1)		RATIOS
		Nominal	Actual	Nominal	Actual	
b_i	mm	150.0	150.0	90.0	90.5	$\beta = 0.6$
h_i	mm	150.0	149.5	90.0	89.5	
t_i	mm	6.3	6.2	6.3	6.2	$b_o/t_o = 23.8$
A	mm^2	3600	3505	2090	2062	
f_y N/mm^2		355	420	355	423	$b_1/t_1 = 14.3$
f_u N/mm^2		490	546	490	530	
Weld a mm		-	-	6.3	6.9	
f_s *N/mm^2		-	392	-	422	

* Based on squash tests

Some results are also provided for joints with $\beta=0.25$ and
1.0.

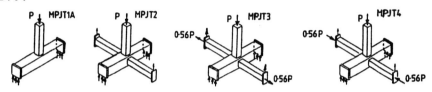

(a) Joint series designation

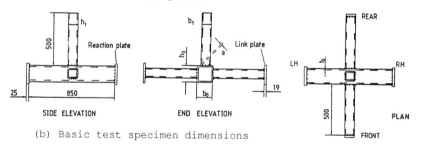

(b) Basic test specimen dimensions

Fig.1. Experimental joint layout

2 Finite Element Model $\beta=0.6$

The joint material properties used are given in Table 1
with the testing arrangement of some of the joint series
shown in Figure.1. Using the finite element package ABAQUS
(1990) with symmetry of the joints, one quarter models
were established. Modelling of the weld in joints such as
these is of paramount importance due to its effect on the
effective branch to chord width ratio β. A preliminary

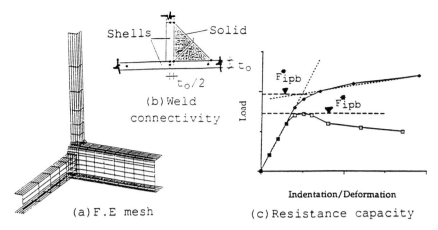

(b) Weld
connectivity

(a) F.E mesh

(c) Resistance capacity

Fig.2. Basis of comparison of experimental and FE results,
with mesh and weld connectivity details

investigation by Crockett et al. (1992) of the effects of
using different shell and solid elements with various mesh
arrangements to model the area local to the weld was
undertaken with both 4 and 8 noded thick shells for the
chord. While it is understood that 8 noded shell elements
are more accurate, 4 noded shells can offer the advantage
of a reduced CPU time and may be used to give satisfactory
results. The most reliable model found utilised 4 noded
thick shell elements to model the brace and chord with 6
noded solid elements to model the weld, details of which
are shown in Figure 2. This model allows for a zero
penetration of the fillet weld between the brace and
chord, thus being only connected through the weld via
multi-point constraints.

The joints in the test series were analysed for this
model with the mesh displayed and comparisons of resistance capacity were made on the basis of the extrapolation
shown in Figure 2c. Two sets of end boundary restraint
conditions have been considered for the OPB, viz completely free, or restrained to deflect approximately the same
amount as the point of attachment to the chord. Comparisons are made with the experimental results in Figure 3 on
the basis of the local joint deflection. In-Plane Brace
(IPB) indentation was taken as the difference between the
axial deflection of the same point on the IPB and that of
the far chord corners, this in-punching being the mode of
failure for this particular joint series. Figure 3 shows
the experimental results for MPJT2 along with the finite
element results with and without a weld. This clearly
displays the requirement for modelling of the weld to
obtain reasonable results. Comparisons are also shown for
MPJT3 & MPJT4, these being the best (0% error) and worst
(-8% error) respectively with regard to the comparison
within the plastic region of the indentation curve.

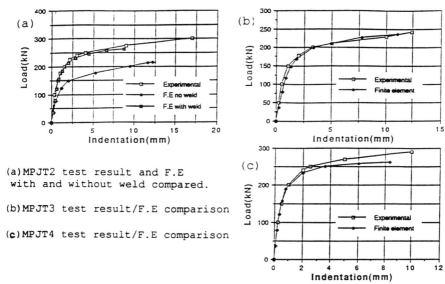

(a) MPJT2 test result and F.E
with and without weld compared.

(b) MPJT3 test result/F.E comparison

(c) MPJT4 test result/F.E comparison

Fig.3. Comparisons between experimental and FE results

Capacity interaction diagrams (Figure 4) have been
produced in each case, based on the capacities derived as
in Figure 2c, for β=0.6 multi-planar T joints. It has also
been used to analyse the capacities for the IPBs in
tension. These particular joints exhibit much increased
membrane action resulting in increasing ultimate strength
long after plasticity. In practice failure would progress
after cracking at the toe of the weld, but due to the
greater complexity in modelling the behaviour, it has not

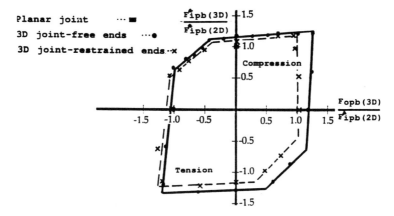

Fig.4. Interaction diagrams for β=0.6

been modelled here. Capacities of such tensile loaded
joints, when determined in the above way should give
resistance capacities below those at which cracking
failure will occur.

3 Finite Element Model $\beta=0.25$

A further study was undertaken to observe the 3D effects
for a lower β ratio. The model for the $\beta=0.25$ joints was
set up in exactly the same way as for the $\beta=0.6$ joints,
the weld model and dimensions also being similar. A full
interaction diagram was not investigated but the effects
of the presence of the OPBs, their method of restraint and
the presence of both tensile and compressive loads in
these braces was examined. A total of seven analyses were
undertaken, the loading and restraint conditions being

Table 2. Interaction effects for 3D joints for $\beta= 0.25$

b_o/t_o=23.8	Planar or multi-planar	Restraint condition	F_{opb}/F_{ipb}	Finite Element F_{ipb} (kN)	Ratio Multi-plnr /planar
A	Planar	---	--	79	1.00
B	Multi-plnr	a	0.0	79	1.00
C	Multi-plnr	a	0.5T	71	0.90
D	Multi-plnr	a	0.5C	80	1.01
E	Multi-plnr	b	0.0	79	1.00
F	Multi-plnr	b	0.5T	69	0.87
G	Multi-plnr	b	0.5C	80	1.01

a=OPBs restrained to remain horizontal, b=OPBs free to rotate

shown in Table 2. Again the mode of failure exhibited was
local deflection of the IPB, the joint capacity being
obtained as previously. The capacity calculated using the
IIW (1989) design strength prediction is 88kN.

4 Finite Element Model $\beta=1.0$

Previous experimental work with planar T joints has shown
that side wall buckling can dominate with near or full
width joints, and is sensitive to chord wall slenderness.
It was considered important to examine the effect of out-
of-plane axial loading also on large β ratios. Initial
investigations were undertaken using both 4 and 8 node
shells to model the brace and chord elements. Due to the
poor ability of 4 noded elements to pick up the buckling
behaviour of the side walls, 8 noded shells have been used

Table 3. FE results for β=1.0

b_o/t_o=23.8	Description	F_{opb}/F_{ipb}	Finite Element F_{ipb} (kN)	Ratio Multi-plnr /planar..
A	Planar	---	989	1.00
B	Multi-planar	0.0	1321	1.34
C	Multi-planar	0.5T	1259	1.27
D	Multi-planar	0.5C	1284	1.30

throughout. A brief investigation of the effects of the presence of the transverse chord/branch fillet weld spanning the top face of the chord was undertaken, using the model displayed earlier (Figure 2b) but the results indicated that its effect on compressive strength was small and hence it was modelled as a butt weld with 'branch' material thickness for subsequent analyses. A typical mesh used to analyse the multi-planar β=1.0 joint is shown in Figure 5a. The side wall butt weld is also included in the model as 8 noded shell elements with nominal chord and base material properties. In order to prevent chord bending failure due to the increased capacity it was thought appropriate to support the lower corners of the chord on roller bearings on a rigid foundation. This is illustrated in Figure 5b along with the general layout and dimensions for the analysed specimen. Yield stress and other material properties were assumed to be the same as those of the previous joints. The loading conditions of the joints and their results are shown in Table 3. Joint resistance capacities quoted for the finite element analyses are those peak loads achieved during the analyses, as shown in Fig-ure 2c.

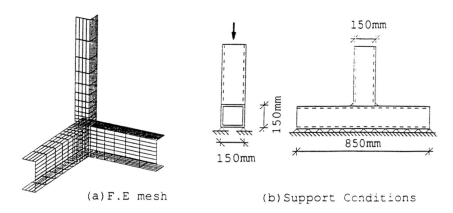

(a) F.E mesh   (b) Support Conditions

Fig. 5. Support arrangement and mesh for FE analysis for full width joints β=1.0.

5 Discussion

The interaction diagrams (Figure 4) display the resistance capacities for the multi-planar $\beta=0.6$ T joints relative to a planar T joint, which is represented by the value 1.0. It can be seen that the presence of the unloaded OPBs raises the ultimate capacity by between 18kN(9%) and 35kN(18%) depending upon the restraint applied to the OPBs. 'Welding in' of the unloaded OPBs increases this capacity. The effect of applying axial forces to these braces in the same sense as that in the IPB can be seen to be almost negligible. This fits in clearly with the yield line approach by Davies et al (1992). It can also be seen that the ultimate capacity as an X joint (i.e only out-of-plane braces loaded) is almost equal to that of a T joint validating the ultimate capacity formulae in the IIW/CIDECT recommendations. When OPB forces of the opposite sense are applied the reduction in capacity can clearly be seen, this again confirming the yield line theory trend. Current design guidelines are all based on uni-planar assumptions and although in most cases the addition of the out-of-plane braces in itself is enough to offset most load effects, this would appear not to be the case when the tension in one plane is of a similar or greater magnitude than the compression in the other plane.

A comparison of the planar result with the unloaded OPB results for the $\beta=0.25$ joints would appear to indicate that the presence of these unloaded OPBs, no matter how restrained, has little effect upon the capacity of the joint. This is likely to be explained by the much smaller stiffening effect on the sidewall exercised by the smaller β ratio. Application of out-of-plane loads however yields similar results to that above, compression having very little effect (apart from a stiffening in the elastic region of the load vs indentation curve), with tension reducing the resistance capacity by 10kN (12%). This has implications for design as the compression IPB, tension OPB joint has a lower capacity than the planar T joint on which the design recommendations are based.

For the $\beta=1.0$ joints, the 'welding in' of the unloaded OPBs has a very significant effect upon the resistance capacity of the joint, raising the capacity by 30% for this particular slenderness. This is associated with the much increased stiffening of the chord sidewalls along the edges of the OPBs, significantly restricting the freedom of the side wall buckling. The effect of applying tensile or compressive loads in the OPBs is much less significant than for lower β ratios. Effects of imperfections in the chord side wall have been ignored in this paper although a finite element study has been undertaken by the authors.

The direction of the joint side-wall buckling is influenced by the sense of the OPB forces- the models without OPBs, with unloaded OPBs and with tensile loaded OPBs deflecting outwards, whilst those joints with compression in the OPBs deflect inwards.

6 Conclusions

FE analyses for three dimensional T joints with chord slenderness of around 24 have been used to determine the variation of in-plane resistance capacity due to the presence and sense of loading of out-of-plane members.Good agreement was obtained with experimental tests when the welds were appropriately modelled for joints with $\beta=0.6$. Based on the resistance criteria chosen, a small increase in strength was achieved solely due to 'welding in' of the out-of-plane members. Subsequent loading in the same sense did not achieve any significant strength enhancement. However reversing the sense reduced the joint resistance below that for the planar joint. This confirms the conclusions of the yield line approach.For low β ratios the same trends were noticeable, but the physical presence of 'welding in' OPBs produced little enhancement. For full width joints 'welding in' of the OPBs produced a significant increase of strength, the magnitude of the OPB forces being much less important.

7 References

Crockett.P. and Davies, G. (1992) 'Modelling of Fillet Welds in RHS' Department of Civil Engineering Research Report No SR92024, Nottingham University, Nottingham,UK.

Davies.G.,Coutie, M.G. Bettison, M. and Morita, K. (1992) Three dimensional Tee Joints in Rectangular Hollow Section, in **Proceedings 3rd Pacific Structural Steel Conference**, Tokyo, Japanese Society of Steel Construction, pp655-662

Davies.G.,Coutie.M.G,and Bettison.M,(1993)'The Behaviour of Three Dimensional RHS Tee Joints Under Axial Branch Loads. **Proceedings 5th International Symposium on Tubular Structures**, Nottingham. E & F.N.Spon, London,

Hibbett,Karlsson & Sorensson,Inc(1989)'**ABAQUS**'**finite Element Package,USA.**

IIW (1989) '**Design Recommendations for Hollow Section Joints-Predominantly Statically Loaded**'. International Institute of Welding Doc XV-701-89.

Paul.J.C,Van Der Valk.C.A.C. and Wardenier.J. (1989),'Static Strength of Cicular Multi-Planar X Joints'. **Tubular Structures 3rd International Symposium**. Elsevier Applied Science.

74   STATIC STRENGTH OF
     MULTIPLANAR K-JOINTS IN
     RECTANGULAR HOLLOW
     SECTIONS: NUMERICAL
     MODELLING

M.A. O'CONNOR
The Steel Construction Institute, Ascot, UK

Abstract
This paper investigates the important variables which
govern the analysis of multiplanar K-joints by direct
comparison of numerical modelling and experimental
results. Conclusions are drawn from the results and
guidance is given on the future modelling of this type
of joint.
Keywords: Multiplanar, Rectangular Hollow Section,
Double K-Joints, Numerical Modelling.

1 Introduction

Multiplanar connections are commonly specified in
tubular structures. However, specific design guidance
for such connections is scant for joints between
circular hollow sections and non-existent for joints
between rectangular hollow sections. To address this
deficiency in the design process, an ECSC research programme is currently underway to investigate the effect
of the multiplanar interactions on the static strength
of rectangular hollow section joints. The programme is
project managed by the Steel Construction Institute and
consists of a series of twenty-four multiplanar tests of
various configurations which are investigated both experimentally and numerically.

   The experimental testing has been carried out by
Nottingham University (T-joints), British Steel plc (K-
joints) and TNO Building and Construction Research/Delft
University of Technology (X-joints). Numerical
modelling is being undertaken by both Delft University
of Technology and the Steel Construction Institute.
This paper concentrates on the Steel Construction Institute's numerical modelling of nine multiplanar K-joint
specimens.

Tubular Structures V. Edited by M.G. Coutie and G. Davies.
© 1993 E & FN Spon, 2–6 Boundary Row, London SE1 8HN. ISBN 0 419 18770 7.

2 Experimental programme

Details of the experimental research on multiplanar K-joint specimens have been outlined elsewhere at this symposium, Yeomans (1993). Briefly, the main parameters investigated were the brace to chord width (β) ratio, the chord width to wall thickness (γ) ratio, and the brace wall thickness to chord wall thickness (τ) ratio, in a total of nine specimens. The loading regime for each specimen was the same with each pair of braces loaded in compression and tension respectively in the same sense as the pair of braces on the adjacent face. Material properties were measured in the longitudinal and transverse directions for each length of material used in the specimens.

3 Numerical analyses

3.1 Purpose
The purpose of the numerical modelling of the multi-planar K-joint specimens is to develop a model, fully calibrated against the experimental results, which is capable of adequately representing the static structural behaviour of multiplanar K-joint specimens. Once developed this model can be used in further parametric studies thus cost effectively increasing the database of results on which to base design guidance.

3.2 Procedure
The models in this study were developed using the LUSAS general purpose finite element analysis package. Two separate models were developed. One model was used to model the specimens with a β ratio significantly less than 1 where the expected failure mode was predominantly chord face deformation and the second model was used for the specimens with a β ratio of 1.0 where the expected failure mode was by chord shear deformation.

The numerical investigation for the K-joints can be split into three stages. The first stage is the model development stage where the effect of using different elements, the modelling of welds and corner radii, etc. on the failure load of the models is investigated. The second stage is correlation of the preferred developed model against all the experimental results. The third stage for modelling of the K-joint specimens is the development and analyses of uniplanar numerical models for comparison with the multiplanar models.

3.3 Modelling details
One of the meshes used for the two developed models is shown in figure 1 for specimen KK06 (β=0.6, γ=15 and

$\tau = 0.8$). This mesh represents half a multiplanar K-joint specimen as a plane of symmetry exists at an angle of 45° through the two corners of the chord member. The semi-loof shell element was used for both the chord and brace members. The semi-loof shell elements have translational degrees of freedom at each node with semi-loof rotations about the loof points on each side of the element. The welds were modelled using solid elements. The solid elements have translational degrees of freedom only at each node. The incompatibility between the solid elements and the bending freedoms of the shells is not important from a structural point of view as the weld elements are only present to ensure the correct load path between the brace and chord elements.

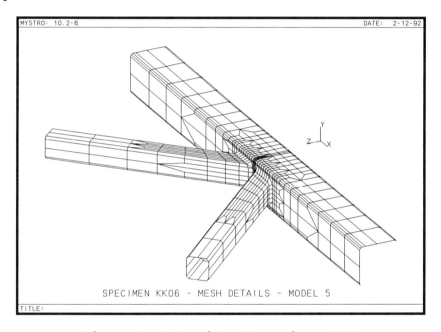

SPECIMEN KK06 - MESH DETAILS - MODEL 5

Fig.1. Mesh details of specimen KK06

The large pinned joints which were welded to the end of each brace and chord member were modelled using rigid beam elements. This enabled an accurate representation of the support conditions at the end of each member. The brace/chord loads and end conditions were applied by transforming the freedoms at the end of the pinned ass-emblies and applying the load and restraints in the appropriate direction. An automatic load incrementation procedure was used for the nonlinear analysis ensuring that the chord and brace loads were applied in the same proportions as the first load increment.

4 Numerical Results

4.1 Deformation criteria

Before the results can be discussed the effect of the choice of deformation criteria and the actual method of deformation measurement on the ultimate failure load need to be considered.

An historically accepted deformation criterion at serviceability is that chord face deformations should not exceed 1% of the chord width. In multiplanar connections it has been discovered that some kind of deformation limit is also required at ultimate limit state as the post yield portion of the load-deformation curve keeps on rising long after the useful capacity of the joint is exceeded (Davies et al. (1993),Liu et al. (1993)). This deformation limit could be any reasonable value as long as consensus is reached in the adoption of such a criterion.

Once it is accepted that a deformation limit to determine the ultimate failure load is needed, in the absence of a definite ultimate failure mode, then the method used to measure chord face deformation needs to be carefully examined. Deformation in the experimental portion of this study is measured by monitoring the change in length of a displacement transducer located between two steel offsets welded to the chord and brace centrelines respectively. Deformation was measured for each brace on both the inner and outer faces of the braces of the specimen. The displacement transducers had to be offset from the chord and brace face for each specimen by varying amounts to avoid the possibility of the transducers clashing. The deformation measured in this way is affected by the amount of the offset as the offset rotates with chord face deformation especially between the braces in the post yield phase of the load-deflection curve.

The effect of this rotation has been studied as part of the numerical modelling. Deformations including and excluding offset rotation for specimen KK06 are plotted against load in figure 2. From this figure it can be seen that offset rotation has a significant effect on the load-deflection curves especially in the post yield phase. The true chord deformation will be closer to the load-deflection curve obtained excluding offset rotation and therefore the ultimate failure load based upon a deflection limit state will be significantly higher than that obtained from the experimental results.

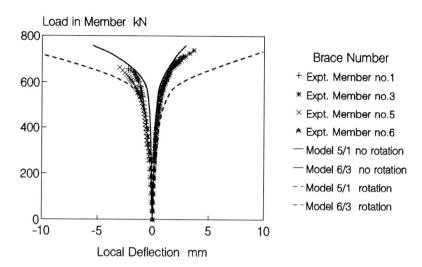

Load in Member kN

Brace Number
+ Expt. Member no.1
* Expt. Member no.3
× Expt. Member no.5
★ Expt. Member no.6
— Model 5/1 no rotation
— Model 6/3 no rotation
-- Model 5/1 rotation
-- Model 6/3 rotation

Local Deflection mm

Inner transducer readings

Fig.2. Comparison of load-deflection curves including
and excluding offset rotation

4.2 Model development stage

The model which gave the closest results to the experimental specimen KK06 included the details outlined
below:
- semi-loof shell elements for brace and chord
- chord and brace radii modelled
- actual chord and brace thickness distribution
- solid elements representing the welds
- actual measured gap between braces
The load-deflection curves obtained for this solid weld
model are compared in figure 2. The curves with offset
rotation included should be compared with the experimentally obtained curves. The curves can be seen to lie on
the safe side of the experimental load-deflection
curves. The effect of different model details on the
load deflection curves were also examined. The effect of
using no weld elements and the effect of using the nominal gap between the braces instead of the actual gap is
shown in figure 3. The effects of using no chord corner
radii, of using thick shell elements instead of semi-
loof shell elements and of using a uniform thickness
distribution instead of the actual distribution is shown
in figure 4. All these results include the effect of
offset rotation for comparison with the experimental
results. It can be seen that the effect of all these
details is to reduce the ultimate load capacity of the

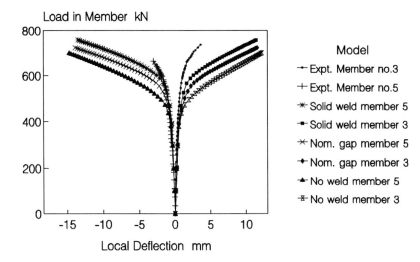

Load in Member kN

Local Deflection mm

Inner transducer readings

Model
— Expt. Member no.3
+ Expt. Member no.5
✳ Solid weld member 5
● Solid weld member 3
✳ Nom. gap member 5
✦ Nom. gap member 3
▲ No weld member 5
☀ No weld member 3

Fig.3. Effect of various modelling details on the load-
       deflection curves obtained - 1

joint model in comparison with the solid weld model with
details as outlined above. The effect of modelling the
welds as shell elements on ultimate load capacity was
also studied. Various shell thicknesses were examined
but all gave results which were stiffer than the model
with solid weld elements and tended to err on the unconservative side in comparison with the experiments.

4.3 Correlation of experimental and numerical models
The comparison between the experimentally and numerically obtained loads at serviceability and ultimate limit
state is given in Table 1 for the first six experimental
specimens. A deformation limit of 5% of the chord width
has been assumed for the ultimate limit state. Numerical
results are given including and excluding the effect of
offset rotation. It can be seen that including the offset rotation reduces the ultimate load by as much as
36%. Direct comparison of the ultimate loads obtained by
numerical modelling and by the experiments is difficult
due to the varied and often complex bridging arrangements adopted in the experimental measurement of chord
deformation. However the loads obtained should fall
between the numerical values obtained by including and
excluding offset rotation. This occurs in all cases
indicating that the numerical modelling is satisfactory.

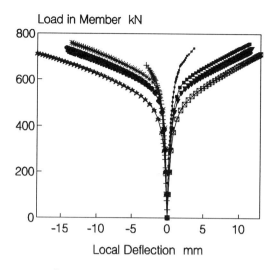

Load in Member kN

Model
⊸ Expt. Member no.3
+ Expt. Member no.5
⁎ Solid weld member 5
⊷ Solid weld member 3
● No radii member 5
▼ No radii member 3
★ Thick shell member 5
⊟ Thick shell member 3
✦ Uni. thick member 5
⊹ Uni. thick member 3

Local Deflection mm

Inner transducer readings

Fig.4. Effect of various modelling details on the load-
       deflection curves obtained - 2

Table 1. Correlation of experimental and numerical
         serviceability and ultimate loads

Spec. Ref No.	Serviceability Load kN 1% b_0			Ultimate Load kN 5% b_0		
	Expt.	Model incl. rot.	Model excl. rot.	Expt.	Model incl. rot.	Model excl. rot
KK01	82	75	88	141	130	174
KK02	140	117	135	193	178	235
KK03	365	300	328	393*	392	450
KK04	188	148	193	268	219	340
KK05	245	230	290	355	315	440
KK06	615	588	654	674*	726	860

    * Failure mode in these tests was by brace buckling
the chord face deformation criterion was not reached.

4.4 Uniplanar numerical modelling

At the time of writing only one uniplanar result is available. This result is for the uniplanar version of specimen KK06. The ultimate load obtained in the uniplanar model was 738kN in comparison with 726kN obtained for the multiplanar model. It is difficult to draw significant conclusions from one result but the multiplanar effect does not appear to be significant for specimen KK06.

5 Conclusions

- The model developed using semi-loof shell elements for the chord and brace members and solid elements for the welds gives good or conservative agreement with the experimental results.
- It is important to carefully consider methods of deformation measurement if deformation criteria are to be used to determine ultimate joint load capacity. The numerical investigation indicates that the method of measurement used in the K-joint experimental study could affect the ultimate loads determined from the experiments.
- Numerical modelling will be used further to determine the multiplanar effect in double K-joint specimens by examining uniplanar models of the specimens studied in the experimental phase of this research programme.

6 Acknowledgements

Thanks is given to ECSC, British Steel and CIDECT for sponsoring the research programme discussed in this paper.

7 References

Davies, G. Coutie, M.G. and Bettison, M. (1993) The behaviour of three dimensional RHS Tee joints under axial branch loads. **5th International Symposium on Tubular Structures** E & F N Spon, London

Liu, D.K. de Koning, C.H.M. Puthli, R.S. and Wardenier, J. (1993) Static strength of multiplanar DX joints in rectangular hollow sections. **5th International Symposium on Tubular Structures** E & F N Spon, London

Yeomans, N. (1993) Rectangular hollow section double K-joints - experimental tests and analysis. **5th International Symposium on Tubular Structures** E & F N Spon, London

75 DEFORMABILITY OF COLD FORMED HEAVY GAUGE RHS – NONLINEAR FEM ANALYSIS ON STRESS AND STRAIN BEHAVIOR

H. KAMURA, S. ITO and H. OKAMOTO
NKK Corporation, Tokyo, Japan

Abstract
FEM statical elasto-plastic analysis is used to present the primary factors in the deformation and stress-strain behavior related to the fracture of cold-formed columns. Cold formed columns are modeled as rectangular hollow sections. General deformational behaviors of the cold formed columns with various forming processes are presented. The relationship between the local stress and strain of the cold formed corner and the stress-strain curves of the cold formed columns is investigated. Numerical examples are presented to illustrate the concentration of stress and strain, the effect of axial force under earthquake and the geometrical configuration factor on the local stress and strain behavior in the cold formed corner welded to a diaphragm plate.
Key Words: Cold-formed Rectangular Hollow Sections, Local Stress and Strain Behavior, Mechanical Property, FEM Statical Elasto-plastic Analysis

1 Introduction

Cold formed rectangular hollow sections (C-RHS below) have come to be widely used to the extent of more than one million metric tons a year for steel buildings in Japan. Most of them are manufactured by cold forming processes such as roll forming and press forming.

In the studies on C-RHS, the influence of work hardening and residual stress on the structural instability such as local and global buckling has been investigated (Kato, 1987, Kimura and Kaneko, 1987, Chan, 1990). So far, however, studies on material instability such as ductile and brittle fracture are very rare, Fujimoto et al., (1977), Hayashi et al., (1987). Even in these studies, only material testing is carried out.

In resent years, as large sized heavy gauge C-RHS are popularized, deformability, and changes in mechanical properties have been investigated by Akiyama et al., (1992), Kuwamura et al., (1993). In this research, heavy sectioned C-RHS manufactured by the press forming process was fractured from ductile cracks which occurred at the welded joint in a case of unsuitable welding conditions.

When C-RHS are used for the structure of high rise buildings, material instability becomes significant as well as structural instability. In the study of this problem, full scale experiments become very important since the similarity law is not satisfied. Test results of full scale experiments, however, are very few, because large sized testing equipment is needed. Moreover, it is generally difficult in full scale experiments to make clear the local stress and strain which affects the criteria of occurrence of ductile cracks. The local stress and strain seem to be affected by the details of welds and the changes of mechanical properties in the bend corner welded to a diaphragm plate. Therefore, analytical procedure comes to be effective on this problem.

Tubular Structures V. Edited by M.G. Coutie and G. Davies.
© 1993 E & FN Spon, 2–6 Boundary Row, London SE1 8HN. ISBN 0 419 18770 7.

In this study, FEM statical elasto-plastic analysis is carried out to investigate the local stress and strain in the bend corner and present the primary factors for the general behaviors of C-RHS produced by various forming processes. The general behavior of C-RHS is analyzed with shell element and local stress and strain is calculated with 3-D solid element. Numerical examples are presented to discuss the concentration of surface stress and strain on the cold formed corner and effects of the axial force on the local stress and strain behavior using the mechanical properties which are assumed from the tensile test results of C-RHS.

2 Analytical model and method

2.1 Mechanical properties of the materials

To determine the stress-strain relation in cold-formed corner and straight part of C-RHS, tensile strength tests were carried out in accordance with JIS Z 2241 by using JIS Z 2201 No.6 test piece. Fig. 1 shows the position from which the test pieces are taken , mechanical properties and their stress-strain relations. This indicates the changes in mechanical properties of the press-bend corner. It is found that the cold forming causes increase in strength and decrease in ductile elongation capacity.

Material I ; The relation of the material which is annealed or not cold-formed is modeled as a bi-linear solid line (hardening slope ; 1.5Es/100).
Material II ; The relation of the outer test piece (solid round-house shaped line) is presented as that of the material which is cold-formed.

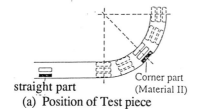

straight part
(a) Position of Test piece

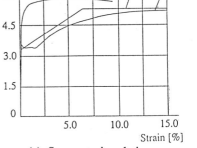

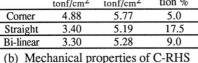

	YP tonf/cm^2	TS tonf/cm^2	Elonga-tion %
Corner	4.88	5.77	5.0
Straight	3.40	5.19	17.5
Bi-linear	3.30	5.28	9.0

(b) Mechanical properties of C-RHS

(c) Stress-strain relation

Fig.1 Stress-strain relations of the materials

2.2 Mechanical properties of the section for each C-RHS model

Three models are assumed as follows: (Cf. Fig.2)

a. Anneal model ; C-RHS which is postheated by normalizing process after cold forming. The property of whole part is assumed to be Material I.
b. Press model ; C-RHS which is produced by press-bending process. Only press-bend corners are cold-formed. The stress-strain relation of the straight part is Material I and that of the corners is Material II.
c. Roll model ; C-RHS which is produced by roll forming process and the whole part of the section is cold-formed. The stress-strain relation in the section is assumed to be Material II.

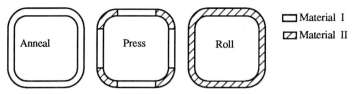

Anneal   Press   Roll

Fig.2 Mechanical properties of the section of each C-RHS model

2.3 Analytical model

The cold-formed columns are modeled as rectangular hollow sections on FEM analysis with isoparametric 4-node shell elements and 8-node 3-dimensional solid elements. The model is 2m long cantilever beam or beam-column with fixed end. The enforced deflection is applied at the other end. The dimensions of the standard model are as follows: Diameter D= 450mm, column thickness t=25mm, the radius diameter of the corner r=75mm (3t). In elasto-plastic incremental FEM analysis without axial force, model is quarter section of the column introducing symmetric and anti-symmetric condition as shown in Fig. 3a. When axial force is applied, model is half section of the column by symmetric condition as shown in Fig. 3b . Further, on the 3-d solid element analysis, a diaphragm plate and welding bead are also modeled in addition to the quarter section of C-RHS.

For 4-node shell elements, the corner part is divided into 5 elements along the circumference and minimum element dimension is 25mm (1.0t, t=thickness of the plate) near the fixed end. For 8-node 3-D solid elements, the number of layers in the thickness direction is three. The corner is divided into 8 elements along the circumference and minimum element dimension is 12mmx8.3mmx8.3mm (1/3t) near the fixed end. The dimension of the welding bead is triangle shape which height is 1/3t and length is 2/3t in the section. Table 1 shows analytical models.

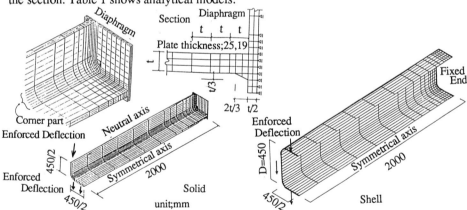

Fig.3a.Analytical models (Without axial force)Fig.3b.Analytical models (With axial force)

2.4 Analytical Method

The computer program of Finite Element Method for nonlinear analysis is ADINA (Automatic Dynamic Nonlinear Analysis), ADINA R&D Inc.(1987), which is generally used for displacement and stress-strain analysis. In nonlinear analysis, the von Mises yield condition, an associated flow rule using the von Mises function and multi-linear hardening law are used on elasto-plasticity model, Bathe, (1982).

Table 1. Analytical models

Analytical model	Model Type	Element Type	Axial Force	Thickness t	Radius Diameter of the corner r
No.1	Anneal	4-node shell	-	25mm	3t = 75mm
No.2	Press	4-node shell	-	25mm	3t = 75mm
No.3	Roll	4-node shell	-	25mm	3t = 75mm
No.4	Anneal	8-node solid	-	25mm	3t = 75mm
No.5	Press	8-node solid	-	25mm	3t = 75mm
No.6	Roll	8-node solid	-	25mm	3t = 75mm
No.7	Anneal	8-node solid	-	25mm	2t = 50mm
No.8	Anneal	8-node solid	-	25mm	4t= 100mm
No.9	Anneal	8-node solid	-	19mm	3t = 57mm
No.10	Anneal	4-node shell	Constant	25mm	3t = 75mm
No.11	Anneal	4-node shell	Compression	25mm	3t = 75mm
No.12	Anneal	4-node shell	Tension	25mm	3t = 75mm

2.5 Axial Force

Fig.4 shows the axial force-deflection relation for model No.10~12. These relations are determined as follows:

1) The axial force assumed as center column is constant, 20% of yield axial force of the column Ny which is estimated ordinary design dead load.
2) The axial force assumed as side column is supposed to be proportional to the horizontal reaction force and becomes 50% of Ny at deflection=8cm. Therefore the axial force is changed from 0.2Ny to 0.5Ny or -0.1 Ny.

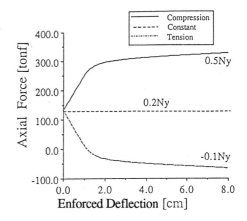

Fig. 4. Axial load pattern

3 Results and discussion

3.1 Load- Deflection Relations

Fig. 5 shows the load-enforced deflection relations due to cold forming processes. Open circles are the results by shell element and solid circles are those of 3-D solid element. Further, solid lines are the result by column deflection curve method which is ordinary used in design procedure. Dotted lines are the experimental results of C-RHS by press-bend and annealed C-RHS, respectively. It is observed that they almost coincide and the strength of as-cold-formed C-RHS is higher than that for annealed C-RHS.

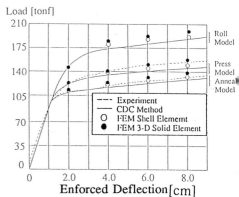

Fig. 5. Load- deflection relations

3.2 Distribution of Surface Stress and Strain

Fig.6 shows the maximum surface principal stress and strain distribution of three types of C-RHS without axial force by 3-D solid element at deflection=40mm (relative storey displacement angle = 1/50). The stress tends to concentrate on the corner in all models.

On the press model which has multi-stress-strain relation (i.e the stress-strain relations are different between in bend corner and in the straight part of the section), the concentration of stress is more remarkable than that of the anneal model or roll model. The maximum surface principal stress is relatively greater than that of the annealing model.

On the anneal model and roll model which have uni-stress-strain relation, the concentrated area of stress distributions not only occur in the corner but also in the flange of the C-RHS. On the other hand, in press model, the concentrated area of stress mainly distributes in the press-bend corner, and the value of maximum principal stress is the largest of the three models.

The distribution of strain for the three models are similar. Though the concentration of strain at the press-bend corner is observed in all models, the value of the maximum principal strain of the anneal model is relatively greater than that of roll and press models. That of roll model is the smallest.

These indicate that the mechanical properties of C-RHS relatively affect the local stress and strain in the press-bend corner. The feature of press model is that both local stress and strain are relatively large compared with the other models.

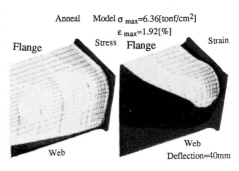

Anneal Model σ_{max}=6.36[tonf/cm^2]
ε_{max}=1.92[%]

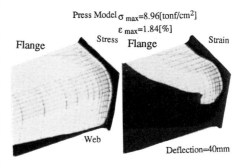

Press Model σ_{max}=8.96[tonf/cm^2]
ε_{max}=1.84[%]

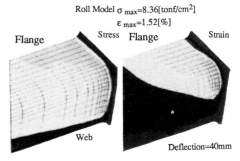

Roll Model σ_{max}=8.36[tonf/cm^2]
ε_{max}=1.52[%]

Fig. 6. Maximum principal stress and strain distribution

3.3 Deflection-Strain Relation

Figs. 7a and 7b show the deflection-maximum principal strain relation due to forming processes at integration point by shell element and 3-D solid element, respectively. These figures indicate that as the deflection becomes larger, the strain exceeds that of 3-D solid element. Using shell elements the strain of the anneal model is the largest, and that of press model and roll model are almost the same level. On the other hand, based on the analysis using solid elements, all models show similar values and the local strain of roll model is always the smallest and that of annealing model is the largest within deflection=8cm.

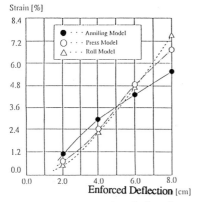

Fig. 7a. Deflection-strain relation due to forming processes (shell elements)

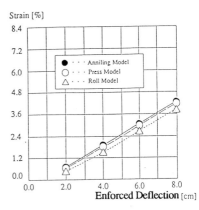

Fig. 7b. Deflection-strain relation due to forming processes (3-D solid elements)

3.4 Strain distribution along circumference of RHS

Fig. 8 shows the distribution of outer surface axial strain along the circumference at near welding bead at deflection=40mm (relative story displacement angle= 1/50) due to forming processes. Concentration of strain which may be caused by the local strain by plate bending moment is observed. This is caused by the effect of Poisson's ratio of the C-RHS section. In elasto-plastic state, as axial elongation, width of the C-RHS becomes smaller. Despite the tension side of the section tends to deform, the end section welded to a diaphragm can not deform on account of restraint of the diaphragm plate. Consequently, the strain in the corner is larger than that of CDC method, because there is large plate bending moment from two directions.

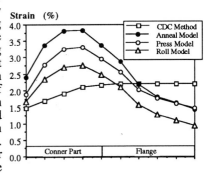

Fig. 8. Distribution of axial strain along the circumference of the column

3.5 Stress and strain state near welding bead

Fig. 9 shows the stress and strain relation at the integration point near welding bead due to forming processes. Solid and dotted lines are the stress and strain relation of material I and material II, respectively. The results of press model and roll model are similar. Though that of annealing model is smaller than the others, all of them are relatively large compared with that of the materials. The multi-stress state of press model is shown in Fig.10. The stress in thickness direction and circumferential direction are relatively large as well as longitudinal stress. And solid circles in Fig.10 are the deviatoric stress. They are relatively small. These indicate multi-stress state is observed near welding bead in the corner.

3.6 Strain distribution in plate thickness

Fig. 11 shows the distribution of axial strain in plate thickness direction of annealing model due to the variation of plate thickness and radius diameter of the cold-formed

762

corner (deflection=40mm). The strain varies linearly from inner surface to outer surface. This indicates that plate bending moment is applied in the corner. The maximum principal stress and strain is shown in Table 2. As the radius diameter of the corner becomes smaller, the axial strain becomes larger. As the thickness of the plate become smaller, the axial strain becomes larger. This increase is caused by the difference of the area of the corner which tries to restrain the deformation of the section.

3.7 Effect on Strain Amplitude due to the Fluctuation of Axial Force

Fig.12 shows the relations between the strain amplitude and deflection due to cold forming process. The three strain amplitudes shown in Table 3 are estimated. $\Delta\varepsilon_\pm$ and $\Delta\varepsilon_{cst}$ assumed to be the strain amplitudes of side and center column, respectively. Table 4 shows the strain amplitude ratio normalizing by $\Delta\varepsilon_0$. These indicate that $\Delta\varepsilon_\pm$ is 1.2 times as much as $\Delta\varepsilon_0$ if the fluctuation of axial force is considered and $\Delta\varepsilon_{cst}$ is almost the same as $\Delta\varepsilon_0$. As the deflection becomes larger, $\Delta\varepsilon_\pm$ and $\Delta\varepsilon_{cst}$ becomes smaller and tend to converge larger than deflection = 60mm.

4 Conclusions

The conclusions of this study can be summarized as follows:

1. Concentration of stress and strain which may be caused by the local plate bending moment is observed in the corner of C-RHS. This concentration is remarkable in the case of C-RHS by press-bend forming compared with annealed C-RHS and rolled C-RHS.
2. The strain of the corner is much larger than the results by column deflection method indicate with Navier's hypothesis. This is causes by the effect of Poisson's ratio of the C-RHS section. The stress and strain at the corner surface near welding bead become lager as the radius diameter of the corner and the plate thickness become smaller.
3. Both large strain and multi-stress state occurs near welding bead in the corner of C-RHS. From this point of view, the C-RHS formed by press-bending is in the most severe condition.
4. The axial strain of side C-RHS column subjected to dead and lateral loading is about 1.2 times as large as that without axial load. The axial strain of the center column is almost the same.

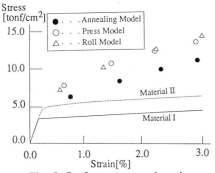

Fig. 9. Surface stress and strain relation near welding bead

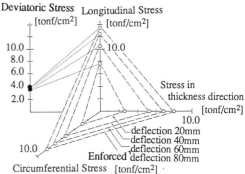

Fig.10. Multi-stress state of press model

Table 2. Maximum principal stress and strain

No. (mm)	σmax kgf/cm²	εmax %
No.4 t=25 r=75	6.36	1.92
No.7 t=25 r=50	7.34	2.25
No.8 t=25 r=100	6.16	1.77
No.9 t=19 r=75	7.38	2.29

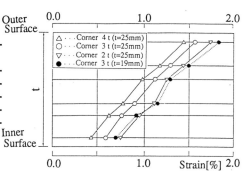

Fig.11 Distribution of axial strain in plate thickness

Table 3. Definition of the strain amplitude

Strain Amp.	Definition
$\Delta\varepsilon\pm$ Outer Column	(Surface strain on Compression Side in Compression Axial Force) - (Surface strain on Tension Side in Tension Axial Force)
$\Delta\varepsilon_{cst}$ Center Column	(Surface strain in Compression Side in case of Constant Axial Force) - (Surface strain in Tension Side in case of Constant Axial Force)
$\Delta\varepsilon_0$	(Surface strain in case of Non Axial Force) x 2

Table 4. Strain amplitude ratio

Model	Deflection	$\Delta\varepsilon\pm$	$\Delta\varepsilon_{cst}$
Annealing	2cm	1.28	0.94
	4cm	1.21	0.95
	6cm	1.20	0.95
	8cm	1.19	0.95
Press	2cm	1.35	0.95
	4cm	1.24	0.92
	6cm	1.22	0.92
	8cm	1.21	0.93
Roll	2cm	1.30	0.99
	4cm	1.23	0.96
	6cm	1.21	0.95
	8cm	1.21	0.95

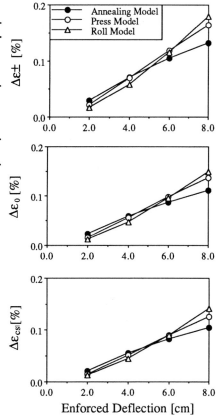

Fig. 12 Strain amplitude-deflection relation

5 References

ADINA R&D Inc.(1987) ADINA Report ARD 87-1 Vol.2, ADINA R&D Inc..

Bathe, K.J. (1982) Finite Element Procedures in Engineering Analysis, Prentice-Hall, Inc..

Chan, S.L. (1990) Strength of cold-formed box columns with coupled local and global buckling, **The Structural Engineer**, Vol.68, No.7, pp.125-132.

Fujimoto, M , Aoki, H , Kitawaki, S (1977) Influence of Plastic Strain History on Mechanical Properties of Structural Metals (Part 1. Influence of Tensile Plastic Strain History), **Journal of Structural and Construction Engineering**, AIJ, No.253, pp.17-26

Hayashi, K , Takano, K , Sasaki, S , Horikawa, H (1987) On Deformation Capacity and Fracture Toughness of Structural Steel Subjected to Pre-strain , **Materials**, Vol.36, No.410, pp. 69-75.

Kato, B. (1987) Deformation Capacities of Tubular Steel Members Governed by Local Buckling , **Journal of Structural and Construction Engineering**, AIJ, No.378, pp.27-36.

Kimura, M , Kaneko , H (1987) Local Buckling Behavior And Evaluation On Width-to-thickness Ratio of Square Steel Tubes, **Journal of Structural and Construction Engineering**, AIJ, No.372, pp.65-70.

Kuwamura, H., Akiyama, H., Yamada, S., Chiu, J.C. (1993) Experiment on the Mechanical Properties and their Improvement of Cold Press-bend Steel Plates, **Journal of Structural and Construction Engineering**, AIJ, No.444, pp.125-134.

AUTHOR INDEX

SUBJECT INDEX

This index has been compiled from the keywords assigned to the individual papers by the authors, edited and extended as appropriate. The numbers refer to the first page number of the relevant paper.

Architecture and Construction in Steel

Edited by **Alan Blanc**, Chartered Architect, **Michael McEvoy,** School of Architecture and of Engineering, University of Westminster and **Roger Plank**, School of Architectural Studies, University of Sheffield, UK

A Steel Construction Institute publication

This book provides a comprehensive guide to the successful use of steel in building and will form a unique source of inspiration and reference for all those concerned with architecture in steel.

Expert contributions have been commissioned by the Steel Construction Institute from 29 leading architects and engineers. They are presented in six sections: history of iron and steel construction, materials, principles of steel framing, steel construction, secondary steel elements, and outstanding contemporary steel architecture. Illustrated with over 1000 photographs, plans, sections and diagrams of buildings from around the world, *Architecture and Construction in Steel* will be an essential source book.

September 1993: 297x210: c.640pp, 550 line illus, 428 halftone illus
Hardback: 0-419-17660-8

E & F N Spon
An imprint of Chapman & Hall

Flexural-Torsional Buckling of Structures

N S Trahair, Challis Professor of Civil Engineering, School of Civil and Mining Engineering, University of Sydney, Australia

This book provides an up-to-date and comprehensive treatment of flexural-torsional buckling, and shows how to design against this mode of failure. It also gives detailed summaries of knowledge on flexural-torsional buckling so that it can be used as a source book by practising engineers, designers and researchers and by advanced students of structural engineering.

Contents: Preface. Units and conversion factors. Glossary of terms. Principal notation. Introduction. Equilibrium, buckling and total potential. Buckling analysis of simple structures. Finite element buckling analysis. Simply supported columns. Restrained columns. Simply supported beams. Restrained beams. Cantilevers. Braced and continuous beams. Beam-columns. Plane frames. Arches and rings. Inelastic buckling. Strength and design of steel members. Special topics. Appendices: in-plane bending; uniform torsion; warping torsion; energy equations for flexural-torsional buckling; differential equilibrium equations for the buckled position. References. Index.

June 1993: 234x156: 384pp, 215 line illus, 4 halftone illus
Hardback: 0-419-18110-5

E & F N Spon
An imprint of Chapman & Hall